TAILINGS AND MINE WASTE '02

PROCEEDINGS OF THE NINTH INTERNATIONAL CONFERENCE ON TAILINGS
AND MINE WASTE/FORT COLLINS/COLORADO/USA/27–30 JANUARY 2002

Tailings and Mine Waste '02

A.A.BALKEMA PUBLISHERS   LISSE / ABINGDON / EXTON (PA) / TOKYO

Published by: A.A.Balkema, a member of Swets & Zeitlinger Publishers
    www.balkema.nl and www.szp.swets.nl

ISBN 90 5809 353 0

Printed in the Netherlands

Table of contents

Preface   IX

Organizing Committee   XI

Rocky Mountain Regional Hazardous Substance Research Center

Remediation of mine waste sites through the Rocky Mountain Regional Hazardous Substance
Research Center (RMRHSRC)
   C.D. Shackelford, S.L. Woods & D.L. Macalady   3
Mining and the environment: The issues that steer the Region 8 Hazardous Substance
Research Center
   T.R. Wildeman & C.D. Shackelford   9
Redox transformations, complexation and soil/sediment interactions of inorganic forms of As and
Se in aquatic environments: Effects of natural organic matter
   D.L. Macalady, D. Ahmann, J. Westall, J. Garbarino & J. Meyer   13
Fate and transport of metals and sediment in surface waters
   P.Y. Julien, B.P. Bledsoe & C.H. Watson   17
Metal removal capabilities of passive bioreactor systems: Effects of organic matter and microbial
population dynamics
   L. Figueroa, D. Ahmann, D. Blowes, K. Carlson, C. Shackelford, S. Woods, N. DuTeau,   21
   K. Reardon & T. Wildeman
Evaluating recovery of stream ecosystems from mining pollution: Integrating biochemical,
population, community and ecosystem indicators
   W.H. Clements & J. Ranville   23

Site characterization

Abandoned quarries surveying
   G. Bonifazi, L. Cutaia, P. Massacci & I. Roselli   27
The effectiveness of single and multiple open standpipe piezometers in monitoring of the pore
pressure regime in tailings dams
   G.J.R. le Roux   35
Monitoring of soil, air, water, and noise at the Lega Dembi Gold Mine
   Y. Binega   39
Variations in composition on SA gold tailings dams
   N.J. Vermeulen, E. Rust & C.R.I. Clayton   45

Preliminary ecological risk assessment for the Elizabeth Mine site, South Strafford, Vermont
  I. Linkov, S. Foster, E. Hathaway & R. Sugatt   53

Impact of acid rock drainage in a discrete catchment area of the former uranium mining site
of Ronneburg (Germany)
  J.W. Geletneky, G. Büchel & M. Paul   67

Hydrochemical investigation at the uranium tailings "Schneckenstein" (Germany)
  T. Naamoun   75

Design, operation, and disposal

Tailings beach slope forecasting – copper tailings
  M. Pinto & S. Barrera   87

Quebrada Honda tailing storage facility conception, design, construction, and operation,
Southern Peru Copper Corporation, Tacna, Peru
  A.H. Gipson, Jr. & H. Walqui Fernandez   93

Retrofitting an HDPE-lined raise over an unlined tailing facility
  A.H. Gipson, H.P. Vos, W.J. Cole & S.R. Aiken   101

Geotechnical considerations

Advanced laboratory compression tests and piezocone measurements for evaluation of time-
dependent consolidation of fine tailings
  U. Barnekow, M. Paul & A.T. Jakubick   113

Prediction of the desiccation and sedimentation behavior of a typical platinum tailings
  J.C.J. Boshoff   121

Paste technology

Planning, design and implementation strategy for thickened tailings and pastes
  F. Sofra & D.V. Boger   129

Effects of settlement and drainage on strength development within mine paste backfill
  T. Belem, M. Benzaazoua, B. Bussière & A.M. Dagenais   139

Use of copper mine tailings as paste backfill material in mining operations – Approach to minimise
land occupation?
  S. Moellerherm & P.N. Martens   149

Pumping paste with a modified centrifugal pump
  A. Sellgren & L. Whitlock   155

Liners, covers, and barriers

Construction and instrumentation of waste rock test covers at Whistle Mine, Ontario, Canada
  B.K. Ayres, M. O'Kane, D. Christensen & L. Lanteigne   163

Field investigation to support the closure design of the Yankee Heap
  P.E. Kowalewski, S. Boyce & R. Buffington   173

Surface water quality

Turnover in pit lakes: I. Observations of three pit lakes in Utah, USA
  D. Castendyk & P. Jewell   181

Turnover in pit lakes: II. Water column stability and anoxia
  P. Jewell & D. Castendyk   189

Groundwater and geochemistry

Long term persistence of cyanide species in mine waste environments
B. Yarar 197

Stabilization / solidification of pyritic mill tailings by induced cementation
J. Ouellet, F. Hassani, S. Somot, S. Shnorhokian & M. Hossein 205

The removal of arsenic from groundwater using permeable reactive materials
J. Bain, L. Spink, D. Blowes & D. Smyth 213

Estimation of the mobility of heavy metals in tailing sediments
T. Naamoun 217

Uranium tailings of "Schneckenstein" (Germany) reservoir of contaminants
T. Naamoun 231

Remediation and reclamation

Abandoned mine site waste repositories, site selection, design and costs
K.L. Ford & M. Walker 245

In situ bioremediation of uranium and other metals in groundwater
W. Lutze, W. Gong & H.E. Nuttall 249

In situ treatment of metals in mine workings and materials
J.M. Harrington 251

Response of plants to oil sand tailings
M.J. Silva, M.A. Naeth, D.S. Chanasyk, K.W. Biggar & D.C. Sego 263

Column leaching test to evaluate the beneficial use of alkaline industrial wastes to mitigate acid mine drainage
I. Doye & J. Duchesne 271

Spatial surface flux boundary model for tailings impoundments
M.E. Rykaart, D.G. Fredlund & G.W. Wilson 283

Radioactivity and risk

The benefits of the risk assessment to the mining industry
T. Alexieva 295

Mobility tracing of radionuclides in the uranium tailings "Schneckenstein" (Germany)
T. Naamoun, D. Degering & D. Hebert 303

Radioactive tailings issues in Kyrgyzstan and Kazakhstan
R.B. Knapp, J.H. Richardson, N. Rosenberg, D.K. Smith, A.F.B. Tompson, A. Sarnogoev, B. Duisebayev & D. Janecky 313

Reprocessing, utilization, and treatment

A brief discussion of the behavior of a loose iron tailing material under undrained triaxial loading
S. Tibana, T.M.P. de Campos & G.P. Bernardes 325

New technologies and approaches

Solar radiation on surfaces of tailings dams – effects of slope and orientation
G.E. Blight 333

Stabilization of uranium in pitwaters using phosphate and coal tailings
S.C. Muller, T. Delaney, R. Ryan, J. Weber, S. Swapp, C. Eggleston & D. Nash 339

The tailings pond at the Milltown Dam, Montana, can be cleaned
G. Ter-Stepanian 349

Consolidation of clay slurries amended with shredded waste plastic
  N. Lozano & L.L. Martinez   351

The use of the powder marble by-product in the raw material for brick ceramic
  G.C. Xavier, F. Saboya Jr. & J. Alexandre   357

Researches concerning the Purolite assimilation for use within the uranium
separation-concentration Resin In Pulp process
  E. Panturu, Gh. Filip, St. Petrescu, D. Georgescu, F. Aurelian & R. Radulescu   361

Cu(II) separation and recovery from mining aqueous systems by flotation (DAF) using
alkylhydroxamic collectors
  L. Stoica, C. Constantin & O. Micu   365

Case histories

Pozo Azul tailings impoundment: Design modifications made to utilize a difficult site
  M.L. Fuller, R.L. Byrd, J. Tagliapietra & K.D. Ball   373

Performance of the Kennecott Utah Copper tailings embankment, Magna, Utah
  J. Pilz & D. Stauffer   379

Performance of vertical wick drains at the Atlas Moab Uranium Mill tailings facility after 1 year
  M.E. Henderson, J. Purdy & T. Delaney   387

Steep Rock Iron Mines: Dredging and draining of Steep Rock Lake and some of the effects after
45 years
  V.A. Sowa   393

Community consultation in the rehabilitation of the South Alligator Valley Uranium Mines
  P.W. Waggitt   403

Remediation of streamside tailings along Silver Bow Creek near Butte, Montana
  W.H. Bucher, G. Fischer, B. Grant, A. Shewman, L. Cawlfield, J.E. Chavez & T. Reilly   411

A rapid response to cleanup – Gilt Edge Superfund Site, South Dakota
  M.J. Gobla   421

A "no action" alternative that worked
  R.B. Meade   427

Mercury spill response, clean-up, and assessment, Choropampa, Peru
  K.J. Esposito, T.A. Shepherd, M.C. Meyer & N.R. Cotts   435

Physical and geochemical characterization of mine rock piles at the Questa mine, New Mexico:
An overview
  S. Shaw, C. Wels, A. Robertson & G. Lorinczi   447

Assessment of store-and-release cover for Questa tailings facility, New Mexico
  C. Wels, S. Fortin & S. Loudon   459

Factors influencing net infiltration into mine rock piles at Questa mine New Mexico
  C. Wels, S. Loudon & S. Fortin   469

ARD production and water vapor transport at the Questa mine
  R. Lefebvre, A. Lamontagne, C. Wels & A.MacG. Robertson   479

Selection of a water balance cover over a barrier cap – a case study of the reclamation of the
Mineral Hill Mine dry tailings facility
  R.J. Frechette & F.W. Bergstrom   489

Hard rock mine closure under modern rules – TVX Mineral Hill Mine case study
  F.W. Bergstrom & R.J. Frechette   495

Evaluation of alternative designs for a water-balance cover over tailings at the Mineral Hill Mine,
Montana, using the EDYS model
  T. McLendon, W.M. Childress, C.L. Coldren, R.J. Frechette & F.W. Bergstrom   505

Author index   519

Preface

This is the ninth annual Tailings and Mine Waste Conference held at Colorado State University in Fort Collins, Colorado. The purpose of these conferences is to provide a forum for discussion and establishment of dialogue among all people in the mining industry and environmental community regarding tailings and mine waste. Previous conferences have been successful in providing opportunities for formal and informal discussion, exhibits by equipment and instrumentation companies, technical exhibits, and general social interaction.

This year's conference includes over 60 papers. These papers address the important issues faced by the mining industry today. These proceedings will provide a record of the discussions at the conference that will remain of value for many years.

Tailings and Mine Waste '02, © 2002 Swets & Zeitlinger, ISBN 90 5809 353 0

Organizing Committee

Organized by the Geotechnical Engineering Program, Department of Civil Engineering, Colorado State University, Fort Collins, Colorado and co-sponsored by the EPA Region 8 Rocky Mountain Regional Hazardous Substance Research Center.

ORGANIZING COMMITTEE

John D. Nelson, Conference Chair	Colorado State University, Fort Collins, Colorado
William A. Cincilla	Golder Paste Technology, Lakewood, Colorado
Cary L. Foulk	M.F.G., Boulder, Colorado
Linda L. Hinshaw	Colorado State University, Fort Collins, Colorado
Victor Ketellaper	US EPA, Denver, Colorado
Daniel D. Overton	Shepherd Miller, Inc., Fort Collins, Colorado
Debora J. Miller	Gannet Fleming, Denver, Colorado
Sean C. Muller	SRK Consulting, Fort Collins, Colorado
Suzanne S. Paschke	U.S.G.S., Denver, Colorado
Charles D. Shackelford	Colorado State University, Fort Collins, Colorado

Rocky Mountain Regional Hazardous Substance Research Center

Remediation of mine waste sites through the Rocky Mountain Regional Hazardous Substance Research Center (RMRHSRC)

C.D. Shackelford & S.L. Woods
Department of Civil Engineering, Colorado State University, Fort Collins, CO 80523, USA

D.L. Macalady
Department of Chemistry and Geochemistry, Colorado School of Mines, Golden, CO 80401, USA

ABSTRACT: The Rocky Mountain Regional Hazardous Substance Research Center (RMRHSRC) for remediation of mine waste sites has recently been formed. The RMRHSRC is funded by the U. S. Environmental Protection Agency (EPA), represents EPA Region 8 states, and consists of a consortium of participants from Colorado State University, Colorado School of Mines, and several academic and non-academic participants from other regions of the U. S and Canada. The research goal of the RMRHSRC is to develop new and improve existing methods or technologies for remediation of mine waste sites that are cost effective and lead to clean ups that are protective of human health and the environment. Also, the activities of the RMRHSRC include training, technology transfer and outreach programs that will focus on the development of new technologies. These outreach programs will provide educational information to allow communities to make informed decisions concerning environmental contamination, and will provide technical assistance to communities and other stakeholders with an ultimate goal of redeveloping brownfields sites.

1 INTRODUCTION

Six states comprise the U. S. Environmental Protection Agency (EPA) Region 8, viz. Colorado, Montana, North Dakota, South Dakota, Utah, and Wyoming. Mining has played a critical role in the socio-economic history of all of these states, and all have environmental problems associated with historic or current mining operations. Of particular concern in this regard is the potential threat to human health and the environment resulting from inactive or abandoned mine lands (AMLs), many of which have yet to be reclaimed in the Region 8 states. In general, these AMLs represent hardrock and non-coal mines located on private, state, and public lands that were actively mined over the past century and a half, but subsequently abandoned prior to the promulgation of environmental regulations (e.g., Clean Water Act, RCRA, CERCLA) enacted since the 1970s.

In response to this concern, the Rocky Mountain Regional Hazardous Substance Research Center (RMRHSRC) for remediation of mine waste sites recently has been designated and will be jointly administered at Colorado State University (CSU) and the Colorado School of Mines (CSM). In addition to the consortium of participants from CSU and CSM, the Center includes several academic and non-academic participants from other regions of the U. S and Canada. This Center is funded by the EPA in the amount of $4 million for an initial period of 5 years, with an additional $1 million in funding being contributed by CSU and CSM. The objective of the Center is to become self-supporting upon completion of this initial 5-year period.

2 CENTER ORGANIZATION

The organizational structure for the RMRHRC is illustrated in Figure 1. The Center consists of two main activities related to research and outreach programs that pertain to remediation of mine waste sites, and a third activity related to quality assurance/quality control (QA/QC) for these programs. The research activities are separated almost equally between the two main participating universities (i.e., CSU and CSM), and the outreach activities include three main components, viz. technical transfer, technical outreach to communities, and technical outreach to brownfields. The research activities are overseen by a Science Advisory Committee (SAC), and the outreach activities are overseen by a Training and Technology Advisory Committee (TTAC).

The membership of the SAC consists of 6-9 technical peers drawn from the public and private sectors and academia. At least one-third of the SAC membership must consist of appropriate personnel from EPA's Regional Offices and Laboratories, and at least another one-third of the membership must be drawn from the academic community. The remaining members may come from industry or other Federal, State, or local governmental units. Appointments to this committee are subject to approval by the EPA project officer. Duties include reviewing

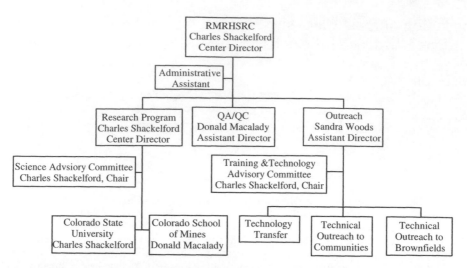

Figure 1. Organizational structure for the Rocky Mountain Regional Hazardous Substance Research Center (RMRHSRC) for remediation of mine waste sites.

the RMRHSRC's research plan, annual development of a list of relevant research topics, preparing recommendations regarding the relevance and technical merit of project proposals, and reviewing ongoing projects. Meetings are held twice a year. Members must be chosen from outside the institutions comprising the HSRC.

Membership of the TTAC includes representatives from relevant EPA Regional Offices, EPA's Office of Emergency Response, academia, and States or localities. Others may be added by the Center Director with approval by the EPA project officer. Duties include annual meetings to recommend outreach activity plans for the next year, review progress, and recommend changes to current year programs.

3 RESEARCH ACTIVITIES

The key issues associated with the environmental impacts resulting from AMLs include: (1) an inadequate ability to rapidly and cost effectively characterize the extent and impacts of the effects of contamination resulting from mining activities; (2) an inadequate ability to accurately characterize the fate and transport of metals and other toxic chemicals from mining sites; (3) a paucity of cost-effective technologies that can clean up mine waste sites; and (4) a need to develop less costly and more rational clean-up strategies. Based on these issues, the goal of RMRHSRC research activities will be to extend our knowledge of the geochemical, biological, hydrological/mineralogical and engineering aspects of environmental problems associated with mining and mine wastes and, based on this knowledge, develop

new or improved methods or technologies that are cost effective and lead to clean ups that protect human health and the environment. The specific research objectives that will be addressed to accomplish the goal of the Center are:
(1) to more rapidly and effectively characterize the extent and impacts of the effects of contamination resulting from mining activities;
(2) to more accurately characterize the fate and transport of metals and other toxic chemicals from mining activities;
(3) to develop cost-effective treatment processes and associated technologies that can effectively clean up mine waste sites; and
(4) to improve our ability to evaluate risk assessments for developing rational clean-up strategies.

To achieve these objectives, Center research will address three main areas of activity with respect to the remediation of mine waste sites: (1) fate and transport; (2) treatment and technologies; and (3) risk assessment. Each of these research activities represents an essential component of the remediation process. For example, fate-and-transport analyses using appropriate modeling approaches typically are required as an integral part of a risk assessment in order to estimate exposure-point concentrations of a given contaminant. Based on these concentrations, a toxicity and risk assessment is performed to determine the cleanup goals that ultimately affect the technology that is implemented for the remediation. Within each of these three research activity areas, both basic and applied research will be included. Mathematical and physical models will be used to better understand the processes being studied and to help extend the results of the basic research to field demonstrations.

3.1 *Research focus areas and approach*

In order to address the three main areas of research activity, the Center research program is divided into five research focus areas, each with a focus group leader, as shown in Figure 2. The focus areas include: (1) site characterization and contaminant transport/transformation; (2) surface water and sediment transport; (3) treatment processes; (4) technologies; and (5) ecological and human health toxicity. The types of contaminant problems and the specific processes to address these problems are identified within this structure. Mathematical and physical modeling will be key components of each of the focus areas. In addition to the research focus areas of the Center, the first-year research projects of the Center also are identified in Figure 2.

To meet the goals of the Center, a multidiscipli-nary group of researchers has been assembled, as shown in Table 1. These researchers have a history of working on complex environmental processes, and taking these processes from the "laboratory to the field". Also shown in Table 1 are the principal investigators (PIs) and Co-PIs for the first year projects. A broad range of multidisciplinary expertise and experience are represented in this group and in their research proposals. A particular strength of this research group is the excellent distribution of Center investigators among each of the three categories, with 12 investigators from CSU, 11 investigators from CSM, and 11 participating investigators from other states within Region 8 and other EPA Regions, and Canada. Also, two of the first year projects (Projects 3 and 4) have investigators from both partnering universities (CSU & CSM) as PIs/Co-PIs.

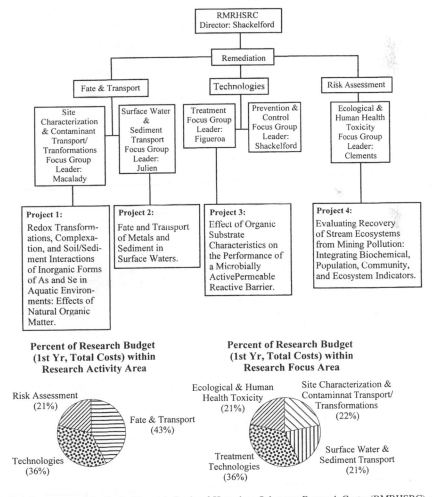

Figure 2. Research focus areas of the Rocky Mountain Regional Hazardous Substance Research Center (RMRHSRC), and first-year research projects.

5

4 OUTREACH PROGRAMS

4.1 *Technology transfer*

The primary goal of the RMRHSRC Training and Technology Transfer Program is to provide effective training and technology transfer resulting in the progression of ideas from the laboratory to application.

The purpose of the Training and Technology Transfer Program is to support the mission of the Center by: (1) promoting organizational linkages, (2) ensuring outreach to industry, communities, and states, (3) facilitating the use of innovative means of information transfer, (4) supporting investigations at the interface of disciplines, (5) exploiting opportunities in science, engineering, and technology where the complexity of the research needs requires the advantages of scope, scale, duration, equipment, and facilities, and (6) capitalizing on diversity through involvement of under-represented groups. The Center will facilitate the progression of laboratory research to field applications by supporting activities that result in idea generation, information transfer, laboratory and pilot-scale testing, field demonstrations and applications.

4.2 *Technical Outreach and Service for Communities (TOSC)*

The goal of the TOSC program is to provide educational resources to help citizens gain a better understanding of the environmental problem, allowing them to make informed decisions and participate more fully in activities affecting their communities. This project will meet the following program objectives: (1) creating technical assistance materials tailored to the identified needs of a community, (2) informing community members about existing technical assistance materials, such as publications, videos, and Web sites, (3) providing technical in-

Table 1. Investigators and participants of the Rocky Mountain Regional Hazardous Substance Research Center.

Investigator/Participant	Discipline and/or Expertise Areas
Colorado State University:	
Rajiv Bhadra	Chemical & Bioresource Engineering
Brian Bledsoe	Sediment transport
Kenneth Carlson	Environmental Engineering
William Clements	Fish and Wildlife Biology
Nancy DuTeau	Microbiology
Pierre Julien	Sediment transport
Kenneth Reardon	Chemical & Bioresource Engineering
Elizabeth Pilon-Smits	Biology
Charles Shackelford	Geoenvironmental Engineering
Chester Watson	Sediment transport
Sandra Woods	Environmental Engineering, outreach
Ray Yang	Environmental Health Sciences
Colorado School of Mines:	
Dianne Ahmann	Environmental microbiology, arsenic geochemistry
Ronald Cohen	Zoology, outreach
Linda Figueroa	Environmental engineering microbiology
Bruce Honeyman	Environmental geochemistry, surface interactions
Tissa Illagansekare	Meso-scale environmental testing, remediation
Junko Munkata Marr	Environmental microbiological engineering
Donald Macalady	Aquatic chemistry, metal/organic interactions, AMD
Harold Olsen	Geotechnical Engineering
James Ranville	Particle size effects, AMD, surface chemistry
Robert Siegrist	Environmental Engineering, remediation technology
Thomas Wildeman	Constructed wetlands for AMD, outreach
Other Investigators/Participants:	
George Aiken (USGS)	Natural organic matter, analytical geochemistry
Katherine Banks (Purdue U.)	Phytoremediation, bioremediation, wastewater treatment
Craig Benson (U. of Wisconsin)	Geoenvironmental Engineering
David Blowes (U. of Waterloo)	Geochemistry, permeable reactive barriers
John Garbarino (USGS)	Analytical chemistry
Joe Meyer (U. of Wyoming)	Environmental toxicology, zoology
Danny Reible (Louisiana State U.)	Environmental transport and mechanics, turbulence
Paul Schwab (Purdue U.)	Chemistry of heavy metals in soil
Otto Stein (Montana State U.)	Environmental Engineering (wetlands)
Richard Wanty (USGS)	Environmental geochemistry, AMD remediation
John Westall (Oregon State U.)	Geochemical models, inorganic and redox chemistry

formation to help community members become active participants in cleanup and environmental development activities, (4) providing independent and credible technical assistance to communities affected by hazardous substance problems, (5) reviewing and interpreting technical documents and other materials for affected communities, and (6) sponsoring workshops, short courses, and other learning experiences to explain basic science and environmental policy related to hazardous substances.

4.3 *Technical Assistance to Brownfields (TAB)*

Brownfields are defined by the Environmental Protection Agency as "abandoned, idled, or under-used industrial and commercial facilities where expansion or redevelopment is complicated by real or perceived environmental contamination." The TAB Program provides technical assistance to communities and other stakeholders with an ultimate goal of redeveloping brownfields sites.

Tailings and Mine Waste '02, © 2002 Swets & Zeitlinger, ISBN 90 5809 353 0

Mining and the environment: The issues that steer the Region 8 Hazardous Substance Research Center

T.R. Wildeman
Department of Chemistry and Geochemistry, Colorado School of Mines, Golden, CO 80401, USA

C.D. Shackelford
Department of Civil Engineering, Colorado State University, Fort Collins, CO 80523, USA

ABSTRACT: Many issues are brought to bear on the treatment of mining wastes. Some of these issues, such as cost, apply to every possible method of treatment. Others such as whether to concentrate on abandoned or active operations are somewhat mutually exclusive. In establishing an action plan for the EPA Region 8 Rocky Mountain Regional Hazardous Substances Research Center, many of these issues were considered. The issues that were considered, the conclusions that were made, and how these conclusions affected the decision of which research projects to fund is the subject of this paper.

1 INTRODUCTION

Analysis of how environmental concerns affect mining reveals a number issues; some of these issues form pairs that are somewhat mutually exclusive whereas other issues span across the whole subject. Cost is an issue that drives all decisions concerning mining and the environment, whereas the question of whether to concentrate efforts on active or abandoned operations can be considered as a somewhat mutually exclusive choice. In any environmental project, there are political, social, scientific, and engineering issues that will have to be considered in the final action plan. Also, there is the question of whether environmental problems brought about by mining are regional, national, or international issues. Finally, the question of whether the major environmental problems concern the atmosphere, water, or the earth is an issue when limited funds are applied to a major problem.

Because we have been awarded funding by the U.S. Environmental Protection Agency (EPA) for a major research Center that concentrates on mining and the environment, i.e., the Rocky Mountain Regional Hazardous Substances Research Center (RMRHSRC), we have had to assess these issues and decide which ones will control the research and outreach activities of the Center. Certainly, our decisions on which areas of this topic of mining and the environment deserve our concentrated efforts can be modified or changed; however, at this time it is necessary to make decisions so that our projects have an appropriate focus. In the case of the outreach and community service efforts of the Center, all hazardous substances will be addressed. However, the EPA specifically requested that this Center, located in

Region 8 of the 10 national EPA regions, concentrate on the environmental issues related to mining wastes. This paper presents these issues and gives our beginning opinions on how they impact on the goals and operation of this new Hazardous Substances Research Center.

2 THE ISSUES

Cost is a dominant issue particularly because of the physical size of some mining wastes. Because many of the sites that have been impacted by organic wastes are U. S. Department of Defense and Department of Energy sites, development of innovative and low cost methods of treatment at these sites has been supported by significant Federal funding initiatives. However, the Federal government is not involved in any of the priority pollutant sites found in Colorado, Utah, Montana, Wyoming, South Dakota, and North Dakota, the states that make up Region 8. So, state and local governments are going to be required to some extent to initiate treatment efforts, and these entities demand a low-cost solution. Also, the record so far shows that high costs paralyze most current attempts at remediation. Consequently, cost limits both the scientific questions that can be answered with respect to a given problem as well as the level of technology that can be implemented to solve the problem.

Although this is a regional center, treatment of environmental problems related to mining is obviously a global issue. Also, some academic, governmental, and private agencies beyond this region and nation have a great track record on working on the treatment of mining wastes. The MEND Program of

Canada and the Acid Drainage Technology Institute of the University of West Virginia are two examples. Also, political restrictions, such as the good Samaritan problem within Colorado, sometimes create hurdles that must be overcome when attempting pilot-scale and full-scale treatment projects within the region. Consequently, in preparing our proposal and now in carrying out the objectives of the RMRHSRC, our approach will be global. At least two of the four research projects have investigators outside of the region. The EPA dictates that outreach activities be coordinated across the EPA Regions. However, for the technology transfer and outreach of this Center to be successful, the Center also must work with other entities outside of Colorado and Region 8 that have already been concentrating on the treatment of mining wastes.

In the case of mining wastes, social and political issues will eventually drive most treatment efforts. However, this Center will concentrate on the scientific and engineering aspects of the problem. The objective will be to make the answers to all the scientific and engineering questions to be clear and complete enough that the social and political issues can be played out on a well-defined scientific and engineering field.

With respect to concentrating on active versus abandoned mining operations, the plan will be to start studies on abandoned operations and then to convince the mining industry that solutions developed through the research activities can be transferred to active operations. Since the Summitville disaster of December 1992, the permitting procedures for all states appear to be so extensive that all contingencies are taken into consideration. In particular, a complete closure plan has to be included in any new mining permit, and companies understand that their liability extends through the closure of the operation. Currently, opening and operating a mine depends on environmental regulations that are based on reasonably good science and engineering. Thus, the Center will begin operations concentrating on abandoned mining lands (AMLs). There also are some issues related to current operations that certainly deserve study, such as whether the models used to predict pit water chemistry are accurate, and how to bring a continuing operation that was started in the 1970's or 80's into compliance with the current regulations that relate to new operations. However, AMLs comprise the majority of the problems in Region 8 and appear to be a more fruitful area for the application of science and engineering for developing inexpensive and innovative treatment strategies. Also, bringing a current operation into compliance with more modern regulations means that treatment is done after contamination has occurred. Consequently, this can be considered to be similar to the treatment of an abandoned operation.

When considering whether the most important problem is contamination of the air, water, or earth, the water and the earth command the most attention. The question of what to do with abandoned tailings and waste rock piles represents the largest physical problem in any of the national priority pollutant sites that are in the EPA list. The problem of acid mine drainage (AMD) contaminates more miles of surface streams than any other type of contamination by hazardous substances. Consequently, these materials will command the research attention of the RMRHSRC.

3 RELATION OF THE ISSUES TO THE RESEARCH PLAN

In the initial year of the Center, four research projects were chosen for funding. How do these research projects relate to the examination of the issues presented above? All the projects concentrate on science and engineering rather than politics and sociology. Also, all of the projects are not tied to a particular mining site, so the results could be transferred to any specific project. In the respect that the transfer of the technology can be to any other site, the issue of focusing on abandoned sites is not obvious. However, all of the projects deal with questions that have to be answered at any abandoned site. An examination of how other issues are involved in each specific project follows.

The first research project deals with the availability and transformations of arsenic and selenium in natural environments where organic matter is an important component. At first glance, this project seems unrelated to mining problems. However, this is a key area where the science is lacking to make the proper decisions on how treatment should proceed. For the heavy metals such as Cu, Zn, Pb, and Cd, principles of geochemistry and microbiology can be applied to treatment situations with a good degree of certainty. For the most part, the metals remain in one positive oxidation state and the precipitation sequence for carbonates, hydroxides, and sulfides is well known. However, for As and Se there are multiple oxidation states. In some of these oxidation states, removal is quite difficult. Furthermore, both these elements can occur in volatile compounds and their fate and transport in a removal system confuses the issue. So, for these two contaminants, fundamental scientific questions about speciation, transformation, and fate in biogeochemical systems have to be answered. Currently, we are designing treatment systems for arsenic and selenium. However, if pressed, we cannot answer questions about the ultimate fate of arsenic and selenium in these systems.

There are thousands of tailings and waste rock piles throughout the Western United States that are the relics of earlier mining operations. The questions that even the casual observer asks are: How are all those piles going to be treated, does every one of those piles have to be treated, and are there really contaminants moving from those piles into the surface and ground water? At present, the answers to the questions of which piles are contributing the most to the contaminant load in a watershed and which contaminants are coming from which pile are not known. In addition, material may move from the pile only during severe storms, and the material that moves may not dissolve but rather travel to a stream as suspended solids. However, upon reaching and entering the stream, the continuous contact of the material with the water may result in dissolution and thereby contribute to the contaminant load in a watershed. The answers to these questions are also unknown. Project 2 on the fate and transport of metals and sediment in surface water addresses these questions. Ultimately, because the cost is so great, treatment of waste piles will only be accomplished when there is a sufficient scientific basis to justify the cost. That basis can only developed when good models for the fate and transport of contaminants in waste pile are developed.

Over the past decade, passive treatment is one method for the removal of contaminants from AMD that has been extensively investigated. Passive treatment meets the criteria of being an innovative and low-cost solution. Because they require no utilities and do not need constant attention, passive treatment systems are especially attractive for treating contaminants in mine drainage at abandoned sites. Because the objective in treatment is to fix the problems at the lowest possible cost, the criteria that are used in the design of these systems have never been thoroughly investigated. Because designs are made using incomplete criteria, failure sometimes occurs and the reasons for failure are not understood. However, enough is understood so that the weak points in the design are known. So, these areas need to be investigated so that passive treatment systems can be made more reliable and efficient. These systems rely on the activity of microbes to produce products such as sulfide and carbonate that form precipitates with the metal contaminants. A key question is just what can be added to the system to increase the activity of the microbes? Another key

question is whether some of the contaminants are toxic to the microbes? Also, there is a nagging question of whether the organic material that is used in these systems causes the dissolution of the contaminants through complexation. In Project 3, these and other questions concerning passive bioreactor systems will be answered so that this promising method of water treatment can be made to be more reliable and efficient.

Finally, there are the skeptics on both sides of a mine treatment project. One side says, "How do you know that you really have to carry out this expensive remediation?" This side especially focuses on waste rock piles, and sometimes argues that removal means destruction of part of our mining heritage. On the other side says that, if you don't remove all the piles and clean all the sediment from the stream, then achieving a natural ecological condition is impossible. Obviously, both sides have to be convinced that a middle ground exists and that there is a good scientific basis for defining what is the middle course of action. Project 4 addresses these specific issues, and deals with the questions of deciding whether a contaminant in a stream is really toxic to the aquatic ecosystem and determining under what conditions that toxicity is more or less severe. If good information is available for answering these questions, then using this data along with the answers to the questions raised in Project 2 can be used to pinpoint what waste rock piles should be removed to provide the greatest benefit to watershed restoration.

4 SUMMARY

Hopefully, the above exposition helps to inform how Rocky Mountain Regional Hazardous Substances Research Center (RMRHSRC) will go about solving the problems associated with the remediation of mining waste sites. Also, the people associated with the RMRHSRC have a strong desire for people to understand the issues that were considered in formulating the initial action plan. To the extent that important issues have been overlooked, the intent of this paper is to solicit responses from the stakeholders so that such issues can be addressed and included with the Center activities in the future.

Tailings and Mine Waste '02, © 2002 Swets & Zeitlinger, ISBN 90 5809 353 0

Redox transformations, complexation and soil/sediment interactions of inorganic forms of As and Se in aquatic environments: Effects of natural organic matter

D.L. Macalady
Department of Chemistry and Geochemistry, Colorado School of Mines, Golden, CO 80401, USA

D. Ahmann
Department of Environmental Science and Engineering, Colorado School of Mines, Golden, CO 80401, USA

J. Westall
Department of Chemistry, Oregon State University, Corvallis, OR 97331, USA

J. Garbarino
United States Geological Survey, USA

J. Meyer
Director of Red Buttes Experimental Station, University of Wyoming, Laramie, WY 82071, USA

ABSTRACT: Waters and suspended sediments bearing inorganic forms of arsenic and selenium occur widely and are especially prevalent in sites impacted by mining. These systems present significant concerns for human and ecosystem health and most, if not all, contain natural organic matter (NOM). The chemical and physical properties of NOM suggest its involvement in several critical processes that affect the behavior of As and Se, including oxidation-reduction reactions, competitive interactions affecting sorption, and formation of metal-organic (bridging) complexes. NOM is expected to alter the mobility, transformations, bioavailability, and toxicity of As and Se, and therefore to play an important role in the design of methods to mitigate contamination problems.

1 INTRODUCTION

We seek to understand and quantify the nature of these interactions and to delineate the environmental conditions under which they are expected to occur. Further, the research will assess the extent to which NOM influences the transport, transformations, bioavailability, and toxicity of As and Se in aquatic systems as a function of system characteristics. To meet these objectives, we will test the following general hypotheses:

- NOM accelerates redox transformations of As and Se, either directly or by acting as an electron shuttle, to an extent determined by properties of the NOM and the system.
- NOM increases the mobility of Se and As through complexation and interference with sorption onto metal oxide surfaces.
- NOM alters As and Se bioavailability and toxicity through its influence on redox reactions, sorption/desorption, and/or aqueous complexation; the net effects represent a balance among these mechanisms.
- Microbial transformations of As and Se are influenced by the abundance and characteristics of the NOM present, as well as the aqueous geochemistry of the system.

2 PRELIMINARY DATA AND WORK IN PROGRESS

We have recently examined several interactions among NOM samples and inorganic arsenic speciation using NOM samples collected from the Rio Negro in Brazil, the South Fork of the Forty-Mile River in Alaska, a stream in the Upper Peninsula of Michigan, USA, monitoring well 6 (MW-6) in a wetland near Leadville, Colorado, the Suwannee River in Georgia, and the Inangahua River in New Zealand.

Investigations of aqueous interactions between NOM and arsenic revealed first that NOM samples can complex extensively with inorganic As; second, that they can cause significant oxidation of As(III), as well as some reduction of As(V); and third, that both complexation and redox activities vary greatly with the origin of the NOM.

NOM, arsenate, and arsenite all sorb strongly onto metal oxide surfaces, and sorption appears to be a primary factor limiting As mobility in natural environments. Because the number of surface sites in a particular environment is finite, these species have significant potential to interfere with each other, altering their respective mobilities in potentially important ways. Additional experiments therefore investigated the competition between NOM and As for

13

adsorption onto hematite surfaces. When hematite was pre-equilibrated with NOM, both arsenate and arsenite effectively displaced the NOM upon their addition. This scenario represents the introduction of arsenic into a pristine environment with moderate NOM content, and indicates that NOM pre-sorbed onto surfaces would not completely obstruct arsenic sorption onto metal oxide surfaces. Conversely, when hematite was pre-equilibrated with arsenate or arsenite, NOM also displaced sorbed arsenic, in addition to reducing As(V) and oxidizing As(III). This result indicates that introduction of NOM into a highly-contaminated environment, as in the construction of a wetland to treat acid mine drainage, might significantly mobilize As sorbed onto metal oxides.

The results of these experiments show that NOM has considerable potential to influence As speciation and mobility in natural environments. Additional support for our hypotheses is found in the recent discovery of a strong correlation between aqueous arsenic concentrations and DOC levels in groundwater well samples from Bangladesh. This observation indicates that further investigation of the effects we have found, with both arsenic and related oxyhydroxides such as those of selenium, is strongly warranted, as it may reveal that these interactions have profound impacts at much larger scales.

Investigations of selenium and arsenic ecotoxicology are also underway at Oregon State University and at Red Buttes Experimental Station, respectively. Selenium impacts are being studied in conjunction with the Kennecott Utah Copper Corporation (KUCC) mining company in Magna, Utah. In 1995, KUCC investigated the possible impacts of Se from their operations on surrounding ecosystems, and completed an initial study characterizing the ecological food webs as well as total Se levels in water, in sediments, and at various trophic levels. Subsequently, a two-year (04/00 - 06/02) field-oriented research project was commissioned with OSU to investigate Se bioavailability and pathways of uptake through the wetland ecosystem.

3 APPROACH

Our experimental approach is to use wellcharacterized NOM samples from various environments within and outside EPA Region 8 and to test these NOM samples for their abilities, under standardized conditions, to affect the processes hypothesized above. Using techniques already developed in our laboratories, we will evaluate the ability of each NOM sample to affect As and Se abiotic and microbial redox transformations, influence surface and aqueous complexation, and to alter bioavailability and toxicity. These data will be used to develop conceptual models of the roles of NOM in As and Se biogeochemical processes, and the models will be used for predictions of As and Se behavior at several field sites. Field work and experiments using field samples will be used to modify the models as necessary and provide recommendations for the inclusion of the roles of NOM in remediation and abatement plans for sites contaminated with As and/or Se.

4 SPECIFIC OBJECTIVES

Because our previous research has shown that the chemical behaviors of NOM samples from different natural sources depend upon sample sources and histories, we will complete the research objectives below in the process of testing the project hypotheses.

- NOM characterization. We will collect water samples from sites within and beyond EPA Region 8 and analyze the waters and associated NOM for pH, alkalinity, conductivity, metals content, anion character and quantity, and total organic and inorganic carbon. Elemental composition, acidity, aromatic content, FTIR spectrum and average molecular weight of the NOM in each sample will also be determined. These samples, as well as selected members of our existing library of well-characterized NOM samples, will be used in subsequent experiments.

- Aqueous complexation. We will determine the ability of each NOM sample to form aqueous complexes with As and Se. For selected NOM samples, we will also measure the effect of additional Fe(III) and/or Al chelated to the NOM on the ability of the NOM to form As and/or Se complexes in solution.

- Sorption competition. We will measure the ability of each NOM sample to compete with As and Se for sorption onto hematite and gibbsite. Through concomitant redox speciation measurements we will estimate the extent to which each NOM sample accelerates redox transformations of As and Se.

- Redox transformations. We will measure the ability of selected NOM samples to accelerate redox transformations in the presence or absence of microorganisms and external electron donors and acceptors.

- Bioavailability and toxicity. We will investigate the influences of selected NOM samples on As and Se bioavailability and toxicity. Toward that end, we will refine analytical methods for determining inorganic and organic Se speciation in waters, sediments, and biota, and we will use these and established techniques to determine As and Se speciation in organisms exposed to these contaminants in laboratory microcosms containing NOM. Results will be interpreted in light of known influences of NOM on As and Se complexation, adsorption, and redox behavior.

- Statistical analysis. We will use techniques such as principle component analysis, ANOVA, and linear regression to reveal correlations among NOM characteristics, aqueous geochemistry, and influences on complexation, sorption, redox reactions, bioavailability, and toxicity of As and Se.
- Conceptual modeling. Finally, we will formulate a conceptual model describing the interactions among NOM, As, and Se that are pertinent for remediation design. We will test this model with experiments from field sites for which predictions of As and/or Se behavior can be formulated and tested based on the site properties and the conceptual model. Iterative processes will enable modification and improvement of the model based on field measurements and observations.

5 EXPECTED RESULTS

This project will generate 3 primary categories of results. The first of these will be a set of quantitative assessments describing chemical interactions of NOM, As, and Se, including the ability of NOM to form aqueous complexes with these metalloids, to interfere with their sorption onto metal oxide surfaces, and to promote their redox transformations. The importance of specific NOM characteristics to these chemical interactions will also be revealed. The second category will be a quantitative description of the ecotoxicological impacts of the chemical NOM/As/Se interactions that are observed, as well as the importance of specific NOM characteristics to these effects, and the third result category will be the conceptual models that are developed from conceptual and statistical evaluation and synthesis of the results of categories 1 and 2. The ultimate expected result is the provision of a body of knowledge that will clearly and directly inform the design of passive treatment systems for acid mine drainage.

Tailings and Mine Waste '02, © 2002 Swets & Zeitlinger, ISBN 90 5809 353 0

Fate and transport of metals and sediment in surface waters

P.Y. Julien, B.P. Bledsoe & C.H. Watson
Department of Civil Engineering, Colorado State University, Fort Collins, CO 80523, USA

ABSTRACT: This study focuses on surface water and sediment transport, with an emphasis on the fate and transport of metals in rivers from mining wastes. The main thrust of this project is to: 1) develop a predictive scientific methodology for watershed rehabilitation strategies and Total Maximum Daily Loads (TMDLs) that address both point and diffuse sources of metal pollution; and 2) improve and develop computer modeling tools for the simulation of point-source and non-point source metals and fine sediment contamination in surface waters. The ultimate goal of the research is to improve our mechanistic understanding of the interaction between heavy metals and fine sediment.

1 INTRODUCTION

Sediment pollution is ubiquitous yet difficult to identify and attenuate. Excessive sediment or "silt-tion" from nonpoint sources impairs water quality in over eighteen percent of surveyed river and stream miles in the United States (EPA 1996, Foster et al. 1995). Water quality degradation via sedimentation is probably even more pervasive than reported due to difficulties associated with monitoring. Due to practical constraints, sediment flux is often measured on weekly, monthly, or even less frequent intervals without regard to the timing of geomorphically important processes and stream flow episodes. Such protocols and the inadequacy of actual sediment transport models limit our ability to characterize how sediment and associated pollutants may induce chronic degradation of aquatic ecosystem health over long periods of time.

The Rocky Mountain streams of Colorado provide an excellent opportunity to investigate heavy metal contamination from historic mining operations and the interactions between metals and fluvial sediments. Heavy metal pollution is recognized as a major environmental problem in Rocky Mountain streams (Clements et al. 2000, Herlihy et al. 1999). The fate and transport of metals in western streams is closely related to the transport of fine sediments in suspension. Contaminants can be stored in sediment deposits. The adsorption properties of metals vary with the size of sediment particles transported in streams and rivers. Modeling contaminant transport can be enhanced when coupled with numerical models that calculate sediment transport by size fractions. Adsorbed metal fate and transport is therefore inextricably linked to the processes of erosion,

sediment transport and sedimentation in the bed and banks of rivers, as well as on the flood plain. The processes of aggradation and degradation in rivers, bank erosion and lateral river migration cause changes in hydraulic geometry of natural rivers (Simons and Li 1982, Kauffmann et al. 1999). This also impacts the biological integrity of the aquatic habitat, particularly the benthic macro-invertebrate population (Lenat et al. 1981, Clements et al. 2000).

The development of watershed rehabilitation strategies and TMDLs for both point and diffuse sources of metal pollution requires improved computer modeling tools for characterizing the fate and transport of metals. Modeling contaminant transport can be enhanced when coupled with numerical models that calculate sediment transport by size fractions. Computer models should support decision makers in the development of watershed and stream rehabilitation strategies.

2 OBJECTIVES

The specific objectives of the research are to:
1) Develop a predictive scientific assessment for improving the effectiveness of watershed rehabilitation strategies and TMDLs that address both point and diffuse sources of metal pollution from mining activities in different stream types and geomorphic contexts; and
2) Improve computer models for the characterization of fate and transport of metals in western streams and rivers. Models will explicitly include interactions between sediment and metal flux, channel type and morphology, and flow regime.

One of the main hypotheses to be tested is the relative importance of riffles and pools in the detention and storage of contaminants and as a potential added mechanism for the dispersion of contaminants in mountain streams. Other hypotheses include the possibility of using intermittent releases for more rapid dispersion of contaminants in mountain streams. With the use of numerical models, one can determine the duration and frequency of intermittent releases required to maintain contamination levels below design levels at specific locations downstream of point sources.

3 BACKGROUND

The Federal Water Pollution Control Act designed the Total Maximum Daily Load (TMDL) to insure that all sources of pollutant loading are accounted for in devising strategies to meet water quality standards. The TMDL represents an estimate of the greatest amount of a specific pollutant that water bodies or streams can receive without violating water quality standards. This estimate includes a margin of safety, a waste load allocation for point sources, and a waste load allocation for non-point sources. Section 303(d) of the Clean Water Act requires states to identify waters that do not or are not expected to meet applicable water quality standards with technology-based controls alone. This identification of water quality-limited waters is presented in a document called the 303(d) List, updated biennially.

The states are required to prioritize these waters, analyze the causes of the water quality problem, and allocate the responsibility for controlling the pollution. This analysis and allocation is called the TMDL Process, and results in the determination of: 1) the amount of a specific pollutant that a water body can receive without violating water quality standards; and 2) the apportionment to the different contributing sources of the pollutant loading. The TMDL approach is a rational method for developing an integrated pollution reduction strategy for point and non-point sources. TMDL development includes five basic steps: 1) select the pollutant to consider; 2) estimate the water body assimilative capacity; 3) identify the contribution of that pollutant from all significant sources; 4) determine the total allowable pollutant load; and 5) allocate the allowable pollution among the sources so that water quality standards can be achieved.

Heavy metals usually refer to metals between atomic number 21 and 84, besides aluminum, selenium and arsenic. Metals are found naturally and their presence is greatly enhanced by mining and industrial activities. The fate of metals is tied to the fate of solid matter. Therefore, the processes of sorption and sediment settling/resuspension significantly affect the fate of metals. It is important to assess the background levels of metals from natural sources besides the anthropogenic sources. Metals are conservative but can be found in different chemical forms that can exhibit different transport and fate besides different toxicity levels. The physical adsorption of metals to solid surfaces requires a quantitative evaluation of the very fine sediment load, also called washload in natural streams.

The concept of washload refers to the quantity of fine sediment not found in large quantities in the bed of a stream (Vanoni, 1975). The washload originates from the detachment of soil particles during rainfall events on the entire watershed. Quantitative predictions of washload do not depend on the hydraulic characteristics of the channel flow but depend on the supply of sediment from upland areas. In mountain streams, the major fraction of the sediment transported most of the time is washload.

The quantitative evaluation of washload is possible on a long-term basis using mean annual quantities of upland erosion form models like the Universal Soil-Loss equation (USLE). The USLE is based on a mean annual value of rainfall erosivity, soil erodibility, topography in terms of runoff length and slope, a cropping-management factor, and a conservation practice factor. Molnar and Julien (1998) developed a computer model for the long-term analysis of washload using Geographic Information System (GIS) data. Washload can also be quantified on a single storm basis from the GIS-based information on the watershed topography, soil type, vegetation type and density, land use and a spatial and temporal distribution of rainfall intensity obtained from a set of raingages or radar data. Julien et al. (1995) developed the model CASC2D for the simulation of surface runoff from rainstorms using GIS and radar data. Recent developments included the calculation of washload with CASC2D-SED by Johnson et al. (2000). A hybrid model to determine long-term estimates of soil erosion from a simulated 10-year sequence of rainfall precipitation has been linked to the WEPP model. The model has been developed for the simulation of the transport of radionuclides at Rocky Flats (unpublished).

Other modeling approaches in streams are based on the processes of advection from flow velocity, turbulent diffusion or mixing in the transversal direction of large rivers and dispersion of material in the downstream direction as a result of the logarithmic velocity profile in the downstream direction and vertical mixing. The determination of mixing and dispersion coefficients from the hydraulic characteristics can be obtained from Fischer et al. (1979) and useful algorithms can be found in Chapra (1997). The calculation of settling and resuspension is performed with the Exner equation along with the settling properties of sediment of different sizes and densities. The properties of coarse and fine sediment are described in Julien (1995). Standard procedures for the calculation of aggradation and degradation for coarse material where sediment transport corresponds to the sediment transport capacity of a river can be found in models like HEC-6, FLUVIAL-11, GSTARS, etc. In the case of fine sediment transport where sediment supply determines the actual sediment transport rate, algorithms that account for the supply and settling properties of sediment for different size fractions are required.

Advection-dispersion models are subjected to the criteria for model stability and consistency, which determine convergence. The Courant number controls the advection process and with the dispersion number, the grid Peclet number controls the nature of the simulation as to whether it is primarily advective or dispersive. Many numerical schemes like the FTBS (forward in time, backward in space) induce numerical dispersion, which should be avoided or controlled. Higher order schemes are possible under the constraint that additional boundary conditions are required, and that the grid Peclet number is sufficiently large to avoid numerical "leakage" in order to satisfy conservation of mass. Other algorithms include the Crank-Nicholson, McCormack, Preissman.

The QUAL2E software package is currently the most widely used computer model for simulating stream-water quality. It is capable of simulating up to 15 constituents in dendritic stream that are well mixed both vertically and laterally. Rooted at the Texas Water Development Board in the 1970s, it has been under development at Camp, Dresser and McKee and updated several times, e.g. by Roesner et al. 1981.

4 METHODS AND APPROACH

The proposed research would focus on the fate and transport of metals attached to fine sediment loads in rivers. The analysis would be two-fold: 1) predictive scientific assessment for developing watershed rehabilitation strategies and TMDLs for both point and diffuse sources of metal pollution from mining activities; and 2) improved computer modeling tools for characterizing fate and transport of metals in western streams and rivers. A predictive scientific assessment is a flexible, changeable mix of small mechanistic models, statistical analyses and expert scientific judgment. The analysis focuses on the simulation of the transport of point sources and non-point sources of fine sediment contaminated with metals in surface waters. The modeling effort will be calibrated and tested with field data at a few sites on the 303(d) List. The resulting tools will be designed for practical application and decision making in the rehabilitation of metal impaired streams in Region 8.

In the first part of the analysis, three to five streams on the 303(d) list will be reviewed and classified according to the main type of pollutant loading. Three faculty members will share their expertise with a graduate student. The purpose is to identify the most common source of metal pollution, point source or diffuse source, impact on stream water quality, impact on stream stability and river bed aggradation and degradation as a possible source or sink of metals and contaminants. Specific sites will be identified in collaboration with EPA and the Water Quality Control Division of the Colorado Department of Public Health and Environment (CDPHE). These selected sites will be subjected to site visits, data acquisition and monitoring for modeling purposes. Specific consideration will be given to point and diffuse sources of contamination, channel geometry, sediment sizes and riparian characteristics, bedload versus suspended load, metal adsorption by size fraction. In-stream characteristics also include geomorphic features such as riffles and pools, the residence time of sediment and contaminants in pools and riffles, backwater areas, bedforms and large sand/gravel bars, and management issues relative to metal deposition and the flushing of contaminated sediments stored in pools, bars, river banks and reservoirs. The impact of periodical sediment flushing on biotic life cycles cannot be understated. The development of management strategies such as seasonal sediment storage and flushing, the use of residual pool volume, and protection of aquatic habitat for endangered species requires a predictive scientific assessment of the sources and fate of the contaminants.

Appropriate computer models need to be identified and/or developed for specific analysis of contaminant transport and loading in streams including TMDLs. As a result, emphasis is being placed on the development and use of computer models for the analysis of TMDLs for heavy metals. Simulation examples of heavy metal/ fine sediment transport including advection, turbulent diffusion, and dispersion of metals/fine sediment for both point and non-point sources at a few sites from the EPA Region 8 will be evaluated.

5 EXPECT RESULTS

This proposed research integrates the fields of stream monitoring and computer modeling of contaminant transport in mountain streams. The two most important expected benefits of this research are: 1) an improved understanding of the mechanics of heavy metal fate and transport in mountain streams; and 2) the development and validation of numerical models for the simulation of advection, mixing and dispersion of fine sediment and heavy metals in mountain streams. Field measurements will be used to calibrate and test numerical models at several sites where the water quality has been altered by mining waste contamination. The proposed study supports the U.S. EPA, the U.S. Bureau of Reclamation, and the State of Colorado in a large-scale restoration project to assess recovery of Rocky Mountain streams from heavy metal pollution. In the case of the Arkansas River, the primary goal is to reduce metal concentrations and to restore a productive brown trout fishery. By integrating the fields of field monitoring and numerical modeling, we aim at the development of strategies for the storage, detention and dispersion of contaminants in mountain streams. Funding is requested to support two graduate students to be supervised by three academic faculty members for a period of two years.

This proposal will assist the States within EPA Region 8 in providing tools for the analysis and allocation of pollutant loading in the TMDL Process. The proposed analysis will focus on metals including Fe, Mn, Zn, Pb, Al, Cu, Cd, Ag, As and sediment. The intent is to provide tools and computer models to quantify the concentration, fluxes and load of pollutants and fine sediment in contaminated streams. The emphasis will be on an assessment of the relative contributions of both point and non-point sources of contaminants. Models that can simulate the load of fine sediments and the processes of advection, turbulent diffusion and longitudinal dispersion are prescribed. Site-specific calibration and testing of models is also contemplated under this proposal. Possible sites for the analysis of heavy metals include Arkansas River, Whitewood Creek, Alamosa River of the Rio Grande River Basin, Clear Creek of the South Platte River Basin, and French Gulch of the Upper Colorado River basin. In terms of fine sediment load, the reach of the North Fork of the Cache la Poudre downstream of Halligan Reservoir is most interesting given the proximity of the site and the tremendous database available for the analysis.

REFERENCES

Chapra, S.C. 1997. *Surface Water-Quality Modeling*, McGraw-Hill, New York.

Clements, W.H., D.M. Carlisle, J.M. Lazorchak, and P.C. Johnson 2000. Heavy metals structure benthic communities in Colorado mountain streams. *Ecological Applications*, 10(2):626-638.

Fischer, H.B., List, E.J., Koh, R.C., Imberger, J., and Boorks, N.H. 1979. *Mixing in inland and coastal waters,* Academic Press, New York.

Foster, I.D., A.M. Gurnell, and B. Webb, eds. 1995. *Sediment and water quality in river catchments*. John Wiley and Sons Ltd. Chichester, England. 473 pp.

Herlihy, A.T., J.M. Lazorchak, F.H. McCormick, D.J. Klemm, M.E. Smith, T. Willingham, and L.P. Parrish 1999. Quantifying the regional effects of mine drainage on stream ecological condition in the Colorado Rockies from probability survey data. *EMAP Symp. of Western Ecological Systems: Status, Issues, and New Approaches,* April 6-8, San Francisco, CA.

Johnson, B.E., Julien, P.Y., D.K., and Watson, C.C. 2000. The two-dimensional upland erosion model CASC2D-SE. *Journal of the American Water Resources Association*, AWRA, 36(1): 31-42.

Julien, P.Y. 1995. *Erosion and Sedimentation*, Cambridge University Press, New York.

Kauffmann, P.R., P. Levine, E.G. Robison, C. Seeliger, and D.V. Peck 1999. *Quantifying physical habitat in wadeable streams.* EPA/620/R-99/003, U.S.E.P.A., Washington, DC.

Lenat, D.R., D.L. Penrose, and K.W. Eagleson 1981. Variable effects of sediment addition on stream benthos. *Hydrobiologia*, 79:187-194.

Molnar, D.K. and Julien, P.Y. 1998.Estimation of upland erosion using GIS, *Journal of coputers and Geosciences*, 24(2): 183-192.

Simons, D.B. and R.M. Li. 1982. *Engineering analysis of fluvial systems. Simons*, Li, and Associates, Inc. Colorado.

USEPA. 1996. *Biological Criteria: Technical guidance for Streams and Small Rivers.* Revised ed. EPA-822-B-96-001, May 1996. US Environmental Protection Agency, Office of Water, Washington, DC.

Vanoni, V.A. 1975. *Sedimentation engineering.* ASCE, 54. New York.

Metal removal capabilities of passive bioreactor systems: Effects of organic matter and microbial population dynamics

L. Figueroa & D. Ahmann
Department of Environmental Science and Engineering, Colorado School of Mines, Golden, CO 80401, USA

D. Blowes
Department of Earth Sciences, University of Waterloo, Waterloo, Ontario, Canada N2L 3G1

K. Carlson, C. Shackelford & S. Woods
Department of Civil Engineering, Colorado State University, Fort Collins, CO 80523, USA

N. DuTeau
Department of Microbiology, Colorado State University, Fort Collins, CO 80523, USA

K. Reardon
Dept. of Chemical and Bioresource Engineering, Colorado State University, Fort Collins, CO 80523, USA

T. Wildeman
Department of Chemistry and Geochemistry, Colorado School of Mines, Golden, CO 80401, USA

ABSTRACT: The ability of passive bioreactor systems (anaerobic wetlands, passive bioreactors and permeable reactive barriers) to remove metals from mine drainage has been demonstrated. However, problems arise with the performance of some passive bioreactor systems. The overall goal of this project is to improve process performance of these systems. This will be done by first evaluating the effect of organic matter characteristics on the types, rate of growth and sustainability of microbial populations. Also, studies will be conducted on the effect of organic matter on metal speciation, complexation, and removal. The project tasks are: 1. Solid phase organic and inorganic material characterization (physical, chemical and microbial), 2. Batch studies on the effect of different PBR mixtures, 3. Bench-scale studies on the effect of selected substrate mixtures, 4. Field sampling of anaerobic wetland and passive bioreactor systems, and 5. Fate and transport modeling. This project will result in improved designs of passive bioreactor systems to achieve target metal concentrations and to protect aquatic and human health.

1 INTRODUCTION

The ability of passive bioreactor systems (anaerobic wetlands, passive bioreactors and permeable reactive barriers) to reduce metals from mine drainage has been demonstrated. However, the success of these systems has been mixed. Problems include cold temperature and stress effects, stable long-term performance, variable effluent metal concentrations, and nitrate effects on performance. Improvements in performance can be achieved by a better understanding of how the organic substrate affects the microbial population distribution and microbial activity, and how the nature of the microbial products, particularly organic products, as well as organic matter produced by degradation affects the fate of metals.

Our hypotheses are:

1 Organic substrate composition affects the microbial community structure and activity and thus the effectiveness of the treatment system,
2 Organic matter produced in anaerobic reactors affects the concentration and speciation of effluent metals,
3 Microbial species will be distributed at different distances along the flow path of anaerobic bench-scales and wetland systems,

4 Microbial population stratification will lead to different metal removal patterns in bench experiments and constructed wetland systems from that observed in batch cultures.

2 OBJECTIVES

The overall goal of this project is to improve process performance of these systems. This will be done by first evaluating the effect of organic matter characteristics on the types, rate of growth and sustainability of microbial populations. Also, we consider that the effect of organic matter on metal speciation, complexation, and removal is an important issue. To achieve this goal, a research plan with the following objectives will be followed:

1 To evaluate the physical, chemical and biological composition of the components used to create the PBR mixtures,
2 To determine if the organic substrate characteristics affect the character and concentration of soluble organic matter and metal speciation and concentration,
3 To determine the variation of microbial populations with time and location,

4 To evaluate the use of mathematical models to relate metal removal and transport to various system parameters.

3 WORK PLANS

The approach will utilize experimental study in batch and bench-scale systems to test the proposed hypotheses and meet the project objectives. The project tasks are:

1 Solid phase organic and inorganic material characterization (physical, chemical and microbial). An easy method of screening organic substrate material will be developed. The results will be used to characterize different organic mixtures used in batch and bench-scale experiments. This task is an important part of testing hypothesis 1 and will yield a better understanding of the function and effectiveness of each organic source element.

2 Batch studies on the effect of different PBR mixtures. These tests will establish the important factors within organic substrate composition that control sulfate reducing bacterial activity. Also, the question of whether certain metals are toxic to sulfate reducing bacteria will be investigated. Depending upon the results of the toxicity question, methods to protect bacteria from toxic metals will be investigated. The results will allow identification of the best mixture of organic and inorganic materials for use in bench-scale experiments.

3 Bench-scale studies on the effect of substrate mixtures and perturbations. The results will be used to test the hypothesis on spatial variability of microbial populations and further examine the hypotheses on organic and inorganic materials to be used. The results of the bench-scale experiments will be used to design field-scale treatment systems to be tested in a subsequent project at Colorado site(s) with low temperature, at high altitude and with variable flow conditions. This part of the project will be done in collaboration with a public or private agency that is attempting to remediate a mine drainage problem.

4 Field sampling of anaerobic wetland and passive bioreactor systems. "Biopsies" of actual bioreactor systems will be made to test the hypothesis on spatial variability of microbial populations. Also, the physical characteristics of these actual reactors will be used to uncover important parameters to be included in the fate and transport models.

5 Fate and transport modeling. Batch test data will be used to generate the thermodynamic parameters necessary for fate and transport models. Data from bench-scale tests will be used to provide relevant parameters for advection-dispersion and kinetic modeling. The models generated from the batch and bench-scale tests will be calibrated using the field-scale systems. The usefulness of the models will be validated by comparison with experimental data from other field-scale systems.

4 EXPECTED RESULTS

This project will result in improved designs of passive bioreactor systems to achieve target metal concentrations to protect aquatic and human health.

Tailings and Mine Waste '02, © 2002 Swets & Zeitlinger, ISBN 90 5809 353 0

Evaluating recovery of stream ecosystems from mining pollution: Integrating biochemical, population, community and ecosystem indicators

W.H. Clements
Department of Fishery & Wildlife Biology, Colorado State University, Fort Collins, Colorado 80523, USA

J. Ranville
Department of Chemistry and Geochemistry, Colorado School of Mines, Golden, Colorado 80401, USA

ABSTRACT: This research will integrate the fields of aquatic toxicology and ecotoxicology to characterize the recovery of a stream ecosystem from mining pollution. The ultimate goals of our research are: i) to improve our mechanistic understanding of ecological responses to heavy metals across several levels of biological organization; and ii) to evaluate indicators of recovery in a metal polluted stream (the Arkansas River) following improvements in water quality. Experiments conducted in stream microcosms will quantify concentration-response relationships between heavy metals and biochemical, and ecosystem-level indicators. To test the hypothesis that these indicators are sensitive to improvements in water quality, we will validate responses in a large-scale 'natural experiment' (sensu Diamond 1986) conducted in the Arkansas River, a metal polluted stream in central Colorado. Metal concentrations in the Arkansas River are expected to decline over the next few years as a result of remediation activities in California Gulch, a U.S. EPA Superfund.

1 INTRODUCTION

Effects of contaminants on aquatic organisms are manifested at all levels of organization, from molecules to ecosystems. Therefore, comprehensive biological assessments of water quality should measure biochemical, population, community, and ecosystem responses simultaneously. The rationale for this approach is that indicators at different levels of biological organization provide different types of information necessary for ecological risk assessment. Effects of contaminants at lower levels of organization (e.g., molecular and biochemical) occur more rapidly and provide an early warning of toxicological effects. In addition, specificity of responses and our mechanistic understanding of contaminant effects are usually greater at lower levels of organization. Finally, some endpoints at lower levels of organization may be linked to exposure (e.g., chemical residues and metabolites) and are direct measures of contaminant bioavailability.

Despite our greater mechanistic understanding of molecular and biochemical indicators, their ecological significance is generally unknown. This is especially true for residue levels and metabolites, which are excellent indicators of exposure but poor indicators of ecological effects. Consequently, some researchers have argued that responses at higher levels of organization (populations, communities, and ecosystems) measured in the field are more ecologically relevant than traditional toxicological indicators. The emerging field of ecotoxicology generally emphasizes effects of contaminants on populations, communities, and ecosystems. However, despite their greater ecological relevance, responses at higher levels of organization are more complex and often lack mechanistic explanations. Thus, establishing a cause-and-effect relationship between stressors and responses at higher levels of organization is not straightforward.

Because it is unlikely that responses at any one particular level of organization will satisfy the mutually exclusive criteria of endpoint specificity, mechanistic understanding, and ecological relevance, an alternative approach in ecological assessments is to measure responses at several levels of organization simultaneously. Establishing causal linkages across levels of biological organization will help establish the ecological relevance of biochemical responses. Understanding the biochemical mechanisms responsible for changes in populations and communities will improve our ability to link observed responses directly to specific contaminants. Thus, one important goal of this research is to characterize effects of heavy metals from abandoned mines on responses at several levels of biological organization.

This research will integrate the fields of aquatic toxicology and ecotoxicology to assess recovery of a Rocky Mountain stream (the Arkansas River, CO) from heavy metal pollution. We will conduct field and experimental stream studies to validate biochemical (bioaccumulation), population (growth, survival, size structure), community (diversity, species composition) and ecosystem (respiration) level indicators of heavy metal pollution. The ultimate goals of our research are to identify indicators of recovery from mining pollution and to improve our mechanistic understanding of responses to heavy metals across levels of biological organization.

2 APPROACH

We will combine experimental stream studies and a long-term field monitoring project to characterize responses of aquatic organisms to heavy metals. During the first year of the study experiments conducted in stream microcosms will quantify concentration-response relationships between heavy metals and biochemical (bioaccumulation), population (mortality, size structure of dominant taxa), community (species diversity, community composition) and ecosystem (respiration) level indicators. To test the hypothesis that these indicators are sensitive to improvements in water quality, we will validate responses in a large-scale 'natural experiment' (sensu Diamond 1986) conducted in the Arkansas River, a metal polluted stream in central Colorado. We have monitored benthic macroinvertebrate communities and assessed water quality in the Arkansas River bi-annually (spring and fall) since 1989 as part of a long-term monitoring program. Metal concentrations in the Arkansas River are expected to decline over the next few years as a result of remediation activi-ties in the Yak Tunnel-California Superfund Site. Data collected before and after the start of remediation project will be used to validate indicators and to test the hypothesis that aquatic organisms have responded to expected improvements in water quality.

3 EXPECTED RESULTS

The most important expected benefits of this research are: i) an improved understanding of the mechanistic linkages among ecological indicators at different levels of biological organization; and ii) development and validation of a suite of indicators that can be used to assess recovery of metal-polluted streams in the Rocky Mountain region. Traditional biological monitoring programs for evaluating water quality and for assessing ecological integrity are seriously limited because of the inability to demonstrate direct cause-and-effect relationships. Our microcosm experiments are designed not only to show causation but to establish concentration-response relationships between heavy metals and a suite of biochemical, population, community, and ecosystem level indicators. We will test these predictions by evaluating indicator responses to improvements in water quality in the Arkansas River following a large-scale remediation program.

Site characterization

Tailings and Mine Waste '02, © 2002 Swets & Zeitlinger, ISBN 90 5809 353 0

Abandoned quarries surveying

G. Bonifazi, L. Cutaia, P. Massacci & I. Roselli
University of Rome "La Sapienza", Department of Chemical, Material, Raw Material and Metallurgical Engineering, Rome, Italy

ABSTRACT: The quarry activities are subjected in Italy to regional regulations that foresee the obligation to restore the site at the end of the activity. Nevertheless, since these regulations have been implemented after the 1980s, several quarries remain without any kind of restoration. However, such sites evolved naturally from a situation of not equilibrium with the surrounding ecosystem, toward a new situation of equilibrium through dynamic processes of soil formation and plants and animals colonization.

The aim of this work is to point out a monitoring methodology in order to determine the state of restoration of abandoned quarries as to survey the land and its evolution in the time and, therefore, to program suitable restoration projects.

The adopted monitoring methodology consists of remote-surveyed images, analyzed through GIS systems, and correlated with data collected in situ. The methodology has been applied to abandoned quarries in the Lazio region in Italy.

The image analysis has allowed selecting two normalized artificial spectral fractions able to effectively discriminate the naked rock (initial state) from the undisturbed vegetation (final state), as well as the evolution from the initial state to another. Moreover, suitable indicators have been extracted for each quarry in order to describe the state of the situ-restoration and soil evolution. The state of restoration of the analyzed quarries results correlated to the value of the two select normalized artificial spectral fractions.

1 INTRODUCTION

The state of restoration surveyed in abandoned quarries can be achieved by human intervention through works of reclamation (artificial restoration) or it can be the result of a natural process (spontaneous restoration). In both cases abandoned quarries evolve from a situation of not equilibrium with the surrounding ecosystem (initial state: before restoration), toward a new situation of equilibrium (final state: after restoration) through dynamic processes of soil formation and plant-, animal- and micro-organism-colonization.

The chemical and physical actions by atmospheric agents such as rain, snow and wind slowly alter and crumble the rock on the surface (weathering processes) forming a layer of debris. The formation of a thin layer of debris (soil) on the surface of the naked rock allows some plants (pioneer species) to take root.

On the other hand, the bio-chemical and physical interactions between these plants and the rock help the further evolution of the layer of soil, which gets thicker and richer in organic compounds.

As the soil evolves, other plants, microorganisms and animals find their proper habitat and enough nutriment to colonize the site (figure 1). The more organisms colonize the soil, the more the soil evolves toward a state of maturity in which it reaches the maximum thickness and organic fraction in equilibrium with the final ecosystem (dynamic series) (Abate, 1999; Orlandi, 1998).

To assess the state of restoration of abandoned quarries, in this work it has been pointed out a monitoring methodology using remote sensing images.

Such methodology has been applied to three abandoned quarries at the base of the south-western

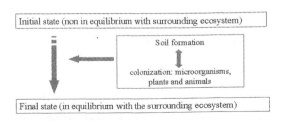

Figure 1. Restoration process.

slope of the Lepini Mountains in Latium, Italy. In all the studied sites the raw material quarried was limestone (Accordi et al., 1967). No kind of reclamation has ever been applied to these quarries, so their state of restoration has been achieved spontaneously.

2 MONITORING METHODOLOGY

Surveying the state of restoration of abandoned quarries practically means monitoring some particular physical and chemical parameters which can be correlated to the development of the ecosystem (soil formation, plants and animal colonization) in the studied area.

The surveyed values of such parameters must be compared to the values representative of the initial state of the restoration process (naked rock) and to the ones representative of the final state (undisturbed surrounding ecosystem). So the initial and the final stadium represent both a reference set for the restoration surveying.

The monitoring techniques used in this work are:
– remote sensing monitoring;
– in situ surveying;
– analysis on soil samples.

The remote sensing monitoring is based on the analysis and processing images by remote sensing. From such images it has been built normalized synthetic bands whose value is related to the physical characteristics of the materials on the earth surface. So it is possible to distinguish the naked rock in unreclaimed quarries from the vegetation sites in the restored quarries (Brivio et al., 1990; Drury, 1987).

The results obtained by imaging have been compared to information coming from in situ surveying and to the results of laboratory analysis on soil samples coming from each studied quarry.

The study of the territory under observation has been supported by the use of a GIS through which it has been possible to geo-referring the images coming from remote sensing and all data related to the studied area (Aronoff, 1989; La Marca, 1998).

3 STUDIED AREA

The quarries studied in this work are situated at the base of the southwestern slope of the Lepini Mountains, in the territories of Sezze, Sermoneta and Bassiano, Latium region, Italy (figure 2).

In this area there are 10 limestone quarries: three of them have been taken into consideration, since other quarries are still working or have been used as dumps or for other uses after their closure.

The studied quarries are identified by the targets A3, A4 and P9.

A quarry still working, identified by the target C7, is assumed as reference for naked rocks.

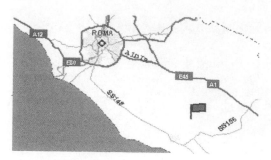

Figure 2. The flag is a landmark for the studied area.

4 REMOTE SENSING MONITORING

Remote sensing is defined as the science of obtaining information about an object, area or phenomenon through the analysis of data acquired by a device that is not in contact with the object, area or phenomenon under investigation.

Remote sensing usually makes use of sensors capable of acquiring data connected to the electromagnetic energy emitted or reflected by the objects being investigated.

In remote sensing of the earth resources electromagnetic energy sensors are placed on airborne or satellite platforms. From such sensors it is possible to obtain digital images of the area under observation. A digital image is a grid of square pixels (or picture elements) in which the brightness is related to the electromagnetic energy acquired by the sensor from the corresponding portion of the earth surface.

The pixel brightness (acquired electromagnetic energy) can be numerically expressed by the Digital Number (or DN) so that a digital image can be considered as a matrix of DN pixel values.

The electromagnetic energy acquired by a remote sensor is the sum of two type of radiation:
– the energy emitted by the earth surface;
– the energy emitted by the sun and reflected by the earth surface toward the sensor.

The energy emitted or reflected depends on the wavelength. The electromagnetic energy emitted by the earth surface is prevailing over the other type of radiation for wavelength $\lambda > 3$ μm.

Since the emitted energy is related to the earth surface's temperature this spectral band (or wavelength range) is called thermal infrared. On the other hand, for wavelength $\lambda < 3$ μm the electromagnetic energy, acquired by the sensor, is almost all reflected by the earth surface. Different materials reflect a different percentage of the incident energy depending on the wavelength (spectral reflectance) so each material has a spectral signature (figure 3).

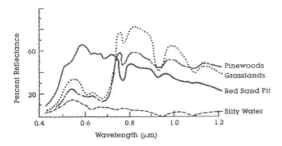

Figure 3. Spectral signatures of some materials on earth surface.

Atmospheric scattering and absorption affect both emitted and reflected energy. Remote sensing is based on recognizing the materials on the earth surface by their different spectral signatures.

4.1 *Landsat 5 TM satellite system*

The images used in this work came from remote sensing system based on Landsat 5 TM (Thematic Mapper). The Landsat 5 satellite rotates at an altitude of 705 km with a near polar, sun synchronous orbit. Image data is acquired nominally at 9:30 a.m. local time with a repeat cycle of 16 days and 14.56 orbits per day. Each acquisition (full scene) covers an area of 185×185 km^2. By the different image devices on this satellite the Thematic Mapper (TM) allows acquisition in seven different spectral bands (Table 1). This sensor has a geometric resolution (dimension of image pixels) of 30 m for the bands 1, 2, 3, 4, 5 and 7 and 120 m for the band 6 (thermal infrared). As regards the radiometric resolution, DN pixel values go from a minimum of 0 to a maximum value of 255 (8 bit digital images).

Table 1. Main characteristics of Landsat 5 TM bands.

Band	Spectrum	Wavelength range (μm)	Mean wavelength λ (μm)
B1	Blue	0.45 - 0.52	0.49
B2	Green	0.52 - 0.60	0.56
B3	Red	0.63 - 0.69	0.66
B4	Near infrared	0.76 - 0.90	0.83
B5	Mid infrared	1.55 - 1.75	1.65
B6	Thermal infrared	10.42 - 12.50	11.46
B7	Mid infrared	2.08 - 2.35	2.22

4.2 *Model calibration*

The calibration model has been achieved by sampling the pixels corresponding to two reference classes of land coverage:
– naked rocks (corresponding to quarry C7 still working);
– undisturbed vegetation.

The reference classes corresponded to the initial state and to the final state of the restoration process. The pixels of the first reference class have been sampled in correspondence to a still working quarry (quarry C7) being representative of the spectral response of the real state before starting the restoration process. Pixels corresponding to undisturbed vegetation sites have been also achieved.

4.2.1 *Spectral profile of reference classes pixels*
The spectral profile of each sampled quarry has been built for each pixel. From figures 4 and 5 it is evident there is no difference in the spectral response related to the band 6, related to thermal infrared.

In the figures 6 and 7 it is also evident a very high DN value in correspondence to band 1 (blue). This effect is caused by the Rayleigh scattering, which is a selective scattering that strongly affects the short wavelength radiation (proportional to λ^{-4}).

Moreover, none of the 7 bands seem to be suitable to characterize the different state of the land surface in correspondence of the different conditions of quarrying. The spectral profiles of each class pixels seem to follow a similar trend.

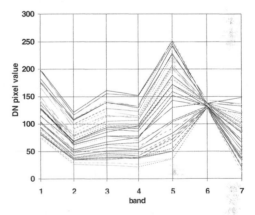

Figure 4. Spectral profiles of pixels related to the quarry C7.

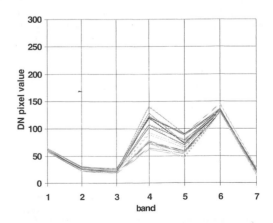

Figure 5. Spectral profiles of pixels sampled in sites characterized by undisturbed vegetation.

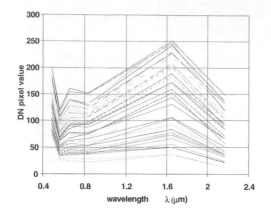

Figure 6. Scaled spectral profiles for λ < 3 μm of pixels related to naked rocks (quarry C7).

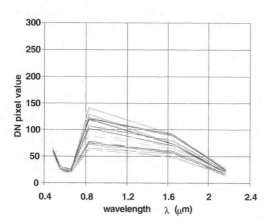

Figure 7. Scaled spectral profiles for λ < 3 μm of the pixels related to undisturbed vegetation sites.

Not considering the emitted energy (band 6) and the selective scattering distortion (band 1) the electromagnetic energy acquired by the sensor is only the energy reflected by the earth surface, which can be expressed as follows:

$$E_R(\lambda) = R(\lambda)I(\lambda) \tag{1}$$

Where:

λ = wavelength;
E_R = reflected electromagnetic energy;
R = reflectance;
I = radiance.

Since the DN value of each pixel is proportional to the electromagnetic energy acquired by the sensor, then, considering the atmospheric non-selective scattering and absorption:

$$DN(\lambda) = cr_s e^{-a_\lambda m} R(\lambda)I(\lambda) \tag{2}$$

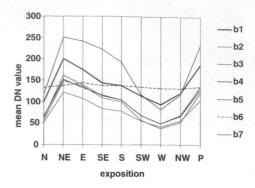

Figure 8. Influence of pixel exposition on DN values related to naked rocks (quarry C7).

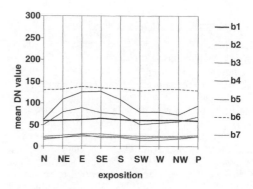

Figure 9. Influence of pixel exposition on DN values related to undisturbed sites.

Where:
DN = digital number;
c = constant;
r_s = view factor;
a_λ = attenuation constant (non-selective scattering and absorption);
m = distance pixel-sensor.

In the figures 8 and 9 it is evident the increase of DN value (except for the bands 1 and 6) for the pixels with exposition from NE to S clockwise in confrontation to those related to the horizontal land surface (P). In fact, pixels exposed to NE-S receive a higher radiance (I), since Landsat 5 TM images of the studied area are taken at about 9:45 a.m.: at this time, in fact, solar radiation comes approximately from ESE.

4.2.2 Normalized synthetic bands
To create a parameter that is not influenced by pixel radiance, but only by the pixel material, it is necessary to build suitable normalized synthetic bands.

Normalized synthetic bands can be built in several ways by finding specific relations between the basic bands.

Considering two different bands with mean wavelength λ_1 and λ_2 among the bands 2, 3, 4, 5 and 7 (related to reflected energy), the ratio between their DN values concerning the same pixel is:

$$\frac{DN\ (\lambda_1)}{DN\ (\lambda_2)} = \frac{e^{-a_\lambda m}\ I\ (\lambda_1)R\ (\lambda_1)}{e^{-a_\lambda m}\ I\ (\lambda_2)R\ (\lambda_2)} \qquad (3)$$

which can be expressed as follows:

$$\frac{DN\ (\lambda_1)}{DN\ (\lambda_2)} = K\ (\lambda_1, \lambda_2)\frac{R\ (\lambda_1)}{R\ (\lambda_2)} \qquad (4)$$

Where
K = coefficient depending only on λ_1 and λ_2.

In fact, a_λ and m are the same for the same pixel and $I(\lambda_1)/I(\lambda_2)$ depends only on the spectral distribution of the solar radiance, which is the same for all pixels (solar radiance spectrum at sea level as in figure 10).

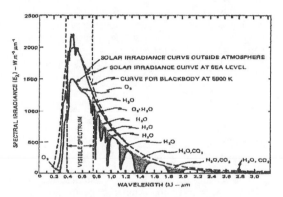

Figure 10. Spectral distribution of solar radiance.

The ratio $DN(\lambda_1)/DN(\lambda_2)$ has been used to build normalized synthetic bands depending only on the chosen basic bands (λ_1 and λ_2) and the spectral reflectance (R), which is a physical property of the materials corresponding to the pixel taken into consideration. In the figures 11 and 12 the values of the normalized synthetic band, obtained dividing the DN value of band 2 by the DN value of band 3 (b2/b3), and the normalized synthetic band, obtained dividing the DN value of band 2 by the mean DN value of all bands (b2/bm), are plotted. Following the same way different synthetic bands are taken into consideration and the related pixel value are calculated for the sampled sites. Among all these normalized synthetic bands it is possible to choose the ones that are more suitable to distinguish the reference classes.

From the figures 11 and 12 it is evident that pixel exposition do not affect the values calculated for the normalized synthetic bands taken into consideration.

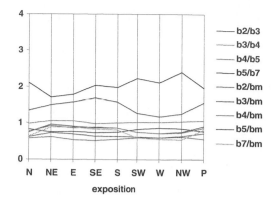

Figure 11. Influence of exposition on the values calculated for the synthetic bands related to naked rocks (quarry C7).

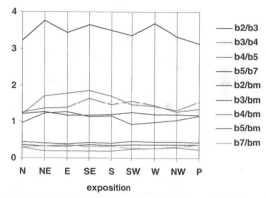

Figure 12. Influence of exposition on the values calculated for the synthetic bands related to undisturbed vegetation sites.

Unsupervised classification has been applied to recognize the synthetic bands suitable to distinguishing the classes related to the different land states.

The bands b3/b4 and b4/b5 seem to give the best results in recognizing naked rocks (quarry C7) from undisturbed vegetation sites (figure 13).

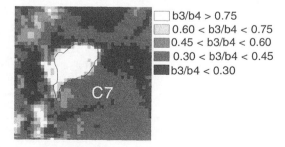

Figure 13. Landsat 5 TM unsupervised classified image of quarry C7.

4.3 *Restoration assessment by remote sensing*

The state of restoration of quarries A3, A4 and P9 can be assessed by comparing their spectral response to that of the reference classes (naked rocks and undisturbed vegetation sites) as shown in figure 14.

The neighbors of spectral response of a pixel to that of the undisturbed vegetation pixel has been assumed as a land state probably covered by vegetation. On the contrary, a pixel response similar to that of naked rocks followed to conclude that it corresponds to a land state with very few plants.

Figure 15 shows the mean values calculated for the synthetic bands of b3/b4 and b4/b5 related to the quarries taken into consideration and for the reference sites taken into consideration (naked rocks and undisturbed vegetation). The barycenter of the points representative of the characters related to the sampled pixels of each site taken into consideration is calculated.

All the barycenters are approximately aligned according to a curve from the site related to naked rocks (state before restoration) to the site related to undisturbed vegetation (state after restoration). From the figure 15 it is evident the restoration hierarchy of quarries taken into consideration: A4 > P9 >A4.

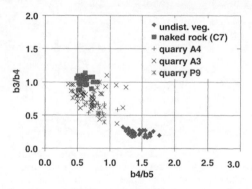

Figure 14. Spectral response of the pixels sampled for the reference sites and for the sites of the considered quarries.

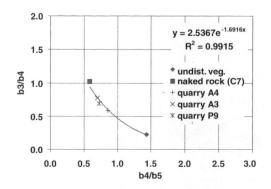

Figure 15. Barycenters of the spectral responses related to the reference sites and to the sites of the considered quarries.

5 IN SITU MONITORING

Monitoring of the quarried sits is carried out following three steps:
- selection of sampling strips representative of the state of restoration of the whole quarry;
- choice of proper parameters suitable to characterize the sampling strips: each parameter has been selected to be indicative of a particular aspect of the restoration process;
- extraction some indices from the adopted parameters:
  - *total restoration indices* to describe the state of restoration for every strips of part of this;
  - *comprehensive restoration index* to describe the state of restoration of the whole quarry.

In this work the *sampling strips* have been assumed 5 m wide and long as necessary to be representative of the whole quarry or of a part of it.

Each *sampling strip* is divided into *square plots* of 5×5 m^2.

In situ monitoring consists in measuring or assessing the value of some suitable parameters referred to the state of vegetation and the state of the soil inside each *square plot*.

In order to assess the state of restoration the following parameters have been evaluated in each sampled *square plot*:
- soil cover (%);
- grass cover (%);
- number of shrubs;
- number of trees;
- number of trees higher than 0.5 m;
- diameter (ϕ) of the tree foliage (m).

Concerning the last parameter, it has been considered the sum of the diameters of all the tree foliages ($\Sigma\phi$) if more than one tree is present in a *square plot*.

As shown in table 2, the data collected for each *square plot* of each *sampling strip* have been compared to the values correspondent to:
- the site state before restoration (naked rocks),
- the site state after restoration (undisturbed vegetation

The value of each parameter relieved for the *site state before restoration* is posed equal to 0, because in this case there isn't any kind vegetation on the ground.

The value of each parameter relieved for the *site state after restoration* has been assessed taking into consideration the undisturbed areas in the surroundings of the monitored quarry.

Normalization of the different parameters has been carried out in order to obtain comparable values: the values of each parameter relieved in situ have been divided by the *maximum value* of same parameter relieved in all the quarried sites taken into consideration.

Table 2. Data collected in situ, referred to the sampling strip S1 of the quarry A3.

SP	SC (%)	GC (%)	NS	NT	NTH	φ (m)
1	70	60	2	0	0	0.9
2	50	40	3	0	1	0.5
3	20	10	0	0	0	0.0
4	30	10	0	0	0	0.0
5	30	10	0	0	0	0.0
6	20	20	0	0	0	0.0
7	60	20	0	0	0	0.0
8	100	90	0	0	0	0.0
9	100	100	0	0	0	0.0
10	100	100	0	0	0	0.0
11	40	40	7	0	6	10.9
12	70	60	0	0	0	0.0
13	40	20	0	0	0	0.0
14	20	10	0	0	0	0.0
15	90	80	0	0	0	0.0

SP: target of sampling plot
SC: soil cover (%)
GC: grass cover (%)
NS: number of shrubs
NT: number of trees
NTH: number of trees higher than 0.5 meters
φ: diameter of foliage (m)

Table 3. Maximum values (normalized values) adopted for calculation of each *restoration index* after in situ-monitoring.

	Partial restoration indices					
	I_1	I_2	I_3	I_4	I_5	I_6
Target	SC	GC	NS	NT	NTH	φ
Normalized value	100%	100%	10	2	12	15 m

SC: oil cover (%)
GC: grass cover (%)
NS: number of shrubs
NT: number of trecs
NTH: number of trees higher than 0.5 meters
φ: diameter of foliagc (m)

The values of the normalized parameters (*partial restoration indices*: I_1, I_2, I_3, I_4, I_5 and I_6), calculated starting from data of table 2 divided by data coming from table 3, are reported in table 4.

Table 4. Normalized values of the *partial restoration indices* obtained for the strip S1 of the quarry A3

Sampl.plot	I_1	I_2	I_3	I_4	I_5	I_6
1	0.70	0.60	0.20	0.00	0.00	0.06
2	0.50	0.40	0.30	0.00	0.08	0.03
3	0.20	0.10	0.00	0.00	0.00	0.00
4	0.30	0.10	0.00	0.00	0.00	0.00
5	0.30	0.10	0.00	0.00	0.00	0.00
6	0.20	0.20	0.00	0.00	0.00	0.00
7	0.60	0.20	0.00	0.00	0.00	0.00
8	1.00	0.90	0.00	0.00	0.00	0.00
9	1.00	1.00	0.00	0.00	0.00	0.00
10	1.00	1.00	0.00	0.00	0.00	0.00
11	0.40	0.40	0.70	0.00	0.50	0.73
12	0.70	0.60	0.00	0.00	0.00	0.00
13	0.40	0.20	0.00	0.00	0.00	0.00
14	0.20	0.10	0.00	0.00	0.00	0.00
15	0.90	0.80	0.00	0.00	0.00	0.00

A *total restoration index* (I_r) has been pointed out for each sampling *square plot*, considering all *partial restoration indices*, by the equation:

$$I_r = \sqrt{\frac{I_1^2 + I_2^2 + I_3^2 + I_4^2 + I_5^2 + I_6^2}{6}} \quad (5)$$

A *comprehensive restoration index* for each quarry (I_{rq}) has been be calculated by the mean value of the *total restoration indices* (I_r), obtained for every sampling *square plots* (table 5).

Figure 16 shows the relation between the normalized synthetics bands (b3/b4 and b4/b5) and the *comprehensive restoration indices* (Irq).

Table 5. Average values of the *comprehensive restoration index* (I_{rq}), obtained from the values related to the *site state before restoration* (naked rocks) and to the *site state after restoration* (corresponding to sites characterized by undisturbed vegetation in the surroundings of the quarries) compared with the values of the two normalized synthetic bands (b3/b4 and b4/b5) related to the same sites.

Class of restoration	I_{rq}	b3/b4	b4/b5
Naked rocks	0.00	1.02	0.58
Quarry A3	0.27	0.77	0.70
Quarry P9	0.42	0.70	0.73
Quarry A4	0.46	0.59	0.85
Undisturbed vegetation	1.00	0.23	1.43

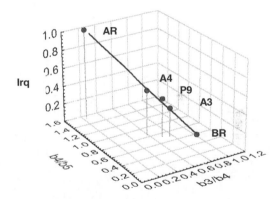

Figure 16. Relation between the values calculated for the normalized synthetic bands (b3/b4 and b4/b5) and that for the *comprehensive restoration indices* (Irq) related to each quarry (A3, A4 and P9) and to the sites before and after restoration (BR and AR respectively).

6 SOIL CHARACTERIZATION

Soil samples coming from the sampled *square plots* have been analyzed taking into consideration:
– soil reactivity (pH);
– organic matter content (%);
– grain size distribution.

At least two soil samples have been taken in each quarry: one representative of a plot showing a scarcely vegetated soil and one representative of a plot related to the most evolved soil in the quarry.

Data, coming from analyses carried out on soil samples, have been related to the I_r values obtained for the same plot as shown in figures 17 and 18.

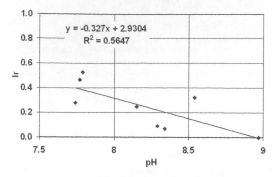

Figure 17. Soil reactivity (pH), after in situ sampling, versus *total restoration index* (Ir) calculated for all the sample plots.

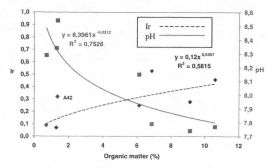

Figure 18. Organic matter content (%) versus soil reactivity (pH) for each sample in situ and *total restoration index* (I_r) obtained for the same plot.

7 CONCLUSIONS

Data coming from monitoring abandoned quarries by remote sensing results well correlated with data coming from soil sampling in situ.

In particular, good results in recognition of the state of the land are obtained taking into consideration data coming from calculation of two *synthetic*

normalized bands and from evaluation of a *comprehensive restoration index* (I_{rq}) pointed out in a suitable way.

Besides, the results of the analysis performed on soil and samples, collected in situ, are well related with the corresponding *total restoration index* (I_r) and, therefore, with the *synthetic normalized bands*.

The tools, designed taking in consideration data coming from both remote sensing and in situ monitoring, result effective to check abandoned quarries and to survey their evolution in the time in order to point out suitable operations for land restoring.

8 ACKNOWLEDGEMENTS

Special thanks have to be addressed to the Corpo Forestale of the Province of Latina (in the Latium Region of Italy) for assistance in land monitoring and to the Company IPT for image acquisition and supplying.

REFERENCES

AA.VV. – Metodi di analisi chimica dei suoli, Ministero delle Politiche Agricole e Forestali, 1993.
Abate I. – Recupero delle cave in ambiente mediterraneo, Quarry & Construction, Sept. 1999;
Accordi B., Angelucci A., Sirna G. – Note illu-strative alla carta geologica d'Italia. Foglio n° 159 (Frosinone) e n° 160 (Cassino), Servizio Geologico d'Italia, 1967;
Aronoff S. – Geographic Information Systems: a management perspective, WDL Publications, 1989;
Brivio P. A., Lechi G. M., Zilioli E. – Il Telerilevamento da aereo e da satellite, Delfino ed., 1990;
Drury S. A. – Image interpretation in geology, Chapman & Hall, 1987;
La Marca F. – L'applicazione di tecniche di telerilevamento all'indagine del territorio, Dispense del Corso di Cave e Recupero Ambientale, Prof. P. Massacci, Università "La Sapienza", Roma, 1998;
Lillesand T. M., Kiefer R. W. – Remote sensing and image interpretation, John Wiley & Sons, 1979;
Massacci P. – Remote Sensing. A critical review, Dispense del Corso di Cave e Recupero Ambientale Università "La Sapienza", Roma, 1999;
Orlandi S. – Caratteristiche del suolo. Appunti del Corso di Idrologia Tecnica, Prof.ssa C. Peiser Siniscalchi, Università "La Sapienza", Roma, 1998;
Sabins F. F. Jr – Remote sensing, principles and interpretation, Freeman, 1986.

Tailings and Mine Waste '02, © 2002 Swets & Zeitlinger, ISBN 90 5809 353 0

The effectiveness of single and multiple open standpipe piezometers in monitoring of the pore pressure regime in tailings dams

G.J.R. le Roux

SRK Consulting, Johannesburg, South Africa

ABSTRACT: The use of open standpipe piezometers, in monitoring of the piezometric levels in tailings dams, is common practice in South Africa. These piezometers are often installed as single piezometers, in a line, to monitor the piezometric level at various cross sections through the tailings dam. Case studies in South Africa have indicated that the true phreatic surface is not always well presented by single open standpipe piezometers. Furthermore open standpipes suffer from a time lag in detecting fluctuations in the piezometric level. This paper discusses the shortcomings of monitoring of the phreatic surface in tailings dams with single standpipe piezometers and concludes with a case study of incorrect 'phreatic levels' at platinum tailings dams in South Africa. It is recommended that multiple piezometers be installed at critical sections.

1 INTRODUCTION

The pore pressure regime in a tailings dam, or any earth dam for that matter, has a significant influence on the stability of the dam wall. Single standpipe piezometers, at various cross-sections around a tailings dam, is commonly used to monitor the pore pressures in the area close to the outer wall. Typically piezometer data is recorded on a monthly basis and used in stability analyses. Once a base case analysis has been carried out, further analyses are usually only carried out once significant changes in the piezometer values are recorded.

This method of monitoring pore pressures has shown shortcomings and a case study is presented to highlight the significant influence of these shortcomings in terms of slope stability of platinum tailings dams.

2 PIEZOMETERS

2.1 *Open standpipe piezometers*

Piezometers are used to monitor pore water pressure and are sometimes referred to as pore pressure cells. They can be grouped into two categories viz. those that have a diaphragm between the transducer and the pore water (pneumatic, vibrating wire, electrical resistance strain gauges) and those without a diaphragm (open standpipe and twin tube hydraulic piezometers).

In South Africa, open standpipe piezometers are commonly used for monitoring of the pore water regime in tailings dams. The convention for the construction of standpipe piezometers is to seal the porous filter element so that the instrument responds only to sub surface water pressure around the filter element and not to fluctuations at other elevations. The water level in the standpipe stabilizes at the piezometric elevation and is determined by sounding with a probe (electrical dipmeter).

Open standpipes are also referred to as Casagrande piezometers, after publication of measurement methods for monitoring pore water pressure during construction of Logan Airport in Boston by Casagrande in 1949 and 1958.

Open standpipe piezometers, when correctly installed, have the following advantages and limitations:

- Advantages: Reliable, long successful performance record, integrity of seal can be monitored, can be used for water sampling, can be used to measure permeability
- Limitations: Long time lag, filter can clog due to repeated water inflow and outflow

It is therefore clear that standpipe piezometers are useful tools in monitoring of the piezometric level in tailings and earth dams, with more advantages than limitations, provided they are correctly installed.

2.2 *Piezometric level*

The general assumption from tailings dam operators and designers, that the water level in a standpipe piezometer reflects the position of the phreatic surface, is incorrect. The water level in an open standpipe piezometer will only represent the phreatic surface if the piezometer tip is located on the phreatic surface or if no-flow conditions exist. For all other tip locations, the piezometric level is indicated and can be lower than the phreatic surface.

The reason for the difference in phreatic and piezometric levels is flow/seepage through the tailings mass. Pore pressures at the piezometer tips are therefore sub-hydrostatic. If no-flow conditions exist, the piezometric level will be the same as the phreatic surface and the pore pressure gradient would be hydrostatic.

In order to use standpipe piezometers to present the phreatic surface, multiple piezometers have to be installed, with the piezometer tip at different elevations for each piezometer.

Figure 1 indicates the effect of multiple piezometers, installed at the same location, parallel to the embankment wall, but with the tips at different elevations, on evaluating the piezometric level. Piezometer A is installed deeper than Piezometer B and the water level inside the open standpipe will rise to a point where the equipotential line and the phreatic surface intersects. Piezometer B is installed slightly shallower, but in line with piezometer A. Water will rise in standpipe B to a level higher than that of A. The difference in level is shown in Figure 1. If a third piezometer is installed at an elevation higher than that of B, no pore pressure readings will register. The tip of piezometer B, as shown in Figure 1 is practically on the phreatic surface.

The difference in piezometric level and true phreatic surface might be as much as several metres, with serious stability implications for the embankment walls under consideration.

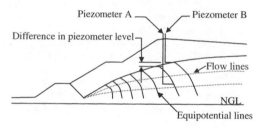

Figure 1. Piezometric level as a function of tip elevation.

3 PIEZOCONE TESTING

The development of instruments capable of measuring porewater pressures and their dissipation in situ, started in the early 1970's with pore pressure probes. The pore pressure probe has been superseded by the piezocone. Different piezocone designs are currently in use with the major differences between the systems the location of the porous filter element in the cone.

To obtain more detailed information of the pore pressure regime in a tailings dam, a piezocone investigation may be conducted. Results from the piezo-

cone investigation can then be compared to piezometer data and an assessment of the reliability of the piezometer data can be made.

The piezocone utilises a pressure transducer to monitor the change in porewater pressure as the cone is pushed into the material to be tested or monitored. As soon as the cone intersects the phreatic surface the transducer responds and pore pressure is recorded.

Depending on the design of the instrument, the piezocone can be used to identify thin layers of different materials.

4 CASE STUDY – PLATINUM TAILINGS DAM IN SOUTH AFRICA

During a routine inspection of one of the Platinum Tailings dams in South Africa, seepage was noted at the toe of one of the dam walls. First reaction was to review the open standpipe piezometer data in order to determine if an increase in piezometric level was recorded. As the piezometer data showed no increase in level, the cause for the seepage was investigated. A piezocone investigation was carried out.

The area under consideration contains two cross section sets of single open standpipe piezometers. Two sets of piezocone tests were carried out adjacent to the standpipe piezometers. Due to access limitations the piezocone tests were only conducted on near horizontal surfaces. No comparison of piezocone results with standpipe piezometers on inclined faces could therefore be made.

The piezocone results were extrapolated to indicate the phreatic surface through that flank of the tailings dam.

4.1 *Piezocone test results*

The results obtained from the piezocone investigation are summarised in Table 1. The depth from tailings surface to the estimated phreatic surface is compared to similar results obtained from the single standpipe piezometers. These results clearly indicate a substantial difference in 'piezometric level' between the two methods of monitoring.

The two sets of results were used in a stability analysis to compare the effects of the difference in pore pressure regime on the stability of the tailings dam wall.

Table 1. Piezocone results vs. piezometer readings

Test No.	Piezocone m	Standpipe Piezometer m	Difference m
1	9.16	13.45	4.29
2	10.80	16.90	6.10
3	5.43	13.62*	8.19

* extrapolated value

4.2 Stability analyses

Limit equilibrium methods were used to analyse a representative section of the tailings dam wall where both piezometric and piezocone levels were known. A typical section through the tailings dam wall is shown in Figure 2. The figure illustrates the difference between the open standpipe piezometer levels and the phreatic surface obtained from piezocone testing.

Due to the fact that the piezometers only provide pore pressures at single points across the tailings dam wall, i.e. at the tip locations, it is strictly speaking not possible to use single standpipe piezometers to monitor the pore pressure regime in the dam wall.

By utilising the data as presented in Table 1, the following factors of safety against overall slope failure were calculated:
- Piezocone level: FoS = 1.498
- Piezometric level:FoS = 1.645

Comparative factors of safety for failure of the bottom slope of the tailings dam are:
- Piezocone level: FoS = 1.335
- Piezometric level:FoS = 1.446

Further analyses indicated that the probability of overall slope failure increases to 50% once the bottom slope fails. The bottom/lower slope is therefore the ruling entity and its factor of safety / probability of failure has to comply with acceptable norms.

The factors of safety were used to assess comparable probabilities of failure using probabilistic theory.

The probabilities of failure estimated are as follows for the bottom/lower slope:
- Piezocone level: PoF = 1.100 %
- Piezometric level: PoF = 0.255 %

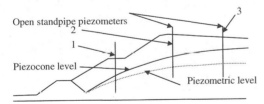

Figure 2. Typical section indicating piezometric and piezocone levels.

4.3 Data interpretation

There is a substantial influence of a change in 'phreatic surface' on the stability of a dam wall. It is therefore important to determine the true phreatic surface in tailings dams, as the piezometric level obtained from single standpipe piezometers could give a false sense of security.

Table 1 shows that the piezometric level and the phreatic surface may differ as much as 8m in elevation. As mentioned before, sub-hydrostatic conditions occur when there is a horizontal flow component in the dam.

The phreatic surface location is determined from processing of the pore pressure results obtained from the piezocone tests. Regression analysis is performed on the representative pore pressure data and the pore pressure gradient is determined. A typical piezocone test result is shown in Figure 3.

The pore pressure gradient for the representative section was approximately 1:6.5, compared to a hydrostatic gradient of 1:9.8. Standpipe piezometers, with the piezometer tip installed below the phreatic surface, will therefore register a water level lower than the phreatic surface. Figure 4 shows the influence of sub-hydrostatic pore pressures on standpipe piezometer readings.

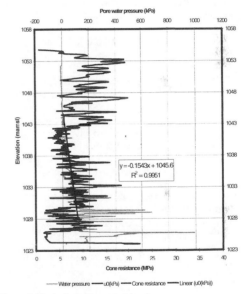

Figure 3. Typical piezocone test result.

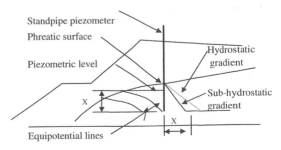

Figure 4. Sub-hydrostatic pore pressures due to seepage.

4.4 Discussion of results

Although both probabilities of failure, related to piezometer and piezocone levels respectively, prove to be unacceptable in terms of acceptable norms, there is nearly an order of magnitude difference between the two results.

The factor of safety reduces by 8% when the phreatic surface, obtained from piezocone testing, is used in the analyses instead of the piezometric level. As the dam rises, and no additional piezometers are installed, the difference between phreatic level and piezometric level will increase. The standpipe piezometers will eventually indicate water levels much lower than the critical level in terms of slope stability. A false sense of security with possible catastrophic consequences may exist.

If more than one standpipe piezometer had been installed at the same location, with the tip at different elevations, the difference between piezometric level and phreatic surface would have been reduced to practically zero when the tip is located on the phreatic surface. The phreatic surface changes however and it would not be possible to install a piezometer with the tip permanently on the phreatic surface.

In order to cater for fluctuations in the phreatic surface, the critical level of the phreatic surface in terms of slope stability needs to be determined. A number of piezometers, say three, at the same location but with their tips at various elevations, need to be installed to cover the range of critical level of the phreatic surface. The registration of pore pressures in the piezometer with its tip on the critical level should act as a warning sign. The stability of the wall would then have to be investigated by conducting stability analyses and piezocone tests to confirm the pore pressure regime in the wall. A decision on mitigating measures would be based on such an investigation.

5 CONCLUSION

The stability of a tailings dam wall is greatly influenced by the pore pressure regime in the dam wall.

Open standpipe piezometers are commonly used to monitor the pore pressure regime. Standpipe piezometers however only measure the pore pressure at the tip of the piezometer, if correctly installed. If seepage occurs, the pore pressure at the piezometer tip will be sub-hydrostatic. The water level in the standpipe piezometer will therefore rise only to the sub-hydrostatic pressure level. This level, commonly known as the piezometric level, can be substantially lower than the true phreatic surface. The difference between the two levels is mainly influenced by the flow regime and the location of the piezometer tip.

The anomaly described can have serious consequences on the stability of the dam wall. It can be overcome, reasonably cost effective, by installing more than one piezometer at the same location but with the piezometer tips at different locations. If three piezometers are installed with their tips located at different elevations, the phreatic surface location can be determined by connecting the three pressure readings and fitting a straight line through it.

A stability analysis needs to be conducted and a critical phreatic surface determined. Registration of pore pressures at that location should act as warning signal and proper investigations, such as piezocone tests, would be carried out.

The case study of a platinum tailings dam in South Africa indicates that an error in factor of safety of 8% was made by assuming that the water level in the single standpipe piezometers represent the phreatic surface. This error can be reduced to practically zero by installing more piezometers (multiple piezometers) at the same location with the tips at different elevations.

REFERENCES

Dunnicliff, J. & Green, G.E. 1988. *Geotechnical Instrumentation for Monitoring Field Performance*. New York: John Wiley & Sons.

Rust, E. 1996. *Geotechnical Investigation and In Situ Testing*, lecture notes at Pretoria University. Pretoria.

Van der Berg, J.P. & Jacobsz, S.W. 1996. *A Method for monitoring the Pore Pressure Regime in Gold Tailings Dams*. Young Geotechnical Engineer Conference, 1996. Botha's Hill, Kwazulu Natal.

Tailings and Mine Waste '02, © 2002 Swets & Zeitlinger, ISBN 90 5809 353 0

Monitoring of soil, air, water, and noise at the Lega Dembi Gold Mine

Yigzaw Binega
Ministry of Mines and Energy, Ethiopia

ABSTRACT: Lega Dembi Gold Mine (LGM) commenced production in 1989 and is currently ranked as the largest gold producer in Ethiopia. In a strategy to improve environmental monitoring plan, monitoring the key characteristics of operations that have a significant impact on the environment – soil, water, air and noise were identified.

Soil and water samples taken from selected locations for analysis was performed to monitor the quality of soil and water respectively. In monitoring air quality at the mine and plant , dust and mercury vapor were the main forms of air pollution identified at the site.

Sampling of surface and ground water was performed in order to monitor the quality of water in the vicinity of the mine. The potential effects of noise on the surrounding are varied and complicated , and the sources of noise from the pit and plant are identified.

1 INTRODUCTION

From the commencement of gold extraction at Lega dembi the government has been concerned in legislation setting standards that have environmental impact on the environment, beside this in order to minimize the environmental impacts of the operations.

The paper is trying to explain the effects on air, water, soil, noise and vibrations at the lega dembi gold mine. While monitoring at the mine and processing plant the effects were identified, some if not all of the following critetia:

- Level and concentration of chemical emissions and their environmental effects,
- Effects of pollutants on water quality ,
- Eeffects of chemical emissions and deposits on soil of site and surrounding.

In addition to this the sources and prevention methods (for air pollution, water pollution, and noise effect at the mine) are explained.

In attempt to further monitor the quality of water, quality of air; soil analysis, and noise ,the Ministry of Mines and Energy is investigating the environmental laws that could be compatable with world mining practice.

The guidelines and standards for industrial pollution control presented by the federal Environmental Protection Authority are provisional and must be periodically reviewed and updated, because the mining sector is dependent for these standards set up by the Authority.

2 DESCRIPTION OF THE LEGA DEMBI DEPOSIT, AND MINING LEGISLATION

2.1 *Description of the Lega Dembi deposit*

The Lega Dembi deposit is located on the right-hand side of lega-dembi valley, 7 km south west of shakisso which is linked to Addis Ababa by a 185 km all-weather and 315 km asphalt roads. Duc south, the deposit area is accessible from shakisso by 11km of all-weather road, maintained by the mine. The Lega dembi gold minc is managcd by thc National Mining Corporation on the behalf of Midroc gold Lega dembi.

The exploration carried out over the past decade by the Ethiopian Mineral Resources Development Corporation (EMRDC), led to the discovery of the Lega dembi vein gold mineralization in the shakisso region. This vein system extends in a north-south direction over a length of about 3 km and has been divided into three sectors: northern, central and southern.

The total reserve of the deposit to the final pit limit level (200m) was 92 tonne. But, at present the company is undertaking extensive underground exploration to a depth of 400m level. At the beginning the pilot plant started to produce 500kg of gold per annum. When the plant started to produce to its full capacity the output is/was 3000kg-4000kg of gold per year.

2.2 Mining legislation

The Mining Law of 1993, 52/1993 in section 26.3 states that the licensee shall ' conduct mining operations in such a manner as to ensure the health and safety of his agents, employees and other persons and to minimize damage or pollution to the environment'. Besides, the Mining Regulations of 1994, No. 182/1994 section 29.3, state that ' the licensee shall notify the Licensing Authority of any occurrence which has resulted in the loss of life or series injury to any person , that may jeopardize any property, the environment or operations and shall immediately take such steps as are necessary to mitigate the impact of such a situation.

At present, the Ministry of Mines and Energy is therefore trying to adopt the environmental laws, that comply with the applicable Ethiopian laws and Regulations consistent with world mining practice and compatible with the long term sustainability of the ecosystems.

3 MONITORING OF SOIL, AIR, WATER, AND NOISE AT LEGA DEMBI GOLD MINE

Different definitions is given to the term monitoring by different authors. By monitoring we understand periodic or continuous surveillance or testing to determine the level of compliance with statutory requirement and/or pollutant levels in various media or in humans, animals or other living things.

3.1 Monitoring of air

Like other forms of pollution , air pollution is an unfortunate by-product of success. Emissions from factories have also traditionally been a major source of pollution like cement-batching plants, power plants , mining and metallurgical industries, etc. But the main forms of air pollution at the mine and plant were identified dust and mercury. Open-pit mining produces a certain amount of carbon monoxide and nitrogen and sulphur oxides from blasting.

Air in which a hazardous substance is supposed to be present should be sampled regularly and frequently, the interval depending on the situation. Both environmental (area sampling and personal sampling should be undertaken). Samples have to be analysed in a well equipped laboratory . Evaluation of the test results is normally done by reference values provided by large international research institutes like those of the European community, the National Institute for occupational safety and Health (NIOSH) in the U.S. A.

The degree of exposure of air quality should be monitored on a regular basis in order to determine whether at the open pit mine and processing operations generate nuisances, both to workers and the surrounding inhabitants. As a result of these operations poor quality could damage the surroundings and may potentially impact land.

The main sources of emissions to air observed at the site are:
- dust from haul roads;
- dust from drilling and blasting operations in the open pit mine,
- exhaust fumes from mobile; concentration and mining plant;
- burning of waste.

The amount of dust at a specific point can be measured with different equipment:
- standard dusty deposit gauge which provides an integrated record of total fall out over the period of exposure,
- directional deposit gauge to measure horizontally borne dust,
- smoke filter.

The amount of dust collected can be measured: all kinds of chemical analysis can be also done on the collected dust sample. Data collected by the company over four years (until 1996) during the operation of the plant showed levels of mercury in the gold room above the eight hour occupational exposure limit is given below in table.1. According to the Ethiopian Environment Protection Authority, the tolerence limits for ambient air pollutants for some substances in the air are illustrated below in table. 2.

Mercury is the most dangerous pollutant emitted by the processing plant; in spite of the steam condensation systems, a small part of the mercury escapes into the air. As a precaution, therefore, the atmosphere in the building containing the amalgamation plant is checked daily.

Table 1. Mercury reading in the gold room over a four year period.

Measurement taken from:	Mg/m^3
Retort door	0.2-0.7
Distiller	0.1-0.33
Hood	0.01-0.2

Table 2. Tollerence limits for ambient air pollutants

Pollutants	Long-term Limits		Short-term Limits	
	Mg/m^3	hours	mg/m^3	min
Mercury	0.003	24	–	–
Lead	0.005	24	0.002	30
Hydrogen Cyanide	0.01	24	–	–
Hydrogen Sulphide	0.008	24	0.008	30
Nitrogen Dioxide	0.085	24	0.085	30
Nitric acid	0.006	24	0.006	30
Silica	0.02	24	5.0	30

The overall policy objective for air quality management at the mine is to achieve as soon as reasonably practicable and to maintain thereafter an acceptable level of air quality, concentration of chemical emissions to safeguard the health and well being of the miners, the surrounding community and their environmental effect.

3.2 *Monitoring of water*

Water (surface and ground) monitoring is one of the important methods supporting the strategy and policy of water resource protection and conservation. The implementation of water quality monitoring results helps to improve the planning, development, protection and management of water resources, to anticipate or reduce threats of water pollution and depletion problems and to enhance anti-pollution policy.

According to "water quality bulletin" journal (1985) – water monitoring has the following objectives:
- to collect, process and analyse background data on water quality and quantity as a baseline for evaluating the current state and for anticipating the changes and trends of the hydrogeological system,
- to provide information for the planning, management and decision – making about water resource development, protection and conservation and for the implementation of legislative and control measures and regulations.

In monitoring the quality of water in the vicinity of the mine, sampling of surface and ground water was done for geochemical analysis purposes.

The impacts on water during mining operations are caused by :
- acid mine drainage,
- surfacewater,
- groundwater,
- tailings from the processing plant.

Acid mine drainage : The major and most significant source of liquid waste in the non-ferrous metal mining industry is acid mine drainage. Acid mine drainage is common in areas where mine openings intersect the water table and where the rocks contain iron sulphides (pyrite and/or pyrrhotite) or, less commonly, certain other sulphides; where such pyretic ores are mined, rainfall leaching of rock waste stockpiles may be responsible for environmental damage. Even though sulphur (mainly pyrite) is ubiquitous in the lega dembi ore, the proportion of sulphides is less than 1% per tonne of ore. As a result acid mine drainage is not considered a possible threat at lega dembi.

Groundwater : The mine does not intersect any significant flowing spring and does not act as a drain to an established water table. The existing springs are located in areas not affected by the workings. The main source of water inflow is torrential rains.

Surfacewater : The mine is located in the upper part of the lega dembi drainage basin, but it does not seriously affect the flow of the run off waters, especially as much of the rain rapidly infiltrates through the rock fissures.

Tailings from the Processing Plant: The tailings from the ore processing are the main source of potential pollution of surface and ground water, especially as a sodium cyanide solution is used to dissolve gold and silver. The cyanidation process has been employed for nearly a century. The chemistry of cyanide solutions is quite complex and it is this complexity which is responsible for its ability to dissolve gold and silver. Other potential contaminants in liquid effluents include process reagents such as acids, flocculants, mercury, sodium hydroxide, etc.

Tailings from a processing plant are composed of a slurry which contains particles of ground material (gangue minerals; minor amounts of valuable minerals) in suspension in water. Consequently, the major liquid waste from a processing plant consists in the water decanted from the tailings pond. The contaminants in tailings pond effluents include suspended solids composed of elements of the ore treated in the plant, heavy metals in solution, thiosalts and chemicals used for the processing.

The risks associated with ore processing is done by monitoring and water treatment. A laboratory built on the site of the ore processing plant provides permanent control on the quality of the tailings water, especially as far as the cyanide concentrations are concerned. The ultimate goal of treatment is to maintain the water quality and to protect the beneficial uses and life forms associated with a receiving system. To protect against surface water – such as digging ditches above the mine so as to divert the run off water , the same system should be used at the edge of the tailings dam.

The effluent limit must comply with the restrictions established by the EPA (Ethiopian Protection Authority) on quantities, rates on concentrations in waste water discharge. In monitoring water and waste water quality, it is important that the effluent characteristics to ascertain the degree of compliance with directives and guidelines of the EPA. The result of water chemical analysis is given in table 3.1 and 3.2 respectively, in which water analysis was done by taking water samples from different places.

Successful control of water pollution is based on the knowledge of: qualities of all waters which may be affected by mining activities; quantities of water required during mining and beneficiation; qualities of process waters after use. Unless a properly designed monitoring programme has carried out, any control measures which are taken are unlikely to result in the desired result at the optimum cost.

Table 3.1. Major constituents (mg/l or ppm)

Elements	B2[*]	B3[*]	B4[*]	D2[*]	U1[*]
Electrical Conductivity (us/cm)	552	863	879	1025	1491
CO_3^{-2}					
HCO_3^{-1}	337	127	220[**]	189	368
Cl^{-1}	29	44	17	83	79
SO_4^{-2}	9	307	58	204	494
NO_3^{-1}	0.89	<.04	.89	39.9	.44
Na	19.4	43	28	94	44
K	3.4	5.6	3.1	26	7.4
Ca	58	73	70	36	236
Mg	26	39	.1	67	42

[*] Sampling codes
[**] Total alkalinity as $CaCO_3$

Table 3.2. Trace constituents (mg/l or ppm)

Elements	B2[*]	B3[*]	B4[*]	D2[*]	U1[*]
Al	<0.1	<0.1	1.7	<0.1	0.5
Ba	3.1	<0.1	1.9	3.2	1.4
Cd	<0.1	<0.1	<0.1	<0.1	<0.1
Cr	<0.1	<0.1	<0.1	<0.1	<0.1
Pb	<0.1	<0.1	<0.1	<0.1	<0.1
Co	<0.1	<0.1	<0.1	3.7	<0.1
Ni	<0.1	0.1	<0.1	2.4	<0.1
Zn	<0.1	<0.1	<0.1	<0.1	<0.1
Cu	<0.1	<0.1	<0.1	1.1	<0.1
Total Fe	0.1	0.3	<0.1	0.2	0.1
Mn	2.1	1.2	<0.1	<0.1	0.4
Ag	<0.1	<0.1	<0.1	<0.1	<0.1
Li	<0.1	<0.1	<0.1	<0.1	<0.1
CN	.02	<0.01	0.02	0.02	<0.01
PH	7.29	7.39	11.43	8.29	7.29

[*] Sampling codes
Source: Solomon Tale and Seyoum Zenebe.

3.3 Monitoring of noise

Noise from the open pit mine is generated by mining equipment (drills, dum trucks, loading shovels, excavators, graders and crushers) and blasting operations. Blasting gives a strong impulse noise, which arises from the air pressure wave generated by detonation of the explosives. This is an unavoidable consequence of blasting. However the noise nuisance caused by blasting can be reduced significantly by appropriate blast design. Noise generated due to mechanical plant operation can be minimized by appropriate maintenance of the plant and fitting of adequate silencers.

Machinery and engines easily produce noise of 85-100 decibel dB(A). Compressed air drilling and pneumatic picks produce not less than 100 dB(A). All noise preventing normal conversation at a close distance (< 50cm) is thought to damage one's hearing. Exposure to noise of 85 dB(A) or more for 8 hours a day over a long period should be regarded as damaging noise. Technical prevention is difficult because machinery and equipment are often used in narrow confined spaces and this produces extra reverberation of the noise.

Basic techniques are available to control noise in mining industry, which tend to:

- Reduce the noise energy generated at source: such measures include :
  o changing a machine for another one that runs slower for example,
  o altering the design, construction or installation,
  o improving maintenance,
  o replacing compressed air powered equipment by electric devices,
  o using properly designed air silencers, resilient mountings.
- Isolate the source by enclosure of noisy fixed equipment,
- Increase the noise absorption between the source and the listener by locating noisy activities as far as possible from areas of potential nuisance and by erecting some form of screening structure between the source and the listener. Screens used in practice include walls, waste banks, trees.

Personal protection like earmuffs and ear defenders is advised where average noise levels of 85 dB(A) are present. It is known that environmental monitoring of noise levels is carried out with sound level meters and personal dose meters. According to noise exposure limits for Ethiopia, it is recommended that daily noise exposure for workers should not exceed 90 decibels dB(A) for an 8-hour working period. The proposed summation formula for estimating equivalent 8-hour noise exposure at the workplace is shown below:

$$D = t_1/T_1 + t_2/T_2 + t_3/T_3 + \ldots + t_n/T_n$$

Where:

D - Daily noise dose which must not exceed unity,
t - Actual exposure time at a given noise level,
T - Permissible exposure time at that level in accordance with table.4,
n - Number of discrete periods of exposure above 90 dB.

Table 4. Noise exposure limits for Ethiopia.

Duration per day, Hour	Permissible Exposure Limit dB
8	90
6	92
4	95
3	97
8	90
6	92
4	95
3	97
2	100
1.5	102
1	105
0.5	110
0.25 or less	115

Note: Exposure to impulsive or impact noise should not exceed 1.
40 dB peak sound pressure level.

Note: 1. Maximum exposure corresponds to D =
1.0 control required for D>1.0
2. Noise levels below 90 dB are not included in the summation.

Potential causes of vibration in mining are different. Vibration from drilling and pick-hammering
implicates the hands and joints of hand, forearm,
arm and shoulder are involved . Exposure to vibration of this type (40 to 300hz) over a long period of
time may give rise to microtraumata, peripherial
nerve stimulation, spasm of the arterioles and eventually disturbed local circulation.

3.4 *Monitoring of soil*

From the point of view of engineering geology the
rocks of the deposit are divided into two major complexes with fundamentally different compositions,
textures and properties. The upper part is composed
of unconsolidated and semi-consolidated rocks. The
underlying rocks are hard metamorphic facies. The
properties of the weathered rocks (unconsolidated)
can potentially result in rock failures or other unfa-
vourable processes (land slips, rocks falls, bench
slope failures, intense erosion) especially at the contact with bed rocks. Soil samples were taken from
selected locations in which soil analysis was performed to monitor the quality of soil and check that
levels of contaminants comply with the emission
levels agreed with the Ministry of Mines and Energy. Recognized international soil standard guide
lines consider toxicity effects on natural ecosystems,
thus the LGM is adopting an "intervention level"
above which pollution represents an unacceptable
risk to water, humans, animals or plants. The analysis of soil result is given in table.5.

4 CONCLUSION

The Mines control department at the Ministry of
Mines and Energy is controlling the gold mine over
all its activities including Environmental issues. Environmental monitoring was done at the lega dembi
gold mine in order to monitor the effects of air, water, soil and noise in the work environment. Thus, by
comparing the obtained results monitored from the
site with those standards used in the country shows
that the situation is still needed to be improved.

Table 5. Result of soil analysis.

Sampling Code	As	Ba	Cd	Co	Cr	Cu	Mn
ss-1	103		0.1	21	175	26	
ss-2	383		0.15	30	289	42	
ss-3	309		0.16	38	122	69	
ss-4	152		0.15	34	56	40	
ss-5	414		0.15	30	289	114	
ss-6	82		0.1	39	58	33	
ss-7	52		0.1	50	57	34	
ss-8	206		0.16	84	90	64	

Sampling Code	Hg	Ni	Pb	Se	V	Zn
ss-1	150	98	13			106
ss-2	0.3	240	16			49
ss-3	0.2	101	68			62
ss-4	0.2	43	22			112
ss-5	0.1	202	74			55
ss-6	0.1	71	11			42
ss-7	0.2	61	10			58
ss-8	0.1	90	14			82

Source: Lega Dembi quarter report.

REFERENCES

BRGM, June 1991. Environmental impact and safety, Addis
    Ababa, Ethiopia.
Golder Associates Audit Report, October 1998. Baseline Envi-
    ronmental Audit Report of the Lega Dembi Gold Mine.
Golder Associates Audit Report, October 1998. Environmental
    Management plan for the Lega Dembi Gold Mine.
Mining Law of 1993, and Mining Regulation of 1994, Addis
    Ababa , Ethiopia.
UNIDO, February 2000. Provisional guidelines and standards
    for industrial pollution control in Ethiopia, Addis Ababa,
    Ethiopia.

Tailings and Mine Waste '02, © 2002 Swets & Zeitlinger, ISBN 90 5809 353 0

Variations in composition on SA gold tailings dams

N.J. Vermeulen & E. Rust
Department of Civil Engineering, University of Pretoria, South Africa

C.R.I. Clayton
Department of Civil Engineering and Environmental Engineering, University of Southampton, United Kingdom

ABSTRACT: Natural soils consist of relatively loose assemblages of discrete mineral and organic particles of various shapes and sizes. Although it is convenient to simplify them as continuous media for the purposes of analysis, properties at particle level may ultimately control their engineering behavior. Gold tailings can be considered a "man-made" soil with properties between those of sand and clay. The soil forming processes on a typical tailings impoundment lead to a highly layered vertical profile, with coarse layers ("sands") alternating with fine layers ("slimes"). In addition to the vertical layering, the horizontal spatial distribution is also highly variable and dependent on the properties of the tailings slurry and the depositional program. In South Africa gold tailings impoundments are almost always constructed using the daywall-nightpan paddock system. In designing a tailings impoundment it is assumed that the more competent coarser material will be deposited near the embankment and the finer material in the central part of the impoundment. This paper will illustrate that although there is a general trend of finer composition towards the central area, a significant amount of fines are trapped in the day-wall and settle on the beach. The composition, i.e. mineralogy and particle properties of both fine and coarse materials has been investigated using electron microscope and x-ray techniques, and is described.

1 INTRODUCTION

The basic properties of the solid phase of a soil that are of importance from a soil mechanics point of view are mineralogy, particle density, size distribution, shape and surface texture. The mineral composition influences a diverse range of characteristics including frictional strength, electrical charge imbalances so important in clay minerals, stiffness, crushing resistance as well as bonding. Particle density on the other hand is often used in phase relationships to calculate saturated densities and void ratios as a function of water content. Grading properties influence relative density, limiting densities and hydraulic conductivity. Particle shape and surface texture are also significant properties in determining the frictional characteristics or shear strength of especially granular soils.

It is remarkable that so little of the routine soil mechanics tests are able to detect these basic properties, considering their importance. It is only grading by way of sieve testing that is able to determine directly the particle size distribution down to approximately 63 μm. The hydrometer sedimentation test relies on the assumption of spherical particle shape that is known to be unrealistic for most soil fines. Other tests such as the Atterberg limits and shrinkage tests are concerned more with the behavioral aspects of a soil under certain moisture conditions, than identifying fundamental properties.

The electron microscope has proven to be quite useful in the field of geosciences. Together with its high resolving power and extremely large depth of field of focus, the Scanning Electron Microscope (SEM) has the ability to render a clear and accurate image of the specimen surface and geometry. In addition the SEM has the ability to perform a qualitative and quantitative spot analysis of the elemental composition of a particle through characteristic x-ray emission (EDS). Powder X-ray Diffraction (XRD) is one of the primary techniques used by mineralogists and solid state chemists to examine the physico-chemical make-up of unknown solids.

This paper discusses briefly the fundamental properties of Witwatersrand gold tailings from South Africa and illustrates some of the spatial variations of these properties on a typical impoundment constructed by the upstream daywall-nightpan paddock system. It will be shown that this method of construction not only leads to entrapment of significant fines in the structural areas of the embankment wall, but also to deposition of coarse material into the semi-permanent pond.

2 PROPERTIES OF GOLD TAILINGS FROM THE LITERATURE

Although tailings literature abound, very little is known about the fundamental properties of this "man-made" material. The following represents a summary of the main findings from a comprehensive literature review:

- *Slurry*: Gold tailings can be classified as a low plasticity, fine, hard and angular rock flour, slurried with process water in a flocculated slightly alkaline state together with soluble salts (Vick, 1983; McPhail & Wagner, 1989). The flocculated state of the slurry promotes low post sedimentation densities, as the material does not readily segregate (Truscott, 1923; Hamel and Gunderson, 1973).
- *Rheology*: The rheology of a gold tailings slurry lies somewhere between a Bingham plastic and a Newtonian fluid (Blight & Bentel, 1983; Blight, 1988; Blight, 1994).
- *Mineralogy*: Quartz is by far the most abundant mineral in gold tailings with small quantities of phyllosilicates as well as pyrites and other sulphides (Donaldson, 1965; Hamel & Gunderson, 1973; Mlynarek et al., 1995). Specific gravity ranges between 2.5 and 3.0 (Pettibone & Kealy, 1971). Oxidation of sulphide minerals in tailings (especially pyrite, FeS_2) results in the production of ferrous and ferric oxides and sulphuric acid (Cowey, 1994). This in turn lowers the pH of the tailings water and may lead to leaching of toxic substances, which become soluble in an acidic environment.
- *Grading*: Gold tailings gradings (Figure 1, fully dispersed) are generally limited to and uniformly distributed in the silt size range with small percentages of fine sand and clay sized particles (Pettibone & Kealy, 1971; Van Zyl, 1993).
- *Particle Shape and Texture*: The sands or coarser fraction of gold tailings range in shape from very angular to sub-rounded (Mittal & Morgenstern, 1975; Lucia et al., 1981; Garga & McKay, 1984; Mlynarek et al., 1995). The fines are invariably angular, sometimes needle shaped, with very sharp edges and resemble shards of broken glass under the microscope (Hamel & Gunderson, 1973). Papageorgiou et al. (1999) mention, in addition to the irregular shapes of the particles, also harsh surface textures.

3 THE WITWATERSRAND GOLD TAILINGS

The Witwatersrand Goldfields discovered in 1884 constitute the largest known deposit of gold in the world. The sediments were laid down between 2.7 and 3 billion years ago in a large basin south of Jo-

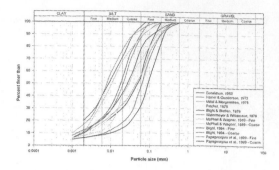

Figure 1. Grading curves from gold tailings literature.

hannesburg and are derived from the surrounding Archaean granite-greenstone terrains (Stanley 1987). The deposits lie in an oval area of approximately 42,000 km^2 in Gauteng, North-West Province and the Freestate Province. Throughout the Witwatersrand, gold ores occur in sheets or reefs originally deposited horizontally under water. The reefs were subsequently covered by material up to thousands of meters deep. Following the consolidation and cementation of these layers, geological movements transformed it into tilted and faulted strata. The thickness of the reefs ranges between a line of grit to several meters, with an average of 300 mm. The sediments were also intersected by dykes and sills of dolerite, diabase and syenite.

The gold bearing reefs can be in the form of either coarse conglomerates or, less frequently, grayish metamorphosed sedimentary rock formations. In the conglomerates, rock pebbles are cemented in a silicate matrix. Pebbles, usually derived from vein quartz, may also consist of quartzite, chert jasper and quartz porphyry and vary in composition, size and color. The matrix consists of pure silica, but also contains minute flakes of muscovite and pyrophyllite as well as visible pyrite and other sulphides. Table 1 summarizes the mineral composition of a typical gold reef on the Witwatersrand.

Table 1. Mineral composition of a typical Witwatersrand gold reef (Stanley, 1987).

Mineral	Abundance
Quartz	70 - 90%
Phyllosilicates (Clays)	10 - 30%
Pyrites	3 - 4%
Other sulphides	1 - 2%
Grains of primary minerals	1 - 2%
Gold	~45 ppm

4 IMPOUNDMENTS IN SOUTH AFRICA

In South Africa, most gold tailings dams are constructed using the upstream, daywall-nightpan pad-

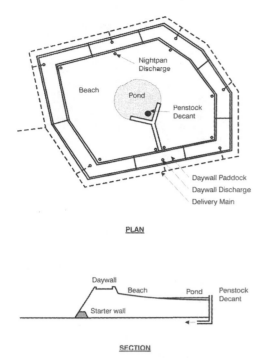

PLAN

SECTION

Figure 2. Layout of a typical South African gold tailings impoundment (McPhail & Wagner, 1989).

dock method, Figure 2, where the entire structure, except for a small initial starter wall, is built with tailings. A typical impoundment is divided into two sections, the embankment or daywall and the interior or nightpan. The daywall is designed to provide sufficient freeboard to retain the accumulated water from deposited tailings and that from the design storm. The daywall is sectioned into paddocks around the perimeter, each paddock being filled from its midpoint by a delivery station. During the day-shift pulp is delivered into these daywall paddocks and distributed by gravity. Excess or supernatant water is decanted into the nightpan with the understanding that most if not all the fines are passed into the nightpan with this water. The need for supervision and close control of the pulp depth makes daywall raising entirely a daytime procedure, hence the name. During the night, tailings are discharged into the nightpan from delivery stations located just inside the daywall. Clear supernatant water is drawn off the next day by penstock decant or barge pump. A natural beach forms in the nightpan from the delivery point towards a semi-permanent pond surrounding the decant facility. Deposition into the paddocks are cycled to allow time for desiccation and consolidation of the embankment material, thus improving its mechanical properties by densification. Deposition and the progression of flow and

sedimentation of the tailings slurry on such an impoundment are well described by Bentel (1981).

Most authors agree that there is a general tendency for decreasing permeability towards the pond as a result of increasing fineness in the material deposited (Jerabek & Hartman, 1965; Kealy & Busch, 1971). However, our observations will show that at any given location on a dam the composition of individual layers can vary significantly over small depths as a function of the properties of the delivery slurry and depositional practices.

5 EXPERIMENTAL WORK

Two tailings impoundments from Vaal River Operations, west of Johannesburg, were investigated. The first, Pay Dam, is the oldest dam on this particular mine and was originally started in the 1940's. It has not been used since the mid 1970's and is currently being reclaimed at approximately 300,000 tons per month. The latter, Mizpah (Figure 6a), was commissioned in 1993 and receives approximately 150,000 tons of tailings per month. The dam was designed for a final height of 60 m with a total surface area of approximately 165 Ha. The average rate of rise is 2.4 meters per year or approximately 200 mm per month, with one deposition cycle taking between 9 and 10 days. Both dams have been constructed with the upstream daywall-nightpan system.

Figure 3. Highly layered nature of a 5 meter exposed tailings profile next to the penstock.

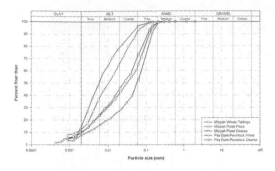

Figure 4. Fully dispersed grading curves from sieve and hydrometer tests.

(a)

(b)

Figure 5. SEM micrographs (a) 150 μm and (b) <10 μm tailings particles.

(a)

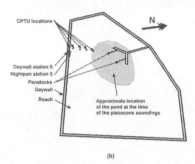

(b)

Figure 6. Site layout at Mizpah.

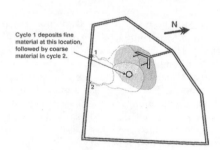

Figure 7. Conceptual model of deposition on a typical SA tailings dam.

Table 2. Experimental program.

Test Type	Test Method	Comments

Specific Gravity
 	
G_s = 2.74 Mg/m³
for all samples
	Density Bottle
	(500 ml Pyknometer)
	BS1377: Part 2: 1990:8.3
		Special care was taken to ensure full de-aeration of the samples (Sherwood, 1970) including:
		    Using a 500 ml pyknometer
		    Using vacuum and stirring to assist de-aeration
		    Placing sample in vacuum overnight.

Grading
Table 3
Figure 4
	Wet Sieving
	BS1377: Part 2: 1990:9.2
	Hydrometer
	BS1377: Part 2: 1990:9.5
		Pre-treatment included:
		    Test for calcareous content (*negative*)
		    Dispersion with Calgon
		    Removal of organic material (H_2O_2)

Imaging *Figure 5*	Scanning Electron Microscope	15 specimens were selected from the whole tailings slurry to be representative of specific size fractions in the tailings grading. Specimens were coated with a thin conductive layer of gold to prevent charge build-up and interference with image quality.
X-ray Analysis *Table 4*	Energy Dispersive Spectrometry X-ray Diffraction	EDS gave the elemental composition of individual particles imaged in the SEM. XRD provided the mineralogical composition of the 5 tailings samples studied.

Reclamation of Pay Dam afforded a rare opportunity to gain access to 5 m deep profiles within the dam, where an open face has been cut by hydro cannon. Figure 3 shows such a profile next to the penstock of this dam.

A second site was investigated on the upper beach area of this dam and was similarly layered (Vermeulen, 2001). No excavations were made on Mizpah, although a cross section of piezocone tests were performed from the daywall, across the beach and into the semi-permanent pond area (Figure 6). Bulk samples of the tailings delivery slurry, representing the whole tailings mix, were collected for laboratory testing. In addition, representative samples of coarse and fine layers at both Pay Dam (from the open faces) and Mizpah (from the surface of the pond area) were also collected.

To investigate the fundamental properties of the tailings a number of standard soil tests as well as electron microscope and x-ray analyses were performed (Table2). Figure 8 shows the piezocone test results.

6 DISCUSSION

6.1 Composition of SA gold tailings

The results of the EDS and XRD analyses were used to determine the mineralogical composition (Table 4) of the tailings samples considered in this study. Based on these limited results the mineralogical composition of Witwatersrand gold tailings is as follows:

- Quartz: 75% ranging from 59% to 83%
- Muscovite: 8% ranging from 7% to 9%
- Pyrophyllite: 5% ranging from 1% to 17%
- Illite: 5% ranging from 3% to 11%
- Small percentages of Clinochlore, Kaolinite and Pyrite.

The coarse particles (sands) are basically pure quartz except for a small percentage of illite clay identified, most probably electro-magnetically attached to the quartz grains. The slimes are predominantly quartz but with significant amounts of pyrophyllite, muscovite and illite clay minerals as well as traces of kaolinite and pyrite. These percentages are also in good agreement with the composition of the Witwatersrand gold reef according to Stanley (1987): Table 1.

The gradings performed on all samples in this study indicated very little, less than 2% per mass, coarser than 200 µm (limit of fine sand) and generally of the order of 10% smaller then 2 µm (clay sized). The remaining material was distributed in the silt and fine sand size ranges as summarized in Table 3. The fine graded samples all have an abundance of fines, which tend to coat and push the coarser particles apart. The behavior of these materials should, therefore, be governed by the fines fraction.

The shape of the particles is fully as important as the size in determining the engineering behavior of a soil. Angularity has a profound influence on engineering behavior. Under load the angular corners break and crush, and particles tend to resist displacement. However, vibration and shock cause loose arrangements of angular bulky grains to be displaced easily. More rounded particles are less resistant to displacement, but can be less likely to crush depending on the surface texture. Tailings sands are bulky particles. The grains imaged on SEM micrographs are highly angular to angular and generally flattened, sometimes elongated with sharp edges. These observations are consistent with the products of rock crushing and grinding. Tailings slime particles are generally flaky grains consisting of disintegrated mica, clay and quartz minerals with very sharp edges. Compared with the bulky sands, the slimes should be more compressible and behave like an intermediate plasticity clay. The fine sample collected from the Pay Dam penstock site consisted mostly of slimes and had a PI of 17%. The slimes particles will be much more susceptible to surface and electromagnetic forces than the body force of gravity, which is the predominant force acting on the sand particles. The fact that the slimes can be flocculated is evidence of the effects of the surface forces. The change in properties from bulky to flaky, i.e.

Table 3. Results from the x-ray analyses on the tailings.

Sample	Mizpah			Pay Dam	
	Whole	Fine	Coarse	Fine	Coarse
Quartz	69	75	83	59	79
Muscovite	8	9	8	7	9
Pyrophyllite	5	4	1	17	5
Illite	6	6	5	11	3
Clinochlore	2	3	2	3	3
Kaolinite	2	2	1	3	1
Gypsum	7				
Pyrite				1	

Table 2. Summary of the grading properties of gold tailings.

Tailings Dam	Description	Median Particle Size D_{50}		CU	CC
		(µm)	Description		
Mizpah	Whole Tailings	30	Coarse silt	28	0.9
	Pond Fines	10	Medium silt	11	0.8
	Pond Coarse	60	Coarse silt	25	2.8
Pay Dam	Penstock Fines	6	Fine Silt	5	0.8
	Penstock Coarse	25	Coarse silt	22	0.9

sand to slime, was identified at approximately 20 μm from the full range of micrographs published by Vermeulen (2001).

Inspection of the electron micrographs revealed the coarser or sand tailings particles to exist either with completely smooth surfaces or rough and irregular surfaces. The surfaces on the smooth sands appear to be the result of splitting and breaking of larger particles in the crushing and grinding processes of cumminution. These surfaces show the typical concave geometry of pure quartz when broken. On the other hand, sand particles with irregular surfaces may have formed by fines attaching themselves to the particles and/or as a result of shattering and chipping of the particle rather that splitting during the reduction process. Individual particles of tailings slimes have very smooth and flat surfaces. However, even dispersed there were some agglomerations and flocks of these flakes, which as a whole present a rough and irregular surface.

6.2 Variations in composition on a typical SA tailings impoundment

Material deposited on a tailings impoundment, whether sub-aerial or sub-aqueous, contain various percentages of sand and slime resulting in fine to coarse layers or mixtures. The spatial distribution of these fine and coarse layers vary significantly within a tailings impoundment as a function of the properties of the slurry (density, water content etc.), depositional practices (discharge method, sequence, duration, etc.) and geometry of the impoundment. Vertically there is almost an alternating sequence of fine and coarse layers throughout the dam, from the embankment and beach areas right into the pond. Horizontally, large areas of coarse or fine material are deposited in depositional plumes. These geometries result from the open ended discharge method where the material runs down the beach in scoured fast flowing channels with material, mainly coarse, deposited at hydraulic jumps. These channels widen into deltas at the pond interface and covers a fairly large area. Coarse material and fine material will be deposited as a function of the energy states. When deposition moves to the next outlet station the geometry of the depositional plume changes and where the previous cycle may have deposited fine material at a specific location, the new cycle may deposit coarse material etc. (Figure 7). Thus, even if there is a general increase in fineness of material towards the decant pond, coarse and fine material are deposited at any specific location as shown by the piezocone test results (Figure 8), and the in-situ profiles (Figure 3). The shaded areas on the piezocone data show the upper and lower trends in cone resistance, where the low values represent fine material and the high val-

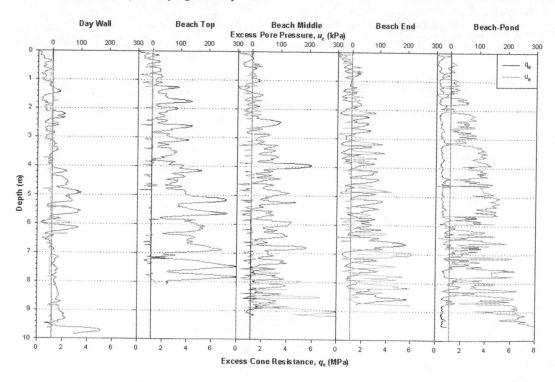

Figure 8. CPTU cross-section through Mizpah.

ues, coarse material. What is worrying though is the fact that the embankment wall profile, constructed by the paddock method, compares more with the pond profiles than the beach profiles. This confirms that large quantities of fine material is trapped within the embankment wall itself.

7 CONCLUSIONS

- This study has shown that tailings consist almost exclusively of tectosilicates (quarts) and sheet-like phyllosilicates (clays), with quartz being by far the most abundant mineral. The coarser tailings sands are predominantly pure silica quartz, with a small percentage of clay minerals. Tailings slimes on the other hand, although still dominated by quartz, contain significant amounts of phyllosilicates (20% muscovite, 15% illite and 20% pyrophyllite, kaolin and clinochlore) as well as traces of pyrite and other sulphides.
- *Specific gravity*: In this study the tailings had a specific gravity of 2.74 Mg/m^3, including the delivery slurry and all other samples collected, irrespective of sampling location.
- *Grading*: Fully dispersed gradings were uniformly distributed in the fine sand and silt-size ranges with approximately 2% coarser than 200 μm, 10% finer than 2 μm and at least 50% slimes. Variations in the gradings between fine and coarse tailings were largely around the median particle size, D_{50}, which ranged between 6 and 60 μm.
- *Particle shape*: The coarser tailings sands consisted of highly angular to sub-rounded bulky but flattened particles. The finer slimes consisted mostly of thin and plate-like particles characteristic of clay minerals.
  Operators of cyclone systems will benefit greatly from a knowledge of the size split between predominantly coarse bulky grains and flaky fines in the delivery slurry, roughly 20 μm in this case.
- *Surface texture*: Surface textures ranged from completely smooth to rough on a micro-scale.
- *Variations in composition on a typical SA tailings dam*: This study has shown that the composition of material varies significantly on an impoundment both with depth and spatial distribution. These variations are the result of the slurry properties, but more importantly the depositional practices and geometry of a dam. Depositional plumes from one outlet station overlap those from other stations and lead to an almost alternating deposition of coarse and fine layers at any specific location. What is surprising is that although there is a general trend of finer material towards the decant pond, significant amounts of fines are trapped in the embankment wall and upper beach areas.

REFERENCES

Bentel, G.M. (1981), Some aspects of the behavior of hydraulic deposited tailings, *Research Project: University of the Witwatersrand*.

Blight, G.E. (1988), Some less familiar aspects of hydraulic fill structures, *Hydraulic Fill Structures, ASCE Geotechnical Special Publication No. 21*, 1000-1027.

Blight, G.E. (1994), Master profile for hydraulic fill tailings beaches, *Proceedings of the Institution of Civil Engineers: Geotechnical Engineering*, 107(1), 27-40.

Blight, G.E., and Bentel, G.M. (1983), The Behavior of mine tailings during hydraulic deposition, *Journal of the South African Institute of Mining and Metallurgy*, 73-86.

Blight, G.E., and Steffen, K.H. (1979), Geotechnics of gold mining waste disposal, *Current Geotechnical Practice in Mine Waste Disposal*, ASCE, New York, 1-53.

Cowey, A. (1994), *Mining and Metallurgy in South Africa - a pictorial history*, MINTEK, 17-36.

Donaldson, G. (1965), The effects of capillary action on the consolidation and shear strength of silt in a hydraulic fill dam, *Proceedings, 6th International Conference on Soil Mechanics and Foundation Engineering*, Montreal, 454-463.

Garga, V.K., and McKay, L.D. (1984), Cyclic triaxial strength of mine tailings, *ASCE, Journal of Geotechnical Engineering*, Vol. 110, No. 8, 1091-1105.

Hamel, J.V., and Gunderson, J.W. (1973), Shear strength of Homestake slimes tailings, *ASCE Journal of the Soil Mechanics and Foundation Engineering Division*, Vol. 99, SM5, 427-431.

Jerabek, F., and Hartman, H. (1965), Investigations of segregation and compressibility in discharged fill slurry, *Transactions of the Society of Mining Engineers*, March, 18-24.

Lucia, P.C., Duncan, J.M., and Seed, H.B. (1981), Summary of research on case histories of flow failures of mine tailings impoundments, *Mine Waste Disposal Technology, US Bureau of Mines Information Circular*, IC8857/1981, 46-53.

McPhail, G.I., and Wagner, J.C. (1989), Disposal of residues, Chapter 11, in G.G. Stanley (ed.), The *Extractive Metallurgy of Gold in South Africa*, The Chamber of Mines of South Africa, Volume 2, 655-707.

Mittal, H.K., and Morgenstern, N.R. (1975), Parameters for the design of Tailings dams, *Canadian Geotechnical Journal*, Vol. 12, 235-261.

Mlynarek, Z, Tschuschke, W., and Lunne, T. (1995), Use of CPT in mine tailings, *Proceedings, International Symposium on Cone Penetration Testing, CPT'95*, Linköping, Vol. 3, 211-226.

Papageorgiou, G.P., Fourie, A.B., and Blight, G.E. (1999), Static liquefaction of Merriespruit gold tailings, *Proceedings, 12th Regional Conference for Africa on Soil Mechanics and Geotechnical Engineering*, Durban, South Africa, 61-72.

Patchet, S. (1977), Fill support systems for deep-level gold mines, *Journal of the South African Institute of Mining and Metallurgy*, Sept., 34-46.

Pettibone, H., and Kealy, D. (1971), Engineering properties of mine tailings, *ASCE, Journal of the Soil Mechanics and Foundations Division*, Vol. 97, No. SM9, 1207-1225.

Sherwood, P.T. (1970), The reproducibility of the results of soil classification and compaction tests, *Road Research Laboratory Report No. LR 339*, Road Research Laboratory, Crowthorne, Berks.

Stanley, G.G. (ed.) (1987), *The Extractive Metallurgy of Gold in South Africa*, The S.A. Institute of Mining and Metallurgy Monograph Series M7, The Chamber of Mines of South Africa, Volume 1 & 2.

Stokes, Sir George G. (1891), *Mathematical and Physical Paper III*, Cambridge University Press.

Truscott, S.J. (1923), *A Text-Book of Ore Dressing*, Macmillan and Co.

Van Zyl, D. (1993), Mine waste disposal, in D.E. Daniel (ed.), *Geotechnical Practice for waste disposal*, Chapman Hall, 269-286.

Vermeulen, N.J. (2001), *The Composition and State of Gold Tailings*, PhD thesis, University of Pretoria, South Africa.

Vick, S.G. (1983), *Planning, design and analysis of tailings dams*, New York: Wiley.

Watermeyer, P., and Williamson, R. (1979), Ergo tailings dam - Cyclone separation applied to a fine grind product, *Proceedings, 2nd International Tailings Symposium*, San Francisco, 369-396.

Tailings and Mine Waste '02, © 2002 Swets & Zeitlinger, ISBN 90 5809 353 0

Preliminary ecological risk assessment for the Elizabeth Mine site, South Strafford, Vermont

I. Linkov & S. Foster
Arthur D. Little, Inc., 20 Acorn Park, Cambridge, MA 02140, U.S.A., linkov.igor@adlittle.com

E. Hathaway & R. Sugatt
USEPA Region I, Boston, MA, U.S.A

ABSTRACT: The Elizabeth Mine, located in South Strafford, Vermont, is one of the oldest and largest hard-rock former metal-sulfide mining sites in New England. The site has been listed on the National Priority List by the United States Environmental Protection Agency (USEPA) and investigations are underway at the site to determine the environmental impact of tailings and waste rock on the surrounding area and downstream receiving waters. This paper presents a preliminary-level ecological risk assessment approach in planning and evaluation to support an early cleanup action. Extensive sampling of surface water, sediments, ground water and soils was conducted to establish the spatial and temporal pattern of contamination in different media. A screening level ecological risk assessment was conducted to identify exposure pathways and contaminants that pose significant risks to ecological receptors. Chemical-specific hazard quotients (HQs) and hazard indices (HIs) were found to be elevated in the immediate vicinity of the source areas, in the Ompompanoosuc River and two affected tributaries. Multiple lines of evidence were used in the evaluation, including surface water and sediment chemistry, surface water and sediment toxicity tests, benthic organism studies, and fish population studies. The measurement endpoints included sediment and water toxicity testing for the site and reference areas, and analyses of composition and structure of the benthic and fish communities in affected sites and selected reference areas. Each line of evidence, independently and in conjunction, confirmed that the site contamination has resulted in significant detrimental effects on ecological receptors. A decrease in chemical concentrations in surface water with distance downstream is correlated with ecosystem recovery, with site-related impacts observed several miles downstream of the source areas. The ecological risk evaluation is critical at a site where the risk basis will be ecological impact rather than human health. The use of ecological risk as a basis for a Non-Time-Critical Removal Action (NTCRA) is not common in the Superfund program. A NTCRA, however, can be an effective way to accomplish rapid risk reduction (ecological or human health).A NTCRA can be implemented at the same time as the Remedial Investigation/Feasibility Study (RI/FS) allowing for prompt risk reduction for the more obvious site hazards while the comprehensive investigation is completed.

1 INTRODUCTION

A Superfund cleanup action must be based upon a finding that there is a current or future potential threat to human health and/or the environment. The human health and ecological risk assessment (HHRA and ERA respectively) process provides a framework for this determination. The risk assessment also helps to identify the contaminants that are causing the impacts, so that cleanup options can be developed. The risk assessment is a critical step in the Superfund process. USEPA does not undertake a cleanup action if the result of the risk assessment indicates that the level of risk to human health or the environment is within USEPA's acceptable risk range. The Superfund cleanup of the Elizabeth Mine site is being implemented as a two phase program. The first phase, or Non-Time Critical Removal Action (NTCRA), consists of an early cleanup action

that targets the acid mine drainage (AMD) generated by the tailings and waste rock. The second, or "Remedial" phase, will involve a comprehensive investigation and subsequent cleanup action of any remaining threats posed by the Site.

Most of the planned NTCRAs at Superfund sites have been conducted based on findings of apparent risks to human health, often deriving from a screening of site concentration data against concentration benchmarks that USEPA considers acceptably safe for long-term human contact. Since the objective of any HHRA is to detect risks to the most sensitive individual in the population, the use of this conservative approach is justified to support a NTCRA decision.

An ERA should generally assess population level effects, rather than effects on individual organisms, except in situations where endangered or threatened species occur. If such species are absent (as in the affected river at the Elizabeth Mine – West Branch of the Ompomponoosuc River - WBOR), establishing that an individual ecological receptor is at risk does not provide a sufficient basis for implementing a NTCRA. For a site with little or no human health risks, a NTCRA decision could not be justified based solely on a screening level ecological risk assessment; more extensive analyses are required.

This paper illustrates the use of an ERA to support a NTCRA decision regarding a former mining site. A two-step ERA evaluation process, consistent with USEPA Guidance (USEPA, 1997), was proposed. First, concentrations of contaminants measured in surface water and sediments were used to identify the Contaminants of Concern (COCs)— chemicals most likely to cause an impact. This data was then used to develop Hazard Quotients (HQs) for each COC; these are equal to the site contaminant concentration divided by the selected safe benchmark concentration for ecological receptors. The HQ represents how many times greater a given sample concentration is than the "safe level".

The second step of the ecological assessment involves the use of biological measures of effects: in this case, fish diversity and abundance, benthic community diversity and abundance, and toxicity tests. Contaminant levels above the numerical water quality criteria do not always result in a significant impact to aquatic organisms; therefore, biological assessments were used to provide a more direct evaluation of any impact from site-derived contaminants.

2 SITE HISTORY AND SETTING

The Elizabeth Mine is located in east-central Vermont, in the Copperas Brook watershed, which drains into the West Branch of the Ompompanoosuc River (WBOR), approximately 6.2 miles (10 kilometers) upstream from its confluence with the East Branch of the Ompompanoosuc River. The Ompompanoosuc River empties into the Connecticut River approximately 9 miles (15 kilometers) downstream from the site (see Figure 1).

Beginning in the early 1800s, the mine's massive sulfide ore body (Beshi-type) was exploited for pyrrhotite to manufacture "copperas", a hydrated iron sulfate salt. By 1830 copper was extracted and processed on site from small amounts of chalcopyrite in the pyrrhotite matrix; by the late 1800's no fewer than 8 separate furnaces were actively smelting copper ore. In 1942 the U.S. government sponsored a revitalization of the mine and construction of a modern flotation plant for copper extraction in support of the war effort. Mining was conducted in two long, narrow open cuts (North Cut and South Cut) as well as underground workings extending over 1 mile in

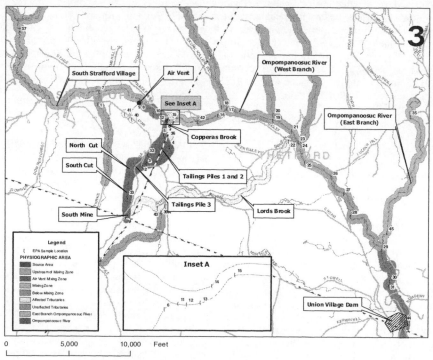

Figure 1. Data groupings by physiographics areas Elizabeth Mine, South Strafford, VT.

54

length. By 1958, more than 100 million pounds of copper were produced over a nearly 150 year life span. As a result of these activities, acid mine drainage (AMD) originating from the tailings, waste rock, and copperas heap leach piles has contaminated local surface water, sediments and soils with metals, including copper, cadmium, aluminum, and zinc.

Waste materials from the mining and milling operations exist today as tailings, waste rock piles, heap-leach piles, and smelter slag at various locations around the mine site. When exposed to natural conditions at the ground surface, these materials break down and release metals and acid contamination. Copperas Brook flows from its headwaters within waste rock piles in the upper portion of the watershed, onto the surface of fine-grained tailing piles from the WWII era operations. Much of this surface water infiltrates into the 35 acre tailings pile(s), exiting at the downstream base as acid mine drainage. Copperas Brook flows over a distance of nearly one mile, before it discharges into the WBOR. An additional source of AMD is a continuous discharge of acidic mine water through an air vent for the underground mine workings. This contaminant source combines with the WBOR upstream of the confluence with Copperas Brook.

A complicated spatial pattern of contamination results from redistribution of contaminants through surface water runoff as well as possible dispersion by wind and redistribution through groundwater. A variety of ecosystems (streams, river, wetlands, and uplands) throughout the mine area complicate the ecological risk framework and exposure pathways.

3 METHODS

3.1 Sampling locations

To assess the extent of environmental impact resulting from the Elizabeth Mine, we collected surface water and sediment samples throughout the impacted and reference watersheds in the mine area. Sample locations are broadly divided into the following nine groupings (see Figure 1):

- Contamination Source Areas (S) include locations within the Copperas Brook watershed and the Air Vent (discharge point for contaminated underground mine pool) prior to discharge into the WBOR;
- Unaffected tributaries to the West Branch of the Ompompanoosuc River (UT) include Sargent Brook, Abbott Brook, Fulton Brook, Jackson Brook, Bloody Brook, and lower Lord Brook;
- Affected tributaries to the West Branch of the Ompompanoosuc River (OR) include upper Lord Brook, two intermittent streams on Mine Road, and an intermittent stream within the Copperas Brook drainage;

- WBOR upstream of Mixing Zone (UMZ) includes the WBOR upstream from the Air Vent and Copperas Brook;
- Air Vent Mixing Zone (AMZ) includes locations within the WBOR between the Air Vent and the confluence with Copperas Brook – approximately 2500 feet in length;
- WBOR Mixing Zone (MZ) includes the section of the WBOR from Copperas Brook confluence to a point approximately 2500 feet downstream;
- WBOR Below Mixing Zone (BMZ) includes the stretch of WBOR between the EBOR/WBOR confluence and USEPA sample location No. 42;
- East Branch of the Ompompanoosuc River (EBOR); and
- WBOR below confluence of EBOR and WBOR.

3.2 Surface water sampling

Surface water samples have been collected at a total of 46 locations throughout the Elizabeth Mine area for use in the ERA. All locations were sampled at least three times in 2000. In April-May 2000, weekly stream sampling was conducted at six locations to evaluate spring runoff metals loading and pH. Monthly sampling was conducted at a subset of locations to assess possible seasonal trends. In addition, episodic sampling was conducted to assess effects from storm events.

3.3 Sediment sampling

Samples of sediment were collected following VTANR (2001) at each surface water sampling location, plus several additional locations, and submitted for metals analysis during two sampling events (June and September/October 2000). In July 2000, 41 locations were sampled for total metals, acid volatile sulfide/simultaneously extracted metals (AVS/SEM), grain size, and total organic carbon (TOC). One location was sampled for cyanide, and five locations were sampled for volatile organic chemicals (VOCs), base neutral and acid extractable organic compounds (BNA), pesticides, and PCBs. (organic compounds were not detected at significant concentrations in sediment). In October 2000, 11 of the 41 locations were sampled for total metals and AVS/SEM.

3.4 Benthic invertebrates

A benthic macroinvertebrate survey was performed in conjunction with the October 2000 surface water and sediment sampling program. A total of 22 sample stations were sampled; 5 for epifauna only; 2 for infauna only; and 15 for combined epifauna/infauna sampling. Epifauna are benthic invertebrates that live on the surface of the sediment substrate. Infauna are benthic invertebrates that live within the sediment. A total of 59 replicate samples have been analyzed to date; the remainder are currently being analyzed. Epifauna samples were collected by placing a 18-inch by 9-inch, 500 μm mesh dip net vertically on the substrate, perpendicular to flow, followed by thoroughly disturbing the substrate within an 18-inch by 18-inch area upstream from the net so that dislodged organisms were washed downstream into the net. The substrate was also rubbed clean of attached organisms. At each station, the substrate was agitated for 30 seconds at four locations, two in fast water and two in slow water, and combined to comprise one replicate sample. Infauna samples were collected using a stainless steel petite Ponar grab sampler, which samples a 6-inch by 6-inch area of the substrate. Taxonomic analysis and bioassessment of the benthic community was conduced in accordance with guidance provided by Vermont Agency of Natural Resources (VTANR, 2001). Epifauna data were analyzed using the following metrics:

- Density
- Taxa richness
- EPT index
- % Oligochaeta
- % modal affinity of orders
- Hilsenhoff Biotic Index
- Pinkham-Person Coefficient of Similarity.

3.5 *Surface water and sediment toxicity tests*

Toxicity tests using aquatic invertabrates and fish were conducted to evaluate the effect of exposure to surface water and sediment from the site on selected organisms. The tests were conducted following procedures detailed in EPA (1994) and EPA (2000). The test organisms included fathead minnow, *Pimephales promelas*; scud, *Hyalella azteca*; bloodworm, *Chironomus tentans*; and water flea, *Ceriodaphnia dubia*. Toxicity tests evaluate cumulative effects of chemicals by introducing healthy organisms to site surface water or sediment for a specific time period. The same types of organisms were exposed to upstream (reference) area surface water or sediment over the same test period as control groups. Two rounds of toxicity testing were performed, corresponding to the June and September sampling events.

4 STEP 1: SCREENING LEVEL ECOLOGICAL RISK ASSESSMENT AND COC IDENTIFICATION

4.1 *Methodology*

Hazard Quotients (HQs) were used to identify COCs and calculate *potential* ecological risks from COCs for each of the nine general site areas (Source Area, Mixing Zone, etc.). This approach uses well-defined and scientifically established chemical-specific exposure levels that have not shown impact to aquatic receptors in studies and which are believed to be non-toxic to the aquatic community. The HQ is the quotient of the site contaminant concentration divided by the non-toxic concentration, and describes how many times greater a given sample concentration is than the "safe level". Safe levels were taken from the numerical Vermont Water Quality Standards (VTWQS, 2000), when available. Several constituents in surface water did not have a VTWQS. For these constituents, USEPA identified a scientifically valid safe level from available literature. There are no Vermont standards for sediment; therefore, all of the safe levels for sediments were from USEPA-accepted sources. Table 1 lists the criteria used to assess each contaminant in surface water and sediment.

The HQ is calculated by dividing the metal concentration (at any given location) by its corresponding, medium-specific benchmark (e.g., VTWQS). Because more than one metal of concern is present in the Elizabeth Mine area media, chemical-specific HQs were summed to estimate total risk to ecological receptors (e.g. fish) or humans. This sum is called the Hazard Index (HI), and is based on the assumption that risk from different metals is additive. In other words, the total potential risk to ecological receptors is the sum of risks posed by individual contaminants. A HQ or HI greater than one indicates potential harm to ecological receptors and that further evaluation using biological measures is warranted.

Table 1. Hazard indices for surface water (metals)

	Value	Reference
Surface Water (ug/L)		
Aluminum (pH 6.5-9.0)	87	EPA, 1999
Barium	3.9	EPA, 1996
Cadmium	2.2	EPA, 1999
Chromium VI	11	EPA, 1999
Cobalt	3	EPA, 1996
Copper	9	EPA, 1999
Cyanide	5.2	EPA, 1999
Iron	1,000	EPA, 1999
Lead	2.5	EPA, 1999
Manganese	80	EPA, 1996
Selenium	5	EPA, 1999
Silver	0.36	Suter and Tsao, 1996
Thallium	12	Suter and Tsao, 1996
Vanadium	19	EPA, 1996
Zinc	120	EPA, 1996
Sediment (mg/kg)		
Aluminum (pH 6.5-9.0)	73160	Jones et al., 1997
Barium	NC	
Cadmium	0.99	MacDonald et al., 2000
Chromium VI	43.4	MacDonald et al., 2000
Cobalt	NC	
Copper	31.6	MacDonald et al., 2000
Iron	20,000	MOE, 1994
Lead	35.8	MacDonald et al., 2000
Manganese	460	MOE, 1994
Selenium	NC	
Silver	1	EPA, 1996
Thallium	NC	
Vanadium	NC	
Zinc	121	MacDonald et al., 2000

The following is a summary of the key findings of the evaluation of chemical data and development of HQs.

4.2 Surface water chemistry

The surface water data collected since April 2000 indicate that 15 contaminants arc detected at concentrations above VTWQS or USEPA criteria: aluminum, barium, cadmium, chromium, cobalt, copper, cyanide, iron, lead, manganese, selenium, silver, thallium, vanadium, and zinc. VTWQS are available for cadmium, chromium, copper, cyanide, iron, lead, selenium, and zinc. USEPA referred to available literature to establish the criteria for aluminum, barium, cobalt, manganese, silver, thallium, and vanadium.

Of these fifteen contaminants, nine appear to be clearly related to the source material (tailings and waste rock), based on concentration and frequency of occurrence in the source area samples: aluminum, cadmium, cobalt, copper, iron, manganese, selenium, silver, and zinc. These contaminants have been designated as COCs. The remaining six contaminants (barium, chromium, cyanide, lead, thallium, and vanadium) were determined to warrant further evaluation as part of the RI/FS to determine if they are truly site-related, based on concerns such as data quality, frequency of occurrence, and naturally occurring background levels.

Figure 2 provides a graphical summary of the surface water samples that exceed applicable criteria for the six COCs. Figure 2 shows that aluminum is the only contaminant that is consistently detected above acceptable criteria at the upstream locations. Figure 2 also shows that the concentration of contaminants in the Source Area and the Mixing Zone Area are frequently greater than the accepted criteria. The percent of samples exceeding accepted criteria drops with distance downstream from the confluence with Copperas Brook.

Figure 3 compares the HQs for the COCs for each area (e.g Source Area, Mixing Zone) to the levels detected upstream of the source area, using HQ ratios. For example, aluminum is detected in the Mixing Zone Area at concentrations 5 times higher than the Upstream of Mixing Zone Area. The HIs for surface water are summarized in Table 2.

Surface water quality within Copperas Brook and the Mixing Zone of the WBOR is clearly impacted by acid mine drainage from the site. The levels of contaminants detected in Copperas Brook and the Mixing Zone are many times the accepted safe levels. Significant impact to the aquatic organisms would be expected as a result of these exceedances. Copper is the only COC to remain significantly above upstream concentrations beyond Union Village Dam at USEPA Location 44, approximately 6

Table 2. Hazard indices for surface water (metals)

Area	Chronic		Acute	
	Avg	Max	Avg	Max
UT	12	47	3	21
UMZ	17	82	2	12
AMZ	29	335	4	48
S	715	5230	161	1483
MZ	64	415	10	77
BMZ	17	69	3	22
AT	15	49	5	21
OR	14	37	2	7

miles downstream of the confluence of Copperas Brook and the WBOR.

4.3 Sediment chemistry

HQs and HIs were calculated for sediments for each of the identified data groups (Source Area, Mixing Zone). The HQs for sediments are summarized in Tables 3.

Aluminum, iron and zinc concentrations in sediments do not display the strong site-related pattern observed for copper. Source Area concentrations for these metals are clearly elevated with respect to the upstream (reference) area; however, the downstream concentrations of iron and zinc are fairly close to upstream concentrations. Hazard Quotients and associated HIs for metals below the confluence of the EBOR and the WBOR are comparable to the Mixing Zone, suggesting that little to modest attenuation of metals contamination in sediment occurs with increasing distance from the Source. It should be noted that the HI for the Air Vent Mixing Zone sediment was comparable to the upstream areas, suggesting that the Air Vent may not represent a significant contaminant source for sediments, or that the Air Vent metals loading is transported downstream due to scour and re-deposition.

The sediment results for the Source and Mixing Zone areas of the WBOR indicate that certain metals are present at concentrations that could have an adverse effect on aquatic and sediment dwelling organisms. Copper levels in sediment, like surface water, continue to be elevated below Union Village Dam, where the copper HQ is five (5).

Table 3. Hazard indices for sediment (metals)

	Chronic		Acute	
	Avg	Max	Avg	Max
UT	4	6	1	3
UMZ	3	3	1	1
AMZ	3	3	0.9	1
S	46	194	12	47
MZ	8	19	2	5
BMZ	6	12	2	4
AT	6	11	2	3
OR	9	14	3	4

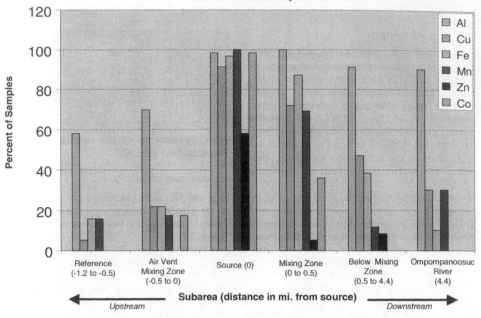

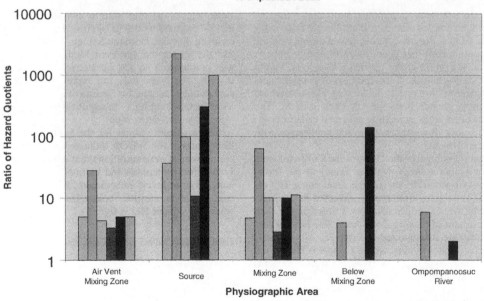

Figure 2. Surface water quality.

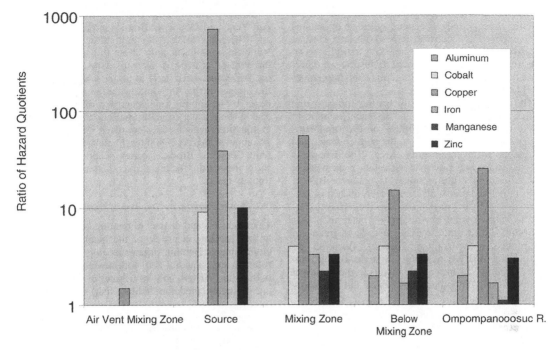

Figure 3. Comparison of sediment hazard quotients for upstream as compared to impacted areas.

4.4 Summary of step 1

Nine contaminants related to the source material were identified as COCs based on their concentration and frequency of occurrence in the Source Area samples: aluminum, cadmium, cobalt, copper, iron, manganese, selenium, silver, and zinc.

The surface water and sediment data document the potential for severe impact to Copperas Brook and a section of the WBOR as a result of the discharges from the Source Areas. All of Copperas Brook and a section of the WBOR fail to meet numerical VTWQS for several metals on numerous sampling occasions.

5 STEP 2: BIOLOGICAL MEASURES OF EFFECT

In addition to the evaluation of the chemical data in Step 1 of the Ecological Evaluation, several lines of biological evidence were examined to determine if significant biological impacts were being observed, including:

- Surface water and sediment toxicity tests: These tests involved the laboratory testing of surface water and sediments collected from the site, using representative organisms. The survival, reproductive success, and growth of these organisms were measured over a period of time to assess the impact, if any, from the site surface water or sediments.

- Benthic Community Surveys: These surveys measured the type and abundance of the bottom-dwelling organisms that often serve as the food for fish communities. The quality of the benthic community is an indicator of the health (in terms of both density and diversity) of the aquatic community.
- Fish abundance and species surveys: These surveys (carried out by the United States Army Corps of Engineers (USACE), and the Vermont Agency of Natural Resources (VTANR) compared density and diversity levels among fish populations to levels expected for the type of river environment represented by the WBOR.

In viewing the risk potential for COCs in the Elizabeth Mine area, these lines of evidence must be "weighted" with respect to the emphasis placed on each. Actual (direct) measures of biological impairment are given greater weight than indirect measures of risk that evolve from the calculation of HIs.

5.1 Surface water and sediment toxicity tests

Toxicity tests were conducted to evaluate the effect of exposure to surface water and sediment from the site on aquatic invertebrates and fish (fathead minnow, Pimephales promelas; scud, Hyalella azteca; bloodworm, Chironomus tentans; and water flea, Ceriodaphnia dubia). Toxicity tests evaluate cumulative effects of chemicals by introducing healthy organisms to site surface water or sediment for a specific time period. The same types of organisms were exposed to upstream (reference) area surface water or sediment over the same test period as control groups. Two rounds of toxicity testing were performed, corresponding to the June and September 2000 USEPA sampling events.

The results of the toxicity testing indicate that Source Area surface water and sediment is toxic to tested organisms. Nearly 100% of the organisms died as a result of exposure to the surface water and sediments from the Source Area. The Copperas Brook surface water was so toxic that even when it was substantially diluted (to levels as low as 10% of the original sample) with clean water, all test organisms died. All test organisms also died from exposure to surface water from sample Locations 8 (Air Vent) and 12 (within WBOR just downstream of confluence with Copperas Brook). Location 13 (within the WBOR, near the Copperas Brook confluence) showed similar toxic results in the sediment toxicity tests. All other areas tested did not show significant impacts.

Toxicity tests of water and sediment between the Air Vent and Copperas Brook (sample locations 9 and 10) indicate that contamination from the Air Vent did not significantly affect the survival, reproduction, or growth of test organisms. Even though the Air Vent water itself is toxic to test organisms, the impact of the Air Vent discharge on WBOR appears to be rapidly dampened by dilution effects. The surface water toxicity tests suggest that impact to aquatic organisms does not reach USEPA location 16, about 4000 feet (3800 feet) below the confluence with Copperas Brook. Figures 4 and 5 graphically show the results for the surface water and sediment toxicity tests.

5.2 Benthic organism community data

Taxa richness and density of benthic organism populations are key measures of the health' of the river environment. Benthic organisms play an important role in the aquatic and sedimentary community within the WBOR and its tributaries. Many of the native aquatic predators (mainly fish) rely on benthic

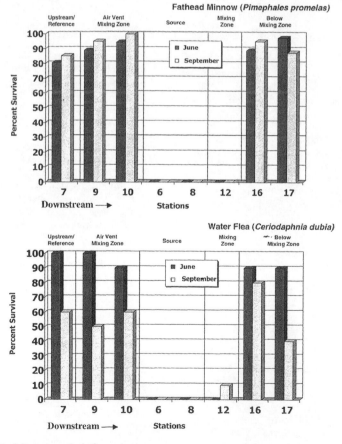

Figure 4. Surface water toxicity test: survival of organisms.

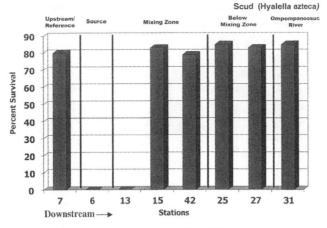

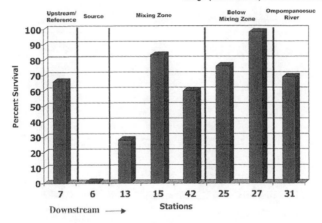

Figure 5. Sediment toxicity test: survival of organisms.

organisms as a food source. Contamination of surface water and sediment adversely affects the community density and diversity of species of benthic organisms present, thereby affecting species that rely on these organisms for food.

Taxa richness and density are severely depressed in Copperas Brook, the Mixing Zone, and the Affected Tributaries. The samples of the benthic community in the Air Vent Mixing Zone are similar in most respects to the upstream (Reference) areas of the Ompompanoosuc River (below EBOR and WBOR confluence) samples. These results indicate that the Air Vent contribution to the WBOR contamination is not significant in terms of biological impact, even though water chemistry results indicate the potential for impacts to the aquatic organisms in this stretch of the river.

The WBOR does not meet VTWQS for all three of the Vermont criteria evaluated for approximately 6 miles downstream of the confluence with Copperas Brook, near USEPA location 44. The WBOR, however, does achieve VTWQS for two of the three

measures by USEPA Location 19, just upstream of Rice's Mills. Figures 6 and 7 show the results for the epifauna survey. Statistical projections (plot of abundance and richness over distance from source) confirm that the VTWQS for all criteria should be met on the stretch of the WBOR near Union Village Dam. The approximate distance downstream of the confluence with Copperas Brook where all state criteria are met in the WBOR is about 6 miles.

5.3 Preliminary assessment of impact on fish communities

Fish density and diversity are key measures in the evaluation and analysis of impacts the Elizabeth Mine on the WBOR and affected tributaries. Studies by the U.S. Army Corps of Engineers (USACE, 1990) and VTANR (1987 and 2000) provide evidence that the Elizabeth Mine has had a severe impact on the fish communities in the WBOR and affected tributaries. In 1990, the USACE studied biomass of forage species fish (black nose dace,

61

Ompompanoosuc River Ecosystem

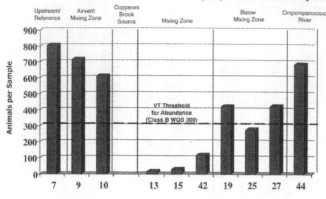

Tributary/Stream Ecosystem

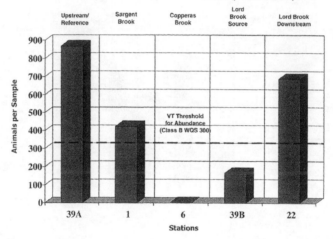

Figure 6. Benthic epifauna density.

longnose dace, slimy sculpin, longnose sucker) in two upstream and two downstream locations in the WBOR (relative to the Copperas Brook confluence). These locations had similar habitats and thus any difference in fish communities can be attributed to the detrimental effects of the mine. The USACE study showed that the biomass of the forage species upstream of the mine (4.5 kg/ha) was more than three times higher than biomass at the downstream locations (1.3 kg/ha).

A VTANR study showed the fish density (i.e. number of fish per unit area) in upstream locations in the WBOR to be almost eight times the density found at locations downstream of Copperas Brook (18.6 fish/100 m^2 upstream vs. 2.5 fish/100 m^2 downstream, VTANR 1987). VTANR calculated an Index of Biotic Integrity (IBI) value for the USACE and VTANR stations. The IBI quantifies the ecological health of the fish community as a whole. Figure 8 (top plot) presents IBI as well as fish den-

sity for the USACE and VTANR data. The IBI in the upstream areas of WBOR is 39 (as compared to the VT threshold values for Class B waters of 29 to 31), whereas the IBI for the WBOR below Copperas Brook is only 9.

A study conducted by the VTANR in the tributaries of WBOR (Langdon, 1987) noted more than a ten-fold reduction of fish density in Lords Brook downstream (1.1 fish/100 m^2) from the South Open Cut source area, as opposed to a stretch of Lord Brook upstream of the South Open Cut. Influence, showed 10.7 fish/100 m^2. No fish were found in Copperas Brook. Sargent Brook shows a fish density (9.9 fish/100 m^2) that is similar to unaffected areas of Lord Brook. The IBI was found to be excellent for the Sargent Brook (45 out of possible 45). Because of the small number of native species (for Lord and Copperas Brooks), the IBI could not be calculated. The study concluded that the low density of fish in these areas implies a toxic impact.

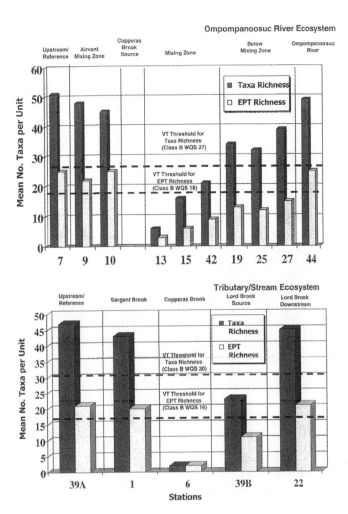

Figure 7. Benthic epifauna diversity.

6 CONCLUSIONS

Multiple lines of chemical and biological evidence regarding impacts to the biological community were evaluated in this study. Figure 9 provides an overall summary of all lines of evidence, indicating the extent of chemical and biological impact to the WBOR watershed from Elizabeth Mine source areas. Assessment of chemical and biological lines of evidence indicates that site contaminants are adversely affecting the aquatic communities.

The biological communities (benthic organisms and fish) are severely impacted in Copperas Brook, the upper reach of Lord Brook, below the South Open Cut, and in the Mixing Zone of the WBOR below Copperas Brook. The biological communities appear to recover to conditions similar to upstream (Reference locations) at some point below Union Village Dam. Surface water and sediment collected from Copperas Brook, the first section of the Mixing Zone, and the Air Vent are toxic to aquatic organisms. Toxicity tests indicate that these impacts are not present below the Mixing Zone.

Collectively, the various lines of evidence suggest that USEPA Location 44, situated six miles downstream from Union Village Dam, represents the best estimate for the location where the WBOR has recovered biologically to a level that is equivalent to upstream and meets the biological component of the VTWQS, even though chemical evidence from surface water sampling indicates the potential for concentrations above numerical VTWQS further downstream. The distance from the Copperas Brook confluence to USEPA Location 44 is approximately 6 miles.

Since all of the lines of evidence show that Copperas Brook and the Mixing Zone are the most severely impacted, it can be inferred that the source

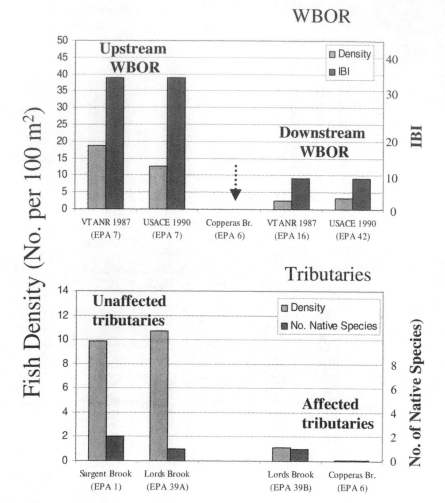

Figure 8. Summary of fish studies.

areas located within the Copperas Brook drainage are the major sources of adverse impacts to the WBOR These impacts firmly support the need for an early cleanup action (NTCRA) to address the principal sources of AMD.

Several studies are underway to further evaluate the extent of environmental damage at the Elizabeth Mine. EPA is evaluating cleanup options designed to restore the WBOR to Vermont biological standards for fresh water rivers. Additional studies are underway to assess impacts to fish populations and terrestrial receptors, including birds and small mammals.

The data collected as part of the Preliminary Ecological Risk Assessment will provide the basis for the proposed NTCRA. This paper illustrates the application of a multiple line of evidence approach to ecological risk assessment for determining the need to implement a NTCRA. The data will also serve as the foundation for the comprehensive Ecological Risk Assessment for the site.

REFERENCES

Jones et al, 1997. ORNL Toxicological Benchmarks for Screening contaminants of Potential Concern for Effects on Sediment Associated Biota: 1997 Revision

Long et al., 1995. Incidence of Adverse Biological Effects within Ranges of Chemical Concentrations in Marine and Estuarine Sediments. Environmental Management, 19 81.

MacDonald et al., 2000 Development and Evaluation of Consensus-Based Sediment Quality Guidelines for Freshwater Ecosystems. Arch. Environm. Toxicol 39:20-31

Ministry of Environment of Ontario (MOE). 1994. Proposed Guidelines for Clean-Up of Contaminated Sites in Ontario.

Suter. and Tsao, 1996. ORNL Toxicological Benchmarks for Screening Potential Contaminants of Concern for Effects on Aquatic Biota: 1996 Revision

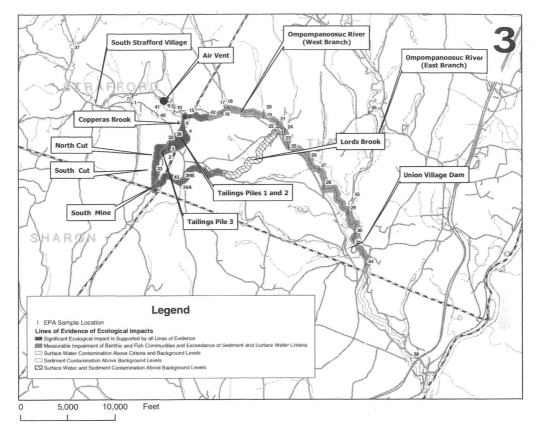

Figure 9. Lines of evidence of ecological impacts, Elizabeth mine, south Strafford, vt.

U.S. Army Corps of Engineers (USACE) 1990. Effects of the Abandoned Elizabeth Copper Mine On Fisheries Resources of the West Branch of the Ompompanoosuc River.

U.S. Environmental Protection Agency (EPA). 1994. Short-Term Methods for Estimating the Chronic Toxicity of Effluents and Receiving Waters to Freshwater Organisms, 3rd edition, EPA/600/4-91/002, July 1994.

U.S. Environmental Protection Agency (EPA). 1997. Ecological Risk Assessment Guidance for Superfund: Process for Designing and Conducting Ecological Risk Assessments.

U.S. Environmental Protection Agency (EPA). 1996. ECO Update: Ecotox Thresholds. Intermittent Bulletin V. 3, No. 2.

U.S. Environmental Protection Agency (EPA). 1999. National Recommended Water Quality Criteria - Correction. EPA Publ. 822-Z-99-001

U.S. Environmental Protection Agency. 2000. Methods for Measuring the Toxicity and Bioaccumulation of Sediment-associated Contaminants with Freshwater Invertebrates, Second Edition, EPA/600/R-99/064, March 2000.

Vermont Agency of Natural Resources (VTANR). 2000. Water Quality Standards, Appendix C: Criteria for the Protection of Human Health and Aquatic Organisms.

VTANR 2001. R. Langdon. Fish Population sampling in West Branch, Ompompanoosuc River and tributaries.

VTWQS, 2000. State of Vermont Water Resources Board, 2000. Vermont Water Quality Standards.

Tailings and Mine Waste '02, © 2002 Swets & Zeitlinger, ISBN 90 5809 353 0

Impact of acid rock drainage in a discrete catchment area of the former uranium mining site of Ronneburg (Germany)

J.W. Geletneky & G. Büchel
Institute of Earth Science, Friedrich-Schiller University Jena, Germany

M. Paul
WISMUT GmbH, Chemnitz, Germany

ABSTRACT: The Ronneburg mining district in the eastern part of Thuringia (Germany) was between 1950 and 1990 one of the largest uranium mining sites in the world. Since 1990 the remediation of the mining site is realized by the Wismut GmbH with financial support of the German government. To reduce the concentrations of radionuclides and heavy metals in ground and surface waters, the remediation activities include two main projects: (1) flooding of the underground mine; (2) backfilling and covering of the former open pit mine with material from the waste rock dumps. The waste rock mainly consists of black shales, metabasaltic rocks and carbonate of Ordovician to Devonian age. These metasedimentary and volcanic rocks contain up to 7 wt% sulfides, 5-9 wt% organic carbon, 30-60 ppm uranium and a series of trace elements. The aim of this case study is to support the application of effective remediation techniques in a catchment area of a small creek, which was effected by acid rock drainage (ARD) generated from a nearby waste rock dump and which is expected to be the most important exfiltration area of flooding waters after the groundwater rise. Since 1997 the influence of highly mineralized, low pH seepage waters on the unsaturated valley sediments as well as on ground and surface waters has been investigated. About 160 drill holes and geophysical investigations (e.g. resistivity measurements) allow a detailed characterization of the present day distribution of different valley sediments. With hydrochemical methods and geochemical modeling, flow paths of seepage water e.g. in the unsaturated valley sediments were detected. The influence of ARD on surface water and groundwater was assessed. Sources and sinks for heavy metals and uranium are revealed. The use of heavy metals and rare earth elements (REE) as tracers, allows the identification of several processes along different flowpaths e.g. dilution and (co)precipitation. These results are completed by a series laboratory investigations.

1 INTRODUCTION

The Ronneburg mining district (Fig. 1) in the eastern part of Thuringia (Germany) was one of the largest uranium mining sites in the world. Between 1950 and 1990 the former SDAG Wismut produced about 216 kt of uranium in the eastern part of Germany. After the USA and Canada the former GDR was the third largest uranium producer in the world (Barthel 1993). About half of the production (113 kt) originated from the underground and the open pit operations near the city of Ronneburg. The legacy By the end of 1990 the mining legacy consisted of (i) a complex underground mine covering an area of 74 square kilometers with 40 shafts and about 3000 km of mine workings, (ii) the Lichtenberg open pit mine with an open volume of about 84 Mio. m³, (iii) 16 waste rock dumps with about 200 Mio m³ of partly acid generating waste rock, and (iv) about 1000 ha of contaminated areas.

Since 1990 the site remediation is carried out by the WISMUT GmbH with financial support of the German government. The remediation projects are mainly aimed at reducing the radiological and chemical exposure of the public and the environment to an acceptable level (Gatzweiler et al. 1997). The most important projects are:

– decontamination of the mine from water pollutants, sealing/plugging of all the shafts and tunnels and backfilling of the uppermost mine levels for stabilisation purposes backfilling of the former open pit mine with material from the waste rock dumps
– flooding the underground mine
– contouring and covering of waste rock dumps which are remediated in-situ
– demolition and decontamination of industrial areas
– water treatment

The flooding of the mine was initiated in 1998 and is expected to be completed between 2003 and 2005. Since there are no open dewatering tunnels within the whole mining field flooding would continue until the natural discharge of the groundwater to the local receiving streams if no active pumping or collection measures are carried out during the final stages of the process. Because of the degree of

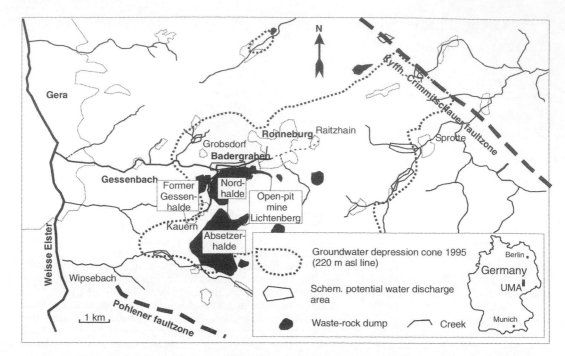

Figure 1. Northern part of the former uranium mining site Ronneburg, Germany (1950-1990) (UMA: uranium mining site Ronneburg).

contamination and the amount of inflow active water treatment is necessary for the southern part of the mine. For that reason a water treatment plant (HDS process) with a basic capacity of 600 m³/h is being built until spring 2002. Contaminants of concern are iron, radionuclides, heavy metals (Ni, Co, Cu) and As.

To control the groundwater rise a 600-mm-pumping well has been drilled which has contact to two of the main levels of the mine. It will be used for water catchment purposes during the implementation of the water treatment plant. Its capacity will be equivalent to the inflow into the mine. Afterwards it is planned to limit the period of pump and treat measures and to continue the flooding process to an optimum level as fast as possible. The overall strategy is to reach a high inundation level to

(i) lower the gradient into the depression cone and to minimise the catchment area of the mine which influences directly the amount of contaminated mine water which has to be treated

(ii) and the contaminant loads which have to be managed

(iii) minimise the thickness of the unsaturated zone which undergoes further AMD generation

(iv) minimise operational costs for water management, water treatment and sludge disposal as a result of (i) and (iii).

For that reason additional water catchment systems have to be installed close to the surface within areas with a high probability for exfiltration of con-

taminated groundwaters into the receiving streams. The first of these areas is the Gessenbach valley, as it is lowest valley with potential contact to the groundwater flow field of interest. For that reason a series of studies have been undertaken to provide a reasonable database for the planning process of water catchment and monitoring systems in the Gessental valley. These studies include geological, geophysical, hydrological and hydrogeological investigations.

The paper describes a part of the results of this field and laboratory work which are gained between 1997 and 2001 in Cupertino between Friedrich-Schiller University Jena and WISMUT GmbH.

2 SITE CHARACTERIZATION

2.1 Geology

The uranium deposit of Ronneburg, Thuringia, Germany, is a strata-controlled, structure bound deposit. It consists of uranium concentrations in small scale brittle structures which form stockworks within or immediately adjacent to carbonaceous, pyritic black shales. The Paleozoic host rocks mainly consist of argillaceous and siliceous black shales with intercalations of dolomitic and phosporite nodule beds (Silurian "Graptolithenschiefer") The main black-shale horizon lies below Ordovician carbonaceous sandy shales and overlies Silurian carbonate rocks. Some Devonian metabasaltic dikes and sills cut the metasedimentary rocks (Dahlkamp 1993). The rocks contain up to 7 wt% sulfides, 5-9 wt% organic carbon, 40-60 ppm uranium and a series of trace elements (Ni, Mo, Zn, Co, Cu, REE).

The intensively folded and faulted, incompetent and competent rocks have a high density small scale brittle structures (fissures, joints, faults). Permian and Tertiary supergene oxidation processes associated with mobilization and precipitation of trace elements created an oxidation and cementation zone. The irregular distribution and size of the ore bodies is controlled by major and minor faults. The uranium ore appeared near the surface in the southern part and down to 1000 m deep in the northern part of the Ronneburg mining district.

2.2 Investigation area

One of the most important drainage systems of the region is the catchment area of the creek "Gessenbach". The Gessental valley is located in the western part of the district between the cities of Ronneburg and Gera. In its upper part near the city of Ronneburg it is influenced by the mining activities. In this part the two waste rock dumps of Gessenhalde and Nordhalde are located. The Gessenbach creek and its tributary Badergraben follow the valley in western direction (Fig. 1, Fig. 4).

The Gessental valley will be one of the main discharge area of minewater in the post-flooding situation for the following reasons:
- The exposed position above the underground mine
- The orographic position (between 240 m and 280 m asl) as the lowest valley in the area
- The geological situation

Parts of the Silurian metasedimentary rocks are expected to have a higher hydraulic conductivity based hydraulic on in-situ tests (Solexperts 1998). In the Eastern part of the Gessental valley these metasedimentary rocks occur in an area of about 400 m. Most Paleozoic rocks in the valley are intensively weathered. The Paleozoic formations are overlain by Quaternary sediments. The Quaternary deposits consist of alluvial sand, silt and mud and redeposited weathered Paleozoic rocks and loess. In these sediments there are several alluvial palaeochannels.

The hydrogeological situation is determined by the groundwater depression cone of the underground mine. South of the Gessental valley two waste rock dumps existed (Fig. 1). The former ore leaching-dump Gessenhalde was transported to the open mine in 1992-1995. Since 1998 27 Mio m³ of waste rock of the dump "Nordhalde" is being relocated into the Lichtenberg open pit mine. This waste rock dump partly consist of acid generating waste rocks. Before relocating highly mineralized and acidic seepage water were diffusely drained into the upper section of the adjacent creeks and Quaternary sediments. So these waters provide information on preferred groundwater flow paths in these sediments.

3 METHODS

Geological, geochemical and hydrogeological studies provide the framework for the characterization of the pre-flooding situation. To characterize the geological situation of the Quaternary sediments about 160 drillholes and additional test pits were made in 2000. Each drill hole was described and sampled. To calculate a the hydraulic conductivity grain size distributions according to DIN 18 123 (1983) were determined for all samples. The fraction below 0.125 mm was analyzed with a particle-analyzer. Because of the uneven grain-size distribution in the samples the calculation of the permeability coefficient was performed according to Beyer (1964). Some samples, especially the fine grained and coarse alluvial sediments were examined using x-ray fluorescence spectrometry (XRS).

Geophysical investigations (e.g. resistivity measurements) were used to characterize different lithological units, to detect the bottom of the valley sediments and to localize groundwater flow paths (Friedrich 1998, Steinhau 1999). The processing of the soundings was performed with the computer programs RS2DINV 2.04 (Loke 1999) and RESI[plus]-2.0 (Interprex 1992).

For the hydrogeochemical investigation a monitoring system with 10 sampling points was installed along the creeks and near the Nordhalde in 1997. At each sampling point the average flow rate, temperature, pH, Redox, conductivity and dissolved oxygen were measured on site. Sampling was performed at least four times a year according to DVWK (1992). HCO_3^- and alkalinity were determined by titration of untreated samples. The samples were filtered in the field (minus on 0.45 μm). A part of the samples was acidified with HNO_3 for cation analyses. Na, K, Ca, Mg, Fe, Mn, Al, Cu, Ni, Zn, Si, Pb, U, Th and REE were determined using atom absorbation (AAS) and/or inductive coupled plasma spectroscopy (ICP-MS). For the determination of Cl, SO_4, NO_3, NO_2, PO_4 ion chromatography and photometry was used.

Groundwater was sampled from two wells with a peristaltic pump. Sample preparation and detection methods were the same as described for the surface water samples.

The computer program PHREEQC 2 (Parkhurst & Appelo 1999) was used to calculate ionic speciation and saturation indexes of mineral phases.

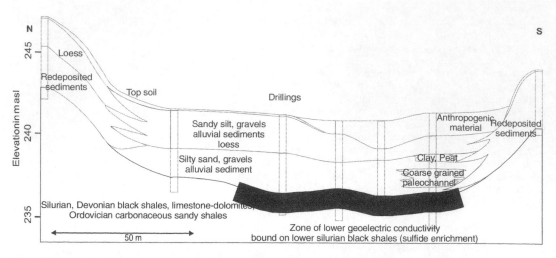

Figure 2. Geoelectrical sounding (resistivity in Ohm × m) with Schlumberger array. Zones of low resistivity are present in mean depths (Friedrich 1998).

4 RESULTS

4.1 *Geology*

Within the investigation area Ordovician, Silurian and Devonian black-shales underlay the alluvial sediments. The top of these metasedimentary rocks is intensively weathered. In the Eastern part of the Gessental valley zones of lower resistivity exist (Fig. 2.). These zones occur only, where lower Silurian black shales are present at the valley bottom (Fig. 3). The lower resistivity of these rocks is assumed to be due to higher sulfide and higher water content of the sediments and/or higher conductivity of the pore and groundwaters, originating from acid seepage from the Nordhalde.

The material of this zone was investigated using XRS. Compared to the alluvial sediments, this material is enriched in e.g. Fe, Al, K, Cr, Ba, U. Compared to the standard rock-chemistry of black shales from Ronneburg, which was derived from 66 samples by Szurowski, 1985, the concentration are similar (Lahl 2001).

The Quaternary sediments consist of alluvial sand, silt, mud and redeposited weathered rocks and loess. At the bottom occur poorly sorted brown to greyish silty sand occurs hosting pebbles with diameters up to 6 cm big. Most of the pebbles consist of quartz, siliceous shales and weathered argillaceous black shales. They are angular to subrounded and show oxidation crusts of limonite and hematite. The particle size analyses show maxima in the fine silt and sand fractions. This suggests that a large amount of redeposited loess is present in this samples. At the western part of the valley coarse grained sediments which are redeposited from the adjacent slopes are common.

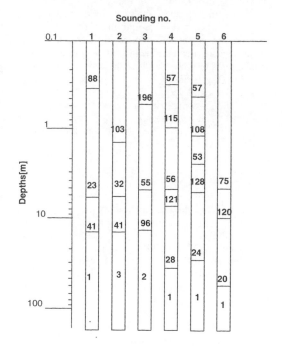

Figure 3. Schematic profile of the Gessental valley.

Paleochannels within the fluvial sediments are present in this areas as well. Some of them are of Quaternary to Holocene age, but others were formed in historic times and can be reconstructed from old maps. They are filled with peat or anthropogenic material and. These channels are expected to be important as preferred flow paths for the post-flooding scenarios.

70

4.2 Hydrology

The hydrological situation of the Gessental valley is dominated by the existence of the Nordhalde and by the groundwater depression cone of the underground mine.

Water balance investigations of the Nordhalde showed that in 1995 over 50 % of the seepage of the waste rock dump discharged directly or indirectly into the underground mine (C&E 1995).

The seepage water in the western part of the Nordhalde (Q4 in Fig. 4, Tab. 1) is highly mineralized, has a low pH (2-3.5) and high redox potentials. This water has enormous concentrations of Fe, Ca, Mg, Al, Mn, SO₄, Si, Cu, Zn, Ni, Cd, Cr, U, REE (Tab. 1). The amount of diffuse discharge at Q4 could only be estimated. Discharge measurements in the "Gessenbach" creek showed that the amount of seepage water flowing into the creek is low (less than 1 l/s).

To investigate the impact of the seepage water of the Nordhalde on the Gessenbach and its tributary, Badergraben sampling was performed regularly at six sampling-locations (G13, G15, G14, G7, G5, B1 in Fig. 4, Tab. 1).

Before both creeks pass the dump the headwaters are strongly influenced by untreated municipal sewage from Ronneburg and Kauern (B1, G13). The concentrations of NH_4, PO_4 and HCO_3 are relatively high. The pH of the water is variable but mostly alkaline (7.5-9). The DO (dissolved oxygen) is low (4-

Table 1. Averaged chemical compositions of water samples from the Gessenbach (G) and the Badergraben (B) creeks and seepage water from the waste rock dump "Nordhalde" (Q) (units in ppm, except for T (°C), pH, Red (mV), Cu, Ni, Co (ppb))

	G5	G7	G14	G15	G13	B1	Q4
T	9.1	9.7	9.7	-	10.7	7.4	10.3
C.	2330	3080	2925	2300	1395	159	12040
pH	6	5	5	6	7	8	3
Red.	324	492	483	383	68	292	567
DO	10.3	11.8	8.9	7.4	3.8	7.3	10.3
Na	69	39	43	72	87	109	27
K	16	13	11	20	25	20	0.4
Ca	185	230	224	207	122	94	950
Mg	175	284	256	201	35	77	1260
Fe	35	105	128	107	0.5	0.5	2393
Mn	7	13	15	6	0.2	0.4	124
Al	-	6	14	0.9	0.2	-	227
NH₄	16	18	19	41	44	18	42
Cl	110	75	78	114	97	137	22
SO₄	1280	1860	1857	1 500	210	308	13270
HCO₃	38	6	47	21	503	371	-
NO₃	11	12	1 4	4	7	8	3
Cu	20	344	263	440	5	5	2120
Ni	1052	2504	2457	1813	13	17	25630
Zn	1	2	2	1	-	-	23
Si	4	7	7	5	5	3	45
Co	87	431	358	140	2	1	8870
U	0.03	0.03	0.04	0.01	-	0.09	0.6
REE	-	-	-	-	-	-	3

5 mg/l). The electric conductivity is low compared to the seepage water (1.5-2 mS/cm). When the Gessenbach creek passes the area of the Nord- and Gessenhalde the water becomes enriched in total dissolved solids (TDS) Fe, Ca, Mg, Al, Mn, SO₄, Si, Cu, Zn, Ni, Cd, Cr, U and REE.

After the unification of Badergraben and Gessenbach the composition of the water changes in flow direction due to mixing and other processes (G5 in Fig. 4). Some of the components are transported as conservative tracers (Mg, Ca, Na, K, Cl, PO₄), others undergo reactive transport (Fe, Mn, Si, Cu, Ni). The concentration of Fe decreases from about 100 mg/l to 10 mg/l over a distance of only 200 m. In the bed of the creek iron precipitation can be observed. The precipitates are flocculated iron-hydroxides and iron coatings. The precipitate consist of iron phases which were supersaturated with respect to the composition of the stream discharge (Tab. 2). Most of the mineral phases were amorphous.

Manganese is also removed from the system. The manganese phases in the geochemical modeling show, however, undersaturation. It is assumed that manganese is removed by (co)precipitation or by adsorption on the surface of the iron-hydroxides (Foos 1997). The precipitate from the creek was extracted after DIN ISO 11466 (1999). The amorphous precipitate shows relatively high concentrations of Cu, Zn, U, Ni, Pb, As, Cr, Co and Mn. These results verify investigations of Wismut GmbH (URST 1992).

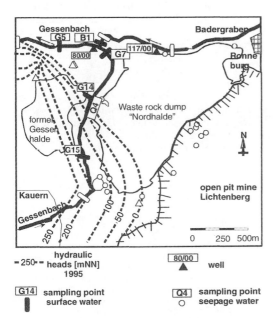

Figure 4. Sampling points of the surface water and groundwater around the waste rock dump "Nordhalde" and hydraulic head of the groundwater depression cone in 1995.

Table 2. Mean saturation index (SI) of selected phases calculated using PHREEQC. SI= log(IAP/K), IAP: ionic activity product, K: equilibrium constant

Phase	Formula	SI
Ferrihydrite	$Fe(OH)_3$	1.8
Ferrite-Ca	$CaFe_2O_4$	1.5
Ferrite-Cu	$CuFe_2O_4$	9.8
Ferrite-Mg	$MgFe_2O_4$	2.1
Ferrite-Zn	$ZnFe_2O_4$	9
Goethite	$FeOOH$	7.1
Hematite	Fe_2O_3	15.1
$CoFe_2O_4$		20.2
Delafossite	$CuFeO_2$	8.8
Jarosite	$KFe_3(SO_4)_2(OH)_6$	6.8
Jarosite-Na	$NaFe_3(SO_4)_2(OH)_6$	4.9
$MnHPO_4$	$MnHPO_4$	2.5
Strengite	$FePO_3 \cdot 2H_2O$	1.5
Trevorite	$NiFe_2O_4$	11.2
Pyrolusite	MnO_2	8.9
Rhodochrosite	$MnCO_3$	-2.3
Manganite	$MnO(OH)$	-5.1
$Mn(OH)_2(am)$		-8.80

4.3 *Hydrogeology*

Most of the alluvial sediments in the Gessental valley are unsaturated with groundwater. However, saturated zones exist which are probably bound to paleochannels. For the coarse alluvial silty sand the coefficient of unconformity is wide spread, mostly between 100 and 600. This results in coefficients of permeability between 3.84×10^{-4} m/s and 2.9×10^{-8}. However, in some areas the values are higher (3.84×10^{-4} m/s to 10^{-5}). These zones are interpreted as paleochannels. The fine grained alluvial and redeposited sediments show lower unconformity (10 to 40) and permeability coefficients between 3.8×10^{-6} m/s and 9.7×10^{-9} m/s.

Two groundwater-wells were drilled in 2000 (117/00, 80/00 in Fig. 4) The groundwater in both wells was sampled. The hydraulic head in well 117/00 was between 8.1 and 8.2 m below surface. However, this well was dry in summer 2000. The water can be classified as Mg-Ca-SO$_4$-water with a pH of 6.4 and an enrichment in heavy metals (Zn, Ni) and REE.

The hydraulic head in well 80/00 varies between 3.3 to 5.3 m (236 to 234 m asl) below surface. The composition of this groundwater can be described as a Mg-Ca-SO$_4$-water with a pH varying between 5.3 to 6.4, enriched in Fe, Zn, Ni, Co, Sr and REE.

The path of the Nordhalde seepage water can be followed into the partly saturated alluvial sediments of the Gessental valley. The chemical composition of the groundwater in the wells shows an influence of acid seepage water, which was originated from the dump because the hydrochemical signatures of REE in the groundwater are comparable with signatures of the seepage. Both show enrichment in the light REE and an anomaly in Gd (Geletneky et al., 2000, Geletneky et al., in prep.).

In well 80/00 oscillating hydraulic heads were measured due to precipitation. It is assumed that this is caused by infiltration of the creek but mainly by seepage water from the Nordhalde. The seepage water flows probably into the valley sediment via discrete flow paths.

5 CONCLUSIONS

Geological, geochemical and hydrogeological investigations provide a framework to understand the-pre-flooding situation of the Gessental valley. The investigations show the detailed structure and composition of the up to 10 m thick Quaternary valley sediments. They consist of alluvial sand, silt and mud and redeposited weathered rocks and loess. The existence of Paleochannels with higher permeability coefficients is assumed. Under these sediments zones of lower electric resistivity exist. These zones are considered to be due to higher sulfide contents or higher degree of mineralization of the pore waters of of the Silurian rocks.

More than 50% of the seepage water of the Nordhalde discharged directly or indirectly into the underground mine. Only a minor part seeps diffusively into the creeks and into the valley sediments. Using the seepage water as tracer, the existence of discrete flow paths within the valley sediments is suggested. This information can be used for the implementation of effective remediation strategies.

The amount of TDS in the surface water increases when the Gessenbach creek passes the dump area. After the unification with the Badergraben creek the chemical composition of the water changes. The amount of TDS decreases rapidly. This is caused by a combination of iron-hydroxide precipitation and dilution.

Remediation carried out by Wismut GmbH will minimize the diffusive discharge of acid seepage water to the surface waters. The Nordhalde as the most important contaminant source will be relocated into the open-pit mine until the end of 2002.

REFERENCES

Barthel, F.H. 1993. Die Urangewinnung auf dem Gebiet der ehemaligen DDR von 1945 bis 1990. *Geol. Jb.* A142: 335-346.

Beyer, W. 1964: Zur Bestimmung der Wasserdurchlässigkeit von Kiesen und Sanden aus der Kornverteilung.-*Wasserwirtschaft-Wassertechnik*:165-169.

C&E 1995: *Wasserhausrechtliche Untersuchungen im Bereich der Nordhalde, Teil I: Untersuchung zum Istzustand.*- C&E Consulting und Engineering GmbH, internal report prepared for Wismut GmbH; Aue.

Dahlkamp, F.J, 1993: *Uranium ore deposits*. Berlin: Springer.

DIN 18 123 1983 *Baugrund-Untersuchung von Bodenproben-Bestimmung der Korngrößenverteilung.*

DIN ISO 11 466 1995: *Bodenbeschaffenheit; Extraktion von in Königswasser löslichen Spurenmetallen.*

DVWK 1992:*Entnahme und Untersuchungsumfang von Grundwasserproben: DK 556.32.001.5 Grundwasseruntersuchungen, DK 543.3.053 Probenahme.- Regeln zur Wasserwirtschaft*, Hamburg, Berlin.

Friedrich, C. 1998: *Kleinräumige geophysikalische Vermessung an Halden im Bereich Ronneburg mit Hilfe der Geoelektrik und des Bodenradars.* Dipl. Thesis., TU Clausthal-Zellerfeld

Gatzweiler, R., Hähne, R., Eckart, M., Meyer, J. & Snagovsky, S. 1997: prognosis of the flooding of uranium mining sites in east Germany with the help of the numerical box-modeling. In Veselic, M & Norton, JP.J. (eds). *Mine water and the environment, Proc. 6ᵗʰ IMWA congr., Bled, 8-12 September 1997.*

Geletneky, J.W., Merten, D., & Büchel, G 2001.: Seepage water from uranium mining dumps in Eastern Thuringia, Germany: A hydrogeochemical study.- *EUG 11, Strasbourg, J.. Conf.. Abs., 6: 44*, Cambridge: Cambridge Publ.

Geletneky, J.W., Merten, D., & Büchel, G in prep: Seepage water from uranium mining dumps: A hydrogeochemical study with special emphasis on REE.

Interprex, 1992: *RESIXᵖˡᵘˢ v 2.0 User`s manual- Resistivity data interpretation software*, Interprex Ltd., Golden, Colorado.

Lahl, K. 2001: *Elutionsversuche an ausgewählten quartären Sedimenten sowie Ockerkalkzersatzmaterial aus dem Gessental bei Ronneburg (ehemaliges ostthüringisches Uranbergnaurevier).*Dipl. Thesis, Univ. Jena.

Loke, M-H. 1999: *Electrical imaging surveys for environmental and engineering studies.*- Penang, Malaysia.

Parkhurst, D.L. & Appelo, C.A.J. 1999: *User`s guide to PHREEQC (version 2).* US Geol. Surv., Water Resour. Inv. Rep. 99-4259.

Solexperts 1997: *Hydrogeologische Testarbeiten bei Ronneburg: Schlussbericht.*, Solexperts AG, internal report prepared for Wismut GmbH.

Steinhau, D. 2000: *Geophysikalische Messungen zur Erkundung der quartären Lagerungsverhältnisse in einem Bergbausanierungsgebiet, Meßgebiet: Talaue des Gessenbaches und des Badergrabens westlich der Ortslage von Ronneburg*, analytec, internal report prepared for Wismut GmbH, Chemnitz-Mittelbach.

URST 1992: *Analyse von Mineralogie, Spurenelementchemismus und Spurenelementfixierung in den Sedimenten der Vorfluter gessenbach, Wipse und Sprotte.*- Auftraggeber: *Umwelt- und Rohstoff-Technologie*, internal report prepared for Wismut GmbH, Greifswald.

Wismut 1998: Exkursionsführer zum Uranbergwerk Ronneburg und zum Aufbereitungsstandort Crossen/Helmsdorf.: In Paul, M & B. Brückner (eds), *Proc. Conf.: Uranium mining and hydrogeology II, 09/1998*, Freiberg.

Tailings and Mine Waste '02, © 2002 Swets & Zeitlinger, ISBN 90 5809 353 0

Hydrochemical investigation at the uranium tailings "Schneckenstein" (Germany)

T. Naamoun
Institute of Applied Physics, T.U. Bergakademie, Freiberg, Germany

ABSTRACT: A hydrochemical investigation was accomplished at the "Schneckenstein" site. At the first five meters of depth, the analysed tailings material show a pH value near 7 and increases slightly with depth. The recorded E_h value ranges from 406 to 430 mV demonstrating an oxidizable medium mainly for iron. In addition, the extracted water can be classified as good mineralised water since the electrical conductivity value lies between 530 and 1014 µs/ cm. Moreover, the content of almost all the main components do not exceed their maximum limit for drinking water recommended by the World Health Organisation. Whereas, the concentrations of Al, As, Cd, Mn, Ni, Pb, and U are considerably high. In the deeper intervals of depth , the pH value increases showing an alkaline medium that may be used as a fingerprint of the alkalinity of leaching procedure. The increase of the pH in the tailing (I) with regard to that in the tailing (II) is probably in relation with the increase of the concentration of chemicals during the mineral processing in order to increase the productivity of the leaching procedure. Furthermore, the sodium content illustrates the contrast between the heap and the treated materials.. The treated material also shows the high mobility of Al, As, Fe, Mn, Ni, Pb and U whereas Ba, Cr, Cu and Mg are low mobile. On the other hand, the DOC content is found very high in most analysed samples. Therefore, an oxygen consumption by organic ligands with time that transforms the tailings environment from oxidizable to reducible is expected.

1 INTRODUCTION

The discharged uranium ore residue in the settling basins in the Schneckenstein site may be considered as a potential source of contamination. It contains material with different physical and chemical properties that influence the ability of elements to migrate and i. e. to participate in the water cycle. To evaluate the environmental contamination risks from the mentioned tailings, a hydrochemical investigation was applied. The current work resumes the most important recorded results as well as it analyses the behaviour of most analysed elements in the studied sites depending on their chemical properties. The work further forecasts the perspectives of their interactions with the tailings environment.

2 A SHORT DESCRIPTION OF THE SITE

The area of investigation is located in the southwest of Saxony, in the Boda valley north of the district of Schneckenstein in the village of Tannenbergsthal/ Vogtland (Figure 1). In the northwest the Boda valley is surrounded by the Runder Hübel mountain (837 m above sea level) and in the southeast by the Kiel mountain (943 m asl). The current surface of the site is located at in altitude of 740 to 815 m above sea level. The mining and tailing sites are situated on the contact between a Biotit Syeno granite and the Biotit-Quart-Serizit schists. Depending on the altitude, an annual precipitation of 960 to 1160 mm is recorded. The mean annual temperature is 5.5 °C.

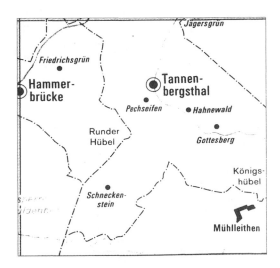

Figure 1. The location of the area of investigation.

3 EXPERIMENTAL

By drilling four boreholes at different depths, four sediment cores were taken at the tailing sites; two boreholes in each tailing respectively (Figure 2). The first two boreholes (GWM 1/ 96; GWM 2 / 96) were drilled down to the granite foundation in Tailing 2 (IAA I). The third borehole (RKS 1/ 96, Tailing 1(IAA II) was sunk to a depth of about eight meters. However, the granite foundation was not reached due to technical problems. The fourth borehole (RKS 2/ 98, Tailing1) is 12 m deep. The cores (diameter 50 mm) were cut into slices of 1 m length and transported in argon filled plastic cylinders to avoid contact with ambient air.

4 LABOUR ACTIVITIES

Two different procedures were used for the chemical water analysis.

4.1 *Pore water extraction conducted by mean of a high pressure device*

Pore water extraction was conducted by mean of a high pressure device presented in Figure 3. Due to the low permeability on the one hand and the limited water content on the other, this procedure was not effective and only 3 samples were analysed. Directly after the extraction process the electric conductivity , the redox potential and the pH values were measured. For the trace elements determination, the water samples were filtered with 0,2μm membrane filter then stabilised with diluted nitric acid (1ml acid/100 ml water) until a pH ~ 2 and finally filled in polyethylene-bottles then stored in a refrigerator. For the measurement process, the ICP-MS equipment was used. In Table 2, the detection limit of each element is given. For the main substance determination, the

Table 1. The detection limits of elements and their error consideration

Element	Detection Limit	Error (2σ)
Floride	0.5 mg/L	0.1 mg/L
Chloride	1 mg/L	0.1 mg/L
Nitrate	0.5mg/L	0.1 mg/L
Sulfate	1 mg/L	0.1 mg/L
Lithium	0.2 mg/L	0.1 mg/L
Natrium	0.5 mg/L	0.1 mg/L
Kalium	0.1 mg/L	0.1 mg/L
Calcium	1 mg/L	0.1 mg/L
Magnesium	1 mg/L	0.1 mg/L

Table 2. The detection limits of different elements by means of ICP AES

Element	Detection limit	Element	Detection limit
Manganese	0.001 mg/L	Silicon	0.044 mg/L
Vanadium	0.007 mg/L	Uranium	0.081 mg/L
Copper	0.018 mg/L	Iron	0.022 mg/L
Molybdenum	0.014 mg/L	Aluminium	0.088 mg/L
Zinc	0.023 mg/L	Magnesium	0.002 mg/L
Titanium	0.017 mg/L	Arsenic	0.321 mg/L
Nickel	0.052 mg/L		

water samples were filtered then stored without stabilisation. The measurement process was accomplished using an ion chromatograph device. The limit of detection of each element is given in Table 1. The iron species were determined with photometry method.

4.2 *Mixed pore water extraction*

The procedure is similar to the first step of the Förstner Salomon extraction procedure described in Salomons & Förstner, (1984) with the difference that in this time the soil material was not crashed. The pH value was measured immediately after the water extraction. The DOC value was determined using an IR-spectrometry with 0.2 mg/L as detection limit and 0.1 mg/L as error (2σ). The arsenic spe-

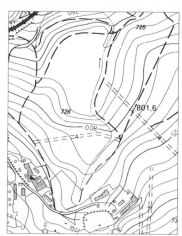

Figure 2. Location of boreholes in the tailing sites.

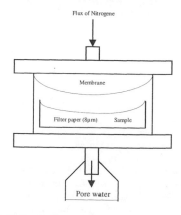

Figure 3. Device for the pore water squeeze.

cies were determined with photometry method. The concentration of trace elements was determined using an ICP AES equipment. The detection limit of each element is given in Table 2. The main elements were detected using an IC device with the same detection limits as sited above (Table 1).

5 RESULTS AND INTERPRETATION

5.1 *Chemistry of the extracted pore water*

5.1.1 *pH-value*

The analysed tailings material have pH value near 7 which increases slightly with depth. It varies between 7.4 and 8.4 with a mean value of 7.9 (Figure 4). These values are nearly to that of the drainage water. The near neutrality of the water samples emphasises the pH buffering that results mostly from the carbonate dissolution as previously mentioned. The increase of pH is probably in relation to the increase of the carbonate content with depth. The recorded values affirm that at first five meters of depth, the analysed water is in contact with only untreated heap material. On the other hand they assure the " low solubility" of most main and trace metals.

5.1.2 *Redox potential*

The E_h value ranges from 406 to 430 mV with a mean value around 420 mV. The recorded values demonstrate an oxidizable medium mainly for iron. According to Wagman et al., (1982), the measured E_h –pH values favour the mobility of arsenic in its $HAsO_4^{2-}$ ionic form. Under the same conditions, Cd and Ni are high mobile onto their Cd^{2+} and Ni^{2+} ionic forms respectively (Wagman et al., 1982). Whereas Co and Cr are immobile and Co tends to form $CoCO_3$ and Co_3O_4 complexes (Garrels & Christ, 1965), while Cr tends to be in its very insoluble Cr_2O_3 form (Wagman et al., 1982). In addition, the same chemical medium favour the presence of copper as well as most of zinc in their immobile

forms. Thus the first mentioned element tends to precipitate onto its immobile CuO compound (Garrels & Christ, 1965). Whereas most of the Zn content must be in its immobile $ZnCO_3$ and ZnO complexes but some zinc may be in its Zn^{2+} ionic form (Wagman et al., 1982).

5.1.3 *Electrical conductivity*

The electrical conductivity value lies between 530 and 1014 µs/ cm with a mean value close to 717 µs/ cm. Thus, after Hütter, (1990), the analysed water samples can be classified as good mineralised groundwater.

5.1.4 *Main components*

Independent on the interval of depth, the content of almost all the main components do not exceed their maximum limit for drinking water recommended by the WHO, (1996). The hydrogen carbonate content lies between 24.4 and 42.7 mg/ l with a mean value of 32.5 mg/ l. The fluoride concentration varies between 1.4 and 1.8 mg/ l with a mean value of 1.6 mg/ l. The total mass of chloride ranges from 3.2 mg/ l to 7.7 mg/ l. The mean value of the nitrate content is around 0.9 mg/ l. The sulphate content reaches its maximum in the first sample and decreases with depth. Its value ranges from 150 to 250 mg/ l with a mean value of about 198 mg/ l. The sodium and potassium contents vary from 8.9 to 19 mg/ l and from 2.7 to 13 mg/ l with mean values of 14 and 7 mg/ l respectively. The calcium and magnesium concentrations range from 40 to 69 mg/ l and from 13 to 30 mg/ l with mean values of 57 and 22 mg/ l respectively.

5.1.5 *Heavy and trace metals*

In almost all analysed samples, the concentrations of Al, As, Cd, Mn, Ni, Pb, and U are considerably high and exceed their maximum limit for drinking water recommended by the WHO, (1996). Although the pH values do not allow Al to be mobile, its content

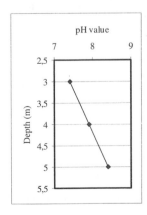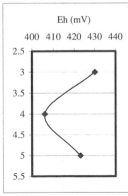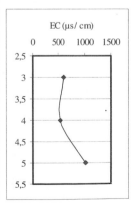

Figure 4. The change with depth of pH, E_h and EC.

is high in all intervals of depth. This is due probably to the presence of some Al amount in form of Al-hydroxide particles in the water solution in spite of the process of filtration. Its concentration varies between 275 and 1030 µg/ l with a mean value of 571 µg/ l. As expected, the arsenic concentration is important lying between 49 and 105 µg/ l with a mean value close to 82 µg/ l. As also expected, the cadmium content is important and decreases with depth varying between 6 and 19 µg/ l with a mean value nearly to 11 µg/ l. The high manganese content demonstrates that the manganese-bearing minerals are contacted by water under "reduced" conditions but may be also by the high bacterial activity. Its concentration lies between 1319 and 4000 µg/ l with a mean value roughly to 2806 µg/ l. In agreement with the above mentioned deductions, the Ni content is relatively high. It decreases with depth varying between 83 and 149 µg/ l with a mean value of 107 µg/ l. Although, under the Chemical conditions of the studied environment lead tends to precipitate onto $PbSO_4$ and $PbCO_3$ complexes, the lead content is slightly high ranging from 6 and 21 µg/ l with a mean value of 11 µg/ l. The uranium content is tremendously high that demonstrates its contact with an oxidizable water that favour its presence mainly in $UO_2(CO_3)_2^{2-}$ high mobile ionic form. Its content increases with depth varying between 55 and 949 µg/ l with a mean value of 423 µg/ l. On the other hand, confirming the recorded chemical conditions that prevail the presence of barium in its immobile $BaSO_4$ compound form, its content is relatively low . Its concentration ranges from 153 µg/ l and 186 µg/ l with a mean value of 173 µg/ l. In confirmation with the chemical conditions above mentioned, the Co

content is low and nearly the same in the different samples and decreases with depth ranging from 37 to 47 µg/ l with a mean value of 42 µg/ l. The Cr content varies between 5 and 17 µg/ l with a mean value of 9 µg/ l. The Cu concentration ranges from 108 to 186 µg/ l with a mean value of 144 µg/ l. These recorded values affirm the expected behaviour of Cu under the above mentioned conditions. The iron concentration decreases with depth and the iron species show its contact with an oxidizable water that it is in agreement with the measured E_h values. Its total content varies between 60 and 150 µg/ l with a mean value of 90 µg/ l. The water solutions do not show any Zn enrichment as expected. Its content varies between 199 µg/ l and 1370 µg/ l with a mean value of 930 µg l.

5.2 Chemistry of the extracted mixed pore water

5.2.1 pH-value

The measured pH value in the four cores show a great similarity between each two boreholes of each tailing. In the borehole N°1 and N°2, the pH value ranges from 6.93 to 9.63 and from 6.93 to 10.07 with mean values of 8.1 and 9.1 respectively (Figure 5). Since In the first five meters of depth of the both mentioned cores, the tailings material is mostly constituted from heap material, the pH value is near neutral. Whereas, in the deeper intervals of depth, its value increases showing an alkaline medium. In the borehole N°3 and N°4, the measured values are remarkably lower than those recorded in the above mentioned cores. The pH value ranges from 6.85 to 7.94 and from 6.18 to 7.68 with mean values of 7.65 and 7.23 respectively. Since in the first two intervals of depth as well as in the next to the last one of the

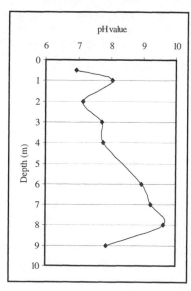 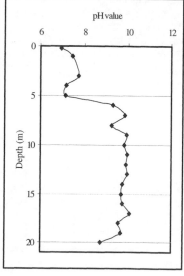

Figure 5. The change with depth of pH in the first two cores.

78

fourth borehole, the tailings material is untreated, thus the recorded pH values are lower than pH 7. Whereas for the other intervals of depth, the measured values are near to pH 8. Since the used deionized water for the shaking process is of pH value ~ 6, thus the real pH values of the analysed tailings material must be higher than the recorded ones. Therefore the alkaline pH mostly in the lower most tailing parts may be used as a fingerprint of the alkalinity of leaching procedure. In addition, the increase of the pH in the tailing (I) with regard to that in the tailing (II) is probably in relation with the increase of the concentration of chemicals during the mineral processing in order to increase the productivity of the leaching procedure.

In addition, it is of importance to point out that by the recorded pH values, arsenic tends to be high mobile mostly in its ionic form $HAsO_4^{2-}$ (Wagman et al., 1982). The same condition prevails the immobility of Ba as most of its amount tends to precipitate mainly onto $BaSO_4$. According to the same scientist, at the same medium, Cr seems to be immobile and most of its amount tends to crystallise onto Cr_2O_3 complex. Depending on the Redox potential of the medium, most of Cu must be in its crystalline phases; native copper, CuO (Wagman et al., 1982) or Cu_2O (Garrels & Christ, 1965) compounds. The recorded pH values favour the precipitation of iron but the E_h value of the medium is the key condition of which compound will appear; $Fe(OH)_3$ (Wagman et al., 1982) or $Fe(OH)_2$ (Winters & Buckley, 1986). According to Wagman et al., (1982), independent on the redox potential of the medium, the mildly alkaline pHs give rise of the solubility of Mg. Whereas, at high pHs; pH> 10, Mg is expected to crystallise

onto $Mg(OH)_2$ compound. According to the same scientist, the E_h value of the medium is considered as the master factor controlling the solubility i.e. the mobility of Mn along with pH value. Other than the high soluble form Mn^{2+} under most E_h conditions and at mildly alkaline pHs, dependent mainly on the E_h value, the immobile Mn forms; MnO_2, MnO, Mn_3O_4 as well as $Mn(OH)_2$ are also probable. Independent on the E_h value of the medium, at most recorded pHs, Ni is immobile and tends to crystallise onto $Ni(OH)_2$ compound (Wagman et al., 1982), but in the case of a stronger alkaline pH; pH>12, Ni demonstrates a high mobility in its ionic form $HNiO_2^-$ (Garrels & Christ, 1965). Under most probable E_h conditions and at the recorded pHs, most of Pb is expected to be in its crystalline form $PbCO_3$ (Wagman et al., 1982). Under oxidizable conditions and at the recorded pH values, uranium is high mobile mainly in its $UO_2(CO_3)_3^{4-}$ ionic form (Langmuir, 1978; Tripathi, 1979). Under the more probable E_h conditions and at the measured pH values, Zn is immobile and most of its amount tends to crystallise onto ZnO compound (Wagman et al., 1982).

5.2.2 Electrical conductivity

Mainly in the second borehole, the EC value is found higher in the both first cores than that in the two other ones. In the borehole N°1, the EC value ranges from 38 to 360 μs/ cm with mean value of 188. While, in the second one it varies between 24 and 800 μs/ cm with mean value of 401 μs/ cm (Figure 6). In the borehole N° 3, its value lies between 56 and 283 μs/ cm with mean value of 151 μs/ cm. Whereas in the fourth one, it varies from 62 and 299 μs/ cm with mean value of 202 μs/ cm. The EC

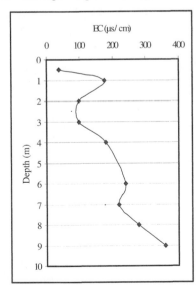
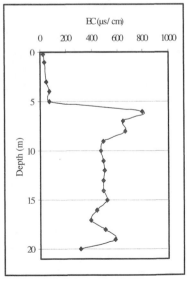

Figure 6. The change with depth of EC in the first two boreholes.

value increases remarkably starting from the fourth meter of depth in the first and the third borehole and from the sixth one in the second core. Whereas in the fourth borehole, the EC value is remarkably low in the next to last interval of depth as expected, but it is higher in the other intervals. The low EC values recorded in the upper most tailings are in relation with the untreated heap material. Whereas, its brutal increase in the deeper intervals of depth as already mentioned is in connection with the presence of wet processed material reach with dissolved solids. Similar to the pH change in the four cores discussed above, the increase of the productivity of the leaching procedure gives rise to the EC in the two first boreholes, mainly in the second one with regard to the other cores.

Comparing between the received pH and EC values of the extracted pore waters from the first intervals of depth of the second core by means of the two different procedures, it will be important to point out that the recorded pH values are nearly the same except the third interval of depth where the pH of the pure pore water is more basic. Whereas the recorded EC values of the mentioned pore water samples are 7 to 15 times higher than in the mixed ones. This effect of water dilution must be taken into consideration in the following part of the interpretation.

5.2.3 *The DOC content*
The extracted DOC content is relatively high in most analysed samples mainly those extracted from the last intervals of depth (Figure 7). This increase of DOC is probably due to the presence of an appreciable quantity of fine roots of plant residue mixed with the tailings sediment. In fact, mainly in the last in-

tervals of depth of the first and the second core, a considerable quantity of vegetable associated with very bad smell was remarkable. This important presence of organic matter conducts the high consumption of oxygen that generates anaerobic conditions in the tailings environment (Cresser et al., 1993). Therefore, metals mainly bound to iron and manganese nodules are exposed to mobilisation.

From the first core, the extracted DOC content lies between 1.2 and 74.8 mg/ l with mean value close to 13.2 mg/ l whereas, from the second one, it varies from 0.8 to 128 mg/ l with mean concentration roughly 8.5 mg/ l and in the third one, it ranges from 1.0 to 1.9 mg/ l.

5.2.4 *Main components*
The extracted hydrogen carbonate content from the first borehole varies from 30.5 to 130.5 mg/ l with mean value roughly to 84 mg/ l. While from the second one, it lies between 12.2 and 256 mg/ l with mean value of 131 mg/ l. Whereas from the third and the fourth cores, it ranges from 24.4 to 97.6 mg/ l and from 12.2 to 73.2 mg/ l with mean values of 67 and 52 mg/ l respectively.

From the first core, the extracted fluoride content varies from 0.2 to 4.5 mg/ l with mean value of 0.9 mg/ l. It lies between 0.4 and 2.4 mg/ l with mean value of 1 mg/ l from the second one. Whereas from the third and the fourth ones, it ranges from 0.2 to 0.8 mg/ l and from 0.3 to 1.7 mg/ l with mean values of 1mg/ l and 0.6 mg/ l respectively.

From the two first cores, the extracted chloride concentration ranges from 0.4 to 7.8 mg/ l and from 0.2 to 24 mg/ l with mean values of 3.2 and 12.2 mg/ l respectively. Whereas from the other ones, it varies

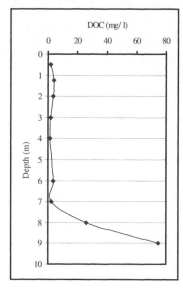 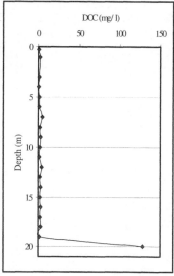

Figure 7. The change with depth of the DOC content in the first two boreholes.

from 0.2 to 0.8 mg/ l and from 0.2 to 29 mg/ l with mean value of 0.8 and 4.3 mg/ l respectively.

From the first borehole, the extracted sulphate content lies between 6.1 and 121 mg/ l with mean value close to 40. From the second one, it varies between 3.3 and 270 mg/ l with mean value roughly 55 mg/ l. Whereas, from the other two boreholes, it ranges from 8.1 to 81 mg/ l and from 4.5 to 83 mg/ l with mean values of 34 and 55 mg/ l respectively.

From the first two cores, the extracted sodium concentration varies from 1.8 to 74 mg/ l and from 0.8 to 180 mg/ l with mean values of 33 and 95 mg/ l respectively. From the third and the fourth boreholes, it ranges from 0.5 to 23 mg/ l and from 0.2 to 37 mg/ l with mean values of 12 and 13 mg/ l respectively.

The extracted potassium content from the first borehole lies between 1.8 and 24 mg/ l with mean value of 10 mg/ l. From the second one, it varies between 1.6 and 23 mg/ l with mean value roughly 6 mg/ l. From the third and the fourth borehole, it ranges from 1.3 to 10 mg/ l and from 0.9 to 8.8 mg/ l with mean values of 4.6 and 2.8 mg/ l respectively.

From the two first boreholes; the first and the second, the calcium concentration ranges from 1.4 to 26.4 mg/ l and from 3.9 to 20.7 mg/ l with mean val-

ues nearly to 9 and 10 mg/ l respectively. From the two other cores, it varies from 8.3 to 28mg/ l and from 10 to 44 mg/ l with mean values close to 18 and 28 mg/ l respectively.

From the first borehole, the magnesium content varies between 0.2 and 0.7 mg/ l with mean value of 1.8 mg/ l. From the second one it lies between 0.4 and 3.7 mg/ l with mean value of 1.7 mg/ l. Whereas from the third and the fourth cores it ranges from 1.4 to 8.6 mg/ l and from 0.1 to 9.5 mg/ l with mean values of 3.6 and 4.7 mg/ l respectively.

As already mentioned the real contents of these components must in many times be higher than that of the recorded ones. However, the recorded EC values in the different analysed samples must be considerably lower than the real values in the tailings pore waters. Therefore the total dissolved solids (TDS) values must be very important. In addition, as expected there is a high correlation between EC and TDS for the four cores. Furthermore, the Piper diagram shows its usefulness mainly for the second borehole by the good illustration of the hydrochemical contrast between the heap material and the treated ones characterised by an extreme sodium content (Figure 8).

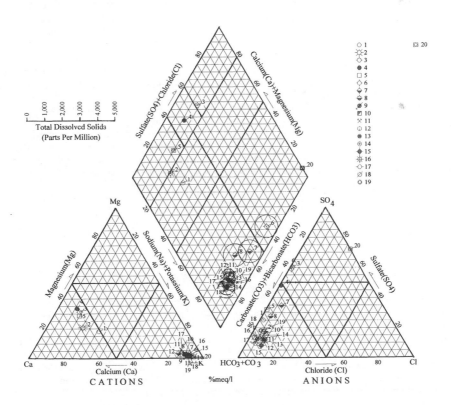

Figure 8. Classification of water quality using Piper diagram; Borehole N°2.

5.2.5 Heavy and trace metals

5.2.5.1 Aluminium

Although most of the total Al is bound to silicates as already demonstrated by means of the selective extraction procedure, independent on the interval of depth, most of the recorded Al concentrations exceed its maximum level recommended by the USEPA, (1991) for drinking water. This is due probably on the one hand to the presence of some Al amount in form of Al-hydroxide particles in the water solution in spite of the process of filtration, on the other hand to the dissolution of clay minerals during the mineral processing that gives rise to the solubility of Al mainly at pH>11.5.

In the samples from the two first cores, the aluminium concentration varies from 0.03 to 13.59 mg/ l and from 0.71 to 40.8 mg/ with mean values of 2.9 and 4.0 mg/ l respectively. Whereas, in the other samples (third borehole) it lies between 0.01 and 1.54 mg/ l with mean value of 0.6. Therefore, Al is a potential pollutant in the prospected environment.

5.2.5.2 Arsenic

Confirming the results of the selective extraction procedure that shows the presence of an important amount of arsenic in the high mobile phases on the one hand and its expected behaviour as already mentioned above on the other hand, its content is tremendously high in almost all analysed samples.

The extracted water samples from the two first boreholes show a total arsenic concentration varying from 13 to 980 µg/ l and from 13 to 2810 µg/ l with mean values of 366 and 1305 µg/ l. Whereas, those extracted from the third borehole demonstrate an arsenic content ranging from 80 to 590 µg/ l with mean value 373 µg/ l. The arsenic species measurements show the dominance of the valence +5 in almost all analysed samples. Therefore the arsenic – bearing minerals are contacted by water under oxidizable conditions.

5.2.5.3 Barium

As expected the Ba content is very low in most extracted water samples. This is due to the fact that most Ba amount is in connection with silicates as already manifested by means of the selective extraction. In addition, with regard to the fact that most recorded pHs are in the alkaline area on the one hand and most of the processed material was under oxidizable conditions mainly during the uranium leaching process on the other hand, therefore most of the non residual Ba tends to be absorbed by nodules and to precipitate with carbonates in form of $BaCO_3$.

In the first and the second borehole, its soluble content varies from 42 to 297 µg/ l and from 44 to 317 µg/ l with mean values of 108 and 119 µg/ l respectively. Whereas, in the third core, it ranges from 61 to 111 µg/ l with mean value close to 80 µg/ l.

5.2.5.4 Chromium

In agreement with the results of the selective extraction procedure ascertaining that more than 90 % of the total Cr content is held in the lattice structure of clay and silicate minerals and only trace amounts of Cr is in relation with the pore water phase on the one hand and with its expected tendency as mentioned above on the other hand, its soluble content is very low in almost all samples. It lies between 0.3 and 40 µg/ l with mean value close to 8 µg/ l. While in the second core, it varies between 0.4 and 75 µg/ l with mean value roughly 8 µg/ l. Whereas in the third borehole, it ranges from 0.3 to 2.7 µg/ l with mean value of 1 µg/ l.

5.2.5.5 Copper

Its soluble content is very low in almost all samples. It varies from 6 to 86 µg/ l and from 8 to 190 µg/ l with mean values of 44 and 63 µg/ l in the first and the second borehole respectively. Whereas in the third borehole it lies between 3 and 13 µg/ l with mean value of 12 µg/ l. In as much as most the non residual fraction of copper is in relation with nodules as already demonstrated by means of the selective extraction, the copper-bearing minerals must be contacted by water under oxidizable conditions. In addition, considering that an appreciable amount of its total mass is in connection with the sulphide/organic phase that can be stable only under reducing conditions, therefore the tailing materials must be in contact with water under "post" oxidizable conditions. Moreover, taking into account the recorded pHs, most of copper tends to be strongly bound to nodules or to crystallise in form of CuO compound (Wagman et al., 1982). Further, the expected E_h-pH conditions of the tailings environment help the stability of carbonates. Therefore the copper associate with carbonate phase is also immobile. Consequently, the measured values of the soluble copper are ascertained and those of the arsenic species are also confirmed.

5.2.5.6 Iron

The soluble iron content is very high. It varies from 0.08 to 23 mg/ l and from 0.1 to 27 mg/ l with mean values of 3.9 and 3.4 mg/ l in the first and the second borehole respectively. Whereas in the third core, it lies between 0.06 and 1.42 mg/ l with mean value of 0.4 mg/ l.

It is of importance to point out that the tailings environment was under oxidizable conditions mainly at the time of the mineral processing. For this reasons most of the non residual iron is found in connection with nodules as demonstrated by mean of the selective extraction procedure. On the other hand, the DOC content is found very high in most analysed samples. Therefore, the high soluble iron content detected in most samples mainly those from the last intervals of depth, is an indicator of an oxygen consumption by organic ligands with time that transforms the tailings environment from oxidizable to reducible.

5.2.5.7 *Magnesium*
Although Mg is well known by its high solubility in a wide range of E_h and pH. The extracted Mg content is found very low in almost all samples. This is due mostly to the alkalinity of the prospected medium that prevails the non dissolution of the Mg bearing minerals. In addition, it asserts the relative high alkalinity of the pure pore water of the tailing materials where Mg tends to crystallise as already mentioned above. Moreover these results are in agreement with those of the selective extraction that assert on the one hand the connection of most residual Mg amount with refractory minerals "clays", on the other hand the presence of an important amount of Mg in association with the stable phase by alkaline pHs "carbonates".

In the first and the second core, the extracted Mg content varies from 0.7 to 5 mg/ l and from 0.2 to 3 mg/ l with mean values of 1.5 and 0.8 mg/ l respectively. Whereas in the third one it lies between 1.1 and 7.9 mg/ l with mean value roughly to 3 mg/ l.

5.2.5.8 *Manganese*
The recorded values of the extracted Mn content exceed 0.5 mg/ l in some samples. In as much as the alkalinity of the prospected medium that do not prevail certainly the high dissolution of Mn and taking into account that the soluble Mn amount in the pure pore water must be higher than the measured ones, thus, these results probably emphasise the reducibility of the tailings environment with regard to Mn. Therefore they also affirm those mentioned above that concern iron. In addition, according to the results of the selective extraction procedure most of the non residual Mn is bound to the moderately reducible phase, however an important amount of Mn is exposed to mobilisation.

In the first two cores, the extracted Mn content varies from 0.04 to 0.4 and from 0.02 to 0.6 mg/ l with mean values of 0.2 and 0.1 mg/ l respectively. From the third one, it ranges from 0.02 to 0.6 mg/ l with mean value of 0.2 mg/ l.

5.2.5.9 *Nickel*
The nickel content is remarkably high in the most extracted samples from the lower part of the tailing sites mainly those from the two first cores. These results probably are the outcome of the alkaline leaching process with pH value higher than pH = 12 where the mobility of nickel is high as already mentioned above. In addition, they demonstrate again the increase of the leaching productivity with the time as already affirmed by means of the gamma and alpha spectrometric measurements. Moreover in view of the fact that a considerable amount of the non

residual Ni is in connection with the sulphide/ organic fraction, therefore, the oxidation of Ni bearing sulphide minerals during the leaching process gives rise to the mobility of the mentioned element. Further, in view of the fact that most amount of the non residual Ni is in association with nodules on the one hand and the process of reducing of the tailings environment on the other hand, Ni is liable to mobilisation with time and it considered as an important pollutant for the prospected areas.

From the two first cores, its total extracted content varies from 1.7 to 38 µg/ l and from 2.2 to 85 µg/ l with mean values roughly 15 and 23 µg/ l respectively. Whereas from the second core, it ranges from 0.7 to 9.6 with mean value around 5 µg/ l.

5.2.5.10 *Lead*
Similarly to Ni, the extracted lead content is relatively high in the most analysed samples mainly those from the lower tailings part. On the other hand, Pb is well known by its high affinity to complexion with organic ligands as well as by its high tendency to form very insoluble sulphide compounds under reducing conditions. Therefore, the current low amount of lead bound to sulphide and organic phase as already emphasised by the selective extraction procedure on the one hand and its current high soluble content on the other hand may be used as a finger print of the oxidation of Pb bearing minerals during the mineral processing and consequently its mobilisation on its $PbOH^+$ ionic form (Wagman et al., 1982). In addition by the same reasons as for Ni, almost the non residual Pb is exposed to mobilisation with time.

From the first and the second core, its extracted content ranges from ~ 0.5 to 85 µg/ l and from ~ 0.5 to 85 µg/ l with mean values roughly 30 and 34 µg/ l respectively. While from the third borehole, it lies between 1 and 16 µg/ l with mean value of 5 µg/ l.

It is noteworthy to point out that the leaching productivity change with time is demonstrated again by the high difference between the extracted Pb from the two first cores with regard to the first one.

5.2.5.11 *Uranium*
As expected, the extracted uranium content is tremendously high in almost all analysed samples. This is due on the one hand to the presence of uranium in its oxidised state and to the alkalinity of the prospected medium on the other hand that prevail for the high mobility of uranium. As far as most of pH values exceed pH 8 in most tailings part thus according to Langmuir, (1978) & Tripathi, (1979), the dissolved uranium is mostly in its mobile $UO_2(CO_3)_3^{4-}$ ionic form. In addition, with regard to the fact that most of the uranium amount is disposed to mobilisation by means of the leaching process, uranium is considered as most potential pollutant for the prospected areas.

From the two first cores, the extracted uranium content varies from 0.04 to 0.39 mg/ l and from 0.02 to 1.65 mg/ l with mean values of 0.19 and 1.03 mg/ l respectively. Whereas from the third one, it lies between 0.34 and 1.61 mg/ l with mean value of 0.9 mg/ l.

5.2.5.12 *Zinc*

The extracted zinc content is relatively low in most samples. From the first and the second borehole, the soluble Zn content ranges from 0.005 to 0.17 mg/ l and from 0.005 to 0.52 mg/ l with mean values of 0.07 and 0.08 mg/ l respectively. Whereas for the third one, it lies between 0.005 and 0.032 mg/ l with mean value of 0.01mg/ l.

The current E_h conditions seems to be still favourable for the stability of both Zn amounts bound to nodules and to carbonates. Further, by pHs of the prospected medium, the Zn amount bound to carbonates is also stable. In addition, it is of importance to point out, that the oxygen consumption by organic substances gives the rise to the stability of Zn amount bound to sulphides. On the other hand, in as much as most of the non residual Zn is attached to nodules, Zn is exposed to mobilisation with time.

6 CONCLUSIONS

The current work affirms the low mobility of most main and trace metals in the upper most tailing parts. On the contrary, the chemical conditions of the same environment favour the high mobility of arsenic as well as of cadmium and nickel. On the other hand, most of analysed metals are high mobile in the lower tailing parts but their present physical conditions restrict their displacement.

REFERENCES

Cresser, M. et al., (1993): Soil chemistry and its applications. Cambridge Environmental Chemistry Series 5. Cambridge University Press. 192 p.

Garrels, R. M. & Christ, C. L., (1965): Minerals, solutions and equilibria. Harper and Rowley, New York, 453 p.

Hütter, L., (1990): Wasser und Wasseruntersuchung. Salle, Frankfurt am Main, Germany.

Langmuir, D., (1978): Uranium solution-mineral equilibria at low temperatures with applications to sedimentary ore deposits. Geochim Cosmochim Acta, 42, 547-570.

Salomons, W. & Förstner, U., (1984): Metals in the hydrocycle. Springer-Verlag, 349 pp.

Tripathi, V. S., (1979): Comments on Uranium solution-mineral equilibria at low temperatures with applications to sedimentary ore deposits. Geochim. Cosmochim. Acta, 43, 1989-90.

USEPA, (1991): The Environmental Protection Agency. National Secondary Drinking Water Regulations. Final Rule. Fed. Reg. 56:20:3526, Jan. 30, 1991.

Wagman, D. D. et al., (1982): The NBS tables of chemical thermodynamic properties. Selected values for inorganic and C1 and C2 organic substances in SI units. J Phys Chem, 11, 2, 392.

WHO, (1996): World Health Organisation, Guidelines for Drinking-Water Quality. http://www.ldb.org/who.htm.

Winters, G. V. & Buckley, D. E., (1986): The influence of dissolved $FeSiO_3(OH)_8^0$ on chemical equilibria in pore waters from deep sea sediments. Geochim Cosmochim Acta, 50, 277-288.

Design, operation, and disposal

Tailings and Mine Waste '02, © 2002 Swets & Zeitlinger, ISBN 90 5809 353 0

Tailings beach slope forecasting – copper tailings

Marcos Pinto
Tailings Dam Operation Supervisor – Collahuasi Mine – Chile

Sergio Barrera
Head of Tailings Area – Arcadis Geotecnica – Chile

ABSTRACT: This article proposes an empirical formula for the determination of the average beach slope value of deposition based on basic background information of the copper tailings slurry. This formula has been calibrated based on measurements taken from several Chilean large conventional copper tailings impoundments. Moreover, an analysis of the values achieved from this formula for other impoundments are included and the correlation reached is discussed. The proposed formula provides a tool to obtain a first estimate of the average slope expected for the type of tailings analyzed.

1 INTRODUCTION

It is well known that tailings discharged as slurry are deposited forming a beach with a slope slightly greater than a horizontal slope.

It is also well known that the slope value is dependent on a series of variables such as grain size distribution, solids content, and flow. For a long time, the slope value was not critical for small impoundments since in the majority of the cases, the tailings were discharged with solids contents between 20 and 45% adopting slope values between 0.2 and 1.0%.

The use of large impoundments was the first real need to try and predict the average slope value of deposition due to the significant impact of the tailings slope on the capacity of the impoundment and the height of the dam walls. Another factor has surfaced recently that reinforces the need to improve the knowledge of the tailings slope: the technology for thickened tailings, which allows for solids contents greater than 60% and, therefore, greater slopes (in many cases greater than 1%).

This article proposes an empirical formula for the determination of the average slope value of deposition based on basic background information of the tailings slurry. This formula has been calibrated based on measurements taken from large conventional tailings impoundments. Moreover, an analysis of the values achieved from this formula for other impoundments are included and the correlation reached is discussed.

2 OBJECTIVE

The first objective of this article is to establish a mathematical model that relates the characteristic variables or parameters of the tailings with the average slope that is developed on the beach (over the pond) for copper tailings in Chilean tailings impoundments.

The second commitment is to establish a model that allows for the measurement or estimate of the trends or influences that represent each of the variables considered.

3 BACKGROUND

The first studies on the slope that hydraulically deposited tailings develop go back to 1973 in Russia and were carried out by Melent'ev et al (1973), indicating that in segregating tailings, the slope is not constant and that the so called "master profile" is formed.

Form the start of his proposal on the benefits of thickened tailings, Robinsky (1979) stated that the main factor affecting the tailings slope was the solids content of the slurry and that this content had little effect under 44%. This relationship was plotted for specific tailings showing that the slope increases with the solids content but a formula that relates these two parameters was never provided.

The concept proposed by Melent'ev et al (1973) was adopted by professor Blight et al (1983 y 1985) in South Africa. The profile of the tailings slope throughout the length of the beach, from the discharge pond to the decant pond, is not constant but rather concave, undergoing modifications due to different factors. The main factor is the segregation (grain size variation due to the sedimentation of coarse particles in the area close to the discharge point). Moreover, this profile can be expressed dimensionless, plotting the elevation of the profile as a fraction of the total descent and the horizontal distance as a fraction of the total length of the beach. Figure 1 and Equation 1 present professor's Blight's proposal, only applicable to tailings deposited over water.

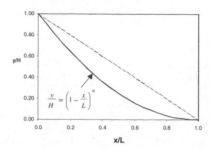

Figure 1.

$$\frac{y}{H} = \left(1 - \frac{x}{L}\right)^N \tag{1}$$

where
H : Elevation difference between the discharge point and the pond
L : Horizontal distance between the discharge point and the pond
x : Distance of a profile point to the discharge
y : Elevation of a profile point from the pond.

Blight also detected that the same tailings deposited at the laboratory or in the field presented or generated the same "master profile" when it was expressed dimensionless. Variations of the main parameters, such as the solids content, grain size, and specific gravity, were investigated in the laboratory to determine the impact of each one of these variables on the "master profile".

One of the problems with the equation (1) is that it does not provide information on the average slope of the deposited tailings. In order to use this equation, the difference in elevation between the head of the beach and the toe, usually the decant pond, and the distance to the pond, which is the same as the average tailings slope, must be known.

The main problem is that the tailings slope determined in laboratory is inversely proportional to the discharge flow (Küpper, 1992), prohibiting the extrapolation of the average slope value for real flows. For this reason, the work done in the laboratory only allows for forecasting the profile shape or

its concavity (N coefficient) but not the average slope.

Different formulas have been proposed to predict the average tailings slope. One of these is based on the shear strength of the slurry and the thickness of the tailings sheet (Williams, 2001)

$$\sin i = \frac{Su}{\gamma d} \tag{2}$$

where
i = slope angle
Su = undrained shear strength
γ = slurry angle
d = sheet depth measured normal to the slope

Other authors have proposed forecasting equations that, in general, are applicable to specific cases. In this regard, Williams, (2001) presents a summary of the studies done in the last 16 years:
— "Williams and Meynink (1986). A semi-empirical formula for channel slopes based on particle fall velocities. Found to be applicable to both river sediment transport and thickened tailings at a single case study but further validation needed.
— De Groote et al (1988) and Winterwerp (1990) working on dykes built from dredged sand in the Netherdlands developed slope equations based on engineering hydraulics.
— Küpper (1991). A semi-empirical formula for beach slope based on the ratio of gravity and inertial forces acting on the particles.
— Morris (1993) employed sediment theory to obtain equations for slope and hydraulic sorting after making simplifying assumptions concerning quasi-steady state flow and wash to bed load transfer. The "constraints" thrown up by the equations were tested against limited case studies with mixed success.
— McPhail (1994) tackled the problem of forecasting both slope and hydraulic sorting by invoking stream power and the principles of entropy. An equation is presented but the stream velocity and channel slope at the head of the beach are needed as inputs in order to forecast performance further down the beach. Since the stream power is proportional to the cube of the velocity, a full-scale operating beach needs to be in existence to obtain reliable input parameters.
— Sofra and Boger (2000) present a paper entitled "Slope Prediction for Thickened Tailings and Pastes" but it is concerned more with run-out from a batch discharge, as could be the case for an underground slope or adit backfill for example, than continuos discharge to an above ground stack."

None of these articles establishes a comprehensive theory or formula to forecast the slope of any tailings hydraulically discharged but explores alternatives that could lead to success and, in any case, provide analytical elements and specific cases.

4 MASTER PROFILE

Taking Equation 1 as a basis and deriving it with respect to x, the slope throughout the beach is obtained:

$$\frac{y'}{H} = \frac{i}{H} = \frac{-N}{L} \times \left(1 - \frac{x}{L}\right)^{N-1} \qquad (3)$$

Evaluating at x=0, the slope developed at the point of discharge (i_o) is obtained

$$\frac{i_o}{H} = \frac{-N}{L} \qquad (4)$$

This evaluation indicates that the N exponent is the quotient between the slope at the discharge and the average slope of deposition, representing a measure of the segregation produced throughout the length of the beach in an impoundment.

$$N = \frac{i_o}{\left(-H/L\right)} = \frac{i_o}{i_M} \qquad (5)$$

This article has the objective of modeling the average slope (i_M) developed on the beach, without including the estimate of the tailings slope under water.

5 NUMERICAL MODEL

5.1 Definitions

P : Tailings production by weight at the time of discharge [KTPD]

C_w : Solids content by weight of slurry [ ]*

d_{50} : Average tailings diameter by weight [μm]

γ_s : Specific weight of solid [t/m^3]

γ_w : Specific weight of water [t/m^3]

$G = \dfrac{\gamma_s}{\gamma_w}$: Specific gravity

A : Model's proportionality coefficient

$\alpha, \beta, \delta, \varepsilon$: Model's exponents

[ ]*   dimensionless parameter.

5.2 Mathematical model

In preparing the model, variables whose average values are easily obtained, including at a conceptual project level that also exhibited low dispersion, were considered.

Other variables, such as viscosity, have average values range widely within a same process and strongly depend on other factors such as temperature.

In preparing this model, the following tailings and slurry characteristic parameters were considered: solids specific gravity (G), slurry's solids content by weight (C_w), tailings dry production at a point per unit of time, and representative grain size (d_{50}).

The adoption of this model considers two basic assumptions:

i..- The average slope is a function of the tailings characteristic parameters previously mentioned

$$i = f(G, C, P, d_{50}) \qquad (6)$$

ii..- And that the influence of each variable can be separated into different functions, such that it can be expressed in the following manner:

$$i = f_1(G) \times f_2(C_w) \times f_3(P) \times f_4(d_{50}) \qquad (7)$$

- Influence of G

The desired tendency for the function associated with this variable is that it should be a function that tends towards zero when the specific gravity tends towards that of water ($f_1(G) \rightarrow 0$, $G \rightarrow 1$)

Therefore, a compatible exponential model would be:

$$f_1(G) = a_1(G-1)^\alpha \qquad (8)$$

- Influence of mass

The model considers that the slope is dependent on the mass per unit of time that flows into the impoundment. This situation can be observed and confirmed by comparing the tendencies from the laboratory (small amount of mass) and that from the real deposition project (large amount of mass)

The mass is expressed as:

$$Mass = P + P \times \left(\frac{1 - C_w}{C_w}\right) = \frac{P}{C_w} \qquad (9)$$

That is, the slope depends on the solids content and production of solids. A function compatible with this consideration is:

$$f_2(C_w) \times f_3(P) = a_2 \times C_w^{\ \beta} \times a_3 \times P^\delta \qquad (10)$$

- Influence of grain size

The following model is considered

$$f_4(d_{50}) = a_4 \times d_{50}^{\ \varepsilon}$$

Finally considering $A = \prod_{i=1}^{4} a_i$ the following model is obtained:

$$i(\%) = A \times (G-1)^\alpha \times C_w^{\ \beta} x P^\delta \times d_{50}^{\ \varepsilon} \qquad (11)$$

6 EXPERIENCE CONSIDERED

There are large tailings impoundments in operation for more than 15 years in Chile, in which a control is kept of the basic parameters and the slope of tailings deposited on the beach. The impoundments considered for this study with their main characteristics are presented in the following table:

Chilean Tailings Impoundments

Name	Owner	P (KTPD)	Start year
Pampa Pabellón	CMDIC	60	1998
Tranque Austral	Codelco Salvador	33	1990
Talabre	Codelco Chuquicamata	165	1951
Las Tórtolas	Disputada	37	1992
El Torito	Disputada	16	1993
Tranque Nº4 [*]	Disputada	13	1967
Carén	Codelco Teniente	110	1986

[*] Ceased operation in 1992
CMDIC: Cía. Minera Doña Inés de Collahuasi
Disputada: Cía Minera Disputada de Las Condes

The following table presents the basic parameters of each impoundment represented by their respective average values:

Impoundment	G	C_w (%)	d_{50} (μm)	Slope (%)
Pampa Pabellón	2.70	50	52	0.50
Tranque Austral	2.70	35	119	0.35
Talabre	2.75	53	70	0.30
Las Tórtolas	2.80	26	75	0.60
El Torito	2.67	20	10	0.30
Tranque Nº4	2.67	33	10	0.60
Carén	2.70	45	51	0.18

7 CALIBRATION

The following model is obtained by performing a regression of the 'least' squares method:

$$i(\%) = 0.0051 \times (G-1)^{17.2} \times \frac{C_w^{1.45}}{P^{0.84}} \times \frac{1}{d_{50}^{0.088}} \qquad (12)$$

that presents a correlation coefficient of **0.95,** which indicates a reasonable adjustment.

As can be seen, the incidence of the grain size (in this model) is not significant. If we consider that this parameter sometimes presents great variability, a forecasting model that does not considers grain size may be presented in the following way:

$$i(\%) = A \times (G-1)^\alpha \times C_w^\beta \, x P^\delta, \quad A = \prod_{i=1}^{3} a_i \qquad (13)$$

Finally the following model is obtained for the average slope as a function of: G, C_w, and P.

$$i(\%) = 0.009 \times (G-1)^{15.6} \times \frac{C_w^{1.38}}{P^{0.86}} \qquad (14)$$

which presents a correlation coefficient of **0.95**, with similar quality of adjustment as in Eq. 12

8 ANALYSIS OF RESULTS

The following table compares the average slope values measured and those estimated by the equation (14).

Impoundment	Average slope (%) Real	Average slope (%) Model	Difference (%)
Pampa Pabellón	0.50	0.40	-20
Tranque Austral	0.35	0.41	+17
Talabre	0.30	0.29	-3
Las Tórtolas	0.60	0.60	0
El Torito	0.30	0.27	-10
Tranque Nº4	0.60	0.64	+6
Carén	0.18	0.21	+16

As can be seen, the results are fairly well adjusted, with maximum differences in the order of 20% between the average slope and the model value, even when the differences in tailings production and solids content present a wide range.

9 CONCLUSIONS

According to the tendencies shown by the model, it can be seen that the deposition slope:
- Is strongly dependant on the solids specific gravity
- Increases with the solids content
- Decreases with the increase in production as indicated by Küpper et al (1992)
- Writing the equation in terms of mass (P/C_w), the following formula is obtained:

$$i(\%) = A \times (G-1)^\alpha \times \frac{C_w^\beta}{P^\delta} = A \times (G-1)^\alpha \times \frac{C_w^{\beta-\delta}}{\left(\dfrac{P}{C_W}\right)^\delta}$$

$$= A \times (G-1)^\alpha \times \frac{C_w^{\beta-\delta}}{(Mass)^\delta} \qquad (15)$$

The previous comments can be described with the general expression:

$$i(\%) = A \times (G-1)^\alpha \times \frac{C_w^\omega}{(Mass)^\delta} \qquad (16)$$

- The exponent of the specific weight is very large. One of the reasons is the narrow range in which data is found (2.67 and 2.80), and that it can not be very numerically stable. For future studies, it would be interesting to analyze the behavior of this expression with heavier tailings. This analysis could lead to other compatible functions that could have the variable G-1 such as:

$$i(\%) = A \times f(G-1) \times \frac{C_w^{\omega}}{(Mass)^{\delta}} \qquad (17)$$

REFERENCES

Blight, G. E. and Bentel, G. M. (1983). "The behaviour of mine tailings during hydraulic deposition". J. S. Afr. Inst. Min. Metall., **83**, 73-86

Blight G. E. Thompson, R. R. and Vorster, K (1985). "Profiles of hydraulic fill tailings beaches". J. S. Afr. Inst. Min. Metall., **85**, 157-161.

De Groot, M. B., Heezen, F. T., Mastbergen, D. R. And Stefess, H. (1988). "Slopes and density of hydraulically placed sands". Proc. ASCE Spec. Conf. On Hydraul. Fill Structures, Denver, Colorado, Aug. 1988, 32-51.

Küpper, A. A. G.(1991). "Design of hydraulic fill". PhD thesis. University of Alberta.

Küpper, A. A. G., Morgenstern, N. R. and Sego, D. C.(1992). "Laboratory tests to study hydraulic fill". Can Geotech. J., **29**, 405-414.

Melent'ev, V. A. Kolpashnikov, N. P. And Volnin, B. A. (1973). "Hydraulic fill structures". Energy. Moscow

Morris, P. H. (1993). "Two-dimensional model for sub-aerial deposition of mine tailings slurry". Trans. Instn Min. Metall. (Sect. A: Min. Industry), **102**, A181-187.

McPhail, G. I. (1994). "Prediction of beaching characteristics of hydraulically placed tailings". PhD thesis, University of the Witwatersrand.

Robinsky, E. I. (1979). "Tailing Disposal by the Thickened Discharge Method for Improved Economy and Environmental Control". Tailing Disposal Today. Vol. 2: Proceedings of the 2nd International Tailings Symposium.

Sofra, F. and Boger, D. V. (2000). "Slope prediction for thickened tailings and pastes". Tailings and Mine Waste'00, Proceedings of the 7th International Conference on Tailings and Mine Waste, Fort Collins, Colorado.

Williams, M. P. A. (2001). Tailings beach slope forecasting – A Review". High Density and Thickened Tailings Conference, Pilanesberg, South Africa

Williams, M. P. A. and Meynink, W. J. C. (1986). "Tailings beach slopes". Proc. Workshop on Mine Tailings Disposal, The University of Queensland, Aug. 1986. 30 p.

Winterwerp, J. C., de Groot, M. B., Mastbergen, D. R. And Verwoert, H. (1990). "Hyperconcentrated sand-water mixture flows over flat bed". J. Hydraul. Eng, **116**, 36-54.

Tailings and Mine Waste '02, © 2002 Swets & Zeitlinger, ISBN 90 5809 353 0

Quebrada Honda tailing storage facility conception, design, construction, and operation, Southern Peru Copper Corporation, Tacna, Peru

A.H. Gipson, Jr.
Knight Piésold and Co.

Henry Walqui Fernandez
Southern Peru Copper Corporation

ABSTRACT: In 1988, Southern Peru Copper Corporation (SPCC) commenced an evaluation for alternatives to tailing deposition in Ite Bay where tailing had been deposited since the Toquepala mine began production in 1959 and the Cuajone mine began production in 1976. The tailing from the two concentrators had been commingled at the Toquepala site and carried in the natural dry stream channel to Ite Bay about 90 km from the concentrator. Tailing storage alternatives considered included deep-ocean deposition or storage in an on-shore structure. In 1994, SPCC implemented stricter environmental policies in accordance with new Peruvian regulations and decided to utilize one of the on-shore structures and halt deposition in Ite Bay by December 26, 1997. To accomplish this, SPCC undertook the investigation, design, construction, and commissioning of the Quebrada Honda tailing storage facility (TSF). The TSF is located in a seismically active area at the northern end of the Atacama Desert, 35 km from the Toquepala mine. The operation currently processes 140,000 tons per day. Ultimately, the TSF will retain 1,100 million tons, have a height of 128 m, and have a 4-km crest length. A 35-m-high starter embankment was constructed of alluvial sands, gravels, and cobbles. The remainder of the downstream-method embankment is being constructed in thirteen 5-m raises from compacted cyclone sands, including a foundation drain.

1 INTRODUCTION

SPCC mines copper ore from two ore bodies located about 90 km north of the town of Tacna in southern Peru. The Toquepala mine is located about 35 km south of the Cuajone mine. Mining at Toquepala commenced in 1959 and at Cuajone in 1976. Both mines are being exploited using open pit mining methods. Each mine has a mill and a concentrator for processing the copper ore and recovering moly also present in the ore body. The facilities are connected by a railroad established as part of the mine facilities. The railroad extends from Cuajone about 35 km south of Toquepala, then west about 90 km to the port town of Ilo. Concentrate is processed at the smelter in Ilo. The copper concentrate is then processed at the smelter and refined at Ilo.

Currently 50,000 dry short tons per day (dstpd) of ore are processed at Toquepala and 90,000 dstpd at Cuajone, bringing the combined daily production to 140,000 dstpd.

Since the beginning of the operations in 1959 and until 1997, tailings produced by the processing operations were deposited in the Pacific Ocean at the Ite Bay, 15 kilometers south of Ilo. SPCC commenced studies leading to the design and construction of the Quebrada Honda TSF in 1988. In December 1997 the TSF became operational, allowing SPCC to stop sending tailings to the Pacific Ocean.

SPCC and a number of contractors completed the investigations and construction of the facility prior to December 26, 1997. SPCC retained Knight Piésold to assist with commissioning the project in 1998.

2 SITE DESCRIPTION

The western portion of Peru is a broad, gently sloping plain rising from sea level to about elevation 1,050 m at Quebrada Honda, 55 km inland. It is informative to note that the term *quebrada* is the Spanish word denoting a canyon or gorge. Honda, Simarrona, and Huacanane are quebradas in the area. Immediately to the east of the site, the foothills of the Peruvian Andes begin to rise steeply. The Toquepala mine is located about 25 km east of Quebrada Honda at an elevation of about 3,000 m. A general site map is shown on Figure 1.

The site itself slopes moderately to the west. A steep ridge of siltstone and sandstone bedrock forms the left abutment of the Quebrada Honda TSF. The main drainage channel is a wide, flat-bottomed riverbed at the base of the left abutment. From the drainage channel, the right abutment rises gradually to the north to a broad, relatively flat plain sloping to the west.

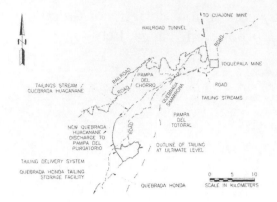

Figure 1. General site map.

3 PROJECT DESCRIPTION

3.1 *Storage requirements and facility life*

The facility has an estimated capacity of 1,100 million tons. The initial design life of the Quebrada Honda TSF was 20 years at a production rate of 72,000 to 125,000 dstpd at an estimated in-place dry density of the tailing of 85 pounds per cubic foot (pcf). In 1998, as a result of a planned increase in production to 146,000 dstpd and a revision in the dry density of the tailing in place to 75 pcf, the design

life was revised to about 14.5 years. The initial estimate of the in-place tailing density was based largely on past experience with other copper tailing. The revised density of 75 pcf was based on laboratory test work, experience with "clayey" tailing, and an estimate of the density of the tailing placed to date. It appears that the clayey component of the tailing has a significant effect on the tailing density.

The projected life spans of the mining operations are 30 years for Toquepala and 30-plus years for Cuajone, respectively.

3.2 *Tailing description*

Tailing from both mills is similar in appearance, grind, mineralogy, and particle size distribution although the Cuajone tailing typically has clayey fines (minus No. 200 sieve size material) and the Toquepala tailing silty fines. This is particularly true when the Cuajone concentrator is milling "soft high clay" ore as different areas of the open pit are mined. The minus No. 200 sieve fraction of the combined tailing stream averages about 53 percent, by weight. The percent solids of the combined tailings stream averages about 53 to 54 percent solids by weight. Tailing gradations are shown on Figure 2.

3.3 *Tailing delivery system*

The tailing from the Toquepala concentrator is conveyed approximately 17 km through an open stream

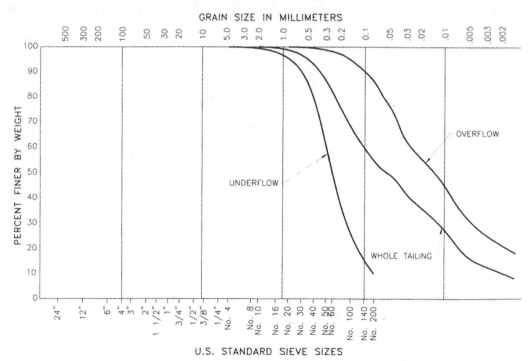

Figure 2. Tailing gradations.

94

channel to Simarrona junction where it is combined with the Cuajone tailing. The Cuajone tailing travels in a concrete-lined channel, approximately 50 km from the concentrator along the railroad line that passes through a series of railroad tunnels to just west of Toquepala where it is discharged into a stream channel that flows to the Simarrona junction. Immediately downstream from the junction of the two tailing streams is a transition structure that diverts the combined stream from the stream channel into a 2-km-long, concrete-lined open channel. The diversion structure has a gate to divert the flow through the Quebrada Simarrona during floods or for maintenance. Overflow from the structure also flows through Quebrada Simarrona to a small storage facility. The Simarrona bypass launder conveys the tailing along the very steep hill slope on the north side of Quebrada Simarrona before crossing a ridge beneath the main access road to Toquepala and discharging into the Quebrada Huacanane. From this discharge point, the tailing flows in another open stream channel for about 16 to 18 km before reaching the first pickup structure of the tailing conveyance system associated with the Quebrada Honda impoundment. At this point, the tailing is diverted from the Quebrada Huacanane drainage to Quebrada Honda. An 8-km concrete launder is used to transport the tailing to Quebrada Honda and 5 km of 42-inch-diameter HDPE pipe is used to carry the tailing on the embankment crest to the ten mobile cyclone stations.

3.4 Stormwater diversion

The Quebrada Honda TSF is located in the Quebrada Honda drainage. The Quebrada Honda drainage itself is a relatively small drainage, sloping moderately from its upper northeast end downward to the southwest. The Quebrada Huacanane drainage, a much larger parallel drainage to the northwest, joins the Quebrada Honda drainage approximately 2 km above the Quebrada Honda embankment. About 4 km before Quebrada Huacanane joins Quebrada Honda, the Quebrada Huacanane bends to the south. Topography at this bend was exceptionally favorable for diverting Quebrada Huacanane drainage to the southwest onto the gently sloping plains to the west, away from the Quebrada Honda drainage. A 35-m-high diversion dam, designed to divert the flood resulting from the 24-hour/500-year storm event, was constructed to divert the stormwater. As Quebrada Honda is normally a dry streambed that is used to transport the tailing slurry, a diversion structure was included in the design to divert the tailing into Quebrada Honda. See Figure 1 for an overview of the facility. A minimum 1-m freeboard was provided for in the early stages to store the runoff from the 24-hour/ 500-year storm event. As the embankment was raised and thus more susceptible to

larger seismic deformations, the minimum freeboard was increased to 2 m and, in the final stages, 3 m.

4 FOUNDATION PREPARATION

The need for foundation preparation was relatively limited. As average annual precipitation in the area approximates 5-mm a year, there was no vegetation to remove. The left abutment was founded on the siltstone and sandstone bedrock, the main portion of the embankment in Quebrada Honda on about 10 m of dense alluvial cobbles and gravels, and the right abutment on older deposits of dense alluvial cobbles and gravels. A number of silt layers on the surface were removed so that the embankment could be founded on the bedrock or alluvium.

5 EMBANKMENT

5.1 Starter embankment

The starter embankment was constructed of compacted alluvium. Its base is at elevation 1,055 m and crest at 1,123 m, giving it a height of 68 m. It was initially designed with a crest at 1,100 m. As a result of needing to develop a water pool on the tailing surface of sufficient size for water reclamation to reduce the percent solids of the tailing from 55 to 35 percent for cycloning, the embankment height was raised during construction. The homogeneous embankment was constructed with a 1.5:1 (horizontal to vertical) upstream slope and a downstream slope of 3:1 to elevation 1,100 m and a 1.5:1 slope above elevation 1,100 m to the crest at elevation 1,122 m. The crest width was about 10 m. The alluvium used to construct the embankment was excavated with a large shovel from borrow areas on the right side of the impoundment and hauled to the starter embankment with mine haul trucks, placed in 1-m lifts, and compacted by controlled routing of the hauling equipment and 10-ton smooth drum vibratory rollers.

5.2 Compacted cyclone sand embankments

The remainder of the downstream method embankment is being constructed with compacted cyclone underflow sands. The upstream slope above the starter embankment and the downstream slope are 3:1. A total of thirteen 5-m raises will be made to raise the embankment to its ultimate crest of 1,188 m for a height of 133 m. The main embankment section is shown on Figure 3.

The cyclone sands for embankment construction are produced at ten cyclone stations spaced evenly along the crest of the embankment During normal operation, seven or eight units operate while the

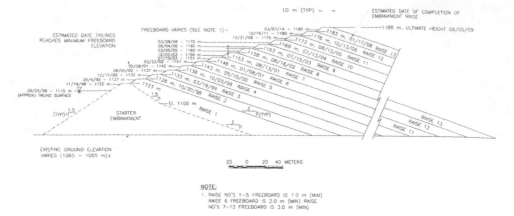

Figure 3. Main embankment section.

other units are kept in standby to allow enough time for the underflow sands to dry.

Each station consists of four Krebs Model DS20B cyclones. Overflow is discharged directly to the TSF. The underflow sands are repulped and passed through Cyclo Wash units at each cyclone station. From there, the underflow slurry is piped to "paddocks" or cells constructed on the top of the advancing raise. The tailing is placed in maximum 1-m lifts, drained, and compacted with a vibratory roller to 98 percent density ASTM D-698. Currently, as the underflow slurry is deposited in the cells, it is spread with a tracked dozer in about 15-cm lifts. A decant is used to remove the supernatant that it is piped to the embankment toe. Figure 4 shows a typical cyclone station and Figure 5 the cells under construction.

The raise schedule required the continuous operation of six cyclone stations to produce and place 514,000 m^3 per month of cyclone underflow sands. Significant operational problems, due to the high clay content of the tailing, were encountered and overcome to achieve the needed production rate. The initial specification for the underflow sands required

the percent fines passing the No. 200 sieve be less than 15 percent. In practice, this did not allow the sands to drain sufficiently in 24 hours so that one lift per day could be constructed to meet the required raise rate construction schedule.

The initial test work was reviewed, and it showed that cyclone simulations performed on the whole tailing feed by Krebs Engineers in January 1995 estimated that the DS29B 20-inch-diameter cyclones would produce an underflow product of which 6 percent would be finer than the No. 400 sieve. In early 1996, Cyclo Wash unites were included to reduce the minus No. 200 sieve fraction in the underflow from over 20 percent to less than 12 percent. SPCC installed the Cyclo Wash units, and when proper feed dilution and cyclone manifold and Cyclo Wash pressures are maintained, the minus No. 200 fraction in the underflow averages less than 10 percent, by weight. This is very good cyclone performance considering the fineness of the minus No. 200 portion of the material and its clayey nature. By observing the performance of the cells, it was learned that if the percent fines was less than 11 percent, the resulting drainage characteristics of the underflow

Figure 4. Typical cyclone station.

Figure 5. Cells under construction.

96

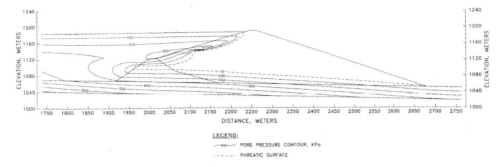

Figure 6. Results of seepage analyses for ultimate embankment.

sands improved significantly, and the desired raise rate could be maintained. In this case, a reduction in the percent fines of 3 to 4 percent had a dramatic effect on the drainage characteristics of the underflow sands.

5.3 Instrumentation

Vibrating wire piezometers were installed in the bedrock beneath the main embankment, in the alluvium beneath the main embankment, and in the compacted cyclone underflow sand portion of the embankment to monitor the pore pressures affecting the structure.

6 STABILITY AND SEEPAGE ANALYSES

One of the key observations made during review of the design work was that the seepage analyses conducted at different times during the design process were in conflict, yielding dramatically different results. The original embankment design alternatives with either a 4:1 or 5:1 downstream slope were predicted to remain unsaturated, while a later 3:1 downstream slope alternative was predicted to become fully saturated unless underdrainage was provided. The reasons for the dramatic shift in the location of the phreatic surface were unclear; documentation was not available to allow a full assessment of the previous analyses. Given the importance of the seepage conditions to the overall stability of the embankment, the analyses were rerun.

6.1 Seepage analyses

The seepage analyses were performed for two embankment configurations at the maximum cross section, one for the embankment at the end of Raise No. 2 and one for the ultimate height embankment. The two-dimensional seepage analyses were based on the finite element method and were performed using the computer program SEEP/W, a commercial finite element code formulated specifically to conduct these types of analyses. The program considers both saturated and unsaturated flow and allows the user to input an unsaturated hydraulic conductivity relationship for each material used in the analysis.

The foundation alluvium consists predominately of well graded sand and gravel with some cobbles and boulders. There were also widely scattered lenses of silt reported. The average amount of fines was about 5 percent. It was characterized as dense to very dense. The results of field infiltration tests and empirical estimates of hydraulic conductivity based on grain size were used to develop the parameters for the analyses. Input parameters for the alluvium utilized in the initial analyses were reviewed and judged to be valid. The input parameters are summarized on Table 1.

The analyses indicate the main embankment will be substantially unsaturated under steady state conditions. The computed phreatic surface within the starter embankment rises to about 8 m during raise No. 2 construction and 25 m at the ultimate configuration above the level of the foundation in the starter embankment. It drops evenly through the underflow sands to exit at the toe of the embankment. The reason that much of the phreatic surface stays close to the upstream face of the embankment and near the foundation is that the impounded tailing has a low hydraulic conductivity relative to the starter embankment, underflow sand, and foundation alluvium.

There is insufficient water flowing from the impoundment to the embankment to saturate the up-

Table 1. Estimated saturated hydraulic conductivities used in seepage analyses

Material Type	K_h (m/s)	K_v (m/s)	K_h/K_v
Alluvium	3.7×10^{-5}	7.0×10^{-6}	5
Starter Embankment	5.4×10^{-5}	5.4×10^{-6}	10
Cyclone Underflow	2.2×10^{-5}	1.1×10^{-5}	2
Cyclone Overflow Slimes:			
0-15 m depth	3.0×10^{-7}	3.0×10^{-8}	10
15-30 m depth	5.0×10^{-8}	5.0×10^{-9}	10
30-45 m depth	4.0×10^{-8}	4.0×10^{-9}	10
45-60 m depth	3.0×10^{-8}	3.0×10^{-9}	10
> 60 m depth	2.0×10^{-8}	2.0×10^{-9}	10
Bedrock	1×10^{-12}	1×10^{-12}	1

stream portion of the embankment and raise the phreatic surface. It should be noted that flow will occur within the unsaturated portions of the embankment; however, the quantity of this flow will be much smaller than that within the saturated zones due to the reduced hydraulic conductivity. Results of the seepage analyses for the ultimate embankment are shown on Figure 6.

Exit gradients were estimated at the toe to evaluate the potential for piping. Based on the flat slope, high exit gradients were not anticipated. The calculated exit gradient was 0.10, which is below the value of 1.0 required to initiate piping. The calculated factor of safety against piping is 5.0. The generally accepted factor of safety for exit gradient ranges from 2.5 to 5 (U.S. Corps of Engineers, 1986), indicating piping is not expected.

However, given (1) the unique nature of the phreatic surface, (2) that seepage would be passing through the unsaturated zone, (3) that water would be entering from the cells being used to construct the raises, (4) that experience tells us actual flows seem to be greater than anticipated, and (5) the relatively little cost of a foundation drain versus its benefits, a foundation drain was recommended and installed. It was also recognized that maintaining a lower phreatic surface would enhance removal of seepage from the cell construction and thus speed drainage and construction.

Piezometer readings to date indicate the phreatic surface is consistent with that developed in the analysis.

6.2 Stability analyses

The downstream slope of the main embankment is the critical slope with respect to stability since the upstream slope will always be buttressed by the tailing. The designer performed a series of undrained triaxial shear tests with pore pressure measurements on compacted samples of the cyclone underflow sand and estimated this material to have an effective friction angle of 35 degrees. Based on the low fines content (9 to 11 percent) of the sand, this appeared to be a reasonable value. The minimum factor of safety against slope instability for the unsaturated 3:1 slope is about 2.0, and the critical failure mode is raveling of the slope surface (infinite shallow failure). For deeper failure surfaces which cut through the slope, the minimum factor of safety is greater than 2.0. The generally accepted minimum factor of safety is 1.5, so the calculated factor of safety is greater and deemed acceptable.

The designer had performed a seismic deformation analysis of the embankment under earthquake loading. The analysis indicated that, at the ultimate height, the embankment would not undergo deformations that would result in the release of tailing. Based on review of this work and the results of the

seepage analysis, it was believed that tailing would not be released from the embankment during design earthquake loading.

7 LESSONS LEARNED

One of the key questions an author needs to ask when considering writing a paper is "Does the paper contribute anything to the general body of knowledge of the profession?" To that end, this section summarizes those experiences described above and a few others not elaborated on above.

7.1 Sand production

A few percent change in the percent fines can have a dramatic effect on the behavior of the underflow, especially when the fines are slightly clayey in nature. In this case, the need to use the Cyclo Wash units to decrease the fines content had a dramatic effect on the construction process and cost. The underflow product changed from moist sand that could likely have been spread by dozers and compacted with smooth drum rollers quickly after production and placement to slurry that required dewatering and draining prior to compaction. This required a much more complicated construction process involved the construction of cells in which to place the tailing to drain and the removal of the slurry water from the cells to the embankment toe. The water is piped to the toe to avoid erosion of the slope. In addition, the tracked dozers used to place the tailing operate in wet conditions, which greatly increases the wear rate of the tracks and thus the operational costs. As more intensive construction is required to construct the cells, a larger equipment fleet and a larger work force are needed, which greatly increases the cost.

Thus, it is extremely important to fully examine the potential variations in materials that will be mined to evaluate their effect on the design, construction, and operation of the facility.

7.2 Foundation drain

A foundation drain had not been included in the initial design. The theoretical seepage analyses indicated the embankment would be stable from a seepage standpoint. However, when the practical aspects were considered, it was judged to be prudent to include a foundation drain. Key to the consideration were (1) despite the unique shape of the phreatic surface and the realization that although the majority of the embankment would not be saturated, there would still be flow through the unsaturated zone; (2) providing for the rapid removal of water seeping into the embankment from raise construction would potentially be beneficial to speeding drainage from the cells and would thus increase the overall raise rate; and (3) although the theoretical analyses indicated suitable factors of safety against piping, the fact that the phreatic surface would daylight on the downstream face just above the toe was discomforting. Given that the cost of the foundation drain installed at the time of construction was relatively small and the potential magnitudes of the resulting problems were large if it was not installed, it was decided to install the foundation drain prior to raise construction.

7.3 Staff training for compacted cyclone sand embankment construction

SPCC has a conscientious and dedicated work force experienced in open pit mining operations. The construction and operation of tailing facilities were new to the operations as previously tailing was deposited in Ite Bay. Although components of the operations appear similar, they each are specialized operations in their own right. SPCC recognized the need to train its workforce specifically for TSF construction. This was accomplished in two parts. The first part involved training personnel to perform the quality control portion of the work and the second training of SPCC personnel in the construction- and operation-related aspects. Training was very successfully accomplished by providing an engineer experienced in construction management and earthworks construction to work with SPCC personnel for a six-month period.

7.4 Water planning and management

One of the key aspects to the successful operation of the facility is water management. At start-up, water was in short supply, so provision was made to use the supernatant to reduce the percent solids in the tailing slurry from 55 percent to 35 percent for cycloning. A barge-mounted pump station was installed in the water pond with a pipeline to the tailing delivery launder on the right abutment. Water pumped from the pond was used to dilute the slurry. Excess water was discharged downstream from the point along the launder where the tailing slurry was diluted with supernatant. Pond size control was critical. A large pond impaired the stability of the facility, and a small pond did not allow sufficient settling time to provide a source of clean water for the pumps. If dirty water was pumped, pump life was dramatically reduced.

8 CREDITS

The authors want to dedicate this paper to Walther Sotomayor who lost his life in an accident at Quebrada Honda. Walther was the tailing facility superintendent, and it was through his dedicated efforts and relentless hard work that the project was successful.

We also want to thank SPCC for permission to prepare and publish this paper.

Tailings and Mine Waste '02, © 2002 Swets & Zeitlinger, ISBN 90 5809 353 0

Retrofitting an HDPE-lined raise over an unlined tailing facility

A.H. Gipson, H.P. Vos & W.J. Cole
Knight Piésold and Co., Denver, Colorado, USA

S.R. Aiken
Knight Piésold Limited, North Bay, Ontario, Canada

ABSTRACT: A gold mine, located in the Middle East, required a modification to its existing tailing storage facility to (1) reduce seepage losses from the existing unlined facility, (2) capture water being lost to seepage for use in a new heap leach operation, and (3) increase capacity. A unique aspect of this project was the founding of the 15-meter-high, HDPE-lined containment over 40 percent of the existing unlined facility. The tailing deposited in the existing facility served as the foundation for the modification and provided a smooth surface for the subsequent geosynthetic liner on both the basin and upstream embankment slopes. Sub-aerial deposition was used as the method of placing the tailing, creating a drained deposit for environmental stewardship, increased capacity, and ease of reclamation. The modified facility has been in operation since 1995. The design, construction, commissioning, and reclamation aspects of the tailing facility modification are discussed.

1 PROJECT OVERVIEW

1.1 *Site history*

The mine began operations in April of 1991. The existing tailing facility was constructed in two stages. Stage 1 involved constructing a series of six dikes connecting topographic ridges to form a rectangular impoundment approximately 1280 meters east to west by 500 meters north to south. Stage 1 was built of mine waste rock with a crest width of approximately 4 meters. The slopes, both upstream and downstream, were constructed at the angle of repose of the waste rock at approximately 1.5H:1V. The Stage 1 dikes had a maximum height of about 4 meters. The Stage 2 raise, which comprised four phases described below, was started in early 1995 with the construction of Phase 1. During this phase, the existing dikes were raised and joined to form a continuous embankment around the entire perimeter. The height of the Phase 1 raise ranged from 4.9 meters on the north to 3.4 meters on the south. The completed Phase 1 perimeter embankment had a maximum height of about 8 meters. The existing Stage 1 facility and Phase 1 of Stage 2 are shown in Figure 1.

1.2 *Site characteristics*

1.2.1 *Topography*
The elevation at the mine site is approximately 1000 meters and the topographic setting is gently undulating, with smoothly rolling ridges and hills.

1.2.2 *Climate*
The climate at the mine site is consistent with the arid climate that covers most of the Middle East. Precipitation amounts average about 70 mm per year with annual evaporation of 2600 mm per year. Average temperatures range from 30 to 45 °C (86 to 113 °F) in the summer to 10 to 30 °C (50 to 86 °F) for the winter months.

Wind is frequent and predominantly from the southeast or northwest. Several times a year, heavy thunderstorms or sandstorms form and wind speeds of 70 to 100 kilometers per hour can be generated.

1.2.3 *Geologic setting*
The mine site is underlain by very-low-grade, metamorphic, laminated sandy sediment. From near surface to a depth of 20 to 40 meters, the sediment often consists of siltstone or fine-grained sandstone/mudstone that generally dips gently, forming low ridges. In the wadis (dry stream beds), the rock is covered with sand or sandy clay, which is about 3 meters thick at the deepest portions, but averages less than 1 meter. Stocks of mainly dioritic composition have intruded the Murdana formation and host sporadic quartz-vein-associated gold mineralization. The region is not considered a seismically active area. No earthquakes have been recorded in the vicinity.

1.3 *Objectives*

The original design concept envisioned the tailing as being self-sealing, such that the tailing fines would

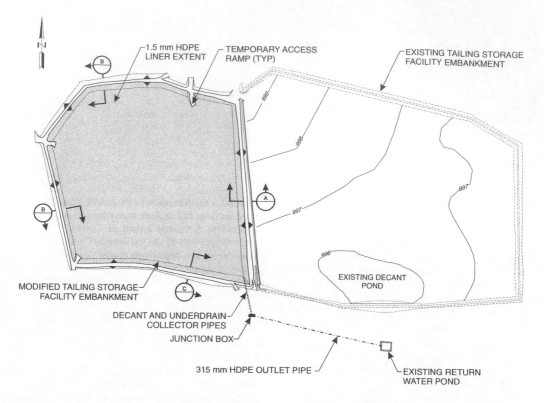

Figure 1. Stage 2-Phase 1 facility superimposed upon the Stage 1 facility.

"seal" the basin and virtually eliminate seepage losses. However, a water balance analysis indicated that about 60 cubic meters per hour of water was being lost to seepage. Consequently, the facility was modified to reduce seepage losses from the existing unlined facility, capture water being lost for use in a new heap leach operation, and increase capacity. This was achieved by constructing a geosynthetic liner on the surface of the existing tailing facility and providing an underdrainage system to recover water. The tailing deposition method was changed from a single-point sub-aqueous discharge method to a thin-layer, rotational, sub-aerial deposition method (drained-and-air-dried) to increase the in situ density of the tailing.

Constructing a geosynthetic liner on the surface of an existing tailing facility is not a common practice and careful consideration had to be given to the effects of settlement (especially differential settlement) and undue strain on the geomembrane liner.

2 ALTERNATIVES EVALUATION

2.1 Preliminary evaluation

Preliminary alternatives were considered based upon the project objectives, site conditions, topography,

and results of geotechnical analyses of collected soil and tailing samples. Four alternatives emerged, consisting of the following design concepts:
- Alternative 1 - raising the western third of the existing facility with a drained-and-air-dried tailing deposition method.
- Alternative 2 – raising the western two-thirds of the existing facility with undrained deposition.
- Alternative 3 – expanding to the north in a new facility, using the north dike of the existing facility with drained-and-air-dried deposition.
- Alternative 4 – expanding farther to the north, also using the north dike of the existing facility.

2.2 Final alternative evaluation

Two variations (Options A and B) of Alternative 1 identified in the preliminary alternatives eval auon were studied further. Detailed volumetr'.s we per-formed for each option, which defined construction stages.
- In Option A, the western end of the existing tailing facility would be raised and a divider dike constructed on the existing tailing. Approximately 40 percent of the existing tailing facility would be covered.
- Option B followed the same concept of raising the western end of the existing tailing facility and

102

using a divider dike. However, approximately 50 percent of the existing facility would be covered.

Operationally, Options A and B were essentially the same. Considering that both options would perform at similar levels, the Owner chose Option A because of its lower capital cost.

3 LABORATORY TESTING

In conjunction with the project requirements and observations of the performance of the current facility, a laboratory testing program was conducted to characterize the tailing and other construction materials. Engineering characteristics of the tailing were defined for use in (1) estimating the required facility size to store the anticipated tailing volume, (2) designing the embankments and foundation, and (3) designing the underdrain pipe system to further consolidate the tailing.

The testing to characterize the material properties included index testing, such as Atterberg limits and grain-size analyses. Moisture content and density tests were also conducted on relatively undisturbed tailing samples to assist with material characterization and capacity estimates.

Tailing settling and drying tests were conducted to develop density versus initial solids content for the three alternative deposition methods (undrained, drained, and drained-and-air-dried). Figure 2 summarizes the tailing characterization test results. Generally, the results indicated a dry density of 1120 kg/m³ for undrained deposition, 1280 kg/m³ for drained deposition, and 1760 kg/m³ for drained-and-air-dried deposition. These densities were achieved at an expected slurry solids content of approximately 40 percent. An in situ tailing dry density (deposited in a drained condition) of 1290 kg/m³ was estimated by obtaining the volume of historic tailing, using the pre-construction topography and the existing tailing surface elevation. This was in excellent agreement with the tailing characterization test results, which predicted a dry density of 1280 kg/m³.

Undrained triaxial shear tests were performed to estimate the shear strengths of the in situ tailing and tailing to be used for construction purposes. The test on the tailing sample remolded to simulate the in situ tailing density resulted in an effective angle of internal friction of 22.9° with zero cohesion. The test on the tailing sample remolded to a dry density representing the anticipated future drained condition resulted in an effective angle of internal friction at 35.5° with cohesion of 6.7 kPa.

Direct shear tests were performed in a modified ⁴-inch-square shear box to estimate the interface friction angle between the tailing and the 1.5-mm-thick high-density-polyethylene (HDPE) geomembrane for use in stability analyses. This is an impor-

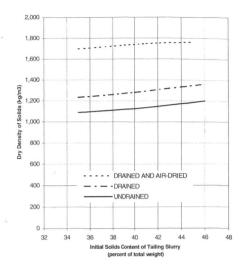

Figure 2. Relationship between dry density and initial solids content.

tant test because the HDPE presents a smooth planer surface on which movements can occur. The stress/strain curves peaked at relatively low strain and then declined. For this reason, a conservative residual or post-peak strength was used for stability analysis and design purposes. Test results on remolded tailing samples showed angles of friction at 21.5° and 24.8° and zero cohesion. Test results using residual strength showed an angle of friction of 18°.

Permeability tests were performed to develop properties of materials for use in the basin underdrain system. The constant-head permeability tests conducted on three tailing samples indicated average permeabilities ranging from 7.6×10^{-5} to 1.3×10^{-5} cm/s.

One-dimensional consolidation tests were performed on tailing samples to estimate settlement of the existing tailing once additional tailing was deposited. From the test data it was concluded that consolidation would be relatively uniform throughout the footprint of the proposed facility but that some differential settlement could occur around the perimeter where tailing thickness varied. These potential issues were addressed in the design.

4 DESIGN

4.1 Design basis

The final alternative evaluation and current and future project needs were reviewed to establish a philosophy to serve as the basis for designs. Included in the modification was an expansion covering approximately 40 percent of the western portion of the existing tailing facility. Upstream and downstream raises were used, and the basin and Phase 1 embankments were lined with a 1.5-mm-thick HDPE synthetic geomembrane overlain by an underdrain collection system. The modification was planned to be a four-phased construction, raising the western, northern, and southern embankments of the existing facility and constructing a divider dike on the existing tailing to form the eastern embankment.

The downstream embankment construction technique was used in the southeast corner where the new decant pond was adjacent to the perimeter embankment. The upstream construction technique was specified for the remainder of the east and south embankments as well as for the northern and western embankments.

A semi-annual construction schedule for raises was selected as commonly used in similar circumstances. Construction every three years was eliminated because it involved relatively large volumes of earthwork to construct perimeter dikes. Even though annual raises resulted in a reduction in the amount of earthwork for each phase, they were also excluded because of the higher mobilization costs involved.

Cumulative capacities of the Stage 2 phased expansions were 1,400,000, 2,470,000, 3,500,000, and 4,500,000 tonnes for Phases 1, 2, 3, and 4, respectively. At the design production rate of 700,000 dry tonnes per year, the phases have capacities for 2.0, 1.5, 1.5, and 1.4 years, respectively. The Stage 2-Phase 1 raise was a maximum of 5.0 meters and the subsequent three raises were approximately 3.5 meters each. The ultimate facility will have a maximum height of 21 meters, or overall increase of about 15 meters over the level of the existing facility, with a crest length of approximately 1600 metersSequential, thin-layer, rotational tailing deposition was accomplished through a header pipe with dropbars. Supernatant fluid exits via a sloping decant structure into an existing HDPE-lined reclaim pond and is ultimately recycled to the plant for reuse.

4.2 Stability analyses

Theoretical stability analyses were performed on representative sections of the proposed embankment to confirm that the facility, as designed, met commonly accepted minimal factors of safety for slope stability. Two sections were selected for evaluation. The first section represented the maximum anticipated section for upstream construction founded on natural soils and rock. This was where the existing embankment was also the highest. The second section, also constructed by the upstream method, was selected because it would be founded on existing tailing. Remaining sections of the embankment were considered to be more stable since they were not as high or were constructed of stronger materials in the downstream direction.

Table 1. Factors of Safety

Description	Factor of Safety	
	Section 6+60	Section 10+50
Downstream Face Circular Failure	1.7	1.7
Downstream Face Block Failure	1.8	1.7
Upstream Face Circular Failure	1.8	1.9

To evaluate stability, three cases were considered for each section. In Case 1, downstream slope stability was evaluated using circular arc analyses. In Case 2, the sliding-block or wedge method was used to evaluate resistance to movement along the tailing/HDPE liner interface. In Case 3, slope stability was evaluated in the upstream direction, where the facility was raised on recently deposited tailing.

Material properties, including density, angle of internal friction, and cohesion were developed for the materials existing or planned for use in construction. The material properties were developed based on experience with similar materials or the results of laboratory tests on samples of tailing and HDPE lining materials. For purposes of analyses, a phreatic surface located 2 meters above the HDPE liner was assumed.

Stability analyses were performed on a personal computer using XSTABL software using the modified Bishop method for circular arc analyses or the modified Rankine for block- or wedge-shaped surfaces. Since the site is located in a non-seismic area, analyses were limited to static cases. Typical results of the stability analysis are summarized in Table 1.

4.3 Foundation preparation

The tailing previously deposited in the existing facility served as the foundation for the facility modification. Foundation preparation was required for two main purposes: (1) preparation of the surface to support equipment for lining construction and (2) to provide a smooth, compacted surface suitable as bedding for the HDPE liner.

4.4 Embankment

Typical embankment sections are shown in Figures 3 and 4.

The southern and eastern embankments of the facility modification were designed for construction using downstream and upstream construction methods. The downstream construction method was used in the area at the decant pond. Both embankments were designed as homogeneous random fill/waste rock structures and were constructed using material provided by the Owner. A 300-mm-thick tailing layer was placed on the upstream face of the embankment as an underliner for the 1.5-mm-thick

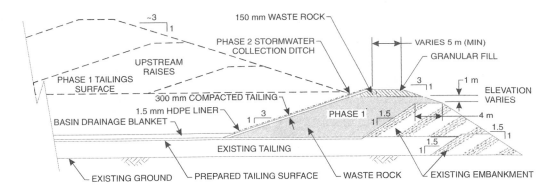

Figure 3. Embankment section B-B, upstream construction.

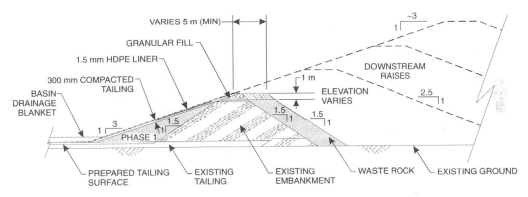

Figure 4. Embankment section C-C, downstream construction.

HDPE liner. The embankments were designed to have a 5-meter-wide crest and 3H:1V upstream and 2.5H:1V downstream slopes. This geometry was selected to provide embankments with sufficiently high factors of safety while attempting to minimize the amount of construction materials and subsequent cost. The 3H:1V upstream slopes allowed placement of the underliner and geosynthetic liner. The use of waste rock was planned in both the southern and eastern embankments to provide suitable starter embankments for future construction expansions.

The northern and western embankments were designed for construction using upstream construction methods. These two embankments were designed as homogeneous earthfill structures constructed with waste rock or tailing material borrowed from within the modified facility or the unused eastern portion of the existing tailing facility beach.

4.5 Liner and underdrainage collection systems

The liner and underdrainage collection systems were designed to work together to collect underdrainage for reuse. A 1.5-mm-thick HDPE geosynthetic was selected for the liner due to its resistance to ultraviolet photodegradation (which was a primary con-

sideration in a desert climate), its ability to be thermally seamed (which allows for positive seam quality control procedures), and its puncture resistance, strength, and workability. The HDPE liner was covered with a 300-mm-thick protective layer of tailing imported from the adjoining existing facility.

Around the perimeter of the facility, where the maximum potential differential settlement was expected, the embankment raises were constructed over previously placed tailing so that the embankment could settle with the liner and alleviate the potential for differential settlements that could impact the geomembrane liner.

The tailing basin was designed with an underdrainage system that overlies the HDPE liner and protective layer. The intent was to provide a collection layer several orders of magnitude more permeable than the tailing to evacuate seepage from the tailing. The underdrainage system also reduces the head on the HDPE liner, thereby decreasing the likelihood of seepage from the facility.

The underdrain consists of a herringbone system of 100-mm-diameter, perforated, corrugated polyethylene tubing (CPT) collection pipes placed at approximately 15 meters on center. These collection

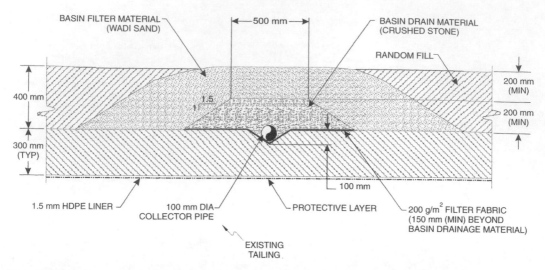

BASIN FILTER MATERIAL (WADI SAND)
500 mm
BASIN DRAIN MATERIAL (CRUSHED STONE)
RANDOM FILL
200 mm (MIN)
400 mm
1.5 / 1
200 mm (MIN)
300 mm (TYP)
100 mm
1.5 mm HDPE LINER
100 mm DIA COLLECTOR PIPE
PROTECTIVE LAYER
200 g/m² FILTER FABRIC (150 mm (MIN) BEYOND BASIN DRAINAGE MATERIAL)
EXISTING TAILING

Figure 5. Typical section through drainage blanket.

pipes flow into 150-mm-diameter collection header pipes. The upstream ends of the perforated CPT collection header pipes change from perforated to solid at the toe of the northern and western embankments, then continue up the embankment face, where they are sealed with end caps and will remain exposed. In the event that the pipes should become clogged or the flow otherwise restricted, the end caps can be removed and fresh water used to rinse the lines. The downstream ends of the collection header pipes are connected to solid HDPE pipes, before passing through the existing embankment, and deliver the seepage to an existing lined collection pond.

The herringbone pattern of pipes is covered with a 200-mm-thick layer of basin drain material covered in turn by a 200-mm-thick layer of basin filter material. The basin filter material acts to reduce movement of fine tailing material into the basin drainage material and, hence, underdrainage collection piping. The exposed areas between the underdrainage pipes were filled with a 400-mm-thick layer of fine random fill to present a uniform surface in the bottom of the tailing facility. A typical section through the basin drainage blanket is shown in Figure 5.

4.6 Tailing distribution system

The tailing from the carbon-in-leach (CIL) plant was designed to be pumped as a slurry to a distribution system running along the northern and western embankments of the modified facility. From the distribution main, the tailing was deposited through dropbars in thin layers, in a controlled rotational sequence, to develop a large, exposed tailing beach sloping toward the reclaim pond in the southeast corner of the facility. The tailing distribution system consists of 14 sections with two dropbars in each

section, for a total of 28 dropbars. The dropbars have 50-mm-diameter holes cut into the overt at 400 mm on center, along their entire length. This allows deposition of tailing without any maintenance to the dropbars. As the tailing level rises, the dropbar deposition holes fill with tailing, and deposition occurs through holes higher on the dropbar. The dropbars are buried in the tailing and new dropbars are installed for future phases of construction. Each of the 14 sections is operated until approximately 100 to 150 mm of tailing have been deposited. The system is then rotated to the next section, allowing the freshly deposited tailing to consolidate, drain, and air dry.

This drained-and-drying method or sub-aerial technique of tailing deposition produces a denser in situ tailing, providing substantial cost savings. Sub-aerial deposition also provides for enhanced liquid/solid separation during deposition to maximize subsequent air drying and consolidation of the tailing while maximizing the liquid recovery from runoff and underdrainage collection. This method fulfills the basic design requirement to maximize the recovery of process liquids from within the tailing storage facility, thereby minimizing the volume needed for storage of tailing solids and reducing the cost of storage. By adopting the sub-aerial deposition method, an increase in tailing density of approximately 36 percent is achieved. Had this method not been adopted, the life of the modified facility would have been 4.7 years compared to the 6.4 years achieved.

4.7 Decant system

The decant structure is located at the southeast corner of the facility and connects to the concrete-encased HDPE outlet pipework for transfer of the decanted liquids from the tailing basin to an existing lined pond.

The decant structure consists of a 450-mm-diameter cold-rolled steel section with a 400-mm-wide slot at the top of the section. The slot is fitted with a slide guide rail, which allows for insertion of 50-mm × 100-mm wooden stop logs that swell to close the section to advancing tailing and allow control of the pond water level.

5 CONSTRUCTION

5.1 *Phase 1 construction*

Phase 1 of the tailing storage facility was completed during 1995 and consisted of raising the western one-third of the existing Stage 1 facility, lining the raised facility with geomembrane, and constructing the underdrain, decant, and a tailing distribution piping systems.

5.1.1 *Earthworks*

Construction activities began in February 1995, with the placement of the working surface for the eastern dike onto the existing tailing surface. The material was placed by end-dumping waste material trucked from the mine and spread using a dozer. Existing tailing was not displaced noticeably during the placement of the eastern dike material, indicating that the tailing provided an adequate foundation for the dike construction. For the northern, southern, and western embankments, a working surface was created in a similar fashion, but the existing tailing offered more support because it was adjacent to the perimeter of the facility. After the first 1-meter-thick lift of construction material had been placed across the tailing surface, additional 0.6-meter-thick lifts were constructed in a similar manner by end-dumping waste material from the current mine operations or waste material borrowed from mine dumps, spreading with a dozer, and compacting, using systematic routing of construction equipment over the construction surface. After embankment fill had been placed, an excavator was used to smooth the upstream face to an even 3H:1V slope. Once the upstream embankment face had been shaped, tailing material was hauled from the existing tailing facility and spread on the upstream face of the embankment using a dozer, after which it was compacted to a smooth, uniform 300-mm-thick layer using a vibratory smooth-drum roller.

Because the surface of the existing tailing within the Phase 1 construction area was soft and very slick, it was initially scarified, using a farm tractor and disk attachment, to remove the layer of crystallized salts and promote drying and increase density so that it would function as subgrade for the geomembrane liner and support construction equipment. Once the tailing had air dried for a few days, it was scarified approximately 150 mm deep or as needed to facilitate further drying. Very soft areas were excavated using an excavator or dozer, dried by working it in the sun, and then replaced. After the tailing had been dried sufficiently, it was smoothed using a grader and compacted using a smooth-drum vibratory roller. This method of drying, smoothing, and compacting the tailing surface achieved a smooth, uniform surface to serve as geomembrane liner subgrade. Final shaping and compacting of the tailing surface was done intermittently as equipment was available and to keep just ahead of geomembrane liner deployment so that the tailing surface was not damaged by construction equipment.

5.1.2 *Liner and underdrainage collection systems*

Prior to liner placement, the subgrade was accepted and signed off by the liner installer. The liner crew consisted of one supervisor, one quality-control technician, four liner technicians and fifteen laborers. The liner was deployed using a loader and spreader bar, and seams were made using double-wedge fusion welding machines. Shorter welds, patches, and welds at the bottom of slopes were performed using extrusion welding machines.

The protective layer component of the basin drainage system over the geomembrane liner was borrowed from the existing tailing facility, using loaders; hauled to the construction site, using haul trucks; end-dumped; and spread by a rubber-tired loader. This produced a reasonably uniform tailing layer between 300 and 600 mm in depth which was further worked using a grader to produce a uniform layer which ranged in thickness between 200 and 300 mm.

Underdrainage installation began by cutting a 100-mm-deep vee-trench into the protective tailing layer using a farm tractor with a specially constructed plow apparatus. Geotextile and underdrainage piping was then installed lengthwise along these vees. Waste rock was crushed on site and used for the 200-mm-thick basin drain material that was placed over the underdrainage piping, using loaders. A farm tractor with a specially constructed attachment was used to strike off the basin drain material above the top of the underdrainage piping. Basin filter material was hauled from the nearby wadi and placed in a 200-mm-thick layer over the basin filter material on the underdrainage pipes, using loaders. Finally, the remaining exposed areas in the bottom of the facility were filled with mine waste to a thickness of 400 mm to complete the underdrainage system.

5.1.3 *Tailing distribution system*

Components of the tailing distribution system were obtained from local suppliers. The Owner prefabricated the dropbar offtakes in its workshop on the mine. This consisted of shortening the outlet leg of the 225-mm-diameter HDPE tee as much as possible before welding on the 225-mm × 100-mm HDPE reducer, and welding on the HDPE flanged stub end and backing flange. The 150-mm-diameter dropbar pipes were cut to the correct length, the tailing discharge holes were drilled, and then the endcap was fitted.

5.1.4 Site-specific limitations

Construction activities were affected by several factors. These included the remote location of the mine site, workforce experience and language barriers, equipment condition and availability, and climate.

The closest major city to the mine site is 750 km away. That resulted in delayed delivery of construction materials and equipment parts on several occasions, slowing construction.

The workforce comprised several nationalities, most of whom were of the Muslim faith. To accommodate these workers, work breaks were taken throughout the day for prayer. Although most of the workforce spoke English fairly well, communication was difficult at times due to the variety of languages spoken. These included Arabic, Swedish, Filipino, Indian, and English. English was, however, the one common language. Occasionally an interpreter was needed. If one was not available, sketches were used to visually communicate the issue at hand.

The technical specifications for the tailing facility modification adhered to U.S. standards. On occasion this created problems with the construction supervisors and workers who weren't accustomed to such exacting standards. Several times during liner deployment and seaming, the owner of the liner installation company was on site pleading his case to lower the destructive testing requirements, saying it was unreasonable and that they never had to do such a large amount of testing in the past. Due to several destructive test failures at the beginning of geosynthetic liner construction, the frequency was never altered.

The overall condition of the earthwork contractor's construction equipment on site was marginal at best. Machinery continually broke down and spare parts' availability was an issue. A lack of working equipment resulted in slow production of the tailing component of the basin drainage system, forcing the Owner to relieve the earthworks contractor of further involvement with the Phase 1 project. Construction was slowed down until another earthworks contractor mobilized to the site and was familiar with the project.

Without question, the most challenging aspect to deal with during construction was the temperature fluctuations during the course of a 24-hour period. Temperatures rose to over 40 °C (104 °F) during the day and cooled off to 25 °C (77 °F). This made it exceedingly difficult for liner deployment and seaming. The HDPE liner used to line the facility is susceptible to extreme temperature fluctuations, and fieldwork had to be carefully planned and timed. Deployment and seaming were completed during the early morning hours before the onset of high temperatures and before the liner began to wrinkle and fold. Afternoons were limited to QA/QC testing and liner repair. "Trampolining" of the liner on the slopes and especially in the corners was common due to the contractor not leaving enough slack in the liner to account for contraction at night. The result was that the liner had to be cut to relieve the tension and then reinstalled and seamed. In addition, because the liner had numerous wrinkles and folds during the heat of the day, it was apparent that the overlying protective layer had to be placed at night when the liner surface was smooth. This required a second shift of workers. At night, visibility was poor, making it difficult to place the protective material on the liner and spread it to the specified thickness. The HDPE liner was torn several times by the loader so the area had to be staked off, the protective layer over the tear removed, and the liner repaired.

The high daytime temperatures also required that the HDPE tailing distribution system piping be installed at night.

The overbearing daytime temperatures also took a toll on the equipment and workers, resulting in numerous breaks due to fatigue and/or machinery breakdowns. Compounding the heat was working on the black surface of the HDPE liner, which, during the heat of the day, had temperatures exceeding 50 °C (122 °F) – not to mention the high winds and resultant sandstorms that exacerbated the situation.

Careful planning and scheduling were required to place the infill material between the underdrainage pipes so that equipment did not damage the pipes and trucks did not have excessively long, circuitous routes to maneuver between the herringbone layout. In some areas, continuous traffic over the protective layer caused the underlying tailing to heave, and alternative routes had to be created.

6 COMMISSIONING

The tailing storage facility was operated by Knight Piésold from August 10, when tailing was first delivered to the facility, until August 15, 1995, when the operation was handed over to the Owner. Commissioning aspects of two key elements of the system, namely, the underdrainage system and tailing distribution system are described below.

6.1 Underdrainage system

The first tailing was delivered to the facility on August 10, 1995, through two dropbars at a time on the north side of the containment. For the first two days, turbid water reported to the junction box from Header 1. During this time the flow slowly decreased, ceasing the night of August 11-12. Tailing distribution was moved to the west side of the facility on August 13 while the cause of the blockage was investigated.

After the tailing discharge points were moved, tailing started covering Headers 2 and 3 and clear flow was noted from both these pipes in the junction box on August 13. As new areas of underdrain were being covered, the discharge from Headers 2 and 3 appeared cloudy for short periods, but the flow cleared up as the tailing was established on top of the underdrain. The fine tailing particles that migrated into the pipes while the filter mechanism was established were flushed out without problems by the flow in the pipes.

It was determined by inspection that the blockage in Header 1 was located approximately 270 meters from the downstream end of the pipe, at the point where laterals from both sides first join the header. Just upstream of the blockage, water was flowing out of the header pipe, indicating that the pipe above this point was not blocked. At the time it was not clear what had caused the blockage, and it was not possible to expose the pipe at the blockage location because of the amount of water in the area. To resolve the problem, sections of Header 1 were exposed, commencing at the outlet end. When the pipe was opened in a number of places, it was found that the pipe was about one-third full of fine tailing. Survey measurements indicated that settlement had occurred in Header 1 pipe near the point where it started going under the embankment. The combination of the settlement in the header line and an excessive amount of tailing entering the pipe, which could not be washed out by the small volume of drainage water, is believed to have caused the blockage. There was too much tailing in the area to accurately identify where the tailing might be entering the system. It was surmised that the ridges of basin filter material, which projected slightly above the general level of the bottom of the containment and mirrored the underlying pattern of the drainage pipes, had created preferential flow paths for the discharged tailing. This led to the scouring of the filter material, allowing direct access for tailing to the coarse filter material and, hence, the drainage pipes.

The final 70 meters of Header 1, from the embankment to the first lateral, were exposed and the grade was corrected. A new pipe was laid and covered with filter material. Flow in the new pipe was confirmed by running water through the line. On the section of line from the 70-meter point as far along as possible towards the blockage (approximately a further 130 to 150 meters), a number of sections of pipe were opened and flow was confirmed. Flow was noted along the full length of the pipe and the areas where the pipe had been exposed by scouring

were reinstated. The Owner continued discharging tailing along the western side of the facility in an attempt to dry out the area around the blockage in Header 1 so that it could be cleared. When it became clear that the area around the blockage was not drying out, the Owner reverted to depositing tailing in the adjacent old tailing facility.

Once the area dried out, 270 meters of pipe between the blockage and the outlet end of Header 1 were exposed. Survey measurements were taken, and the elevations revealed that sections of the pipe were at shallow grades with isolated low points. As mentioned above, an excessive amount of tailing entered the pipe, and since the slope on the pipe was not intended to be steep enough to scour tailing out of the line at low flow volumes, the blockages had occurred again.

The header pipe over the 270-meter section was again removed, cleaned, and re-laid to a grade of 0.25 percent. Seven of the lateral drains could not be reconnected once the header was realigned, and these were abandoned. A continuous stream of ±25 cubic meters per hour of reclaim water was then flushed through the header from the upstream end to clear the line and, on September 6, 1995, tailing was again deposited in the new facility. Subsequently, once a layer of fresh tailing had been established over the pipes, all headers continued to function as designed.

During commissioning, deposited tailing, although limited in quantity, displayed the anticipated drying, shrinking, and cracking pattern envisaged in the design, indicating that with rotational deposition in about 100-mm lifts and 14-day drying periods, the facility will function as intended.

6.2 *Tailing distribution system*

The distribution pipework and dropbars functioned as intended, and the planned rotation system of operating two dropbars at the same time achieved the planned tailing distribution. The dropbars served their function to reduce energy in the flow stream, and laminar flow developed at the discharge points as desired.

7 SUBSEQUENT RAISES

In addition to the Phase 1 raise, the tailing facility has undergone three additional phases of development, as originally intended, to progressively increase the capacity of the facility. The facility is currently undergoing the Phase 4 (and final) raise, using mine waste rock to construct the embankment. The southern and eastern embankments are being constructed using the downstream method, and the upstream face is HDPE lined. This allows for a larger decant pond to develop, if required. Once complete, the Phase 4 raise will provide sufficient capacity through the remaining two years' life of the mine.

The actual slopes of the embankment are steeper and the crest wider than those in the original design. A cross section was developed on the south side of the facility and reviewed static stability, using two primary failure modes (rotational failures and sliding block failures) through the embankment slope. The section analyzed had a crest elevation at the planned ultimate height of the dam. The results of static stability analyses on the existing geometry yielded a minimum factor of safety of approximately 0.9 for both the rotational and sliding block failures through the ultimate planned embankment height. Due to the assumed good strength characteristics of the foundations, the critical failures were all generally shallow. The factor of safety is expected to increase once a planned safety berm is in place and before the Phase 4 development is complete. To achieve the post closure factor of safety of 1.5, the overall embankment slope will require regrading to a maximum 2H:1V. The pseudostatic analyses show that a seismic coefficient of 0.15g would reduce the factor of safety to just above one.

8 FUTURE RECLAMATION

The goals of reclamation for the tailing facility are to reduce or eliminate dust from the tailing surface and manage the contained water to create a zero-discharge facility. At the cessation of the CIL process, the tailing pipeline, reclaim-water return line, and associated pumps and equipment will be dismantled and removed. Buried pipelines will be purged, cut off, and capped and will remain in place. Surface lines will be purged, dismantled, and removed for salvage or disposal in the new industrial waste disposal site. The existing pond decant structure will be plugged.

Without input from the tailing, precipitation and water pumped to the facility from a nearby project will be the only remaining sources of water contained in the facility. The reclamation design allows a 0.8-meter freeboard at closure. This is expected to be sufficient to contain water from a maximum rainfall event at the pond and from the nearby project. Under normal conditions, evaporation will regulate the water level in the pond.

As a result of the sub-aerial deposition and the dry climate, tailing exposed above the pond water level will quickly dry out to form a beach. In moist climates, vegetation is used to control dust from tailing ponds. Since vegetation is not practical in this arid location, a layer of granular materials, such as detoxified heap leach rock, will be used to provide a suitable cover. The proposed design places a 300-mm-thick layer of granular material over the entire exposed tailing beach. This granular material will extend to cover any exposed HDPE liner to protect it from ultra-violet radiation. The tailing dam embankments will be regraded to a 2H:1V slope to provide a 1.5 factor of safety for closure.

9 CONCLUSION

The design concepts and parameters developed to modify an unlined tailing storage facility, by constructing an HDPE-lined tailing storage facility over previously placed tailing, proved in practice to be excellent. Seepage was recovered for use in the heap leach operation as anticipated in the design. The facility provides the capacity required by the Owner as a result of the higher densities achieved by the application of the sub-aerial deposition technique and underdrainage system. Subsequent raises were undertaken at minimal cost, and the upstream raises were placed over deposited tailing as envisioned. Reclamation of the facility should prove to be simple and economical.

Geotechnical considerations

Tailings and Mine Waste '02, © 2002 Swets & Zeitlinger, ISBN 90 5809 353 0

Advanced laboratory compression tests and piezocone measurements for evaluation of time-dependent consolidation of fine tailings

U. Barnekow, M. Paul & A.T. Jakubick
WISMUT GmbH, Technical Services, Chemnitz, Germany

ABSTRACT: The paper presents the progress achieved from 1995 to 2001 in evaluating time-dependent consolidation of fine tailings. Results are being applied to the decommissioning of large uranium tailings ponds at WISMUT, in particular at Helmsdorf and Culmitzsch A tailings ponds both located in Southeastern Germany. Uranium tailings ponds of the former Soviet-German Wismut company in Eastern Germany cover nearly 6 km^2 and contain about 150×10^6 m^3 of tailings. Time-dependent consolidation behaviour of thick fine tailings is of critical importance for ongoing in-situ-decommissioning including the dewatering by technical means and covering the tailings surfaces. Based on historic data and on first geotechnical data gained from drillings and soundings the actual consolidation state was recalculated for each tailings pond. For this the disposal period was backcalculated using a one-dimensional model applying the non-linear finite consolidation theory (Schiffman et al. 1984). To gain input parameter needed for such modeling a new advanced laboratory compression test was developed to determine the fine tailings consolidation behaviour characterized by the functions effective stress vs. void ratio and permeability vs. void ratio. A new cone penetration test including a piezocone was adopted to measure the hydraulic conditions in situ and the consolidation state of the fine tailings. Pore pressure gauges were installed in fine tailings in situ to measure the consolidation progress continuously. Data from the new laboratory tests and field measurements were used for a second run of the time-dependent consolidation modeling with respect to the covering of the fine tailings during remediation. Modeling was calibrated by newly collected geotechnical lab and field data. Results from both modeling and geotechnical investigations/measurements are in agreement with fine tailings consolidation behaviour predicted by the non-linear finite strain consolidation theory. In addition the paper presents the use of different one- and two-dimensional time-dependent consolidation models for cost-effective designing of the cover, for designing relevant construction elements of the final cover and for controlling and monitoring the construction progress of the cover on the tailings ponds.

1 INTRODUCTION

In Eastern Germany uranium mining and milling by the former Soviet-German WISMUT company lasted from 1946 to 1990. The total amount of uranium produced was 220,000 metric tons. Uranium ores were milled and processed by acid or soda-alkaline leaching mainly at two sites: Seelingstädt south of Gera (Thuringia) and Crossen near Zwickau (Saxony). The wastes from the hydrometallurgical uranium extraction processes were discharged into large tailings ponds covering a total area of nearly 6 km^2 and containing 150×10^6 m^3 of tailings. The Culmitzsch and Trünzig tailings ponds belong to the Seelingstädt site, whereas the tailings ponds Helmsdorf, Dänkritz 1 and Dänkritz 2 are hosting the tailings from the Crossen mill. Acid milling pulps were neutralized before discharging into the ponds. The tailings themselves as well as the pore and seepage waters contain considerable amounts of radionuclides. Seepage waters usually show high salt concentrations. In addition the tailings solids and in particular the seepage and pond water of the Helmsdorf pond are contaminated by arsenic.

All of WISMUT´s tailings ponds are located in the immediate vicinity of villages. The Helmsdorf tailings pond for example (area 2 km^2; vol. 50×10^6 m^3) is located near the villages of Oberrothenbach and Crossen (5000 inhabitants) as well as the city of Zwickau (120,000 inhabitants).

During the disposal period coarse and fine grained tailings materials settled and consolidated by their self weight in the tailings ponds due to discharge pattern. Subaerial sandy tailings beaches developed in the vicinity of the discharge pipes. In the distal area cohesive fine tailings of low permeability settled and consolidated below the water table. Self-weight consolidation is observed currently ongoing.

The so-called intermediate zone located in between the area of coarse and fine tailings is characterized by an interlayering of permeable coarse tailings layers and cohesive, low permeable fine tailings layers.

The time-dependent consolidation behaviour of fine tailings with thicknesses of several tens of meters is of critical importance for the ongoing in-situ-decommissioning including the dewatering by technical means and for the covering of the tailings surfaces.

2 OVERALL REMEDIATION CONCEPT FOR WISMUT´S TAILINGS PONDS

Remediation of uranium tailings ponds started immediately after German reunification in 1990 with defense measures against acute risks, complex environmental investigations and the preparation of first site specific remediation concepts. As a first defense measure subaerial tailings beaches were interim covered to guarantee dust control and to distinctly lower radon exhalation rates from the tailings surfaces.

Pond water of all tailings ponds except tailings ponds Helmsdorf and Culmitzsch A was expelled until 2000. The Helmsdorf tailings pond currently contains about 2.6×10^6 m^3, the Culmitzsch A tailings pond about 600,000 m^3 of pond water. Pond water and seepage water catched are treated before discharge into the receiving streams. Lowering the pond water table led to increasing subaerial areas of tailings surfaces to be progressively interim covered.

Based on extended investigation programs WISMUT has decided to prepare for the remediation option of dry decommissioning in-situ with partial dewatering by technical means for all of its tailings ponds. This preferred option was admitted for each of the tailings ponds by the governmental authorities responsible in the mid 90´s. Decommissioning techniques for dry option following the acute defense measures comprise:

- the removal and treatment of pond water
- interim covering including dewatering of fine tailings by technical means
- reshaping the tailings dams
- contouring of the interim covered tailings surfaces
- final covering and landscaping including revegetation.

From a geotechnical point of view interim covering the poorly consolidated fine tailings surfaces currently ongoing is the decisive step of the whole decommissioning. Preceding the interim covering of the fine tailings drying-out of subaerial fine tailings surfaces is used to initially improve trafficability.

Interim covering measures on fine tailings surfaces start with the placement of a geotextile, geogrid and/or combined geomaterials like drainmats on dried tailings crust. Technical dewatering is enhanced by stitching vertical wick drains into the tailings. Loading of tailings surface is done by placing ahead thin earthen layers using common earthwork equipment like small dozers or hydraulic excavators. Tailings pond Trünzig A (67 ha) was interim covered until 1995, tailings pond Dänkritz 1 (27 ha) until 2000. The tailings pond Trünzig B (48 ha) will be wholly interim covered this year.

Reshaping the dams is needed to guarantee long-term stability and long-term erosion control of the dams and their future final cover. First dam reshaping started at Tailings Pond Trünzig A this year using hydraulic excavators and scrapers. The reshaped tailings dam is covered with a 0.5 m-layer of earthen material to avoid erosion and dust blowing of sandy tailings during the remediation period. Reshaping the main dam at the Helmsdorf tailings pond (max. height 58 m; length 1700 m) is foreseen to be started in early 2002.

Contouring of the tailings surfaces within the pond area follows the interim covering progressively. Contouring the pond area prepares for later final covering. It shall create a long term stable surface contour to ensure future surface runoff on the final cover and to ensure the functionality of the final cover itself. For this total settlement portions, spatial different settlement portions and time-dependent settlement rates, especially of the cohesive uranium fine tailings are of critical importance. For example the tailings pond Culmitzsch A which is located in a former open pit covers an area of about 1.6 km^2 from which 0.6 km^2 are covered by homogeneous up to 57 m thick fine tailings. Currently ongoing annual settlements in a range from 0.2 m/year to 0.5 m/year are measured on fine tailings surfaces below water table. Due to loading by only 2 m of interim cover total settlement portions from to 4 m to 7 m must be assumed. Additional loading during final covering will lead to additional settlements developing over an estimated time period of several decades. To handle this problem during ongoing remediation consolidation of fine tailings it is foreseen to enhance fine tailings consolidation using deep wick drains in those pond areas that are sensitive to the design of the future surface contour and to the functionality of the final cover. Currently WISMUT prepares for speeding up the consolidation of max. 33 m thick fine tailings in the fine tailings area of the tailings pond Trünzig A. For this purpose an embankment fill shall be placed combined with stitching wick drains down to a maximum depth of 30 m. This remediation step is currently under permission process and shall start next year. Contouring works on the tailings beaches of Trünzig A began this year.

Final covering is still in the permitting process for several of the tailings ponds. The final covers shall provide long-term stability of the reshaped dams and contoured tailings. They shall control infiltration and adon exhalation and have to guarantee erosion control and revegetation. In addition surface runoff due to heavy rainfalls must be controlled by the design of the final cover and by runoff catchment construc-

The contouring step has in particular a huge potential for cost reduction. Cost-optimization of contouring means minimization of cut and fill material volumes needed. I.e. reducing the average thickness of earthen cover layers by only 1m on an area of 50 ha means a reduction of approximately 3.3 million EUR. The surface contour has to be designed accordingly. In addition the design of the final cover depends on the spatial distribution of settlement portions developing during and after remediation phase.

The aim is to reach a minimum volume of earthen contouring materials to be placed on the pond area with respect to the design and long-term stability of the final cover. To solve this task it is of critical importance to know exactly the time-dependent settlement rates due to successive surcharge loading during ongoing remediation over the long term. The consolidation behaviour of the different fine tailings in the tailings ponds, the spatial distribution of geotechnical and hydraulic tailings properties as well as the hydraulic conditions in the tailings ponds and their surroundings must be well understood. The progress achieved by WISMUT on these works since 1995 is presented in the following chapters.

Preparing for the remediation a three-dimensional structural model was established for each tailings pond based on geodetic data, on historical data and on the results of the extended drilling and piezocone measurement programs. The structural models were prepared using the Software package EarthVision (Dynamic Graphics). They contain all geodetic data of the surfaces and subsurfaces of the tailings and dams, of the mine waste dumps, of the topographic data in the surrounding area and of relevant geologic layers in the underground.

Geotechnical data gained from geotechnical investigations were analysed by statistic methods to derive their spatial distribution within the tailings layers in the pond. Piezocone tests (= cone penetration test incl. piezocone) were carried out. Data gained from piezocone tests were used to develop the three dimensional distribution of the different tailings layers within the tailings body as well as to check the as-made tailings dam construction.

Based on this a zoning of each tailings pond into sandy beach zones, intermediate zone of interlayering fine and coarse tailings and fine tailings zone was prepared. Different tailings zones identified were divided up into subzones with respect to different hydraulic conditions to prepare for later one-dimensional consolidation modeling of each of these subzones seperately applying non-linear finite strain consolidation theory (Schiffman et al. 1984).

3 METHODOLOGY OF COMPRESSION TESTS AND PIEZOCONE TESTS

3.1 *Newly developed compression tests*

First modeling of the disposal history using computer codes applying one-dimensional finite strain theory was carried out to evaluate the current consolidation state of the fines. Resulting from this first modeling it became clear that data gap on the fine tailings consolidation behaviour had to be closed. Conventional compression testing according to German regulation DIN 18135E was found unsuitable to successfully carry out compression tests on pulpy fine slimes. The task to be solved was to construct an improved compression test apparatus and to develop a methodology for deriving all input data needed for non-linear finite strain consolidation calculation. The recently developed automatic oedometer test apparatus KD 314 S is presented in Fig. 1 below.

Geotechnical input functions, which shall be derived, are the relationships of void ratio vs. effective stress and of permeability coefficient vs. void ratio.

The automatic oedometer test apparatus KD 314 S was jointly developed by WISMUT and WILLE Geotechnik, and optimised for pulpy fine uranium tailings. The apparatus consists of an oedometer cell, a consolidation press with electromechanical driving mechanism and a PC for test steering (see Fig.1). The oedometer test allows software-controlled testing and continuous measuring of all parameters of

Figure 1. Automatic compression test apparatus KD 314 S.

interest which are related to the material behaviour of consolidating fine-grained mill tailings. Usually samples of 200 mm diameter and up to 200 mm height are tested. The ratio of height vs. diameter of the sample varies during testing from 1: 1 to 1: 5. This has to be taken into account for evaluation of the test results. Parameters measured during testing include settlement value, surcharge pressure (load) on top and base pressure at the bottom of the sample as well as pore pressure. Based on these parameters it is possible to eliminate the effect of the geometry of the tested sample. Tailings samples are loaded gently with respect to non-linear consolidation behaviour of the fine tailings. Usually one starts with a first step of less than 1 kPa surcharge pressure. Stepwise loading is done with respect to suitable covering steps, i.e. in steps of 2, 4, 8, 12, 20, 50, 100, 200, 300 kPa of surcharge load. Each loading step is carried out until full primary consolidation. Secondary consolidation can be measured as well. During each step pore pressure is measured continuously with time with an accuracy of 0.1 kPa.

3.2 Piezocone tests

Piezocone tests or cone penetration tests including a piezocone for pore pressure measurement (CPTU) were adopted for weak or pulpy fine tailings. Such piezocone tests are currently used for the following purposes:
1. To determine the profile of the different tailings layers with depth at a given location in order to derive the spatial distribution of different tailings layers in the tailings pond. Based on the piezocone test results a detailed zoning of sandy beach zones, intermediate zones and fine tailings zones is carried out.
2. To determine the hydraulic conditions with depth by dissipation testing. For this the piezocone is stopped at a given depth and time-dependent pore pressure variation is measured continuously. Static pore pressure within the tailings body and in the underground layers is measured step-wise with depth. From such results actual consolidation state can be derived. In addition permeability at a given depth can be evaluated successfully in low permeable fine tailings layers using dissipation test results.
3. According to common practice geotechnical parameters are derived standardwise for specific purposes if needed.

Piezocones used measure tip pressure, sleeve friction and pore pressure u_2 immediately above the tip. The results gained from older equipment showed such piezocone tests unsuitable for weak or pulpy fine tailings. Pore pressure gauges used were not able to measure small absolute pore pressures with accuracy needed for evaluating the static in-situ pore pressures with depth, in particular in the most interesting shallow depths.

In addition the friction ratio typically used to identify the different soil layers with depth does not work in fine tailings.

Therefore a complete new system was constructed. The new piezocone is installed on a small crawler vehicle having a soil pressure below 7 kPa. This crawler is able to drive on pulpy fine tailings surfaces covered only by geotextile and geogrid. Wooden planks are used if needed. The new piezocone is also used to carry out measurements from a floating platform fixed in the fine tailings underground. Maximum sounding depth is 50 m below surface. Soil classification of the tailings layers with depth is done automatically by the evaluation software using the Bq-value according to the Robertson-formula as follows:

$$Bq = (u_2-u_0)/(q_T-p_o)$$
u_2 = dynamic pore pressure behind the tip
u_0 = static pore pressure
q_T = tip pressure
p_o = surcharge pressure from overlying layers.

The soil profile detected by the piezocone test is calibrated with borehole results. Actual static pore pressure is measured stepwise with depth by dissipation testing. Static pore pressure measurement is also calibrated with pore pressure gauges installed in situ in boreholes. Permeability can be derived from dissipation test data for low permeable tailings layers. Beyond this electrical conductivity can be measured using electrodes installed according to the Wenner system. Electrical conductivity data are usually used to detect salty tailings pore water.

4 RESULTS

4.1 Geotechnical characterization of fine tailings

Extensive drilling and sampling programs were carried out on the tailings ponds of WISMUT in the past. Drillings, ram core soundings including liner sampling and corer sampling were carried out from the pond water table, from tailings surfaces and from interim cover surfaces. Liner samples were gained from drillings over the entire tailings thickness. Corer-sampling was carried out in the uppermost 4 m below tailings surface.

Some tailings samples were investigated for their mineralogical content using X-ray diffractometry. A typical mineralogical composition of soda-alkaline fine tailings is: muscovite/illite 40%...50%, quartz 20%, chlorite 15%; feldspar 7%. Soda-alkaline tailings contain usually more than 4%...5% carbonates. Acid tailings contain less than 4%...5% carbonates. Due to neutralization of acid mill pulps using lime milk gypsum appears in these tailings. Organic carbon content can vary in a range of < 1 % up to 7% in the fine tailings.

Tailings samples are usually investigated for their soil physical properties. Cohesive fine tailings can be characterized typically as silts or clays of moderate or high plasticity. Consistency under field conditions is mainly weak or pulpy. Typical geotechnical properties of fine tailings from acid leaching and soda alkaline leaching are presented below in Table 1.

Table 1. Geotechnical parameters of fine tailings

Parameter	TP Helmsdorf soda-alkaline leaching	TP Culmitzsch A acid leaching
Soil group (DIN 18196)	clay, silt of moderate to high plasticity (TA, UA, TM)	silt, clay of high plasticity (UA, TA)
water content	0.45 ... > 1.2	0.50 ... > 1.8
liquid limit	0.50 ... 0.60	0.60 ... 0.75
plasticity limit	0.20 ... 0.24	0.22 ... 0.30
void ratio e	1.5 ... > 3.5	2 ... > 5
compression index C_C	0.4 ... 0.55	0.6 ... 0.85
Permeability k_f	1*E-7...1*E-9	2*E-7...5*E-10

4.2 Results of the compression tests

Based on the compression test measurement results the compression curve (void ratio vs. effective stress) and the permeability coefficient vs. void ratio relationship of fine tailings is determined. An example for the evaluation results of a compression test carried out on a typical fine tailings sample gained from the fine tailings zone of tailings pond Culmitzsch A is presented in Fig. 2 and Fig. 3 presented below. Fig. 2 shows a typical curve of void-ratio vs. effective stress (load: 0.5...250 kPa). Fig. 3 presents the permeability coefficient vs. void ratio relationships. The upper curve in the diagram (Fig. 3) was derived from conventional evaluation of time-settlement curve according to the Terzaghi consolidation theory. This curve represents a filter velocity vs. void ratio relationship. The lower curve is derived from the time-dependent pore pressure decay and represents the permeability coefficient vs. void ratio relationship.

During each loading step one observes the (normalized) pore pressure decay to be distinctly slower than the (normalized) settlement with time. This is in agreement with consolidation behaviour predicted by non-linear finite strain consolidation theory (Schiffman et al. 1984).

The results from the newly developed compression tests, as presented above, are used as input parameter relationships in the consolidation modeling applying the non-linear finite strain consolidation

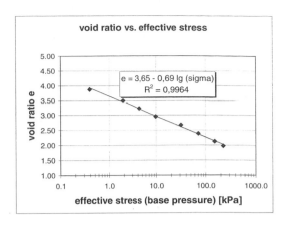

Figure 2. Compression test on sample no. CA-Wa3 from southern fine tailings zone of Culmitzsch A tailings pond: void ratio vs. effective stress curve.

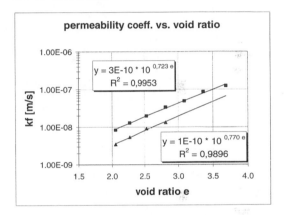

Figure 3. Compression test on sample no. CA-Wa3 from southern fine tailings zone of Tailings pond Culmitzsch A: permeability – void ratio relationship.

theory (Schiffman et al. 1984). Using the newly developed compression test permeability coefficients in a range from 5×10^{-11} m/s up to 3×10^{-6} m/s were successfully measured in the past.

4.3 Piezocone test results

Fig. 4 presents a tailings layer profile with depth located on the Culmitzsch A tailings pond. In the diagram one can see the tip pressure and the calculated Bq-value with depth. The layering derived is presented on the left hand side. The sounding is located in an intermediate zone next to the fine tailings zone. Hydraulic conditions in this area of the tailings pond are strongly influenced by pumping from 10 deep wells dewatering the beach zone. Therefore dissipation testing was done to evaluate hydraulic conditions in local aquifer layers seeping laterally. Accu-

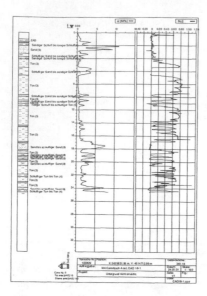

Figure 4. Results from piezocone tests.

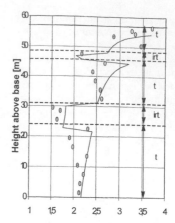

Figure 5. Void ratio – depth profile (borehole SBA 6) observed in borehole samples (points) and modeled using FS-Consol (line).

racy of the pore pressure gauge is 1 kPa. In tailings pond of Culmitzsch A significant excess pore pressures were measured using piezocone tests as well as by pore pressure gauges installed in situ measuring continuously pore pressure since 1996. Excess pore pressure was detected in homogeneous fine tailings of usually about more than 50 m thickness.

5 CONSOLIDATION MODELING AND MODELING RESULTS

The first run of one dimensional modeling of the disposal period including interim covering was carried out using codes like ACCUMV or FSConsol (GWP Software) both applying the non-linear finite strain theory. This was done for all of WISMUT´s tailings ponds from 1995 until 1999.

Fig. 5 presents a typical result of the first modeling step recalculating the actual consolidation state. In the diagram the modeled void ratio depth relationship is compared with void ratios measured in borehole samples gained from different depths. The borehole is located in the fine tailings zone in the tailings pond of Culmitzsch A but near to intermediate zone.

Input and calibration data for the consolidation model are:

- detailed data evaluation of annual discharge: filling time, filling rates, solid density and density of the discharged slurry
- initial void ratio estimated from slurry density and in situ void ratio observed near tailings surface

- in-situ borehole profile data for void ratio – depth relationship (void ratio - effective stress and so-called initial void ratio e_0)
- compression test data for void ratio vs. effective stress relationship and void ratio vs. permeability relationship
- current tailings height (thickness)
- annual settlement rates measured each year
- continuous pore pressure measurements in situ using pore pressure gauges installed in boreholes to get (static) pore pressure with depth and time dependent varying of pore pressures due to progressive dewatering of the whole tailings pond
- piezocone measurements including dissipation testing to evaluate the spatial distribution of (static) pore pressures in the tailings at a given time.

The second modeling step is carried out not only for disposal period and interim covering but also for the contouring and final covering. This has already been done for some of the tailings ponds like e.g. tailings pond Helmsdorf. Data of tailings consolidation behaviour gained from newly developed compression test were used as input data for this second step of consolidation modeling at the Helmsdorf tailings pond. The consolidation modeling is calibrated with static in situ pore pressures gained from piezocone data and pore pressure gauges installed in situ.

The third step is the calculation of the spatial distribution of total settlement portions on the tailings pond. For this purpose one uses the three dimensional structural model of each tailings pond prepared beforehand, as presented above. Resulting from the second model run one gets for each homogeneous tailings subzone total settlement portions due to load and due to tailings thickness. Due to the design of the contoured surface of the tailings pond the thickness of the cover layers varies spatially on the tailings surface. Spatial distribution of total settlement portions are then calculated using the three dimensional structural model (EarthVision software (Dynamic Graphic Inc.)). Such works are currently ongoing and will be prepared for all of the tailings ponds of WISMUT within the next years. First results gained for the tailings pond of Helmsdorf are presented in Fig. 6.

To model the time-dependent consolidation of fine tailings enhanced by using vertical wick drains the two-dimensional radialsymmetric code CONSOL-2D was developed at WISMUT with support of Technical University Chemnitz. The code developed applies the non-linear finite strain theory (Schiffman et al. 1984).

WISMUT currently prepares for speeding up the settlement rates in certain areas of the fine tailings zone in the tailings pond of Trünzig A (67 ha; vol 13×10^6 m^3)) by using deep wick drains and additionally placing a temporary embankment fill on the tailings surface. The aim is to gain most of the settlement portions during the remediation foreseen to be finalized within a time period of 5 years. Such enhancement of time-dependent consolidation must be sophistically monitored. Time-dependent settlement rates and total settlement portions gained within the 5 year period as well as final settlement portions under the placed final cover must be predicted by consolidation modeling and monitored during and after execution of the construction works. For this purpose the consolidation model CONSOL-2D is calibrated with measured lab and field data.

6 CONCLUSIONS

In the last years the consolidation modeling applying the non-linear finite strain consolidation theory (Schiffman et al. 1984) was improved with respect to the specific needs of the remediation of large uranium mill tailings ponds.

All input parameters on tailings consolidation properties needed for consolidation modeling are now measured using the newly developed compression test called KD 314 S using proved standard procedures.

Data on hydraulic conditions in fine tailings needed are measured using new piezocone test technique and evaluation software adopted and improved for fine tailings.

New consolidation software applying two-dimensional radialsymmetric consolidation implementing the non-linear finite strain consolidation theory (Schiffman et al. 1984) was developed in the past and applied successfully for e.g. the embankment fills placed to enhance tailings consolidation.

During the ongoing remediation progress the actual state of the works on the consolidation modeling on fine tailings will proceed in the same manner for

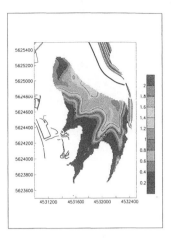

Figure 6. Helmsdorf tailings pond: Spatial distribution of settlement portions under cover load in the fine tailings zone derived from non-linear finite strain consolidation modeling.

all the other tailings ponds at WISMUT as well. After having characterized the three-dimensional internal structure of each tailings pond using the new piezocone test this piezocone test will be used in the future more and more to monitor the time-dependent consolidation progress in needed accuracy.

7 ACKNOWLEDGEMENTS

We would like to thank Dr Andy MacG Robertson and Dr Christoph Wels both Robertson GeoConsultants Inc., Vancouver, Canada for their support during the implementation of consolidation modeling on fine tailings at WISMUT during the last years and for their suggestions on the improvement of the modeling and field investigations on WISMUT´s tailings ponds.

In addition we would like to thank WISMUT´s Remediation Unit Ronneburg, / Project Management "Decommissioning of Tailings Ponds" for their support and hereunder especially Mr Jürgen Müller responsible for piezocone investigations for his support and commitment on pushing forward the field investigation techniques to be applied.

At last we would like to thank Mr Thomas Mehlhorn from geotechnical laboratory of the Engineering Department for his sophisticated R&D-works on the oedometer test techniques.

REFERENCES

Schiffman, R. L.; Pane, V. & Gibson, R. E. (1984): The theory of one-dimensional consolidation of saturated clays, IV. An overview of non-linear finite strain sedimentation and consolidation, in Yong, R. N. & Townsend, F. C. (editors): Sedimentation Consolidation Models – Prediction and Validation, American Society of Civil Engineers, New York 1984.

Prediction of the desiccation and sedimentation behavior of a typical platinum tailings

J.C.J. Boshoff

SRK Consulting, Johannesburg, South Africa

ABSTRACT: A one-dimensional model of the sedimentation and desiccation of high-clay-content, subaerially deposited tailings is described. The approach is semi-empirical and predicts average density and water content of a layer of tailings during sedimentation, first stage evaporation, varying rainfall, and variation of initial height and solids content. The method employs both geomechanical and soil physics principles and allows for varying evaporation, varying rainfall, and variation of initial height and solids content. The model allows the effects of different initial layer depths, to be compared with total drying time and resulting final dry density.

1 INTRODUCTION

The recent increasing demand to increase the rate of deposition on existing platinum tailings dams, has highlighted the importance of understanding the sedimentation and desiccation behavior of sub-aerially deposited tailings.

The rate of deposition on a tailings dam must be sufficiently slow such that there is a sufficient degree of dissipation of excess pore pressures induced by the raising of the tailings dam. As tailings are deposited, water is drawn of the dam, thus allowing the solids to settle out. These layers of solids are then subjected to the process of consolidation and settling as excess pore pressures are dissipated. After deposition, the thin tailings layers are allowed to desiccate through evaporation. The successful dewatering by desiccation increases density and shear strength and reduces compressibility.

The upstream method of tailings dam construction which predominates in South Africa, therefore necessitates that particular attention be paid to the marginal rate of rise of the tailings dam crest and that the behavior of the tailings material after deposition be well understood.

A one dimensional model that can accurately predict the post depositional behavior of tailings undergoing sedimentation and desiccation induced consolidation was developed at the School of Civil Engineering, University of New South Wales in 1992. The approach is semi-empirical and predicts average density and water content of a layer of tailings during sedimentation, first-stage evaporation, and second-stage evaporation, during which desiccation occurs. The method employs both geomechanical and soil physics principles and allows for varying evaporation, varying rainfall, and variation of initial height and solids content. It is upon this model that this investigation is based. The model allows the effects of different initial layer depths, different initial water contents and different available evaporation potentials to be compared with total drying time and resulting final dry density.

2 MODELING OF TAILINGS SEDIMENTATION, DESICCATING AND CONSOLIDATION

Analysis of tailings desiccation has been undertaken in a fairly simplistic manner (Guerra, 1978; Vick, 1983). A better approach is to employ soil physics principles to predict water movement within the soil profile. The application of such theory, and an empirical engineering approach, allow modelling that can predict rates of sedimentation and desiccation of tailings.

2.1 *Method for prediction of sedimentation and desiccation*

The approach taken by Swarbrick and Fell in predicting sedimentation and desiccation is based upon the following assumptions:
- the layer is assumed to be of uniform water content and density;
- sedimentation is not affected by the rate and amount of desiccation;
- evaporation is analyzed as a two-stage process;
- all supernatant water is reclaimed;
- drainage is not significant.

2.1.1 Sedimentation

This model component covers the settling of flocs to a state where particle-to-particle interaction becomes substantial and significant effective stresses have been developed (Yong, 1984). The most suitable approach in terms of experimental observations for the tailings used in the research is based upon solids volume concentration of the form (Thomas, 1964).

$$v = v_n 10^{\left(K_0 \frac{v_s}{h} \right)} \tag{2.1}$$

Here

v = the settling velocity of the interface between settled material and clarified water for a given height of settled solids in min/day

H = height of settled solids in millimeters

v_0 = the unburdened average settling velocity (i.e. Stokes)

V_s = the total volume expressed as a height of solid material (in millimeters)

k_o = a material constant.

Both v_0 and k_o are determined by regression analysis from sedimentation tests for a particular tailings.

While (2.1) predicts the rate of settlement for a given tailings, it does not predict when sedimentation ceases or becomes insignificant. The final height, H_f, at which sedimentation ceases, is a function of effective stress, and is mainly dependant on V_s. Noting that H_f must approach zero as V_s does, then a suitable relationship between the two is:

$$H_f = K_h \left(V_s \right)^{\phi} \tag{2.2}$$

The material constraints K_h and $\emptyset$ are determined by regression analysis from laboratory column test results.

2.1.2 First stage of evaporation

Analysis of the first stage of evaporation (Covey, 1963; Gardner and Hillel, 1962) generally involves an expression relating the water content of the soil to the evaporation potential ep (mm/day) and the depth of the soil profile at the commencement of stage-one evaporation Hs (mm).

Gardner and Hillel (1963), developed an expression for the average water content within a profile at the end of stage-one evaporation. Assuming this relationship is applicable to tailings, the following equation was derived:

$$w_1 = \frac{H_s}{k_1 G_s V_s} \ln \left(1 + \frac{e_p k_1 H_s}{2 D_o} \right) \tag{2.3}$$

Here

w_1 = the average gravimetric water content at the end of stage-one evaporation (M_w/M_s) and is found from e_p and H_s.

Having obtained w_1, the duration of the first stage of evaporation is found knowing e_p using

$$t_1 = \frac{(w_s - w_1) G_s V_s}{e_p} \tag{2.4}$$

e_p can be obtained from pan evaporation (Pruitt, 1996; Howell et al., 1983). The parameters D_o and k_1 in (2.3) by definition, are from the approximate relationship used relating diffusivity and water content.

2.1.3 Second stage of evaporation

Observations have shown that the cumulative evaporation during the second stage is approximately proportional to the square root of time. As noted previously, as the applied evaporation potential decreases, the duration of stage one, t_1, increases. This is evident from theory because the rate of increase in w_1 from (2.3) is not as great as the increase in t_1 due to (2.4), hence t_1 increases. As a result of this increase, the rate at which the second-stage rate falls will be less pronounced.

Gardner and Hillel's (1962) work involved a comparison of observed cumulative second –stage evaporation for different rates of available potential evaporation with Gardner's theoretical solution with an infinitely high evaporation rate (Gardner, 1959). They conclude that the cumulative evaporation during the second stage of evaporation follows that of the theoretical solution for an infinitely high rate of potential evaporation after some point on the theoretical curve. This point depends upon the available potential evaporation. This only applies when cumulative evaporation (CE) is expressed as a normalized water loss. In Figure 2.1 the normalized water loss has been plotted against time for a particular value of e_p. Gardner's theoretical solution for $e_p = \infty$ will describe the rate of water loss during the second stage of its origin with regard to commencement of state one evaporation, i.e. t_o and e_o, is known.

As the second stage is approximated by assuming that cumulative evaporation is proportional to the square root of time, then a general relationship is proposed.

$$t = \left(\frac{E - E_0}{b_s} \right)^2 + t_0 \tag{2.5}$$

Where

t = time since start of evaporation

E = the normalized cumulative water loss

E_o and t_o = fitting parameter shown in Figure 2.1

b_s = the rate of change of E with respect to the square root of $(t - t_o)$, usually called sorptivity.

2.1.4 Calculation of saturation

One of the critical parameters needed for design purposes is the average density of the deposited layer during desiccation. While the bleed rate from sedimentation exceeds the available evaporation, the tailings remain saturated, and calculation of density is straightforward. Once evaporation exceeds the bleed rate, desaturation begins, and an estimation of saturation is required. If the equivalent saturated water content, w_{sat}, is known, then the level of saturation would simply be w/w_{sat}. In deforming soils however, as S decreases, the total volume decreases, which lessens the decrease in S. S may therefore be approximated by the bulk density divided by the equivalent bulk density assuming full saturation. This may be expressed in terms of w by

$$S = \left(\frac{1 + \dfrac{1}{w_{sat}}}{1 + \dfrac{1}{w}} \right) \quad (2.6)$$

After some point during desiccation, no more shrinkage can occur and w_{sat} will not change. The value of S during desiccation after this point will be entirely dependent upon w.

During the desiccation process, it is necessary to determine w_{sat}, the most suitable approach being linear interpolation between an upper and lower bound. The upper bound for w_{sat} is the saturated water content after full sedimentation has occurred.

This is termed w_{ub} and is calculated by

$$w_{ub} = w_i - \frac{H_i - H_f}{G_s V_s} \quad (2.7)$$

The lower bound is dependent upon the final dry density of the profile, which reaches a maximum after some period during desiccation. Letting ρ_d denoting dry density at water content w and letting ρ_w denote the density of water, then

$$S = \left[\frac{1 + \dfrac{1}{w_{sat}}}{1 + \dfrac{1}{w}} \right] = \left[\frac{G_s w \rho_d}{G_s \rho_w - \rho_d} \right] \quad (2.8)$$

Rearrangement gives

$$w_{sat} = \frac{G_s \rho_w - \rho_d}{G_s \left[(1+w) \rho_d - \rho_w \right] + \rho_d} \quad (2.9)$$

For the lower bound, we let $w = 0$, $\rho_d = \rho_{d\,max}$ (maximum dry density), and $w_{sat} = w_{lb}$ at this point. Substitution of these into (2.9) yields (2.10).

$$w_{lb} = \frac{\left(G_s \rho_w - \rho_{d\,max} \right)}{G_s \left(\rho_{d\,max} - \rho_w \right) + \rho_{d\,max}} \quad (2.10)$$

Having obtained the two extreme values of w_{sat}, linear interpolation between the upper and lower bounds using the previous minimum value of saturation (S_{min}) is used to estimate S.

If sedimentation is still occurring during evaporation, which is usually the case, w_{sat} should be determined using the height predicted by sedimentation.

2.1.5 Explicit model procedure

All of the model parameters required (i.e. G_s, v_o, k_o, k_h, k_1, ϕ, D_o, b_s and ρ_{dmax}) are determined by the methods described in Section 2.2 for a specific type of tailings and used as input parameters for the model.

2.1.6 Rainfall

Rainfall events are modeled on a daily basis. w is not allowed to rewet above the current value of w_{sat}, assuming S_{min} does not increase. This is represented algebraically by

$$w_{rewet} = \max \left(w + \frac{V_{rain}}{G_s V_s} \right) \quad or \quad w_{sat} \quad (2.11)$$

where V_{rain} represents the volume of rain per unit area (in millimetres) of rainfall for that particular period.

2.2 Explicit model procedure

The specific steps required to predict critical initial settlement and desiccation is shown below:
- Step 1: Calculate V_s from H_i (mm), w_i, and G_s

$$V_s = \frac{H_i}{(1 + G_s w_i)} \quad (2.12)$$

- Step 2: Using V_s calculate H_f (mm) from (2.2)
- Step 3A: Initiate saturation calculations – Determine upper and lower bounds for w_{sat} from (2.7) and (2.10). Set $S_{min} = 1$.
- Step 3B: Sedimentation only - ΔH is greater than e_p – Using a time step, (Δt), of 1 min, find the drop in height ΔH for the current value of H using (2.1) i.e.

$$\Delta H = \Delta t v_0 10^{(k_0 V_s / H)} \quad (2.13)$$

and hence the new height is found from $H = H - \Delta H$. The new w is found from

$$w = w - \frac{\Delta H}{G_s V_s} \quad (2.14)$$

and saturation, S, equals 1. A time step of 1 min gives the best results in terms of computational effort and accuracy. Repeat step 3 until e_p exceeds ΔH or H reaches H_f. Set H_s, and w_s to the current values of H and w and find V_{ws} using

$$V_{ws} = w_s G_s V_s \quad (2.15)$$

During steps 4 and 5, a separate variable, H_{set}, is required to record the height of settled solids over time due to pure sedimentation. Hence, initially, H_{set} k = H_s.

123

- Step 4: Sedimentation and first-stage evaporation – Using a suggested Δt of 1 hr, find ΔH using (2.13), hence $H_{set} = H_{set} - \Delta H$. Calculate new w and w_{sat}.

$$w = w - \frac{e_p \Delta t}{G_s V_s} \qquad (2.16)$$

$$w_{sat} = \max \ of \ \left[\frac{1}{G_s}\left(\frac{H_{set}}{V_s} - 1 \right) \right] \ and \ (w) \qquad (2.17)$$

Using (2.6), an estimate of S is determined and hence height. The true height is found from

$$H = V_s \left(\frac{wG_s}{S} + 1 \right) \qquad (2.18)$$

This ensures that allowance is made for any layer shrinkage due to evaporation. Repeat step 4 until w is less than w_l or H_{set} reach H_f. Once this occurs, stage one ends so t_l is known. Then find t_o using (2.5) and parameter b_s.

- Step 5: Sedimentation and second-stage evaporation – Using a suggested Δt of 1 day, find ΔH using (2.13) hence, $H_{set} = H_{set} - \Delta H$. Calculate W_{sat} from (2.17) and new w from

$$w = w - \frac{w_s b_s \Delta t}{2\sqrt{t - t_o}} \qquad (2.19)$$

Determine H using (2.18). Repeat step 5 until desired final water content is reached or H_{set} reaches H_f.

- Step 6: Once H_{set} reaches H_f in steps 3, 4, and 5, sedimentation ceases. Steps 4, then 5 are repeated ignoring the use of (2.13), (2.17), and (2.18). The value of w_{sat} is found by linear interpolation with respect to $S_{min.}$.

$$w_{sat} = \max \left[w_f + \left(w_r - w_f \right) * S_{min} \right] \ and \ (w) \quad (2.20)$$

And S is found by (2.6). A time step of one day is suggested.

- Step 7: Having obtained water content and saturation, calculate other parameters as required, using:

$$\rho_d = \frac{SG_s \rho_w}{S + G_s w} \qquad (2.21)$$

and the corresponding estimated height is

$$H = \frac{G_s V_s \rho_w}{\rho_d} \qquad (2.22)$$

3 EXPERIMENTAL PROGRAM

Sedimentation and desiccation of tailings were monitored and analyzed as separate phenomena in the laboratory. This was in order to establish material dependent parameters for each phenomenon independently. The theoretical approach used in this study assumes that sedimentation and desiccation occur simultaneously.

3.1 *Test apparatus*

The process of sedimentation was mainly studied through the use of short (500ml) column tests, which provided useful information about the general settling behaviour of the platinum tailings. The analysis of tailings desiccation was undertaken using the same column with settled material. The freewater water was drained off before drying commenced.

3.2 *Field experiments*

A method of operation, which may be described as a hybrid of the paddock method and the spigot method, is employed on this particular dam.

Settlements of the hydraulically deposited tailings were measured by means of level pegs that were installed close to the outer paddock wall, in the middle of the paddock and close to the inner paddock wall. The settlement was measured over a period of approximately four months. Evaporation was measured by means of an A-pan.

3.3 *Tailings type*

The ore body at this particular mine, located in the Bushveld Igneous Complex in the northern province of South Africa is characterised by sepentinised zones within the pyroxenite host rock. The serpentinised zones are variably high in talcs and calc silicates. The specific gravity of this mineral at 2,24 to 2,3 is low especially in comparison to the host rock, which has a specific gravity in excess of 3.11. As a consequence of this low specific gravity the mineral tends to "float" on the tailings stream surface and forms fine layers wherever it settles out.

3.4 *Summary of model parameters*

The parameters obtained from the sedimentation and desiccation tests and a summary of the model input parameters are given in Table 3.1.

A comparison between the values obtained from field experiments and laboratory experiments can be seen in Table 3.2 below. The intermediate results for moisture content and dry density are presented in Figures 3.1 and 3.2 respectively.

3.5 *Discussion*

From Figures 3.1 and 3.2 it is evident that the model has proven capable of predicting the field behaviour of the platinum tailings layer with great accuracy. The model slightly under predicted both moisture content and dry density mainly because of the following reasons:

Table 3.1. Summary of model parameters

Parameter	Unit	Value
V_0	mm/day	4528
k_0	-	-6.245
D_0	mm^2/day	500
k_h	-	4.24
k_l	-	15
ϕ	-	0.95
b_s	-	0.087
e_p	mm/day	50
H_i	Mm	500
V_s	Mm	144.37
H_f	Mm	477.33

Table 3.2. Predicted vs actual properties

	Field (Actual)	Laboratory (Predicted)
Settled Height (mm)	317	312,79
Moisture content (%)	0,147	0,146
Dry density (kg/m^3)	1631	1420,7

- freewater is drawn off towards the penstock almost immediately after deposition because of the beach angle
- evaporation and the dry climate out in the field has a major impact on the density and rate of consolidation of the tailings material
- tests in the field were conducted on the outer shell of the tailings dam which contains coarse material (sands) and is free draining. No drainage were allowed in the laboratory.

4 CONCLUSIONS

Tailings deposition using the sub-aerial technique and underdrainage to achieve liquid-solid separation, can lead to significant benefits in the construction of tailings storage facilities, and in the control of seepage. In practical terms this requires the removal of all excess water from the tailings facility by minimising as far as possible any surface ponding, and by the provision of an extensive underdrainage system. Potential benefits resulting from the use of these concepts can be summarised as follows:
- Significantly greater unit weights are achieved in the stored tailings resulting in better utilisation of the storage facility.
- The drained nature of the tailings and the removal of surface ponding adjacent to facility embankments, enable major portions of the structural elements of the facility to be constructed out of tailings material by so-called "upstream" construction, thereby effecting significant economies.
- At decommissioning of the facility, the tailings are fully drained and consolidated allowing immediate construction of a surface seal and the elimination of any long-term seepage.
- The drained nature of the tailings increases the resistance to liquefaction under seismic loading,

and the elimination of surface ponding reduces the possibility of adverse consequences that could result form liquefaction induced embankment deformations.

The laboratory experiments show that small scale testing can produce model parameters of sufficient accuracy and that settlement and desiccation may be predicted from column-settling tests and a two-stage drying model. The model has proven capable of predicting field behaviour with sufficient accuracy. The flexibility of the model, in terms of input and climatic conditions, covers a wide range of alternatives, which promotes optimum design and operation. The predicted water content for a given layer of tailings is a useful variable, with which other important soil characteristics may be correlated.

Natural Moisture Content

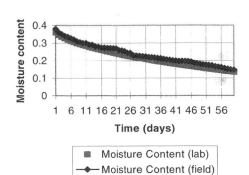

Figure 3.1. Moisture content versus time (predicted and actual).

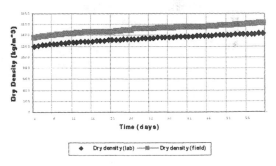

Figure 3.2. Dry density versus time (predicted and actual).

REFERENCES

Swarbrich, G.E. and Fell, R. 1991. Prediction of Improvement of Tailings Properties by Desiccation
Swarbrich, G.E. 1992. Transient Unsaturated Consolidation in Desiccating Mine Tailings
Vick, S.G. 1990. Planning Design and Analysis of Tailings Dams

Paste technology

Planning, design and implementation strategy for thickened tailings and pastes

Fiona Sofra
Rheological Consulting Services Pry Ltd., The University of Melbourne, Australia

David V. Boger
Particulate Fluids Processing Center, The University of Melbourne, Australia

ABSTRACT: The use of thickened tailings and pastes stems from the aim to design tailings disposal sites to suit their surrounding environment, rather than manipulating the environment to accommodate the tailings. Thickened tailings and pastes usually exhibit non-Newtonian behaviour, so determination of the disposal plant operating conditions requires a thorough understanding of the rheological characteristics of the material. The implementation and optimisation of dry disposal methods involves three concurrent and interdependent rheological studies to determine i) the concentration required to achieve the optimum spreading and drying characteristics of the tailings once deposited ii) the optimum conditions for pipeline transport and iii) the feasibility of dewatering the slurry to the required concentration.

The purpose of this paper is to provide an informed background to ensure the appropriate rheological issues are addressed as a part of the overall tailings management strategy.

1 INTRODUCTION

The main criterion for the success of a tailings management system is to implement and operate a safe and environmentally responsible system at a minimum cost. Exploitation and manipulation of the transport and deposition characteristics of the tailings are vital to ensure that the appropriate disposal system is operated successfully.

A dry disposal system is selected and designed to minimise the pumping energy required whilst delivering a material at the velocity, yield stress and viscosity needed for the desired slope and spreading characteristics. The purpose of this chapter is to elucidate the significance of rheological characterisation for the planning, design, operation and optimisation of dry disposal systems including dry stacking, thickened tailings disposal and paste backfill.

2 SUMMARY FLOWSHEET FOR SELECTION, DESIGN AND TESTING FOR DRY DISPOSAL IMPLEMENTATION

Understanding the material rheology is essential in selecting and designing a dry disposal system to minimise pipeline transport costs whilst delivering a material with the properties required for deposition. Furthermore, insight into the operational parameters that alter the tailings' rheology can be exploited to ensure optimal operation of the system.

Several material and operational parameters at may be manipulated to provide the desired characteristics of the deposited tailings. Material parameters include the solids concentration, viscosity and yield stress. Operational parameters include the flow rate, determined by the pipe diameter and throughput and the extent of shear to which the tailings are subjected. The extent of shear is determined by the pump type, the flow regime (laminar or turbulent) and the pipeline length and configuration.

The deposition and pipeline flow requirements need to be considered together using an iterative process to optimise the system for economical operation whilst maintaining the integrity of the chosen disposal method. Figure 1 summarises the rheology based decisions, considerations and test work required for the proposed implementation strategy in flow chart form.

3 THE IMPORTANCE OF RHEOLOGY IN A DRY DISPOSAL IMPLEMENTATION STRATEGY

By considering tailings rheology at all points during the design phase, implementation strategies can be developed for new operations or existing systems optimised to reduce operating costs and/or the impact on the surrounding environment.

The strategy proposed in this chapter looks at designing the tailings to suit the surrounding environ

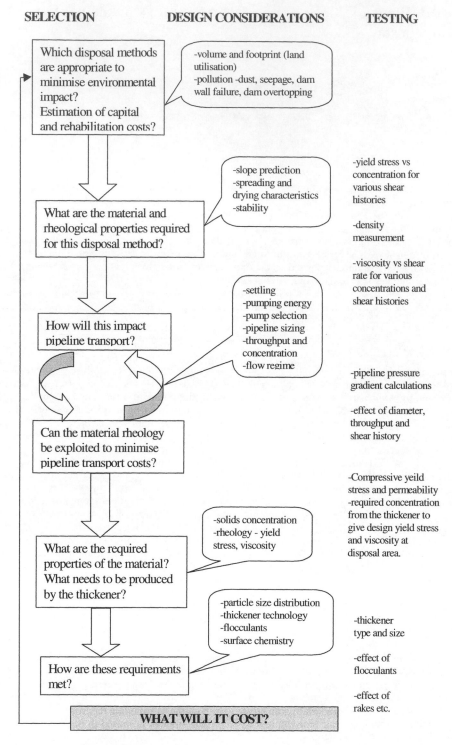

Figure 1. Planning, design and testing flow sheet.

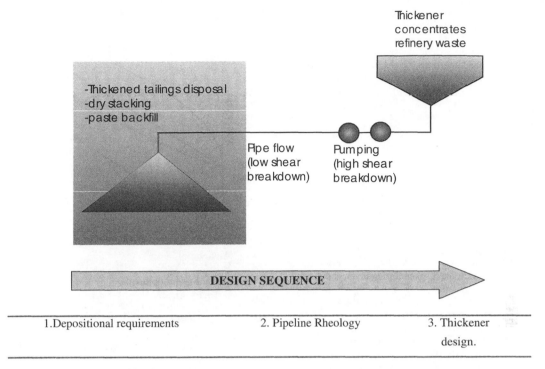

Figure 2. Suggested approach for determination of tailings disposal system and requirements.

ment, with consideration being given to the choice and operation of the disposal method. Figure 2 briefly summarises the steps involved in determining a disposal system. It is important to note that the design sequence begins at the disposal point and works upstream to the thickening stage.

Point 1 in the design sequence represents the choice of the disposal method and the rheological properties required to achieve the desired footprint and shape of the final deposit (depositional requirements). Point 2 refers to the pumping and pipeline conditions needed for optimal transport whilst ensuring the tailings reach the disposal site with the rheological properties necessary for the chosen disposal method (pipeline requirements).

Once the depositional and pipeline issues have been quantified and optimised on paper, it is possible to determine the thickening requirements represented by point 3 on Figure 2. The thickener should be designed and operated to produce the required tailings' properties via determination of thickener size and geometry, and control of flocculation and pre treatment requirements. If the thickener requirements cannot be achieved then the process is repeated.

For clarity, the design sequence is depicted in Figure 2 as a linear sequence. However, in reality the disposal system and thickener will be designed using an iterative approach in order to optimise the entire disposal operation.

4 IMPORTANT RHEOLOGICAL CONCEPTS

Rheology is the study of the deformation and flow of matter. In terms of fluid flow, materials may be classified as either Newtonian or non-Newtonian fluids. The viscosity (η) of a fluid is defined as the ratio of the shear stress (τ) to the shear rate $(\dot{\gamma})$, as shown in Equation 1. In many flows, the shear rate is equivalent to the gradient in velocity.

$$\eta = \frac{\tau}{\dot{\gamma}} \tag{1}$$

Inelastic Newtonian fluids exhibit a linear relationship between the applied shear stress and the shear rate, as shown in curve A in Figure 3. Flow is initiated as soon as a shear stress is applied. The linear relationship between the shear stress and the shear rate indicates a constant viscosity.

Concentrated mineral tailings often display non-Newtonian flow behaviour in that they possess a yield stress. The yield stress (τ_y) is the critical shear stress that must be exceeded before irreversible deformation and flow can occur. For applied stresses below the yield stress, the particle network of the suspension deforms elastically, with complete strain recovery upon removal of the stress. Once the yield stress is exceeded, the suspension exhibits viscous liquid behaviour where the viscosity is usually a function of the shear rate. Yield stress behaviour is shown by the constitutive relationships of Equations 2 and 3.

$$\gamma = 0 \qquad \tau < \tau_y \qquad (2)$$

$$\tau = \tau_y + \eta\left(\dot{\gamma}\right)\dot{\gamma} \qquad \tau > \tau_y \qquad (3)$$

Curve B on Figure 3 shows a yield stress followed by a linear shear stress - shear rate relationship, commonly known as Bingham behaviour. Although not a true viscosity according to Equation 1, the gradient of this line is referred to as the Bingham plastic viscosity.

In addition to yield stress behaviour, the viscosity of the material may vary with shear rate. As the shear rate is increased, pseudoplastic or shear thinning materials exhibit a decrease in the viscosity (curve C, Figure 3). Dilatant or shear thickening materials exhibit an increase in viscosity with increasing shear rate (curve D, Figure 3). Dilatant behaviour, although relatively rare, is sometimes observed in mineral suspensions. For the different fluid categories, various empirical flow models are used to describe the flow behaviour. The most commonly used Equations are the Ostwald-De Waele model for shear thickening or shear thinning materials, the Bingham model for yield stress materials and the Herschel Bulkley model for yield stress, shear thinning or yield stress, shear thickening materials.

Ostwald-De Waele power law model;

$$\tau = K\dot{\gamma}^n \qquad (4)$$

Bingham Model;

$$\tau = \tau_B + \eta_B \dot{\gamma} \qquad (5)$$

Herschel Bulkley Model;

$$\tau = \tau_{HB} + K\dot{\gamma}^n \qquad (6)$$

The shear thinning nature of industrial tailings, such as red mud in the alumina industry, is attributed to the alignment of particles or flocs. An increase in the shear rate from rest results in instantaneous alignment of particles in the direction of shear, therefore providing a lower resistance to flow. As such, the suspension will show a decreasing viscosity with increasing shear rate.

A further deviation from Newtonian flow behaviour is the result of time dependence or thixotropy. Thixotropy is the result of structural breakdown under shear and manifests itself as a decrease in the viscosity and yield stress with time for a given, con-

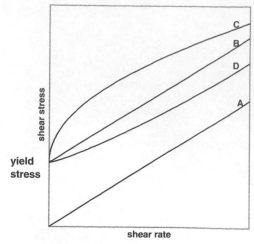

A Newtonian
B Bingham (yield-constant viscosity)
C Yield-pseudoplastic
D Yield-dilatant

Figure 3. Typical flow curves.

stant shear rate. As time of shear increases, the rate of breakdown will decrease, as fewer structural bonds are available for breakdown. At the same time, structural reformation can be taking place and the rate of this process increases with time of shear due to the increasing number of bonding sites available.

A state of dynamic equilibrium, where the breakdown and reformation rates are equal, is possible, however this state is not always achieved in industrial applications due to the extended shearing times often required. The material in pipeflow, for example, may be in a partially sheared state, where the shear stress-shear rate behaviour is still changing with the time of shear. Problems may arise when using flow curves (shear stress – shear rate data) generated in a laboratory environment for pumping energy prediction, due to the difficulties in ensuring that the material is in the same structural state as in the process pipeline.

5 RHEOLOGICAL REQUIREMENTS FOR THE CHOSEN DISPOSAL METHOD

5.1 Depositional requirements - Point 1, Figure 2

Having determined that a thickened tailings or paste disposal system is desirable, it is necessary to consider the rheological properties to a) determine the operating requirements of the disposal system and b) estimate the operating costs. For thickened tailings and pastes, an understanding of the tailings rheology is prerequisite to quantifying the deposition, pipeline transport and thickening behaviour of the dewatered tailings.

The essential depositional requirement for dry disposal is to deliver tailings to the disposal point with a yield stress sufficient to support the largest particles and ensure a homogeneous suspension where segregation does not occur. Differential segregation upon deposition will result in a non-uniform deposit slope and inefficient use of the disposal area. The relationship between the solids concentration and the yield stress, must be determined to indicate the minimum concentration necessary to obtain this yield stress.

The rheological characterisation of concentrated mineral suspensions requires specialised equipment and techniques. A significant amount of work on the measurement of the yield stress of mineral suspensions has been completed by many workers. From this work, novel and simplified measurement techniques have resulted. The vane-shear instrument and technique allows direct and accurate determination of the yield stress from a single point measurement (Nguyen, 1983, Nguyen and Boger, 1983, 1985). Many workers worldwide have adopted the vane-shear method and confirmed its applicability for all types of yield stress materials (Yoshimura et al., 1987, James et al., 1987, Avramidis and Turian, 1991, Liddell and Boger, 1996, Pashias, 1997).

In an attempt to further simplify yield stress measurement, the 'slump test' has been modified to accurately evaluate the yield stress of mineral suspensions (Showalter and Christensen, 1998, Pashias et al., 1996). This technique has been typically used to determine the flow characteristics of fresh concrete and is important in mine stope fill. The slump test is conducted using only a cylinder and a ruler, eliminating the need for sophisticated equipment and allowing easy, on-site yield stress measurement by plant operators.

The flow properties of concentrated mineral suspensions vary significantly with solids concentration and type; however, a number of common characteristics have been observed for concentrated suspensions in general. The strong dependence of the rheology on concentration is exemplified in Figure 4. This figure shows the yield stress as a function of concentration for a number of industrial slurries. Although the relationships vary for the different materials, all materials exhibit an exponential rise in the yield stress with concentration. Furthermore, for all materials the yield stress begins to rise rapidly beyond 80 to 100 Pa, regardless of the concentration.

The presence of a yield stress is essential for dry stacking and thickened tailings disposal to ensure that the material comes to rest at the required angle for stability and maximum storage capacity. An adequate yield stress also ensures that particle size segregation does not occur and that the final tailings stack will be homogeneous.

The concentration at which the yield stress begins to rise rapidly is significant when optimising pumping energy requirements. The yield stress must be sufficient to prevent solids deposition, but not so high that start-up problems will be encountered.

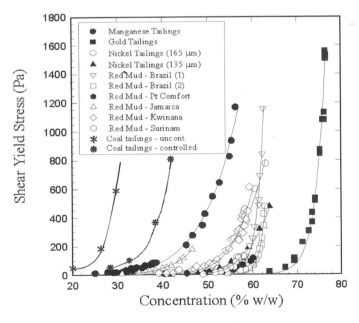

Figure 4. Yield stress as a function of concentration for a number of mineral tailings.

For thixotropic materials, shear history must be taken into account when determining the yield stress - concentration relationship. During the design phase, it is unlikely that the extent of structural breakdown during pipeline transport will be known. As such, the yield stress - concentration relationship must be determined for structural states ranging from the initial to the equilibrium state. From this data, the concentration range required to produce the desired yield stress can be determined.

The viscosity and yield stress must be known in order to determine the spreading characteristics and to predict the angle of repose (slope) formed upon deposition. The slope obtainable for a given material enables the disposal area to be sized. At this point, the design material characteristics may be varied to produce a steeper slope and reduce the footprint, or a shallower slope to increase the capacity and lifetime of the disposal area.

The determination of the design slope value sets the rheological properties required of the tailings when the material reaches the disposal site (point 1 on Figure 2).

5.2 *Pipeline transport requirements - Point 2, Figure 2*

In manipulating the rheological properties of the material to achieve the required slope, it is important that their effect on the pipeline transport requirements also be considered. The yield stress determines the minimum pressure differential required for flow. A combination of the yield stress and viscosity will dictate the frictional pressure losses expected in the pipeline. Calculation of the pressure gradient for various combinations of pipe diameter, flowrate and throughput must be undertaken for the design viscosity and yield stress.

Measurement of fundamental flow behaviour may be undertaken using a capillary rheometer (Want et al., 1982, Nguyen and Boger, 1983, Leong, 1987). The capillary rheometer generates shear stress-shear rate data for the determination of pumping energy requirements in addition to describing the influence of thixotropy and pseudoplasticity.

Figure 5 shows the decrease in viscosity with increasing shear rate (shear thinning) for a number of concentrations of red mud.

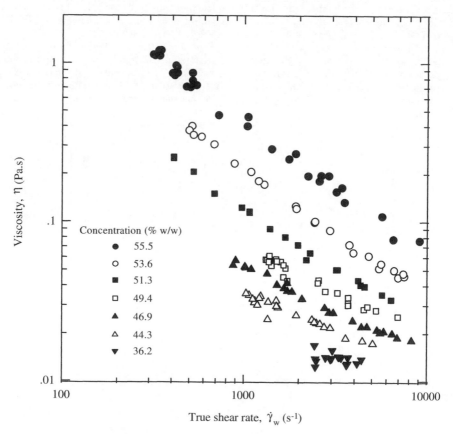

Figure 5. Red mud; equilibrium viscosity vs. shear rate for various solids concentrations.

For thixotropic materials, the amount of pipeline transportation may alter the tailings' viscosity and yield stress between the thickening stage and the disposal area. It is vital to understand these changes in order to ensure that the material reaches the disposal area with the yield stress and viscosity required to attain the desired deposition characteristics. Point 2 on Figure 2 refers to the study of the pipeline flow and represents the necessity for understanding the changes in the tailings' rheology for the proposed pipeline conditions.

It is not sufficient to determine the concentration corresponding to the required viscosity and yield stress at the disposal site, and operating the thickener to produce this concentration of tailings. Should the tailings undergo structural breakdown during pipeline transport, the rheological properties will differ between the thickener underflow and the point of disposal. Consequently, the disposal area will not attain the desired slope. Determination of the shear history dependence, completed during prediction of the deposition characteristics, will indicate the maximum reduction in the yield stress and viscosity expected.

To determine the effect of shear history on the flow properties, the suspension is sheared (by mix-ing) in between measurement of the shear stress-shear rate behaviour by capillary rheometry. Typical flow data for 47-wt % bauxite tailings samples (red mud) from alumina production are shown in Figure 6.

Capillary rheometry data in Figure 6 illustrate a yield stress material that is strongly thixotropic and shear thinning. Although knowledge of the shear thinning behaviour is imperative, for many materials the time dependent nature, which results in thixotropic behaviour, has a more significant influence on the flow properties.

Combined with the deposition requirements, quantifying the expected changes in the rheological characteristics for the proposed pumping and pipeline conditions from the point of thickener discharge to disposal will set the requirements of the thickener underflow, point 3, Figure 2. The concentration and rheological properties required from the tailings produced by the thickener will dictate thickener operation to ensure provision of tailings suitable for the disposal system selected.

5.3 Thickening requirements - compression rheology - Point 3, Figure 2

Prior to deposition, the slurry must be thickened to the concentration found in shear rheology tests to

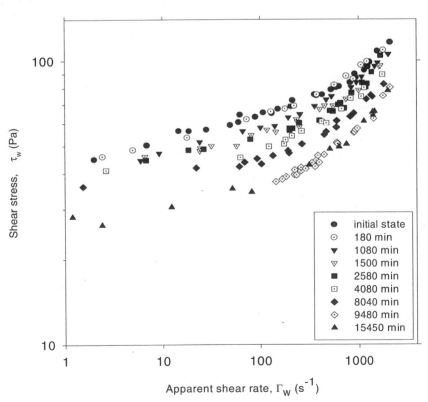

Figure 6. Shear stress vs shear rate variation with shear history.

provide the desired deposition characteristics whilst ensuring that the material is still easily pumpable. Thickening also ensures that little or no particle size segregation occurs during deposition (Robinsky, 1978; Wood and McDonald, 1986; Williams, 1992).

Segregation is inhibited due to an increase in the viscosity and yield stress with increased solids concentration and results in the uniform gradient required for dry disposal methods. The observed uniformity is desirable as it leads to improved stability while minimising tailings storage volume, which facilitates easy site rehabilitation.

Thickening is usually achieved using gravity thickeners with an appropriate polymeric flocculant. Choice of the flocculant and operating conditions within the thickener is determined by examining the shear, compression and permeability characteristics of the suspension.

The shear yield stress (the shear stress required for irreversible flow to occur) is important in the determination of piping and deposition behaviour. Likewise, the compressive yield stress, coupled with the permeability of the material, provide information regarding the feasibility of dewatering the residue to the concentration required for dry disposal. The compressive stress is the stress required for irreversible compression of the network structure and the permeability dictates the rate of dewatering and accounts for the hydrodynamic interactions between falling particles in a suspension.

In a conventional gravity thickener, a continuous network structure is formed through the aggregation of particles or flocs containing interparticle water. The compressive yield stress, beyond which the transmitted network pressure of the overlying structure will cause the collapse of the structure and syn-

eresis of liquor, is concentration dependant (Buscall and White, 1987). By increasing the compression zone depth, the material at the bottom of the thickener will be subjected to a greater applied pressure and will consolidate to a higher concentration until the compressive yield stress is equal to the increased stress.

An understanding of the compression zone depth required to overcome the compressive yield stress at a given concentration, and the variation in this relationship with factors such as flocculant dosage and shear history will facilitate optimisation of thickener performance.

Figure 7 shows typical results for red mud. In the unflocculated state, a conventional thickener with negligible compression zone will yield an underflow concentration of 38wt%. By increasing the sediment height to 5m, an underflow concentration of over 52wt% is possible and flocculating with 145ppm flocculant can further increase this to 58wt%. Whilst this sounds straight forward, these numbers represent the solids concentration that could be achieved if the thickener was operated at a throughput rate that allowed full dewatering to occur. In practice, this rarely happens and it is the permeability of the solids that dictates the underflow solids concentration achieved. Since flocculation generally increases the permeability, the most permeable solids usually achieve the highest solids output.

Permeability is influenced by the liquor viscosity, the flocculation conditions and the solids concentration. The liquor viscosity varies with temperature and dissolved solids concentration and affects the rate at which the liquid can move upward through the particulate matter. Flocculation conditions (type, dose and addition method) affect the suspension

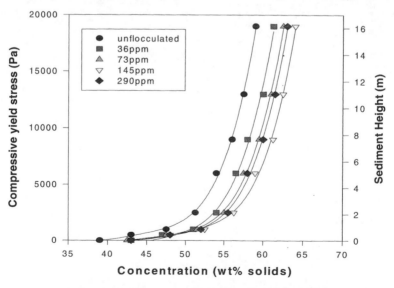

Figure 7. Compressive yield stress vs concentration for flocculated and unflocculated red mud.

structure formed, the amount of entrapped interfloc liquid and the ease of liquid permeating through the structure. For a given set of process conditions, ie temperature, dissolved solids concentration and flocculation conditions, permeability is a function of the solids volume fraction.

A portable apparatus that is capable of determining the compressive yield stress and the permeability has been developed and built at the University of Melbourne to allow optimisation of the thickening process (de Kretser et al, 1999).

6 CONCLUSION

Due to the complex rheology of mineral tailings under shear and compressive forces, knowledge of the rheological properties allows the design of a disposal scheme that takes best advantage of the material behaviour without affecting upstream plant performance. Furthermore, an insight into how to change or manipulate the rheology may facilitate waste disposal in a manner that is more environmentally or economically favourable than conventional methods.

The implementation and optimisation of dry disposal methods involves three concurrent and interdependent rheological studies to determine i) the concentration required to achieve the optimum spreading and drying characteristics of the tailings once deposited ii) the optimum conditions for pipeline transport and iii) the feasibility of dewatering the slurry to the required concentration. The technical methods outlined in this paper provide the rheological characterisation required to complete these studies.

As environmental factors translate into economic issues, the push for minimising waste production using dry disposal methods is gaining popularity. The use of rheological information is of high importance in evaluating the spreading characteristics, the transportation and the dewatering of thickened mineral tailings. The principles outlined in the examples given in this paper may be applied to many industries encompassing a wide range of waste materials.

REFERENCES

Avramidis, K.S, Turian, R.M 1991. Yield stress of Laterite suspensions. *J.Colloid Interface Sci.* 143:54-68.

Buscall, R., White, L.R., 1987. On the consolidation of concentrated suspensions, I. The theory of sedimentation, *J.Chem.Soc.Faraday Trans.* 1(83):873-891.

de Kretser, R.G., Scales, P.J, Aziz, A.A.A., 1999, Optimising the surface chemistry of suspensions for dewatering in mineral processing. *Rheology in the mineral Industry II.*

James, A.E., Williams, D.J.A, Williams, P.R. 1987. Direct measurement of static yield properties of cohesive suspensions. *Rheol. Acta* 26:6489-6492.

Leong Y.K. 1988. *Rheology of modified and unmodified Victorian brown coal suspensions.* PhD. Thesis, The University of Melbourne.

Liddell, P., Boger, D.V. 1996. Yield stress measurement with the vane. *J Non-Newtonian Fluid Mech.* 63:235-261.

Nguyen, Q.D., 1983. *Rheology of Concentrated Bauxite Residue*, PhD Thesis, University of Melbourne.

Nguyen, Q.D., Boger, D.V. 1983. Yield stress measurement in concentrated suspensions. *Journal of Rheology*, 27(4): 321-349.

Nguyen, Q.D., Boger, D.V. 1985. Direct Yield Stress Measurement with the Vane Method, *Journal of Rheology*, 29(3):335-347.

Pashias, N; Boger, D.V; Summers, J; Glennister, D.J. 1996. A Fifty Cent Rheometer for Yield Stress Measurement. *Journal of Rheology*, 40(6):1179-1189.

Pashias, N. 1997, *The Characterisation of Bauxite Suspensions in Shear and Compression*, PhD Thesis, The University of Melbourne.

Robinsky, E.I, 1978, 'Tailings Disposal by the Thickened Discharge method for Improved Economy and Environmental Control', *Tailings Disposal Today, Proceedings 2nd Internat. Tailings Symp.*:75-95.

Showalter, W.R., Christensen, G. 1998. Toward a rationalization of the slump test for fresh concrete: comparisons of calculations and experiments. *J.Rheol*, 42(4):865-870.

Want, F.M., Colombera, P.M., Nguyen, Q.D., Boger, D.V 1982. Pipeline design for the transport of high density bauxite residue slurries. *Proc. 8th Int Conf. Hydraulic-Transport of Solids in Pipes*, Johannesburg; 242-262.

Williams, M.P.A. 1992. 'Australian Experience With The Central Thickened Tailings Discharge Method for Tailings Disposal', *Environmental Issues and Waste Management in Energy and Minerals Production*; 567-577, Balkema, Rotterdam.

Wood, K.R., McDonald, G.W. 1986 'Design and Operation of Thickened Tailings Disposal System at Les Mines Selbaie', *CIM Bulletin*, 79(895):47-51.

Yoshimura, A.S., Prud'homme, R.K. Princen, H.M. Kiss, A.D, 1987. A comparison of techniques for measuring yield stresses. *J. Rheol.*

Tailings and Mine Waste '02, © 2002 Swets & Zeitlinger, ISBN 90 5809 353 0

Effects of settlement and drainage on strength development within mine paste backfill

T. Belem, M. Benzaazoua, B. Bussière & A.M. Dagenais
Département des Sciences Appliquées, Unité de recherche et de service en technologie minérale
Université du Québec en Abitibi-Témiscamingue, Rouyn-Noranda, Québec, Canada

ABSTRACT: Paste backfill mixture with binder was poured and cast into rigid PVC moulds especially designed to permit load application to the paste in both drained and undrained conditions. After 28 days of curing, uniaxial compression tests and microstructure analysis through the measurements of specific surface area were accomplished to investigate the mechanical strength development within paste backfill samples. The results of the study showed that external loading leads to an increase in compressive strength (UCS) when paste backfill sample is drained, but a reduction of UCS when paste backfill sample is undrained.

1 INTRODUCTION

In the past, the Canadian mine wastes were stockpiled in tailings ponds designed for this specific purpose. However, when such tailings contains sulfide minerals, these react with oxygen rainwater to form acid mine drainage (AMD). This acidified water may also lead to the leaching of heavy metals which can contaminate the environment. However, because of recent environmental legislation, mining companies have the obligation and duty to manage properly the tailings in order to replace the environmental impact.

In order to achieve this legislation, a new type of backfill was then developed, namely the paste backfill, to replace the traditional hydraulic backfill. This new technology reuse the full size fraction of the tailings in a mixture with a binding agent (e.g. ordinary Portland cement, hydraulic minerals, silica fume, etc) and water (e.g. mine process water, rainwater, lake water, etc.) to fill underground mine excavations. The use of paste backfill became an increasingly common practice in the hard rock mines worldwide (Barsoti 1978; Stone 1993; Bodi et al., 1996) and in Canada since the 1990 (Viles and Davis 1989; Landriault 1992 & 1995; Landriault and Lidkea 1993; Landriault and Tenbergen 1995; Landriault et al., 1997; Naylor et al., 1997). Among the numerous advantages attributed to the use of paste backfill one can cite (i) a significant tonnage reduction of tailings to store on surface infrastructures (Hassani and Archibald 1998), (ii) the stability of underground excavations and consequently, an increase in extracted ore body (Mitchell 1989a &b; Lamos and Clark 1989; Lawrence 1992; Petrolito et al., 1998; Ouellet et al., 1998a & b; Benzaazoua et al., 1999a & b; Bernier et al., 1999; Benzaazoua and

Belem 2000; Belem et al., 2000), (iii) a reduction in mine operating costs (Hassani and Bois 1992; Hassani and Archibald 1998) and (iv) an increased in safety for miners.

One of the difficulties in using paste backfill is that it is a complex material, in continuous evolution since its preparation in the backfill plant, and through the different stages like its underground delivery by pipelines to the disposal point, its setting in the stope until its short-term, mid-term and long-term hardening in the excavations. Due to this complexity, paste backfill is the object of numerous studies by various research centers and teams. Despite this, many questions remain unanswered concerning physical, chemical and mechanical properties of paste backfill. The object of this study is to understand the effects of settlement and drainage on strength development within *in situ* paste backfill. The basis for this study was observations made by ground control engineers that the mechanical strength of core samples taken from certain *in situ* paste backfill was greater (sometimes by a factor of 2) than the one coming from samples of the same paste backfill poured into plastic cylinder moulds and intended to the quality control follow-up. It clearly follows that the differences in mechanical strength between the paste backfill moulded into the plastic cylinders and the *in situ* core samples coming from the same paste backfill mixtures are likely due to a difference in the mechanical strength development within the paste backfill during curing time (e.g. hydration mode of the binding reagents). Among the numerous probable hypotheses explaining these differences one can cite: (i) *in situ* exudation or vertical drainage of part of the water used for the paste backfill preparation which favors a rapid and massive formation of binder hydrates (e.g. saturation of hydration reactions), (ii) combination of drainage and settlement of the *in situ* paste backfill mass and, (iii) confining pressures exerted by the excavation walls on the *in situ* paste backfill. Because all the laboratory studies and quality control follow-up are carried on cylindrical plastic moulds of paste backfill, the direct application of such results to mine designs requires that one work with high factor of safety, implying an increase in backfill operating costs.

The purpose of this paper is to identify the probable causes of the differences in mechanical strength between the *in situ* and experimental paste backfills. It is a preliminary and prospective study of the combined effects of paste backfill drainage and settlement on its mechanical strength development. To this end, tailings from a Canadian hard rock mine were sampled for the paste backfill mixtures preparation using 5 wt.% of a binder agent made up of 50:50 of ordinary Portland cement (Type I) and sulfate-resistant Portland cement (Type V). The resulting paste backfill was then poured and cast into rigid PVC moulds especially designed to permit different load application to the paste. Two types of moulds were used: moulds with a perforated bottom to favor drainage and non-perforated moulds to prevent it. Uniaxial compression tests and microstructure analysis and specific surface area measurements were done to investigate the changes in physical properties of paste backfill samples.

2 EXPERIMENTAL PROCEDURES

2.1 Tailings characterization

Tailings from a Canadian underground rock mine (mine M1) was sampled after having been filtered for the backfill preparation. Cyanides present in the tailings were removed by the SO_2-Air method prior to the filtration process. The resultant tailings contain about 60 wt.% sulfides, essentially pyrite, and thus giving it a specific gravity, G_s, of 3.78. Particle size distribution was determined using a *Malvern®* *Mastersizer* laser granulometer under humid conditions.

2.2 Load application device

To reach the objectives of this study, a device was designed and manufactured that allows load application on the paste backfill mass during the curing process. Figure 1 presents a schematic representation of this device. Two types of experiment set-up with this device were used: (i) the perforated bottom set-up was used to asses the effects of drainage (D-test) and, (ii) the non-perforated bottom set-up for simulating undrained conditions (U-test). Control moulds (perforated and non-perforated bottom) were also used in order to understand the effect of the applied

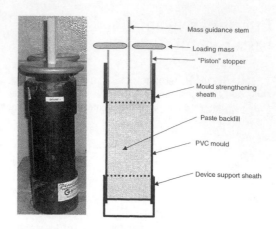

Figure 1. Load application device.

loads on the mechanical strength development within the backfill.

Moulds average diameter measured 10.125 cm and were 24 cm in height. As shown in Table 1, three different loads were applied in the drained and undrained tests: 10 lbs, 20 lbs and 50 lbs. It should be noted that a first load of 5 lbs was applied after 3 days of curing, followed by the addition of 5 lbs per day until the target load is reached. This procedure simulates a sequential backfilling of 0.12 m of paste backfill each day. Thus, the total duration of loading was 2 days for D-test 1 and U-test 1, 4 days for D-test 2 and U-test 2, and 10 days for D-test 3 and U-test 3. Table 1 also presents the values of the loads applied to paste backfill in terms of mass (lbs) and weight (N) and the corresponding simulated backfill column height.

Table 1. Applied loads and simulated backfill column height for drained (D-test) and undrained (U-test) tests.

Applied load (lbs)	Applied weight (N)	Applied pressure (kPa)	Column height (m)
0	0	0	0
10	44.5	5.5	0.23
20	89.0	11.1	0.47
50	222.5	27.6	1.18

2.3 Paste backfill preparation

For the present study, the chosen paste mixture recipe corresponds to that already used by the mine M1 where the tailings were sampled. This mixture recipe consists of a binder agent made up of 50% ordinary Portland cement (Type 10 or Type I) and 50% sulfate resistant Portland cement (Type 50 or Type V). The mixture contains 5 wt.% binder. The mixing water used was municipal water.

Tailings, binder agent and water were mixed and kneaded in a 3-speed cement mixer for about 15

Photo 1. Load device set-up placed in the humidity chamber.

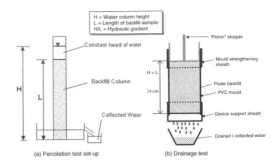

Figure 2. Drainage test compared to the percolation test.

minutes in order to obtain a homogenous paste. The mixture was calculated so that the final solid mass percentage, C_w, was 78% and the moisture content, w, of 28%. This corresponds to a solid volume percentage, C_v, of 66%, a water-to-cement ratio (w/c) of 6 and a water-to-solids ratio (w/s) of 0.28.

The resulting paste backfill was then poured into the different devices for both D-tests and U-tests (see Fig. 2.) according to B.N.Q. 2622-913 standard method and modified by the URSTM Laboratory (procedure PE4 EG-04). This consists of pouring three layers of approximately equal volume, each layer being consolidated by pounding it 25 times with a rounded end stainless steel rod of 9.5 mm in diameter. These blows were evenly distributed over the entire section of each mould. Finally, all samples were stored and conditioned in a humidity chamber at about 70% humidity with an average temperature of 24° for a total curing period of 28 days (see Photo 1). These conditions are similar to those observed in the concerned mine M1.

2.4 Paste backfill drainage measurement

Measurement of paste backfill drainage may be done either by weighing each sample set-up or by collecting the percolated water, at regular time intervals. In the present study, drainage was measured by using the second method, at a time interval of 30 minutes. The first measure was taken 1 hour after molding of the paste. Figure 2b presents the set-up for drainage measurements used in this study, in comparison to the classical percolation test set-up (Fig. 2a). The fundamental difference between these two tests set-up being that in the case of the percolation test an external supply of water is necessary to maintain a constant water level (Fig. 2b).

In the drainage test the starting height of the water column (H) is equal to the height of the backfill column (L). The percolation rate or the drainage rate, v (cm/s) of the paste backfill is given by the following equation:

$$v = \frac{q}{A} \tag{1}$$

where $q = V/t$ is the volume of drained water (V) at the time interval t; $A =$ mould section area.

2.5 Uniaxial compression tests

After 28 days of curing, all the devices were dismantled (see Photo 2). Uniaxial compression test were done on the recovered samples to determine their mechanical strength. This test consists of placing a sample between two plates of a mechanical press and applying an axial force until the sample breaks. The stress corresponding to this maximum force is the uniaxial compressive strength (UCS). The compression tests were carried out using a rigid mechanical press MTS 10/GL with a loading capacity of 50 kN and a minimum deformation rate of 0.0001 mm/min. The applied force to the sample was measured by a pressure cell while the displacement was measured by a magnetic induction displacement sensor. The accuracy of the force measurement is about 1%. The shape factor (height-to-diameter ratio) of the backfill samples for these tests was of 2.

2.6 Specific surface area measurements

The specific surface area, S_s, is a good indicator of the fineness of backfill particles. This fineness depends on both the fineness of the initial tailings and that of the hydrates formed. That is only S_s can be used to characterize the microstructure of the paste backfill. To be more precise, the specific surface area of a material includes the external and internal surfaces. This measure includes also all of the surface irregularities at the molecular scale level as well as the surface represented by pores in the solid matrix.

As such, the value of the specific surface area allow us to determine the influence of the load appli-

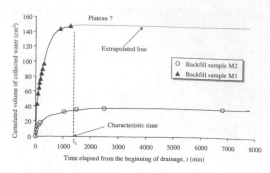

Figure 3. Variation in collected water with time elapsed from the beginning of drainage.

Photo 2. Dismantling of the device after 28 days of curing.

cation on the strength development within the paste backfill. By definition, S_s is the ratio of the sum of the surface areas to the mass of the sample (m²/g or m²/kg). The specific surface area measurements were performed by B.E.T method (Brunauer, Emmett and Teller) which consists of an argon adsorption using *Micromeritics® Gemini* surface analyzer.

3 RESULTS

3.1 *Paste backfill drainage and percolation*

3.1.1 *Paste backfill drainage*
As mentioned before, paste backfill drainage was measured by collecting the drained water at a 30 minutes time interval. Drainage period lasted 21 hrs and the total cumulative volume of water collected was about 147 cm³. Figure 3 presents the evolution of collected water as a function of time for the paste backfill sample M1 compared to the data obtained from the paste backfill sample M2 (from another mine not presented in this study). This diagram shows that the quantity of collected water increases gradually with time but seems to plateau after a characteristic time $t_c = 21$ hrs.

One can conclude that if this type of paste backfill is placed in mine excavation the resulting drainage will be minimal and almost inexistent after 24 hrs of curing. The validity of this observation is supported by the fact that recuperation of water was continued for 5 days after obtaining a plateau. Also to be noted is the fact that the drainage capacity of backfill sample M1 (present study) is 4 times greater

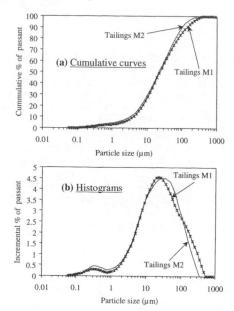

Figure 4. Grain size distribution of tailings samples M1 & M2.

than the one for backfill sample M2. This can be explained in part by the difference in the particle size distribution of these two types of tailings. But in addition to the tailings particle size distribution, the type of binder agent used plays probably a certain role in drainage that has not been identified in this study.

Figure 4 presents the particle size distribution curves of the tailings samples M1 and M2 used for the preparation of paste backfill samples M1 and M2. It should be noted that despite the strong similarity between the particle size distribution of these two types of tailings, the particles of tailings sample M1 are slightly coarser than the one of tailings sample M2. This difference is evident from Figure 5 for parameters D_{60} and D_{90}.

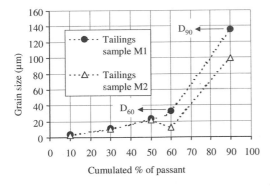

Figure 5. Grain size vs cumulative % of passant of tailings samples M1 & M2.

The two curves on Figure 3 are well fitted by the following general regression equation:

$$V(t) = \frac{V_m \bullet t}{a + t} \qquad (2)$$

where t (min) = time elapsed from the beginning of drainage; V_m (cm³) = maximum volume of collected water at the end of drainage; a (min) = constant corresponding to the drainage characteristic time.

Results from regression gave V_m = 172.523 cm³ and a = 212.899 min for the backfill sample M1; and V_m = 38.6 cm³ and a = 153.303 min for the backfill sample M2 with a correlation coefficient R = 0.99. Although, the maximum total volume of collected water at the end of the test was 147.48 cm³ for the backfill sample M1 and 37.75 cm³ for the backfill sample M2. Authors suggest that this equation could be used for different types of paste backfill by calibrating V_m and the constant a with regard to the particle size distribution of different types of tailings. V_m should depend on the particle size distribution of the initial tailings, particularly the parameters D_{60} and D_{90} (large fraction) as well as the coefficient of uniformity, C_u, and the void ratio (e).

3.1.2 Paste backfill percolation rate

Data of the paste backfill drainage (see §3.1.1) were converted into percolation rates (or drainage velocity), v (cm/s) using Equation 1, and plotted on Figure 6 as a function of time elapsed from the beginning of drainage. As it could be anticipated from the cumulative volume of drained water (Fig. 3), the drainage velocity (or the percolation rate) decreases gradually with time. Under certain conditions, the drainage velocity may be directly related to the coefficient of permeability or hydraulic conductivity k (cm/s). In fact, these two parameters are equivalent if the hydraulic gradient (H/L) used when measuring the percolation rate is 1, when the tortuosity of the porous medium formed by the paste backfill is neglected (Thomas 1966; Hassani and Archibald 1998). These

observations are of greater concern for hydraulic backfill.

The percolationn data of backfill samples M1 (present study) ad M2 (Fig. 6) was best fitted by the regression curves given by the following equation:

$$v = 10^5 \times \left(v_0 + a \times t\right)^{-\frac{1}{b}} \qquad (3)$$

where t (min) = time elapsed from the beginning of drainage; v_0 (cm/s) = percolation rate at the drainage start; a and b = material constants. From the regression curves, v_0 = 0.014737064 cm/s, a = 0.00022823277 and b = 1.3143908 with R = 0.999 for the backfill sample M1 and v_0 = 0.065354171 cm/s, a = 0.00187135 and b = 1.1711637 with R = 0.999 for the backfill sample M2.

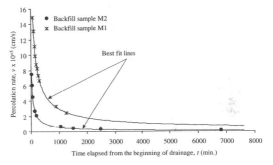

Figure 6. Drainage velocity vs time elapsed.

Table 2. Geochemical data of collected water from drained paste backfill.

Time elapsed (min)	pH	Eh	Electric coductivity (μmohs)
60	12.07	181.35	8910
90	12.06	188.95	8110
120	12.06	140.30	9645
150	12.02	127.40	9545
180	11.99	123.15	9800

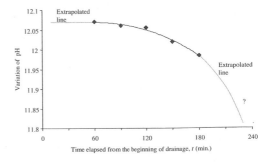

Figure 7. Variation in pH of collected water with time.

3.2 Chemistry and geochemistry of collected water

Water drained from the mould was collected for pH, oxydo-reduction potential (Eh) and electric conductivity measurements using a selective electrode. Water was then acidified for chemical analysis by ICP ES (Inductively coupled plasma emission spectroscopy) in order to determine its main ionic elements: calcium (Ca), sulfur (S) and sulfate (SO_4^{--}). Calcium is mainly related to the presence of a binder agent such as Portland cement. The presence of Ca in the collected water can indicate a loss of cement during backfill drainage and in which proportion. Another good indicator of the presence of cement in the collected water is the pH. If the pH is neutral to acidic ($5 \leq pH \leq 7$), the collected water does not contain cement. A basic pH ($pH \geq 12$) indicates the presence of cement in collected water. The oxydo-reduction potential can also indicate Ca presence in water.

Table 2 contains the results of the geochemical analysis of collected water from the drained paste backfill samples after a period of 3 hrs. These results show that a basic pH for this time period confirm the presence of Ca (and thus of cement) in the collected water (Fig. 7). This indicates that the drainage leads to the loss of a certain amount of cement.

ICP ES results indicate that the drained water contained 837 ppm of Ca, 13870 ppm of SO_4^{--} and 575 ppm of Na. The total volume of drained water was of 0.14748 L, so this quantity contains 123.44 mg of Ca, 2045.55 mg of SO_4^{-} and 84.80 mg of Na. Knowing that each mould contained approximately 208 g of cements (Types 10 and 50 Portland cements) and that 100 g of anhydride binder agent contains 50 g of Ca, then each paste backfill mould contains 104 g of Ca. A small calculation shows that there is a loss of 0.12 % of cement from the paste backfill.

3.3 Geotechnical properties of loaded paste backfill

Tables 3a and 3b contain the physical and geotechnical parameters obtained from the drained (D-test) and undrained (U-test) paste backfill samples after a curing period of 28 days. These parameters are: the moisture content ($w\%$) and volumetric water content (θ), the solid mass percentage ($C_w\%$) and water percentage ($E\%$), the density (ρ) and dry density (ρ_d), solid particles density (ρ_s), the unit weight (γ) and dry unit weight (γ_d), the degree of saturation ($S_r\%$), the porosity (n) and the voids ratio (e).

These parameters indicate a slight variation due to both water leaching and the load application. It is also evident that the voids ratio (e) of the drained control samples is lower than the one of the undrained control samples. These results also show that e slightly decreases with load increases in the case of the drained paste backfill (D-test) and slightly increases with the load increases for the undrained samples (U-test). In this last case, it ap-

Table 3a. Physical and geotechnical parameters of drained and undrained paste backfill.

	Moisture content w (%)	Volumetric water cont. θ	Solid percentage C_w (%)	Water proportion E (%)	Density ρ (g/cm^3)	Dry density ρ_d (g/cm^3)
D-test (control)	18.9	0.40	84	16	2.55	2.14
D-1@10 lbs	20.6	0.41	83	17	2.40	1.99
D-2@20 lbs	19.8	0.40	83	17	2.40	2.00
D-3@50 lbs	20.1	0.40	83	17	2.42	2.01
U-test (control)	21.5	0.43	82	18	2.44	2.01
U-1@10 lbs	21.1	0.42	83	17	2.40	1.98
U-2@20 lbs	21.5	0.42	82	18	2.39	1.96
U-3@50 lbs	21.3	0.41	82	18	2.36	1.95

Table 3b. Physical and geotechnical parameters of paste backfill (continued).

	Saturated unit weight γ (kN/m^3)	Dry unit weight γ_d (kN/m^3)	Degree of saturation S_r (%)	Porosity n	Voids ratio e	Solids density ρ_s (g/cm^3)
D-test (control)	25.0	21.0	99	0.41	0.69	3.62
D-1@10 lbs	23.6	19.6	92	0.45	0.81	3.61
D-2@20 lbs	23.5	19.6	89	0.44	0.80	3.60
D-3@50 lbs	23.7	19.7	91	0.44	0.79	3.61
U-test (control)	24.0	19.7	98	0.44	0.79	3.61
U-1@10 lbs	23.5	19.4	93	0.45	0.82	3.60
U-2@20 lbs	23.4	19.3	92	0.46	0.84	3.62
U-3@50 lbs	23.2	19.1	90	0.46	0.85	3.60

pears that the load increased the pore water pressure within the backfill which prevented its skeleton from settling. Instead, this pore water pressure favored a gradual relaxation of its physical structure which allowed an increase in the amount of empty pore. This can also be seen by the lower unit weight value of the undrained backfill compared to the one of the drained backfill sample.

3.4 Mechanical behavior of loaded paste backfill

3.4.1 Compressive strength
Applied loads on the backfill and the backfill column height simulated by the addition of loads are presented in Table 1 in terms of mass (lb), weight (N and kN) as well as the corresponding applied pressure (kPa). The simulated height h corresponds to the ratio of the applied weight to the unit weight γ of the paste backfill which equals to 24 kN/m^3.

Results of uniaxial compression tests are shown in Table 4 for both drained and undrained paste backfill samples, loaded and unloaded control samples. Figure 8 shows the variation in paste backfill compressive strength as a function of applied load.

The first major observation is that compressive strength of the drained samples is always higher than the one of the undrained samples. It can also be observed that the UCS of drained backfill samples increase with increasing load while the UCS of undrained backfill samples decreases with increasing load (Fig. 8).

Table 4. Geochemical data of collected water from drained paste backfill.

Load mass (lbs)	Load pressure (kN)	UCS drained sample (kPa)	UCS undrained sample (kPa)
0	0	1 141	735
10	5.5	1 000	845
20	11.1	1 048	835
50	27.6	1 097	763

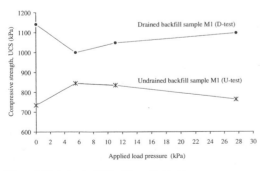

Figure 8. Variation in UCS with applied load pressure.

Load applications on the drained backfill samples lead to a more rapid drainage which favors the formation of hydrates caused by the saturation of hydration reaction, thus increasing mechanical strength. For the undrainded paste backfill, load applications increase pore water pressure which can cause the break of cement bonds, thus favoring a reduction in mechanical strength. This reduction can also be due to the inhibition of the hydration reaction by the surplus of water in the backfill.

3.4.2 Stress-strain behavior of loaded paste backfill
Figures 9 and 10 present the stress-strain curves of the drained and undrained paste backfill samples, respectively. These figures show that when cured under loading conditions, this type of paste backfill exhibits an elastoplastic behavior, whether drained or not. These figures also show clearly the stress-strain curve of the control drained backfill sample is above the loaded samples curves. For each type of sample, these curves show that the loads do not have much influence on the deformability of the backfill (linear ascending part of the curves).

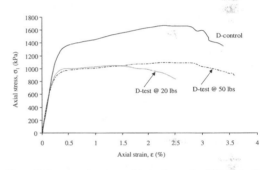

Figure 9. Stress-strain curves of drained paste backfill (D-test).

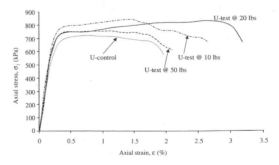

Figure 10. Stress-strain curves of undrained backfill (U-test).

3.5 Microstructural analysis of loaded paste backfill

Table 5 presents the values of specific surface area (S_s) of drained and undrained paste backfill samples.

145

Table 5. Geochemical data of collected water from drained paste backfill.

Load mass (lbs)	Load pressure (kN)	S_s (m²/kg) drained sample (kPa)	S_s (m²/kg) undrained sample (kPa)
0	0	6547.2	6952.6
10	5.5	6668.0	7672.6
20	11.1	7076.5	7274.1
50	27.6	6416.9	6959.3

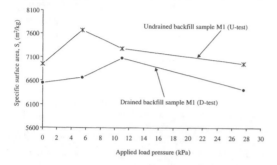

Figure 11. Specific surface area as a function of applied load pressure.

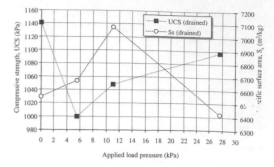

Figure 12. Variation in UCS and S_s with applied pressure for drained samples.

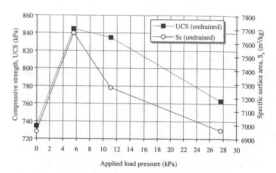

Figure 13. Variation in UCS and S_s with applied pressure for undrained samples.

This parameter provides indirect information about the microstructure of paste backfill during the course of its curing. Indeed, by describing the degree of particle fineness, this parameter takes into account the binder agent hydrates formed. Consequently, the specific surface area can be related to the compressive strength of paste backfill. However, when comparing the specific surface area values of both drained and undrained samples, there is a difference in favor of the undrained backfill samples.

Figure 11 is a graphical representation of data in Table 5. One can note that contrary to the mechanical strength (Figure 8), the specific surface area of drained backfill is always lower than the one for undrained backfill. For the undrained backfill, S_s increases from the unload state (control) up until the application of the first 10 lbs, 5 days after moulding of the samples. The specific surface area then decreases after the application of the 20 lbs and the 50 lbs loads. For the drained backfill, however, S_s increases slightly from the unload state to the point when the 10 lbs load was added and continuing to do so with the application of the 20 lbs load before decreasing under the application of the 50 lbs load.

4 DISCUSSION

Figures 12 and 13 show the variation in compressive strength of paste backfill and S_s as a function of applied load pressure for both drained and undrained conditions, respectively.

Previous work on undrained paste backfill (Benzaazoua et al. 1999) has provided evidence for the existence of a proportionality relationship between the specific surface area, S_s, and the binder agent proportion used in the paste backfill and consequently, its UCS value.

From the curves of Figure 12, there does not seem to exist a proportionality relationship between S_s and compressive strength for the drained backfill samples. As a matter of fact, the highest UCS value is not associated to the highest S_s. From the drained water analysis, we know that there was not any loss of solid anhydrous cement or tailings particles. Moreover, assuming that drainage diminished the amount of pore water in the backfill, saturation of the hydration reaction of the binder agent should be favored, thus precipitating a significant amount of hydrates. These hydrates should generate strong cement bonds within the backfill and consequently, strong compressive strength (UCS). However, on Figure 12, the UCS varies in opposite direction of S_s. This has yet to be understood.

In contrast, the curves for the undrained backfill on Figure 13 suggest that the S_s is proportional to the compressive strength. Indeed, the greatest value of the specific surface area is associated to the highest value of compressive strength.

5 CONCLUSION

This preliminary study has demonstrated that load applications have a significant influence on the strength development within drained and undrained paste backfill samples. Loading leads to an increase in mechanical strength when the paste backfill is drained, but a reduction of UCS when the paste backfill sample is undrained.

In the case of drained paste backfill samples, the chemistry of collected water showed a loss of a small proportion of anhydride cement (0.12 %). The increase in mechanical strength is probably due to both the settlement of particles and the massive formation of hydrates. In the case of undrained backfill samples, the reduction in mechanical strength due to load application may be explained by the development of pore water pressure within the paste backfill. Microstructure analysis via S_s seems to confirm these results, but further research is needed.

6 ACKNOWLEDGMENTS

This research was supported by the Fond de l'Université du Quebec en Abitibi-Témiscamingue (FUQAT) under grant N3-2052397. The authors gratefully acknowledge their support. The authors would also like to thank our mining partner, Agnico Eagle (Laronde Division) and Aur-Novicourt-Teck (Louvicourt Mine) for their collaboration in the completion of this work. Also, thanks to Hugues Bordeleau, technician at URSTM, for performing the experiments.

REFERENCES

Barsotti, C. 1978. The evolution of fill mining at the Ontario Division of Inco Metals. *In* Proceedings of the 12th Canadian Rock Mechanics symposium, Mining with Backfill, CIM special volume **19**: 37–41.

Belem T., Bussière B. & Benzaazoua M. (2001). The effect of microstructural evolution on the physical properties of paste backfill. *In* Proceedings of Tailings and Mine Waste'01, January 16-19, Fort Collins, Colorado, A.A. Balkema, Rotterdam, pp. 365–374.

Belem T., Benzaazoua M. & Bussière B. (2000). Mechanical behaviour of cemented paste backfill. *In* Proceedings of the Canadian Geotechnical Society Conference "*Geotechnical Engineering at the dawn of the third millennium*", 15-18 oct. 2000, Montréal, **1**: 373–380. ISBN 0-920505-15-5.

Benzaazoua, M., and Belem, T. 2000. Optimization of sulfide-rich paste backfill mixtures for increasing long-term strength and stability. *In* Proceedings of the 5th Conference on Clean Technology for Mining Industry. *Edited by* M. Sanchez, S. Castro and F. Vergara, Santiago, Chile, **II**: 947–957.

Benzaazoua M., Belem T. and Jolette D. 1999. Investigation de la stabilité chimique et de son impact sur la qualité des remblais miniers cimentés. Rapport d'activité de recherche IRSST, 152p. + 100p. annexes.

Benzaazoua, M., Ouellet, J., Servant, S., Newman, P., and Verburg, R. 1999. Cementitious backfill with high sulfur content: Physical, chemical and mineralogical characterization. Cement and Concrete Research, **29**: 719–725.

Bernier, R.L., Li, M.G., and Moerman, A. 1999. Effects of tailings and binder geochemistry on the physical strength of paste backfill. Sudburry'99, Mining and the environment II. *Edited by* N. Goldsack, P. Belzile, Yearwood and G. Hall. **3**: 1113–1122.

Bodi, L., Hunt G. and Lahnalampi T. 1996. Development and shear strength parameters of paste backfill. *In* Proceedings of the 3rd International Conference on Tailings and Mine Waste'96, Fort Collins, Colorado, 16-19 January. A.A. Balkema, Rotterdam, pp. 169–178.

BNQ 2622-911. Confection et mûrissement en laboratoire d'éprouvettes de béton destinées aux essais de compression et de flexion. Norme d'essai, 1973.

Das B.M. 1983. Advanced in soil mechanics. McGraw Hill Ltd., New York.

Hassani, F., and Archibald, J. 1998. *Mine backfill, CD-ROM* 263p.

Hassani, F.P. and Bois, D. 1992. Economic and technical feasibility for backfill design in Quebec underground mines. Final report 1/2, Canada-Quebec Mineral Development Agreement, Research and Development in Quebec Mines. Contract No. EADM 1989-1992, File No. 71226002.

Lawrence, C.D. 1992. The influence of binder type on sulfate resistance. Cement and Concrete Research, **22**: 1047–1058.

Lamos, A.W. and Clark, I.H. 1989. The influence of material composition and sample geometry on the strenght of cemented backfill. Innovation in Mining Backfill Technology. *Edited by* Hassani et al. eds, A.A. Balkema, Rotterdam, pp. 89–94.

Landriault, D.A. 1992. Paste fill at Inco. In Proceeding of the 5th International Symposium on Mining with backfill, Johannesburg, South Africa, September.

Landriault, D.A. 1995. Paste backfill mix design for Canadian underground hard rock mining. In *Proceedings of the 97th Annual General Meeting of the C.I.M. Rock Mechanics and Strata Control Session,* Halifax, Nova Scotia, May 14-18.

Landriault, D.A. and Lidkea, W. 1993. Paste fill and high density slurry fill. In Proceedings of the International Congress on Mine Design, Queens University, Kingston, Ontario, Canada, August.

Landriault, D.A. and Tenbergen, R. 1995. The present state of paste fill in Canadian underground mining. In Proceedings of the 97th Annual Meeting of the CIM Rock Mechanics and Stata Control Session, Halifax, Nova Scotia, May 14-18.

Landriault D.A., Verburg, R., Cincilla, W. and Welch, D. 1997. Paste technology for underground backfill and surface tailings disposal applications. *Short course notes,* Canadian Institute of Mining and Metallurgy, Technical workshop – april 27, 1997, Vancouver, British Columbia, Canada, 120p.

Mitchell, R.J. 1989a. Stability of cemented tailings mine backfill. *In* Proceedings of Computer and physical modelling in geotechnical engineering. *Edited by* Balasubramaniam et al., A.A. Balkema, Rotterdam, pp. 501–507.

Mitchell 1989b. Model studies on the stability of confined fills. *Canadian Geotechnical Journal*, **26**: 210–216.

Naylor, J., Farmery, R.A. and Tenbergen, R.A. 1997. Paste backfill at the Macassa mine with flash paste production in a paste production and storage mechanism. In Proceedings of the 29th annual meeting of the Canadian Mineral Processors (division of the CIM), Ottawa, Ontario, 21-23 january, pp. 408–420.

Ouellet, J., Benzaazoua, M., and Servant, S. 1998. Mechanical, mineralogical and chemical characterisation of paste backfill. *In* Proceedings of the 4th International Conference on Tailings and Mine Waste, A.A. Balkema, Rotterdam, pp. 139–146.

Ouellet, J., Bidwell, T.J., and Servant, S. 1998. Physical and mechanical characterisation of paste backfill by laboratory and *in situ* testing. *In* Proceeding in the 6th International Symposium on Mining with Backfill, MINEFILL'98. *Edited by* M. Bloss (Austr. Inst. Min. Metal.), Brisbane, Australia, pp. 249–l253.

Petrolito, J., Anderson, R.M., and Pigdon, S.P. 1998. The strength of backfills stabilised with calcined gypsum. *in* Minefill'98. Proceedings of the 6th International Symposium on Mining with Backfill. *Edited by* M. Bloss (Austr. Inst. Min. Metal.), Brisbane, Australia, pp. 83–85.

Stone, D.M.R. 1993. The optimization of mix designs for cemented rockfill. *In* Minefill'93. Proceedings of 5th International Symposium on Mining with Backfill. Johannesbourg, SAIMM, pp. 249–253.

Strömberg, B. 1997. Weathering kinetics of sulphidic mining waste: an assessment of geochemical process in the Aitik Mining Waste Rock Deposits. AFR-Report 159, Department of Chemistry, Inorganic Chemistry, Royal Institue of Technology, Stockholm, Sweden.

Thomas, E.G. 1966. The important properties of hydraulic fill, with particular reference to mechanized cut and fill mining operations. *Proc. Australian Inst. Min. Metall.* 220:1–20.

Viles, R.F., Davis, R.T.H. and Boily, M.S. 1989. New material technologies applied in mining with backfill. *In* Innovation in Mining Backfill Technology. *Edited by* F. Hassani et al., A.A. Balkema, Rotterdam, p p. 95–101.

Weaver, W.S., and Luka, R. 1970. Laboratory studies of cement-stabilized mine tailings. Canadian Mining and Metallurgical Bulletin, **64**(701) 988–1001.

Use of copper mine tailings as paste backfill material in mining operations – Approach to minimise land occupation?

S. Moellerherm & P.N. Martens

Institute of Mining Engineering I, Aachen University of Technology, Germany

ABSTRACT: In modern society copper is widely recognised as an important raw material. In 2000 total copper production amounted to 13,2 Mio t and it is expected to increase in the next years. On the other hand copper ore mining has severe impacts on the environment like acid mine drainage or land use due to tailings disposal. A possible solution to minimise these effects could be the application of mining methods with paste backfill. In a project funded by the German Federal Institute of Geosciences and Natural Resources the Institute of Mining Engineering at the Aachen University of Technology, Germany, modelled a conversion of ore production from mining methods without backfilling either underground or open pit to a mining method with backfill and analysed the impact on copper market and world market price.

The paper gives a short review on copper mining. Then a description of the paste backfill plant model and the resultant backfill mining method will be given. A cost estimation for the solution and the different influences on the environment are discussed with regard to the effect on the world copper price and world demand.

1 INTRODUCTION

Copper is mainly used in Germany in electro technical and electronical consumer products and in the construction industry. Total world copper production amounted to 13.2 Mt in 2000. The World Bank estimates an annual demand increase in primary copper of 2,2% despite growing recycling rates. It is therefore an essential metal for society's economical development. Copper mining is in many developing countries like Chile, Peru, Indonesia or Papua New Guinea a major contributor to their national GDP. Figure 1 shows the global distribution of underground and open pit copper ore production in 1998.

In Asia, North- and South America open pit excavation is the favourite mining method whereas underground mining dominates in Europe and in Australia. As mining undoubtedly influences the natural environment future challenges of copper ore mining are minimization of land use in area and duration, proper land rehabilitation as well as tailings management. One possible solution to avoid these negative impacts could be the application of underground mining with backfilling tailings compounds. This approach would definitely increase the production cost. For this background the paper describes the influence of such an "ecologically satisfying" copper ore mining on world market prize and demand.

2 STATE OF THE ART OF COPPER MINING

In the project 261 copper ore mines accounting for 98% of world copper production have been examined. In 1998 total ore production amounted to 2.066 billion t, 81% from open pit and 19% from underground mining without in-situ leaching operations. In the open pit mines 2.5 billion t of overburden were hauled. Table 1 presents the global distribution of overburden and ore production.

A review of applied underground mining methods reveals the following results (Figure 2).

Dominating mining methods are block caving and room and pillar mining due to low costs. However, in some operations backfilling is still applied like in Louvicourt, Myra Falls, Kidd Creek (all in Canada) or in Neves Corvo, Portugal.

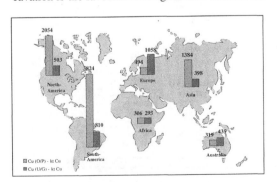

Figure 1. Distribution of copper ore production.

Table 1. Overburden and ore production

continent	overburden (O/P) Mt	ore (O/P) Mt	ore (U/G) Mt
North-America	646	664	57
South-America	1052	587	97
Europe	178	92	93
Africa	81	62	55
Australia	156	61	24
Asia	422	200	74

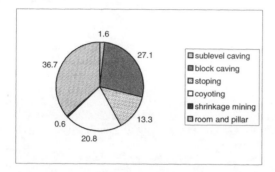

Figure 2. Percentage share of underground mining methods.

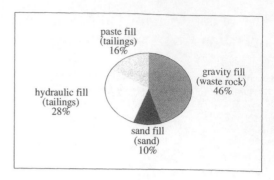

Figure 4. Share of backfill techniques in copper mines.

Figure 4 shows the share of the different backfill techniques.

Almost 50% of the mines use gravity backfill because of the easy and sufficient availability of overburden and waste rock due to a combined open pit underground operation. The other reason is the use of additives like smelter slag (e.g. Norilsk in Russia).

Sand as an additive is used in cases where this compound material can be obtained with low cost.

Figure 5 presents the average backfill mixture applied in copper ore mining.

Table 2 contains the significant parameters of the three backfill techniques.

As table 2 indicates paste backfill has two main advantages: 90% of the cavity will be filled by the material and it contains tailings from the processing plant. This could decrease the land demand for tailings disposal. Furthermore it is at the moment theBest Available Technology. Therefore it was decided to use a paste backfill plant for modelling an "ecologically satisfying" operation.

3 MODELLING THE "ECOLOGICALLY SATISFYING" COPPER MINING

3.1 Backfill plant model

One major aspect of the project was the design of a backfill operation meeting the requirements of an "ecologically satisfying" mining like less land need and use of tailings for backfill. The sketch summarizes present backfill techniques (Figure 3). The drawing bases on information obtained by literature review, mine visits, statistics and company reports.

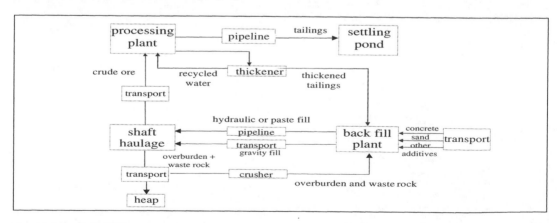

Figure 3. Scheme of backfill process.

150

Table 2. Parameters of backfill techniques

parameter	gravity fill	hydraulic fill	paste fill
solids content	75-90%	50-75%	65-85%
water content	7-11%	25-35%	12-25%
concrete content	3-12%	3-15%	2-5%
compressive strength	1-20 MPa	0,5-3,0 MPa	1-5 MPa
density	1,8-2,5 t/m3	1,4-2,3 t/m³	2,1-2,35 t/m³
fill grade (cavity utilisation)	~ 80%	~85% (tailings backfill) ~60% sand backfill)	~90%
solid	overburden, waste rock (coarse grained)	fine- and coarse grained tailings, sand	fine- and silt grained tailings
examples	Kidd Creek, Norilsk	Neves Corvo, Myra Falls	Louvicourt, Brunswick

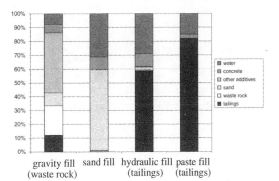

Figure 5. Average backfill mixture.

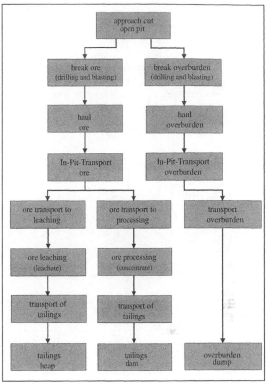

Figure 6. Process steps in copper open pit.

3.2 Method of cost estimation

The cost estimation is an important aspect to determine the effect of the "ecologically satisfying" copper ore mining on world copper prize because production cost would rise to a certain extent. It is aimed to obtain a representative cost assessment for three scenarios and then transfer the results on a global scale.

The cost estimation itself bases on the US-Bureau of Mines Cost Estimating Handbook. The handbook splits operating cost into labour cost, supply cost (energy, explosives, bits and miscellaneous) and material cost (spare parts, diesel oil, lubricants, tyres and miscellaneous). Based on a process chain analyses for open pit and underground mines (Figure 6 and Figure 7) the cost are estimated for each process step.

Because the Cost Estimating Handbook only allows to assess projects in the USA for the year 1984 it is necessary to introduce specific correction factors:

- Correction factor for investment goods

  Different cost indices have been used as the Chemical Engineering Plant Cost Index for thickeners and filters, the Marshall & Swift Equipment Cost Index and the Construction Cost Index for shaft sinking.
- Correction factor for supply goods and material
  For the cost correction of energy, diesel, lubricant and blasting devices the index "Fuel and Energy" of the World Bank is an appropriate mean.
- Correction factor for productivity and labour
  The handbook contains for each process step the average labour cost. The cost can be transferred to various countries with indices provided by the World Bank. Productivity is another interesting aspect. Because the productivity varies in different countries and the different mining methods a productivity correction factor was created and applied in different geographical regions.

The correction factor for capital investment is the last factor applied in the project. With all these factors and information one can assess the production cost of the "ecologically satisfying" copper mining. In the next step three scenarios have been conducted.

151

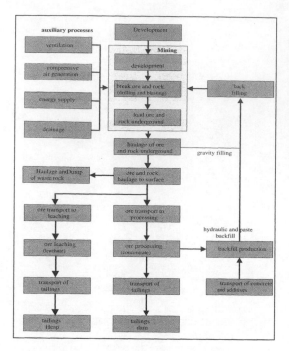

Figure 7. Process steps in underground copper mining.

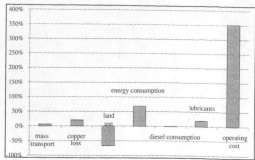

Figure 8. Effect of the conversion on ecological parameters and cost for scenario A.

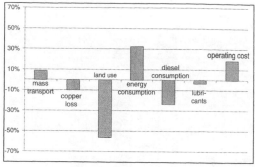

Figure 9. Effect of the conversion on ecological parameters and cost for scenario B.

4 SCENARIOS

4.1 Scenario A

In scenario A three cases have been analysed:

- Mining method without backfilling to mining method with tailings paste fill
- Mining method with waste rock gravity fill and composites to tailings paste fill
- Mining method with sand fill to tailings paste fill.

This scenario comprises 107 mine operations representing 11% of total copper ore production. Figure 8 summarizes the results:

The increase in mass transport relates mainly to the concrete transport. The mining losses will decrease due to a better ore recovery. New land requirement will significantly decrease because less land will be needed for tailings disposal. Operating cost will in average increase up to 0,13 $/t Cu.

4.2 Scenario B

Scenario B includes 11 mines with block caving representing 4% of copper ore production. A conversion will lead to the following results (Figure 9).

Land use will decrease to 67% and operating cost will significantly increase from 7,5 $/t ore to 34 $/t ore.

4.3 Scenario C

This scenario can be transferred to 95 operations accounting for 78% of world ore production. It contains the conversion from open pit to underground mining. Figure 10 demonstrates the effects.

Mass transport and land use will both be minimized due to less overburden transport and dump. On the other hand ore loss, energy consumption and cost will increase.

5 DISCUSSION

The effect of the conversion on world market prize show the following two figures. The LME market prize in 1998 was in average 0,75 $/lb Cu (Figure 11 and Figure 12).

After the transition of all underground mines to paste backfill total copper production would drop down from 9 Mio t to 8 Mio t.

In case all mines – open pit and underground – would use paste backfill the production would fall down to less than 3 Mio t. If the copper demand would stay stable the prize would rise to 2,5 $/lb Cu.

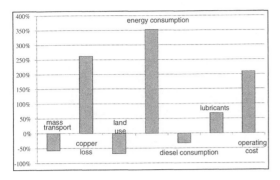

Figure 10. Effect of the conversion on ecological parameters and cost for scenario C.

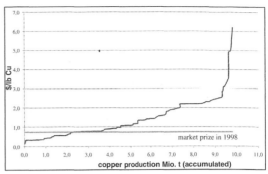

Figure 12. Effect on the market after the conversion of all mines.

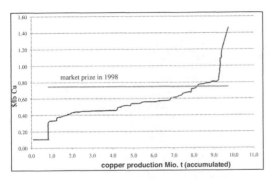

Figure 11. Effect on the market after underground mines to paste backfill operation.

6 CONCLUSION

The application of backfilling could contribute to a minimization of land use on the other hand energy consumption and operating cost will rise. The change and increase in energy would obviously lead to other ecological impacts like increase of Greenhouse Gas Emissions. These effects on copper prize have not been analysed in the project nor the possibility to substitute copper by other materials. Effects of the conversion on the local and national economy and society have not been investigated, either. Furthermore a detailed analyses at the locations could not been conducted and wasn't intended, either. In many cases a change-over from open pit to underground would not be applicable instead the mine would be closed.

7 ACKNOWLEDGEMENT

This project was conducted with a financial grant by the German Federal Institute of Geosciences and Natural Resources.

Tailings and Mine Waste '02, © 2002 Swets & Zeitlinger, ISBN 90 5809 353 0

Pumping paste with a modified centrifugal pump

A. Sellgren
Lulea University of Technology, Sweden

L. Whitlock
GIW Industries Inc., U.S.A.

ABSTRACT: There is today an interest to apply paste technology for surface disposal in the base metal, gold mining, and oil sand sectors. Pipeline loop experiments at the GIW Hydraulic Testing Laboratory with an open shrouded and augered centrifugal pump (impeller diameter 0.3 m) showed that a simulated paste product could be pumped at yield stresses well over 200 Pa. The feasibility of different pumping systems is briefly discussed.

1 INTRODUCTION

There is an increased interest throughout the world in the treatment and processing of tailings from mineral beneficiation plants in order to reduce the volumes and sizes of tailing dams. The reasons are generally a combination of environmental and economical considerations.

The economical feasibility of a high degree of thickening is determined by the longterm impact in the disposal area. The local conditions in the disposal area strongly determine the feasibility and economical effectiveness of different degrees of thickening.

The concept of Thickened Tailings Disposal, TTD, means that the tailings are thickened more deeply at the concentrator, conveyed to the disposal area by pipeline, and discharged from an elevated or man-made position (Robinsky,2000). The principal aims are to reduce the need for large dams and settlement ponds and to better utilize available space. The thickened tailings, considered here, might be self-supporting enough to attain, for example, a 1 and 5° slope when discharged at solids concentrations by mass of about 50 and 60%, respectively.

With the higher angle more volume can be filled per unit surface area for a constant dam height. With AMD-generating tailings multi-layer capping may be needed. The reduced cost for dams, capping, and decommission can be considerable over the life of the mine. A smaller area also means less evaporation, which is important in areas of water shortage. Less water in the disposal area also concerns dam safety . Some reported tailings dam failures and incidences may be related to large amounts of water in the disposal area.

The typical angles of up to 5° in the normal TTD-concept can be increased to 10° or more when using paste, (Landriault, 2001).

A tailings mixture that is produced and handled with an extremely high solids concentration is often called paste. The term paste has mainly been associated with backfilling in underground mines with underground transportation in pipelines and boreholes. A small amount of water means less use of binder in order to obtain high strength. The amount of particles smaller than 20 µm normally exceeds 15 % an-with the average particle size may varying from 20 to 100 µm. The solids concentration by volume must normally exceed 50 % in order to obtain the properties that give virtually no drainage of water or segregation of particles.

There is currently an interest to apply paste technology for surface disposal in the base metal, gold mining, and oil sand sectors. In 2001, several plants have been commissioned for paste disposal of tailings on the surface or a combination of surface disposal and underground backfill production (Landriault, 2001).

In order to get the high solids concentrations for paste conventional thickeners are combined with mechanical (filter) dewatering or deep tank thickeners. The transportation of the highly thickened slurry is normally carried out with a positive displacement type of pumps.

Centrifugal slurry pumps can be used for solids concentrations by volume of up to 50 % when pumping sands, (Addie et al.1993, Berg et al. 1991).

The thickening of tailings to high solids concentrations often means that there is an intermediate area where both centrifugal and positive displacement pumps are feasible. For example, centrifugal pumps are used to feed positive displacement pumps in pipeline transportation systems for highly concentrated tailings slurries. Centrifugal pumps can be characterised as being cost-effective for large flow rates and low to moderate working pressures; positive displacement pumps are generally cost-effective for small flow rates and high pressures.

The objective here is to present experimental results with a modified centrifugal pump when pumping a simulated paste product.

2 CHARACTERISATION AND EXPERIMENTS

Pastes and thickened tailing slurries behave in a highly non-Newtonian way and exhibit often a yield stress, meaning that they behave as a solid until sufficient force is applied. The yield stress is defined as the minimum stress required causing the solid-liquid mixture to flow.

The experimental work was carried out at the Hydraulic Testing Laboratory, GIW Industries Inc., U.S.A. Here, slurry pipeline hydraulics and pump performances can be investigated in loops with pipe diameters of up to 0.5 m and pipeline lengths of up to 200 m. The experiments in this case were carried out in a pipeline-loop system with pipe diameter of 0.075 m.

A phosphate clay slurry with a slurry S.G. of 1.11 was used to simulate paste. Adding fine sand with an average size of 135 microns increased the consistency. The rheological properties and the pumping characteristics of the slurries were determined from differential pressure drop and flow rate measurements.

The pump was an open shrouded and augered GIW LCC-type (3-vane, all-metal) centrifugal pump with an impeller diameter of 0.3 m, see Figure 1.

The slump test using a standard slump cone is used in the concrete industry to measure the flow properties of "workable" fresh concrete. This method has been adopted in the mining industry to determine the optimum properties of pastes. The inexpensive, reliable, and simple test method has been modified for pastefill mix designs, see for example (Kuganathan, 2001).

Recent techniques with high solids concentrations for the disposal of tailings suspension are critically dependent on the yield stress. The slump test as a means of directly measuring the shear stress with a cylinder (piece of pipe) was presented by Pashias and Boger (1996).

Slump measurement with a cylinder of a height and diameter of 0.1 m each was carried out here for the highest solids concentration investigated, see Figure 2.

The yield stress corresponding to the slump of about 0.025 m in Figure 2 is 350 to 400 Pa, following the evaluation procedure given by Pashias and Boger (1996).

Figure 1. Open shrouded and augered centrifugal pump (impeller diameter = 0.3 m).

Figure 2. Slump test adapted in a 0.1 m-standing pipe for measurement of the yield stress. Slurry S.G. = 1.67.

The yield stress values in this range are suitable for paste surface disposal of, for example, gold tailings (Boger,2000). The solids concentration by mass for the gold tailings at that time was about 75 %. Suitable properties were also reported for a similar product (Tennbergen, 2000) for a yield stress of about 200 Pa based on slump and pipeline pressure requirement data.

In summary, the clay-sand mixture with a slurry specific gravity of up to 1.67 (Figure 2) may well simulate the yield property of a tailings paste.

3 RESULTS

The rheological properties of the clay-sand mixtures were determined from the pipeline experiments. Calculated shear stress versus the rheological scaling parameter 8V/D (V= velocity, D=diameter of pipe) are shown in for only phosphate and a clay-sand mixture with S.G.= 1.60 Figure 3.

Sand was then added at 8V/D=400 up to a mixture S.G. of 1.67-corresponding to a shear stress value of about 360 Pa. The slope of the curve for

S.G.=1.62 is about 0.2 in Figure 3. Rheograms were then constructed for S.G.-values of 1.60 and 1.67, following the procedures described in, for example Wilson et al. (1997). It was assumed that the slope for S.G.=1.67 remained the same (0.2) as for 1.60. The rheograms with the results represented by a Bingham model are shown in Figure 4.

When pumping slurries with centrifugal pumps the relative reduction of the water head and efficiency for a constant flow rate and rotary speed may be defined by the ratios and factors shown in Figure 5. Figure 6 shows how the pump head and efficiency were lowered by the highly viscous clay-sand mixture.

When the pump in Figure 6 was operating in the best efficiency region (0.015-0.020 m^3/s), the reductions in head and efficiency were about 10 and 15 %, respectively. However, the performance became very sensitive to small variations in the mixture S.G. for lower flow rates. It can be seen in the figure that the pump can produce head fairly well for S.G-values of 1.62 to 1.63. However, it cannot maintain the head at a S.G. of 1.67, when an unstable head curve is created.

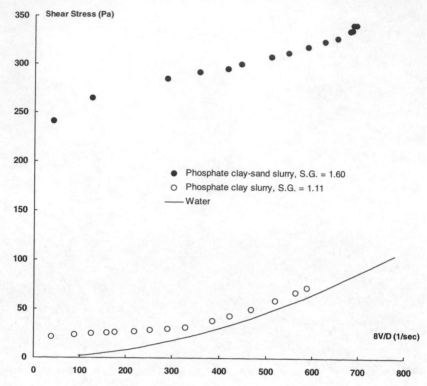

Figure 3. Shear stress versus the rheological scaling parameter.

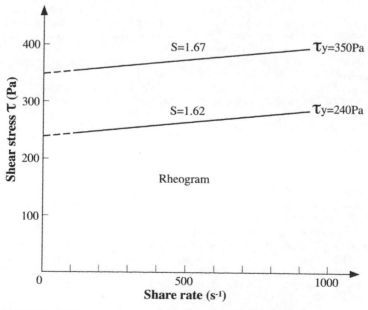

Figure 4. Rheogram with estimated yield stresses. S denotes the specific gravity of the mixture.

4 DISCUSSION AND CONCLUSIONS

The pipeline friction losses are very sensitive to changes in the slurry S.G. for this type of slurries flowing laminarily. For example, the increase in S.G. from 1.62 to 1.67 (3 %) in Figure 5 corresponds to an increase in pump head or pressure requirement of about 40 % in a horizontal pipeline.

The sensitivity of variations in the consistency of the mixture means that the mixture S.G. must be effectively controlled when using a centrifugal pump. Furthermore, the results in Figure 6 also show that there is a limit in S.G. for which the operation should be within the best efficiency region in order to avoid complex situations with an unstable head curve.

Thickening to very high solids concentrations requires an effective control of slurry densities and flow rates which is needed with centrifugal pumps. Using positive displacement pumps instead of centrifugal pumps generally implies costs that are ten times greater (Cowper, 1999).

Experiences from test at the GIW Hydraulic Laboratory with regular closed impeller slurry pumps has sometimes indicated similar instability problems, Figure 6, for slurries with yield stresses of about 100 Pa, (Sellgren et al. 1999a, Sellgren et al. 1999b). The open shroud and augered impeller seems to have been the determining factor for pumping in excess of 200 Pa without the problem of an unstable head curve for lower flow rates.

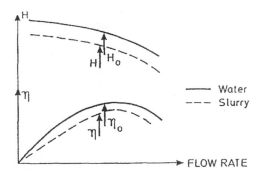

Head ratio: $HR = H/H_0$
Head reduction factor: $R_H = 1 - HR$
Efficiency ratio: $ER = \eta/\eta_0$
Efficiency reduction factor: $R_\eta = 1 - ER$

Figure 5. Sketch defining the reduction in head and efficiency of a centrifugal pump pumping solid-water mixture.

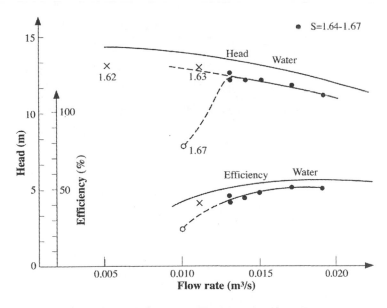

Figure 6. The effect of slurry on the pump head and efficiency at different slurry specific gravities.

159

REFERENCES

Addie, G. and Hammer J. 1993. Pipeline head loss of a settling slurry at concentrations up to 49 % by volume, *Hydrotransport 12, Brügge, Belgium, 28-30 September.*

Boger D.V. 2000. Yield stress measurements, Proceedings, *Transportations and Deposition of Thickened Tailings Seminar*, October 23, 24, University of Alberta, Alberta, Canada.

Berg C.H. van den, Vercruijsse P: and Broek M. Van den 1999. The hydraulic transport of highly concentrated sand-water mixtures using large pumps and pipeline diameters. *Hydrotransport 14, Maastricth, The Netherlands, 8-10 September.*

Cowper, T. 1999. Rheology, a key factor in pumping and Slope Disposal of high density mine tailings. Proceedings, *Rheology in the Mineral Industry II, HI*, U.S.A., March.

Kuganathan, K. 2001. A method to design optimum pastefill mixes through flow cone and mini slump cone testing. Proceedings, *Minefill 2001 Sept. 17-19*, Seattle, U.S.A.

Landriault, D., Welch, D. and Frostiak, J. 2001. Bulyanhulu Mine: Blended paste backfill and surface paste deposition. *The state of the art in paste technology, Proceedings, Minefill 2001, Sept. 17-19*, Seattle, U.S.A.

Pashias and Boger .1996. A fifty cent rheometer for yield stress measurements. *Journal of Rheology, 40(6).*

Robinsky, E. (1999): Thickened tailings disposal in the mining industry, Quebecor Printpak, Toronto.

Sellgren, A. and Addie, G. and Yu W.C. 1999a. Effects of non-Newtonian mineral suspensions on the performance of centrifugal pumps, *Journal of Mineral Processing and Extraction metallurgy Review, vol. 20, pp 239-249.*

Sellgren, A. and Addie, G.R. and Juzwiak, J.H. 1999b. Factors involved in the pumping on non-settling slurries with centrifugal pumps. Proceedings, *Rheology in the mineral industry II*, E.J. Wasp, Ed., Kahuku, Oahu, Hawaii, U.S.A.

Tenbergen, R.A. 2000. Paste dewatering techniques and paste plant circuit design. *Proceedings, Tailings and Mine Waste 00*, Balkema, Rotterdam.

Wilson, K.C., Addie, G.R., Sellgren. A. and Clift R. 1997. *Slurry transport using centrifugal pumps*, Blackie Academic and Professional, London, U.K.

Liners, covers, and barriers

Tailings and Mine Waste '02, © 2002 Swets & Zeitlinger, ISBN 90 5809 353 0

Construction and instrumentation of waste rock test covers at Whistle Mine, Ontario, Canada

B.K. Ayres, M. O'Kane & D. Christensen
O'Kane Consultants Inc., Saskatoon, Saskatchewan, Canada

L. Lanteigne
Inco Limited, Copper Cliff, Ontario, Canada

ABSTRACT: Inco Ltd. is currently in the process of decommissioning the Whistle Mine site located in northern Ontario, Canada. Closure of the mine site includes relocation of waste rock from two waste rock piles into an open pit, and construction of an engineered dry cover system over the potentially acid-generating waste material. The final contoured surface of the backfilled pit will have a 20% slope. Site-specific field information is required on the construction feasibility and potential performance of dry cover systems prior to finalizing the design of the full-scale cover system.

Three experimental dry cover systems were constructed over acid-generating waste rock at Whistle Mine in the fall of 2000. Each test cover plot has a different barrier layer overlain by a protective layer of non-compacted soil. The three barrier layers being evaluated are a geosynthetic clay liner (GCL), a compacted sand-bentonite mixture, and a compacted local silt/trace clay material. A monitoring system was installed to evaluate field performance of the test covers, which includes continuous monitoring of climatic parameters, gaseous oxygen / carbon dioxide concentrations and moisture / temperature conditions within the cover and waste materials. The quantity of net percolation through each test cover is also being monitored. A waste rock platform with a 20% slope was constructed to support the test cover systems, as well as a seepage collection system to prevent contamination of the local groundwater system.

This paper describes the construction of the test cover plots, the waste rock platform and seepage collection system, and installation of the field performance monitoring equipment.

1 INTRODUCTION

Inco Limited, Ontario Division (Inco) has finalized the design of the closure works for the Whistle Mine site. The Whistle Mine site is located approximately 30 km north of Sudbury, Ontario, Canada (Fig. 1). A closure plan for the mine site was reviewed and approved by the Ministry of Northern Development and Mines, Ontario in 1998. The closure plan includes relocation of waste rock from two waste rock piles and all acid-generating waste rock used for the various site roads into the pit. The final surface of the filled pit will be contoured once all of the designated materials have been placed, and will possess a constant 20% slope as a result of the natural relief adjacent to the pit. An engineered soil cover system will subsequently be constructed over the backfilled waste material, covering a surface area of approximately 9.7 ha.

The cover will be a critical aspect of the work, as it must be an effective barrier against oxygen diffusion, water infiltration, and deep tree-root penetration over the long term for maximum environmental protection. Field information is required on the construction feasibility and potential performance of alternate dry cover systems prior to finalizing the design of the full-scale cover system.

Figure 1. Location of Whistle Mine in Ontario, Canada.

A test cover program for acid-generating waste rock was initiated at Whistle Mine in the fall of 2000. Three test cover plots with alternate designs, each approximately 12 m wide and 24 m long, were constructed on a waste rock platform with a 20% slope. A seepage collection system was installed at the study site to prevent potential acidic drainage from the waste rock platform from entering the local groundwater system. Monitoring equipment, including automated net percolation collection and monitoring systems (i.e. lysimeters), was installed to evaluate the field performance of the various test covers for a minimum of two years.

This paper describes the construction of the test cover plots and the waste rock platform / seepage collection system, as well as installation of the field performance monitoring equipment. Issues related to construction of the various barrier layers in a full-scale application are also discussed.

2 BACKGROUND

2.1 Mine history / site description

The Whistle Mine orebody, which was originally discovered in 1897, was developed as an open pit mine in 1988. Mining and the production of waste rock at Whistle Mine occurred between 1988-1991 and 1994-1998. Approximately 6.4 million tonnes of waste rock was produced during these periods and stored in two surface stockpiles adjacent to the pit. The waste rock is composed of approximately 80% mafic norite, which has an average sulphide content of 3% (DeVos et al. 1997). The mine site is currently being decommissioned, which includes relocation of all waste rock to the open pit. Backfilling of the open pit should be completed in early 2002.

The mine site is part of the Post Creek watershed, an area of approximately 5400 ha, which drains into Lake Wanapitei, only 3 km east of the mine. The area immediately surrounding the mine site is undeveloped wilderness. Bedrock outcrops are frequent and typically form hills that rise up to 50 m above the surrounding areas (DeVos et al. 1997). A thin discontinuous blanket of glacial till covers the bedrock. Whistle Mine is situated in a semi-arid environment; the mean annual precipitation and potential evaporation for the region is approximately 870 mm and 520 mm, respectively. Approximately 30% of the annual precipitation occurs as snow.

2.2 Dry cover systems

The application of a dry cover system over reactive waste rock is becoming a common technique for preventing and controlling acid rock drainage following closure of a mine site or waste storage facility (MEND 2001). The objectives of dry cover systems are to minimize the influx of water and provide an oxygen diffusion barrier to minimize the influx of oxygen. Apart from these functions, dry covers are expected to be resistant to erosion and provide support for vegetation. Dry covers can be simple or complex, ranging from a single layer of earthen material to several layers of different material types, including non-reactive tailings and/or waste rock, geosynthetic materials, and oxygen consuming organic materials (MEND 2001).

In general, the design of a dry cover system for acid-generating waste rock must consider the performance of the dry cover on both horizontal and sloped surfaces, the internal hydraulic and geochemical performance of the waste material, and the potential influence of basal flow (MEND 2001).

2.3 Study objectives

The objectives of the Whistle Mine test cover program are:

1 To evaluate the "relative" field performance of three different test cover plots in response to varying site climatic conditions, as well as the "absolute" field performance of each test cover compared to a control (i.e. uncovered) test plot;

2 To collect accurate field performance data for calibration and subsequent validation of numerical models used for cover design; and

3 To evaluate cover construction techniques and in particular, gain some insight into potential quality assurance / quality control (QA/QC) difficulties associated with construction of the barrier layer.

3 DESIGN OF TEST COVER PLOTS

Three test cover systems are being evaluated in this study for potential placement over the backfilled open pit at Whistle Mine (Fig. 2). Each test cover system has a barrier layer to limit the ingress of atmospheric oxygen and meteoric water. The barrier layer for Test Plot #1 is a geosynthetic clay liner (GCL). Test Plot #2 has a 0.45 m thick sand-bentonite mixture barrier layer, which has a sodium bentonite content of 8% on a dry mass basis. Test Plot #3 has a 0.60 m thick barrier layer of a silt/trace clay material, obtained from a nearby borrow pit.

Each of the test cover systems has a 0.9 m thick layer of relatively well-graded material over the barrier layer. This layer protects the integrity of the barrier layer and provides a medium for the growth of vegetation. In addition, this layer will reduce the percolation of meteoric waters to the underlying waste through storage and release of moisture to the atmosphere as a result of evapotranspiration.

Different methodologies were used to determine the design thickness of the various test cover layers. Laboratory and numerical modeling investigations were completed in order to determine the optimum

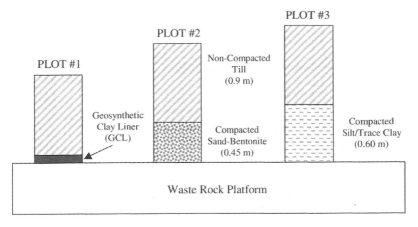

PLOT #3

PLOT #2

PLOT #1

Non-Compacted
Till
(0.9 m)

Geosynthetic
Clay Liner
(GCL)

Compacted
Sand-Bentonite
(0.45 m)

Compacted
Silt/Trace Clay
(0.60 m)

Waste Rock Platform

Figure 2. Schematic diagram of the Whistle Mine waste rock test cover plots.

thickness for the barrier layers in Test Plots #2 and #3 (KP 1998). The GCL being evaluated in this study is Bentomat® ST, which is a prefabricated product consisting of a layer of sodium bentonite (<0.1 m thick) between woven and non-woven geotextiles needle-punched together. Although preliminary cover design modeling results showed a 0.5 m thick protective layer was adequate for the Whistle Mine test cover systems, the design thickness of this layer was increased to 0.9 m to provide greater protection for the barrier layers against deep-root penetration and freeze-thaw effects (KP 1998).

Saturated-unsaturated flow numerical modeling was completed for this project in order to determine the optimum dimensions for the cover test plots. The software package SEEP/W (Geo-Slope International 1995) was configured for steady-state, two-dimensional seepage and used for all simulations. A critical component in the design of the cover dimensions was to ensure that the cover was large enough so that any potential flow around the test plots would not be intercepted by the lysimeter. In addition, the cover had to be large enough so that lateral moisture movement under the cover due to liquid and vapour gradients or slope effects would be minimized. The final dimensions of the cover that met these criteria had a top surface of approximately 24 m long and 12 m wide with 2H:1V side-slopes.

A waste rock platform located in an undisturbed area was required for this study because of the ongoing open pit backfilling operation. A continuous 20% slope at the existing waste rock piles would not be available for the duration of the test cover project. A seepage collection system was also designed for this project to prevent potential acidic drainage from the waste rock platform from entering the local groundwater system. A fourth test plot with no cover (Test Plot #4) was also established on the platform.

4 CONSTRUCTION OF STUDY COMPONENTS

Construction of the Whistle Mine test cover plots and associated infrastructure occurred in September and October 2000. Figures 3 and 4 show the major components of the test cover project. The study site is located in a cleared area on the Whistle Mine property. OKC (2001) provide complete details on the construction of the various study components. A summary of the construction details is provided below.

4.1 Seepage collection system

The seepage collection system has two components; namely, the waste rock containment area and the seepage collection pond. The waste rock containment area has a length and width of approximately 90 m and 45 m, respectively, and a slope of about 1.7% from the north to south perimeter. The south perimeter of the containment area is also the north perimeter of a lined pond, which collects all surface runoff and seepage from the containment area. The waste rock containment area is comprised of an earthen foundation, a 0.3 m thick sand sub-base layer, a geosynthetic liner and finally, an overlying sand drainage layer. The geosynthetic liner consisted of one 20-mil thick panel of low-liner density polyethylene (LLDPE). A 0.3 m containment berm was constructed around the perimeter to ensure all runoff and seepage from the area reports to the seepage collection pond. The sand drainage layer was a minimum thickness of 0.5 m to protect the liner from damage during placement of the waste rock.

The seepage collection pond is located immediately south of the containment area and has a length of 90 m and width of 10 m. The pond is 1.5 m deep and was designed to retain runoff from the 24-hour design storm event with a 1:100 year return period

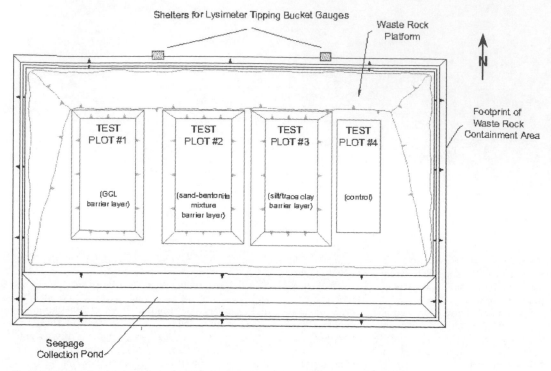

Figure 3. Plan view of the major components of the Whistle Mine test cover project (after OKC 2001).

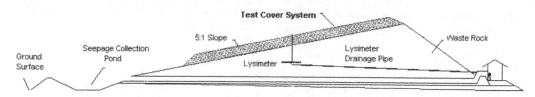

Figure 4. Cross-section of the major components of the Whistle Mine test cover project (after OKC 2001).

(113 mm), assuming the pond is less than one-third full prior to the storm event. One 30-mil thick panel of LLDPE was installed to provide containment of runoff and seepage waters.

4.2 Waste rock platform

A platform of waste rock with a 20% slope was required to support the three test cover plots and one control test plot. This resulted in the waste rock platform having a footprint of approximately 85 m by 40 m and a height of 6.0 m. The volume of waste rock in the platform is approximately 12,000 m³.

The waste rock platform was constructed in three stages. The first stage involved placing and grading a single lift of waste rock, with an average thickness of 0.5 m, up to the design elevation of the bottom of the lysimeter tanks. Once the lysimeter tanks and drainage pipes were installed, waste rock haulage to the test area resumed and subsequent placement in

0.5 m thick lifts. The third and final stage consisted of placing and grading waste rock to achieve the desired final elevation of the 20% slope. Four 50-ton haul trucks were used to place waste rock in the containment area, and a D8N and D3 bulldozer were used for rough and final grading, respectively. The surface of the 5H:1V slope was compacted with an 84" wide smooth-drum vibratory roller to provide a smooth foundation for the test plots. Waste rock samples were collected throughout construction for geotechnical and/or geochemical laboratory testing.

4.3 Test cover plot construction

The waste rock test cover plots at Whistle Mine were constructed in a similar manner, with the exception of the barrier layers. The protective layer in each test cover, which consisted of a well-graded soil material obtained from a nearby borrow pit, was placed in a single, 0.9 m thick non-compacted lift

166

Figure 5. Moisture conditioning and mixing the first lift of the sand-bentonite mixture barrier layer (from OKC 2001).

following construction of the barrier layer. A 25-ton haul truck and a D3 bulldozer were used to haul, place and grade the protective layer material. Heavy equipment was not allowed to operate directly on the barrier layer in order to protect its integrity.

A seed mixture was applied to the surface of each test cover, including the control test plot, following construction of the protective cover layer and installation of the near surface field monitoring equipment. A truck-mounted hydroseeder was used to apply the seed mixture to the entire test plot area.

Details related to construction of each barrier layer are provided below.

4.3.1 Test plot #1 barrier (GCL)

The GCL was shipped from the factory to the site in two rolls, each 38 m long by 4.5 m wide. Each roll was hauled to the top of the waste rock platform with a forklift, and subsequently unrolled and placed directly on the prepared smooth waste rock surface. A small anchor trench, approximately 0.2 m deep, was excavated along the entire north perimeter of the test plot to prevent the GCL from moving during placement of the overlying protective layer material. A strip of sodium bentonite was placed along all overlapping seams to provide a watertight seal at these locations.

4.3.2 Test plot #2 barrier (sand-bentonite mixture)

The sand-bentonite mixture being evaluated in this study consists of relatively uniform sand with a bentonite content of 8% on a dry mass basis. The ben-

tonite used in the mixture is Envirogel® 12, which was supplied by Wyo-Ben, Inc. out of Montana, USA. This product is a granular sodium bentonite containing particles ranging in size from fine sand to clay. The bentonite was delivered to the test site in 3000 lb canvas bags on semi-trailer flatbed units.

Construction of the compacted sand-bentonite mixture barrier layer was completed in two 0.23 m (9") thick lifts. Slightly different construction techniques were used for each lift. Construction of the first lift involved placing the host material within the footprint of the test plot and subsequently mixing in the bentonite and moisture conditioning the mixture directly on the waste rock platform (Fig. 5). Mixing and moisture conditioning the sand-bentonite mixture for the final lift occurred in a cleared area immediately west of the waste rock platform. The prepared mixture was then hauled, placed and graded on top of the first lift.

The following methodology was applied to both lifts of the barrier layer for mixing, moisture conditioning and compacting the sand-bentonite mixture:
- The bentonite host material was placed and graded, and then bentonite was applied evenly to the lift surface corresponding to the design bentonite content.
- The bentonite was mixed thoroughly into the host material with a pulvi-mixer prior to moisture conditioning.
- The sand-bentonite mixture was moisture conditioned to achieve a moisture content corresponding to 2% wet of the optimum moisture content

167

(OMC) for the mixture, based on a standard Proctor compaction test.
- A smooth-drum vibratory roller was used to compact each lift following moisture conditioning.

4.3.3 Test plot #3 barrier (silt/trace clay)

The compacted silt/trace clay barrier layer was constructed in three 0.20 m (8") thick lifts. A loader was used to place and grade the barrier layer material. Each lift was moisture conditioned to achieve a moisture content corresponding to 2% wet of the optimum moisture content (OMC), based on a standard Proctor compaction test. A smooth-drum vibratory roller was used to compact each lift following moisture conditioning.

4.4 Evaluation of test cover construction techniques

Construction techniques used for the test covers were evaluated and assessed as to their potential application for construction of a cover system on the backfilled open pit at Whistle Mine. This component of the study focused on gaining some insight into potential quality assurance / quality control (QA/QC) difficulties associated with construction of the barrier layer. Discussions are provided below on the installation of the GCL and construction of a sand-bentonite mixture barrier layer.

4.4.1 Installation of the GCL

Installation of the GCL barrier layer was a relatively simple task. This was reflected in the brief amount of time required to cover the surface area of the test plot (three hours). The force of gravity (i.e. the 20% slope) played a key role in the simplicity of unrolling and installing the GCL on the waste rock surface. The use of a spreader bar, inserted through the core of the GCL roll, attached to the bucket of an excavator or loader may be required for a full-scale installation. Nevertheless, it is expected a GCL could be installed on the entire surface of the backfilled open pit in a relatively short period of time.

The GCL was installed directly on the surface of the waste rock platform following compaction of the 20% slope. Compaction of the waste rock surface provided a relatively smooth foundation for the GCL. However, a visual inspection of the GCL following installation revealed that puncturing of the GCL could potentially occur during placement of the non-compacted cover material. This issue must be addressed if GCL is to be used in the design of the full-scale waste rock cover system. It is anticipated that a geotextile, installed between the GCL and underlying waste rock, would provide the necessary cushioning to prevent puncturing of the GCL.

4.4.2 Construction of sand-bentonite barrier layer

The sand-bentonite mixture for the barrier layer in Test Plot #2 was prepared using an in-situ or field batch technique. The primary reason for preparing the mixture adjacent to the waste rock platform for the second lift was due to space constraints on the platform and concerns for the safety of equipment operators. However, this would not be an issue for construction on the backfilled open pit. In summary, both techniques would be suitable for full-scale cover system construction.

Another technique that has been used at other sites for preparation of sand-bentonite mixtures involves the use of a pug mill or batch plant. The bentonite host material, along with appropriate quantities of bentonite and water, are mixed in the batch plant and subsequently transferred to a stockpile or directly to haul trucks via a conveyor belt. This technique as well as the in-situ or field batch technique generally produce a similar product; however, due to differences in rotation speed, batch plant mixing may not produce as well of a mixed product. Precipitation can affect both techniques; however, batch plant mixers have more difficulty working in moist conditions because of their lack of drying capability.

The key criteria for placing and compacting the sand-bentonite mixture material, as well as the silt/trace clay material, was moisture content, as opposed to dry density. One of the design objectives of the Whistle Mine test cover systems is to reduce the ingress of atmospheric oxygen to the underlying waste rock by maintaining the barrier layer at or near saturation. Therefore, placing and compacting the barrier layer materials above the OMC, as opposed to below or at the OMC, benefits this particular cover system design objective. In addition, a minimum saturated hydraulic conductivity will occur wet of the OMC for typical silt/trace clay soils and sand-bentonite mixtures (see Lambe 1958, Mitchell et al. 1965, Haug & Wong 1992). Nonetheless, the dry density of each compacted lift achieved in the field was between 90 and 95% of the standard Proctor maximum density.

5 INSTALLATION OF FIELD PERFORMANCE MONITORING SYSTEM

A monitoring system was installed in order to evaluate the field performance of the various test covers constructed on Whistle Mine acid-generating waste rock. The objectives of the field performance monitoring system are to:

1 Develop an understanding for key processes and characteristics that control performance;
2 Obtain a water balance for each of the test cover systems and the control test plot;
3 Develop credibility and confidence with respect to performance of the proposed cover system from a closure perspective; and

4 Develop a database with which to calibrate the cover system design using numerical modelling tools.

The test cover field monitoring program was designed to quantify as many parameters as possible influencing the performance of a sloped cover system (Fig. 6). The critical parameters being measured are the net percolation of meteoric water and the ingress of atmospheric oxygen into the underlying waste material. Net percolation is a component of the water balance for the cover system, and is related to the other water balance components as follows:

$$PERC = \Delta S + D_r + NSI \tag{1}$$

$$NSI = PPT - AET - RO \tag{2}$$

where PERC = net percolation into the waste material; ΔS = change in moisture storage within the cover layers; D_r = lateral drainage or percolation within the cover layers; NSI = net surface infiltration; PPT = precipitation; AET = actual evapotranspiration; and RO = runoff.

Table 1 summarizes the field instrumentation that was installed in each of the test plots in the fall of 2000. OKC (2001) provide complete details on the installation of the various field monitoring components. Installation of the major monitoring components is summarized below.

5.1 Net percolation collection & monitoring system

A net percolation collection and monitoring system (lysimeter) was installed at each test plot for monitoring the quantity and quality of percolating water through each test plot. Each lysimeter is comprised of a net percolation collection tank, in-situ moisture monitoring system, underdrain system and a net percolation monitoring system. Installation of each of these components is described briefly below.

The net percolation collection tank, which is the main component of the lysimeter, consists of a large vertical storage tank. Two-dimensional saturated-unsaturated flow numerical modeling was carried out to determine the optimum tank dimensions and tank position within the test plot, based on design criteria outlined in Bews et al. (1997). The results of steady-state simulations utilizing SEEP/W (Geo-Slope International 1995) dictated the installation of 2.4 m diameter tanks, 2.3 m high, at approximately the centre point of each test plot. Figure 7 shows the location of the four lysimeter tanks. The top of the lysimeter tanks, which were cut to have a 20% slope, were positioned immediately below the waste rock / barrier layer interface. A thin layer of drainage sand (<0.1 m) was placed in the bottom of each tank prior to backfilling with waste rock. Samples of the drainage sand and waste rock backfill were collected for future laboratory characterization.

An in-situ moisture monitoring system was installed in each of the net percolation collection tanks for monitoring temporal and spatial changes in moisture storage in the waste rock backfill. The in-situ moisture monitoring system selected for this project is the Diviner 2000, a product manufactured and distributed by Sentek Pty Ltd. of Adelaide, Australia. The Diviner 2000 consists of one sensor on a shaft with an automatic depth sensor (i.e. the probe) and pre-installed access tubes. Insertion of the probe into an access tube provides an immediate profile of soil moisture as a function of depth.

The underdrain component of the lysimeter consists of a 51 mm diameter PVC pipe that extends from the base of the net percolation collection tank to a point just above the monitoring system. The underdrain pipes have a downward slope ranging between 1% and 2% to allow gravity flow of net percolation waters from the tanks. Portions of the underdrain pipes were covered with insulation to minimize the potential for freezing of percolating waters draining during the winter months. A water trap oxygen barrier was installed at the end of each

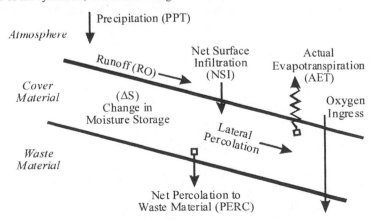

Figure 6. Parameters affecting the field performance of a sloped cover system (from MEND 2001).

169

Table 1. Summary of field instrumentation installed in the Whistle Mine waste rock test cover plots.

Instrumentation (parameters measured)	Test Plot #1 – GCL barrier	Test Plot #2 – Sand-bentonite barrier	Test Plot #3 – Silt/ trace clay barrier	Test Plot #4 – Control
Lysimeter (net percolation)	1 – at base of cover	1 – at base of cover	1 – at base of cover	1 – at surface
Campbell Scientific 229-L thermal conductivity sensor (matric suction & temperature)	7 – in N/C layer 2 – in waste rock	7 – in N/C layer 4 – in barrier layer 2 – in waste rock	7 – in N/C layer 4 – in barrier layer 2 – in waste rock	4 – in waste rock
Campbell Scientific CS615-L frequency domain reflectometer (volumetric water content)	7 – in N/C layer 2 – in waste rock	7 – in N/C layer 4 – in barrier layer 2 – in waste rock	7 – in N/C layer 4 – in barrier layer 2 – in waste rock	4 – in waste rock
Gas sampling port (O_2 & CO_2 gas concentrations)	1 – above barrier in N/C layer 1 – below barrier in waste rock 1 – at depth in waste rock	1 – above barrier in N/C layer 1 – below barrier in waste rock 1 – at depth in waste rock	1 – above barrier in N/C layer 1 – below barrier in waste rock 1 – at depth in waste rock	3 – in waste rock
Surface runoff and sub-surface collection & monitoring systems	Preliminary work completed in 2000 (installation to be completed at a later date)			NA
Meteorological station – installed on Test Plot #2 (air temperature, relative humidity, wind speed & direction, net radiation, barometric pressure, rainfall and snowfall)				

NA = not applicable; N/C = non-compacted

Figure 7. Looking northwest at the Whistle Mine test plot area showing the location of the four lysimeter tanks (from OKC 2001).

underdrain pipe to prevent oxygen from entering the underdrain system and oxidizing the waste material in the net percolation collection tank. Ethylene glycol was placed in the oxygen barrier, as opposed to water, to prevent freezing of the oxygen barrier during the winter months. The solution in the water trap oxygen barrier can be changed-out at any time with distilled water to facilitate the collection of representative seepage waters for chemical analysis.

The lysimeter monitoring systems are comprised of a flow meter to automatically record the time and quantity of water discharged from each net percolation collection tank, and a sample bucket to collect net percolation waters for chemical analysis. The flow meter is simply a tipping bucket rain gauge, connected to an automated data acquisition system (DAS). The flow meters and sample buckets are housed in sheds located at the north perimeter of the test plot area.

5.2 In-situ moisture / temperature sensors

Sensors to measure in-situ matric suction, temperature and volumetric water content in the test cover materials and waste rock were installed in all four plots. Model 229 thermal conductivity sensors, supplied by Campbell Scientific Canada Corporation, were selected for monitoring in-situ matric suction and temperature in the various test plot materials. Model CS615 frequency domain reflectometer (FDR) probes, also supplied by Campbell Scientific Canada Corporation, were installed to measure in-situ volumetric water content in the various test plot materials. Where possible, these sensors were installed into the face of undisturbed material.

These sensors were installed in a single instrumentation nest located approximately 1.5 m up-gradient of each lysimeter tank. Sensors installed in Test Plots #1 and #2 were connected to an automated DAS located on Test Plot #2, while sensors installed in Test Plots #3 and #4 were connected to an automated DAS located on Test Plot #3. Each DAS consists of a datalogger and multiplexer, housed in an environmentally sealed enclosure, powered by a rechargeable battery / solar panel system. In-situ moisture and temperature measurements are currently being collected every six hours.

5.3 Gaseous O_2/CO_2 monitoring system

Three sampling ports were installed within the three test cover systems; one 0.05 m above the barrier layer, one 0.05 m below the barrier layer, and one about 0.6 m below the barrier layer. Data collected from the sampling ports just above and below the barrier layers will be used to assess the effectiveness of the barrier layers in reducing the ingress of atmospheric oxygen to the underlying waste material. Three sampling ports were also installed at various depths within the waste rock of the control test plot.

The sampling ports were connected to a gas analyzing system, which consist of an O_2/CO_2 analyzer, a 12-point sequencer and a condensate trap panel. The gaseous concentration analyzer is capable of measuring O_2 and CO_2 concentrations in the range of $0 - 25\%$ and $0 - 10\%$, respectively. The O_2/CO_2 analyzer and 12-point sequencer were connected to a DAS to automatically record gas concentrations every twelve hours.

6 SUMMARY AND CONCLUSIONS

The construction and instrumentation of experimental dry cover systems for acid-generating waste rock at Whistle Mine in Ontario, Canada were reviewed. Three cover system alternatives were constructed on a specially designed waste rock platform possessing a 20% slope. Each test cover system has a barrier layer to limit the ingress of atmospheric oxygen and meteoric water, and an overlying non-compacted layer to protect the integrity of the barrier layer and provide a medium for the growth of vegetation. The barrier layers being evaluated in this study are a geosynthetic clay liner (GCL), a 0.45 m thick compacted sand-bentonite mixture and a 0.6 m compacted silt/trace clay material. Techniques used to construct each of the barrier layers were assessed in terms of their potential application for construction of the full-scale cover system.

A state-of-the-art monitoring system was installed to assess field performance of the test cover systems during all seasons of the year. The system includes continuous monitoring of various climatic parameters, gaseous oxygen / carbon dioxide concentrations and moisture / temperature conditions within the cover and waste materials, and the quantity of net percolation through each test cover.

Collection of test plot field data commenced in November 2000. Data collected over the next two years will be used for calibration and subsequent validation of numerical models used for cover design. The field calibrated model will be a key tool for determining the optimum cover design for the Whistle Mine backfilled open pit.

REFERENCES

Bews, B.E., O'Kane, M.A., Wilson, G.W., Williams, D. & Currey, N. 1997. The design of a low flux cover system, including lysimeters, for acid generating waste rock in semiarid environments. *Proceedings of the Fourth International Conference on Acid Rock Drainage, Vancouver, BC, May 31-June 6, 1997*: 747-762.

DeVos, K.J., Pettit, C., Martin, J., Knapp, R.A. & Jansons, K.J. 1997. Whistle Mine waste rock study: Volume I. MEND Project 1.41.4.

Geo-Slope International Ltd., 1995. SEEP/W User's Manual. Geo-Slope International Ltd., Calgary, AB.

Haug, M.D. & Wong, L.C. 1992. Impact of molding water content on hydraulic conductivity of compacted sand-bentonite. *Canadian Geotechnical Journal* 29: 253-262.

Knight Piesold (KP). 1998. Design report for closure, Inco Ltd., Ontario Division, Whistle Mine. Volume I – Report and Volume II – Appendices. Report No. D2350/1 prepared for Inco Ltd.

Lambe, T.W. 1958. The engineering behavior of compacted clay. *ASCE Journal of the Soil Mechanics and Foundations Division* 84: 1655-1 to 1655-35.

MEND, 2001. Dry covers. In G.A. Tremblay & C.M. Hogan (eds), *MEND Manual: Volume 4 – Prevention and Control*: 155-232. MEND Project 5.4.2d.

Mitchell, J.K., Hooper, D.R., & Campanella, R.G., 1965. Permeability of compacted clay. *ASCE Journal of the Soil Mechanics and Foundations Division*, 91: 41-65.

O'Kane Consultants Inc. (OKC). 2001. As-built report for the Inco Ltd., Whistle Mine acid-generating waste rock test cover plots. Report No. 647-1 prepared for Inco Ltd.

Field investigation to support the closure design of the Yankee Heap

P.E. Kowalewski
Olsson Associates; Lakewood, Colorado

S. Boyce
SRK Consulting; Elko, Nevada

R. Buffington
Bald Mountain Mine; Elko, Nevada

ABSTRACT: A field investigation and laboratory-testing program was developed to aid in the closure design of the gold heap leach pad at the Yankee Mine near Ely, Nevada. The field investigation included rotosonic drilling of three boreholes through the heap to develop material and moisture profiles through the heap, assess the physical properties of the heap leached ore, and estimate the fluid inventory at closure to aid in the development of a solution disposal plan. In addition, local soils were sampled and tested for use in the design of the reclamation soil cover for the heap. The field- and laboratory-testing program included: in-situ density, in-situ moisture content, grain size distribution, saturated permeability, and the determination of the soil water characteristic curves. The data obtained through the testing program were used to estimate the residual solution inventory that would have to be handled at closure and was used to establish boundary conditions in finite element modeling of the proposed reclamation soil cover. The results of the field and laboratory investigation showed the heap to be very well drained, which led to a reduction in the overall design and cost of the solution disposal facilities at the site.

1 INTRODUCTION

The Yankee Mine is located in White Pine County, Nevada, about 70 miles northwest of Ely, Nevada. The 6 million ton heap leach pad was operated from 1993 through 1999, and gold was removed from the host ore utilizing conventional cyanide leaching technology. The heap leach pad covers a plan area of about 34 acres, and the ¾-inch minus crushed ore was placed in 20-foot lifts via radial stacker to an ultimate height of over 120 feet.

The design and construction of the pad included a synthetic, high-density polyethylene (HDPE), liner system to contain and control pregnant solution flow for gold extraction and protection of groundwater from cyanide and metal-bearing fluids. After the operational life of the pad is complete, the pad and liner system act to collect and contain meteoric water along with residual cyanide and metal-containing spent ore, and deliver those fluids to the drainage system located above the synthetic liner. The meteoric waters are altered during contact with the spent ore,.resulting in an observed (and previously predicted) increase in arsenic within the effluent at this particular site. Additionally, in the near term, nitrate and other conservative constituents may be elevated as a result of evapoconcentration of fluids within the operationally closed circuit, and through the degradation of cyanide to nitrate within the solution.

In order to limit the potential for these chemical species to mobilize in the natural environment, several strategies have been developed. Of primary import to any of the scenarios is the effective long-term minimization of the meteoric water inputs to the system. Three alternatives were reviewed for applicability at this site:

- Synthetic membrane encapsulation;
- Low-permeability clay cap; and
- "Store and release" (ET) soil cover.

The first option was dismissed as infeasible, given the requirements of system longevity and the regulatory stipulation to provide for a beneficial post-mining land use (rangeland). Capping with low-permeability clays was also considered but rejected because of the erosion and dessication potential in a semi-arid environment. The sporadic nature of precipitation in the area and the low annual rainfall do not generally support the successful implementation of a low-permeability clay cap for this site.

The single alternative that meets the above-mentioned criteria (longevity, post-mining land use, and low flux of meteoric waters) is the "store and release" soil cover.

2 FIELD INVESTIGATION

To appropriately design the cover for the Yankee heap, data representing the physical nature of the heap material was gathered through an intensive field investigation.

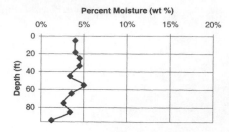

Figure 1. Moisture profile for YPRS-1.

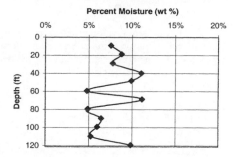

Figure 2. Moisture profile for YPRS-2.

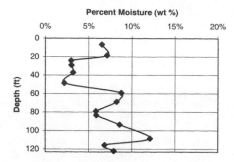

Figure 3. Moisture profile for YPRS-3.

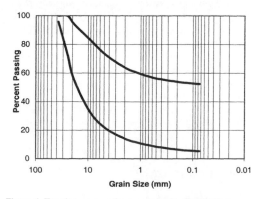

Figure 4. Envelope curves for ore grain size distribution.

Potential soil cover areas were identified based on the geomorphology of the local alluvial materials. Additionally, soil had been stockpiled by the operator to facilitate the reclamation and revegetation of the heap and adjacent facilities. Bulk samples were collected at each of the potential borrow sites for later evaluation of physical properties. It was expected that the soils would be similar within the spatial limits of the site, although that assessment would be tested by the results of laboratory procedures.

In order to collect representative data from the heap profile, three boreholes were drilled to within approximately 20 feet of the synthetic liner using a rotosonic drill. The rotosonic drilling method results in a continuous core of relatively undisturbed sample measuring five inches in diameter. The method provides for accurate retrieval of in-place bulk density, moisture content (see Figures 1 through 3), and gradation analysis, which provide the basis for further analytical testing of reconstituted samples.

3 LABORATORY TESTING

Samples obtained during the drilling program were subjected to laboratory testing to determine the grain size distribution, saturated permeability, and the soil water characteristic curve (SWCC) of the ore and the cover soil. The SWCC relates a soil's moisture content to the applied matric suction.

3.1 Ore testing

3.1.1 Grain size distribution

Ten samples of ore were tested to determine the range of grain size distribution in the ore. The results show the ore classifies as poorly graded gravel with silt and sand (GP-GM) to clayey gravel with sand (GC). Grain size testing was used to direct further testing of the ore. Figure 4 presents the range of grain size distributions determined for the collected ore samples.

3.1.2 Permeability testing

Four samples of the ore were subjected to constant head hydraulic conductivity tests. The results of the analyses show very little variation in saturated permeability values, with a minimum value of 1.5×10^{-2} cm/sec and a maximum value of 3.5×10^{-2} cm/sec. The average saturated permeability of the ore was determined to be 2.8×10^{-2} cm/sec. One of the most important conclusions made from the field and laboratory investigation of permeability was that there was no apparent decrease in permeability with depth. The results of the permeability testing are presented in Table 1.

Table 1. Summary of ore saturated permeability testing

Sample	Depth (ft)	Saturated Permeability (cm/sec)
YPRS-2	9	3.5×10^{-2}
YPRS-2	68.5	1.5×10^{-2}
YPRS-3	29	3.5×10^{-2}
YPRS-3	108.5	3.3×10^{-2}
	Geometric Avg	2.8×10^{-2}

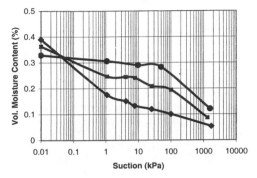

Figure 5. SWCCs determined for ore.

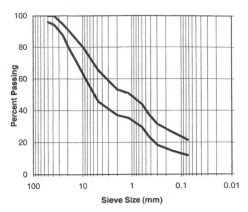

Figure 6. Envelope curves for cover soil grain size distribution.

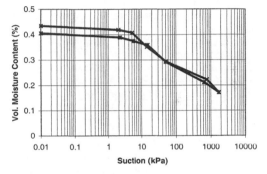

Figure 7. Cover soil soil water characteristic curves.

3.1.3 Ore soil water characteristic curves

Several samples of ore were tested to determine their soil water characteristic curves (SWCC). The SWCCs of the ore provide information regarding the draindown characteristics of the ore. Three-, five-, and seven-point characteristic curves were determined for several different samples of ore. Samples were selected for testing based on grain size distribution and the in-situ moisture profiling completed during the field investigation. Figure 5 shows the SWCCs determined for the ore.

3.2 Cover soils testing

Samples of soils available for use in the closure cover were tested in the laboratory to determine grain size distribution, maximum dry density, optimum moisture content, saturated permeability, and SWCC.

3.2.1 Grain size distribution

Results of the grain size analyses show the borrow soils classify as a silty sand to a silty sand with gravel (SM). Figure 6 presents the envelope curves for the cover soils.

3.2.2 Maximum dry density and optimum moisture

Proctor analyses were conducted to gain an understanding of the soil density that will likely be achieved when the cover soils are excavated from the borrow source and placed on the heap. The analyses show a maximum dry density of approximately 98 pcf and an optimum moisture content in the range of 12-14% for the cover soils.

3.2.3 Permeability testing

Results of falling head hydraulic conductivity testing of the proposed cover soil show a saturated permeability of 1.2×10^{-5} cm/sec when the soils are compacted in the range of 90% of optimum dry density. The 90% compaction is the likely range obtained using vehicle traffic only for compaction.

3.2.4 Soil water characteristic curves

Two samples of the cover soils were tested to determine their soil water characteristic curves (SWCC). The SWCC provides the relationship between a soil's moisture content and an applied matric suction. This relationship is unique for each type of soil and is dependent upon the grain size distribution of the tested material. Figure 7 shows the SWCCs determined for the cover soils sampled during the field investigation.

4 APPLICATION OF FIELD DATA

The data gathered during the field and laboratory investigations were used to conduct draindown analyses for the heap and to design and evaluate the performance of the proposed cover design. The draindown analyses were completed to estimate both the total draindown volume and the expected flow rates.

4.1 Draindown analyses

The potential short-term draindown of solution currently stored in the pore spaces within the heap was evaluated using a two-step approach. The first step included using a numerical model, FEFLOW, to assess the time rate of draindown from the ore, while the second step provided a hand-calculated estimate of the volume of solution using the measured ore properties and calculating matric potential in the heap.

As discussed in Section 2, three borings were completed through the heap. Samples of the ore were obtained from the borings and in-situ density and moisture content were determined. Using the FEFLOW code, the anticipated time for draindown from leaching moisture to residual moisture was on the order of 60 to 135 days. The model also showed that a maximum draindown volume of 100,000 gallons could be expected. The numerical approach using the hand-calculation showed a maximum draindown volume of 250,000 gallons. The higher estimate of 250,000 gallons of solution was used as a conservative estimate of heap draindown when developing approaches to handling the draindown solution at the site.

4.2 Cover design and evaluation

SoilCover Version 5.2 (GeoAnalysis 2000 Ltd., 2000) was used to assess the performance of several different cover configurations proposed for the Yankee heap. Three different heap profiles were modeled to assess the performance of the different soil cover configurations. To complete the modeling of the ore columns, the ore was grouped into eight (8) different units and assigned representative soil properties. The initial moisture contents used in the model were obtained during the field investigation (see Section 2).

4.2.1 Cover evaluation

Several cover profiles, including a no cover option, were evaluated on each of the three ore columns. Cover thicknesses of 12-, 18-, 24-, and 36-inches were evaluated using the SoilCover model. In most of the cases a maximum rooting depth of 18 inches was used as a conservative estimate of rooting depth. The rooting depth was never specified to a depth that would allow roots to penetrate into the ore, as it is

not believed that this would happen at the site due to the poor water holding properties of the ore. In a few cases (24- and 36-inch covers only), the rooting depth was increased to 24 inches to investigate the sensitivity of the rooting depth on the amount of meteoric water percolating through the cover. In all of the cases, a "poor" quality vegetation (leaf area index = 1.0) was assumed to be growing on the surface of the cover, which is representative of the vegetation that is expected at the site.

4.2.2 Evaluation results

The results of the analyses show that constructing a cover over the exposed ore will significantly reduce the amount of meteoric water entering the heap. Left uncovered, approximately 0.6 to 2.2 inches of meteoric water will infiltrate into the heap on an annual basis. This is equivalent to an average annual drainage rate of 0.03 to 0.11 gallons per minute per acre (gpm/acre). The addition of a soil cover will reduce the meteoric infiltration rate to a range of 0.02 to 0.10 inch annually, which is equivalent to an annual average drainage rate of approximately 0.001 to 0.005 gpm/acre.

The results show that an 18-inch cover will reduce the infiltration of meteoric water to approximately 0.03 inch annually (0.002 gpm/acre). Construction of a thicker cover (24- or 36-inch) will only provide a marginal improvement over the 18-inch cover (reduction in annual infiltration to approximately 0.001 gpm/acre) and therefore is not considered to be warranted. Construction of a much thicker cover (in excess of 36 inches) could potentially lead to an increase in infiltration into the heap, as the potential increase in hydraulic head could force more water out of the cover and into the heap.

4.2.3 Sensitivity analyses

Several sensitivity analyses were completed to assess the potential impacts of the assumed rooting depth on infiltration, the effect of the permeability of the cover soil on infiltration, and the effect of a "wet" (125% of normal) year on infiltration. The effect of the assumed rooting depth was analyzed on all three ore profiles, while the effects of increasing the permeability of the cover soil and including a "wet" year were only analyzed using the 18-inch cover on ore column YPRS-3. YPRS-3 was used because the modeling showed the underlying ore column transmitted flow the most rapidly out of the three columns, most likely due to the moisture conditions identified during the field investigation.

In both the 24- and 36-inch cover cases, increasing the rooting depth from 18 to 24 inches showed no significant (<0.01 inch) reduction in the annual infiltration rate. For the cases involving YPRS-3 only, when the saturated permeability of the cover soil was increased by one-half an order of magnitude (from 1.2×10^{-5} cm/sec to 5×10^{-5} cm/sec) the amount of average annual infiltration into the ore increased from 0.02 inch to 0.09 inch, which is equivalent to an increase of 0.0036 gpm/acre. Increasing the saturated permeability of the cover soil by an order of magnitude (from 1.2×10^{-5} cm/sec to 1×10^{-4} cm/sec) led to an increase in average annual infiltration from 0.02 inch to 0.10 inch, which is equivalent to an increase of 0.0041 gpm/acre. The fact that the amount of infiltration is not increasing by an order of magnitude when the saturated permeability of the cover is increased by an order of magnitude is due to the fact that the cover is operating in an unsaturated condition for most of, if not all of the year.

The case that was run that included a "wet" year in the 5-year precipitation cycle showed an increase in annual infiltration from 0.02 inch to 0.05 inch. It is important to note that the "wet" year increased annual precipitation by approximately 3.3 inches, and yet resulted in an increase in average annual infiltration of only 0.03 inch. The amount of precipitation infiltrating into the ore is not higher in the wet year cycle because the plants and the atmosphere have excess "capacity" to remove water from the cover soil profile. The amount of water being removed through plant transpiration and soil surface evaporation is being limited by the availability of water in the profile. Therefore, as the amount of precipitation is marginally increased, the cover system has the capacity to remove this additional water.

5 BENEFITS OF FIELD INVESTIGATION

The field investigation of the Yankee heap leach pad indicated that the facility was draining more readily than had been anticipated. The information provided by the data gathered during the investigation resulted in a more effective cover design and improved the options for short-term and intermediate-term solution removal.

By optimizing a soil cover to reduce flux into and through the heap, each of the requisite goals are more likely to be satisfied. Constructing a cover too thin at the site would result in greater flows, shorter lifespan, and lower revegetative success. Conversely, creating a cover thicker than required would result in an excess of collateral surface disturbance, increased expense of pad closure, and the real possibility that the cover would be less effective given the rooting depths of native vegetative species at the site.

The final result of this field investigation is to provide increased accuracy in the final cover design, thereby improving the chances of successful decrease in fluid flux through the heap, and minimization of the potential to adversely affect the local environment.

5.1 *Solution inventory*

The field investigation provided information relative to the predicted short-term and intermediate-term solution inventory. This information provided the operator with an assessment of the costs associated with handling and removing solution inventories prior to final closure of the facility.

Contrary to our early assumptions about short-term draindown flows, the investigation indicated that no significant allowance was required for the short-term solution handling. This resulted in savings of considerable expense associated with the handling of the fluids, and the permitting time and expense required to authorize various short-term solution removal scenarios.

5.2 *Material properties*

The field investigation conducted for the closure of the Yankee heap provided valuable information regarding the material properties of the ore and the soils proposed for use in the cover. The determination of in-situ ore densities and moisture contents provided initial conditions for numerical modeling efforts.

While operational information provided anecdotal evidence that the Yankee heap was well-drained, the data provided by the drilling confirmed this belief. In addition, grain-size data gathered during the drilling showed material segregation, which could be correlated with the moisture contents of the ore. As anticipated, the finer grained ore retained a higher amount of moisture, while the coarser grained ore retained less moisture.

5.3 *Design of solution disposal facilities*

Long-term solution handling requires the use of several designed features. As other facets of the field investigation had indicated that the fluids predicted to flow from the heap would be properly reintroduced to the unsaturated subsurface without potential to impact the regional groundwater, the central component to solution handling was a subsurface infiltration field.

A long-term infiltration site with an area on the order of two acres was designed to spread and reintroduce fluids to the subsurface, based on the chemical relationships determined in a related investigation. As the cover model predicts very low flow rates from the closed heap, the splitting of this fluid into numerous pipes for infiltration becomes problematic.

To equalize and split flows, an engineered system was designed which utilized a large synthetic-lined pond, backfilled with granular media, a sanitary dosing siphon, several splitting boxes, and the infiltration field.

The pond design required that an inflow header system be created at a level about 5 feet from the bottom of the 20-foot deep pond. The pond is designed to be backfilled with granular media with a saturated permeability greater than 1×10^{-4} cm/sec and less than 1×10^{-2} cm/sec. The outflow manifold is designed to be constructed at least 10 feet above the inflow system, and will collect the flow and route it to the dosing siphon. Given the very large surface area of the pond and the low flow rates into the pond, this upflow system will maintain very low vertical fluid velocities.

The flow from the pond will be received by a dosing siphon that collects 1,100 gallons of fluid before flushing it at a rate of 50 gpm to the infiltration field. By increasing the instantaneous flow rate, the splitting of the fluid into 10 pipes within the infiltration area becomes hydraulically more stable and significantly less influenced by slight elevation differences as a result of construction or settling of materials.

The resulting system provides the following benefits:

1. Control of fluids for future routing options, including the potential to utilize water as a permanent wildlife water resource in an area of scarce water resources; and
2. Even dispersal of the fluids over a greater area.

6 CONCLUSIONS

The field investigation of the Yankee heap leach facility provided numerous opportunities to optimize the final, closed configuration of the heap. Optimization of the design led to significant cost savings from decreased solution handling requirements, construction requirements, and permitting efforts. The investigation was effective in significantly reducing costs, providing for a benefit to cost ratio which significantly exceeds 1.0 in the short term, helps to ensure that the proper long-term closure options lead to an optimized closure cost, and a design which will protect the local environment.

REFERENCES

EarthInfo (1994). NCDC Summary of the Day Climatic Database; West1 1994; Boulder, Colorado.

Fredlund, D. G. and Rahardjo, H., (1993) Soil Mechanics for Unsaturated Soils. John Wiley and Son's, Inc., New York.

SoilCover, 2000. Unsaturated Soils Group, Department of Civil Engineering, University of Saskatchewan, Saskatoon, Canada.

van Genuchten, M.Th., Leij, F.J., and S.R. Yates (1991). The RETC Code for Quantifying the Hydraulic Functions of Unsaturated Soils. United States Environmental Protection Agency. EPA/600/2-91/065. Washington, D.C.

WASY GmbH (2000); FEFLOW Version 4.8 Finite Element Subsurface Flow & Transport Simulation System; WASY Institute for Water Resources Planning and Systems Research Limited, Berlin, Germany.

Surface water quality

Tailings and Mine Waste '02, © 2002 Swets & Zeitlinger, ISBN 90 5809 353 0

Turnover in pit lakes: I. Observations of three pit lakes in Utah, USA

Devin Castendyk
School of Environmental and Marine Sciences, The University of Auckland, Auckland, New Zealand

Paul Jewell
Department of Geology and Geophysics, University of Utah, Salt Lake City, Utah, USA

ABSTRACT: Turnover events influence the chemistry of pit lake water by circulating dissolved oxygen and controlling redox conditions at depth. To investigate the role of lake morphology on turnover and lake chemistry, three pit lakes that varied only in morphology were studied. Duncan, Blackhawk and Blowout lakes in Utah, have corresponding depths of 9, 26.5, and 71 m. Temperature, dissolved oxygen, conductivity, and water chemistry data were collected from each lake between 1997 and 1998. Results indicate holomictic behavior in the largest lakes due to the absence of strong vertical density gradients, while Duncan Lake is believed to circulate throughout the year owing to its shallow depth. Blackhawk and Duncan lakes behave similarly to other pit lakes of equivalent morphology, while Blowout Lake is unique among pit lakes with very high relative depths. The presence or absence of vertical density gradients may be more influential toward meromictic behavior than morphology alone.

1 INTRODUCTION

Pit lakes have the potential to become environmental hazards to adjacent water resources and to organisms utilizing the lake surface. As the practice of open pit mining becomes more prevalent, pit lake abundance will increase proportionally in the future (Miller et al. 1996). The possibility of environmental damage creates a need to understand the geochemical evolution of pit lake waters in an effort to predict lake chemistry before mining begins. Ultimate pit lake chemistry results from a combination of factors including the flux of solutes into and out of lake water, geochemical processes within the lake, microbial activity within the lake, and the physical behavior of the lake itself. This paper examines the physical behavior of pit lakes with respect to lake mixing, an event known as lake turnover.

Holomictic pit lakes exhibit complete turnover at least once a year. In the process of turnover, dissolved oxygen is distributed throughout the water column. During the summer and winter, surface water temperature and density are significantly different from the underlying water body, creating thermal stratification. For holomictic lakes, the surface layer is known as the epilimnion, while the deep layer is known as the hypolimnion. Oxygen is not supplied to the hypolimnion under stratified conditions. Shallow eutrophic lakes with limited hypolimnion volumes may become depleted in dissolved oxygen at depth due to organic decay. In deep oligotrophic lakes, the large volume of the hypolimnion may behave as a reservoir for dissolved oxygen over time

(Jewell 1992). Not only is oxygen available for organic decay, excess oxygen is available for the oxidation of sulfide minerals contained in the wall rock of the lake basin. Sulfide oxidation increases acidity and ion concentrations. The abundance of sulfides in precious metal deposits combined with the large hypolimnion volumes of pit lakes suggests subaqueous sulfide oxidation can affect pit lake chemistry. One example where this occurs is Yerington Lake, Nevada. Jewell (1999) found elevated levels of selenium in pit lake water to be the direct result of sulfide oxidation within the lake as apposed to groundwater flux.

In contrast, meromictic pit lakes exhibit incomplete turnover resulting in a permanently isolated layer at the base of the lake. For meromictic lakes, the surface layer is known as the mixolimnion, while the isolated layer is known as the monimolimnion. The monimolimnion does not have a source for dissolved oxygen and low redox conditions prevail. Redox conditions strongly influence water chemistry, including the redox state of aqueous species, the solubility of sulfide minerals, and the precipitation or dissolution of iron oxide species (Miller et al. 1996, Stumm & Morgan 1996, Davis & Eary 1997). Monimolimnion waters often display greater density than mixolimnion water. This is a product of increased solute concentrations, identified by changes in conductivity, total dissolved solids, or salinity.

The depth and frequency of turnover dictates whether a lake is holomictic or meromictic. To understand pit lake water chemistry, it is necessary to understand the factors that limit the turnover capacity of pit lakes. Most research in this area has focused on the morphology of pit lakes. Pit lakes tend to have a small surface area relative to the maximum depth. Several authors have suggested that these morphologic relationships predispose pit lakes to meromictic behavior (Lyons et al. 1994, Levy et al. 1995, Doyle & Runnells 1997). Other factors limiting turnover capacity include density gradients and the topography of the catchment area. Density gradients can result from low-density surface water input above high-density lake water, or high-density groundwater input below low-density surface water. The topography of the catchment area, such as high pit walls surrounding a pit lake, can reduce the wind energy applied at the lake surface, resulting in meromictic behavior.

In this paper, we investigate three pit lakes that differ primarily in morphology. The lakes are situated immediately adjacent to one another and can be assumed to have identical climate conditions, similar wall rock mineralogy, and similar input groundwater chemistry. Limnologic and geochemical results are compared with data from 10 pit lakes in North America to identify relationships between pit lake characteristics and turnover.

2 STUDY AREA

Blackhawk, Blowout, and Duncan pit lakes are situated on the southeast slope of Iron Mountain, in the Iron Springs Mining District of Utah (Fig. 1). Blackhawk Lake has a long axis equal to 184 m, a short axis equal to 111 m, a surface area of 16,000 m^2, and a maximum depth of 26.5 m. The largest lake, Blowout Lake, has a long axis equal to 224 m, a short axis equal to 192 m, a surface area of 33,800 m^2, and a maximum depth of 71 m. Duncan Lake is the smallest of the three, with a long axis equal to 132 m, a short axis equal to 63 m, and a surface area of 6,500 m^2. The average depth of Duncan Lake is 3 m, although one narrow region extends to 8 m.

The age and geology of the Iron Mountain pit lakes contributed to the observed lake chemistry. Open pit mining began in the mid 1930s and concluded in the late 1960s (Stegen 1979). Lakes received groundwater recharge for approximately 30 years prior to study, suggesting lake levels were near to steady state conditions when observed. This was confirmed by negligible changes in lake levels over the one-year study period. Lake basin wall-rock was predominantly limestone, with replacement deposits of iron oxide minerals, magnetite (Fe_3O_4) and hematite (Fe_2O_3) (Mackin 1968, Bullock 1970, Rowley & Barker 1978). Sulfide levels are negligible with the exception of Duncan Lake which has a sulfur concentration of 1% or greater (Bullock 1970). Small grains of pyrite (FeS_2) were noted in hand samples of iron ore from the wall-rock of Duncan Lake. The

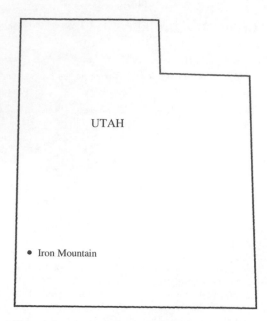

Figure 1. Iron Mountain location map.

abundance of carbonate minerals relative to sulfide minerals predisposed these lakes to alkaline conditions.

Four additional pit lakes are located in the district. Pinto and Burke lakes are proximal to the three lakes studied on the southeast slope of Iron Mountain. Comstock Lake is situated on the northeast slope of Iron Mountain. Desert Mound Lake is situated on the southwest slope of Granite Peak, approximately 10 km northeast of Iron Mountain. The abundance and morphologic variability of pit lakes in the Iron Springs Mining District, makes this an ideal location to study the effect of morphology on pit lake turnover.

3 METHODS

Six sampling trips were made to Iron Mountain at approximately two-month intervals, between September 1997 and July 1998. Blackhawk and Blowout lakes were sampled at five-meter intervals throughout the water column, while Duncan Lake was only sampled at the surface. A 1.5 L PVC lake sampler from Science Source was used to retrieve samples from a given depth. Dissolved oxygen (mg/L), temperature (°C), pH, and conductivity (μmho/cm) were measured at the surface, in this sequence. A Yellow Springs Institute (YSI) dissolved oxygen probe, model 50B, was used to measure dissolved oxygen and temperature. A Beckman Φ 12 pH/ISE meter was used to measure pH. Conductivity was measured using a YSI conductivity probe, model 33.

Three water samples were decanted from the lake sampler for cation, anion, and alkalinity analysis at the University of Utah in Salt Lake City. Cation and anion samples were filtered through a 45 μm membrane. The cation sample was acidified 1:5 with concentrated nitric acid. An unfiltered sample was collected for alkalinity analysis. All samples were packed in ice following collection, and refrigerated until analyzed. Major cations (Na^+, K^+, Ca^{2+}, and Mg^{2+}) were measured with a Perkin-Elmer 630 Atomic Adsorption Spectrophotometer. Major anions (F^-, Cl^-, NO_3^-, PO_4^{3-}, and SO_4^{2-}) were measured with a Dionex DX-100 Ion Chromatograph. Carbonate alkalinity ($CaCO_3$) was measured using the standard electrometric titration method described by Fishman & Friedman (1989). After analysis, the geochemical speciation program MINTEQA2 (Alison et al. 1991) was used to calculate potentially over-saturated mineral species and species charge balance for each data set.

4 LIMNOLOGY RESULTS

Temperature, dissolved oxygen and conductivity profiles are shown for Blackhawk Lake (Fig. 2) and Blowout Lake (Fig. 3). Summer, fall, winter and spring profiles were measured in early October 1997, early November 1997, mid February 1998, and early April 1998, respectively.

4.1 Blackhawk Lake

The summer temperature profile for Blackhawk Lake indicates thermal stratification, with an epilimnion extending from the surface to approxi-mately 10 m depth, a transitional layer, or metalimnion extending from approximately 10 to 20 m depth, underlain by a hypolimnion (Fig. 2). This apparent stratification is non-existent in the fall profile, suggesting a fall turnover event may have occurred. In the winter profile, thermal stratification is suggested by the increase in temperature with depth. This is likely to be a reflection of the frozen lake surface observed at the time of measurement. Temperatures are nearly homogeneous in the spring profile, suggesting a spring turnover event may have occurred.

The summer dissolved oxygen profile is similar to a positive heterograde curve described by Wetzel (1983), with a metalimnetic oxygen maximum occurring at approximately 15 m (Fig. 2). Oxygen maximums in the metalimnion are the product of two factors: the reduced solubility oxygen in the epilimnion due to warm summer temperatures, and oxygen consumption within the hypolimnion resulting from organic decay (Wetzel 1983). The depletion of oxygen with the hypolimnion may be indicative of the limited oxygen supply within a small hypolimnion volume, as described by Jewell (1992). In the fall profile, dissolved oxygen appears to be constant throughout the water column, supporting the occurrence of a fall turnover event. Compared to the summer profile, a progressive increase in dissolved oxygen is noted between the fall, winter, and spring profiles. This is possibly the result of progressively cooler water circulating. Given the temperature dependence of oxygen solubility, cooler waters will transport more dissolved oxygen than warmer waters. Spring turnover is suggested by elevated dissolved oxygen concentrations and the relatively uniformity of dissolved oxygen with depth.

Fall and spring conductivity profiles are nearly constant with depth, while the winter profile appears to indicate density stratification (Fig. 2). Since conductivity is reflective of solute concentrations, there appears to be minimal change in solute concentra-

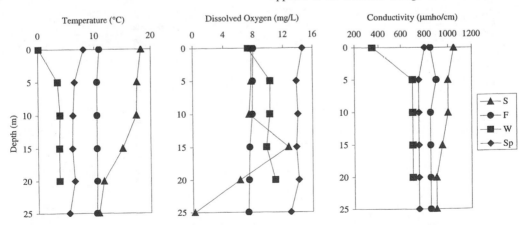

Figure 2. Seasonal temperature, dissolved oxygen and conductivity profiles for Blackhawk Lake. The summer (S), fall (F), winter (W), and spring (Sp) profiles were meas ured in early October 1997, early November 1997, mid February 1998, and early April 1998, respectively.

tion with depth during the fall and spring. Strong solute concentration gradients are indicative of strong density gradients that can inhibit turnover. Therefore, the fall and spring conductivity profiles may indicate the absence of strong chemical density gradients, a precondition for turnover. A slight decrease in conductivity with depth is noted in the summer profile. This is possibly the product of the evapoconcentration of surface waters yielding elevated solute concentrations and conductivity in the epilimnion. The winter conductivity profile may reflect ice covering the lake surface at the time of measurement, resulting in low surface water conductivity.

On the basis of these data, we believe that Blackhawk Lake undergoes turnover twice a year, during the spring and fall.

4.2 Blowout Lake

The summer temperature profile in Blowout Lake suggests the existence of thermal stratification, with an epilimnion layer extending to approximately 15 m, a metalimnion layer between approximately 15 and 25 m, and a hypolimnion layer below 25 m (Fig. 3). This apparent stratification does not occur in the winter profile, which displays relatively uniform water temperature with depth. Similarly, the spring temperature profile is nearly constant.

In meromictic lakes, dissolved oxygen in the monimolimnion decreases over time. The presence of dissolved oxygen in Blowout Lake below the previously defined metalimnion at approximately 15 m, strongly suggests that Blowout Lake exhibits holomictic behavior (Fig. 3). A positive heterograde curve as described by Wetzel (1983) is shown in the summer profile, with a metalimnetic maximum at approximately 15 m. In this profile, dissolved oxy-

gen decreases between approximately 15 m and 35 m, and appears to be constant below approximately 35 m. Dissolved oxygen in the hypolimnion during summer may reflect the capacity of large hypolimnion to behave as oxygen reservoir, as suggested by Jewell (1992). Compared to the summer profile, dissolved oxygen concentrations are reduced in the winter profile below approximately 15 m. Above this depth, the winter profile may show an increase in dissolved oxygen due to lower surface water temperatures and the temperature dependency of oxygen solubility. In the spring profile, mean dissolved oxygen concentrations are 1 to 2 mg/L greater than the winter profile, below a depth of approximately 25 m. The spring profile also shows less variability with depth than the summer and winter profiles. These data may indicate that a turnover event has occurred in the spring.

With the exception of the surface water measurement in the spring profile, conductivity profiles appear to exhibit uniformity with depth throughout the study period (Fig. 3). Given the relationship between conductivity and solute concentrations, it is unlikely that strong differences in solute concentration existed in the water column during the spring, with the exception of surface water. Therefore, it is likely that strong density gradients existed during the spring, which could impede turnover. The slight rise in conductivity in the spring profile may be a product of the evapoconcentration of surface water. Conductivity values are approximately 150 μmho/cm greater in the summer profile than in the winter or spring profiles. To reduce conductivity at depth, the addition of dilute groundwater or surface water is required. Dilution could be a result of circulating water from the surface during turnover.

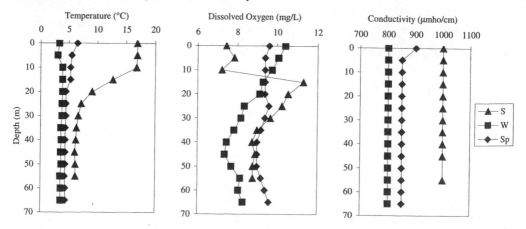

Figure 3. Seasonal temperature, dissolved oxygen and conductivity profiles for Blowout Lake. The summer (S), winter (W), and spring (Sp) profiles were measured in early October 1997, mid February 1998, and early April 1998, respectively.

The dissolved oxygen data suggest Blowout Lake exhibited one turnover event in the spring. Other data support this observation.

4.3 Duncan Lake

A detailed limnologic analysis of Duncan Lake was not conducted. With the exception of a narrow pocket, the average depth of the lake was 3 m. Given the average depth of the lake, it was assumed that the lake circulated throughout the year.

5 WATER CHEMISTRY

Chemical data for Blackhawk, Blowout, and Duncan lakes collected in July 1998, are presented (Tab. 1). Solute profiles for Blackhawk and Blowout lakes showed minimal variability with depth. For this reason, the value presented represents the average value for a given parameter across all lake depths. Speciated charge balances were calculated using the average value. Since the speciated charge balance did not exceed the general tolerance level of 10%, these data are a reasonable representation of average lake chemistry.

Data from Duncan Lake were derived from a single sample collected at the lake surface.

5.1 Blackhawk Lake

Blackhawk Lake was found to have a slightly alkaline solution pH, as speculated from the composition of basin wall-rock (Tab. 1). Conductivity levels were low, reflecting an equally low concentration of total dissolved solids (TDS). Calcium (Ca^{2+}) and magnesium (Mg^{2+}) were the dominant cations, showing nearly equivalent concentrations, while chloride (Cl^-) was the dominant anion. Sulfate (SO_4^{2-}) levels were low, suggesting minimal sulfide oxidation. Geochemical modeling with MINTEQA2 indicated super-saturated conditions with respect to limestone ($CaCO_3$) and aragonite, a limestone polymorph.

5.2 Blowout Lake

Blowout Lake was found to have a similar chemistry to Blackhawk Lake, and identical super-saturated mineral species as predicted by geochemical modeling (Tab. 1). The principle difference was that calcium concentrations exceeded magnesium concentration by approximately 25 mg/L, defining calcium as the dominant cation.

Blowout Lake exhibits a slightly great concentration of all constituents than Blackhawk Lake. This may be the result of dilution or hydrology. The catchment area of both lakes is nearly equivalent, while the volume of Blowout Lake is significantly greater. During precipitation events, both lakes should receive an equal volume of freshwater runoff. This could dilute the solute concentrations of Black-

Table 1. Iron Mountain Water Chemistry, July 1998.

Parameter (mg/L)	Blackhawk	Blowout	Duncan
Temperature (°C)	15.70	11.40	21.30
pH	8.33	8.21	8.59
Conductivity (μmho/cm)	950.00	1000.00	8000.00
Na^+	31.80	54.00	328.00
K^+	2.60	4.00	51.00
Mg^{2+}	63.20	68.30	805.00
Ca^{2+}	65.70	91.30	411.00
F^-	0.39	0.50	-
Cl^-	185.00	213.00	1730.00
NO_3^-	0.65	2.09	-
PO_4^{3-}	-	-	-
SO_4^{2-}	106.00	158.00	3120.00
Carbonate Alkalinity ($CaCO_3$)	139.00	171.00	77.00
Total Dissolved Solids	622.00	799.00	6539.00
Speciated Charge Balance (%)	1.70	0.60	7.70

hawk Lake with respect to Blowout Lake. Alternatively, the flux of solutes into and out of each lake may be different as a result of variable groundwater flow entering and leaving each lake.

5.3 Duncan Lake

Duncan Lake is significantly more concentrated than Blackhawk and Blowout lakes (Tab. 1). Conductivity and TDS are nearly one order of magnitude greater than either lake. The lake has a solution pH of 8.6, slightly more basic than the other lakes. Carbonate alkalinity is lower, at 77 mg/L. The dominant cation is magnesium, while the dominant anion is sulfate.

Elevated sulfate is possibly the product of sulfide oxidation, given that sulfide minerals are more abundant in Duncan Lake than the other locations (Bullock 1970). Reduced carbonate alkalinity may result from the neutralization of acidity generated by sulfide oxidation.

Geochemical modeling indicates the lake solution was supersaturated with respect to limestone, aragonite, and gypsum ($CaSO_4 \cdot 2H_2O$). Precipitation of gypsum could account for greater magnesium concentrations relative to calcium.

The overall concentrated water chemistry of Duncan Lake is believed to be a product of higher evaporation rates resulting from the shallow average depth of the lake (i.e. 3 m). This may be a direct effect of morphology on lake chemistry.

6 PIT LAKE COMPARISONS

To further explore the occurrence of turnover in pit lakes, data from Iron Mountain pit lakes were compared with morphologic, limnologic and geochemical data form 10 pit lakes in North America (Tab. 2). Lakes are listed in order of increasing relative depth. Relative depth (z_r) expresses the maximum depth (z_m) of a lake as a percentage of mean lake diameter (Hutchinson 1957):

$$z_r = \frac{50 \cdot z_m \cdot \sqrt{\pi}}{\sqrt{A_0}} \qquad (1)$$

where A_0 represents the surface area of the lake. Relative depth has been frequently used to compare the morphology of lakes and to account for limnologic behavior (Walker and Likens 1975, Anderson et al. 1985, Levy et al. 1995, Doyle & Runnells 1997). Lake turnover status, maximum conductivity (μmho/cm) and maximum TDS (mg/L) are presented following relative depth data. We attempt to relate each of the Iron Mountain pit lakes to a pit lake previously described in the literature.

6.1 Blackhawk, Yerington and Aurora Lakes

Blackhawk Lake appears to have similar characteristics to Yerington and Aurora lakes, in Nevada (Tab. 2). These lakes have a relative depth between 17% and 18%, exhibit summer stratification, are holomictic, and have TDS levels of approximately 500 to 600 mg/L, based on available data (Price et al. 1995, Miller et al. 1996, Atkins et al. 1997, Jewell 1999). The maximum depth and surface area of Blackhawk Lake is similar to the maximum depth and surface area of Aurora Lake. All three lakes are situated in the Great Basin and are subject to similar arid-temperate climatic conditions.

Despite having a moderately high relative depth, turnover may be possible in these lakes due to the absence of strong salinity gradients between lake layers, as reflected by the low TDS value reported. The absence of strong salinity density gradients enable wind energy applied to lake surfaces during periods of uniform water temperature to completely mix the water column. Jewell (1999) cited the low TDS and minimal surface water input to Yerington Lake as the factors enabling lake turnover.

As a product of seasonal turnover, dissolved oxygen is distributed throughout the water column in Blackhawk, Yerington, and Aurora lakes (Atkins et al. 1997, Jewell 1999).

6.2 Blowout Lake

Blowout Lake appears to be unique compared to the other pit lakes with very high relative depth (Tab. 2). In theory, the high relative depth of Blowout Lake should inhibit the turnover capacity of the lake, resulting in meromictic behavior. Past research on Spenceville, Gunnar and Berkeley lakes suggests that high relative depth pit lakes are predisposed to meromictic behavior due to the inability of wind energy to circulate the water column (Levy et al. 1995, Doyle & Runnells 1997). While all of the meromictic lakes identified have a high relative depth between 20% and 40%, Blowout Lake is holomictic and has a relative depth of 35%.

One possible explanation for holomictic behavior in Blowout Lake is the absence of strong vertical density gradients, as indicated by low conductivity and low TDS concentrations. The limited data provided for meromictic pit lakes show conductivity levels approximately 1000 to 11,000 μmho/cm greater than Blowout Lake, and TDS levels 200 to 4800 mg/L greater. This could reflect the presence

Table 2. Morphology, limnology and chemistry of some existing pit lakes in North America.

Pit Lake	Location	Maximum Depth (m)	Surface Area (km²)	Relative Depth (%)	Turnover Behavior	Maximum Conductivity (μmho/cm)	Maximum TDS (mg/L)
Duncan	Utah	8	0.0065	9	Holomictic	8000	6539
B-Zone*	Saskatchewan	55	0.2900	9	Holomictic	-	-
Boss**	Nevada	7	0.0025	12	Holomictic	-	12,000
Yerington**	Nevada	109	0.3100	17	Holomictic	-	631
Blackhawk	Utah	26	0.0160	18	Holomictic	950	622
Aurora**	Nevada	20	0.0100	18	Holomictic	-	491
D Pit*	Saskatchewan	26	0.0160	18	Holomictic	-	-
Brenda***	British Colombia	140	0.3800	20	Meromictic	12,000	-
Island Copper	British Columbia	380	1.9000	24	Meromictic	-	-
Spenceville	California	17	0.0020	34	Meromictic	12,000	-
Blowout	Utah	71	0.0340	35	Holomictic	1000	799
Gunnar**	Saskatchewan	110	0.0700	37	Meromictic	2000	1050
Berkeley**	Montana	242	0.2900	40	Meromictic	7000	5000

* Doyle & Runnells 1997
** Price et al. 1995, Miller et al. 1996, Atkins et al. 1997, Jewell 1999
*** Stevens & Lawrence 1998
**** Fisher & Lawrence 2000
***** Levy et al. 1995
****** Tones 1982, Doyle & Runnells 1997
******* Davis & Ashenberg 1989, Doyle & Runnells 1997

of strong vertical density gradients in the meromictic lakes shown. The presence of strong vertical density gradients contributes toward meromictic behavior in Brenda, Island Copper, Spenceville, Gunnar, and Berkeley lakes, while the absence of strong vertical density gradients could enable turnover in Blowout Lake.

Lakes presented with relative depth greater than 20% are situated in different North American locations and are subject to variable climate conditions (Tab. 2). This is likely to have an influence upon resulting lake chemistry.

It is important to note that morphology ratios, such as relative depth or the Petersen Scaling Parameter employed by Lyons et al. (1994), do not reflect the presence of vertical density gradients that can lead to meromictic behavior. While morphology ratios are useful tools for measuring the general tendency of lakes to turnover and remain oxygenated, the presence or absence of vertical density gradients may cause pit lakes to behave differently than predictions based on morphologic ratios alone. Blowout Lake is one example were this occurs.

6.3 Duncan and Boss Lake

Duncan Lake appears to have similar characteristics to the Boss Lake in Nevada (Tab. 2). Both lakes have a maximum depth between 7 and 8 m and a surface area in the order of one thousandth of a square kilometer. Resulting relative depths are between 9% and 12%. Water chemistry shows elevated concentrations of TDS ranging from 6500 mg/L in Duncan Lake to 12,000 mg/L in Boss Lake. Both lakes are situated in the Great Basin, and subject to similar arid-temperate climatic conditions.

Studies by Atkins et al. (1997) show that Boss Lake circulates throughout the year and does not display summer stratification. While a detailed limnologic study of Duncan Lake was not conducted, the shallow, maximum depth of Duncan Lake is thought to allow for similar circulation behavior. Given the shallow depth of both lakes, wind energy applied at the lake surface may physically mix the entire water column throughout the year. High dissolved oxygen concentrations in Boss Lake were attributed to constant circulation (Atkins et al. 1997).

Elevated conductivity and TDS levels are believed to be the product of high evaporation rates, resulting from lake morphology and regional climate. Water temperatures increase more rapidly in shallow lakes than deep lakes, contributing to the rate of evaporation. Both lakes are situated in an arid climate zone, where net evaporation exceeds net precipitation.

7 DISCUSSION & CONCLUSIONS

Limnologic observations of three pit lakes in Utah that varied only in depth, identify holomictic behavior in both a medium (i.e. Blackhawk) and a large (i.e. Blowout) pit lake. The smallest lake (i.e. Duncan) was believed to undergo constant turnover as a product of shallow average depth. Water chemistry data for Blackhawk and Blowout lakes showed these lakes are chemically similar, while Duncan Lake is concentrated. Elevated concentrations in Duncan Lake are believed to be the product of shallow lake morphology.

When compared with other existing pit lakes, Blackhawk Lake appears to be similar to Yerington and Aurora lakes in Nevada, which exhibit similar morphology and holomictic behavior. Blowout Lake is unique among pit lakes because it exhibits high relative depth and holomictic behavior. The absence of strong vertical density gradients is suspected to allow turnover in this lake, whereas other lakes with strong density gradients become meromictic. Vertical density gradients are not reflected by morphology ratios, such as relative depth. Duncan Lake is similar to Boss Lake, in Nevada. Both lakes are shallow, which may promote the elevated water concentrations observed.

In moderate to deep pit lakes, morphology appears to have a limited influence upon turnover behavior. The presence or absence of vertical density gradients may exert a stronger influence. Turnover and lake chemistry are strongly influenced by morphology in shallow lakes, which allow perpetual turnover and promote elevated concentrations.

8 ACKNOWLEDGMENTS

Special thanks are extended to Melissa Mitchell, Greg Waite, Scott Tangenberg, and Chuck Williamson for their assistance in the field.

REFERENCES

Allison, J.D., Brown, D.S. & Novo-Gradac, K.J. 1991. *MINTEQA2/PRODEFA2, a geochemical assessment model for environmental systems: version 3.0 user's manual.* Athens, Georgia: U.S. Environmental Protection Agency.

Anderson, R.Y., Dean, W.E., Bradbury, J.P. & Love, D. 1985. *Meromictic lakes and varved lake sediments in North America (U.S. Geological Survey bulletin 1607).* Washington D.C.: United States Government Printing Office.

Atkins, D., Kempton, J.H., Martin, T. & Maley, P. 1997. Limnologic conditions in three existing Nevada pit lakes: observations and modelling using CE_QUAL_W2. *Proceedings of the Fourth International Conference on Acid Rock Drainage* 2: 697-713.

Bullock, K.C. 1970. *Iron Deposits of Utah, Utah Geological and Mineralogical Survey Bulletin 88*: 23-62. Salt Lake City, Utah: Utah Geological and Mineralogical Survey.

Davis, A. & Ashenberg, D. 1989. The aqueous geochemistry of the Berkeley Pit, Butte, Montana, U.S.A. *Applied Geochemistry* 4: 23-36.

Davis, A. & Eary, L.E. 1997. Pit lake water quality in the western United States: an analysis of chemogenetic trends. *Mining Engineering* 6: 98-102.

Doyle, G.A. & Runnells, D.D. 1997. Physical limnology of existing mine pit lakes. *Mining Engineering* 12: 76-80.

Fisher, T.S.R. & Lawrence, G.A 2000. Observations at the upper halocline of the Island Copper pit lake. *Fifth International Symposium on Stratified Flows* 1: 413-418.

Fishman, M. & Friedman, L. (eds.) 1989. Alkalinity, electrometric titration. In *Techniques of water-resource investigation of the United States Geological Survey: Methods for determination of inorganic substances in water and fluvial sediments*: 55-56. Washington D.C.: United States Department of the Interior.

Hutchinson, G.E. 1957. *A Treatise on Limnology*. New York: Wiley and Sons.

Jewell, P.W. 1992. Hydrodynamic controls of anoxia in shallow lakes. In J.K. Whelan & J.W. Farrington (eds), *Organic matter: production, accumulation, and preservation in recent and ancient sediments*: 201-228. New York: Columbia University Press.

Jewell, P.W. 1999. *Stratification and geochemical trends in the Yerington pit mine lake, Lyon County, Nevada (unpublished report)*. Reno, Nevada: Nevada Division of Environmental Protection.

Levy, D.B., Custis, K.H., Casey, W.H. & Rock, P.A. 1995. Geochemistry and physical limnology of an acidic pit lake. In *Tailings and Mine Waste 1995*: 479-489. Rotterdam:Balkema.

Lyons, W.B., Doyle, G.A., Petersen, R.C. & Swanson, E.E. 1994. The limnology of future pit lakes in Nevada: the importance of shape. In *Tailings and Mine Waste 1994*: 245-248. Rotterdam:Balkema.

Mackin, J.H. 1968. Iron ore deposits of the Iron Springs District, southwest Utah. In *Ore Deposits of the United States, 1933-1967*: 992-1019. New York: The American Institute of Mining, Metallurgical, and Petroleum Engineers, Inc.

Miller, G.C., Lyons, W.B. & Davis, A. 1996. Understanding the water quality of pit lakes. *Environmental Science and Technology* 30(3): 118A-123A.

Price, J.G., Shevenell, L., Henry, C.D., Rigby, J.G., Christensen, L., Lechler, P.J., Desilets, M., Fields, R., Driesner, D., Durbin, W. & Lombardo, W. 1995. *Water quality at inactive and abandoned mines in Nevada, Report of cooperative project among state agencies, Open-File Report 95-4*. Reno, Nevada: Nevada Bureau of Mines and Geology.

Rowley, P.D. & Barker, D.S. 1978. Geology of the Iron Springs Mining District, Utah. In *Guidebook to Mineral Deposits of Southwestern Utah*: 49-58. Salt Lake City, Utah: Utah Geological Association.

Stegen, R.J. 1979. *Utah Mineral Occurrence Files – Page Ranch Quadrangles*. Salt Lake City, Utah: Economic Division of the Utah Geological Survey.

Stevens, C.L. & Lawrence, G.A. 1998. Stability and meromixis in a water-filled mine pit. *Limnology and Oceanography* 43(5): 946-954.

Stumm, W. & Morgan, J.J. 1996. *Aquatic Chemistry*. New York: Wiley and Sons.

Tones, P.I. 1982. *Limnological and fisheries investigations of the flooded open pit at the Gunnar uranium mine, Saskatchewan Research Council Report C-803-10-E-82*. Saskatchewan: Saskatchewan Research Council.

Walker, K.F. & Likens, G.E. 1975. Meromixis and a reconsidered typology of lake circulation patterns. *Verhandlungen Internationale Vereinigung fur theoretische und angewandte Limnologie* 19: 442-458.

Wetzel, R. 1983. *Limnology*. Philadelphia: Saunders College Publishing.

Tailings and Mine Waste '02, © 2002 Swets & Zeitlinger, ISBN 90 5809 353 0

Turnover in pit lakes: II. Water column stability and anoxia

Paul Jewell
Department of Geology and Geophysics, University of Utah, Salt Lake City, Utah, USA

Devin Castendyk
School of Environmental and Marine Sciences, The University of Auckland, Auckland, New Zealand

ABSTRACT: The tendency of pit mine lakes to permanently stratify and become anoxic has tremendous influence on the accumulation of trace metals and thus the environmental liability of mine owners and operators. Most pit mine lakes have a large depth/surface area ratio commonly referred to as the relative depth (z_r). This, plus the fact that pit walls tend to shelter the lake from wind stress, means that pit lakes do not mix as easily as natural lakes and therefore have a tendency to stratify. In order to become permanently stratified, the density contrast between the surface and deep waters of a lake must be sufficiently high that wind-induced shear stresses cannot mix the water column. The gradient Richardson number (Ri), defined as the ratio of vertical density gradient to the square of the vertical velocity gradient, is typically used to quantify these two competing forms of energy. A water column is permanently stratified when the Ri > 0.25. This is more likely occur when z_r is large because wind energy is less effective at penetrating the water column. Detailed measurements of currents and density at the Yerington, Nevada pit mine lake confirms the generalized application of the Richardson number criteria. Numerical models of lakes also show that intermediate depth lakes (10–30 m) lakes are more likely to become seasonally anoxic while shallow lakes (< 10 m) are neither permanently nor seasonally anoxic.

1 INTRODUCTION

A number of environmental concerns surround pit mine lakes. Perhaps the most important is chemical evolution of water within the pit. During initial stages of filling, the pit is considered to be terminal whereby ground water flows into the lake, but not out. As the lake level begins to approach that of the surround potentiometric water surface, the lake takes on a "flow through" character in which groundwater upgradient of the lake flows in and evolved pit lake water becomes a source of dissolved constituents for water flowing down the hydraulic gradient and away from the lake. The nature of pit lake water as a function of time is therefore of considerable concern with respect to groundwater quality in the surrounding area. Additional concerns come from the pit lake itself which despite efforts to the contrary is often used as habitat and drinking water by wildlife and birds.

The hydrology of any lake can be described within the context of a simple model in which water enters by groundwater or surface water inflow and leaves by groundwater or evaporation. Chemical constituents enter or leave the lake in the same manner with additional sources and sinks from dissolution of minerals due to water-rock interactions in the pit lake walls; adsorption/desorption onto clays, Fe-oxides, and organic matter; and the formation of solid phases which become sequestered in sediment at the pit lake bottom. The formation of many solid mineral phases can be enhanced or retarded by the redox state of the water column.

The ability to predict the redox state of pit waters is crucial to any prognostic model of lake water quality. The solubility of many metals and colloidal iron is strongly redox dependent. For instance, it has long been recognized that most sulfide minerals are insoluble in low redox state waters (Stumm & Morgan, 1996) meaning that the dissolution of sulfides in these environments is minimal. On the other hand, colloidal iron strongly absorbs selenium and arsenic in oxygenated environments and will dissolve and release these elements in anoxic waters (Davis & Eary, 1997). Redox conditions in the water column therefore play a critical role in determining which suite of trace metals may end up in solution when ground water interacts with ore-bearing rocks.

Surface waters of lakes tend to be oxygenated due to the exchange of gases with the atmosphere. In the deeper portion of lakes, oxygen is consumed by organic matter produced by photosynthesis at the surface or the oxidation of sulfide minerals at depth. The replenishment of oxygen to these deep waters is dependent on (1) solar heat flux which warms the upper water and tends to stratify the water column, (2) vertical solute gradients which also stratify the water, and (3) wind shear stress which tends to mix the water column. The redox state of the deep water of lakes is dependent on the two counteracting effects of vertical density gradients and velocity shear. In the past, it has been believed that lakes with large depth/area ratio (known as relative depth, r_z) are less prone to deep water mixing due the more limited effects of wind-induced currents. Pit lakes tend to have very large relative depths (Lyons et al., 1994; Doyle & Runnels, 1997). While this ratio is a useful general measure of the tendency of lakes to mix and remain oxygenated, this approach neglects the complex development of vertical density gradients which prevent deep hypolimnonic water from mixing with the overlying epilimnion (Castendyk & Jewell, 2002). In addition to pit lake geometry, the presence of high walls tends to shelter the lake surface from winds and thus is a potentially important factor for inhibiting water column mixing.

This paper outlines the general theoretical framework for water column stability and dissolved oxygen in pit lakes and then places that framework within the context of existing pit lake data and a detailed study of the Yerington, Nevada pit lake (Fig. 1).

1.1 Methods

Data on the hydrology, climatology, and geochemistry of the Yerington mine were collected between May, 1998 and May, 1999 and incorporated with existing data from previous studies of the lake. The combined data sets were then used to construct simple hydrodynamic and geochemical models of the lake waters.

A portable meteorological station was deployed near the lake surface on the southern access road to the lake. The station sampled wind direction and speed and air temperature every hour. The data were stored internally and downloaded during visits to the lake which took place at intervals of 6-7 weeks. A continuous meteorological record was collected for the times periods of May 2 to September 18, 1998 and from December 12, 1998 to May 12, 1999. The gap in the meteorological record toward the end of 1998 was the result of technical problems in data collection.

Figure 1. Location map of the Yerington pit lake.

Data on water velocity and temperature were collected with an S-4 current meter manufactured by Applied Microsystems. The S-4 is an electromagnetic current meter which produces a small electrical field during operation. Moving water disturbs the field and generates the measurement. This type of current meter is capable of measuring very low velocities (~1 cm/s) of the type expected in pit mine lakes. For this project, the probe was retrofitted with thermistor with an accuracy of ±0.01°C in order to test the validity of numerical models of temperature.

The probe was deployed at a 10 m depth in the middle of the lake and programmed to sample the water over a 10 minute period which was then recorded at a variety of time intervals. The 10 m depth was intended to correspond to the approximate depth of the thermocline. The probe was recovered and the data downloaded at the same time as the portable meteorological station. At the time of these visits, the probe was reprogrammed to significantly decrease the sampling interval (to the order of seconds) and a vertical profile of water column temperature was determined.

2 WATER COLUMN STABILITY

Previous studies of water column stability in pit lakes have applied lake surface area/depth ratio calculated as the Peterson Scaling Parameter (PSP) (Lyons et al., 1994) or relative depth (r_z) (Doyle & Runnels, 1997; Castendyk & Jewell, 2002). The underlying assumption of these parameters is that deep lakes are less prone to wind shear forcing and thus more likely to stratify. While these ratios have broad applicability to natural freshwater lakes, geothermal heat flux (primarily from geothermal waters), evaporation losses, and groundwater outflow are also recognized as key variables for understanding pit lake stratification (Lyons et al., 1994). Of these three variables, geothermal heat flow is the most difficult to predict. Detailed knowledge of bottom heat flux into a lake is site specific and will not be considered in the analysis here. Net groundwater flux to a lake is difficult to determine without detailed knowledge of local hydraulic conductivity and head gradients. The role of groundwater in lake stratification is important when the solute concentration of the groundwater is sufficiently greater than surface water flowing into lake so that wind shear will not mix the lake when the vertical thermal gradient reaches zero (during the late fall for temperate latitude lakes). The third variable, high seasonal evaporation, should increase the density of surface waters and thus promote mixing during high wind events.

At the most basic physical level, the tendency of a water mass to become stratified is dependent on the vertical density gradient (which encompasses the three variables mentioned by Lyons et al., 1994) and water velocity shear brought about by wind forcing (which is implied but not specifically considered in PSP and r_z analysis). All of these variables can be conveniently expressed as the gradient Richardson number, the ratio of the density gradient (which tends to stabilize a water column) and the velocity shear gradient (which tends to destabilize it). If the gradient Richardson number exceeds 0.25, the water column is believed to be stable (Turner, 1973).

$$Ri = \frac{-g\dfrac{\partial \rho}{\partial z}}{\rho\left(\dfrac{\partial V}{\rho z}\right)^2} \qquad (1)$$

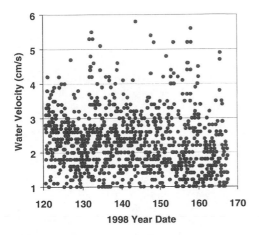

Figure 2. Water velocity at 10 m depth in the Yerington lake measured between approximately May 1 and June 18, 1998 (Jewell, 1999).

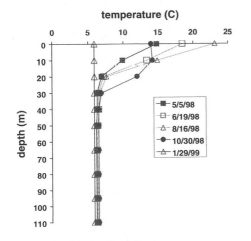

Figure 3. Water column temperature at Yerington between May, 1998 and January, 1999.

In Equation (1), V represents velocity and ρ is density.

2.1 Application to pit the Yerington pit lake

The gradient Richardson number can be used to make general determinations of the tendency of Yerington mine to mix or remain permanently stratified. The velocity gradient at Yerington was determined by taking the maximum measured velocity at 10 m (6 cm/s), at which time the water column would be most unstable (Fig. 2), over the shortest vertical distance at which the water velocity was probably close to zero (assumed to be the point where the vertical temperature gradient is zero). This depth is approximately 30 m (Fig. 3). Under these conditions $\partial V / \partial z$ is ~ 0.06 ms^{-1}/30 m = 0.002 s^{-1}. The denominator of the Richardson number is therefore 0.004 kg m^{-3}s^{-2}. In order to maintain a stable water column $\partial \rho / \partial z$ must be approximately [(0.004 kg m^{-3}s^{-2})*(0.25)]/9.8 ms^{-2} = 10^{-4} kg/m^3 m^{-1} or approximately 0.1 mg/L m^{-1}. This would correspond to a 3 mg/L change in TDS over a 30 m hypolimnion depth.

Although this density gradient seems modest, it would require considerable freshwater inflow to the lake during seasonal thermal stratification. If the surface area of Yerington is ~ 0.3 km^2 and TDS ~ 600 mg/L (Castendyk and Jewell, 2002), decreasing the hypolimnic TDS by 3 mg/L would require influx of ~ 5 x 10^4 m^3 of water during seasonal stratification. This would be in addition to calculated net seasonal evaporative losses from the lake surface of ~ 10^5 m^3 (Jewell, 1999). In other words, surface runoff into the lake would have to be significantly higher than calculated net evaporation in this very arid setting. The small surface catchment at Yerington and many other pit mine lakes in the arid Great Basin simply does not favor these conditions.

2.2 General observations of pit mine lakes

Application of the gradient Richardson number to pit mine lakes is limited by a lack of detailed velocity records. Even so, insight can be gained by considering analysis of published TDS and depth data from other pit mine lakes (Table 1). In general permanently stratified (meromictic) pit lakes are deep and have high TDS (Berkeley, Brenda, Gunnar). Shallow, high TDS lakes (Duncan, Boss) do not stratify since wind mixing homogenizes the water column. Rather, these lakes tend to become relatively saline due to evapoconcentration. Spenceville is a relatively shallow, permanently anoxic lake. This can be explained by the fact that a large amount of surface water runoff (1.35 x 10^4 m^3/yr) entered a small surface area (2000 m^2), high TDS lake over a very short period of time (2 years) (Levy et al., 1997).

Table 1. Summary of some existing pit lakes in North America

Pit Lake	Location	Maximum Depth (m)	Deep water Dissolved oxygen	Turnover Status	Conductivity (μmho/cm)	TDS (mg/L)
Duncan*	Utah	8	Oxygenated	Holomictic	8000	6539
Boss**	Nevada	7	Oxygenated	Holomictic	-	12,000
Yerington**	Nevada	109	Oxygenated	Holomictic	-	631 *******
Blowout*	Utah	71	Oxygenated	Holomictic	1000	799
Blackhawk*	Utah	26	Oxygenated	Holomictic	950	622
Aurora**	Nevada	20	Seasonally anoxic	Holomictic	-	491 ********
Brenda***	British Colombia	140	Permanently anoxic	Meromictic	8000-12,000	-
Spenceville****	California	17	Permanently anoxic	Meromictic	4000-12,000	-
Gunnar*****	Saskatchewan	110	Permanently anoxic	Meromictic	300-2000	175-1050
Berkeley******	Montana	242	Permanently anoxic	Meromictic	4000-7000	2000-5000

* Castendyk, 1999
** Atkins et al. 1997
*** Stevens & Lawrence 1998
**** Levy et al. 1995
***** Tones 1982
****** Davis & Ashenberg 1989
******* Miller et al. 1996
******** Price et al. 1995

3 DISSOLVED OXYGEN

Dissolved oxygen in the deep portions of lakes depends on stability of the overlying water column, the amount of biological productivity, amount of sulfide minerals in the wall rock, and water column depth. Permanent stratification will lead to deep water anoxia even when biological activity or oxidation of sulfides are modest. Under certain circumstances, seasonal stratification will produce anoxia and metal remobilization in the deepest portions of lakes. These conditions are examined here within the context of data from the Yerington mine and previously published numerical studies of lakes.

Dissolved oxygen concentration is a function of temperature and dissolved salts and is therefore most easily examined within the context of the their relative saturation in water. Gas solubility equations (Weiss, 1970) were used to determine oxygen solubility for a given temperature at Yerington. Measured oxygen concentrations could then be examined within the framework of saturation concentrations (Fig. 4). High temperature (>10°C) lake samples near the surface are generally saturated or supersaturated with respect to oxygen, a condition that is commonly observed in photosynthetically active lake and ocean waters (e.g., Broecker & Peng, 1982). Hypolimnotic waters are depleted in DO by 1-2 mg/L as a result of oxidation of organic matter sinking from the photic zone. The fact that deep water oxygen depletion was not greater in 1998-99 than it was in previous sampling years (Kempton, 1996) suggests that the lake turns over on an annual basis in addition to the overturn event observed in late 1998 (Jewell, 1999).

Oxygen in different holomictic pit mine lakes exhibits different seasonal behavior condition. For instance, the Yerington pit lake does not become sea-sonally anoxic whereas the much shallower Aurora pit lake does (Table 1). While Aurora has a some-what higher nitrate concentration (4-5 ppm vs. 1-3 ppm at Yerington) and thus higher biological productivity, the difference probably is due to the much shallower depth at Aurora (20 m vs. 110 m at Yerington) and can be explained within the context of a numerical modeling study of lake water column and dissolved oxygen (Jewell, 1992). The model used in this study was a 1-dimensional version of the Princeton Ocean Model (POM), a primitive equation model that has been widely used to study circulation in lakes, estuaries, and shallow coastal oceans. POM

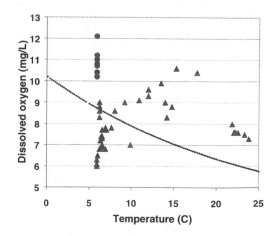

Figure 4. Temperature-dissolved oxygen data from the Yerington pit lake. Triangles represent data from Jewell (1999) and Kempton (1996) during thermal stratification while circles represent samples taken during lake overturn in January, 1999. The solid line represents calculated oxygen concentrations (Weiss, 1970).

192

incorporates a sophisticated turbulence closure scheme to predict vertical eddy viscosity and diffusivity and has a long track record in accurately predicting the physical, chemical, and biological characteristics of lakes and shallow marine settings (see reference list at http://www.aos.princeton.edu/WWWPUBLIC/htdocs.pom). For this study the hydrodynamic model was coupled to a simple biogeochemical model of dissolved and particulate nutrients and dissolved oxygen. Surface forcing of the model was accomplished using observed temperature and wind forcing sufficient to produce the observed thermocline depth.

In holomictic lakes, spring overturn provides the hypolimnion with a large dissolved oxygen reservoir capable of oxidizing the seasonal influx of organic matter from the epilimnion. For a fixed surface productivity, deep lakes (> 40 m) have less of a tendency to become anoxic than lakes which are between 10 and 40 m deep (Jewell, 1992) (Fig. 5). An intermediate depth lake such as Aurora exhibits seasonal stratification while a deeper lake such as Yerington does not.

4 CONCLUSIONS

Permanent stratification and anoxia in pit mine lakes is most likely when (1) the concentration of dissolved solutes in groundwater greatly exceeds that of inflowing surface water (2) surface water inflow is relatively high and (3) the lake is relatively deep. In order for a water body to become permanently stratified, vertical density gradients from freshwater surface inflow during seasonal thermal stratification must be sufficiently strong large that winds will not mix the waters once the lake becomes thermally homogenous in the fall and/or spring. Detailed analysis of the Yerington pit lake shows that these conditions are highly unlikely to occur and that the ·lake will continue to turn over on a seasonal basis. In fact, the magnitude of seasonal evaporation is such that the TDS of surface waters is greater than the TDS of deep waters during thermal stratification (Kempton, 1996; Jewell, 1999).

Seasonal oxygen depletion in pit mine lakes is a complex phenomenon that depends on biological productivity, oxygen consumption by sulfide oxidation, and lake depth. All other factors being equal, very shallow and very deep lakes tend to remain oxygenated during seasonal stratification while moderate depth lakes (10-40 m) are most likely to become anoxic.

5 ACKNOWLEDGMENTS

Financial support for the project was provided by Arimetco, Inc. Funds for detailed study of the sele-

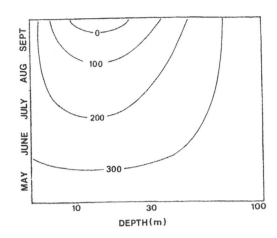

Figure 5. Bottom water oxygen concentrations (μmol/L) during seasonal stratification as a function of depth. 1 mg/L of oxygen is approximately equal to 43 (μmol/L). Results are from numerical simulations of lakes assuming a fixed biological productivity and wind shear (Jewell, 1992) and show that seasonal anoxia is most likely in lakes that are 10-25 m deep.

nium in sulfides was provided by the University of Utah Undergraduate Research Program in a grant to Melissa Mitchell. Field assistance at Yerington was provided by Ben Passey, Gary Robertson, Alfanso Rios, and Melissa Mitchell. A special thanks goes to Dennis Dalton and Joe Sawyer of Arimetco for sharing their knowledge of the Yerington pit lake during the course of this project.

REFERENCES

Atkins, D., Kempton, J. H., Martin, T., & Maley, P. 1997. Limnological conditions in three existing Nevada pit lakes: Observations and modeling using CE-QUAL-W2. *Fourth International Conference on Acid Rock Drainage, Proceedings Vol. II*: 697-713.

Broecker W. S. and Peng, T.-H. 1982. *Tracers in the sea.* New York, Eldigio Press.

Castendyk, D. N. 1999. *Chemical, hydrologic, and limnologic interactions at three pit lakes in the Iron Springs Mining District, Utah.* Unpublished M.S. thesis, University of Utah.

Castendyk, D. N. & Jewell, P. W. 2002. Turnover in pit lakes. I: Theory and observations. *Tailings and Mine Waste '96.* Rotterdam: Balkema.

Davis, A. & Ashenberg, D. 1989. The aqueous geochemistry of the Berkeley Pit, Butte, Montana, U.S.A. *Applied Geochemistry* (4) 23-26.

Davis, A. & Eary, L. E. 1997. Pit lake water quality in the western United States: an analysis of chemogenetic trends: *Mining Engineering,* 6, 98-102.

Doyle, G. A. & Runnells, D. D. 1997. Physical limnology of existing mine pit lakes. *Mining Engineering,* 12, 76-80.

Jewell, P. W., 1992, Hydrodynamic controls of anoxia in shallow lakes. In J. K. Whelan and J. W. Farrington (eds), *Organic matter: production, accumulation, and preservation in recent and ancient sediments.* New York, Columbia University Press.

Jewell, P. W. 1999. *Stratification and geochemical trends in the Yerington pit mine lake, Lyon County, Nevada*: unpublished report to the Nevada Division of Environmental Protection, 28 p.

Kempton, H. 1996. Report from PTI Environmental Services to Santa Fe Pacific Gold Corporation on chemical composition, limnology, and ecology of three existing Nevada pit lakes (PTI project No. CA1Q0601).

Levy, D. B. & Custis, K. H., Casey, W. H., and Rock, P.A. 1997. The aqueous geochemistry of the abandoned Spenceville copper pit, Nevada County Nevada. *Journal of Environmental Quality*, 26, 233-243.

Lyons, B. W., Doyle, G. A., Petersen, R. C. & Swanson, E. E. 1994. The limnology of future pit lakes in Nevada: the importance of shape. *Tailings and Mine Waste '94*. Rotterdam: Balkema Press.

Miller, G.C., Lyons, W.B. & Davis, A. 1996. Understanding the water quality of pit Lakes. *Environmental Science & Technology* (30): 118-123.

Price, J. G., Shevenell, L., Henry C. D., Rigby J. G., Christensen, L., Lecher, P. J., Desilets, M., Fields, R., Driesner, D., Durbin W., & Lombardo, W. 1995. Water quality at inactive and abandoned mines in Nevada, Report of a cooperative project among state agencies. *Nevada Bureau of Mines and Geology Open-File Report 95-4*.

Stevens, C.L. & Lawrence, G.A. 1998. Stability and meromixis in a water-filled mine pit. *Limnology and Oceanography* 43(5): 946-954.

Stumm, W. & Morgan. 1996. *Aquatic chemistry*. New York, Wiley and Sons.

Tones, P.I. 1982. Limnological and fisheries investigations of the flooded open pit at the Gunnar uranium mine (Saskatchewan Research Council Report C-803-10-E-82). Saskatchewan, Canada: Saskatchewan Research Council.

Turner, J. S. 1973. *Buoyancy effects in fluids*. Cambridge: Cambridge University Press.

Weiss, R. F. 1970. The solubility of nitrogen, oxygen, and argon in water and seawater. *Deep Sea Research* 17, 721-735.

Groundwater and geochemistry

Tailings and Mine Waste '02, © 2002 Swets & Zeitlinger, ISBN 90 5809 353 0

Long term persistence of cyanide species in mine waste environments

B. Yarar
Colorado School of Mines, Mining Engineering Department, Golden, CO 80401, USA

ABSTRACT: World headlines in the media, in recent years, have included tailings-dam failures and sigrufi-cant cyanide-spills. Presently, it is recognized that cyanides are essential for precious metal recovery from low grade resources with a substantial contribution to the world economy. However, simultaneously, search for alternative lixiviants is in progress, together with widespread, global anti-cyanide activism Health effects of cyanide compounds other than free-cyanide are often insufficiently addressed or remain not required for reporting to regulatory agencies. This paper addresses cyanidation-extraction byproducts and their long-term fates in the ecosystem Data show that a number of currently accepted cyanide abatement approaches can create long-lasting products that degrade at very low rates. It also underscores the need for research on the numerous long-term ecological and health effects of metal-cyanide compounds and complexes including their carcinogenic potentials.

1 INTRODUCTION

In 1988 about 2.46 metric tons of gold were produced in the world, and more than 90% of this was by cyanidation. The remaining 10% or so can be accounted for by gravity separations and such technologies (USGS, 1999; von Michaelis, 1984). Based on reaction (1) the molecular ratio of cyanideto-gold consumption is 2:1 which works out as 0.498 kg of NaCN for each kilogram of gold recovered.

$$Au + CN^- \Rightarrow Au(CN)_2^- \qquad (1)$$

Considering that some of the cyanide is recycledand some of it is lost to cyanicides, that is, cyanide-consuming-non-gold entities, the amount of sodium cyanide handled in the world for gold recovery comes out to be some 1000 metric tons per year. These numbers are summarized in Table 1 which excludes unrecorded numbers, likely, utilized in some developing countries.

Cyanide usage is not confined to the gold recovery industries but also includes metal recycling and metal plating and finishing enterprises. Notably, cyanide as NaCN and KCN, is a ubiquitous chemical used in the metallurgical industries since 1887. Flotation-concentration of Cu-Zn ores are also common users of cyarude as a modifying agent to enhance the recoveries and grades of products. Cyanides are also components of pesticide formulations and their degradation products and are generated by numerous plants and organisms as well (Towill, 1978; Huiatt, 1983). Yet, their high toxicity is firmly under control during usage. Control of pH and safe handling practices appear to be sufficient requirements for safety in the workplace. Figure 1 is the speciation-diagram for dissolved HCN gas whcrc it is sccn that at pH >10 essentially all of the cyanide is in the ionic (CN⁻) form.

Table 1. World gold production using cyanide; *basis for actual consumption is 0.4 kg NaCN/kg Au.

Gold production		kg, NaCN equivalent needed	
	kg	Stoichiometric	Actual*
World production (including US production	2,460,000	1225080	984,000
US production	366,000	182268	146400
US production by cyanidation	341,000	169818	136400
US cyanidation, in tanks and closed containers	238,000	118524	95200
US cyanidation in heaps and dumps	103,000	51294	41200
NaCN Usage (kg /kg Au)			
Stoichiometric kg NaCN/kg gold	0.498		
Usage in Canada (average kg NaCN/kg Au)	0.450		
Usage in S-Africa (average kg NaCN/kg Au)	0.280		
Usage in Free world (average kg NaCN/kg Au)	0.400		

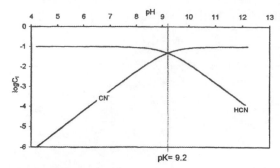

Figure 1. Speciation diagram for HCN in water (NaCN equivalent: 5 g /L).

Under these conditions, in the plant, cyanide is not the "ogre" that it has been made to be, in recent years.

In this paper, the primary focus is on the use of cyanides as lixiviants in the recovery of gold from natural resources. In this context, primary problems associated with it arise, mostly, after usage as a lixiviant (leaching-agent), in tailings ponds, active and abandoned heaps and at disposal after usage as a laboratory chemical and pilot plant material.

Numerous high profile and some secondary spills in the last decade have brought cyanide and its apparent perils to the limelight. A few of these are collected in Table 2. In all of the cases cited here, extensive efforts have been made for the remediation of the immediate damage to the environment with results considered satisfactory, while in some others such as the Summitville Mine in Colorado large sums of dollar have been dedicated towards remedia tion efforts.

Table 2. Some examples of recent high-profile cyanide accidents in the world.

Date	Accident
1992	Summitville, Colorado, USA; leaking mine abandoned, now a superfund site
1995	Omi gold mine, Guyana dam ruptures about 3.5 million tons of toxic waters are spilled with tailings.
1997	Nevada, gold quarry mine leach pad collapse releases 980 million tons of cyanide bearing slurry spills
1998	Kyrgyzstan-Issyk-Kul lake receives 1.7 tons of solid sodium cyanide from a truck overturn accident
1998	Homestake mining operation; Black-hills, S.D , USA 6 tons of cyanide-bearing tailings are lost to the environment.
2000	Baia Mare gold Mine, Romania, 100,000 tons of slurry is released into the rivers Lapus, and followed to Somes, Tizsa, and the Danube, from dam failure
2000	Papua, New Guinea, Dome Resources Freight helicopter spills some 100 kg of sodium cyanide during transportation

2 ENVIRONMENTAL CONTAMINATION

Potential mechanisms of contamination in mining and metallurgy related areas include the following:
1) During transportation of sodium cyanide to site of use
2) Tailings dam failures
3) Heap and dump failures
4) Seepage from heaps dumps and impoundments and
5) Illegal usage.

Especially in the cases of dam failures and seepage processes, cyanides take a secondary seat to metals such as Pb, Hg, Zn, Cu etc. which gain special notoriety and exacerbate abatement and control costs. The Summitville Colorado, Superfund site is a case in point where the current effort includes extensive attention to heaw metals. The cost of cleanup for this site has been estimated to be in excess of $100 million.

The primary concern, in addition to the elimination of the immediate effects of such spills is the need to provide answers to a number of questions: i.e.: "what happens to the toxic material?"; "is it really rendered harmless and destroyed permanently?" and "where does it all go?"

3 ANALYTICAL REPORTING METHODS

Cyanide compounds associated with mining and metallurgy can be broadly grouped, (Scott, 1981) as shown in Table 3. The strength of the metal-cyanide complexes on the other hand is better expressed in terms of their stability constants, that is, in essence, their resistance to dissociation under ambient conditions. These values (recalculated after Scott, 1981 and Beck, 1987) are compiled in Table 4, where it is clearly observed that gold, cobalt silver and iron complexes are thermodynamically highly stable against dissociation.

For compliance with regulatory agency requirements (in the USA) cyanides are reported in three forms:

a) Free cyanide, which includes HCN and CN^-
b) WAD (weak-acid-dissociable) cyanide and
c) Total cyanide

It is also customary to determine and report metals such as Zn, Cu, Ag, Hg, Mn, Fe etc. together with pH and TDS (total dissolved solids).

As can be seen from Table 5, these techniques do not permit the reporting of all species, among them cyanate (CNO^-) and thiocyanate (SCN^-) which are evidently not-negligible sources of toxic materials. Species missed by these techniques are considered innocuous although there is ample evidence indicat-

Table 3. Examples of cyanide species in groups as used in assay and reporting procedures

Type	Examples
Free cyanide	CN^- , HCN
Simple cyanide compounds (some partly soluble)	NaCN , KCN, $Ca(CN)_2$, $Zn(CN)_2$, $Cu(CN)_2$, $Ni(CN)_2$ Ag(CN) ,
Cyanide complexes	Weak complexes: $Zn(CN)_4^{2-}$ $Cd(CN)^{3-}$ Moderately strong complexes: $Cu(CN)_2^-$, $NI(CN)_4^{4-}$, Strong complexes: $Fe(CN)_6^{4-}$, $Co(CN)_6^{4-}$, $Au(CN)_2^-$ $Ag(CN)_2^-$

Table 4. Stability constants of some metal-cyanide complexes.

Species	$\log_{10}$ Ks
$Au(CN^-)_4$	56
$Au(CN^-)_2$	3 7.65
$Co(CN)_6^{4-}$	5 0
$Ag(CN)_2^-$	39.1
$Fe(CN)_6^{4-}$	35.4
$Cu(CN)_3^{2-}$	29.2
$Ni(CN)_4^{2-}$	26
$Cu(CN)_2^-$	19.95
$Zn(CN)_4^{2-}$	19.01
$Cd(CN)_4^{2-}$	18.93
$AgCN_{(s)}$	13.8

Table 5. Cyanide types determined for reporting and species they may miss.

Assay type	Reports	Test conditions	Misses
Free Cyanide	CN^-, HCN	alkaline pH	Most compounds and complexes of Ni, Co, Au, and PGM
WAD-Cyanide	free CN, HCN, compounds and complexes that break up at mild pH	$pH \cong 4\text{-}4.5$	Cyanates, SCN^-, complexes of Co, Ni, Au and Fe
Total Cyanide	Compounds and complexes that break up at $pH \leq 1$ in hot water	$pH \leq 1$ in hot water	Cyanates, SCN^-, complexes of Co, Ni, or PGM

ing that microorganisms, certain plants, crustaceans and other aquatic organisms such as fish and frogs are highly sensitive to their toxic effects.

Accumulation of cyanides is also known to occurin some plants and also fish. These include numerous complexes and SCN^-. Furthermore, not only fatal toxicity, but toxicity that reduces the agility of these organisms as well as their reproductive cycles should not be overlooked. Sometimes, low levels of ingested toxic concentrations, become lethal when temperature or oxygen content or heavy metals as well as nitrates and ammonia concentration in the aqueous environment change. Beside heavy metals such as Pb, Hg, Cu, Au, Co etc., sulfide species aris-

ing from the presence of sulfide minerals (e.g.; sulfide, polythionates, thio-sulfate and sulfate) can create thiocyanate (SCN^-) which is a reactive and persistent species of cyanide.

Persistent cyanide compounds present two types of hazard:

1: They are constant sources of cyanide emission and
2: They act as sources of heavy metals which would otherwise have been diluted as they travel away from the source.

The stabilities of metal cyanides and complexes constitute a "hazard" because in effect, they act as repositories of cyanide until an opportunity arises for its release. Taking for example $Zn(CN)_4^{2-}$ with a dissociation constant of 1.02×10^{-19}, on its own in equilibrium with alkaline water it is an emitter of small concentrations of CN^-. In contact with a sulfide containing solution, in the same environment however, it is liable to the formation of ZnS since the solubility product of ZnS equals 1.2×10^{-28} (Weast, 1979). The equilibrium thermodynamics of this system is clearly in favor of the displacement of CN^- by S^{2-}. Ion-sensitive electrode measurements conducted at the author's laboratory show that such reactions occur at high rates.

4 WHAT HAPPENS TO CYANIDE

Cyanides are disseminated and modified by two major routes: (1) Natural attenuation and (2) planned treatment procedures.

4.1 Natural attenuation and breakup

Processes such as evaporation, wind action, oxidation by oxygen from air, acidulation by CO_2/H_2CO_3; bacterial action, uv-radiation; dilution and attenuation by rocks and minerals, including adsorptions, precipitations and catalytic phenomena are operative natural processes. CO_2 which results in HCN generation at shallow depths of disposal lagoons is the primary operator, accounting for about 90 of all cyanide loss.

Yet, the sum of these mechanisms has only limited effect on the total rate of natural decyanidation of process waters in tailings ponds. Even then, the fmal products beside HCN, CNO⁻, SCN⁻ or NH_3 are not environmentally-friendly. Furthermore, many of those processes that involve rocks minerals and dissolved ionic species create precipitates and adsorbed species that are longerlasting than the original free-cyanide if it were to be treated.

On the optimistic side, it appears that natural degradations in abandoned heaps can be utilized as a preliminary de-cyanidation stage followed by chemical destruction rinses of the residual cyanide. This is a promising prospect in the exploitation of natural processes to this end. Indeed, some test data indicate that an 84000-ton heap, after 3 months of abandonment retains about 11.5% of the applied cyanide. The calculated amount of total cyanide retained in the pores of this heap is almost 4.4 tons. What is more interesting is that after another 18 months, some 85% of the cyanide present had been naturallydegraded (Huiatt 1983).

These natural processes and their products are compiled in Table 6. Thiocyanate, cyanate and metal cyanides in this table are the persistent ones which act, essentially, as constant reservoirs and emitters of cyanides and metal-ions.

Table 6. Natural cyanide attenuation processes and products.

ATMOSPHERE
Operators and reactants: Air currents, HCN, O_2, CO_2 H_2O, uv-radiation, air-bome-micro-organisms
Processes and Products: Dilution and dispersion processes and decompositions giving NH_3, HCOO⁻ and H_2 CO_3

WATER ENVIRONMENTS
Operators and reactants: HCN, CN-, O_2, Metals (e.g.: Ni^{2+}, Cu^{2+}, $Fe^{2+,3+}$, etc.), S^{2-}, microorganisms, mineral catalysts.
Processes and Products: Hydrolysis "HCOO⁻, NH_4^+
Precipitations and complexations, e.g. Ni $(CN)_4^-$, $Cu(CN)_2^-$ $Fe(CN)_6^{3,4-}$ SCN⁻ and degradations by uv-radiation e.g. CN⁻, CNO⁻, NH_3, HCO_3^- NO_3^- etc.

SOIL ENVIRONMENTS
Operators and reactants: HCN, CN⁻, H_2O, sulfur, micro-organisms and enzymes, metals, mineral catalysts
Processes and Products: Products, as in water-environments also numerous biodegradation products including NH_3, CO_2, CH_4, SCN⁻ and $Me(CN)_x^-$ species

---

4.2 Planned treatment procedures

The printed literature contains numerous cyanide removal and/or destruction methods (Davuyst, 1991; Botz, 1998; Robbins, 1996). These include :
(1) Physical separations, (2) biological procedures and (3) chemical destruction methods.

4.2.1 Physical separations
Membrane filtration, electro and reverse osmosis, use of zeolites and ion exchange resins, or indeed adsorption on active carbon are typical methods of physical separation.

Ion exchange or reverse osmosis may be used as "polishing steps" for treated water. The problem is: they are complex and costly to operate and generate their own saline solutions which need special arrangements for disposal. Presently active carbon is a potential alternative when all else fails. Its additional advantage is that it removes ions beside free cyanide and by-products such as Pb, Cu, Hg and also WAD cyanides.

Admixture of metal salts or metal oxides, hydroxides or composite ores such as the iron-rich bauxite with contaminated solutions can remove cyanide by the formation of $K_{3,4}Fe(CN)_6$, $Cu_2Fe(CN)_6$ or $Zn_2Fe(CN)_6$ followed by filtration. Such substances during impoundment however, can decompose by the influence of solar radiation, bacterial action or catalytic effects and emit HCN.

Nevertheless, these hybrid processes have not yet been demonstrated at industrial scale.

4.2.2 Biological oxidation
Micro-organism based oxidations can be represented by the following simplified equations:

$$CN⁻ + 1/2 \ O_2 \ [\text{enzymatic catalysis}] \Rightarrow CNO⁻ \qquad (2)$$

$$CNO⁻ + 3H_2O \ [\text{enzymes}] \Rightarrow NH_4^+ + HCO_3^- + OH \ (3)$$

Use of bacteria such as "Pseudomonas Pseudoalkaligenes" or "Bacillus Pumilus" isolated from mine-waters are known to degrade cyanide (Arps, 1994). One plant in the USA uses bacteria to remove cyanide and some of the heavy metals from the effluents of a metallurgical plant. In this practice a biological treatment by the "attached growth method" uses rotating biological contactors to facilitate the removal of cyanide, thiocyanate and some toxic metals from effluent water. The primary problems with this approach relate to the necessity of controlling temperature, bacterial nutrients such as phosphate and difficulty to maintain reproducible strains of bacteria.

4.2.3 Chemical breakup methods
Cyanide can be converted to other chemical compounds by numerous methods. A number of these are given in Table 7. Four widely-used industrial methods include the following: (1) Alkalichlorination, (2) Hydrogen peroxide treatment, (3) SO_2 / air oxidation and (4) Volatilization from acidified effluent solutions or slurries. Brief discussions of these methods are given below.

4.2.3.1 Ozone treatment
Ozone gas, is a strong oxidant and functions as shown in equations 4-6. It is used to a limited extent.

$$CN⁻ + O_3 \Rightarrow CNO⁻ + O_2 \qquad (4)$$

Table 7. Some methods used for cyanide abatement.

	Eftluent treatment method(s)	
Cyanide destruction by oxidation	Cyanide destruction by photolytic methods	HCN vaporization by reduced solution pH
Direct radiation	Electrical potential	Hydrolysis, distillation
With Ozone	Electrical potential with chlorine in medium	AVR
With hydrogen peroxide		SO_2/air
With catalysts	Chlorine/hypochlorite	Caro's acid
	Oxygen	
	Ozone	
	Hydrogen peroxide	
	Hydrogen peroxide plus Cu^{2+}	
	Hydrogen peroxide plus Castone (proprietary reagent)	

$$3\ CN^- + O_3 \Rightarrow 3\ OCN^- \qquad (5)$$

$$2CNO^- + 3O_3 + H_2O \Rightarrow 2HCO_3^- + N_2 + 3\ O_2 \qquad (6)$$

One property of ozone is that after the initial formation of SCN- in sulfur containing media, it generates HCN from thiocyanate; it needs to be used at pH<11 and plus, it is a costly substance to utilize at the industrial quantities needed.

4.2.3.2 Alkali chlorination
This is a well-established method that uses chlorine gas (Cl_2), sodium hypo-chloride (NaOCl) or calcium hypochlorite $Ca(OCl)_2$.

The basic reactions in this approach consist of a number of steps summarized in equations (7) and (8).

$$2CN^- + 5Cl_2 + 10\ OH^- \Rightarrow 10\ Cl^- + 2\ HCO_3^- + N_2 + 4\ H_2O \qquad (7)$$

$$CN^- + NOCl^- \Rightarrow CNO^- + Cl^- \qquad (8)$$

The problem with hypochlorite usage is that it generates (CNO⁻) which is objectionable. Application of chlorine gas needs provisions for the capture of fugitive gas while sodium and calcium hypochlorites are water-soluble powders.

4.2.3.3 Hydrogen peroxide
H_2O_2 as cyanide destructant, acts as follows :

$$CN^- + H_2O_2 \Rightarrow CNO^- + H_2O \qquad (9)$$

$$CNO^- + 2\ H_2O \Rightarrow NH_4^+ + CO_3^{2-} \qquad (10)$$

Reaction (9) needs high pH, (which is already present in gold processing effluents) and Cu^{2+} ions to proceed, while reaction (10) occurs at slightly acidic aqueous environments, without need for cupric ions.

An alternative H_2O_2-based reagent is commercially available (DuPont) and primarily consists of a formulation of hydrogen peroxide with formaldehyde (HCHO) with some proprietary additives. The mixing of reagents is conducted at about 50 °C and the net reaction is given as:

$$HCN + HCHO + H_2O_2 \Rightarrow$$
$$3\ NH_3.5HCNO.\ 5H_2(OH)CONH_2 \qquad (11)$$

The problem is that, not all of the CN⁻ is removed from the medium. Additional treatment technologies are needed for the satisfactory de-cyanidation of effluents.

4.2.3.4 SO_2 / air oxidation
The use of sulfur dioxide-air mixtures in the presence of Cu^{2+} as catalyst forms the basis of the INCO (International Nickel Company) process (Robbins, 1996) where the reaction is represented as follows:

$$CN^- + SO_2 + O_2 + H_2O \Rightarrow OCN^- + H_2SO_4 \qquad (12)$$

One of the arguments made against alkali chlorination (and in favor of the INCO process) is that if the effluent slurries contain iron arising from pyrite or pyrrhotite, ferro and ferricyanides are formed. These solids in turn can generate HCN when exposed to solar radiation at solid disposal sites. The INCO process is said to be among the ones that generate the least of harmful compounds, though, it may generate large quantities of $CaSO_4$ - rich sludge which adds to the cost of the process. A simplified flowsheet for the application of this technology is given in Figure 2.

There are more than 30 plants that use the INCO process in North America; worldwide, more than 50 projects are reported to have licensed this technology. The approach has been shown to be applicable

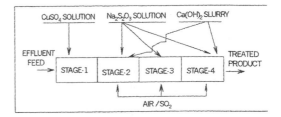

Figure 2. The INCO-SO_2/Air process (schematic). pH=10, CaO: 2-4 g/g cyanide, (SO_2/Air): 2-10/100, SO_2: 3-6 g/g cyanide; T=5-60°C, Cu^{2+} = 0-50 mg/L

to solutions, slurries and abandoned heaps (as heap rinse).

4.2.3.5 *Volatilization from acidic solutions*

The HCN speciation diagram given in Figure 1 shows that at pH<9.3, HCN gas is the predominating hydrolytic species. Thus the HCN gas can be driven off from the aqueous medium by distillation, or displacement by sparged air. Then HCN in the gaseous mixture can be captured by an alkaline solution such as NaOH or $Ca(OH)_2$ in a subsequent process step. This is in principle, forms the basis of AVR (acidification-vaporization-re-adsorption) technologies. A simplified flowsheet for the use of this approach is given in Figure 3.

One potential drawback relates to the safety of operators of such plants because HCN gas which is highly toxic is present in concentrated form in the plant pipeline system though, no incident has been

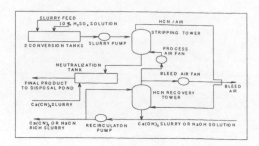

Figure 3. Scheme for the partial recycling of cyanide by the Cyanisorb technology.

recorded in some 8 plants that have operated using this technology. A synopsis of treatment technologies with critical commentary are summarized in Table 8.

Table 8. Cyanide abatement and destruction technologies for CN^- - containing effluents.

Method	Primary Mechanism(s)	Demonstrated scale	Advantages	Disadvantage
Dialysis, E-osmosis	Membranes	Lab	Future Potential	Cost; Maintenance
Ion Exchange Resins	Ion-exchange	Lab.; pilot	Potential for cyanide recycling	Cost; Maintenance
Metal salts and/or Mineral powder addition	Precipitation, Adsorptions; Catalysis	Small scale and pilot	Capture of ions and suspended solids	Product solids removal and disposal; materials volume
Active carbon contact	Physical & chemical uptake; Catalysis	Lab. and small scale	Widely effective on ions and solids	Cost, Fouling of carbon; catalytic-byproducts
Ion or precipitate flotation	Surface chemistry of foams and solids	Lab. demonstration	Potential for treatment of large volumes	Large scale demonstration and; supplementary techniques needed
Direct and With O_3, H_2O_2 or Sensitizing Solids: ZnO, TiO_2	UV-and catalysis-induced redox	Commercially available small-scale	Demonstrated technology	Cost, LTV radiation does not penetrate water, special reactor design needs.
Electrical potential	Oxidation	Lab and large pilot	Demonstrated technology, can be combined with O_2, O_3, Cl_2, or OCl^-	Potential for poisonous by products
Ozone	Oxidation of CN^-	Industrial	Cost -effective; minimal amount of harmful end-products	More effective at low pH, while CN-effluents are alkaline
Hydrogen peroxide	Oxidation of CN^-	Industrial, can be used with accelerators and catalysts	Cost -effective; proven technology	Cost of pH control when used with acids; Can produce nitrite and nitrate as byproducts
SO_2	Oxidation	Industrial	Proven technology can be accelerated with Cu^{2+}	pH-control needed; sludge formation and handling is problematic
Chlorine (Cl_2) and hypochlorite (OCl^-)	Oxidation of CN^-	Industrial	Versatile, usable at small and large scale	Minimal
H_2SO_4	HCN-generation	Pilot	Allows recycling of CN^-	
Bio-oxidation	Metabolic and enzymatic oxidation and metabolic adsorption	Industrial with some strains and pilot or lab. - scale with others	Partly proven technology, promising potential; partial removal of metal ions also occurs.	pH and nutrient control may be needed; biomass and bacterial strain control may create problems

5 CYANIDE DESTRUCTION ECONOMICS

Two motivations that form the basis for cyanide abatement and control are obvious: 1) Recycling of a valuable resource and 2) Destruction of a toxic component added to nature by man.

One of the alternatives for the minimization of cyanide emission is to reduce its use whenever possible. For example in flotation technology where cyanide is used as a selective-depressant for certain minerals such as copper or say, chromiumcontaining minerals sodium oxalate or sodium thiosulfate have been shown to be usable. Similarly, in the recycling of scrap metal, or exposed photographic film, thiosulfate or nitric acid solutions can be readily used as alternative lixiviants.

If the economics of cyanide usage in a given precious metal recovery approach is considered successful, the problem then boils down to the toxicity of cyanide and its compounds and the resulting management of this situation. The toxicity problem can conceptually, be solved by three approaches two of which were cited above. "Replacement of cyanide by alternative lixiviants" would be an added method of cyanide abatement. (Yarar 1993; Yarar, 1999).

The cost of conventional destruction technologies can be up to $1.50 per ton of ore treated. Recycling, provided by the AVR technology is an obvious abatement technique which is only partial at the present.

6 CONCLUSIONS

From the study of the published literature on cyanidation-based metal extraction technologics and their cyanide-based products, the following conclusions are reached:

1. Analytical data reporting is not as best it could be because it omits numerous cyanide complexes and compounds.
2. The assumption that free cyanide, in nature, breaks down completely to CO_2 or nitrated compounds, is not always the case. Compounds with heavy metals do form and resist destruction in natural environments for long periods of time. Cyanide compounds assumed to be "destroyed" or "not-present" are in fact, in place and are continually emitting harmful/toxic components.
3. Presently, a universal panacea to all cyanide-related problems does not exist although numerous technological approaches are available in the marketplace and most commercialized cyanide destruction processes claim superiority to competitors, in some form or another.
4. Cyanide usage is not about to cease any time soon, nor is there need for that although research is needed for a less controversial and equally-effective alternative lixiviant.
5. Numerous long-term ecological and health effects of metal-cyanide compounds and complexes including their carcinogenic potentials are still open to study.

REFERENCES

Arps P. J. &. Nelson, M. G . 1994. Cyanide degradation by a bacterium isolated from mine wastewater: pseudomonas pseudoalcaligenes (UA7), *SME Annual Mtg. Albuquerque, N.M, Feb. Preprint No.: 94-162.*

Beck M. Y. 1987. Critical survey of stability constants of cyano complexes. *Pure and Appl. Chem. 59(12), 1703-20.*

Botz, M. C.&. Mudder, T. I. 1998. Cyanide recovery for silver leaching operations, application of CCD-AVR circuits, *Randol Silver Forum, Denver, Colorado: 5.*

Davuyst, E. A. & Robbins, (1991) G. cyanide pollution control-the INCO process. *Randol-Cairns Conf. Proceedings, Randol. Itl. Ltd. Golden, CO, Colorado, 145-147.*

Huiatt, J. L. ct. al. 1983. Proc. workshop on cyanide from mineral processing, *Utah MMRS Institute, Salt Lake City Utah.*

Robbins, G. H. 1996. Historical development of the INCO SO_2 / air cyanide destruction process. *CIM Bull., 89, (1003) 63-69.*

Scott, J. S. & J. Ingles, 1981. Removal of Cyarude From Gold Mill Effluents: *Proc., Can. Mineral Processors 13th Ann. Mtg., Jan. 1981, Ottawa, Ontario.*

Tovill, L. E. et. al. 1978. Reviews of the environmental effects of pollutants: V.-Cyanide, *EPAl600/I -78-027, US-EPA, Cincinnati, Ohio.*

USGS, 1999. *Mineral industry surveys:* Gold, 1998 Annual review. Pittsburhg, PA.

von Michaelis, 1984. Role of cyanide in gold and silver recovery. *Cyanide in the environment, D. van Zyl (ed), CSU-Fort Collins CO, pp.51-64.*

Weast, R. C. (ed), 1979. Handbook of physics and chemistry. *CRC Press, Boca Raton Florida.*

Yarar, B. 1993. Alternatives to cyanide in the extraction of gold. Precious Metals Economics and Refming Technology, J. P. Rosso, (ed) IPMI, Pensicole, Florida.

Yarar, B. 1999. Alternatives for cyanide in precious extraction and methodologies for the destruction of environmental cyanide. *Proc. International Meeting CAECO-99 (Central Asia Ecology-99) K. Karimov (ed.). Tamga, Issyk Kul, Kyrgyzstan (in print).*

Stabilization / solidification of pyritic mill tailings by induced cementation

J. Ouellet, F. Hassani, S. Somot & S. Shnorhokian
McGill University, Mining and Metallurgical Engineering Department, Montreal, Quebec, Canada

M. Hossein
PMK Group, Kenilworth, New Jersey, USA

ABSTRACT: A preliminary experimental phase was conducted to test lime-fly ash binders in stabilizing-solidifying sulfide-rich tailings from mines in Canada. Treated and cemented tailings samples were subjected to mechanical tests alongside mineralogical characterizations. Metal and contaminant leachability from the samples was also checked using TCLP solutions. Results show that secondary minerals are responsible for solidifying samples and decreasing their permeability values. On the other hand, the geochemical stability of cemented tailings is enhanced with increasing curing periods, resulting from the precipitation of insoluble (in acid rain, in groundwater, and in acetic acid solutions) secondary minerals. It is deduced that these are hydroxy-minerals since an alkaline environment is required for their formation. They have not been identified yet but the phases are composed of Fe, Cu, Zn, Mg, and sometimes Al, with a certain part probably consisting of oxy-hydroxides. However, since the chemistry of pore water from the cemented samples is specific and evolves over time, these could be sulfate, carbonate and alumino-silicate complex phases as well. Although minerals that immobilize metals are not well defined yet, results indicate the capability of lime-fly ash binders in immobilizing heavy metals and in enhancing mechanical resistance to compressive forces and freeze-thaw cycles over a period of at least 660 days.

1 INTRODUCTION

Both the mechanical and chemical stability of mill tailings are of concern when looking for long-term management solutions, specifically with respect to the natural processes of erosion and weathering. If the mechanical and geochemical stability of tailings can be established, they could potentially be reused as backfill, thus minimizing both ore extraction and waste management costs. Therefore, promoting the long-term cementation of mill tailings is an economical management solution that makes use of these characteristics. However, pyrite-rich mill tailings are the main source of acid mine drainage, which can initiate the leaching of contaminants such as heavy metals. Acidity and sulfated waters also induce the deterioration of consolidated materials. Metal release can be controlled by maintaining a neutral to alkaline pH. The addition of lime is the most widely used method to maintain a neutral pH in mill tailings, whereas lime-fly ash binders have been used to treat the largest percentage of industrial waste in the United States (Conner 1990). In this paper, the testing of such a method on Canadian sulfide-rich mill tailings is presented.

Cementation is based on the formation of secondary minerals. Where these minerals were observed to precipitate (Marcus & Sangrey 1982, Thomson & Heggen 1982, Mcsweeney & Madison 1988, Blowes et al. 1991, Sevenitrane et al. 1995, Tassé et al. 1997) , a hardening of tailings was noted as occurring naturally in specific favorable conditions (surface crust, semi-arid climate, increase of pressure or temperature). The minerals responsible for cementation in concrete are identified to be mainly calcium silicate hydrates (CSH), and possibly ettringite (Minnick 1967, Ouelle et al. 1998). It is acknowledged that ettringite plays a major role in early (first hours to first days) strength development in concrete (e.g. Smith 1993, (NEVILLE 1996). However, its precipitation at later stages when the stiffness of the cemented material is already established contributes to deterioration (e.g. Neville 1996). On another hand, several studies have shown that the addition of lime-fly ash to wastes was responsible in decreasing contaminant mobility The mineralogical hosts of contaminants still remain a matter of debate. Some researchers have tried to quantify and identify the potential of ettringite to fix metals (Bensted & Varma 1971, Chen & Mehta 1982, Hassett et al. 1989, Mohamed et al. 1999, Perkins & Palmer 2000). Ouellet et al. (1998) investigated the effects of chemical alterations on the overall strength performance of paste backfill in a case study. A series of minerals resulting from ettringite destabilization were identified as a result of change from an alkaline to acidic environment (Myneni et al. 1998). The potential of metal immobilization by lime-fly ash binders in pyrite-rich tailings was investigated by Shnor-hokian (1996) and Hossein (1999). Hossein (1999) also reported favorable geotechnical characteristics in the treated tailings over long periods of curing. In this paper, the preliminary results of the durability tests obtained by these authors are discussed in relation to the mineralogical and geochemical characteristics of the tailings.

2 MATERIAL AND METHODS

2.1 Tailings used as a matrix and additives

The tailings samples studied (S1, S2) resulted from a flotation process aimed at the separation of Cu-Zn from the ore gangue. These samples come from the Canadian volcanic-associated massive sulfide deposits of Norbec (Quebec) and Mattabi (Ontario). Detailed reviews of mine sites mineralogy, geological contexts (e.g. Wolf 1976, Franklin 1995), and of geochemistry (e.g. Barrie et al. 1993) can be found in the literature. Tailings samples (S) and lime (L) were mixed in various proportions in order to reach a sufficiently alkaline pH. Then, fly ash type C (C) was added to these mixes in various proportions. Fly ash C is a mixture of fine crystals and of amorphous phases, and is generated from the burning of sub-bituminous or lignite coals. Mineralogically, it is mainly composed of quartz, mullite, hematite, magnetite, carbon and an amorphous component (Berry, 1976). As mentioned in several references on concrete, elements needed for the early formation of sulfo-aluminates and calcium silicate hydrates (CSH) are supplied by fly ashes (Shikami 1956, Minnick 1967, Carles-Gibergues & Vaquier 1971). The calcium-rich composition of fly ash C (Table 1) is well suited for mixing with potentially acid generating tailings.

In various mixes made up of tailings S1, S2, lime, and fly ash C, aluminum (A) was added in an aluminum nitrite solution, at a concentration of 110 mg/l, with the aim of promoting ettringite formation (Hossein, 1999). Water was added to all mixes in the ratio of solid:water = 0.5. This ratio presented a compromise amongst various goals including ettringite formation, fast drying and workability of the paste (Hossein, 1999). After an initial dry mixing of the ingredients, water was added and the samples were allowed to mellow for 24 hours at 90% relative humidity and 25°C. They were then compacted into specially designed molds (diameter of 30 mm and height of 60 mm), wrapped in plastic sheeting, and cured in a humid room. Mineralogical parameters were evaluated after various periods of aging.

2.2 Durability tests

2.2.1 Leaching tests

A modified Toxicity Characteristic Leaching Procedure (TCLP) test was used for leaching the various samples. A small amount from each sample (2g) was combined with 40 ml of acetic acid (with a pH of 3 for most samples, and with a pH of 5 for the two control samples S1 and S2) and subjected to mechanical shaking for 15 hours. The solution was then vacuum filtered using a 0.45 μm pore size acetate filter, and the resulting leachate was stabilized with nitric acid after taking pH measurements. For the first sample, S1 and S1L4C10A (composed of S1 + 4% Lime + 10% fly ash C + 110 mg/l aluminum) were leached after 3, 35, 90, 140, 360, and 660 days of curing. Leaching tests were conducted on sample S2 mixed with various proportions of lime (L2 to L7) and of fly ash (C0 to C20) after curing periods of 1, 14, 35, 70, 140 and 360 days. Shnorhokian (1996) and Hossein (1999) investigated the effects of lime and fly ash, as well as that of long term curing of samples, on metal leachability in detail.

2.2.2 Unconfined Compressive Strength (UCS)

UCS tests were conducted according to ASTM D2166, and consisted of pressure being applied at the top of a cylindrical sample 60 mm in height and 30 mm in diameter. The piston rate ranged between 0.5 to 1.5 mm/min so that maximum loading time did not exceed 15 minutes. UCS tests were performed on samples S1L4C10A, S2, S2L5, S2L5C10, and S2L5C10A after various periods of curing.

2.2.3 Permeability (Hydraulic conductivity)

For the permeability tests, samples were compacted into special molds 60 mm in height and 30 mm in diameter. They were then thoroughly permeated with water prior to testing in order to minimize the effect of trapped air pockets within the samples. Based on the expected low permeability results, the falling head test method was used instead of the constant head one. Permeability tests were conducted on samples S1, S1L4C10A, S2, S2L5, S2L5C10, and S2L5C10A after 28 days of curing.

2.2.4 Freezing/thawing cycles

A single cycle of the freeze/thaw procedure consisted of freezing a sample for 24 hours at -10°C and then thawing it for another 24 hours at 25°C in a humid room. UCS and permeability tests were conducted on selected samples after 6 or 12 cycles on samples that had been cured for 28 days. Freeze/thaw procedures were performed on samples S1L4C10A and S2L5C10A only, after 28 days of curing.

2.3 Analytical techniques

Leachates were analyzed for Ca, Al, Mg, Cr, Cu, Fe, Pb, and Zn using flame atomic absorption spectrometry (AAS) and for dissolved sulfates by ion chromatography. All concentrations in leachates were multiplied by 20. The geochemical composition of untreated tailings samples was analyzed by X-ray fluorescence spectrometry (XRF), and the mineralogical composition of treated tailings was studied both by X-ray diffraction (XRD) and scanning electron microscopy (SEM) equipped with an energy dispersive spectrometer (EDS).

3 RESULTS AND DISCUSSION

3.1 Mineralogy and geochemistry

The physical characteristics and geochemistry of the samples from the two mines are presented in Table 1. The mineralogy and geochemistry of the control samples S1 and S2 reflect various ore paragenesis and host-rocks composition.

However, in both samples the gangue mineral consists mainly of quartz, and the major sulfide mineral is pyrite. S2 is characterized by its high chlorite amount when compared to S1. Other minerals present in S2 include a pyroxene and residual Cu-As oxidized minerals. Minor amounts of feldspars, carbonates and kaolinite are also found in the samples. Based on XRD spectra, it can be noted that pyrite dissolves quickly when lime is added to S1 samples whereas it remains stable in S2 ones. When lime and water are added to tailings, air could have been introduced to the samples. This could explain the dissolution of sulfides at this stage in S1. The procedure has been the same for S2, but pyrite has not been dissolved.

This suggests that pyrite is protected from oxidation in sample S2. The initial Eh value of S2 is higher than S1 (Table 1). It is possible that pyrite was coated with Fe-oxy-hydroxides prior to mixing,

and that this protected it from further oxidation that has been observed in other tailings samples (e.g. Nicholson et al. 1990). This is also in agreement with the visual aspects of the sample, which shows evidences of oxidation (light brown color). Quartz reacts in all samples when sufficient lime is added to tailings (>1% in S1, >4% in S2). This is not surprising since silica dissolves easily under alkaline conditions. The supply of silica and other compounds brought about by the alteration of inherited minerals are sources of formation for secondary ones, as long as they are not leached out easily from the sample. The manner in which secondary minerals grow and in which chemical bonds are created between original and secondary minerals will enhance the mechanical cohesion of the samples. Different amounts of lime are needed in each sample in order to increase the pH sufficiently, based on the initial pH of the sample (Table 1). In both cases, lime in excess of the optimum value was added to ensure that the pH stays high for a long period of time. Chlorite is seen to react in sample S1, and further reactions with inherited minerals occur when fly ash is added to the mix. For example, pyrite, chlorite and pyroxene are seen to react in S2. As pyrite dissolves after the addition of lime and fly ash, and not lime alone, it is thought that reactions between fly ash and pore water were responsible for the dissolution of pyrite coatings first, followed by that of pyrite. This could be explained by a change in pore water composition resulting from new interactions with minerals.

Results obtained by Shnorhokian (1996) show that in the presence of lime (3%), ettringite and gypsum are both precipitated progressively in sample S1 from day 1 until day 35. In S2, the early precipitation of ettringite (after 1 day) is followed by its decomposition in favor of gypsum precipitation, with the latter dominating after 35 days of curing. Whereas the pH changes from 10.90 to 10.48 in S1L3, it drops from 10.22 to 8.64 in S2L3, which explains the dissolution of ettringite in this case. Similar conclusions can be drawn from Hossein's (1999) experiments with various mixes including varying amounts of lime and fly ash C. Mineralogical characterizations were performed on S2L5 and S2L5C10 after 28 days of curing and where both pH values are higher than 10.5. After 35 days of curing, the pH values are 10.63 and 10.85, respectively. If we compare the intensity ratios of gypsum (Gy; d=7.56 Å) to ettringite (Et; d=9.73 Å) peaks for these two samples, we obtain Gy/Et=2.9 for S2L5, and Gy/Et=1.5 for S2L5C10. Since the addition of fly ash C introduces additional alkaline products into the system, ettringite formation is promoted by it. Ettringite dissolution provides elements for sulfate-rich pore water that result in further precipitation of gypsum and Al-phases, which are stable in more acidic conditions as described by various authors (e.g. Myneni et al. 1998). In the S2L5C10 sample,

Table 1. Initial geochemistry and characteristics of tailings and fly ash (C)

Sample	Units	Norbec	Mattabi	C
PH	unit	3.26	2.46	12.16
Eh	mV	255	410	-
Moisture	%	9.20	8.13	-
SiO_2	%	-	-	53.3
Al_2O_3	%	14.77	6.6	23.63
Fe_2O_3	%	9.91	41.07	4.4
MgO	%	4.81	1.31	1.15
CaO	%	1.06	0	15.81
Na_2O	%	-	-	3.03
As	ppm	192	903	-
Cd	ppm	20	25	-
Cr	ppm	15	0	-
Cu	ppm	968	1767	-
Pb	ppm	516	4015	-
Zn	ppm	2937	13644	-

and after 360 days of curing, SEM observations show that ettringite crystals were smaller and that the amount of gypsum has increased when compared to samples cured for 3 to 180 days. To prevent this type of ettringite destabilization, Al was supplied to the lime-fly ash samples in these experiments in the form of an Al-nitrite solution (A=110 mg/l). The pH value was also maintained at a sufficiently high level an increase in lime percentage so that ettringite stability could be achieved after 660 days of curing. However, even under such improved conditions, ettringite shows beginnings of destabilization (smaller crystals, dispersed crystals).

SEM characterization at various stages of the evolution of treated samples shows large amounts of euhedral crystals – which were identified as ettringite and gypsum (EDS and XRD) – filling the voids between the original minerals in a progressive way. After 28 days of curing, ettringite crystals can be seen in the shape of a dense mat. Only minor amorphous precipitates are observed on the SEM pictures after 180 days of curing. Based on a comparison to the ones presented by Myneni *et al.* (1998), these precipitates could be Al-hydroxides and Al-hydroxy-sulfates.

3.2 *Leaching experiments*

Leaching experiments demonstrate the effects of additives and of aging on the concentrations of elements leached with an acidic solution, which dissolves minerals that are not stable under such conditions. Acetic acid is known to easily dissolve carbonates such as calcite or dolomite after 15 hours of contact (Gupta & Chen 1975), to partly dissolve amorphous Fe-Mn oxy-hydroxides (Chester & Hugues 1967). It is also known to extract exchangeable ions from clays (Ray *et al.* 1957), and zeolites decompose in this reagent as well (Loeppert & Suarez 1996). Amorphous Ca-Al-hydroxy sulfate known to be stable in acidic conditions (Myneni *et al.* 1998) but ettringite is dissolved. Gypsum is also dissolved in the acetic acid solution as L/S=20. However, crystalline oxides and hydroxides, silicates, sulfides are not dissolved. Therefore, element concentrations in the leachates partly reflect the amount of soluble minerals within each sample. If the amount of ettringite in the sample is high, then dissolved Al, Ca, and SO_4 concentrations should also be theoretically (in absence of chemical precipitations within the leachate) high in the leachate. If ettringite is responsible in trapping metals in its structure, then these metals should be released into the leachate simultaneously with Al, Ca, and SO_4. They will not precipitate as hydroxides or be re-adsorbed since the pH would be acidic. Moreover, acetate is a complexing agent that contributes to maintaining dissolved metals within the solution. Another limitation in interpreting these tests is the lack of selectivity of acetic

solution. Indeed, if amorphous Al-hydroxides and gypsum are present in samples with ettringite, they will dissolve and release Al, Ca, SO4, and their associated metal phases. Both ettringite and amorphous oxy-hydroxides are potential hosts for metals. When a mix of these minerals is present in samples, the TCLP method is inadequate to define exactly which phase was responsible for trapping the metals. However, if these ettringite and amorphous oxyhydroxides are present in the samples and only low released metals can be detected in leachates, it means that the minerals are not responsible for major metal retention.

3.2.1 *Effect of lime addition on TCLP results*

As expected, the more lime is added, the higher the pH of tailings sample S2L5 becomes, and Fe, Mg, Cu and Zn are less easily leached with the acetic solution (Fig.1).

The decrease in metal concentrations in the leachate is large, especially for Cu and Fe. When the system becomes alkaline, many dissolved elements probably precipitate as microcrystalline oxyhydroxides. This is in good agreement with Fe, Mg, Cu, Zn hydroxide solubilities.

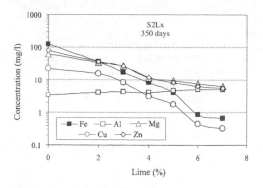

Figure 1. Dissolved elements in TCLP leachates for S2Lx (350 days of curing) as a function of lime content.

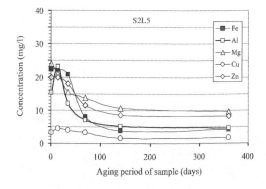

Figure 2. Dissolved elements in TCLP leachates for S2L5 as a function of the aging period.

However, the exact nature of these precipitates has to be determined. They could be complex phases since the chemical composition of pore water in treated tailings is complex.

In these leachates, Al concentrations depend upon the solubilities and dissolution rates of Al-soluble phases that can be weathered, inherited Al-silicates and/or secondary Al-hydroxy phases. The pH of the sample S2L5 after 360 days curing is 10, and ettringite is not observed. In these aged samples, elemental concentrations are low and concentration variations are not significant, taking into account analytical errors. Moreover, when looking at the composition of leachates after various periods of aging for S2L5 (Fig.2), it seems clear that Al concentrations after 360 days are related more to the solubility limit of secondary Al-minerals rather than the dissolution rate of original Al-silicates. After 14 days of aging of S2L5, the pH of the sample is 11.62 and a large concentration of Al is released in the leachate. This is in agreement with the XRD identification of ettringite in this sample after 28 days of aging.

3.2.2 Effect of fly ash C addition on TCLP results

Various proportions of fly ash C are added to S2L5, and they are leached after 360 days of curing. As fly ash proportions increase, all element concentrations in leachates decrease slightly along with an increase in pH values. Fly ash addition itself has a very limited effect on the geochemical stabilization of these lime-treated samples, with the main controlling factor being the slight pH increase due to calcium content in the fly ash.

3.2.3 Effect of aging of mixes with or without fly-ash on TCLP results

Without any fly ash addition in S2L5, all the monitored (except Ca and SO_4) element concentrations that have been leached are observed to decrease significantly after 70 days of curing (Fig.2), and remain stable even after that period. This implies that the amount of insoluble secondary minerals (crystalline oxy-hydroxides or hydroxy-sulfates) that precipitate in the tailings samples increases with time during the early stages of their evolution. Low, constant concentrations of released elements after 70 days could be the result of equilibrium between pore water and stable secondary phases or due to the presence of secondary carbonates.

Indeed, no special precautions were taken in the experimental procedures to prevent carbonate formation. Under alkaline conditions, those materials are highly sensitive to carbonation. It would mean that over a period of 360 days, treated tailings are geochemically stabilized in a satisfied way.

When fly ash is added to samples (Fig.3), leaching results are similar to the previous ones.

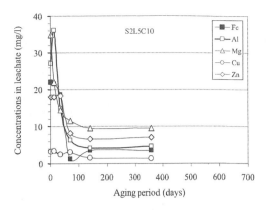

Figure 3. Dissolved elements in TCLP leachates for S2L5C10 as a function of the aging period.

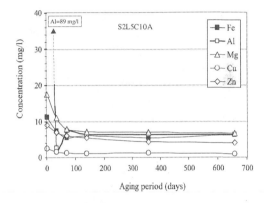

Figure 4. Dissolved elements in TCLP leachates for S2L5C10A as a function of the aging period.

However, after one day of curing, leached Al and Mg concentrations are higher in S2L5C10 than in S2L5. The ratio of Gy/Et calculated previously indicates that the amount of ettringite is higher in S2L5C10 than in S2L5. This is in agreement with higher concentrations of Al in leachate after 14 days of aging and a pH higher than 11. In these samples, and after 360 days of curing, SEM observations show that ettringite crystals were smaller and the amount of gypsum increases, compared to samples aged between 3 and 180 days.

In the long term, leaching test results (Fig.4) obtained from S2L5C10A are similar to the ones presented above with no Al-nitrite addition. Results were also similar between samples S2L5C10A and S1L4C10A. Differences in concentrations of leached Fe, Cu, and Zn are not significant, taking analytical errors into consideration.

The in-situ evolution of these materials also depends upon external parameters such as water infiltration. In order to investigate this aspect, Hossein (1999) performed leaching tests using groundwater and simulated acid rain on Al-treated samples. Leachate concentrations of the samples remained low for the elements analyzed even after 660 days of curing.

3.3 Resistance to shear

Tests on S1 indicate an increase in compressive strength with longer periods of curing (Fig.5). After 28 days of curing, values in excess of 800 kPa were observed and the presence of ettringite was verified through XRD analysis. Values exceeded 1200 kPa after 90 days of curing and remained constant even after almost a year.

As expected, tests on S2 did not show any significant changes in compressive strength in the control sample (Fig.5). Moderate strength gain was observed in the sample treated with 5% lime only, which is the minimum amount needed to raise the pH of tailings S2 to about 11. A value of about 137 kPa was reached after 14 days of curing, followed by disintegration and loss of strength. XRD analysis on this mix showed the presence of gypsum after 28 days of curing, which could account for the loss in strength after a certain period of time. The addition

of fly ash has resulted in a strength value of 90 kPa after one day and about 300 kPa after seven days in S2L5C10. XRD analysis indicated stronger peaks of ettringite in these samples, with slightly reduced peaks of gypsum.

Sample S2L5C10A gave the best results amongst the various S2 mixes. Strengths of 700 kPa were observed after 28 days of curing, which is the minimum strength value recommended by the US EPA for reusability of treated waste material. XRD analysis of these samples showed a strong presence of ettringite. Strength values surpassed the 1000 kPa mark after 90 days of curing, and remained so even after 360 days.

3.4 Permeability

Tests on S1 samples indicate a lower value for permeability in the treated mix than in the control sample. (Fig.6). Whereas the latter gave a value of 3.6×10^{-4} cm/s, the mix that was treated with lime, fly ash and aluminum gave a value of 1.16×10^{-6} cm/s after 28 days of curing. This result is also well above the US EPA recommendation for stabilized waste destined for land disposal, which stipulates a minimum value of 10^{-5} cm/s for k (hydraulic conductivity). Untreated mixes of S2 gave an average value of 3.5×10^{-4} cm/s (Fig.6). Treatment with lime only produced a conductivity of 2.5×10^{-4} cm/s. Permeability values for S2L5C10 and S2L5C10A are significantly lower at 3.2×10^{-5} cm/s and 1.96×10^{-6} cm/s, respectively.

The roles of ettringite and various pozzolanic reactions between lime and fly ash are clearly seen to enhance the physical properties of the treated tailings.

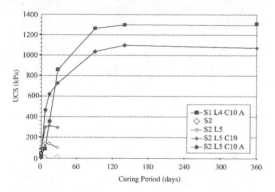

Figure 5. Evolution of unconfined compressive strength (UCS) of samples S1 and S2, with or without additives (L,C, and A) as a function of curing time.

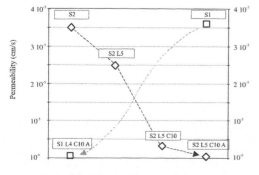

Figure 6. Permeability evolution when additives (L, C and A) are mixed with S1 and S2 samples.

3.5 Effect of freeze/thaw cycles

The results of the freeze/thaw cycles are presented in Table 2. In the case of S1, UCS values were down from 855 kPa to 674 kPa after 6 cycles and 544 kPa after 12 cycles. Conductivity values increased from about 10^{-6} cm/s to 5×10^{-6} cm/s and 1×10^{-5} cm/s after 6 and 12 cycles, respectively.

Table 2. Permeability test results (k in cm/s) on samples 1 and 2

	Normal Treatment	Freeze/Thaw 6 cycles	Freeze/Thaw 12 cycles
S1	3.6×10^{-4}	-	-
S1L4C10A	1.2×10^{-6}	4.8×10^{-6}	1.0×10^{-5}
S2	3.5×10^{-4}	-	-
S2L5	2.5×10^{-4}	-	-
S2L5C10	3.2×10^{-5}	-	-
S2L5C10A	2.0×10^{-6}	4.5×10^{-6}	9.1×10^{-6}

As for S2, and where the mix was treated with lime, fly ash, and aluminum, compressive strength was reduced to 676 kPa after 6 cycles and to 512 kPa after 12 cycles from an initial value 727 kPa.

Permeability values for the same sample came down from $2*10^{-6}$ cm/s to $4.5*10^{-6}$ cm/s after 6 cycles, and $9.1*10^{-6}$ cm/s after 12 cycles. Visual inspections of the samples after the freeze/thaw cycles did not indicate any loss of physical integrity or deterioration.

4 CONCLUSIONS

After curing, lime and lime-fly ash treated tailings were geochemically stabilized for a period of at least 360 days. When Al is added to these samples, the stability was seen to extend further to a total period of at least 660 days. The presence of ettringite and sample stability is related mainly to the pH of pore water in the tailings, since aluminum, sulfates, and calcium are not the limiting factors in these tailings. Gypsum formation occurs independently of pH values. The calcium- and sulfate-rich pore water is over-saturated with gypsum. Ettringite is not responsible for the geochemical stabilization of Fe, Cu, Zn, Mg, and Al in these materials. The formation of other secondary minerals and overall stability depend on the pH and pore water composition, resulting in the entrapment of metals.

Secondary minerals also enhanced the geotechnical parameters of the treated tailings, such as decreasing the overall permeability and as solidifying them. They were also seen to be more resistant to freeze/thaw cycles compared to untreated samples. The mechanical strength of these materials is related to the stability of ettringite.

These various results suggest that potential low cost solutions exist where tailings can be both mechanically and geochemically stabilized. The main parameter that needs further research in the future is improving the stabilization of original sulfide minerals in the tailings, as they were observed to generate acidity when present even in small amounts and after 660 days of aging.

5 ACKNOWLEDGMENTS

All tests and analyses were performed at McGill University. The authors would like to thank NSERC for its financial support and the contribution of the various mining companies involved.

REFERENCES

Barrie, C. T., Ludden, J. N. & Green T. H. 1993. Geochemistry of volcanic rocks associated with Cu-Zn and Ni-Cu deposits in the Abitibi Subprovince. *Economic Geology* **88**, pp. 1341-1358.

Bensted, J. & Varma, S. P. 1971. Studies of ettringite and its derivatives. *Cement Technology* **2** / 3, pp. 73-76.

Berry, E. E. 1976. Fly ash for use in concrete. Part I- A critical review of the chemical, physical and pozzolanic properties of fly ash. **76-25**, CANMET.

Blowes, D. W., Reardon, E. J., Jambor, J. L. & Cherry J. A. 1991. The formation and potential importance of cemented layers in inactive sulfide mine tailings. *Geochimica Cosmochimica Acta* **55**, pp. 965-978.

Carles-Gibergues, A. & Vaquier, A. 1971. Comportement pseudo-pozzolanique d'une cendre volante. *Matériaux Constr.* **6**, pp. 142-148.

Chen, S. S. & Mehta, P. K. 1982. Zeta potential and surface area measurements on ettringite. *Cement and Concrete Research* **12**, pp. 257-259.

Chester, R. & Hugues, M. J. 1967. A chemical technique for the separation of ferromanganese minerals, carbonate minerals and adsorbed trace elements from pelagic sediments. *Chemical Geology* **2**, pp. 249-262.

Conner, J. R. 1990. Chemical fixation and solidification of hazardous wastes. p. -692. New York, Van Nostrand Reinhold.

Franklin, J. M. 1995. Volcanic-associated massive sulfide deposits. In *Mineral Deposit Modeling. Kirkham, R. V., Sinclair, W. D., Thorpe, R. I., and Duke, J. M.* **2**, pp. 315-334. Geological Association of Canada. GAC Special Paper 40.

Freyssinet, P., Piantone, P., Azaroual, M., Itard, Y., Clozel, B., Baubron, J. C., Hau, J. M., Guyonnet, D., Guillou-Frottier, L., Pillard, F. & Jezequel, P. 1998. Evolution chimique et mineralogique des machefers d'incineration d'ordures menageres au cours de la maturation. **280**, p. -146. Orleans,France, BRGM. Eau-Aménagement-Environnement.

Gupta, S. K. & Chen, K. Y. 1975. Partitioning of trace metals in selective chemical fractions of nearshore sediments. *Environ. Lett.* **10** / 2, pp. 129-158.

Hassett, D. J., Pflughoeft-Hassett, D. F., Kumarathasan, P. & McCarthy, G. J. 1989. Ettringite as an agent for the fixation of hazardous oxyanions. pp. 471-481. University of Wisconsin-Madison, Department of Engineering Professional Development. Twelfth Annual Madison Waste Conference, Sept. 20-21.

Hossein, M. 1999. Role of ettringite formation in the stabilization/solidification of sulfide-bearing mine waste. pp. 1-355. McGill University.

Loeppert, R. H. & Suarez, D. L. 1996. Carbonate and gypsum. In *Methods of Soils Analysis: Part 3- Chemical Methods. Bigham, J. M.* / 15, pp. 437-474. Madison, Soil Science Society of America. SSSA Book Series 5.

Marcus, D. & Sangrey, D. A. 1982. Uranium mill tailings stabilization with additives. **IAEA-SM-262/49**, pp. 449-467. Vienna, IAEA. Management of Wastes from Uranium Mining and Milling.

McSweeney, K. & Madison, F. W. 1988. Formation of a cemented subsurface horizon in sulfidic minewaste. *Journal of Environmental Quality* **17** / 2, pp. 256-262.

Minnick, L. J. 1967. Reactions of hydrated lime with pulverized coal fly ash. Bureau of Mines information circular 8348. Proc. Fly Ash Utilization Conference.

Mohamed, A. M. O., Boily, J. F., Hossein, M. & Hassani, F. P. 1999. Ettringite formation in lime-remediated mine tailings: I. Thermodynamic modeling. *CIM Bulletin* **88** / 995, pp. 69-75.

Myneni, S. C. B., Traina, S. J. & Logan, T. J. 1998. Ettringite solubility and geochemistry of the $Ca(OH)_2-Al_2(SO_4)_3-H_2O$ system at 1 atm pressure and 298K. *Chemical Geology* **148**, pp. 1-19.

Neville, A. M. 1996. Properties of Concrete. **4e**, pp. 1-844. London, John Wiley & Sons, Inc.

Nicholson, R. V., Gillham, R. W. & Reardon, E. J. 1990. Pyrite oxidation in carbonate-buffered solution: 2- Rate control by oxide coatings. *Geochimica Cosmochimica Acta* **54**, pp. 395-402.

Ouellet, J., Benzaazoua, M. & Servant, S. 1998. Mechanical, mineralogical, and chemical characterization of a paste backfill. Proceedings of Tailings and Mine Waste '98, pp. 139-146.

Perkins, R. B. & Palmer, C. D. 2000. Solubility of (Ca6[Al(OH)6]2(CrO4)3 . 26 H2O), the chromate analog of ettringite; 5-75°C. *Applied Geochemistry* **15**, pp. 1203-1218.

Ray, S., Gault, H. R. & Dodd, C. G. 1957. The separation of clay minerals from carbonate rocks. *American Mineralogist* **42**, pp. 681-686.

Sevenitrane, H. N., Newson, T. A., Fahey, M. & Fujiyasu, Y. 1995. Some factors influencing the consolidation behavior of mine tailings. **1**, pp. 459-464. Hiroshima, Japan, Balkema. International Symposium on Compression and consolidation of clayey soils.

Shikami, G. 1956. On pozzolanic reactions of fly-ash. pp. 221-227. Proc. Japan Cement Eng. Assoc.X.

Shnorhokian, S. 1996. Immobilization of heavy metals in lime-fly ash cementitious binders. p. -116. McGill University.

Smith, R. L. 1993. Fly ash for use in the stabilization of industrial wastes. pp. 58-72. New York, ASCE. Fly Ash for Soil improvement.

Tassé, N., Germain, D., Dufour C. & Tremblay R. 1997. Hardpan formation in the Canadian Malartic mine tailings: implication for the reclamation of the abandoned impoundment. **4**, pp. 1797-1812. Vancouver (BC).

Thomson, B. M. & Heggen, R. J. 1982. Water quality and hydrologic impacts of disposal of uranium mill tailings by backfilling. pp. 373-384. Vienna, IAEA. Management of wastes from uranium mining and milling. Proceedings of a symposium, Albuquerque, IAEA-OCDE.

Wolf, K. H. 1976. Handbook of Strata-Bound and Stratifor-more Deposits: I-Principles and General Studies.

Tailings and Mine Waste '02, © 2002 Swets & Zeitlinger, ISBN 90 5809 353 0

The removal of arsenic from groundwater using permeable reactive materials

Jeff Bain, Laura Spink, David Blowes & David Smyth
Department of Earth Sciences, University of Waterloo, Waterloo, Ontario, N2L 3G1

ABSTRACT: The use of permeable reactive barriers (PRB) for the treatment of contaminated groundwater is gaining interest as an alternative to conventional (pump and treat) methods. Laboratory experiments investigating the potential for treatment of arsenic contaminated groundwater through the use of potential reactive barrier materials were conducted. For this investigation, arsenic contaminated groundwater collected from a mine located in Ontario, Canada, was passed through laboratory columns of 100% zero valent iron and a mixture containing 20% zero valent iron. During the experiments, the influent groundwater contained between 5 and 15 mg/L of dissolved arsenic. Flow through the columns (0.2 to 0.4 pore volumes/day; PV/day or ~25 to 65 m/a) was several times the groundwater velocity measured at the mine site where the water was collected (10 m/a). Results indicate that the reactive mixture is capable of removing the dissolved arsenic to below current drinking water limits (0.05 mg/L) for considerably more than 90 pore volumes, the current progress of the experiment. After treating 90 PV of water containing an average of 10 mg/L As, concentrations exceeding 0.1 mg/L of As migrate less than 5 cm into the reactive material and concentrations exceeding 0.05 mg/L do not move more than 12 cm into the reactive material. In terms of a field application (e.g. 1 m thick) in situ PRB installation, this represents the equivalent of years or decades of treatment before replacement.

1 INTRODUCTION

Tailings and waste rock are the principal wastes derived from the mining and milling of ore bodies for the recovery of gold. At some mining properties, water infiltrating the surface of the tailings becomes contaminated with arsenic and heavy metals released from the tailings. Research conducted at a mine site in Ontario suggests that, as a result of dissolution of arsenic-bearing iron oxides contained within the tailings, arsenic (As) is released to the tailings porewater and subsequently, the underlying groundwater aquifer (McCreadie et al., 1999). Because of the large reservoir of As stored in the tailings impoundment, the potential impact on groundwater downgradient from the mine tailings and possible remediation methods are being investigated by the company.

The traditional approach for groundwater protection is to pump the groundwater and treat it at surface. Research conducted at the mine site indicates that movement of dissolved As in the aquifer may be attenuated partly by natural processes, including adsorption and reductive precipitation (Ross et al., 1999). These mechanisms are, however, not expected to prevent off-site migration of As. In situ permeable reactive barriers, designed to simulate enhanced natural attenuation processes, and installed in the path of the As-bearing groundwater, are being evaluated as an alternative method for preventing off-site migration of the As (McRae et al., 1999b). The University of Waterloo holds several international patents, dating back to 1994, for the use of zero valent iron in in situ permeable reactive barrier installations for the treatment of electroactive metals contained in groundwater (Canadian Pat. # 2,062,204; U.S. Pat. # 5,514,279).

This paper describes some of the findings of recent investigations, conducted by University of Waterloo researchers, into the remediation of As-bearing water using permeable reactive materials. The paper also describes an ongoing laboratory investigation that has two objectives. One objective is to evaluate and compare the As treatment performance of a reactive mixture containing 20% zero valent iron (20Fe) to a reactive medium composed 100% of zero valent iron (100Fe). Mixtures containing less than 100% zero valent iron have the potential to provide effective As removal, but at lower cost than a 100Fe reactive medium. Another objective is to evaluate the performance of both reactive media for the treatment of groundwater collected from a contaminated property.

2 BACKGROUND

The ingestion of arsenic may lead to serious health effects, such as those currently observed in Bangladesh (Dhar et al., 1997). Arsenic is a redox sensitive metal that occurs in groundwater in two valence states, arsenite (AsIII) and arsenate (AsV). In its most reduced form, As0, arsenic is insoluble. Under the reducing and neutral to acidic pH conditions typical of most natural groundwater, As(III), generally in the form H_3AsO_3 (arsenious acid), is dominant. Under more oxidizing conditions, As(V) is dominant in forms including H_3AsO_4, $H_2AsO_4^-$, $HAsO_4^{2-}$ and AsO_4^{3-} (Farrell, et al., 2001; Hug et al., 2001) Under neutral pH conditions, arsenite is more mobile than arsenate in groundwater. Arsenite is also more toxic than arsenate (Korte & Fernando, 1991; Farrell, et al., 2001; Hug et al., 2001). In a reduced state, arsenite may form sparingly soluble precipitates such as As_2S_3 or FeAsS, and may be co-precipitated with ferric hydroxides.

Due to its sensitivity to redox conditions, one option to control the mobility of arsenic is to manipulate the redox conditions of the medium that the As-bearing groundwater flows through. Zero valent iron is a strong reductant, and has been found suitable for reducing the oxidized form of many dissolved metals (Blowes et al., 1997; Gu et al., 1998; McRae et al., 1999b; Blowes et al., 2000; Morrison et al., 2001; Su & Puls, 2001). McRae (1999a,b) conducted a set of laboratory batch and column experiments evaluating the potential for the treatment of As bearing water with mixtures containing zero valent iron.

The laboratory batch and column experiments initiated by McRae in 1996 indicated that reactive mixtures containing zero valent iron remove dissolved As from water containing 1 to 1.8 mg/L of As (1:1 ratio of AsIII to AsV) to less than 0.018 mg/L. Removal occurred primarily during the first 2 hours of the experiment, with treatment to below 0.05 mg/L within 3 hours. Analysis of reactive materials from the batch experiments by energy dispersive X-ray analysis (EDX) and X-ray photoelectron spectroscopy (XPS) suggests that the As(III) and As(V) were removed from solution by adsorption and also by coprecipitation with goethite (FeOOH) at the surface of the iron particles, after reduction of As(V) to As(III) by the zero valent iron (McRae, 1999a). In the column experiments, water containing approximately 1 to 1.6 mg/L As (1:1 molar ratio of As(III) to As(V)) has been treated to below 0.05 mg/L for more than 1200 pore volumes (Figure 1).

3 EXPERIMENTAL PROCEDURE

Laboratory columns investigating the removal of As from groundwater under flowing conditions were initiated by Spink (2001). The experiment consists of two columns packed with permeable reactive material, through which As-bearing (5-15 mg/L As) groundwater collected from an aquifer at a mine site

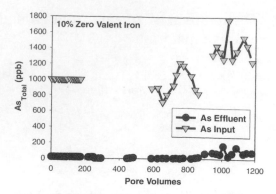

Figure 1. Removal of As from water containing 1 to 1.8 mg/L As (after McRae, 1999a)

in northern Ontario is pumped. Two permeable reactive materials are being evaluated: one composed 100% of zero valent iron (100Fe) and the other a mixture containing 20% zero valent iron (20Fe).

Two Plexiglas™ columns 5 cm in diameter and 41.8 cm in length were filled with the reactive media, and were operated in an upflow manner. Because of differences in the packing density and the porosity of the two reactive media, the velocity is higher in 20% Fe column than in the 100% Fe column. The porewater flow in the columns is 3 to 5 times more rapid than groundwater flow velocities at the mine site. At this velocity, residence time in the columns is ~2.5 days in the 20Fe column and ~6 days in the 100Fe column.

The groundwater used in the experiments contains moderate levels of several dissolved species including SO_4, Cl, Ca, Mg and Fe, a condition not present in the testing conducted by McRae (1999). Table 1 describes the major ion chemistry of the input water, which was fairly consistent during the experiment. The combined interaction of these species with the zero valent iron is not well known. Favourable or adverse effects on the performance of the zero valent iron in its treatment of As-bearing groundwater or in the permeability of the reactive materials should become evident from these experiments. The water was collected on a regular basis (~monthly) from piezometers installed at the mine site.

Sampling ports installed at 2-3 cm intervals facilitated measurement of porewater chemistry along the length of the column at regular periods during the experiment. Influent and effluent water samples were also collected for analysis. Samples were collected and immediately analysed for pH, Eh and alkalinity, while additional sample was collected and filtered to 0.45 μm. Cation samples were acidified. A suite of cations was analysed by inductively coupled plasma-mass spectrometry (ICP-MS) and separate As samples were analysed by hydride genera-

Table 1. Typical chemistry of input water for the column experiments. Concentrations are in mg/L.

Parameter	Value
pH	7.2
Eh (mV)	329
CO_3	62
As_{Tot}	10
SO_4	562
Ca	99
Fe	0.4
K	22
Mg	64
Mn	0.4
Na	133
Si	3
Zn	0.1
Cl	73

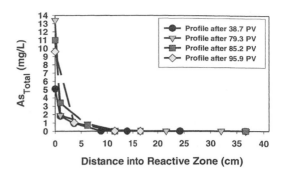

Figure 2. Arsenic concentration profile in the 20% Fe column at selected times.

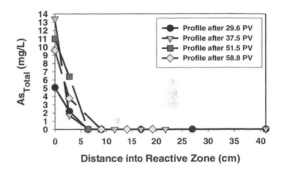

Figure 3. Arsenic concentration profile in the 100% Fe column at selected times.

tion and atomic absorption (HG-AA) at the University of Waterloo or by graphite furnace atomic absorption (GF-AA) at Philip Analytical Laboratories in Mississauga.

4 RESULTS

Figures 2 and 3 illustrate the treatment of As in the 20Fe and 100Fe columns, respectively. It is evident from these figures that at this stage of the investigation, As contained in the groundwater is removed at similar rates, if not faster, in the column with the mixture containing 20% zero valent iron, noting that the residence time in this column is less than in the 100Fe column. In both cases, however, the majority of As removal, occurs within the first 5 cm travel distance into the reactive material. In this distance, input As concentrations of 5 to 15 mg/L decrease to ~1 mg/L. Over the next 5 cm travel distance in the columns the As concentration decreases further to values below detection limits, which were between 5 and 20 ppb depending on the laboratory and analysis procedure used.

At this time, the rate of movement of As in the columns cannot be quantified. At a minimum, the rate of movement and the potential treatment capacity of the reactive materials can be based upon the total movement observed. In both columns, water containing As concentrations above the maximum drinking water limit has not progressed more than 12 cm distance into the reactive material. With this observation, the 20Fe mixture treats ~8 pore volumes of water per centimeter of column length, and the 100Fe treats ~5 pore volumes of water per centimeter of column length. These calculations should be considered conservative, however, because the position of the As front has been stable for several tens of pore volumes. With continued operation of the columns, estimates of the potential longevity of the reactive materials will be improved. An evaluation of the mechanisms responsible for As removal in

both columns will be conducted at a later stage in the experiment. An evaluation of the mechanisms responsible for As removal in both columns will be conducted at a later stage in the experiment.

5 SUMMARY

Porewater, containing 5-15 mg/L As, collected from the aquifer at a mine site, has been treated to less than 0.005 mg/L As in laboratory column experiments operated at flow rates several times greater than the groundwater velocity at the mine site. Two reactive mixtures are being evaluated in the column experiments. Early results, with the treatment of 58 to 90 pore volumes of water, suggest that an effective barrier can be constructed of either reactive material.

REFERENCES

Blowes, D.W., Ptacek, C.J., Benner, S.G., McRae, C.W.T., Bennett, T.A. & Puls, R.W. 2000. Treatment of inorganic contaminants using permeable reactive barriers. Journal of Contaminant Hydrology. 45: 123-137.

Blowes, D.W., Ptacek, C.J. & Jambor, J.L. 1997. In-situ remediation of Cr(VI)-contaminated groundwater using permeable reactive walls: Laboratory studies. Environmental Science & Technology. 31(12): 3348-3357.

Dhar, R., Biswas, B, Samanta, G., Mandal, B, Chakrabortic, D., Roy, S., Jafar, A., Islam, A., Ara, G., Kabir, S., Khan, A.W., Ahmed, S.A. & Hadi, S.A., 1997. Groundwater arsenic calamity in Bangladesh. Current Science (India). 73: 48-59.

Farrell, J., Wang, J, O'Day, P., & Conklin, M., 2001. Electrochemical and spectroscopic study of arsenate removal from water using zero-valent iron media, Environmental Science & Technology. 35(10): 2026-2032.

Gu, B, Liang, L., Dickey, M.J., Yin, X. & Dai, S. 1998. Reductive precipitation of uranium(VI) by zero-valent iron, Environmental Science & Technology. 32(21): 3366-3373.

Hug, S.J., Canonica, L, Wegelin, M., Gechter, D. & Von Gunten, U. 2001. Solar oxidation and removal of arsenic at circumneutral pH in iron containing waters, Environmental Science & Technology. 35(10): 2114-2121.

Korte, N.E. & Fernando, Q. 1991. A review of arsenic(III) in groundwater. Critical Review in Environmental Control. 91(1): 1-39.

McCreadie, H., Blowes, D.W., Ptacek, C.J. & Jambor, J.L. 2000. Influence of reduction reactions and solid-phase composition on porewater concentrations of arsenic, Environmental Science & Technology. 34(15): 3159-3166.

McRae, C.W.T. 1999a. Evaluation of reactive materials for in situ treatment of arsenic(III), arsenic(V) and selenium(VI) using permeable reactive barriers: Laboratory Study. M.Sc. thesis, Department of Earth Sciences, University of Waterloo. 160 p.

McRae, C.W.T., Blowes, D.W & Ptacek, C.J. 1999b. In situ removal of arsenic from groundwater using permeable reactive barriers: A laboratory study. In: Sudbury '99 - Mining and the Environment II. Edited by D. Goldsack, N. Belzile, P. Yearwood and G. Hall. September 13-17, 1999, Sudbury, Ontario, Canada. 601-609.

Morrison, S.J., Metzler, D.R. & Carpenter, C.E. 2001. Uranium precipitation in a permeable reactive barrier by progressive irreversible dissolution of zerovalent iron, Environmental Science & Technology. 35(2): 385-390.

Ross, C.S., Bain, J.G. & Blowes, D.W. 1999. Transport and attenuation of arsenic from a gold mine tailings impoundment. In: Sudbury '99 - Mining and the Environment II. Edited by D. Goldsack, N. Belzile, P. Yearwood and G. Hall. September 13-17, 1999, Sudbury, Ontario, Canada. 745-754.

Spink, L.E. 2001. Confidential. B.Sc. thesis, Department of Earth Sciences, University of Waterloo. 56 p.

Su, C. & Puls, R.W. 2001. Arsenate and arsenite removal by zerovalent iron: kinetics, redox transformation, and implications for in situ groundwater remediation, Environmental Science & Technology. 35(7): 1487-1492.

Estimation of the mobility of heavy metals in tailing sediments

T. Naamoun
Institute of Applied Physics, T.U. Bergakademie, Freiberg, Germany

ABSTRACT: Realistic evaluation of the environmental contamination risks from uranium tailings requires an understanding of ways of binding or of the specific chemical forms of trace metals in the tailing materials and consequently their mobility and participation in the water cycle. For this purpose, samples were taken from the uranium tailings of Schneckenstein and were analysed by means of a seven-steps sequential chemical extraction. For the residual fraction, ICP-AES was used for the analysis of fifteen elements (Ba, Cr, Zn, Co, Cd, As, Mn, Fe, Pb, Ni, Al, V, Cu, U, and Th). For the other separation fractions, ICP-MS was used for all elements except for Fe where flame adsorption spectroscopy (FAAS) was utilised. The data analysis shows the near immobility of Al as well as of chromium and vanadium under all weathering conditions of the tailings environment. Whereas thorium and iron seems to be also low mobile, certain amounts of their total contents are exposed to mobilisation under reducing conditions. In addition, the most mass of Cu as well as of Mg, Pb and Zn are low mobile, while considerable amounts of their whole concentrations are exposed to mobilisation under the same above mentioned conditions. Nevertheless, certain portions are high mobile. Moreover, although the most amounts of elements such as As, Co, Mn, and Ni are inactive, considerable amounts are mobile under anoxic conditions, whereas little ones are high mobile. Furthermore, of their part Cd, Ba and U are considerably mobile but the anaerobic conditions increase the bioavailability of the last two mentioned elements.

1 INTRODUCTION

Besides processed ores from the Schneckenstein area (Germany), uranium ores from Zobes, Niederschlema, Oberschlema and the area of Thuringia were treated using physical (gravitational and radiometric) and chemical mainly soda alkaline procedure in the uranium processing plant called "32 unit". Thus the resulting tailings will contain material that reacts as a sink of contaminants. Many attempts were made to specify the binding forms of trace metals in different natural environments by means of different selective extraction procedures. Thus, the current work is an attempt with the aim of the determination of the mobility of many metals in the tailing environment.

2 EXPERIMENTAL

Sediment samples characterised with their difference in mineral contents and metal concentrations were collected from the second bore hole at different depth intervals (0,5-1,5m; 3,5-4,5m; 5,5-6,5m; 6,5-7,5m; 8,5-9,5m; 11,5-12,5m; 14,5-15,5m; 17,5-18,5m; 19,5-20,2m). In order to avoid contamination, the soil sample was taken with a polyethylene spoon from the middle of each sample container. The samples were freeze-dried. They were subsequently ground in agate mortar until a grain-size $\leq$ 63µm, homogenised and stored until needed.

The leaching procedure and reagents are shown schematically in Figure 1. After the same DIN (DIN 38414 / 7) as in the last fraction presented in the diagram sited above, an aqua-regia unlocks for the entire samples from each depth interval was accomplished. Due to the low concentration of cadmium and thorium (less than their detection limits) and due to inaccurate uranium determination with .X-ray fluorescence analysis, a HF-unlock was accomplished for the entire samples. The procedure is as the following:

Into a well cleaned beaker, from each freeze dried sample, between 0.1and 0.2g of well pulverised soil material was filled. 2.5ml of HNO3 with 5 ml HF were added to it. Without exhausting, the sample was two times heated; for one hour at temperature of 50 °C, then for half an hour at 100 °C. Thereafter, it was evaporated to the dry with exhausting. After the cooling of the sample, the same volume of both acids was added to it and the same cited procedure was repeated. Then the sample was lifted by adding 2 ml of concentrated HNO3 and shortly heated. Finally , by adding an amount of approximately 10 ml

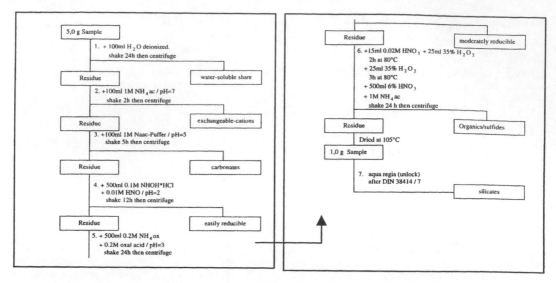

Figure 1. The sequential extraction- schema.

Table 1. The detection limits of each element in different fractions.

Limit of detection(µg/l)	As	Ba	Cd	Co	Cr	Cu	Fe	Mn	Ni	Pb	Th	U	V	Zn
Pore water	10	1	1	1	1	1	100	1	1	1	1	1	1	10
Exchangeable	50	5	5	5	5	5	100	5	5	5	5	5	5	50
Carbonatic	100	10	10	10	10	10	100	10	10	10	10	10	10	100
Easily reducible	100	10	10	10	10	10	100	10	10	10	10	10	10	100
Moderately reducible	100	10	10	10	10	10	1000	10	10	10	10	10	10	100
Sulfide./Org.	50	5	5	5	5	5	100	5	5	5	5	5	5	50
Aqua-regia unlock	100	10	0.1	10	10	10	100	10	100	50	10	50	10	10
HF-unlock			0.1								1	0.1		

of deionized water and heating, the concentrated precipitation was diluted. The resulting solution was filled in special bulb afterwards conserved.

To minimise losses of solid material, the selective extraction was conducted in centrifuge tubes polypropylene. Between each successive extraction, separation was effected by centrifuging at 4000 rpm. The supernatant was removed with a pipette and filled in polyethylene-bottles then analysed for trace metals, whereas the residue was washed with ~ 10 ml of deionized water. After centrifugation for 45 min, this second supernatant was discarded. The volume of rinse water used was kept to a minimum to avoid excessive solubilization of solid material, particularly organic matter. All reagents used in this work was of analytical grade. All glassware used for the analysis was previously soaked in weak nitric acid and rinsed with deionized water.

3 TRACE METAL ANALYSIS

The analytical techniques used to determine the metal concentration in the extracts were selected for compatibility for particular elements and their concentrations and include: FAAS (flame atomic absorption spectrometry, ICP-OES (inductively coupled plasma optical emission spectrometry), ICP-AES (inductively coupled plasma atom emission spectrometry), ICP-MS (inductively coupled plasma mass spectrometry). For all concerned elements(Al, As, Ba, Cd, Co, Cr, Cu, Fe, Mg, Mn, Ni, Pb, Th, U, V, Zn), the metal concentrations in the aqua- regia extracts were measured with ICP-AES, where as ICP-MS was used for other fractions. The elements Al and Mg were determined by ICP-OES. FAAS was appropriate for Fe determination. The detection limits of each element in different fractions are given in Table 1.

3.1 Error consideration

The calculated mass balance between the sum of the seven fractions of the sequential extraction and the aqua –regia unlock of the hole sample for each element, point out the dependence of the error of measurements at the element it self and at the depth interval on the other hand. For the majority of elements, the relative error did not exceed 20%, but for Cd and

Table 2. The selective extraction; error consideration.

Depth	Al	As	Ba	Cd	Co	Cr	Cu	Fe	Mg	Mn	Ni	Pb	Th	U	V	Zn
Interval	%	%	%	%	%	%	%	%	%	%	%	%	%	%	%	%
0,5-1,5m	3	4	5	56	39	5	2	8	6	6	12	34	8	14	1	9
3,5-4,5m	5	3	2	42	6	10	20	10	10	3	6	23	9	25	7	12
5,5-6,5m	4	5	10	9	2	11	0.4	20	5	6	3	8	36	14	10	23
6,5-7,5m	16	11	15	13	3	22	2	23	14	2	3	0.4	47	17	23	28
8,5-9,5m	4	8	15	24	2	18	3	18	13	0.7	1	4	54	11	13	25
11,5-12,5m	14	11	17	39	4	20	8	17	9	0.3	0.5	16	35	18	22	16
14,5-15,5m	9	6	11	19	1	6	11	15	7	5	10	2	31	19	14	21
17,5-18,5m	12	10	17	18	2	17	4	9	3	0.4	10	14	21	1	13	25
19,5-20,2m	8	10	20	62	4	2	5	7	4	14	44	43	26	25	9	51

Th was important and attained 60% for Cd in the last depth interval. Probably, the error is a consequence of the low concentration of the two elements in most fractions (Table 2).

4 RESULTS AND INTERPRETATION

4.1 *Aluminium*

In the analysed samples, the total Al concentration lies between 172000 and 230000 ppm. Since aluminium combines with silica to form aluminosilicates and due to its insolubility over much of the natural pH range which promote its retention in weathering products of low solubility i.e. its very low mobility in weathering processes, most of the Al content is found in association with the lithogenous fraction varying from 98 to 99 % from its above mentioned bulk concentration. Moreover due to the fact that most of Al is detected only by means of X-ray fluorescence analysis (Figure 2), this element seems to be found in clay forms.

Besides, according to Vance et al., (1996), Al tends to be complex with natural organic substances (i.e., biochemical, humic substances, dissolved organic carbon, etc) in the weathering and neogenesis of Al-bearing minerals. Thus, only trace amount from its non residual fraction ranging between ~ 0 to 2 % is in connection with the sulfid./organic phase and this may be due to the scarcity of organic species in the prospected environment as demonstrated by the low amount of TOC in most tailing parts. Further, Al has not any ability to form sulphide complexes although its high tendency to combine with sulfates in acidic medium.

Furthermore, the neglected amount of the non residual Al connected with the carbonate phase ranging from ~ 0.5 to 2 % is an affirmation of the non affinity of the mentioned component for the sited element.

Moreover, because only in the pH range from 4.2 to 5, Al ions can be found with high fractions in the exchangeable cations and since the pH of the studied environment is expected to be mostly in the alkaline range, the non residual Al amount associated with the exchangeable phase is negligible and it lies between ~ 0.01 to 0.13 %.

In addition, most of the non residual Al is found in relation with the nodular hydrogenous fraction to be ranging from 96 to 98 %. This association may be an outcome of the abrasion of a little amount of alumino- iron minerals such as hornblende and chlorite under the effect of the mineral processing that causes the oxidation of iron i.e. its fixation to the new formed compounds and the dissolution of some silica without any attack to Al.

Its noteworthy to point out that the closeness of Al to silicates especially clays and its near absence in the labile phases is a useful tool for the affirmation of the alkalinity of the used leaching procedure and of the studied environment itself. These findings support the well known behaviour of aluminium in an alkaline range.

4.2 *Arsenic*

In the treated sediments, the whole arsenic content varies between 170 and 690 ppm and it is found that about 29-66 % from the mentioned concentration is held in the crystal lattice with only a minor part in association with clays as is demonstrated by the total unlock and X-ray fluorescence analysis (Figure 2). In addition, according to the findings of Onishi & Sandell, (1955), the most arsenic of the mentioned phase is bound to silicate minerals such as quartz, feldspars and hornblende (amphibole). Moreover, according to the same researchers, the relative high As concentration in these minerals results from the ease with which this element substitutes for Si, Al, or Fe in their crystal lattices.

Besides, although the chalcophilic tendencies of this element to form sulphide complexes under anaerobic conditions as well as the presence of many As-sulphide minerals such as arsenopyrite, realgar, orpiment and arsenolite in the most uranium deposits of the Ore Mountains, not any association of this element with the sulfid./organic phase is detected and this may be due to the dissolution of most As-sulphide compounds including those mentioned minerals by the influence of the mineral processing mainly under the flux of oxygen attack. In addition

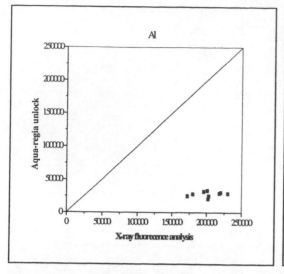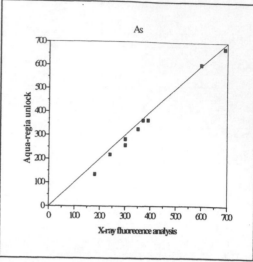

Figure 2. The comparison between the two procedures; aqua regia unlock and X- ray F. A.

As demonstrates a low affinity to organic species with regard to other chemical components.

Further, since As has a high affinity to Mn oxides as well as to Fe oxides and hydroxides (Fordham & Norrish, 1979), the most important of its non residual amount is found in connection with the nodular hydrogenous fraction ranging between 75 to 98 % with the dominance of the moderately reducible phase and these findings are in agreement with those of Jacobs & al., (1970) confirming the strong adsorption of this element mainly to the amorphous iron oxide.

Furthermore, only a trace amount of As ranging from 0 to 2 % from its non residual fraction is in relation to the carbonatic phase, and these results support those sited by Yan-Chu, (1994) and this may be due to the very low affinity of As to carbonates with regard to other components mainly at pH less than 11.

Although As is able to be exchanged on clay minerals in its anionic form AsO_4^{3-} replacing OH^- ion, a negligible amount of its non residual fraction is associated with the exchangeable phase and this may be due to the scarcity of the mentioned As anion which may only be abundant enough at high pHs exceeding pH 12.

On the other hand, an important amount of its non residual fraction is found in connection with the pore water phase ranging from ~ 0.6 to 14 % and this may be due to the increase of the solubility of As with the increasing of pH (Yan-Chu, 1994) mainly in alkaline range.

It is noteworthy to point out that most As amount from its former sulphide complexes (minerals) is dislocated to the nodular forms mainly under the influence of the mineral processing causing a high loss

of sulphur in its sulphate form as well as the formation of iron oxides and hydroxides that are considered as the best scavenger of As under oxidising conditions.

4.3 Barium

In the treated tailing materials, the bulk barium content varies from 390 to 3700 ppm. Causing by the substitution of Ba^{2+} for K^+ due to the nearly identical ion sizes, Ba is distributed among a number of silicate structures and geochemically mainly associated with potash feldspars and micas mainly muscovite and biotite but also with hornblende and even with clays such as chlorite, thus ranging from 43 to 92 % from the mentioned concentration of lithogenous origin. Moreover, the results of both procedures (aqua regia unlock and X-ray F.A) show that the minor amount of this fraction is in connection with clays.

Besides, although the low amount of TOC in the tailings material, a little amount of the non residual Ba is in relation with the sulfid./ organic phase ranges between ~ 3 to 12 %. Thus these results demonstrate the high adsorption of Ba by organic substances and maintain those of Goldberg, (1961).

Further, ranging from 29 to 66 % from its non residual amount is in connection with the nodular hydrogenous fraction and these findings support those of Duval & Kurbatov, (1952) that assert the considerable affinity of hydrous ferric oxide to this element as well as of Puchelt (1967) that affirm also its high association with manganese hydroxide. In addition, possibly the former associated Ba with for example biotite is already dislocated to the freshly precipitated Fe and Mn hydroxides mainly during the min-

220

eral processing that provokes the abrasion of many minerals including the mentioned ones.

Furthermore, relatively a considerable amount of the non residual Ba is in association with the carbonatic phase ranging from ~ 7 to 22 %. These findings are in agreement with the well known chemical attitude of this element because it has a great tendency to precipitate with carbonates as $BaCO_3$ at high pHs exceeding pH 11.6. In addition the alkaline uranium leaching procedure provides the favourable medium for this kind of association.

On the other hand, since clay exchange sites and humus show a high cation exchange selectivity for Ba^{2+} over more strongly hydrating cations such as Ca^{2+} and Mg^{2+} (Mc Bride, 1994), the large majority of the non residual Ba is attached to the exchangeable phase varying between ~ 15 to 42 %.

So as Ba is immobile over most of the Eh-pH field of water, only a trace amount from its non residual fraction is linked with the pore water phase and it ranges from ~ 0.1 to 1 %. In addition, it is noteworthy to point out that except the above mentioned Ba compound forms another fraction from its total content must be in precipitation with sulphate ion forming the mineral barite since on one hand this form is the most stabile one under all weathering conditions. On the other hand the sited mineral is identified by means of the mineral separation by heavy liquids.

4.4 Cadmium

The total Cd content ranges from 0.23 to 6.8 ppm in the analysed sediments. Between 5 to 100 % of the mentioned concentration is of lithogenous origin and Cd seems to be associated with its host minerals such as: hornblende and also chlorite.

Besides, although the absence of any connection of this element with the sulfid./ organic phase in three samples and due to the chalcophilic property of Cd that it means its high tendency to coprecipitate with sulphur under reducing conditions and high availability of HS⁻ as well as to its appreciable ability to form complexes with humic and fulvic matter an important amount of the non residual Cd is in relation with the mentioned phase varying from 16 to 69 %. Likewise, the Cd amount bound to the present phase shows a high positive correlation with the total sulphur content measured by means of X-ray fluorescence analysis (Figure3), and this is in close conformity with those of Gupta & Chen, (1975).

Further, although the appreciable affinity of Fe oxides for Cd chiefly at high pHs, almost not any association of this element with the nodular hydrogenous fraction is identified and this may be due to its higher tendency for complexation with other species than Fe and Mn oxides and hydroxides.

On the other hand, although the absence of Cd in the carbonatic phase in two samples, as much as the sited element in its ionic form Cd^{2+} tends to coprecipitate with $CaCO_3$ (McBride, 1994) or precipitate as $CdCO_3$ above pH 7, a high amount of its non residual fraction is found in connection with the mentioned phase ranging from ~ 11 to 35 % in the other sediments. Thus, it seems that the Cd prefers to form complexes with carbonates better than with Fe and Mn oxides and hydroxides.

Furthermore, except the last sample where not any Cd is bound to the exchangeable phase, a considerable amount of its non residual fraction is connected with the mentioned binding form varying from ~ 19 to 100 %. These findings are in agreement with those of Andersson, (1975) affirming the sig-

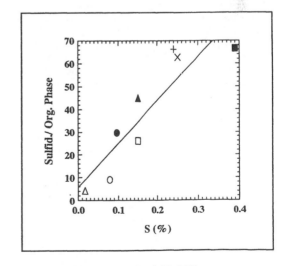

Figure 3. The relationship between sulphur & the Cd; Cu amounts bound to the Sul./ Org. Phase (Cr = 0.86; 0.90).

nificance of Cd in the sited mobile fraction and this is due to the high selectivity of clays for Cd^{2+}.

In addition, since at neutral to alkaline medium Cd tends to precipitate in sulfide minerals or as $CdCO_3$ as well as to co-precipitate with $CaCO_3$, its solubility under the sited conditions is low and therefore, there is no Cd in connection with the pore water phase.

4.5 Cobalt

The bulk concentration of Co in the analysed samples ranges from 26 to 130 ppm. Between 33 to 98 % of the total Co is held in the lattice structure of minerals with a large majority in association with clays.

Besides, due to the high ability of Co for complexation by both humic and fulvic organic matter (Nriagu & Coker, 1980) as well as to its tendency to precipitate with sulphur under reducing conditions forming CoS precipitate, an appreciable amount of the non residual Co is in association with the sulfid./organic phase varying from ~ 7 to 100 %.

Further, the most amount of the non residual Co is in connection with the nodular hydrogenous fraction ranging between ~ 0 to 92 %. This is due to the strong affinity of this element for Fe/Mn oxides especially for Mn nodules (Cronan, 1976). In addition, the tailings environment favours the association of this element with micro nodules since it tends to be completely bound to these complexes at alkaline medium by chemisorption and co-precipitation (McBride, 1994).

Furthermore, because the alkaline uranium leaching procedure provokes the formation of a stable carbonate precipitate (spherocobaltite; $CoCO_3$) and the tailings environment provides its stability a relatively important amount from the non residual Co is in relation with the carbonatic phase ranging from ~ 0 to 13 %.

In addition, according to Kharkar et al., (1968), Co demonstrates a high exchangeability mainly for clays, thus a little amount of its non residual fraction is in connection with the exchangeable phase varying between ~ 0 to 5 %.

Moreover, only a trace amount of Co is in association with the pore water phase and this may be due to the high tendency of Co to complexation under the chemical conditions of the studied areas.

4.6 Chromium

The bulk Cr concentration in the analysed samples varies between 64 and 130 ppm. Inasmuch as many silicates such as hornblende, chlorite and micas are considered as host minerals for Cr, ranging from 89 to 95 % of the mentioned content is held in the lattice structure of clay and silicate minerals.

Further, although the low TOC as well as the total sulphur concentrations in the whole tailings material,

it is found that between ~ 0 to 55 % of its non residual fraction is associated with the biogenic fraction. This is may be due mostly to its high affinity to strongly complexation with organic matter (humic substances) in its immobile cation Cr^{3+} form as well as to its ability to form sulfide complexes under reducing conditions.

Furthermore, as Cr(III) can strongly bind to particles essentially Fe/Mn oxides, from 45 to 100 % of the authigenic fraction is in corporation with the nodular hydrogenous fraction, where the most bulk is in connection with the moderately reducible phase and these findings are in agreement with those of Johnson & al., (1992) confirming the higher closeness of the Cr cycling to that of Fe than that of Mn.

Besides, because Cr prefers to form stable complexes with ligands and to be adsorbed on surface sites, there is no identified Cr association with the carbonatic as well as with the exchangeable phases in most samples.

In addition, as in soils Cr is found predominantly in its insoluble forms and due to its high tendency to form compounds with humus and Fe/Mn oxides as already mentioned, only trace amounts of Cr is in relation with the pore water phase.

Moreover, it is of importance to point out that most occurring Cr form is its very slow reacting trivalent oxide and independent on the chemical medium, any other forms will tend to be converted to the mentioned one. Therefore, Cr is expected to have a very low reactivity under all weathering conditions of the tailings environment.

4.7 Copper

The total Cu concentration of the analysed samples varies between 56 to 670 ppm. The lithogenous contribution to the mentioned content ranges from 16 to 61 %. The large majority of Cu associated with the residual fraction is bound mostly to silicates such as hornblende and minor amount is held in the lattice structure of clays such as chlorite.

Besides, because of its chalcophilic property i.e. its high tendency to associate with sulphur by precipitation and co-precipitation under reducing conditions to form very insoluble sulfide minerals as well as its strong affinity to organic ligands specially for humic type material, a considerable amount of the non residual Cu is in connection with the sulfid./organic phase varying from ~ 9 to 66 %. In addition, the total sulphur content is in a high positive correlation with the Cu amount in the present phase (Figure 3) and this is in close agreement with the finding of Gupta & Chen, (1975). Likewise, it is noteworthy to point out that the present association between Cu and sulphur is mostly due to the presence of sulfide minerals such as: chalcopyrite, covellite, bornite and chalcocite in most uranium deposits of the Ore mountain. Moreover their absence in the

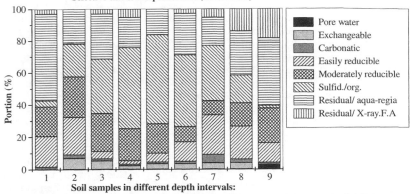

Figure 4. Partition of copper in different phases.

tailings material may be due to the dissolution of the most of their amounts during the mineral processing since a negligible amount of chalcopyrite is already detected but only by mode of mineral separation by means of heavy liquids.

Further, between 28 to 87 % of the non residual Cu is in relation with the nodular hydrogenous fraction and this due to the strong adsorption of Cu^{2+} by Fe oxides (McKenzie, 1980), hydrous oxides and oxyhydroxides (Millward & Moore, 1982) as well as by Mn oxides (Pickering, 1979) so as the pH is raised on one hand and to the expected alkalinity of the chemical tailings medium that provides the high association of Cu with these compounds on the other hand. In addition, it is of importance to point out that except the mentioned complexes, the Al oxides and hydroxides are also important adsorbents for Cu at high pHs (McBride, 1994). Therefore, an appreciable Cu amount is probably bound to the last sited composites.

Furthermore, ranging from 1 to 7 % of the non residual Cu is found in connection with the carbonate phase. These findings are in agreement with those of Salomons & Förstner, (1980) that confirm this type of binding.

Because the milling process during the uranium leaching increases the breakdown of primary minerals accompanied by the retention of metals particularly of Cu in situ by the fine fractions which holds them in a more readily exchangeable form under poor drained conditions, a certain amount of its non residual fraction is joined with the exchangeable phase and it varies between ~ 1 to 9 %.

In addition, only a trace amount of Cu is specified in the pore water phase and this is due to its high tendency to be adsorbed by colloidal material largely at high pHs as already mentioned except the last sample where ~ 6 % of its non residual fraction is identified in the sited phase and this may be due to the presence of a considerable DOC concentration in the last depth interval that increases the soluble copper in the form of Cu^{2+}- organic complexes at the sited chemical conditions.

4.8 Iron

The whole Fe content in the sample sediments varies between 53000 to 69000 ppm. Since Fe is a greater constituent of a large variety of silicates including hornblende, chlorite and biotite, the big majority of the mentioned concentration is of lithogenous origin ranging from 84 to 94 %.

Further, although the high tendency of Fe to form sulphide complexes and the sufficient spread of iron sulphide minerals such as pyrite, chalcopyrite and arsenopyrite in the most mineral deposits of the ore mountains including the uranium ones on one hand. On the other hand, its appreciable ability to form complexes with humus, almost not any Fe is detected in the sulfid./ organic phase. This may be due to the dissolution of most sulphide minerals during the mineral processing since the soda alkaline procedure accompanied with a high aeration not prevails their stability as mentioned in the third paragraph. Further, the mentioned conditions also strips Fe^{3+} from organic matter complexes that are stable only at very low pH.

Furthermore, the association of approximately 100 % of the non residual Fe with the nodular hydrogenous fraction with the dominance of the moderately reducible phase, may be due to the formation of Fe oxides and hydroxides chiefly after the abrasion of most biotite and the dissolution of nearly all non silicate iron mineral contents during the uranium leaching processing.

Moreover, although the presence of an appreciable amount of carbonates as well as the alkalinity of the chemical medium of the tailings environment that the provide the formation of Fe carbonate mineral siderite, only a trace amount of Fe in the carbonatic phase was detected. This may be due to its very low solubility mainly during and after the mineral processing as well as to its high tendency to form its proper oxides and hydroxides with regard to other compound forms.

In addition, only a trace amount of the non residual Fe is detected in the pore water phase and this is due to its low solubility over the Eh-pH conditions of the prospected sites and to its general tendency for complexation.

4.9 Magnesium

In the treated samples, the bulk concentration of magnesium ranges between 14300 to 28900 ppm. Since Mg is an important constituent of a large number of common rock forming silicates including those identified in the tailings sediments such as, hornblende and chlorite, the most amount of the sited concentration is of lithogenous origin fluctuates between 75 to 96 %. In addition, its total contents show that it is mostly bound to clays and it seems that the most Mg portion is associated with chlorite since this mineral is considered as one of the most important rock-forming minerals in sediments containing this element.

On the other hand, Mg has an impressive ability to form complexes with organic species, although the low amount of TOC in the analysed sediments, an appreciable amount from its non residual fraction is found in connection with the sulfid./organic phase varying from ~ 4 to 35 %.

Besides, because Mg is also frequently associated with carbonates in forms of the minerals dolomite and Mg-calcite (calcite in which some of the Ca is replaced by Mg) predominantly at high pH exceeding pH 8 where they show an extreme stability and due to the expected alkalinity of the analysed sediments as well as to the presence of the first mentioned mineral with a low amount in the tailings material, relatively a considerable mass of the non residual Mg is found in the carbonatic phase ranging from ~ 0 to 22 % and may be is totally connected with the sited above mineral.

Further, although Mg occupies an important amount from the total exchange capacity of the negatively charged silicate minerals where it is held as an exchangeable cation and the presence of enough of broken bonds around the edges of the silica-alumina units of clays and nonclay minerals mainly after the mineral processing that rises the exchange capacity of the tailings material, only a little amount of its total content is connected with the exchangeable phase ranging from ~ 0.3 to 2 %. This

may be due mostly to the pH of the medium since it is considered as the most important factor among others controlling on one hand the ion exchange capacity of any mineral or sediment and on the other hand the exchangeable ion itself. Moreover, in such environment (alkaline) as the prospected areas, according to Bolt et al., (1976), Na^+ is the dominant exchangeable ion.

Furthermore, the most amount of the non residual Mg is found in association with the nodular hydrogenous fraction ranging from 56 to 76 %. This may be due to the substitution of the two elements Fe and Mg in their ionic forms due to the great similarity between them.

In addition, a trace amount of the total Mg is detected in the pore water phase varying between ~ 0.02 to 0.4 %. This may be due to the alkalinity of the studied medium mainly at pHs above pH 8 where Mg tends to precipitate with carbonates or to form $Mg(OH)_2$ compound.

It is important to point out that, the low amount of exchangeable Mg is on one hand a good witness for the alkalinity of the studied environment and on the other hand it supports the findings of Bolt & al., (1976).

4.10 Manganese

The total manganese content in the analysed sediment varies between 630 to 1900 ppm. Due to the replacement of Fe and Mg by Mn in many silicates, a range of 43 to 92 % of the mentioned concentration is of lithogenous origin. In addition, the total determined concentrations show that the majority of the residual Mn is in association with clays (chlorite).

Further, due to the restricted affinity of Mn for sulphur at pHs between pH 8 to pH 10 and under anaerobic conditions to form MnS precipitate as well as to its tendency to bound to humic type material mostly at pH >6 (McBride, 1994), a certain amount of the non residual Mn is in connection with the sulfid./organic phase ranging between ~ 2 to 12 %.

Furthermore, the large majority of the non residual Mn is in relation with the nodular hydrogenous fraction ranging from 58 to 77 %. This is due to its high ability to form its own oxides chiefly as MnO_2 form under aerated conditions on one hand and this is demonstrated by its grand association with the easily reducible phase. On the other hand, it tends to be strongly adsorbed on iron oxides and oxyhydroxides (Millward & Moore, 1982) largely at alkaline medium, therefore a considerable amount of its nodular fraction is in relation with the moderately reducible phase.

Moreover, a relatively important amount of the non residual Mn is bound to carbonates varying between ~ 7 to 20 %. This is due to its high ability to precipitate as $MnCO_3$ at high pHs (8<pH<12) and under reducing conditions on one hand and to its tendency to replace Ca and chiefly Mg in calcite and dolomite respectively on the other hand.

In addition, a considerable amount of the non residual Mn is interconnected with the exchangeable phase ranging from ~ 6 to 19 %. This may be due to the increase of the cation-exchange capacity of the processed material after the mineral grinding on the one hand, as well as to the cation exchange selectivity for Mn^{2+} over Mg^{2+} and Ca^{2+} due the alkalinity of the tailings environment on the other hand.

Besides, because the alkalinity of the chemical milieu of the prospected sites (pH>6), Mn tends to form complexes with organic matter, oxides and silicates (McBride, 1994) and therefore the decrease of its solubility and only a trace amount of the non residual Mn is in relation with the pore water phase.

4.11 Nickel

The bulk Ni concentration of the treated sediment varies between 24 to 300 ppm. The total unlocks affirm the big majority of the mentioned content ranging from 25 to 74 % of lithogenous origin and is mostly held in the lattice structure of its host silicate minerals such as hornblende and micas (muscovite) and a little mass is also in association with clays (chlorite).

Further, because Ni tends to incorporate into sulphides by reduction of oxidising sulphate waters by the action of sulphate-reducing bacteria (Salomons & Förstner, 1984) as well as its high ability for bioaccumulation in humus and for bonding to "softer" organic ligands containing nitrogen and sulphur (McBride, 1994), a considerable amount of the non residual Ni is found in connection with the sulfid./organic phase varying from ~ 7 to 20 %.

Furthermore, between 67 to 92 % of the non residual Ni is in relation with the nodular hydrogenous fraction and this is due to the high affinity of Fe and Mn oxides to the mentioned element on the one hand, and to the alkalinity of the studied chemical environment on the other hand, since according to McBride, (1994) the chemisorption on these oxides is favourable above pH 6.

Besides, an appreciable amount of the non residual Ni is bound to carbonatic phase ranging from ~ 0 to 11 %. These results are in agreement with those of Stover & al., (1976) asserting the important association of this element with the mentioned phase.

In addition, the exchangeable Ni is favourable only at low pHs, so that only a low amount of its non residual fraction is bound to the exchangeable phase varying from ~ 0 to 7 %.

Furthermore, because its solubility decreases markedly at high pHs and its mobility becomes very low in neutral to alkaline medium only a trace amount of Ni was detected in a soluble form in the pore water extract.

4.12 Lead

In the analysed tailings material, the whole Pb content varies between 40 to 190 ppm. Due to the similarity in ionic radius of Pb^{2+} and K^+ a large amount of Pb is incorporated in silicates such as potash feldspars and micas where the first sited ion replaces the second one. Thus ranging from 33 to 84 % of the mentioned bulk content is in association with the lithogenous fraction with the dominance of the non clays part.

Although the high affinity of organic substances chiefly humic substances to complexation with Pb (McBride, 1994) as well as the extreme tendency of this element to form very insoluble sulfide compounds under reducing conditions due to its chalcophilic property, there is no detected Pb bound to the mentioned chemical components in the four samples and only a certain amount of the non residual Pb is associated with the sulfid./organic phase ranging between ~ 1 and 14 %. This may be due to the low content of humus in the tailings material and also to the alkalinity of the studied chemical medium that favour the detachment of Pb-organic complexes (McBride, 1994) as well as to the dissolution of most Pb -sulfide minerals during the mineral processing accompanied by a high loss of sulphur as demonstrated by its low total content in the bulk analysed sediment.

Further, due to the high affinity of Fe and Mn oxides and hydroxides for Pb mostly at high pHs as well as of the expected alkalinity of the studied chemical milieu, between 81 to 100 % of the non residual fraction is found in connection with the nodular hydrogenous fraction.

Furthermore, as the high pH favour the precipitation of Pb with carbonate (Salomons & Förstner, 1984), an appreciable amount of the non residual Pb ranging from ~ 0 to 14 % is in connection with the carbonate phase and these findings are in agreement with those of Salomons & Förstner, (1980) confirming the association of a considerable amount of this element with the last sited phase.

Besides, owing to the difficult displacement of Pb as an exchangeable form from clays, only very low amounts of the non residual Pb is in association the exchangeable phase not exceeding 1 % for all samples.

In addition, because lead is the least mobile heavy metal in soils, especially under reducing or non-acid conditions (McBride, 1994), only a trace amount of the non residual Pb is found in connection with the pore water phase.

4.13 Thorium

The gross thorium content varies in the treated samples between 12.7 and 20.5 ppm and as is expected not any Th is detected in the pore water phase since

Th is well known for its extremely low solubility in natural waters.

Besides, although the high tendency of Th to form complexes with sulphates chiefly at pHs below pH 5 (Boyle, 1982) as well as to be adsorbed almost completely to organic matter above pH 7, there is no detected Th in the sulphid./organic phase. This may be due to firstly to the expected alkalinity of the prospected sites and on the other hand to the low organic content in the studied areas as demonstrated by the low TOC content in the most tailings environment including the analysed samples.

Further, no detected Th in the carbonatic phase that may be due to the low amount of carbonates in the studied area mainly after the mineral processing on one hand and to its low affinity for the mentioned complexes on the other hand (Boyle, 1982).

Furthermore, Th has a high ability to deposit in clays and organics by adsorption and precipitation mechanisms but not by the ion exchange ones. Therefore, there is no detected Th in the exchangeable phase.

Moreover, since an appreciable amount of Th is frequently associated with silicates including quartz, feldspars (Boyle, 1982), and micas (mainly in inclusion) by direct cation substitution in the silicate lattice as well as by adsorption in lattice defects, the most amount of the Th content is found to be in connection with the lithogenous fraction ranging from 73 to 92 %. Further, an appreciable amount of Th was probably bound to the initial hematite content, then after the mineral processing apparently the low mobile Th was fixed to the freshly formed hydrous ferric oxides. Therefore, the other Th amount is found totally in relation with the moderately reducible phase.

It is noteworthy to point out that the mentioned element is largely attached to clays and partly to re-

sistate minerals since the most amount of Th is dissolved by the total HF unlock.

4.14 *Uranium*

In the analysed samples the whole uranium content ranges from 11.8 to 225.6 ppm. Between 24 to 57 % of its total mass is associated with the lithogenous fraction (Figure 5) and U is probably held together with Y mostly to feldspars but also to chlorite, hornblende and even biotite by different mechanisms such as the isomorphous substitution in the lattice or the concentration in lattice defects as well as the adsorption along crystal imperfections and grain borders since U tends to replace Y in some minerals including the above mentioned ones due to certain energetic similarities (ionic size, charge, electronegativities, etc.) between them and the high correlation between the above mentioned elements that is already revealed in the preceding paragraph, asserts this position. Moreover, the most U amount is dissolved by the aqua regia unlock and this approves that U is not largely attached to clays but primarily clutched in resistate minerals on the one hand and the greater ease of its mobilisation than that of Th on the other hand, and these findings are supported by those of Pliler & Adams, (1962).

Besides, in as much as U is found to be strongly enriched in certain organic sediments, particularly those formed from humic substances (Boyle, 1982) as these complexes are able to absorb U intensively from waters essentially in reducing environments (Armands & Landergren, 1960) as well as it may be precipitated by reduction from its hexavalent state in sulfide-rich sediment (Van Wambeke, 1971), thus a significant amount of the mentioned element is in association with the sulfid./organic phase varying between ~ 5 to 42 % from its non residual fraction.

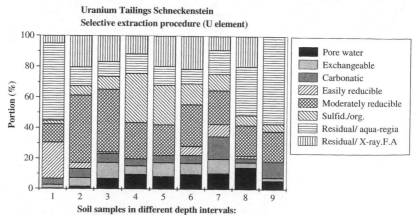

Figure 5. Partition of uranium in different phases.

Further, because an appreciable amount of U is often bound to mineral biotite (Boyle, 1982) as well as in reducing environments it is frequently in connection with hematite (Van Wambeke, 1971). Besides, it tends to be adsorbed and coprecipitate by hydrous ferric oxide and manganese dioxide fractions (Boyle, 1982), a considerable mass from its non residual fraction ranging between 30 to 80 % is found in relation with the nodular hydrogenous fraction with the dominance of the moderately reducible phase and this may be due to the widely spread iron oxides and hydroxides resulting from the abrasion and dissolution of iron minerals such as biotite and hematite during the mineral processing.

Furthermore, since uranium forms very stable complexes with carbonate ion in forms of $UO_2(CO_3)_3^{4-}$ and $UO_2(CO_3)_2^{2-}$ ion complexes at pHs equal or above pH 8, then by the degassing of CO_2 in the atmosphere it tends to coprecipitate in the form of $UO_2CO_3^0$ with other minerals (Gascoyne, M., 1992) and due to the expected alkalinity of the studied environment providing the mentioned processes, a considerable amount of its non residual fraction ranging from ~ 6 to 24 % is intimately associated with the carbonate phase.

In addition, an important amount of the non residual U varying from ~ 5 to 14 % is in connection with the exchangeable phase and this may be due to the ability of U to substitute for calcium in some minerals over all the already identified in the tailings material by the polarisation and cathodoluminescence microscopy of the accessory mineral apatite.

So as the appearance of chemical components such as: sulphate, chloride, nitrate, carbonate, phosphate, or humic matter in the interstitial waters reveal that U contents in the mentioned phase depending mainly on the pH of the concerned medium, these sited constituents play the decisive role in its presence in the naming phase. Further, although their scarcity in the studied areas and due to the expected alkalinity of the studied environment, the carbonate compounds may be considered as the most important among the above mentioned complexes. Therefore, the non residual U portion connected with the pore water phase that it ranges from ~ 0.5 to 29 %, is also in the carbonate ionic forms. Moreover, it is remarkable the low amount of U in the sited phase in the first two samples may be in relation with the distinguished pH of the milieu in such depth intervals.

4.15   Vanadium

In the treated samples, the whole vanadium content varies between 93 to 560 ppm. Since large amounts of V exist as a substitution constituent in muscovite mica as well as in silicate clays including chlorite, the lithogenous contribution to the mentioned concentration ranges from 93 to 96 %. Besides, by means of the aqua regia unlock as well as X-ray F.A, the large majority of V seems to be bound to the sited clay type.

Because V tends to form sulphide complexes as well as a high affinity to humus to form organic complexes in soils (Vinogradov, 1959), an appreciable amount from its non residual fraction ranging between ~ 0 to 19 % is in connection with the sulfid./organic phase, although the low amount of TOC as well as of the total sulphur in the different sediments.

On the other hand, the most non residual V is in corporation with the nodular hydrogenous fraction varying between 77 to 99 % with a dominance of the moderately reducible phase. Thus, these findings are in agreement with the well known chemical behaviour of V inasmuch as the mineral processing provides a well aerated environment where this element presents a high ability to substitute readily for Fe^{3+} in minerals such as Fe oxides (McBride, 1994).

In addition, due to the unimportant affinity for exchangeability as well as for precipitation with carbonates, not any V is detected in both phases carbonatic and exchangeable.

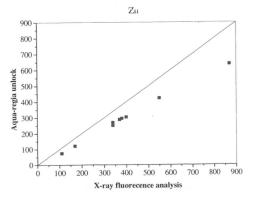

Figure 6. The comparison between the two procedures; aqua regia unlock & X- ray F. A.

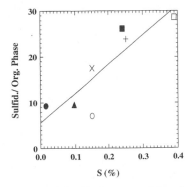

Figure 7. The relationship between sulphur & the Zn amount bound to the Sul./ Org. Phase (Cr = 0.87).

Further, because V shows a high solubility over the most Eh-pH space, an impressive amount from its non residual fraction is associated with the pore water phase ranging from ~ 1 to 13 %.

It is noteworthy to point out that there is a good agreement between the results of the mineral determination confirming the high resistance of muscovite with regard to other minerals and the high amount of the immobile and bound to silicates V.

4.16 *Zinc*

The total Zn content varies between 110 to 870 ppm in the analysed samples. Ranging from 48 to 76 % of the mentioned concentration is of lithogenous origin and approximately equally shared between silicates such as hornblende as well as micas and clays such as chlorite (Figure 6).

Further, except the first sample where not any association of Zn with the sulfid./ organic phase is identified, between ~ 15 to 52 % of its non residual fraction is connected with the sited binding form and this is due on the one hand to its chalcophilic property that is to say, its high closeness for coprecipitation with sulphur under reducing conditions, on the other hand to its high tendency to form complexes with organic matter mostly humic type material (Nissenbaum & Swaine, 1976). In addition, the bulk Zn associated with the mentioned phase is in a high positive correlation with the whole sulphur content (Figure 7) and this finding confirms again the results of Gupta & Chen, (1975).

Furthermore, between 41 to 95 % of the non residual Zn is attached to the nodular hydrogenous fraction. These results are in agreement with those of McKenzi, (1980) demonstrating the appreciable affinity of Fe oxides for this element as well as with

those of Millward & Moore, (1982) affirming its high adsorption by iron oxyhydroxide and also with those of Pickering, (1979) asserting its certain attraction to Mn oxides mostly at alkaline medium.

In addition, excluding the first sample where not any Zn is bound to carbonates and because at mildly alkaline pHs, the sited element tends to form $ZnCO_3$ (BROOKINS, 1988) compound therefore ranging from ~ 2 to 9 % of its non residual fraction is in relation with the carbonatic phase. These findings are in close conformity with those of Salomons & Förstner, (1980) approving this kind of binding.

Besides, since Zn is able to be held in exchangeable forms on clays and organic matter (McBride, 1994), a little amount of the non residual Zn is connected with the exchangeable phase varying from ~ 2 to 8 % and these results are in accordance with those of Stover & al., (1976) confirming this type of association.

On the other hand only a trace amount of the non residual Zn is in relation with the pore water phase and this is due to its tendency to form complexes with different chemical species under the chemical conditions of the tailings environment.

5 CONCLUSIONS

The relative mobility of each element by means of the aqua regia unlock is illustrated graphically in Figure 8. From the data analysis mentioned above it is clear nearly the immobility of Al as well as of chromium and vanadium under all weathering conditions of the tailings environment. While, thorium and iron seems to be also low mobile but certain amounts of their total contents are exposed to mobi-

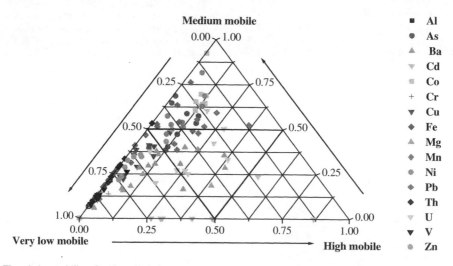

Figure 8. The relative mobility of each studied element.

lisation under reducing conditions. In addition, the most mass of Cu as well as of Mg, Pb and Zn are low mobile, while considerable amounts of their whole concentrations are exposed to mobilisation under the same above mentioned conditions. Nevertheless, certain portions are high mobile. Moreover, although the most amounts of elements such as As, Co, Mn, and Ni are inactive, considerable amounts are mobile under anoxic conditions, whereas little ones are high mobile. Furthermore, of their part Cd, Ba and U are considerably mobile but the anaerobic conditions increase the bioavailability of the last two mentioned elements.

Considering from neutral to alkaline pHs for the studied areas, elements such as Al, Ba, Co ,Cr ,Cu, Ni, Th and Zn have the lowest tendency to mobile. While Fe, Pb and Mn are low mobile. In addition, at the same medium As and Cd are relatively and moderately mobile, respectively. Moreover, the presence of carbonates enhance the mobility of uranium at such pHs. Whereas, Mg and V are high mobile at the same conditions. Therefore the oxygen activity plays the decisive role in the mobility of most studied elements in the prospected areas.

REFERENCES

Andersson, A., (1975): Relative efficiency of nine different soil extractants. Swed. J. Agric. Res., **5**, 125-135.

Armands, G. & Landergren, S., (1960): Geochemical prospecting for uranium in northern Sweden: The enrichment of uranium in peat; in: Noe-Nygaard, A. & al. (eds.). Genetic problems of uranium and thorium deposits, 21st Inter. Geol. Cong., Copenhagen, p. 51-66.

Boyle, R. W., (1982): Geochemical prospecting for thorium and uranium deposits, Develop. Econ. Geol., **16**. Elsevier, Amsterdam.

Cronan, D. S., (1976): Manganese nodules and other ferromanganese oxide deposits. In : Riley, J. P. & Chester, R. (ed.). Chemical Oceanography. Academic, London. New York, **5**, 217-263.

Duval, J. E. & Kurbatov, M. H., (1952): The adsorption of cobalt and barium ions by hydrous ferric oxide at equilibrium. J. Phys. Chem, **56**, 982.

Fordham, A. W. & Norrish, K., (1979): Arsenate- 74 uptake by components of several acidic soils and its implications for phosphate retention. Aust. J. Soil Res. **17**, 307-316.

Gascoyne, M., (1992): Geochemistry of the actinides and their daughters. In: Ivanovich, M & Harmon, R. S. (eds.). Uranium- series disequilibrium. Clarendon Press. Oxford. 910 pp.

Goldberg, E. D., (1961): Chemistry in the Oceans. In: Sears, M. & Washington, D. C. (ed.): Oceanography. Amer. Assoc. Advanc. Sci., **67**, 583-597.

Gupta, S. K. & Chen, Y. K. (1975): Partitioning of trace metals in selective fractions on nearshore sediments, Environ. Lett. Vol. 10, pp. 129-158.

Jacobs, L. W. & al., (1970): Arsenic sorption by soil. Soil Sci. Soc. Am. J. 34, 750-754.

Kharkar, D. P. & al., (1968): Stream supply of dissolved silver, molibdenum, antimony, selenium, chromium, cobalt, rubidium and cesium to the oceans. Geochim. Cosmochim. Acta, **32**, 285-298.

Mc Bride, M. B., (1994): Environmental chemistry of soils. 406 pages, Oxford University Press, New York, USA.

McKenzie, R: M., (1980): The adsorption of lead and other heavy metals on oxides of manganese and iron. Aust. J. Soil Res. **18**, 61-73.

Millward, G. E. & Moore, R.M., (1982): The adsorption of Cu, Mn and Zn by iron oxyhydroxides in model estuarine solution. Water Res. **16**, 981-985.

Onishi, H. & Sandell, E. B., (1955): Geochemistry of arsenic. Geochimica & cosmochimica acta, **7**, 1.

Pickering, W. F., (1979): Copper retention by soil/sediment components. In.: Nriagu, J. O. (ed). Copper in the environment. Part I. Ecological cycling. John Wiley, New York, pp 217-253.

Pliler, R. & Adams, J.A.S., (1962): The distribution of thorium and uranium in a Pennsylvanian weathering profile, Geochim. Cosmochim. Acta, **26**, 1137-46.

Puchelt, H., (1967): Zur Geochemie des Bariums im exogenen Zyklus. Sitzungsber. Heidelb. Akad. Wiss. Math.-nat. Kl. 4. Abh.

Salomons, W. & Förstner, U. (1980) Environ Technol Lett Vol. 1 pp. 506-517.

Salomons, W. & Förstner, U., (1984): Metals in the hydrocycle. Springer-Verlag, 349 pp.

Stover, R. C. & al., (1976): J. Water Pollut. Control Fed., Vol. 48, pp. 2165-2175.

Vance, G. F. & al., (1996): Environmental chemistry of aluminium- organic complexes.

Yan-Chu, H., (1994): Arsenic distribution in soils, in: Nriagu, J. O. (ed): Arsenic in the environment.

Part I: Cycling and characterisation. John Wiley & Sons, Inc.

Van Wambeke, L., (1971): The geology of uranium and thorium. In: Lesmo, R. (ed). Report of the session Part II (1969), The geology of uranium and thorium.E.N.I. – Scuola Enrico Mattei (Italia).

Vinogradov, A. P., (1959): Geochemistry of rare and dispersed elements in soils. Chapman & Hall, London.

Tailings and Mine Waste '02, © 2002 Swets & Zeitlinger, ISBN 90 5809 353 0

Uranium tailings of "Schneckenstein" (Germany) reservoir of contaminants

T. Naamoun

Institute of Applied Physics, T.U. Bergakademie, Freiberg, Germany

ABSTRACT: The beginning of any assessment of contaminant risks from waste tailing requires the determination and evaluation of contaminants. In addition, the estimation of their mobility needs the specification of their relation to the present minerals in the studied environment. For this task, the concentration of a wide range of main and trace elements were determined by means of the X-ray fluorescence analysis. Further, due to their geochemical importance, the non-metals carbon, nitrogen and sulphur contents were defined. Moreover, the ignition loss was also determined. The recorded results show that independently of the weathering conditions of the studied areas, Na, Ca, S, Sr as well as Mg will exhibit a high mobility. Whereas, under the expected weathering conditions of the tailings environment, a very low mobility of Al, Ba, Bi, Ce, Cr, Fe, Ga, La, Pb, Mn, Nb, P, K, Rb, Si, Sn, Ti, W, Y and Zr is forecasted. Moreover, Co, Cu, Zn and Ni tend to be low mobile, but the presence of an oxidising agent may considerably enhance their mobility and the mobility of uranium.

1 INTRODUCTION

Besides the processed ores from the Schneckenstein area, uranium ores from Zobes, Niederschlema, Oberschlema and the area of Thuringia were treated using physical (gravitational and radiometric) and chemical (acid and mainly soda alkaline) in the uranium processing plant called "32 unit". Due to the alkalinity of the leaching procedure, an apparent increase of silicates mainly of quartz and muscovite together with the presence of high resistant clays such as chlorite and kaolinite are recorded. Whereas, sulphide minerals are roughly absent. While an appreciable amount of carbonates (calcite and dolomite) is also measured. the mineral constituents of the studied areas were identified by means of the X-ray mineral phase analysis as well as by the Polarisation together with the cathodoluminescence microscopy. Then, the mineral separation by means of heavy liquids of many samples was also used. The current work tries to combine these elements and minerals and therefore estimate their mobility under the expected chemical conditions of the prospected areas.

2 LOCATION AND SHORT DESCRIPTION OF THE SITE

The Schneckenstein site is approximately 3 km from the village of Tannenbergsthal/county of Vogtland, southwest of Saxony (Figure 1). The surface of the site is situated at an altitude of 740 to 815 m above sea level. In addition, the area of investigation is located on the Southwest border of the Eibenstock granite that could be considered as biotite-syenogranite and covered with a weathered surface layer. Southwest of the investigation area follows the contact zone with quartz-schist.

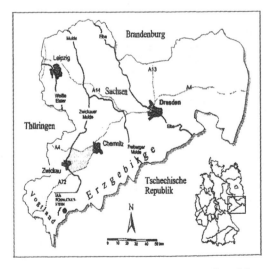

Figure 1. The area of investigation - Uranium tailings Schneckenstein.

3 EXPERIMENTAL

The research was focused on four sediment cores. These cores were taken at the tailing sites by drilling four boreholes at different depths. Two boreholes were dug in each tailing (Figure 2). The first and second boreholes (GWM 1/ 96 and GWM 2 / 96) were drilled down to the granite foundation in Tailing 2 (IAA I). The third borehole (RKS 1/ 96, Tailing 1(IAA II) was sunk to a depth of about eight meters but the granite foundation was not reached due to technical limitations. 12 m deep of the first tailing was sunk for the fourth borehole (RKS 2/ 98). The cores (diameter 50 mm) were cut into 1 m long slices and transported in argon filled plastic cylinders to prevent ambient air contact.

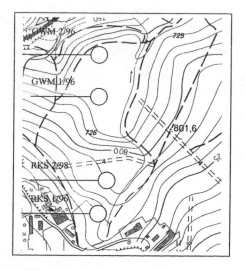

Figure 2. Location of boreholes in the tailing sites.

Table 1. The detection limit of elements and their error consideration.

Element	Limit of detection	Error
Na, Mg, Al,Si, P, S, Cl, K,Ca	0,02-0,1 %	3-5 %
Ti,V, Cr, Mn, Fe	0,01%	2-3 %
Co, Ni, Cu, Zn, Ga, Ge, As, Se, Br	30 ppm	5-10 %
Rb, Sr, Y, Zr, Nb, Mo	20 ppm	3-10 %
Ag, Cd, In, Sn, Sb, Te, I	60 ppm	5-10 %
Ba, Ta, W, Re, Os, Ir, Pt, Au, Hg, Ti, Pb, Bi, Th, U	80 ppm	5-10 %
REE	100-200 ppm	5-10%

4 LABOUR ACTIVITIES (EXPERIMENTAL)

4.1 *X-ray fluorescence analysis*

The X-ray fluorescence analysis is considered as a standard method of element analysis offering a good accuracy for major elements and detection limits in the region of 1ppm. (Table 1).

Approximately 5g of well pulverised soil material was weighed in a shuttle then about 0,5g of wax was added. Using a vibrator, the soil material together with wax was mixed. Around two spoons of Bor-acid were filled and roughly spread in the compressor, afterwards the soil sample was dispersed in the middle then compressed. After the expiration of the default compress-time, the prepared tablet was ejected.

The measurement was accomplished using an ARL-Spectrometer 9400 XRF. For the evaluation of the counting, the computer program UNIQUANT II was used.

4.2 *The non-metals carbon, nitrogen and sulphur*

The determination of the concentration of the three elements carbon, nitrogen and sulphur in soil samples took place at the elementary-analyser Vario EL fully automatic. The detection limit lie here with approximately 0,1% C, 0,01% N and 0,02 % S. The control of the measurement took place with duplicate determination with the standard-test BHA.

From each sample that is used in the x-ray mineral phase analysis, approximately 15 mg became weighed into a shuttle from tin-foil. 45 mg of Wolfram(VI)-oxide was added to it. The subsequent folding of the shuttle took place thereby expulsing the presence of any more air.

4.3 *Determination of the ignition loss*

The ignition loss of a soil sample V_{gl}, is defined as the mass-loss Δm_{gl} during the glowing process divided by the initial dried mass m_d.

$$V_{gl} = \Delta m_{gl} / m_d = (m_d - m_{gl}) / m_d \qquad (1)$$

where: m_{gl} is the mass of the soil sample before the glowing process.

The soil samples were dried at 105 °C of temperature in a cupboard until a constant mass was obtained, then they were grounded in agate mortar until fine grain sizes were achieved.

An empty porcelain crucible was glowed at 550° C of temperature for approximately 20 min in a grouch-oven. Then it was cooled in an exsiccator before its weight was taken.

Approximately 15g (m_d) of prepared soil material for each sample was filled in the porcelain crucible and glowed at 550 °C for about 2 hours then cooled in an exsiccator before its weight was fixated. This procedure was repeated until a constant mass was obtained (m_{gl}) and the ignition loss was determined using the formula mentioned above. The method was subjected to standards set out in DIN 18 128.

5 RESULTS AND INTERPRETATION

5.1 *Arsenic*

The bulk As mass varies from 140 to 950 ppm. Whereas its mean concentrations are found to be 384.8 and 580.5 ppm in the first (II) and the second (I) studied areas respectively. Both values exceed hundred times the average of the earth's crust As concentration (3 ppm) estimated by Leonard (1991) as well as the global contents (Vinogradov,1954) and that of the Ore mountains (Palchen et al., 1987). Moreover, they exceed As soil Clarke abundance ratios, but are between seven and seventeen times the average As concentration in the soil material covering the most important rock varieties from the Schlema- Alberoda area. On the other hand, the results from the SDAG WISMUT (1989) archive show even an extreme increase in its concentration in most samples.

Concerning its expected compound forms, on one hand the high concentration of arsenic is characteristic of nearly all sulphide deposits and probably is associated with the typical arsenic minerals as arsenopyrite, löllingite as well as the native arsenic and arsenides (skutterudite, chloanthite and safflorite). In addition, pyrite, goethite and hematite have been found in the earlier described uranium deposits as well. On the other hand, the absence of most of these minerals in the tailing environment may be in relation to their scarcity in the mother mineral deposits but it can be caused by the uranium leaching procedures since the new chemical conditions enhance the destruction (dissolution) of many minerals and probably provoke the appearance of others (secondary, tertiary as well as quaternary minerals).

With regards to the As and its relationship to other elements, it is noteworthy to point out the frequent association of Fe with both elements chromium and bismuth in many geochemical environments as well as with As, which causes the apparent high correlation of the last mentioned element with these elements (Cr in the first two boreholes and Bi in the other cores) in the areas studied although chemically they do not occur together in any association (Figure 3 and 4). The absence of any expected correlation with Cu may be an outcome of the chemistry of processing as these two elements react differently at any given condition.

In connection with the mineralogy separation, it was noticed that there is an important amount of As

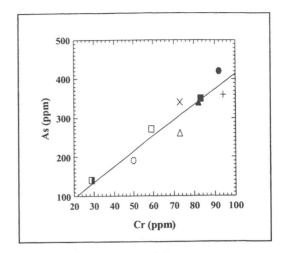

Figure 3. The relationship between Cr & As in the first core ($C_r = 0.95$).

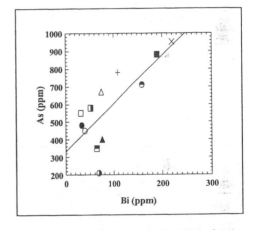

Figure 4. The relationship between As & Bi in the fourth core ($C_r = 0.77$).

bound to the heavy fraction and this strong enrichment of arsenic can be explained by the high adsorptive capacity of FeO(OH) accumulation dominating the same fraction of this element (Salomons & Forstner, 1984). On the other hand, the loss of its concentration after the mineral separation in most samples demonstrated in the end result balance confirms the prediction that on the one hand the medium mobility of As at all pHs as well as under oxidizing conditions on the other hand, and its low mobility under reducing ones (Reimann & Caritat, 1998; Herrmann & Rothemeyer, 1998) are maintained.

5.2 *Barium*

The total Ba content varies between 230 and 4800 ppm. Besides, its mean concentrations are found to be 1602 in the first site (II) and 1932 ppm in the second one. Both values exceed three times that of the Ba soil Clark abundance ratio for the entire Ore Mountains and are even higher for the Schlema Alberoda area where the Ba average concentration is around 81 ppm.

Since the concentration of the typical barium mineral "barite" is not important in most tailings, the most amount of Ba seems to be bound mainly to micas and feldspars (Roy, 1965; 167; Gay & Roy, 1968), even biotite (McBride, 1994) and perhaps clays (Salomons & Forstner, 1984) as well. In fact, the Ba^{2+} cation has a high tendency to change K^+ ion into the minerals mentioned due to the similarity in ionic radius between these ions.

With regard to the expected behaviour of this element in the tailings environment, on the one hand, the low mobility of this element in most weathering conditions is well known (Reimann & Caritat, 1998; McBride, 1994; Brookins, 1988). On the other hand, the end results balance shows a low loss of barium during the mineral separation despite the hard chemicals used demonstrates the low mobility of this element and confirm its predicted behaviour.

5.3 Chromium

The whole chromium concentration ranges from 29 to 160 ppm. Its average content is around 80 ppm in the first prospected area. Whereas in the second one it attains 105 ppm. Both mean values are not far from the average earth crust chromium concentration given by Gauglhofer & Bianchi (1991) as well as by Barnhart (1997), but are approximately twice lower than the Cr soil Clarke abundance ratio for the entire Ore Mountains estimated by Palchen et al., (1987) and even four times lower than the mean value of Cr in the test samples measured by SDAG WISMUT (1989).

With regard to its relation to minerals, as the typical chromium minerals "chromite" as well as "crocoite" were not identified neither in the initial uranium deposits nor in the tailings material, chromium is probably concentrated mostly in clays and also in hornblende because of it frequently replaces iron and mainly aluminium due to the resemblance between Cr^{3+}, Al^{3+} and Fe^{3+} in their chemical properties and ionic sizes (Wedepohl, 1974). Accordingly, the mineral separation is focussed on the equal distribution of this element in the two mineral fractions and the end results balance shows its negligible loss during the chemical processing. Besides, chromium is well known for its dominant stable "trivalent oxide" that occurs in nature. Moreover, it has low to very low mobility in all weathering conditions (Andrew-Jones, 1968; Reimann & Caritat, 1998). Therefore, chromium is not considered as a potential pollutant for the prospected sites.

5.4 Cobalt

The bulk Co content varies from 14 to 150 ppm. In addition, in the two tailings, its mean concentrations are 69 and 85 ppm respectively. These mean values are almost ten times higher than the Co soil Clark abundance ratio for the entire Ore Mountains estimated by Palchen et al., (1987) and four times the average cobalt content in the earth's bulk crust given by Schrauzer (1991). Moreover, according to Vinogradov (1954) and depending on the main rock varieties covering the Schlema Alberoda area, the average Co concentrations fluctuate between twice and four times lower than the values mentioned. In addition these mentioned results are considerably lower than these received from the SDAG WISMUT, (1989) archive.

Concerning the Co compound forms in the tailing sites is mostly bound to the heavy minerals since in the heavy fraction its concentrations are many times higher than those of the light one on one hand. On the other hand, this metal preferentially associates with Fe and Mn oxides because of chemisorption and co-precipitation. In addition, the typical Co minerals are absent in the tailings environment. Therefore, Co is probably bound to its host minerals as hornblende, chlorite as well as biotite.

With regard to the relationship of Co to Ni, their resemblance in many of their physical and chemical properties is well known. Besides, they are often regarded as a geochemical pair exhibiting near-ideal substitution for each other. Accordingly they occur together in a variety of minerals. Therefore, the good correlation of Co with Ni in all boreholes on the one hand confirms the well known chemical behaviour of these elements as well as the well known Ni-Co association characterising the uranium deposits of the Ore Mountains (Figure 5 and 6). On the other

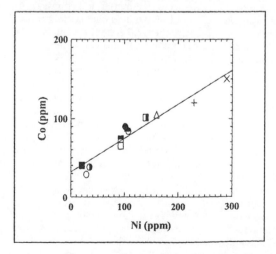

Figure 5. The relationship between Co & Ni in the fourth core ($C_r = 0.96$).

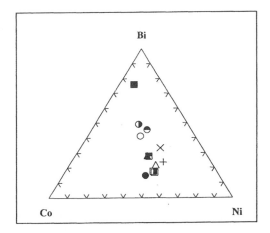

Figure 6. The relationship between Co, Bi & Ni in the fourth core.

hand, it shows that the new chemical conditions did not enhance their mobility but prevail their stability.

In connection with the expected mobility of Co, it is important to notify its easy solubility during weathering. Moreover, it does not form residual silicate minerals. According to Andrew- Jones (1968), at reducing as well as at neutral to alkaline conditions, a very low mobility is expected while at oxidising conditions a medium one, but with high mobility at acidic environment is more probable.

5.5 Copper

The gross Cu content ranges from 24 to 670 ppm. On the other hand, its mean concentrations are close to 339 ppm in the first tailings and 367 ppm in the other one. The two mean values mentioned are about fifteen times that of the Cu soil Clarke abundance ratio for the entire Ore Mountains evaluated by Palchen et al., (1987) as well as the average earth's bulk crust copper concentration after Vinogradov (1954) and roughly ten times that of the Cu soil average concentration covering the Schlema Alberoda area independently of its rock forming varieties. Compared with the SDAG WISMUT (1989) archive results, the Cu mean values found are considerably lower again.

Concerning the relation of copper to minerals and its probable compound forms in the present tailings environment, it is noteworthy to point the wide spreading of some typical copper minerals such as chalcopyrite, bornite, copperpyrite and native copper as well in the initial mineralogy of the uranium deposits to which the tailings materials belong. Besides, not any of the minerals mentioned above was identified. Only negligible amount of chalcopyrite was detected after the mineral separation by means of heavy liquids. Therefore, it will be possible to suppose that the new chemical conditions had provoked on the one hand the disappearance and the dissolution of these minerals, on the other hand the redistribution of copper according to the new environmental conditions. Moreover, most of the colloidal materials of soils (oxides of Mn, Al, and Fe, silicate clays, and humus) tend to absorb Cu^{2+} strongly, and increasingly so as the pH is raised (McBride, 1994). Furthermore, according to Wedepohl (1974), high concentrations of copper may be adsorbed on clay minerals, iron oxides and perhaps also organic matter. Consequently, a considerable amount of copper is expected to be bound to clays such as chlorite, hornblende, micas (biotite and muscovite) as well as organics, proportional to its affinity to these minerals. On the other hand, the end results balance of the mineral separation shows a high increase in copper concentrations after chemical processing accompanied by the mineral loss mainly from the light fraction. Therefore, its most amount is probably bound to iron complexes considered as the main compound form of the heavy mineral fractions.

According to Reimann & Caritat (1998), at neutral to alkaline pH as well as under reducing conditions, a very low mobility is expected whereas under oxidising conditions it is medium. Besides, it is of importance to point out that under high pH conditions the mobility of copper may be significant, e. g. in Cu^{2+}-organic complexes. Therefore in the expected chemical medium of high pH values and in the presence of organics, copper can be considered as an important pollutant for the tailings environment.

5.6 Iron

In its oxide form, the total Fe content varies from 2.5 to 12.6 %. Whereas its mean concentrations are 6 and 8.6 % (4.2 and 6 % of Fe) in the areas (II) and (I) respectively. According to Palchen et al., (1987), these two mean values surpass twice the iron soil Clarke abundance ratio for the entire Ore Mountains that is about 3.2 %.

From the geochemical point of view, iron is considered to be one of the most abundant elements in the earth's crust which exhibit variable valence states. Thus important redox reactions occur. During weathering, by reacting with atmospheric O_2, ferrous minerals transform into ferric oxides. In the presence of sufficient organic carbon in sediments, ferric oxides tend to reduce to ferrous compounds by bacterial activity. The presence of a sufficient amount of dissolved sulfide, results in the formation of iron sulfide , while at low amounts and with the presence of dissolved carbonate, siderite may occur. In the case when both dissolved sulfide and carbonate are low and sufficient silica is abundant, iron silicates are the dominant iron complex.

Concerning the relationship of Fe to the mineral contents discussed above, it is noteworthy to point out the presence of Fe as mineral association mainly in chlorite form since a high negative correlation between Fe_2O_3 and SiO_2 occurs in four boreholes. The good correlation between iron and magnesium in the third and the fourth boreholes may be in relation with the presence of chlorite, hornblende as well as cordierite . The presence of biotite may be demonstrated with the correlation of Fe_2O_3 to K_2O in the third borehole. A negligible amount of iron may be in connection with the mineral pyrite. Since the most amount of Fe is probably bound to silicates, its mobility depends on the dissolution rates of these parent minerals. On one hand, in silicates the solubility of iron is considerably high only at low pH values (pH<5). On the other hand, the dominant presence of silicates tends to increase and neutralise the pH of the tailings environment, so the mobility of iron in the studied area is probably low to very low.

5.7 *Lead*

The whole Pb content ranges from 34 to 370 ppm. Furthermore, its mean concentrations are 141 in the first site and 197 ppm in the other one. These two mean values are found to be more than five times that of the global Pb Clarke abundance ratio after Vinogradov (1962) and even exceed ten times the lead soil Clarke abundance ratio for the entire Ore Mountains (Palchen et al., 1987) as well as the average earth's bulk crust lead concentration estimated by Wedepohl (1995) and by Lide (1996). In addition, depending on the rock forming varieties covering the Schlema Alberoda area, the average Pb concentration varies between two and five times below the values mentioned above. Compared with the Pb mean value measured by the SDAG WISMUT (1987), they are notably lower.

Besides, a part of Pb was probably bound to its typical minerals "galena and cerussite" as they were identified in the whole mineralogy of the Niederschlema-Alberoda district. Furthermore, their absence in the present tailings material may be due to their dissolution under the chemical effect of uranium processing but also due to their low concentration in the initial uranium deposits. Moreover, since Pb tends to form complexes with organic matter and tends to be absorbed mainly on Fe-hydroxides and clays, it can precipitate with carbonates, hydroxides and phosphates especially at high pHs. Lead seems to be bound mostly to its host minerals as potassium – feldspars, kaolinite and even biotite, mainly after uranium processing depending on the affinity of Pb to the mentioned forms as well as the chemical conditions of the tailings sites. Therefore it is probably distributed throughout these forms.

Furthermore, at all weathering conditions, Pb has a low mobility chiefly under reducing ones where it becomes very low mobile to immobile (Reimann & Caritat,1998; Andrew-Jones, 1968). Thus independent of the weathering conditions of the tailings sites, Pb is expected to be low mobile.

5.8 *Magnesium*

In its oxide form, the total Mg content varies between 0.5 and 3.2 %. Moreover, its mean concentration is equal to 1.8 % in the first tailing. Whereas in the second one it is slightly more reaching 1.9 %. These indicate mean values which exceed twice the magnesium soil Clarke abundance ratio for the entire Ore Mountains (Palchen et al., 1987) but are close to the average earth's crust Mg concentration (Aikawa, 1991).

At the same time, MgO is found to be in a high correlation with CaO, Na_2O, K_2O, SiO_2 as well as Fe_2O_3 in most tailings (Figure 7) which is in agreement with the mineral determination indicating that the Mg is probably associated with chlorite, hornblende, dolomite and a negligible amount of biotite as well as cordierite.

Concerning its geochemical behaviour, Mg is known to be highly soluble under acidic to mildly alkaline conditions. According to Reimann & Caritat (1998) as well as Andrew-Jones (1968), Mg exhibits a high mobility in all weathering conditions, therefore independent of the chemical environment of the tailing sites, a high mobility of Mg is expected.

5.9 *Manganese*

The bulk MnO concentration ranges from 0.03 to 0.26 %. While, its mean contents are 0.13 and 0.17 % (0.1 and 0.13% Mn) in the tailings (II) and (I), respectively. These mean values mentioned lie in

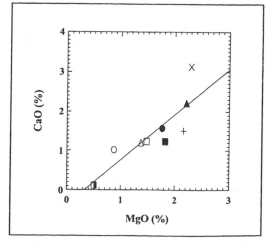

Figure 7. The relationship between Ca & Mg in the first core ($C_r = 0.85$).

236

the range of the Mn soil average concentrations for the most rock forming varieties covering the Schlema Alberoda area and do not extremely exceed the average earth's crust concentration (Schiele, 1991) as well as the Mn soil Clarke abundance ratio for the entire Ore Mountains (Palchen et al., 1987). On the other hand, the SDAG WISMUT (1987) archive demonstrates an apparent excess in its results.

Due to some similarities in crystal chemical properties of Mn^{2+} with Fe^{2+}, Mg^{2+} and Ca^{2+}, manganese substitutes ferrous iron, magnesium as well as calcium in its minerals. Furthermore, since the mineral determination prove the absence of typical manganese minerals, Mn is probably bound to hornblende, biotite (Leelanandam, 1970; Greenland et al., 1968; Mohsen & Brownlow, 1971), chlorite, cordierite, clays (Butler, 1954) and perhaps also dolomite (Yaalon & al., 1972). On the other hand, the high association of MnO with other ions such as Na_2O (Figure 8) in the borehole N°1, with SiO_2 in the second, with K_2O, CaO, MgO, SiO_2, Al_2O_3, Fe_2O_3 in the third and with SiO_2, MgO as well as Fe_2O_3 in the fourth is probable in relation to the presence of mineral varieties such as biotite and mainly hornblende.

Concerning the probable mobility of Mn in the prospected sites, a low mobility in acidic and neutral to alkaline pH range as well as under reducing chemical conditions is expected whereas under oxidizing conditions low to very low mobility is more probable.

5.10 Nickel

The total Ni content varies from 13 to 320 ppm. In addition its mean concentration is 103 ppm in the first prospected site, whereas in the second one it is more important reaching 125 ppm. These two mean values are not far from the soil average Ni concen-

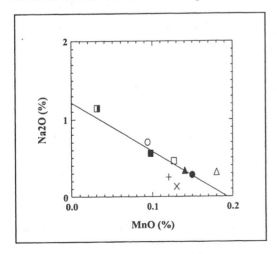

Figure 8. The relationship between Na & Mn in the first core ($C_r = -0.86$).

trations for almost all rock forming varieties covering the Schlema Alberoda area as well as the average earth's crust nickel concentration estimated by Taylor & McLennan (1995) but more than twice the global soil Clarke Ni abundance ratio (Vinogradov, 1954) as well as that of the entire Ore Mountains (Palc hen et al., 1987). Again, the value measured by SDAG WISMUT (1987) shows an extreme excess in the Ni concentrations.

Besides, nickel is probably bound to its host minerals as hornblende, chlorite, muscovite and possibly even pyrite, chalcopyrite and biotite. Furthermore, above pH ~ 6, nickel tends to be absorbed on oxides "Mn and Fe", noncrystalline aluminosilicates, and layer silicate clays, whereas at lower pHs, exchangeable and soluble Ni^{2+} are favourable. Therefore, acid rain has a pronounced tendency to mobilise nickel from soil and to increase nickel concentrations in groundwaters (Sunderman & Oskarsson, 1991). Moreover, Ni favours the formation of complexes with organic matter especially organic ligands containing nitrogen and sulphur. In addition, under reducing conditions as well as at neutral to alkaline pHs, a very low mobility of Ni is expected especially in the presence of sulphides that restrict its mobility. Furthermore, under oxidising conditions a medium mobility of this element is more probable, whereas at acidic mediums a high mobility may be certain.

5.11 Uranium

The whole uranium concentration varies from 25 to 820 ppm. Furthermore, its mean contents are 126 in the first site (II) and 300 ppm in the second one. These mean values are nearly the results of the SDAG WISMUT (1987) but surpass many times the average earth crust's uranium concentration estimated by Burkart (1991) and even more the soil average uranium concentration of the Schlema Alberoda area is close to 3 ppm.

In addition to uraninite, uranium forms various minerals that consist of carbonates, phosphates, vanadates, silicates, sulfates, etc., commonly in combination with uranium plus some other cations. The presence of such uranium minerals as uranophane, uranopilite, uranospinite and zippeite in the whole mineralogy of the "mother" uranium deposits is considered as examples of these associations. On the other hand, their absence in the whole mineralogy of the tailing sites may be in relation to the chemical processing causing the oxidation of uranium. Therefore its redistribution and its appearance on several other compounds results in the new chemical medium with undetectable very low concentrations. Furthermore, depending on the pH of the medium, uranium is able to form different ionic complexes and the presence of reduced iron as well as sulphur tend to limit its mobility by reducing U^{6+} to U^{4+}.

Moreover, except under reducing conditions where uranium's mobility is very low, under other conditions it demonstrates a high to very high mobility. Accordingly, the amount of carbonates in the tailings sites as well as sulphur and iron play a decisive role in the prediction of the mobility of uranium. On the other hand, uranium is in good correlation with many elements. Moreover, the good correlation with Co as well as with Ni (Figure 9 and 10) may be in relation to the initial uranium associations in the uranium gangue and perhaps it proves that the residual uranium is in its initial low soluble form of uranium oxide (uraninite).

5.12 Zinc

The bulk Zn amount lies between 54 and 950 ppm. In addition, its mean concentrations are around 425 ppm in the tailing (II) and 502 ppm in the second one. Notwithstanding these mean values exceed seven times the average earth crust's zinc concentration estimated by Ohnesorge & Wilhelm (1991) as well as the soil Clarke Zn abundance ratio of the entire Ore Mountains (Palchen et al., 1987). Nevertheless, they exceed the soil Zn average concentrations more than four times that of the main rock varieties covering the Schlema- Alberoda area. In contrast to this, they are remarkably lower than the results of the SDAG WISMUT (1989) archive.

Besides, Zn was probably bound to sphalerite since this mineral was identified in the whole mineralogy of Zobes, Schmirchau as well as Niederschlema-Alberoda uranium deposits and the absence of this mineral in the tailings material is perhaps due to its scarcity in the initial uranium accumulation but it can be due also to its dissolution under the chemical effect during the uranium leaching. Moreover, since Zn is able to replace other divalent cations and preferentially Fe^{2+} instead of Mg^{2+} in many minerals, Zn is probably in relation to its host minerals such as hornblende, chlorite and also biotite.

Furthermore, Zn is expected to be very mobile only under oxidising conditions and at acid mediums. However, under reducing conditions as well as at neutral to alkaline pHs its immobile character prevails. In addition, the presence of clays (chlorite and kaolinite) restrict the mobility of Zn and is considered as a geochemical barrier of this element in the tailings environment.

5.13 Other elements

Most amounts of Al are expected to be bound to silicates whereas Bi is probably in relation mainly with sulphide minerals. The current work show that Ce is probably attached to feldspars while Ga is associated with Al in the common minerals. Whereas Rb is probably essentially incorporated in muscovite and potash feldspars. Elements such as La, Nb, P, K and Y seem to be bound to their host minerals. The first

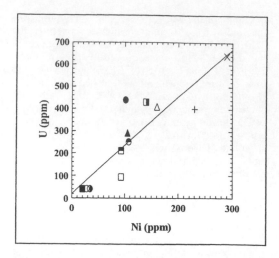

Figure 9. The relationship between U & Ni in the fourth core ($C_r = 0.89$).

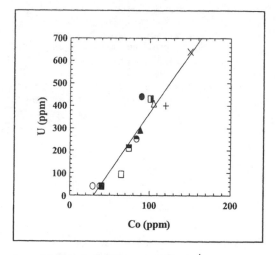

Figure 10. The relationship between U & Co in the fouth core ($C_r = 0.96$).

one is probably mostly incorporated with feldspars and even biotite. Whereas, the second one is in association mostly with hornblende, chlorite, rutile and perhaps even biotite. The third one is in connection with feldspars, muscovite, hornblende and apatite. Potassium seems to be mainly in relation with potash feldspar and even with muscovite whereas Yttrium is probably bound to feldspars and even biotite and apatite.

5.14 The non-metals carbon, nitrogen and sulphur

The concentration of nitrogen is below its detection limit for most tailings material except in some depth intervals where it is in the range of the bulk continental crust nitrogen concentration given by Rei-

238

mann & Caritat (1998). In addition, the nitrogen is probably bound to organic compounds since it correlates with carbon in most tailings (Figure 11).

Moreover, the coincidence of the maximum values of nitrogen with that of carbon are probably in connection with the Culmitsch sedimentary uranium deposit that it is rich in organic material. Besides, the high positive correlation between carbon and vanadium (Figure 12) in the second and the third borehole may also be in relation with the presence of an important amount of soil material from the location mentioned, since an association of vanadium to carbon in such deposits is already recognised (Meinecke, 1973). Furthermore, the concentration of carbon is low in most tailings environment and does not exceed the bulk continental crust carbon concentration

given by Reimann & Caritat (1998).

Concerning the presence of sulphur in the areas studied, it is important to point out that although the little excess of its concentration in some depth intervals with regard to the bulk continental crust sulphur concentration given by Reimann & Caritat (1998), it is considered very low with regard to its concentration in some sediments according to the same authors. On the other hand, some of its amount may be associated with organic compounds since it is found in a high correlation with carbon (figure 13) in the tailing (II).

5.15 *The ignition loss*

Since the tailings material samples were glowed at 550 °C, the main components are lost in form of

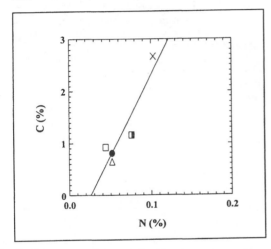

Figure 11. The relationship between N & C in the first core ($C_r = 0.92$).

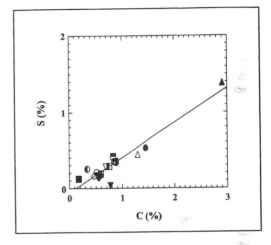

Figure 13. The relationship between S & C in the second core ($C_r = 0.95$).

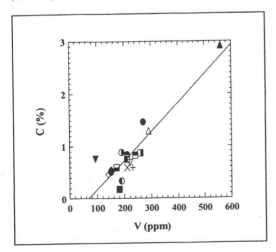

Figure 12. The relationship between C & V in the second core ($C_r = 0.90$).

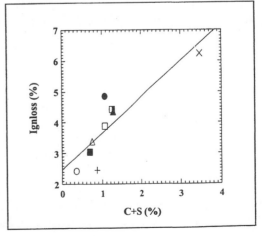

Figure 14. The relationship between C+S & Ignloss in the first core ($C_r = 0.86$).

H2O, CO2 and N2 compounds from organic substances and perhaps also sulphur in the form of S⁻ while the nitrogen in the form of NO3/NH4 can be omitted as the concentration of this element in the whole tailings is negligible.

The high correlation between the ignition loss and the sum of carbon and sulphur (figure 14) amount in the soil samples from each borehole confirms the high dependence of the ignition loss on the carbon and sulphur amounts.

6 CONCLUSIONS

It is of importance to point out that, most of the elements analysed above are expected to exhibit very low mobility in the tailings sites mainly under the anticipated weathering conditions of the area studied. Among these the following element can be mentioned; Al, Ba, Bi, Ce, Cr, Fe, Ga, La, Pb, Mn, Nb, P, K, Rb, Si, Sn, Ti, W, Y and Zr.

On the other hand, independently of the weathering conditions of the area studied Na, Ca, S, Sr as well as Mg will exhibit a high mobility. It is noteworthy to point out that their present concentrations are considerably lower than their initial ones since a high amount of them is probably lost during mineral processing.

Although, Co, Cu, Zn and Ni tend to be low mobile at the expected chemical environment of the tailings, the presence of an oxidising agent may considerably enhance their mobility. At the same condi-

tions it means that at expected high pH values as well as under oxidazing conditions, As is considered to be an important pollutant. On the other hand it is of importance to point out that, the presence of chromium side by side with arsenic may restrict the mobility of the element last mentioned mainly in the case where chromium is reduced to its trivalent form. Thus, this condition causes the fixation of arsenic in the tailings material with chromium. Therefore, the resistance of these two elements considerably increases and consequently they remain in place.

The presence of the Ni-Co association on the one hand and that of Bi-Co-Ni on the other one under the present conditions of the tailings sites as it was mentioned above may be considered as a useful tool to ensure the chemistry of processing. As the strong acid pHs increase the mobility of Co as well as of Ni, the association mentioned probably tends to be disturbed in this case. Therefore, the mostly used soda alkaline procedure is also confirmed.

As uranium exhibits a high mobility under most weathering conditions except the reducing one, the presence of oxidising agents plays a decisive role in the mobility of this element in the area studied. Besides, its association with Y as it is indicated by its high correlation with this element, may be due to its incorporation into silicates. Moreover, the mineral separation by means of heavy liquids confirms this deduction since it proves the presence of most amount of uranium in the heavy fraction.

The concentration change with the depth for all elements mentioned is probably in relation to the

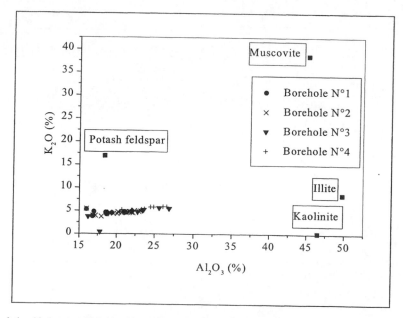

Figure 15. The relationship between K & Al oxides and its mineralogical significances.

uranium deposit type since on the one hand the expected mobility of most elements is not important as already mentioned, on the other hand the low permeability of most tailings material as well as the age of the tailings do not favour a significant displacement of contaminants.

Furthermore, it is of importance to analyse the relationship between the potassium and aluminium oxides since it can be considered as a useful tool for the control of the stability of any minerals as well as the dissociation of some elements bound to them under certain chemical conditions. In fact, on the one hand Figure 15 shows that all presented points represent one cluster located between potash feldspar and kaolinite which is closer to feldspar but is farther from muscovite and illite minerals. On the other hand, the mineralogy of the tailings sites shows an apparent excess in muscovite well known for its high resistance under all weathering conditions, i.e. chemical mediums and the absence of illite. Likewise the chemistry of processing preserves the concentrations of both potassium and aluminium well known for their very low mobility in all environments in the alkaline sector. Consequently, their relationship tends to be maintained. Therefore the graph presented proves on the one hand that all samples belong to the same region; in our case it is the Ore Mountains. On the other hand it proves the abrasion of an important amount of potash feldspar even more of illite. Thus the complete disappearance of illite and a decrease of the potash feldspar concentration result in an apparent increase of muscovite.

REFERENCES

Aikawa, J.K., (1991): Magnesium, in: MERIAN, E. (ed.): Metals & their compounds in the environment. VCH-Verlag.

Andrew-Jones, D.A, (1968): Mineral Ind. Bull. 11, 1.

Barnhart, J., (1997): Chromium chemistry and implications for environmental fate and toxicity, in: J. of soil contamination. Vol. 6, Nr. 6, Dragun, J., (ed.). Chromium in soil: perspectives in chemistry, health, and environmental regulation.

Brookins, D.G., (1988): Eh-pH diagrams for geochemistry. 175 pages. Springer-Verlag.

Burkart, W., (1991): Uranium, Thorium, and dacay products, in: Merian, E. (ed.): Metals & their compounds in the environment. VCH-Verlag.

Butler, J. R., (1954): The geochemistry and mineralogy of rock weathering. (2) The Nordmarka area, Oslo. Geochim. Cosmochim. Acta, 6, 268.

Gauglhofer, J. & Bianchi, V., (1991): Chromium, in: Merian, E. (ed.): Metals & their compounds in the environment. VCH-Verlag.

Gay, P. & Roy, N. N., (1968): The mineralogy of the potassium-barium feldspar series. III. Subsolidus relationships. Mineral. Mag. 36, 914.

Greenland, L. P. et al., (1968): Distribution of manganese between coexisting biotite and hornblende. Geochim. Cosmochim. Acta, 32, 1149.

Herrmann, A. G. & Röthemeyer, H., (1998): Langfristig sichere Deponien. Springer-Verlag, 466 pp.

Leelanandam, C., (1970): Chemical mineralogy of hornblendes and biotites from the charnockitic rocks of Kondapalli, India. J. Petrol. 11, 475.

Leonard, A., (1991): Arsenic, in: Merian, E. (ed.): Metals & their compounds in the environment. VCH-Verlag.

Lide, D. R., (editor-in-chief) (1996): CRC handbook of chemistry and physics. 77th edition, 1996-1997. CRC Press, Boca Raton, USA.

Mc Bride, M. B., (1994): Environmental chemistry of soils. 406 pages, Oxford University Press, New York, USA.

Meinecke, G., (1973): Zur geochemie des vanadiums. Clausthaler hefte zur lagerstättenkunde und geochemie der mineralischen rohstoffe. Gebrüder Borntraeger, Berlin, Stuttgart, 90 pages.

Mohsen, L. A. & Brownlow, A. H., (1971): Abundance and distribution of manganese in the western part of the Philipsburg batholith, Montana. Econ. Geol. 66, 611.

Ohnesorge, F. K. & Wilhelm, M., (1991): Zinc, in: Merian, E. (ed.): Metals & their compounds in the environment. VCH-Verlag.

Pälchen et al., (1987): Regionale Clarkewerte-Möglichkeiten und grenzen ihrer Anwendung am Beispiel des Erzgebirge. Chem. Erde 47. 1-2, p. 1-17.

Reimann, C. & Caritat, P. D., (1998): Chemical elements in the environment. 397 pages. Springer-Verlag.

Roy, N. N., (1965): The mineralogy of the potassium-barium feldspar series. I. The determination of the optical properties of natural members. Mineral. Mag. 35, 508.

Roy, N. N., (1967): The mineralogy of the potassium-barium feldspar series. II. Studies on hydrothermally synthesized members. Mineral. Mag. 36, 43.

Salomons, W. & Förstner, U., (1984): Metals in the hydrocycle. Springer-Verlag, 349 pp.

SDAG WISMUT (1989): Teilbericht: Strahlenexposition in den Aufbereitungsbetrieben und Beprobung szechen der SAG/ SDAG Wismut.

Schrauzer, G. N., (1991): Cobalt, in: Merian, E. (ed.): Metals & their compounds in the environment. VCH-Verlag.

Sunderman, F. W. & Oskarsson, A. and JR., (1991): Nickel, in: in: Merian, E. (ed.): Metals & their compounds in the environment. VCH-Verlag.

Taylor, S. R. & McLennan, S. M., (1995): The geochemical evolution of the continental crust. Reviews of geophysics, 33, pp. 241-265.

Vinogradov, A. P., (1954): Geochemie seltener und nur in Spuren vorhandener chemischer Elemente im Boden. Akademie-Verl. Berlin.

Vinogradov, A. P., (1962): Die Durchschnittsgehalte der chemischen Elemente in den Hauptarten der Eruptivgesteine (russ). Geochimija 7, p. 555-571.

Wedepohl, K. H., (executive editor), (1974): Handbook of geochemistry. Springer-Verlag, Berlin. 5 volumes.

Wedepohl, K. H., (1995): The composition of the continental crust. Geochimica & cosmochimica acta, 59, pp.12117-1232.

Yaalon, D. H. et al., (1972): Distribution and reorganisation of manganese in three catenas of Mediterranean soils. Geoderma, 7, 71.

Remediation and reclamation

Abandoned mine site waste repositories, site selection, design and costs

K.L. Ford
National Science and Technology Center, Bureau of Land Management, Denver, CO, USA

M. Walker
United States Forest Service, Lakewood, CO, USA

ABSTRACT: The Bureau of Land Management (BLM) is conducting priority cleanups of tailings and waste rock dumps at abandoned mine sites. Typically, these sites contain tailings piles, cyanide heaps and rock dumps that were historically constructed in or near drainages and are releasing pollutants into watersheds. BLM selects the most environmentally suitable site for removal of mining waste and placement into repositories and to comply with regulations. A repository is an engineered disposal cell similar to a landfill. Repositories have different design features based on site-specific conditions and use of water balance models. This paper also provides a case-study for screening and selection of a repository site using Geographical Information Systems (GIS) at the Ute-Ulay site in Colorado.

1 INTRODUCTION

The Bureau of Land Management (BLM) administers over 250 million acres of public lands in twelve western states of the USA. There are a large, but unknown number of abandoned hard rock mining sites on BLM lands. Over 9,000 such sites are in BLM's inventory. The National Science and Technology Center (NSTC) is BLM's technical support center for investigation and cleanup of abandoned mines and hazardous waste sites. Within the constraints of available funding, NSTC has investigated hundreds of AML sites and has assisted in the cleanup of 15-20 of the worst sites identified.

Typically, these sites involve old milling sites where ores were crushed and milled using mercury amalgamation, flotation, or cyanide leaching and thus there are often tailings or heap material remaining at the site. These wastes contain high concentrations of metals, cyanide and often acidity and were often placed in drainages with tailing dams. When the dams were no longer maintained, they often breached and serve as an ongoing source of contamination to surface waters. In addition to tailings, other mine features include waste rock, open shafts and adits, flowing adits with or without acid rock drainage and millsites. The mining wastes are typically handled as Superfund wastes under the National Contingency Plan 40 CFR 300.415. The primary focus of this paper is management of the historic tailings in mine waste repositories: site selection, design and cost.

2 SITE SELECTION

BLM is required by policy to locate repositories in the most environmentally suitable locations. These locations are evaluated using the following criteria:

- BLM ownership
- out of the 100 year floodplain and wetlands
- away from shallow groundwater
- in an area of generally flat topography
- within a reasonable haul distance
- away from cultural features and threatened and endangered species habitat
- away from other geologic hazards.

The NSTC used ArcView Geographical Information Systems (GIS) to evaluate potential repository locations in the Henson Creek, Colorado watershed. Data on each of the above criteria were obtained, weighted as to suitability and entered into ArcView as separate layers. The composite view is shown in Figure 1. The most suitable areas are shaded darkest. This area is a steep, high altitude valley and few areas were found to be suitable. This method was very helpful in identifying potential repository areas.

3 DESIGN

Designs depend on site conditions, especially the characteristics of the waste, the climate, hydrology, and topography of the site. These factors are evaluated during site characterization. Waste characteristics are investigated via total metal and leachable

GIS Analysis Results

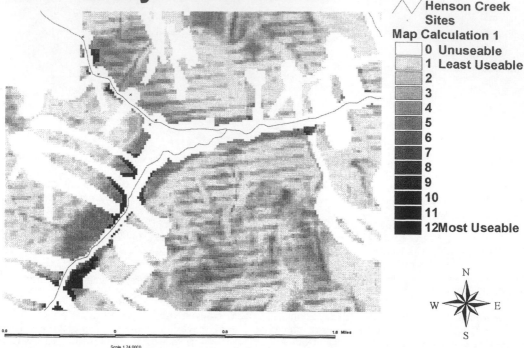

Henson Creek
Sites
Map Calculation 1
- 0 Unuseable
- 1 Least Useable
- 2
- 3
- 4
- 5
- 6
- 7
- 8
- 9
- 10
- 11
- 12 Most Useable

Scale 1 24,0000

Figure 1. Repository suitability analysis for Henson Creek Watershed.

metal concentrations, and acid-base accounting and permeability. Repository locations are investigated with drilling or test pits to determine depth to groundwater and to collect samples for grain size, Proctor, permeability, and agronomic analyses. Floodplain determinations are made from existing flood insurance maps or by watershed modeling.

BLM lands are mostly arid with less than 15 inches or precipitation annually and are subject to long, hot summers with high evapotranspiration. Some BLM lands are wetter and more mountainous with cooler temperatures. For convenience in discussing design, we have classified these climatic conditions into four categories:

1. <12 inches precipitation, no shallow groundwater
2. <12 inches precipitation, shallow groundwater
3. >12 inches precipitation, no shallow groundwater
4. >12 inches precipitation, shallow groundwater.

BLM uses the Hydrologic Evaluation of Landfill Performance (HELP) model version 3.07 or other models such as SoilCover to determine infiltration through the cap material and to investigate the utility of different cap configurations. Generally, for category 1 in our semi-arid environments, 24 inches of suitable soil cover with a hydraulic conductivity of 10^{-5} cm/sec or less will show no infiltration through the mine waste. In marginally wet conditions (12-15

inches/year) or if the tailings are very acid generating or leachable, a capillary barrier comprised of 6-12 inches of gravel in conjunction with the soil cover will preclude infiltration. For highly acidic tailings, deep amendment with lime has been used to stabilize the tailings. For wetter areas (categories 3 and 4), an impermeable liner is required to prevent infiltration. When shallow groundwater (less than 15' below the waste) is encountered, a bottom liner may be required in conjunction with a top liner.

Figure 2. Belle Eldridge Repository Deadwood, SD.

246

4 COST OF MINE WASTE REPOSITORIES

BLM is interested in evaluating the cost of repositories to ensure that government funding is reasonably spent and to help predict costs during the planning stages of cleanup. There are costs associated with site characterization, design, construction oversight, permits, operation and maintenance. This section is concerned only with the direct capital costs of constructing the repository, and does not include categories such as stream restoration or reclamation of removed areas.

Construction costs include mobilization, excavation, haul and placement of mine waste, cap placement, revegetation and runon controls. Haul distance is an important cost factor and the costs presented below are all for less than four miles round trip. Since 1998, BLM has designed or constructed 13 repositories in South Dakota, Montana, Idaho, Colorado, Utah, Nevada, Arizona, Oregon and Washington for which the data in this paper has been compiled. Costs shown below have been estimated using industry sources such as R.S. Means Environmental Cost Data or acquired from bidding and construction cost data. Based on costs, these have been subdivided into two categories:
1. repositories with simple soil covers or simple soil covers with capillary barriers,
2. lined repositories.

For unlined repositories, BLM's experience is that the total volume of wastes to be placed in the repository is the key factor that can predict cost (2001 dollars). Smaller sites require a higher cost per cubic yard disposed than larger sites due to economies of scale. Linear regression was used to investigate the relationship between volume and cost. Because the relationship is not linear, the natural logarithm of the volume was regressed with the cost per cubic yard and the following equation was determined:

Cost= -1.98 x ln(volume) + 31 (unlined sites)

This relationship has a R^2 of 0.6, indicating that more data points are needed to satisfy normal statistical confidence. However, this information is helpful in preliminary project planning and cost estimating.

This relationship seems to hold for volumes up to approximately 500,000 cubic yards. While data from additional larger sites are needed to better predict the cost, costs are expected to level out around $5-6 dollars/cubic yard.

Table 1. Provides the mean unit costs/cubic yard disposed by repository category.

	Unlined Repositories (n=8)	Lined Repositories (n=5)
Mobilization	$0.42	$1.83
Prepare Repository	$0.67	$3.09
Excavate, haul, place, compact	$4.10	$10.69
Construct soil cap	$1.59	$6.81
Revegetation	$0.33	$1.41
Runon controls and fencing	$0.13	$2.35
Sum	$7.23	$26.18

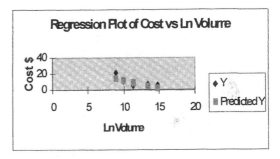

Figure 3. Regression plot: unlined repositories.

To date, too few sites have been completed with liners to estimate the costs using regression. The 2001 mean cost per cubic yard is $26.18 for lined sites.

5 CONCLUSIONS

A primary strategy to help remediate abandoned mine tailings sites in stream channels, BLM is constructing mine waste repositories. The repositories are selected using environmental criteria to avoid flooding and groundwater conditions, cultural features, endangered species habitat, and such practical criteria as haul distance, availability of soil borrow and site slope. Various designs are used ranging from a simple soil cover in arid environments to an engineered cap with capillary barrier or geosynthetic liners, depending on precipitation and characteristics of the waste.

In situ bioremediation of uranium and other metals in groundwater

W. Lutze, W. Gong & H.E. Nuttall

The University of New Mexico, Albuquerque, NM 87131, USA

ABSTRACT: Aquifers near uranium mines may be contaminated with uranium and other metals. Under certain conditions, indigenous microbes can be used to clean up the groundwater. In this case, the microbes control chemical and bio-geochemical reactions such as adsorption, precipitation, and immobilization of radioactive and hazardous metals. In the present study we have selected groundwater from the site of the former uranium mine at Königstein, Saxony, Germany. The mine will be flooded in the near future. The flooding water (termed G4) will be contaminated with uranium and heavy metals. The pH will be <3 because uranium was mined by acid leaching of the sandstone host rock. Fractions of the contaminated water may leak into an adjacent freshwater aquifer (termed G3). In this event, the G3 aquifer must be decontaminated. The depth of about 200 m below the surface prevents construction of a permeable barrier. Conventional pump-and-treat is traditionally the most expensive and least reliable process. Therefore, in situ bioremediation has been considered.

Lutze et al. (Lutze, W., H.E. Nuttall, and G. Kieszig, Microbially Mediated Reduction and Immobilization of Uranium in Groundwater at Konigstein, Session E12, Bioremediation of Metals and Radionuclides, In Situ and On-Site Bioremediation, The Sixth International Symposium, June 4-7, 2001, San Diego, California) have conducted experiments in the laboratory with groundwater and sandstone from the site to establish a scientific database for the potential application of bioremediation. These authors have shown that the dissolved U(VI) species were reduced to U(IV) with the help of indigenous sulfate reducing microbes. U(IV) is extremely insoluble in water and precipitated as uraninite (UO_2). In this way uranium was immobilized within the sandstone. In these experiments, mixtures of sandstone and contaminated G3 water were amended with sodium lactate and sodium trimetaphosphate in a closed system (no access of oxygen). Within three weeks enough sulfate reducing microbes had grown to reduce the most abundant species in the water (sulfate and suspended ferrioxihydroxide). The reaction products, Fe(II) and HS^-, formed FeS_{1-x}. U(VI) was reduced simultaneously. Independent of the particular conditions at Koenigstein, one can assume that sulfate and iron are much more abundant in a groundwater/rock system, than the contaminant uranium. Therefore, FeS_{1-x} will be present in large excess over UO_2. Thermodynamically FeS_{1-x} is less stable against oxygen than UO_2. These two facts make FeS_{1-x} an efficient redox buffer preventing UO_2 from oxidative dissolution, a process likely to

occur when bioremediation is discontinued and dissolved oxygen has access to immobilized uranium. The effectiveness of FeS_{1-x} as a redox buffer was demonstrated experimentally by leaching sandstone columns in which UO_2 was precipitated. The leachant was uncontaminated oxygen-saturated groundwater (G3). Insignificant concentrations of U(VI) were detected in the eluate during passage of 15 pore volumes of water through the column. The dissolved oxygen reacted with FeS_{1-x} to form an equivalent amount of sulfate.

FeS_{1-x} occurs naturally and is known as the mineral mackinawite. Mackinawite forms in nature under reducing conditions (Eh<400 mV) in a microbially mediated reaction. Mackinawite can be amorphous or crystalline, depending on environmental conditions. The mineral is known to occur as solid solution with other metals, particularly nickel.

In this paper, we report on the fate of heavy metal contaminants upon precipitation of mackinawite. Several solutions containing metal contaminants (Mo, Pb, Ni, Cd, Zn, Co, Mn, As) were amended in serum bottles and sealed to prevent access of oxygen. Reaction progress was monitored observing the change of the color of the sandstone. In the course of three months the sandstone had turned black. Previous experiments with uranium had shown that the reaction was complete after this time, because the sulfate had been quantitatively reduced to sulfide. Samples of the solution were taken and analyzed. Generally, metals such as Mo, Pb, Ni, Cd, Zn, Co, Mn, and As, were found in solution at concentrations of 1 μg/L after mackinawite formed. Initial concentrations were usually up to three orders of magnitude higher. Only Mn had increased in comparison with its initial concentration. This is commonly observed under these conditions and is most likely due to reduction of Mn(IV) in MnO_2 to soluble Mn(II). MnO_2 is a minor constituent in sandstone.

Samples of mackinawite were taken from the serum bottles under reducing conditions and analyzed by transmission electron microscopy, energy-**dispersive** X-ray spectroscopy (EDS) and selected area electron diffraction. Results of this study are shown in figure 1. The micrograph shows that mackinawite is present in the form of very small particles (a few nm). The second phase (squares) is lead sulfide (PbS). PbS is perfectly crystalline whereas mackinawite is still in its early state of crystallization as determined by electron diffraction. PbS is the only other phase that formed besides mackinawite. EDS analysis of mackinawite showed that it contained several of the other metals such as Ni, Cd, Zn, Co, and As. These elements may be in solid solution with mackinawite though this has not been shown directly. Mo was not found. It is interesting to note that As was removed from solution and found in mackinawite. The details of this process still need to be worked out. Column experiments analogous to those described above for uranium were conducted and revealed that metal concentrations were still at 1 μg/L after elution with 15 pore volumes of uncontaminated, oxygen-saturated water. Translating the conditions used to elute columns, i.e. using the groundwater flow rate at Koenigstein, yields a calculated period of at least 450 years for which no release is expected from the immobilized material. Hence, bioremediation may be an effective method to immobilize and stabilize uranium and other toxic metals for long periods of time.

100.00 nm

Figure 1. TEM micrograph showing tiny particles of mackinawite (FeS_{1-x}) and lead sulfide (PbS).

Tailings and Mine Waste '02, © 2002 Swets & Zeitlinger, ISBN 90 5809 353 0

In situ treatment of metals in mine workings and materials

J.M. Harrington
Shepherd Miller, Inc., Fort Collins, Colorado, USA

ABSTRACT: Contact of oxygen contained in air and water with mining materials can increase the solubility of metals. In heaps leached by cyanide, metals can also be made soluble through complexation with cyanide. During closure, water in heaps, and water collected in mine workings and pit lakes may require treatment to remove these metals. In situ microbiological treatment to create reductive conditions and to precipitate metals as sulfides or elemental metal has been applied at several sites with good success. Treatment by adding organic carbon to stimulate in situ microbial reduction has been successful in removing arsenic, cadmium, chromium, copper, iron, lead, manganese, mercury, nickel, selenium, silver, tin, uranium, and zinc to a solid phase. Closure practices can affect the success of in situ treatment at mining sites, and affect the stability of treated materials. This paper defines factors that determine the cost and permanence of in situ treatment.

1 INTRODUCTION

1.1 *Problem description*

Most mining operations remove materials from the ground that were, prior to being mined, in a chemically reduced state. The processing operation in many cases directly oxidizes the materials (such as in gold and copper leaching); in other cases, the materials are contained in waste piles in a manner that allow oxidation reactions to occur over time. In an oxidized state, many metals are soluble in water that travels through these materials, and these metals will as a consequence be present in seeps from the materials. The oxidation of elements in mined materials and the movement of water through these materials is the cause of acid rock drainage (ARD) and can also be the cause of the release of such elements as arsenic, chromium, selenium, and uranium in neutral pH drainage. The closure of mining operations may require the chemical stabilization of these mined materials to prevent these materials from being a source of environmental contamination. Returning the mined materials to the reduced state that they were in prior to mining makes sense as the method that will leave these materials in a form most chemically similar to the form they were prior to being mined.

1.2 *Green World Science® process*

The Green World Science® (GWS) process is defined by the addition of a reductant into metal-containing materials to prevent metals and other constituents from becoming mobile, and to remove from solution constituents that are already soluble.

The reductant may chemically react with the mine materials to remove oxygen and reduce oxidized materials, or the reduced compounds may be added that are microbially oxidized, and the mine materials reduced, by microbial enzymes. (GWS currently owns several patents on this process including US 6,196,765, US 5,710,361, and US 5,362,715 and others allowed and pending.) Organic carbon is the typical reductant utilized in this process; however, the addition of reduced gases into vadose materials is also proposed.

This paper presents data from six applications of the GWS process. The applications include two gold heap leach facilities (Yankee Mine and Coeur Rochester Mine), mine workings in an adit (Mike Horse Mine), a backfilled pit (Beal Mountain Mine), a pit lake (Sweetwater Uranium Mine), and groundwater beneath a test gypsum impoundment (J.R. Simplot phosphate plant). Except for the tailings impoundment, all data presented are for full-scale in situ treatment. There have been more than 20 other applications of the GWS process.

2 IN-PAD HEAP LEACH TREATMENT

2.1 *Yankee mine*

The Yankee mine is a gold heap leach operation. The heap contains approximately 7 million tons of leached ore on one pad. At the close of leaching, GWS added over 185 tons of organic carbon and other nutrients into the barren pond during solution recycle and evaporation. The primary constituents that required treatment were WAD cyanide, nitrate, aluminum, copper, mercury, silver, selenium, and zinc. Figure 1 shows results from a meteoric water mobility procedure (MWMP) test performed on leached materials prior to treatment, and compares these with results from MWMP on samples removed after treatment. (The MWMP procedure mixes an equal weight of synthetic rainwater with the mine materials for 24 hours; afterwards the water is extracted and the constituent concentrations in the water are reported.) The results of the MWMP prior to treatment are an average of five samples, and post-treatment results are an average from six samples.

The concentrations of constituents of concern in MWMP extracted solution all decreased in the treated samples. The concentration of WAD cyanide and cyanide-complexed metals Cu, Hg, Ni, and Zn all decreased by several orders of magnitude. Section 2.3 discusses the mechanisms that account for these changes.

2.2 Coeur Rochester mine

Coeur Rochester mine is located in Nevada and has several heaps on separate pads each with contained solutions ponds. The heap leach operations primarily recover silver and some gold. GWS added 2,100 tons of organic carbon and other nutrients through trenches located on the Stage 1 pad; no prior rinsing of this pad had occurred. The constituents being treated included pH, nitrate, sulfate, and WAD cyanide, arsenic, copper, mercury, nickel, and zinc. Figure 2 compares the MWMP results from untreated and treated areas of the pad. The treated areas were tested after 1 year of treatment; however, the results shown are partial treatment results, as the process will require close to two years to complete.

2.3 In situ treatment processes in gold and silver heap leach facilities

The processes that account for the decrease of soluble constituents in heap leach pads are described below:
1) Off-gassing of dissolved constituents: In the treatment process, the added organic carbon is converted to carbon dioxide, which partitions between the gas phase and the interstitial water depending on the interstitial solution pH. Nitrate is converted to nitrogen gas; an organic carbon-dependent process termed denitrification (Narkis et al, 1979).
2) Degradation of weak acid dissociable (WAD) cyanide complexes: Two primary mechanisms account for the microbial removal of cyanide. Detoxification—Cyanide is an inhibitor of metabolic processes, and to prevent this inhibition, cells have specific enzymes capable of converting cyanide to less toxic forms, including cyanate and ammonia. Example enzymes are cyanidase and cyanide hydratase (Raybuck, 1992). Nitrogen Assimilation—Cyanide is also a good nitrogen source for cellular growth, and for this reason is incorporated into biomass during the growth phase of microorganisms (Raef et al, 1977). The removal of free cyanide from solution and the degradation of cyanide complexes with relatively low dissociation constants (which includes most of the WAD cyanide forms) occur because of these two cellular processes.

The removal of copper, mercury, nickel, silver, and zinc from solution is caused by the degradation of cyanide, and the consequent removal of this complexing agent that had maintained these metals in solution. In many instances, the degradation of cyanide is not itself sufficient to meet water quality standards for heap leach or tailings drainage, to allow discharge off containment, and further metal removal

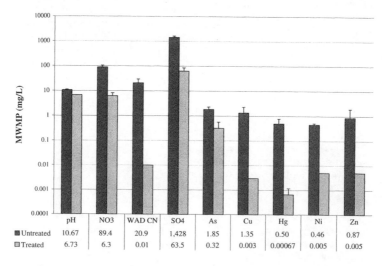

Figure 1. Yankee heap leach prior to and after treatment MWMP comparison.

252

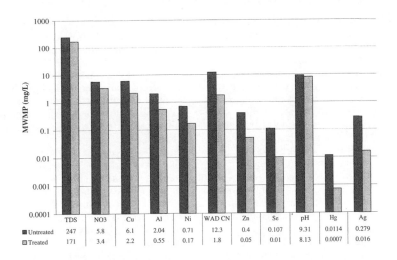

	TDS	NO3	Cu	Al	Ni	WAD CN	Zn	Se	pH	Hg	Ag
■ Untreated	247	5.8	6.1	2.04	0.71	12.3	0.4	0.107	9.31	0.0114	0.279
▫ Treated	171	3.4	2.2	0.55	0.17	1.8	0.05	0.01	8.13	0.0007	0.016

Figure 2. Coeur Rochester Pad 1 treated and untreated MWMP comparison.

by other processes may be required. This can be performed by in situ microbial treatment such as sulfate and metal reduction (see items 3 and 4 below). In other instances, the high pH of the interstitial solutions, created by lime addition prior to leaching, will be sufficient to remove the metals by precipitation in the form of metal hydroxides.

3) Metal Reduction: Several metals are less soluble in their reduced forms. Examples of this are selenium, which is quite insoluble in its elemental form Se^0 and soluble in its oxidized selenate (Se VI) form; chromium, which is quite insoluble as $Cr(OH)_3$ (Cr III) and soluble in the oxidized Cr (VI) form; and uranium, which is quite insoluble in its UO_2 (U (IV)) form and soluble in its oxidized U (VI) form. Cellular metabolic processes oxidize the added organic carbon and enzymatically reduce these metals (Oremland et al, 1999). These metals may also be indirectly reduced by contact with other reduced ions that were created by microbial reduction such as ferrous iron and sulfide (Wiclinga et al, 2001).

4) Sulfate Reduction: The oxidation of organic carbon and the reduction of sulfate is a coupled metabolic process that is performed by microorganisms. Sulfide is a strong reductant, and reacts with both soluble metal and metals in minerals. In some instances, this reaction leads to the formation of insoluble metal sulfide precipitates (including sulfides of As, Cd, Cu, Fe, Hg, Ni, and Zn). In other instances, the presence of sulfide can lead to a temporary increase in metal concentrations (such as the release of iron and manganese into solution from the reductive dissolution of oxide minerals). The amount of metal release can be controlled by managing such parameters as the application rate of organic carbon, providing sufficient time for reactions to come to completion, and by controlling the pH of the treat-

ment solutions. Sulfide is also an efficient reductant of oxidized chromium, selenium, and uranium; the reaction of sulfide with these elements can lead to their immobilization in a sulfide phase.

5) Adjustment of the pH: The microbial oxidation of organic carbon leads to changes in the solution pH by creating carbon dioxide and bicarbonate. In low pH solutions, the pH increases as a result of acidity consumption during iron and sulfate reduction and the formation of bicarbonate (see reactions listed in Table 1). In high-pH solutions, the creation of carbon dioxide leads to the consumption of hydroxide, which results in lower interstitial solution pH. The change of solution pH affects several metals and metalloids whose solubility is pH sensitive. For example, in oxidized conditions, iron, when present, is principally responsible for removing arsenic from solution, and iron-arsenic compounds are least soluble around pH 7.5. In reduced conditions, sulfide can form arsenic sulfides, and arsenic sulfides are least soluble below pH 7. In high pH solutions, soluble arsenic sulfides (AsS⁻) can form and may not precipitate; however, if iron is present in high concentrations arsenopyrite may form even in solutions with pH >7. Thus, if arsenic removal is desired in low pH solutions, the formation of arsenic sulfides should be attempted, and if removal is desired in high pH conditions, formation of iron arsenate compounds should be attempted.

2.4 Factors that affect the application of in situ treatment in heap closure

Regulatory Framework: Regulations that control heap closure call for geochemical stabilization of heap materials so that a closed heap does not cause degradation of waters of the State (*e.g.* Nevada Administrative Code 445A). Some US states classify mine wastes based on extraction tests of the heap materials, allowing far less stringent cover to be placed or unregulated solution discharge for heaps that pass extraction tests. In situ treatment should be considered in cases where heap water quality contains constituents of concern that are amenable to in-heap treatment. In cases where the salt concentration is high, discharge of even treated heap solutions may not be feasible because of TDS restrictions. However, removal of certain constituents within the heap may allow for alternative discharge approaches, such as non-discharging wetlands or infiltration systems, where untreated heap solutions may cause ecological toxicity in the wetland or certain constituents may not otherwise be attenuated.

Timing of Treatment: Microbial treatment of a heap leach facility may begin as soon as cyanide addition ceases. It may be advantageous to utilize the rinse period to enhance treatment coverage, as well as to use this period to achieve water quality goals that will reduce the time required to achieve closure after metal recovery has ceased. While in some situations silver recovery may decrease because silver-cyanide complexes can be degraded by microbial enzymes, gold recovery should not decrease because microorganisms cannot degrade gold-cyanide complexes. Some in situ treatment situations have shown an increase of gold content in the doré because other metal-cyanide complexes have been removed from solution, and hence do not compete with gold sorption to carbon.

The evaporative concentration of salts during leaching is a function of the climate, the solution application method, and the length of time that solutions are recycled. Closure of many heaps now involves the reduction of solution inventories by recycling barren solution over the pad until water concentrations are decreased or the cyanide concentrations decrease by natural oxidation or by solar reactions. The decision to evaporatively concentrate heap solutions in many instances dictates the closure plan, and should not be done without considering the implications to the revegetation of the heaps and the long-term cost of heap water disposal. In situ treatment during the final stage of leaching, *i.e.*, the rinsing residual gold-cyanide complexes, may remove the cyanide and metals so that discharge is possible, or at least there will be a wider range of heap drainage disposal options. As an alternative to evaporation that creates a salty discharge, in situ treatment will leave a relatively fresh water solution suitable for watering vegetation or infiltration.

Factors affecting cost and specificity of treatment: There is a predictable order in which microorganisms will transform constituents present in a heap after organic carbon is added. Table 1 summarizes many of the relevant reactions that may occur in a heap leach pad. These microbial reactions occur in natural systems in a predictable order based on the energy (ΔG) of the reaction (Zehnder and Stumm, 1988). Briefly, the ΔG of these redox reactions is a function of the relative strength of the oxidant (*e.g.* oxygen is a stronger oxidant than is sulfate), and the relative concentration of the oxidants and reductants (Pauling, 1988). For instance, a solution with 1 mg/L nitrate and 100 mg/L selenate may favor selenium reducing organisms over nitrate-reducing organisms even though the ΔG of nitrate reduction is greater than the ΔG of selenium reduction.

These reactions show why, in heap leach closure, much of the oxygen that is present in the heap interstices must be consumed prior to significant amounts of denitrification or sulfate reduction. Thus in a heap in which selenium is the target constituent, the cost of treatment will be a factor of the amount of organic carbon required to 1) reduce all of the oxygen initially present in the heap, 2) reduce the oxygen that will diffuse into the heap during treatment, 3) denitrify the nitrate, and 4) reduce the selenium. The cost of the organic carbon to reduce the selenium alone may be less than 1% of the total organic carbon requirement. Table 1 shows the constituents that must be evaluated to determine the total amount of reductant that must be added to treat water in situ.

For in situ sulfate reductive treatment, the amount of sulfate reduction required to remove a particular metal is a function of the reactivity of sulfide with the metals. For example, in a treatment performed in an adit, nearly complete removal of arsenic (50 mg/L to less than 1 mg/L) and zinc (20 mg/L to less than 1 mg/L) was achieved without affecting the iron in solution (600 mg/L). This occurred because zinc and arsenic sulfides, in the particular pH environment of that adit, were less soluble than iron sulfides. In our experience a typical order of metal sulfide removal has been Pb> Zn> Cu> As> Cd> Fe> Mn. This order may change based on factors such as solution pH or the presence of other complexing agents such as some organics (humics + fulvics) that may also affect the precipitation of metals (Lehman and Mills, 1994).

Table 1. Redox reactions catalyzed by microorganisms.

Oxygen Consumption:
$$O_2 + (CH_2O) \rightarrow CO_2 + H_2O \quad (1)$$
Denitrification:
$$4NO_3^- + 5(CH_2O) + 4 H^+ \rightarrow 2 N_2 + 5 CO_2 + 7 H_2O \quad (2)$$
Iron Reduction:
$$4 FeOOH + (CH_2O) + 7H^+ \rightarrow 4 Fe^{2+} + HCO_3^- + 6 H_2O \quad (3)$$
Selenium Reduction:
$$2 HSeO_3^- + 2(CH_2O) + 2H^+ \rightarrow 2 Se^0 + 2 CO_2 + 4 H_2O \quad (4)$$
Uranium Reduction:
$$2(CH_2O) + 2UO_2(CO_3)_2^{2-} + 2H_2O \rightarrow 2UO_2 + 5HCO_3^- + H^+ \quad (5)$$
Sulfate Reduction:
$$SO_4^{2-} + 2 (CH_2O) + H^+ \rightarrow HS^- + 2 CO_2 + H_2O \quad (6)$$
Note: (CH_2O) is the empirical formula for carbohydrates, e.g. glucose is $C_6H_{12}O_6$.

254

3 BACKFILLED PIT IN SITU TREATMENT

3.1 *Beal Mountain mine*

Beal Mountain mine is a closed gold mine located in Montana. As part of the closure, the mine pit was backfilled with "clean" rock, and seepage that was expected to fill the pit was routed through the pit and into German Gulch located downgradient of the pit. As the pit was filled with rock, seepage from the pit wall and spring precipitation created a pond in the middle of the pit. The pit pond contained nitrate at concentrations (8.1 mg/L) higher than would be allowed to discharge into the gulch (1.0 mg/L). GWS added approximately 20,000 lbs. of nutrients into the pit pond that was located at the center of the pit. As the water elevation rose in the pit, the nitrate concentration in the pond was monitored until the pit was completely filled. Figure 3 shows nitrate concentrations until the pit pond began to drain out of the pit.

After the pit backfilling was completed, two wells installed in the pit backfill were monitored for a suite of parameters. Nitrate concentrations in the pit wells initially were approximately 2 mg/L and have continued to decrease to less than 1 mg/L, and in one well, to concentrations below detection. In addition to the reduction of nitrate, selenium reduction occurred. The two wells initially contained 42 and 47 ppb selenium. As shown in Figure 4, after two years the concentration of selenium has decreased to between 2.0 and 3.0 ppb. In the same time frame, no decrease has been observed for sulfate and other TDS constituents, making an explanation of washout or dilution improbable.

3.2 *In-pit treatment processes*

The treatment of the Beal Mountain backfilled pit illustrates several factors that affect the success of in situ treatment in mine waste. Wells installed on either side of where the pond was located showed the effect from the biological treatment, even though carbon was added only to the pond at the center of the pit. The organic carbon was effectively distributed into the backfill because the carbon was added while the pit was still filling, allowing the carbon to be distributed with the rising water. In general, addition of the reductant to mine wastes as they are placed enhances the distribution and coverage of the treatment process.

Organic carbon was added at a rate sufficient to reduce only the nitrate calculated to be in all of the water in the pit. This rate was calculated to include dissolved oxygen that, it was assumed, was saturated in all of the pit waters. Selenium was reduced because this assumption caused a slight over-treatment of the pit. No measurable sulfate reduction occurred in the pit because the amount of the over-treatment was sufficient only to treat the selenium, but oxidants below selenium, such as sulfate, were not reduced because the carbon had been used up.

At the Beal Mountain backfilled pit, the addition of organic carbon alone was sufficient to remove all of the oxygen present in the pit waters, to denitrify all of the nitrate present in the pit waters, and then to additionally react with selenium. If additional organic carbon had been added, sulfate reduction would have also occurred. In another backfilled pit, organic carbon has been injected at concentrations sufficient to remove more than 99 percent of the sul-

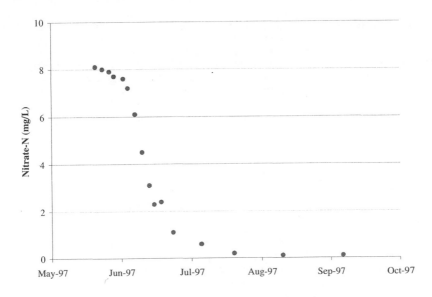

Figure 3. Beal Mountain pit pond nitrate in situ treatment.

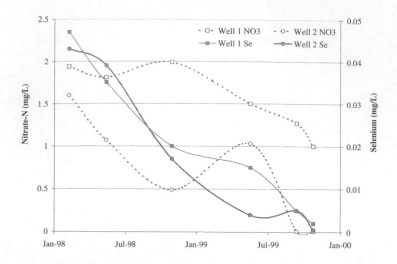

Figure 4. Beal Mountain nitrate and selenium in backfill wells after treatment.

fate, from 1,500 mg/L to less than 1.0 mg/L. In none of these cases was it necessary to inject anything other than organic carbon; rather, injection of organic carbon alone stimulated the organisms naturally present to perform the reaction that was most energetically favorable in that environment. Section 7.2 discusses the abundance and distribution of microbial species that participate in these reactions.

4 PIT LAKE TREATMENT

4.1 Sweetwater Uranium mine

Sweetwater Uranium mine is located in Wyoming. Uranium ore and overburden rock had been removed from pits that were dewatered during mining. Approximately 5 million tons of overburden rock was placed back in the pit on one side of the pit. After mining ceased in 1983, the pits were allowed to backfill, and the resultant lake was 125 feet at is deepest, and averaged 65 feet in depth over 60 acres (total volume in the lake was approximately 1.25 billion gallons). After the water level had reached static conditions (with evaporation being the only outflow from the pit) uranium (8.4 mg/L) and selenium (0.45 mg/L) were the only constituents that were present above the site closure standards (5 mg/L and 0.05 mg/L, respectively). GWS added 1.1 million pounds of organic carbon and other nutrients to the pit beginning on October 19, 1999. Figure 5 shows the results of the treatment. In six weeks, the selenium had been precipitated such that the dissolved selenium concentration was less than 0.05 mg/L. In 14 weeks, the uranium had been precipitated such that the dissolved uranium concentration was less than 5.0 mg/L. The lake has now remained below standards for both of these elements for two years.

4.2 Pit lake treatment processes and issues

The reactions stimulated in the pit are the same as those stimulated in the heap leach pads and in the backfilled pit. Selenium reduction occurred prior to uranium, consistent with the thermodynamically predicted order described in Section 2.3. There was 1.1 mg/L nitrate in the pit that was removed in the first two weeks of the treatment, also consistent with the energetics of denitrification as compared with selenium reduction.

As with all in situ treatment processes, managing oxygen in the treatment was the highest cost item. More than 80% of the added carbon was consumed in the reduction of oxygen that was present in the pit at the beginning of the treatment, and the oxygen that continued to diffuse into the pit during treatment. For this reason, we chose to perform the treatment of the lake over the wintertime, to take advantage of an ice layer would form over the pit and minimize the mixing and diffusion of oxygen into the lake. While timing the treatment to occur in the winter may have slightly slowed the microbial reaction rate, the organic carbon demand was significantly reduced and strongly reducing conditions were more easily achieved under the ice layer.

A common fallacy believed about microbial treatment is that microorganisms cannot function in cold water at high rates. The water temperature in the Sweetwater pit at the beginning of the treatment was 9°C, and all of the uranium precipitation occurred when the bulk of the pit was approximately 3°C. We have treated other sites when the water temperature was 0.5°C. Microorganisms that are specifically adapted to function in cold temperatures (psychrophilic or psychrotrophic organisms) can function at metabolic rates at near freezing temperatures at least half as fast as as other microbes

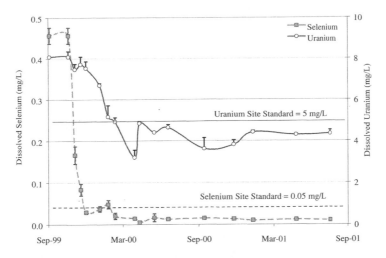

Figure 5. Sweetwater pit lake dissolved selenium and uranium after in situ treatment.

function at 10°C (Knoblauch *et al*, 1999). Psychrophilic cells have a lipid bylayer that allows transport across the cell membrane at cold temperatures, and enzymes with a specific structure to remain flexible in near-freezing water. Our experience is that cold environments do not lack microbes capable of performing the desired metabolic functions at sufficient rates, but rather that the reactants (specifically organic carbon or other reductants) are not present in the cold environments, and must be supplied to allow microbes to be active. The treatment at Sweetwater shows that effective in situ treatment can be achieved in the wintertime. It also illustrates one of the benefits of using organisms that are already present in that system. The fact that the bottom of the lake is 4°C all year means that the lake had 15 years of selection for psychrotrophic organisms that had been exposed to selenium and uranium.

5 UNDERGROUND MINE WORKINGS

5.1 *Mike Horse mine*

The Mike Horse mine is located in Montana. Mike Horse mine is part of a complex of mine workings that together are called the Upper Blackfoot Mining Complex (Anderson and Hansen, 1999). The main Mike Horse adit was plugged and water allowed to pool in the workings in 1996. The average residence time in the pit is 3 months. The fluctuation of the water level in the adit behind the bulkhead on an annual basis is tied to the spring snowmelt. With each rise in water level, additional metals are released, making the flow and loading from the mine workings greatest in the spring and early summer. GWS added 18,000 lbs. of organic carbon and nutrients to the mine workings in October 1996, and added an-

other 40,000 lbs. of organic carbon and nutrients in July 1997.

With each addition of organic carbon, the redox potential in the mine workings decreased down to – 100 mV. The ferrous iron ratio to total iron started at approximately 0.50 (equal concentration) and went to close to 1.0 (only ferrous iron). During this stage of iron reduction, the pH of the mine pool increased and the total iron concentration rose. After iron reduction was complete, metals began to precipitate. Figure 6 shows the results for cadmium and copper. The concentration of both of these metals rapidly decreased, reaching 0.018-mg/L copper and 0.006-mg/L cadmium. At the beginning of treatment, copper was 3.0 mg/L and cadmium was 0.25 mg/L.

The spring flush brought a load of new metals into the mine workings, overcoming the reducing conditions in the mine. However, the reducing conditions and metal removal was rapidly achieved again when new organic carbon was added. It may be possible to build up sufficient "buffering" capacity of reduced compounds (specifically sulfide) to prevent the mine workings from becoming oxidized in the spring and early summer.

5.2 *In-mine treatment processes*

The collection of water behind bulkheads is a common practice in the closure of mine workings. The collected water may then be treated in wetlands or other passive treatment system, or in active water treatment plants. The pre-treatment of water behind the bulkhead offers several advantages to treatment outside of the mine. First, the mine workings offer residence time in which reactions can occur. Second, treatment outside of the mine will ultimately necessitate sludge disposal, while in situ treatment can use the mine workings as the repository of the treatment sludge. Third, treatment in the mine can

257

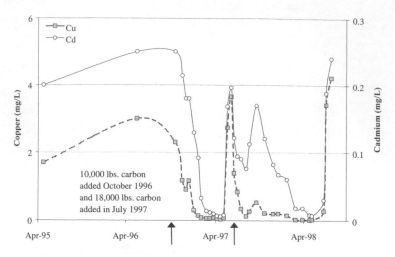

Figure 6. Mike Horse in situ adit treatment results for cadmium and copper.

sludge. Third, treatment in the mine can reduce the leaching of ores that contain sulfide by removing ferric iron from solution. The leaching of sulfides present in rubblized rock in the mine workings and the wall rock by ferric iron can be a significant part of the overall metal load generated in the mine. Removal of ferric iron, both by iron reduction and by precipitation of iron by sulfide, will remove the possibility of leaching reactions from occurring in the mine workings.

The principal reactions that will be useful in the treatment of mine workings are iron reduction and sulfate reduction. Iron reduction raises the pH by consuming acidity during the reaction (reaction 3 in Table 1), and converts ferric iron to ferrous iron, which removes the possibility of additional sulfide oxidation by ferric iron. Sulfate reduction also can reduce ferric iron to ferrous iron by the chemical reaction of aqueous sulfide with ferric iron, producing ferrous iron and elemental sulfur. Sulfate reduction is primarily useful because it can be used to remove metals by the precipitation of metal sulfides. Metals sulfides are often insoluble compounds. The combination of preventing ferric iron leaching in the mine workings, and the in situ precipitation of metals provides significant cost savings compared to any treatment outside of the mine workings.

6 IN SITU GROUNDWATER TREATMENT BENEATH TAILINGS IMPOUNDMENTS

6.1 *J. R. Simplot don plant tailings*

The J. R. Simplot phosphogypsum tailings impoundment is located in Idaho. The reaction of sulfuric acid with phosphate ore produces a tailing solution that is very low pH and typically contains high concentrations of soluble metals. The Simplot tailings

contained arsenic that was the most mobile of all the metals in the tailings (least attenuated by the soils beneath the tailings). A test tailings impoundment (1.5 acres) was constructed with a subsoil layer, and a drainage layer beneath the subsoil. Tailings solution was then placed on the subsoil and the arsenic and phosphate concentrations in the drain were monitored. When the arsenic concentration in the drain layer exceeded the arsenic standard (0.05 mg/L) organic carbon began to be added to the tailings solution. This continued for over 400 days, and the drain concentrations of both arsenic and phosphate were monitored. Figure 7 shows these data.

Arsenic was removed from solution within one pore volume after carbon began to be added. This continued for more than ten additional pore volumes. This occurred even though phosphate concentrations continued to rise throughout the experiment, which indicated that arsenic was not being removed because of a sorption reaction. The redox potential after carbon was added dropped from +415 to –280 mV, and stayed at that potential for as long as organic carbon continued to be added. Aqueous sulfide was measured in the drain at approximately 0.3 mg/L for much of the study, indicating that the sulfide was reacting with the arsenic, precipitating arsenic sulfides. That sulfide was the cause of the arsenic removal was confirmed by the post-experimental soil extraction tests, which showed approximately 90% of the arsenic to be bound up in an operationally-defined arsenic sulfide phase.

7 IN SITU TREATMENT APPLICATION NOTES

7.1 *Complementary closure practices*

The range of sites that can be effectively treated by in situ microbial treatment include underground

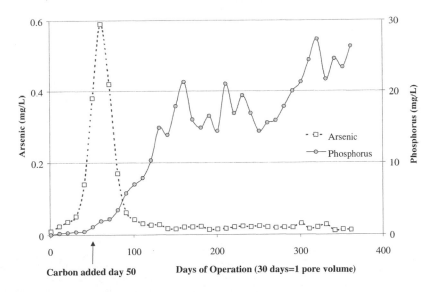

Figure 7. Simplot test gypsum tailings impoundment drain concentrations of arsenic and phosphate.

mine workings, mine pit lakes, backfilled mine pits, heap leach facilities, tailings impoundment ponds, groundwater beneath tailing impoundments, and waste rock facilities. Because oxygen must be removed prior to the treatment of most other constituents, any closure practice that limits the amount of oxygen contact with the mine workings or materials will decrease the cost and increase the effectiveness of the in situ treatment. The installation of a soil cover on a heap, a bulkhead in a mine adit, or the flooding of a mine pit all will reduce the oxygen contact with mining materials. There are two fundamental ways that oxygen contact is minimized: replacement of a gas phase with water (relative oxygen content of oxygen saturated water *vs.* air is 4 x 10^{-5}), and reduction of gas flux through vadose materials by reduction in permeability, typically by installation of a cover.

The effect of a cover on in situ treatment of vadose materials is dual: by limiting recharge, movement of water is deceased, and by limiting oxygen diffusion into the soil. All covers are meant to reduce water infiltration, but not all covers equally reduce gas diffusion. Reduction of gas movement through covers can be achieved by increasing the saturation of a particular layer, or by providing a layer with reduced permeability. Reduction of water infiltration, however, can affect the treatment by reducing the movement of treatment solutions. Thus the timing of the cover placement, after the treatment solutions are adequately applied, is important to increase the coverage and decrease the time for the overall treatment to occur.

Flooding of mine workings or materials is widely practices as an approach to limit the oxidation of sul-

fides. In some cases, it would not be feasible to keep mine workings dewatered. As noted above, it is better to add the organic carbon to the mine pool prior to flooding to get better coverage of the treatment solution. However, it is our experience that recirculation of organic carbon in a mine pool is possible and that these reactions can be stimulated in a uniform manner using this approach.

The amount of treatment that will be required can also be affected by how rapid the flooding is allowed to occur. The oxidation of sulfides can be slowed or prevented by flooding, and so the sooner after exposure the mine workings or materials are flooded, the lower the constituent load will be that may require treatment. We have proposed filling mine workings with an oxygen-excluding gas such as carbon dioxide to prevent sulfide oxidation reactions when water filling is occurring, or in instances where filling with water is not possible.

7.2 Microbial versatility, diversity and distribution

Many microorganisms can utilize a variety of oxidants in the metabolism of organic carbon. Thus, one organism stimulated to perform denitrification by the addition of organic carbon will utilize selenium after the removal of nitrate from solution (reference). In some cases, this metabolic versatility for a single organism covers almost every oxidant listed in Table 1 (Knight and Blakemore, 1998). In other cases, after nitrate has been denitrified, other organisms capable of selenium reduction grow and, from a metabolic perspective, become the dominant organism. The diversity of microorganisms in natural systems is such that in no case that we have yet encountered have the target organisms not been already present and capable to perform the desired reaction (reference). Even in systems where there have been extreme pH conditions, or where toxic constituents have been present, microorganisms have been present that are capable of performing any of the reactions shown in Table 1. The only cases where adding microorganisms proved desirable was when the time necessary to grow the microbes in situ was longer than was available, such as when a pond had to be treated and drained in a short period of time. It is important to remember that microbial growth times in natural systems may be in the order of days or weeks for one single cell to double (references).

7.3 Treatment permanence

In the same way that flooding and covers minimize the potential for sulfide oxidation in untreated mine materials and workings, these same approaches will act to enhance the longevity of an in situ treated mine facility. As can be seen in Figure 6, the Mike Horse mine workings could be easily treated during periods of low recharge, and the treatment effectiveness was lost during periods of high recharge. Slowing recharge will increase the residence time for the microbial reactions to occur.

Maintenance doses of organic carbon can be added to systems to maintain them in a reduced condition. In a reduced state, materials removed during in situ treatment (e.g. metal sulfides, elemental selenium) will remain stable (references). This organic carbon addition can be automated by coupling the addition rate of organic carbon to a redox potential probe.

Removal of metals by sulfate reduction often requires only a small amount of organic carbon, far less than the amount necessary to reduce the total sulfate in solution. However, treatment of such systems can be performed so that excess organic carbon is added, and excess sulfide is created. This excess sulfide may stay in solution as aqueous sulfide, or may precipitate as elemental sulfur. The "bank" of reduced sulfur compounds created can act to maintain the mine materials and workings in a reduced state far after the organic carbon is used up. Elemental sulfur can be microbially disproportionated to produce 3 moles of sulfide and 1 mole of sulfate from 4 moles of elemental sulfur. The sulfide thus produced will react with oxidized metals and precipitate them just as would microbially-produced sulfides from sulfate reduction.

In systems open to the sun, organic carbon can be produced and added to the system through photosynthesis. In the Sweetwater pit, algae were stimulated to create approximately 60,000 lbs. of organic carbon per year. This rate of organic carbon production is sufficient to maintain the reduced selenium and uranium in their insoluble reduced form. Vegetation also produces organic carbon that is deposited to the soil in two forms, through "leakage" of photosynthetic sugars from the roots, to decomposition of dead plant material (roots, stems, leaves). A well-vegetated cover can reduce the oxygen diffusion into the covered materials by consuming oxygen by the decomposition of organic matter.

The burial of constituents removed by in situ treatment will act to minimize the potential for remobilization. In the Sweetwater pit, for instance, the burial of the selenium and uranium removed in one season occurs at a rate of ½ cm/annual lake cycle. This burial though sedimentation is the equivalent of adding a cover layer, because oxygen or oxygenated compounds must diffuse through this layer to react with the reduced compounds removed in the treatment.

The aging of reduced precipitates also acts to minimize the potential for remobilization. Elemental selenium, for instance, when newly formed can be oxidized much more easily than after the precipitate is aged (reference). Pyritized metals are also considered to not be easily mobilized as compared with amorphous sulfide precipitates (reference). The burial and aging of precipitates together reduce the potential for remobilization of metals in many in situ treated mine workings.

7.4 Conclusions

The examples presented in this paper illustrate that the in situ treatment of mine materials and workings is a technology with a strong base of application experiences. The range of sites treated show that there is potential for this technology to be applied at a wide variety of sites. The scientific basis for the technology is now understood to a level that allows a good understanding of the application potential at a particular site, and so that potential pitfalls may be avoided.

REFERENCES

Anderson, R., and B. Hansen. 1999. Mine waste and water management at the Upper Blackfoot Mining Complex. In *Tailings and Mine Waste '99*: 715-723 Rotterdam: Balkema.

Knight, V., and R. Blakemore. 1998. Reduction of diverse electron acceptors by Aeromonas hydrophilia. *Archives of Microbiology*, 169: 239-248.

Knoblauch, C., B. Jorgensen, and J. Harder. 1999. Community size and metabolic rates of psychrophilic sulfate-reducing bacteria in arctic marine sediments. *Applied and Environmental Microbiology*, 65: 4230-4233.

Lehman, R.M. and A. Mills. 1994. Field evidence for copper mobilization by dissolved organic matter. *Water Research*, 28: 2487-2497.

Narkis, N., M. Rebhun, and C. Sheindorf. 1979. Denitrification at various carbon to nitrogen ratios. *Water Research*, 13: 91-98.

Oremland, R., J. Blum, A. Bindi, P. Dowdle, M. Herbel, and J. Stolz. 1999. Simultaneous reduction of nitrate and selenate by cell suspensions of selenium-respiring bacteria. *Applied and Environmental Microbiology*, 65: 4385-4392.

Pauling, L. 1988. General Chemistry. New York: Dover Publications.

Raybuck, S. 1992. Microbes and microbial enzymes for cyanide degradation. *Biodegradation* 3: 3-18.

Raef, S., W. Characklis, M. Kessick, and C. Ward. 1977. Fate of cyanide and related compounds in aerobic microbial systems—II, microbial degradation. *Water Research* 11: 485-492.

Wielinga, B., M. Mizuba, C. Hansel, and S. Fendorf. 2001. Iron promoted reduction of chromate by dissimilatory iron-reducing bacteria. *Environmental Science and Technology*, 35: 522 527.

Response of plants to oil sand tailings

M.J. Silva
The Bioengineering Group, Inc., Salem, MA, USA

M.A. Naeth & D.S. Chanasyk
Department of Renewable Resources, University of Alberta, Alberta, Canada

K.W. Biggar & D.C. Sego
Department of Civil and Environmental Engineering, University of Alberta, Alberta, Canada

ABSTRACT: A greenhouse experiment was conducted to evaluate the response of five agronomic species on initially high water content composite tailings (CT) from Alberta oil sands operated by Syncrude Canada Ltd. The CT mixture having an initial solids content of approximately 65% was placed in lysimeters with no drainage at the bottom to prevent water loss other than via evapotranspiration. The species selected for testing were Altai wildrye (<u>Elymus angustus</u>), Creeping foxtail (<u>Alopecurus arundinaceus</u>), Red top (<u>Agostis stolonifera</u>), Reed canarygrass (<u>Phalaris arundinacea</u>), and Streambank wheatgrass (<u>Agopyron riparian</u>). The electrical conductivity, pH and sodium adsorption ratio of the CT mixture were 3.1 dS/m, 7.7 and 8.2, respectively. All plants survived after an eleven-week period. The six grasses used in the experiment are recommended for revegetation of CT.

1 INTRODUCTION

The Athabasca deposit which has an average thickness of 38 m is the largest of four oil sands deposits located in Alberta, Canada and is the only deposit in the province that can be recovered through surface mining. The Athabasca oil sands of Northern Alberta have been commercially mined by Syncrude Canada Limited and Suncor Inc. since 1978 and 1967, respectively. The two major types of material generated by the oil sands mining and extraction process include overburden and tailings. The overburden consists of all materials lying above the economically minable oil sands. Tailings are a byproduct of the oil sand extraction process. After extracting the bitumen, a slurry waste consisting of residual bitumen, water, sand, silt and fine clay particles is hydraulically transported and stored within surface tailings ponds. Without chemical treatment prior to deposition, the fast-settling sand particles segregates from the slurry upon deposition at the edge of the tailings ponds while the finer fraction accumulates in the center of the pond. Currently, there are approximately 400 million m^3 of mature fine tailings (MFT) at a gravimetric water content of 233% in storage and future predictions estimate that 1 billion m^3 will require storage and future reclamation by 2020 if current discharge methods continue (Liu et al. 1994). The major environmental issues associated with the sludge ponds are their instability and incapability of supporting the weight of animal or machine for a long period of time. Reclamation of these tailings to a desired dry landscape will not be possible until the surface of the deposit be capable of supporting human traffic. The slow rate of consolidation of the existing fine tails is compounded by the continuous addition of new fine tails from the extraction process.

Syncrude Canada Ltd. is currently evaluating a technique for solidifying wet slurries that consists in the addition of phosphogypsum ($CaSO_4$ $2H_2O$) to a mixture of tailings cyclone underflow and MFT. This technique produces a nonsegregating tailings stream known as composite tailings or consolidated tailings (CT). The full evaluation of this technique must proceed in conjunction with the development of reclamation options such that a suitable long-term waste management disposal program can be implemented.

Plant species for reclamation of tailings must adapt to the particular chemical and physical conditions of the growth medium, and to the macro- and microclimates. Species lists must be developed on a regional basis to take general climatic effect into account, but additional screening programs will be necessary to determine the response of proposed species to special soil conditions. As suggested by Ripley et al. (1978) the ultimate selection of species must remain site specific; that is, the choice will have to be made at each individual mining site based on greenhouse experiments and/or field trials. Experimental results by Johnson et al. (1993) and Silva et al. (1998) have shown that oil sands tailings are not phytotoxic. Selected plants should tolerate extremely adverse conditions characteristic of oil sands tailings. Of particular interest will be a tolerance to high pH, water logging, residual bitumen, and a short growing season. This paper is part of a research whose main objective is to study and evaluate the strength enhancement mechanism of suitable plant species growing on initially high water content tailings. This paper presents the results of a greenhouse experiment conducted to evaluate the response of five agronomic species on composite tailings from Alberta oil sands operated by Syncrude Canada Ltd. These plant species proved the most viable in CT as a result of greenhouse experiments conducted to identify the most suitable plant species for dewatering and reclamation of CT (Silva et al. 1998).

2 MATERIALS AND METHOD

2.1 *Composite tailings*

Composite Tailings (CT) was prepared by mixing tailings sand, Mature Fine Tailings (MFT), tailings pond water and gypsum, which were provided by Syncrude Canada Ltd. The amount of gypsum added was approximately 1200 ppm and the proportion of sand, MFT and water was such that the CT mixture had 20% fines and an initial solids content (dry weight of a sample divided by its total weight) of approximately 65%. The mixture was made up in several batches using a 0.22-m^3 cement mixer. This size of the mixer was chosen to make enough volume of CT to fill one lysimeter at a time and obtain a homogenous CT mixture in each lysimeter.

It is worth to mention here that when the CT mixture is transferred from the mixer to the lysimeter, some of the water is trapped in the pore spaces. This entrapped water is under an excess pressure, which is caused by the weight of the CT. Some of the water will flow out of the deposit relieving this excess pressure. This process is known as consolidation. The greatest excess pressure is located at the bottom of the lysimeter and it takes a considerable amount of time for the water to travel from the bottom to the surface of the deposit to relieve this pressure. Figure 1 shows a schematic diagram of the lysimeters used in the experiment. This setup had the purpose to accelerate the consolidation process.

Forty five-gallon drums were lined with a plastic bag and had a 75-mm-thick saturated coarse sand placed at the bottom to act as a filter. One end of a plastic tube, wrapped with geotextile, was inserted into the filter and the other end was located at the top of the lysimeter. This system created double drainage, which accelerated the self-weight consolidation of the CT deposit. The lysimeters had a diameter of 57.2 cm and a height of 84.0 cm. Aluminum tubes with a diameter of 50 mm were installed in the center of each lysimeter to allow access for a neutron probe to monitor the solids content within the tailings. The lysimeters were filled with CT to a depth of about 0.8 m and self-weight consolidation was allowed to take place. Any expressed water was siphoned off and additional CT was added to restore the initial level. The plastic tubes were removed from the lysimeters four days later when no more drainage was noticed. Six lysimeters were instrumented with five mini-TDR probes each built at the University of Alberta. The mini-TDR probes measured solids content in a smaller zone compared with that of the neutron probe. The neutron probe and the TDR probes were specially calibrated to measure solids content of the CT in the lysimeters (Silva 1999).

The CT mixture consists of about 80% sand, 10% silt and 10% clay (Qui & Sego 1998). Values of nutrient concentrations, pH and electrical conductivity (EC) for the CT mixture used in this study are presented in Table 1. Nitrate, phosphate and potassium levels were deficient for plant growth, whereas sulfate was at optimal level. Levels of iron, boron and manganese were adequate for plant growth. Zinc and copper levels were deficient. Chloride was in excess. The mean value of Sodium Adsorption Ratio (SAR) was 8.2 ± 1.2, which is less than 13 (Miller & Donahue 1990); therefore CT can be classified as a non-sodic soil. Based on the combined values of SAR and EC, CT can be classified as a normal soil (non-saline and non-sodic).

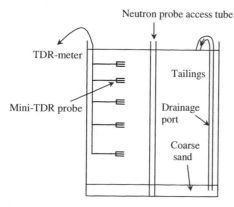

Figure 1. Lysimeter layout.

Table 1. Chemical composition of CT

Analysis	CT
pH	7.7 ± 0.3[*]
E.C. (dS/m)	3.1 ± 0.3
Nitrate (ppm)	1.0 ± 0.0
Phosphate (ppm)	4.0 ± 2.9
Potassium (ppm)	43 ± 3
Sulfate (ppm)	>20
Ca (ppm)	437 ± 40
Na (ppm)	681 ± 93
Mg (ppm)	65 ± 7
Fe (ppm)	17 ± 0.8
Cu (ppm)	0.13 ± 0.0
Zn (ppm)	0.3 ± 0.0
B (ppm)	3.0 ± 0.1
Mn (ppm)	5.4 ± 0.4
Cl (ppm)	>50

[*] Means (n = 4) ± standard deviation

2.2 Plant material and growth room conditions

The plant species used in the experiment were: Altai wildrye (Elymus angustus), Creeping foxtail (Alopecurus arundinaceus), Red top (Agostis stolonifera), Reed canarygrass (Phalaris arundinacea), and Streambank wheatgrass (Agopyron riparian). Plants were started from seeds in root trainers and transplanted to the lysimeters after 4 weeks with the equivalent to 64 plants/m^2. Three lysimeters were used for each species and three were left unplanted as a control. The lysimeters were placed in a growth room in a completely randomized design. Air temperature and hours of light per day were set at 22 °C and 15 hours, respectively, simulating a typical growing environment in Fort McMurray. The average maximum and minimum temperatures recorded were 24.7 and 20.6 °C, respectively. The average relative humidity was 52.9%. Water was added weekly to simulate average precipitation at Syncrude Canada's Mildred lake site from June through August. Normal values of precipitation for this location are 63.9 mm for June, 79.1 mm for July and 71.8 mm for August.

Fertilizer was added cautiously to prevent the total solute load from exceeding the salinity tolerance of the plants. To avoid over-fertilization the amount of fertilizer was based on optimum levels of macronutrients required by agronomic species. The fertilizer was added to give an equivalent of 150 kg N/hectare, 80 kg P_2O_5/hectare, and 105 kg K_2O/hectare.

2.3 Measurements

Solids contents were measured weekly via the neutron probe and TDR probes to obtain information on the water lost due to evapotranspiration in the planted lysimeters and evaporation in the controls. Observation of stress symptoms, survival and tillering were conducted chronologically. Number of tillers, number of leaves, plant height, and leaf area were determined at regular intervals throughout the experiment.

The leaf area was determined on five different dates using the following procedure. The leaves of a subsample from each lysimeter were carefully traced on paper. The leaf area was then measured using a planimeter. The total leaf area was determined by multiplying the area of the subsample by the total number of leaves. These leaf areas were then used to calculate the leaf area index (LAI), which is defined as the area of one side of leaves per unit of soil surface (Jensen et al. 1990).

After ten weeks of plant growth, reed canarygrass and creeping foxtail showed symptoms of water stress. Solids content measurements indicated that indeed the tailings deposit was near the wilting point, which is about 96% solids. This condition dic-

tated the end of the test series. The plants were then cut to measure the dry shoot biomass.

At the end of the experiment samples of tailings were taken every 10 cm to measure the solids content profiles. Root densities were measured at six different depths by inserting rings of known volume. The roots were washed and oven dried to measure the dry root biomass profile. The total root biomass from each lysimeter was then calculated by extrapolation. EC and pH profiles were measured on only six lysimeters, which were chosen at random from each treatment.

2.4 Statistical analyses

Statistical testing of biomass production by species within a treatment was performed using single factor analyis of variance (ANOVA) followed by multiple comparison testing utilizing Duncan's multiple range test as described in Little & Hills (1972).

3 EXPERIMENTAL RESULTS

Mean height growth over time is shown in Figure 2. Creeping foxtail and reed canarygrass demonstrated a high degree of tolerance of the harsh growing conditions in the tailings. There is a general pattern in plant height increase with time. However, it appears that the growth of streambank wheatgrass was dramatically delayed and maximum height was reduced. Transplant shock likely contributed to this poor growth.

Values of LAI (Figure 3) for each plant species were calculated from measured leaf areas. The time used in the horizontal axis is from the date of germination, which in average occurred about 7 days after seeding. The ability of a plant canopy to intercept energy and subsequently to transpire varies with LAI. As an emerging seedling develops leaves and its roots permeate the soil, the plant increases in effectiveness as a conduit for transferring water from the soil profile to the atmosphere. The ground is considered completely covered when the LAI reaches the value of 2.7. At this value the plant canopy becomes also complete (Ritchie 1974).

Water was lost from the lysimeters through evapotranspiration or by evaporation alone in the case of the controls. The cumulative water loss from the lysimeters is shown in Figure 4. On Day 28 the seedlings, which were grown in root trainers, were transplanted to the lysimeters. For the first two weeks after transplanting, there was little difference in the rate of water loss. When the creeping foxtail and reed canarygrass plants increased in height and LAI, their evapotranspiration rates increased over the controls and other species. Streambank wheatgrass lost less water than the controls until Day 85. Because of its poor physiological state, they trans-

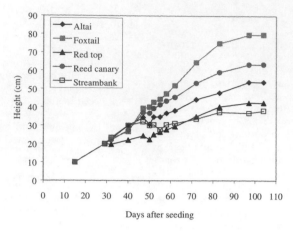

Figure 2. Temporal variations in plant height.

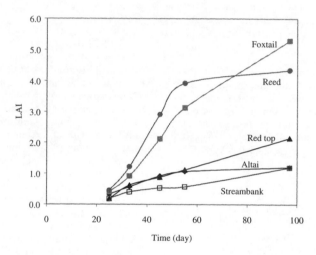

Figure 3. Temporal variations in leaf area index (LAI).

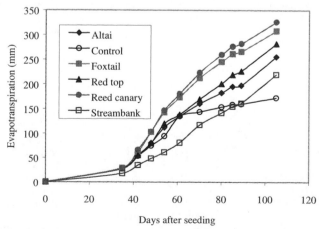

Figure 4. Cumulative evapotranspiration.

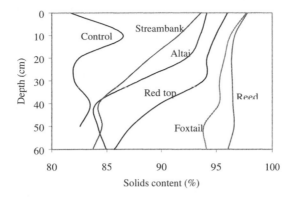

Figure 5. Mean measured solids content profiles at the end of the tests.

pired little water, and the presence of the leaf cover on the surface of the tailings in the lysimeter reduced the evaporative losses.

The cumulative evapotranspiration at the end of the greenhouse experiment is directly reflected in the final solids content profile (comparing Figures 4 and 5). As water was transpired from the surface, a moisture gradient was established. Reed canarygrass and creeping foxtail significantly lowered the moisture content of the CT mixture. Roots reached the bottom of the lysimeters and may have gone further if the containers had been deeper. In these lysimeters the moisture contents were close to the wilting point and the plants started showing signs of water stress. This condition dictated the termination of the experiment. Pure evaporation (control) did not cause any significant dewatering of the CT compared to the plants. Salt accumulations and the presence of a film of bitumen at the surface of the control lysimeters inhibited further evaporation.

Creeping foxtail and reed canarygrass produced the highest shoot and root dry weights (Table 2), which were not significantly different between them, but were significantly different from the other species. Altai wildrye and streambank wheatgrass generated the lowest shoot and root dry weights. Creeping foxtail and streambank wheatgrass produced the highest root to shoot ratio (0.93), followed by reed canarygrass (0.75).

The root biomass profile is presented in Figure 6. These curves are the average of six root density profiles measured within each planted treatment. A

semi-logarithmic scale was used to provide a better visualization of root biomass differences. Reed canarygrass produced the highest root biomass distribution. There appears to be a direct relationship between root biomass and root water uptake, which can be seen by comparing Figures 5 and 6.

The patterns of the electrical conductivity (EC) were similar in the different treatments (Figure 7). EC was higher at the surface and ranged from 2.2 to 3.9 dS/m. Salt accumulations were more evident in the control lysimeters where a white film of salt was present at the surface. The pH was uniform along the tailings profile with a constant value of 8.1 (Figure 8).

Assessments of the plant response were conducted at seven different dates using a symptom scale, which was based on the degree of plant healthiness (Table 3). Altai wildrye and streambank wheatgrass presented signs of stress since transplanting. Approximately 25% of older leaves of streambank wheatgrass slowly turned chlorotic then started to die on Day 57. Altai wildrye presented a poor growth; small necrotic spots on many leaf tips were evident during the initial stage of the experiment. However, both Altai wildrye and streambank wheatgrass displayed a strong propensity for regrowth following initial leaf loss. Creeping foxtail and reed canarygrass presented a healthy growth during the whole experiment; plants started showing signs of water stress after Day 65. The water stress symptom was more notorious on reed canarygrass, leaf tips were curled and dying, many first leaves were dying as well. Later solids content measurements con-

Table 2. Shoot and root dry weights

Species	Shoot dry weight (g)	Root dry weight (g)	Root : shoot ratio
Altai wildrye	37.7 a[*]	24.5 a	0.65
Creeping foxtail	120.0 b	111.4 b	0.93
Red top	86.7 c	37.3 a	0.43
Reed canarygrass	134.3 b	100.8 b	0.75
Streambank wheatgrass	26.9 a	24.9 a	0.93

[*]Means (n = 3) within a column followed by a common letter are not significantly different at the 5% level.

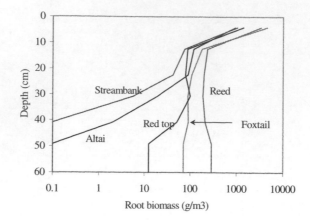

Figure 6. Root biomass profile.

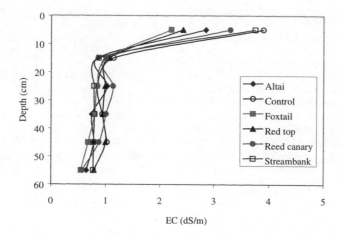

Figure 7. Electrical conductivity profile.

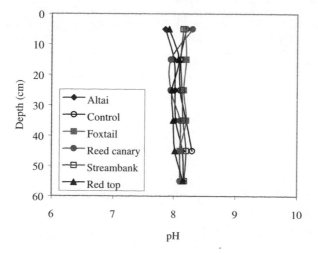

Figure 8. pH profile.

Table 3. Summary of plant behavior assessment.

Plant species	Symptom scale						
	Day 38[*]	Day 41	Day 48	Day 57	Day 65	Day 83	Day 90
Altai wildrye	1.5	1.5	2	2	2	2	2.5
Creeping foxtail	1	1	1	1	1.5	1.5	2
Red top	1	1.5	1.5	1.5	1.5	2	2
Reed canarygrass	1	1	1	1.5	2	2	2
Streambank wheatgrass	1.5	1.5	2	2.5	2.5	2.5	3

[*]Day number is the number of days after seeding

Symptom scale is based on degree of plant health
1 Very healthy, lush, a few older leaves dying, maybe a few tips browning
2 Fairly healthy, many first leaves dying, some symptoms evident, tips dying, a bit of chlorosis
3 Looking stressed, dry leaves, rolling leaves, chlorosis and necrosis very evident, tips curled and dead, perhaps stunted
4 Very stressed, dry, dying tillers
Note: dying refers to mortality

firmed that indeed the soil was close to the wilting point and not sufficient water was available to the plants.

4 DISCUSSION

Five agronomic species were selected to evaluate their response on composite tailings from Alberta oil sands operated by Syncrude Canada Ltd. All plants survived after an eleven-week period. Creeping foxtail, red top and reed canarygrass plants showed signs of healthy growth, whereas Altai wildrye and streambank wheargrass did not grow well in the tailings in the initial period. These plants were shocked in the transplant and their growth was stunted at the beginning. Stress symptoms were more evident in streambank wheatgrass. It appears that the level of metabolic activity in the plants was reduced. This was accomplished by reducing their biomass through shedding their leaves, or by going into a dormant stage. The stress was observed in many plants, which had dry leaves with tips curled and dead. Low plant height increase, low leaf area index, low evapotranspiration and low shoot and root biomass are evidences of poor plant growth. However, both Altai wildrye and streambank wheatgrass displayed a propensity for regrowth. This capacity can be important attribute for species used in reclamation, and may reduce replanting cost (Fedkenheuer et al. 1980). Macyk et al. (1998) reported that streambank wheatgrass presented good growth and initial establishment in tailings sands. They concluded that this plant is suitable for revegetation on the Syncrude tailings dike, especially for the purpose of erosion control.

Creeping foxtail and reed canarygrass had the highest evapotranspiration rates, the highest above and below ground biomasses, the deeper roots and the lowest stress symptoms. The same can be said about the plant heights and leaf area indexes. These plant parameters are linked to the healthy plant growth and physiological conditions.

Red top plants had a fairly healthy growth during the whole period. However, at the end of the experiment the plants showed signs of water stress. Their roots were not as deep as those of creeping foxtail and reed canarygrass plants and the stress was caused by moisture deficiency at the surface of the tailings. The results indicate that red top is a short plant with shallow roots.

The results obtained clearly confirm that CT is not phytotoxic and can be used as a medium for plant growth. This conclusion is supported by the low EC and SAR, which class CT as a normal soil.

5 CONCLUSIONS

A greenhouse experimental program was conducted to evaluate the response of five agronomic species on high initially water content tailings. Based on values of Sodium Adsorption Ratio (SAR) and Electrical Conductivity (EC) composite tailings (CT) can be classified as normal soils (non-sodic and non-saline). Healthy plant growth confirmed that CT is not phytotoxic and suitable plants can be used to implement future reclamation activities when the impoundment reaches full capacity.

In conclusion, reed canarygrass, creeping foxtail and red top plants proved to be the best candidates for future field research in CT in the initial stage when there is still free water at the surface of the tailings. Altai wildrye and streambank wheatgrass plants are not tolerant to waterlogged conditions, however they are also good candidates for future field research in CT with no free water at the surface.

6 ACKNOWLEDGEMENT

The authors acknowledge the financial support of the National Science and Engineering Research Council of Canada and the first author thanks Luscar Ltd. for their Graduate Scholarship in Environmental Engineering. Special thanks is extended to Christine Hereygers, Steve Gamble and Gerry Cyre, whose technical assistance was extremely valuable. Syncrude Canada Ltd. Supplied the tailings used in this study.

REFERENCES

Fedkenheuer, A.W., Heacock, H.M., and Lewis, D.L. 1980. Early performance of native shrubs and trees planted on amended Athabasca oil sand tailings. Reclamation Review, **3**: 47-55.

Johnson, R. L., Bork, P., Allen, E. A. D., James, W. H., and Koverny, L. 1993. Oil sands sludge dewatering by freeze-thaw and evapotranspiration. Alberta Conservation and Reclamation Council Report No. RRTAC 93-8. ISBN 0-7732-6042-0.

Jensen, M.E., Burman, R.D., and Allen, R.G., eds. 1990. Evapotranspiration and irrigation water requirements. Manual of practice No. 70, ASCE. New York.

Little, T.M., and Hills, F.J. 1972. Statistical methods in agricultural research. Agricultural extension. University of California.

Lui, Y., Caughill, D., Scott, J.D., Burns, R. 1994. Consolidation of Suncor nonsegregating tailings. Proceedings, 47th Canadian Geothecnical Conference, Halifax, Nova Scotia, September 21-23, 1994, pp. 504-513.

Miller, R.W. and Donahue, R.L. 1990. Soils. An introduction to soils and plant growth. Sixth edition. Prentice Hall.

Qiu, Y. and Sego, D. C. 1998. Engineering properties of mine tailings. Proceedings, 51st Canadian Geotechnical Conference, Edmonton, Alberta, October 4-7, 1998. Vol. 1. pp. 149-154.

Ritchie, J.T. 1972. Model for predicting evaporation from a row crop with incomplete cover. Water Resources Research, **8**: 1204-1213.

Silva, M.J. 1999. Plant dewatering of tailings. Ph.D. thesis. University of Alberta. Edmonton, Alberta.

Silva, M. J., Naeth, M. A., Biggar, K. W., Chanasyk, D. S., and Sego, D. C. 1998b. Plant selection for dewatering and reclamation of tailings. Proceedings, 15th Annual National Meeting of the American Society for Surface Mining and Reclamation, St. Louis, Missouri, May 17-21, 1998. pp. 104-117.

Tailings and Mine Waste '02, © 2002 Swets & Zeitlinger, ISBN 90 5809 353 0

Column leaching test to evaluate the beneficial use of alkaline industrial wastes to mitigate acid mine drainage

I. Doye & J. Duchesne
Departement de Geologie et de Génie Géologique, Université Laval, Sainte-Foy, Qc., Canada

ABSTRACT: This paper presents laboratory investigation to evaluate the capacity of basic residues to inhibit acid mine drainage. Column tests have been set up with unoxidized tailings, oxidized tailings, waste rock, cement kiln dust, and red mud bauxite. Columns were also performed by leaching oxidized tailings mixed with 10 and 50% alkaline materials. Finally, column tests have been carried out to simulate layered co-mingling, covers, liners, and co-disposal of waste rock and fine tailings. The pH results obtained show that the addition of alkaline materials permits to keep near neutral conditions in the fine material layer. No significant difference in pH values was obtained with mixture of 10 and 50% alkaline material. When the fine-grained layer overlays reactive waste rock, a delay was observed before achievement of near neutral pH. In the cases where the fine-grained layer are below or mixed with the waste rocks, the near neutral pH values are directly reached.

1 INTRODUCTION

Acid mine drainage (AMD) is the most serious environmental problem facing the mining industry. AMD is formed when sulfide minerals in waste rocks or tailings are oxidized in the presence of water and oxygen to form highly acidic, sulfate and metals-rich drainage. Layered co-mingling of waste rock with compacted fine tailings, covers with capillary barriers effect, and co-disposal of waste rock and fine tailings are some of the methods used to reduce the formation of acid mine drainage. These methods shield waste rocks from oxygen by limiting gas flux by diffusion. One of the other techniques used to mitigate AMD is the addition of alkaline materials which serve to either neutralize the acid generated or retard or stop the oxidation of pyrite. Higher alkalinities also help control bacteria and restrict solubility of ferric iron, which are both known to accelerate acid generation.

Recent studies have indicated that certain types of alkaline amendments can successfully control AMD from pyritic spoil and refuse (Skousen et al. 1998, Brady et al. 1990, Burnett et al. 1995, Perry & Bradly 1995, Rich & Hutchinson 1990, Rose et al. 1995, Wiram & Naumann 1995). Limestone is one of the least expensive and most readily available source of alkalinity. However, limestone has a limited solubility and can only raise the pH of a system to approximately 8.3. Lime (CaO) and portlandite $(Ca(OH)_2)$ are examples of materials that can increase water pH above 8.3 but are fairly expensive materials. Some industrial wastes produced in large

volumes present very interesting alkaline properties. Cement kiln dust (CKD) produced by the Portland cement industry and red mud bauxite (RMB) issued from the Bayer process used in the aluminum industry are examples of such materials and were selected in this study. Both materials are industrial residues produced in large volumes, with a strongly basic character, and a fine and uniform particle size distribution. The addition of alkaline materials can be done in different ways to mitigate AMD. According to Skousen et al. 1998, the use of alkaline addition can be divided into several categories, including:

1- blended with potentially acid-producing material to either neutralize the acid, retard, or prevent the oxidation of pyrite;

2- incorporated as stratified layers at specific intervals within the backfill or spoil;

3- applied as trenches or funnels to create alkaline groundwater conduits;

4- applied on or near the surface to create an alkaline wetting front;

5- applied as a chemical cap to create a hard pan (either on the surface or the floor of the mine).

The purpose of this study is to evaluate the effect of CKD and RMB as alkaline addition on leachate compositions when set up with acid-producing tailings or waste rock. Leaching column tests have been carried out in order to study the role of alkaline addition when set up to simulate layered co-mingling, covers, liners, and co-disposal of waste rock and fine tailings. The efficiency of the neutralization and the water quality of the leachates from columns were monitored over time. Clearly, the potential to recycle alkaline industrial wastes to prevent the formation of AMD at its source is one of the most sustainable solutions to the problem.

2 MATERIALS AND METHODS

2.1 *Mine reactive tailings (RT)*

The mine reactive tailings considered in this study come from the Aldermac mining site, Abitibi Temiscamingue, Quebec, Canada. The massive sulfide deposits of the Aldermac site was worked from 1932 to 1943 for copper and gold. The sulfides are hosted by a mafic to felsic volcanic sequence cut by probably synvolcanic sills of gabbro and quartz-feldspar porphyry (Barrett, 1991).

Tailings from Aldermac mine are net acid generators. Indeed, leachates issued from the site present pH values varying from 2.3 to 3.1. Concentrations of iron (453 mg/L), zinc (11 mg/L) and copper (0.6 mg/L) in the leachate do not respect directive 019 of the Ministère de l'environnement du Québec whose maximum acceptable concentrations are 3, 0.5, and 0.3 mg/L for Fe, Zn, and Cu respectively (MENVIQ, 1989). Iron content and pH do not respect US regulation according to the technology-based effluent limitations for active mining discharges set forth in 40 CFR Part 434.

Table 1 presents chemical composition of the materials. The RT which come from massive sulfide ore deposits are rich in Fe, S, Cu, Pb, and Zn. Mine tailings, with a 20 μm median grain-size, are mainly composed of K-feldspar $((K,Na)AlSi_3O_8)$, quartz (SiO_2), pyrite (FeS_2), muscovite $(K, Ca, Na)(Al, Mg, Fe)_2(Si,Al)_4O_{10}(OH)_2)$, chlorite $(Mg_5Al_2Si_3O_{10}(OH)_8)$, chalcopyrite $(CuFeS_2)$ (1%), sphalerite (ZnS) (<1%), K-jarosite $(KFe_3(SO_4)_2(OH)_6)$, and iron oxyhydroxides as ferrihydrite $(Fe(OH)_3)$, and goethite $(FeOOH)$. Mineralogical composition of the tailings was determined by x-ray diffraction (XRD), optical microscopy, and scanning electron microscopy (SEM).

The quantity of pyrite is high and estimated at 12.3% by mass. Acid base accounting modified (ABA) test was conducted on tailings (Lawrence 1990) and net neutralization potential of -418.0 kg $CaCO_3/t$ was measured. This test was chosen for reducing the overestimation of the NP value measured by the Sobek method. Basing the AP on the sulfide-sulfur content assumes that sulfur present as sulfate is not acid producing (e.g., sulfate minerals such as gypsum).

2.2 *Waste rock (WR)*

The waste rock used in this study come from the South Dump of La Mine Doyon, Abitibi, Quebec, Canada. La Mine Doyon South waste rock dump is one of the largest acid generating dump in Eastern Canada (Gélinas et al., 1994). The waste rock in the south dump is a highly altered chloritic-sericitic schist. The schist is mainly composed of quartz, chlorite, muscovite (or sericite), illite $(K_xAl_2(Si_{4-x}Al_xO_{10})(OH)_2)$, jarosite $(KFe_3(SO_4)_2OH_6)$, pyrite (FeS_2), and gypsum $(CaSO_4.2H_2O)$ (Gélinas et al., 1994). The chemical composition of this waste rock is given in table 1. The high SiO_2 content is due to the host rock which is a felsic schist. The Al content is in agreement with a mineralogy dominated by phyllosilicates such as chlorite and sericite. The quantity of pyrite calculated according to the total chemical analysis was 7.0 wt %. Calcite concentration was evaluated to 2.25 wt %. Net neutralization potential of –50.1 kg $CaCO_3/t$ was measured on the Doyon waste rock sample (Fortin et al., 2000).

2.3 *Cement kiln dust (CKD)*

CKD is a waste residue generated from the manufacture of Portland cement. It is a powder composed principally of micron-sized particles collected from electrostatic precipitators during the high temperature production of cement clinker. The sample is a fine powder with a 10 μm median grain-size particle. The CKD used in this study were issued from St. Lawrence cement, Joliette, Quebec, Canada. The chemical composition of the CKD is presented in Table 1. This CKD is rich in calcium, silicon, and potassium oxides. CKD presents a high natural alkalinity and a considerable high sulfate content. The CKD contains also trace elements such as Cr, Pb, and Zn which come from fuel and raw materials used in the kiln.

Crystalline phases detected by XRD were principally calcite $(CaCO_3)$, lime (CaO), quartz, anhydrite $(CaSO_4)$, arcanite (K_2SO_4), with some portlandite, and sylvite (KCl). Some of these phases such as lime, arcanite and sylvite are highly soluble in water. Into contact with water, these phases dissolve completely or precipitate under more stable secondary phases. The XRD pattern for the water-reacted CKD sample showed the presence of portlandite, calcite, quartz, gypsum $(CaSO_4.2H_2O)$, and ettringite $(Ca_6Al_2(SO_4)_3(OH)_{12}.26H_2O)$. Duchesne & Reardon (1998) present a comprehensive study of the leaching behavior of this material.

2.4 *Red mud bauxite (RMB)*

RMB is a clay-size waste issued from the aluminum production. The RMB is produced when the bauxite ore undergoes the Bayer process, in which caustic soda is added to bauxite, producing alumina and muddy red residue. The RMB used in this study was produced at the ALCAN Vaudreuil plant, Arvida, Quebec, Canada. The chemical composition of the RMB is presented in table 1. The sample is composed mainly of alumina, iron, and sodium (caustic soda) oxides, with small quantities of silicon, titanium, and calcium oxides. The high Cr concentration results from chromites which are residual phases in the profiles of laterite and which are very resistant in the alteration process. Into contact with water, RMB presents a very high pH value. The main crystalline phases detected by XRD were hematite (Fe_2O_3), goethite (FeOOH), gibbsite ($Al(OH)_3$), boehmite (AlOOH), anatase (TiO_2), rutile (TiO_2), sodalite ($Na_4Al_3Si_3O_{12}Cl$), katoite ($Ca_3Al_2(SiO_4)(OH)_{12}$), and lime (CaO).

Table 1. pH values, NNP, and element concentrations of waste rock (based on Savoie et al., 1991), reactive tailings (RT), cement kiln dust (CKD) and red mud bauxite (RMB).

Elements	Waste rock	RT	CKD	RMB
pH paste	4.2	2,3	12.6	11.4
NNP*	-50.1	-418.0	425.5	91.6
	mg/L	mg/L	mg/L	mg/L
Al	47 366	37 720	17 350	97 180
Ca	11 721	10 007	294 000	29 740
Fe	41 899	170 500	14 340	265 100
K	10 293	1 329	34 360	2 573
Mg	22 197	6 031	7 237	905
Mn	6 196	410	286	85.1
Na	4 080	16 320	3 041	58 010
Si	303 852	231 900	54 590	62 070
S tot	37 400	123 000	-	-
SO$_4$	-	24 000	89 000	14 800
Co	26	158	4	4
Cr	-	141	61	633
Cu	45	1 083	10	9
Pb	15	114	344	43
Zn	49	145	708	20

*NNP: Net Neutralization Potential (kg CaCO$_3$/t).

3 METHOD

3.1 Column leaching tests

Test columns have been set up to evaluate the effect of alkaline additives on the geochemistry of drainage water. The first series of columns consists of controls where each material is leached individually. A second set of columns present the effect of alkaline additive proportioning when mixed with the reactive tailings. Finally, the third series of columns presents various cases where alkaline additions are used to prevent the AMD from waste rock.

Samples were placed into polycarbonate column (8.5-cm diameter, 40-cm height) open to the atmosphere at the top, and close at the bottom with a filter (stainless 316, 100 mesh) and a drain for sample collection. Columns were fed by the top with 500 ml of deionized water two times per week. This implies two wet/dry cycles per week. Flushes of water were repeated for 35 weeks. To avoid preferential flow, a layer of Ottawa sand (ASTM-190) was disposed over the materials. In the same way, a layer of sand was placed under the materials in order to prevent the obstruction of the drains. The permeability of the fine material layer was increased by addition of Ot-

tawa sand to tailings and alkaline additive at a ratio by mass of 1:2. Ottawa sand was selected because it is an inert material composed mainly of silica dioxide (over 99%). The increase in the permeability of fine materials makes it possible to accelerate the tests in order to obtain geochemical data in a reasonable time. The material was compacted in 3 layers using 30 knocks of compactor per layers. The compactor had a mass of 1386 g and its height of fall was 42 cm.

3.2 Monitoring chemical composition

The parameters monitored after each flush of water include leachate volume, pH, and electrical conductance. Acidity, alkalinity, sulfate, Al, Ca, Cu, Fe, K, Mg, Na, Si, and Zn concentrations were determined for the first four flushes and at every 4 weeks thereafter. Solution samples were filtered through 0.45 μm membrane filters. Each sample was split into two portions. One portion was acidified with 1.0 ml of 1:1 HCl for cation analyses on a Perkin Elmer Analyst 100 Atomic Absorption Spectrophotometer. The unacidified portion was reserved for sulfate analyses on a Dionex DX-100 ionic chromatograph. All solutions were kept at 4 °C until analysis. Solid samples were characterized by combination of inductively coupled plasma emission mass spectrometer (ICP-MS) and X-ray fluorescence spectroscopy technologies (XRF). The solids were dissolved by a lithium metaborate/tetraborate fusion technique. Total sulfur and sulfate concentrations were analysed by infrared technology.

3.2.1 First series (control columns)
The first set of columns consists of controls where each material is leached individually. Figure 1A presents the set up of the columns where a single layer of 3.5 cm of material is placed between two layer of non reactive sand. The materials used in these columns are unoxidized tailings (UT), reactive tailings (RT), CKD, RMB, and a mixture of CKD and RMB 1:1 by mass.

3.2.2 Second series (alkaline additives mixed with reactive tailings)
A second set of columns present the effect of alkaline additive proportioning when mixed with the reactive tailings. Six columns were set up with addition of 10% and 50% by mass of alkaline additive. Figure 1A presents the set up of these columns containing mixtures of tailings with CKD, RMB or blend of CKD and RMB (1:1).

3.2.3 Third series (alkaline additives used to prevent AMD from waste rock)
The third series of columns was set up in order to evaluate the role of the mixture of alkaline additives and reactive tailings on the reduction of AMD from

waste rock. Nine columns were set up according to figure 1.

The first column containing only Doyon waste rock was the reference column (Figure 1B). In the second column, reactive tailings layer was disposed above waste rock (Figure 1D.1,2). Then, six settings of various materials were considered to reproduce the cases of layered co-mingling of waste rock with compacted fine materials, covers or liners with fine materials, and co-disposal of waste rock and fine materials (Figure 1B, 1C, and 1D). The fine materials were mixtures of reactive tailings with 10% mass fraction of CKD or CKD:RMB. One column was set up with a CKD layer overlaying reactive tailings, then waste rock without homogenisation (Figure 1D.1**). The last column includes a layer of reactive tailings mixed with 10% mass fraction of CKD overlaying a double mass of waste rock (Figure 1D*).

4 RESULTS AND DISCUSSION

4.1 First series of columns (control columns)

4.1.1 pH measurements

Figure 2 shows pH values of controls as a function of the volume of water passing through columns. Initial pH values were strongly acidic for the unoxidized tailings (UT) and reactive tailings (RT), with values around 3.4 (UT) and 2.4 (RT). pH of tailings raised to reach 4.5 and 3.3 at the end of experiment.

The pH level of CKD is established at 12.7 early during experiment. Dissolution of pure lime by water can account for a pH of 12.4 (saturation with respect to portlandite). The substantially higher observed pH is due to the combined effect of the dissolution of primary arcanite coupled with the subsequent precipitation of ettringite. At the end of the experiment, pH of the CKD layer decreases to 11.6.

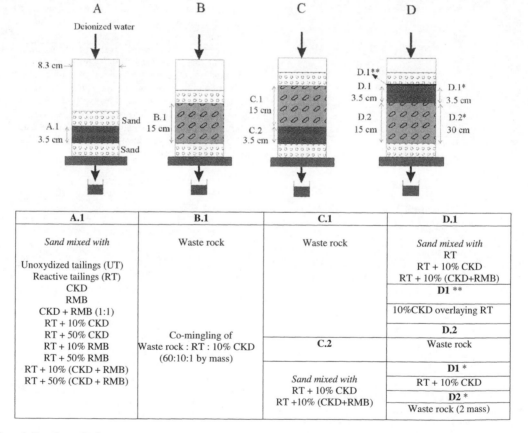

Figure 1. Experimental columns.

274

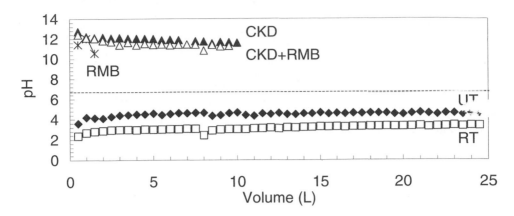

Figure 2. Curves of pH evolution of the control columns.

This pH decrease is the result of the portlandite depletion and the subsequent slow dissolution of silica-rich glass of CKD particles.

The RMB presents at the beginning of the experiment a high alkaline pH corresponding to the leaching of the caustic solution (NaOH) trapped in the red mud. Leaching of the column with RMB was stopped after only 3 flushes because of the slow drainage of the material due to its weak permeability. In a previous study (Doye, 1999), RMB was leached with a water:RMB ratio of 4:1 during 8 weeks. The pH remained over 10 during all the duration of the test due to the presence of hydroxides. The pH values of the mixture of CKD+RMB follows the same trend as the CKD sample but with slightly lower pH values due to a dilution effect given by the contribution of the RMB.

4.1.2 *Chemical composition of the leachates*
Figure 3a, 3b, and 3c presents the values for Ca, Fe, and SO_4^{2-} concentrations as a function of the volume of water passing through columns. The follow-up of the columns was done over a period of 20 weeks with 2 water supplies of 500 ml each week. One can see on figure 2 that certain columns took more time to drain. It is the case for column containing CKD and RMB. These materials have a low permeability.

After the first flush, leachates recovered from tailings samples were rich in iron and sulfates. Iron concentrations were above 1000 mg/L and fall nearly 100 mg/L after the second flush. At the end of the experimentation, iron concentrations vary between 10 and 50 mg/L. The evolution of SO_4^{2-} follows the same trend as iron concentrations with values of 16 000 and 4 000 mg/L for the unoxidized and reactive tailings after the first flush and 15 and 60 mg/L at the end of the experiment. Iron and sulfate are some products of the oxidation of the pyrite. The concentration of Ca found in leachates of RT and UT results from the dissolution of calcite. In the

Abitibi area which is characterized by a weak metamorphic facies, calcite appears as an alteration product of plagioclase feldspars. The Ca concentration values reach 200 mg/L and 300 mg/L for RT and UT respectively and fall below 1 mg/L after 12 flushes. Aluminosilicate minerals dissolution also contribute base cations (Ca, Fe).

The first flush of the CKD sample was rich in sulfate and calcium with concentrations of 2 500 and 1 000 mg/L respectively. Sulfate concentrations come initially from the dissolution of arcanite then in the long run of the ettringite. The first calcium concentrations coming mainly from the dissolution of portlandite are followed by dissolution of calcite. Iron content was very low in solution given that it is the form of stable oxyhydroxides.

Leaching of RMB shows high sulfate concentrations which come from the secondary precipitation of gypsum during the storage of the materials. The sulfate concentrations become less important after the second flush. A very small quantity of iron is lixivied from red mud in spite of the very significant iron content of the RMB. Indeed, iron present in RMB is mainly in the form of hematite and goethite. These phases are insoluble under the pH conditions of the test. Ca is almost non-existent in the leachate of RMB.

While the mass of CKD is decreased in half in the mixture CKD+RMB, the concentration of Ca is ten times as small after the first flush. Given that Ca generated by the RMB is unimportant, this one results only from carbonates, oxides, and hydroxides of calcium found in the CKD. The addition of RMB to the CKD contributes to the decrease of the rate of dissolution of the calcium containing phases or to the precipitation of secondary minerals containing calcium while maintaining alkaline pH conditions similar to those generated by the sample containing only CKD. The sulfate concentration measured of 7500 mg/L is excessively high in the mixture of

275

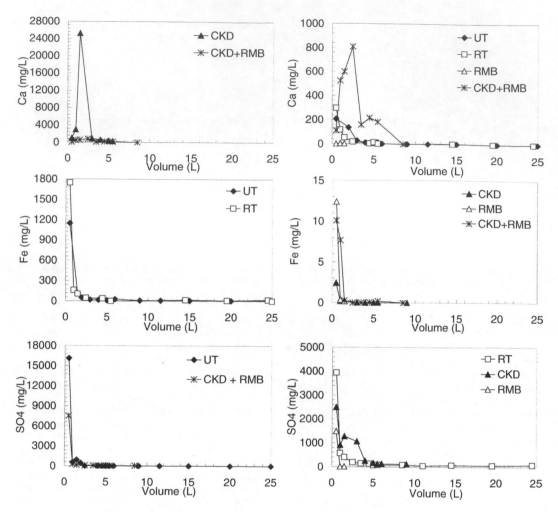

Figure 3. Evolution curves of pH, Ca, Fe, and SO_4^{2-} of the second test

RMB+CKD with regard to the values for CKD and RMB leached separately. The sulfate concentration rapidly decreases to 150 mg/L after the second flush. This value is below the level expected according to the contribution in SO_4 of both materials.

4.2 Second series of columns (alkaline additives mixed with reactive tailings)

4.2.1 pH measurements

Figure 4a shows evolution of pH in the case where alkaline additives are mixed with reactive tailings.

Generally, a rapid stabilization of pH values is noted except for the addition of 10% RMB. Without alkaline additive, pH values of RT vary between 2 and 3 over time. With addition of 10 and 50% CKD, pH increase above 7 and stay stabilized near 7 after 24 L of deionized water. The addition of 50% CKD does not increase the pH at value higher than that

obtained with 10% CKD. RT sample mixed with 10% RMB shows an initial pH value of 4.5 which increases thereafter to 6 for finally going down again. However, addition of 50% RMB to RT shows a pH which is stabilized at value around 8 for the entire period of experiment. The RT mixed with 10% (CKD:RMB, 1:1) follows the same evolution than sample with addition of 10 and 50% CKD. The pH of sample with 50% (CKD:RMB, 1:1) stabilized at value around 9.

For economic reasons, the optimal choice for the third serie of tests will be the use of 10% alkaline material. Addition of 10% alkaline material seems to be sufficient and effective to assure neutral pH conditions over a long period of time.

It is interesting to note that the average yearly precipitation of the Abitibi area (Quebec) represents 860 mm/year. A volume of 25 L of water passing through the columns corresponds to 5.3 years of precipitation. This value does not take into account the run off and the evapotranspiration occurring on site. We thus think that the column tests represent one period of time much longer than 5 years considering the real conditions of exposure and the true nature of the materials (permeability...).

4.2.2 Chemical composition of the leachates

Concentrations of Ca, Fe, and SO_4 are presented in figures 4b, 4c, and 4d respectively. The concentration of these tree elements is already discused for the RT sample in section 4.1.2.

With addition of 10% CKD, the fast dissolution of lime and portlandite minerals involves 700 mg/L of Ca in solution which increase after two flushes to 1 800 mg/L. After the dissolution of the readily soluble phases, the rate of calcium is decreased and it is controlled by the dissolution of calcite maintaining a neutral pH. The conditions of neutralization involve the precipitation of secondary minerals such as hydroxides which participate themself in the neutralization and in the preservation of the alkalinity.

The RT sample mixed with 50% CKD generates 500 mg/L of Ca after the first flush and 780 mg/L after the third one. Then Ca concentrations decrease to less than 40 mg/L. In spite of the fact that mass of CKD added to the RT is five times as important, Ca values remain lower with regard to those with 10% CKD until the end of the test. The addition of 50% CKD induces a slower rate of dissolution of calcite than with 10% CKD. At the beginning of the test, the mixtures with 10% and 50% CKD present alkalinity values of 100 and 120 mg $CaCO_3$/L respectively. This last value is not as high as expected given the high mass of CKD used.

The concentration of Ca in leachates obtained from samples with 10 % and 50% RMB results only from the calcite coming from the alteration of plagioclase feldspars present in the reactive tailings.

Any difference is observed for the values of Ca concentrations between the addition of 10 and 50% (CKD+RMB) to RT. The concentration of Ca increases from 500 mg/L to 800 mg/L and then decreases to value lower than 30 mg/L after a volume of water of 5L. The leaching of Ca in those samples is mainly due to the contribution of CKD.

Generally, the concentration of the oxidation products (Fe, SO_4) in samples with alkaline additives decreases progressively along the test. Iron is particulary reduced with regard to the amount generated by RT. The addition of alkaline material decreases iron values below 1 mg/L except for the mixture with 10 % RMB where the concentration is reduced to 66 mg/L.

The mixtures with 10 and 50 % RMB show a sulfate concentration superior to 3000 mg/L which is as high as the value obtained with the RT sample after the first flush. Sulfate concentration decreases to 500 mg/L as of the second flush. Sulfates which are present in the RT are almost completely solubilized during the first lixiviation.

On the other hand, the use of 10 and 50 % CKD allowed to reduce the sulfate concentrations of half after the first flush if compared with the sample containing only RT in spite of the presence of arcanite in the CKD. The sulfate reduction is more progressive with these samples in which the gypsum and the ettringite can be formed and control the sulfate concentrations in solution. This phenomenon is slightly eased with the use of the mixture of CKD and RMB.

Evolution of Al, K, Mg, Na, Si, Cu, and Zn are not illustred in this paper. Aluminium, copper, and zinc are observed in measurable concentrations in the column with reactive tailings. With the adding of alkaline materials, Al, Cu, and Zn concentrations decrease under detection limits except for the column with 10% RMB. With 10 and 50% CKD, concentrations of K increase by 500 and 1000 times respectively after the first flush of water compared to reactive tailings. After the second flush, concentrations of K are only 7 times higher. Arcanite is a mineral phase highly soluble in water and generates potassium ions essentially during the first flush of water. Samples with 10 and 50% RMB show high initial Na concentrations coming from the residual sodium hydroxide used in the Bayer process. After the second flush, Na concentrations in theses samples are 10 and 50 times the concentration found in the reactive tailing sample. The alkaline material addition presents any change concerning the magnesium concentrations.

4.3 Third series of columns (alkaline additives used to prevent AMD from waste rock)

4.3.1 pH measurements

The pH values of the third series of columns are showed in figure 5a. The column of reference which only consists of waste rock (WR) shows a pH between 2 and 3 for the duration of the test. The same range of pH was obtained for the column with reactive tailings overlaying waste rock (RT/WR) and that with a CKD layer overlaying reactive tailings, then waste rock without homogenization (10%CKD/RT/WR). It should be recalled that the purpose of the study was to determine the geochemical behavior of the layers of fine materials amended with basic materials and that in order to accelerate the reactions, the fine materials were mixed with nonreactive sand in order to increase the permeability. Therefore here, no layer acts like a capillary barrier able to remain at a high degree of saturation in order to prevent oxidation. The addition of sand makes it possible to drain the layers in order to evaluate their geochemical behavior.

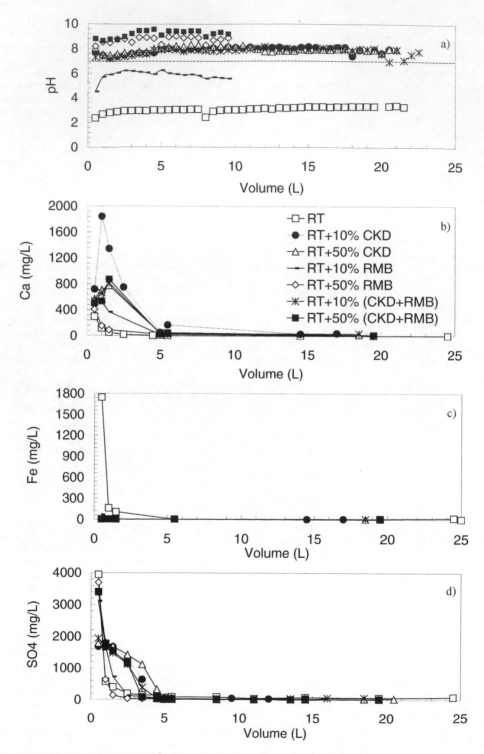

Figure 4. Evolution curves of pH, Ca, Fe, SO_4^{2-} of the second test.

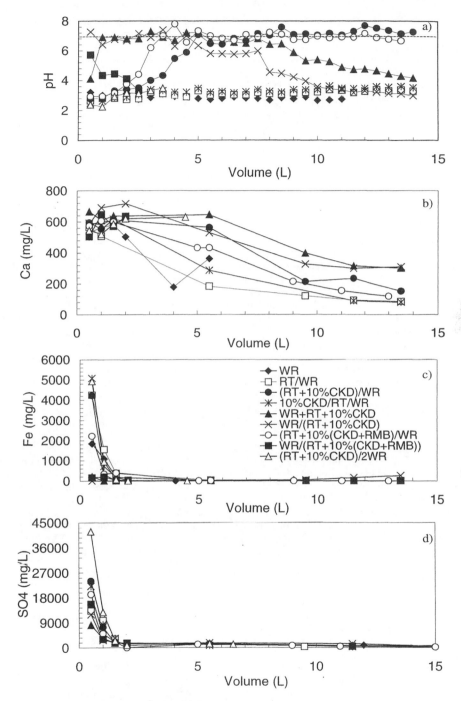

Figure 5. Evolution curves of pH, Ca, Fe, SO_4^{2-} of the third test.

The co-disposal of waste rock with reactive tailings (WR+RT+10%CKD) gives an acidic pH after the first flush followed by an increase to near pH 7 after the second flush. The pH of this sample stabilized around this value until a volume of 8 L of water passed through the column. From that moment, the pH started to go down towards more acidic values.

The evolution of the pH of the layer of RT+ 10%CKD is completely different according to its disposition. When this layer is located over the waste rock like a cover (RT+10%CKD/WR), the pH raises appreciably to 7 after the passage of a volume of water of 5 L and stabilized around this value afterward. The elimination of the products of oxidation accumulated in waste rock causes this late increase of the pH. For the case where the layer is located under the waste rock, like a liner (WR/RT+ 10%CKD), the pH evolves between 6 and 7 at the beginning of the test and then decreases to reach acidic values after a volume of water of 5 L. The products of oxidation accumulated in waste rock are directly neutralized by the alkaline layer. These results suggest that it would be interesting to have these two types of layers simultaneously in order to obtain a neutralization in the short run due mainly to the liner and to provide a longer-term neutralization starting from the cover.

When the fine material layer composed of a mixture of CKD and RMB is located over the waste rock (RT+10%(CKD+RMB)/WR), the pH value raises faster to pH 7 than the use of 10%CKD only. Drainage pH remains close to 7 over time. This result shows a better short term neutralization for this mixture due to the caustic alkalinity readily available from the RMB at the beginning of the test. When this layer of fine materials is situated below the waste rock (WR/RT+10%(CKD+RMB)), the initial pH is near 6 and decreases rapidly to acidic values.

According to the first results, the pH of the column with a double mass of waste rock (RT+10%CKD/2 WR) tends to go up. It seems to have a delay in the increase of pH compared to the similar column with 2 times less waste rock (RT+10%CKD/WR).

4.3.2 Chemical composition of the leachates

Concentrations of Ca, Fe, and SO_4 are showed in figure 5b, 5c, and 5d respectively. Generally, concentrations of Ca for all columns were comprised between 500 mg/L and 700 mg/L after the first flush of water. Like discussed in section 4.1.2, Ca present in leachates from the columns with WR (529 mg/L) and RT/WR (563 mg/L) comes from calcite which is a secondary product of the plagioclase feldspars alteration. Calcium comes also from gypsum stocked in tailings. Afterward, calcium concentrations from WR and RT/WR rapidly decrease.

Leachates from columns containing CKD present the highest calcium concentrations measured after the first flush of water. It is also noticed that the Ca concentrations decrease much less quickly with subsequent flushes in the presence of alkaline materials mainly of CKD. The column with 10%CKD/RT/WR does not followed this general trend and seems to evolve as WR and RT/WR. Indeed, the Ca concentration of this column is rapidly depleted after only 3 flushes.

The column representing the co-disposal of tailings and waste rock (WR+RT+10%CKD) presents a high Ca concentration (665 mg/L) after the first flush. The Ca content results from the dissolution of calcium carbonates, calcium oxides, and calcium hydroxides present in the CKD and from the calcite present in the tailings and mine waste. It is noticed that this Ca concentration remains relatively stable during the passage of 5.5 L of water through the column and that during that time near neutral pH conditions are maintained. In general, the pH begins to decrease when the Ca concentration in solution falls below 500 mg/L.

When the fine material layer with CKD is placed under the waste rock like a liner (RT+10%CKD), high Ca concentrations are reached for at least the 4 first flushes. Ca is available in solution compared with the same layer placed over the waste rock (RT+10%CKD/WR) where precipitation of gypsum and ettringite can occur.

When RMB are included in the fine material layer (RT+10%(CKD+RMB)/WR), Ca concentrations in solution are generally weaker than that measured with only CKD due to a dilution effect considering that no calcium is available from RMB.

Generally, concentrations of iron decrease with time but not in the same way for all columns. Reactive tailings (RT) data presented in the second series of columns showed iron concentration near 1800 mg/L after the first flush of water. About the same concentration was leached from the waste rock (WR). The columns showing the cover up of waste rock by reactive tailings (RT/WR) presents one of the highest iron concentration with value reaching 4200 mg/L. Among the other samples presenting very high iron values we find the case showing a CKD layer overlaying reactive tailings and then waste rock without homogenization (10%CKD/RT/WR), the column with a double mass of waste rock (RT+ 10%CKD)/2WR), and the column with mixture of reactive tailings with 10%CKD overlying waste rock (RT+10%CKD/WR).

In short, we notice that after the first flush, all the columns whose base consists of waste rock present the highest iron concentrations. This first flush leaches iron already present and accessible from corrosion processes in the waste rock. The subsequent flushes show iron concentrations much weaker for the columns containing alkaline additions. For example, the sample RT+10%CKD/WR presents iron concentration in solution of 310 mg/L after the second flush, while it is still at 1550 mg/L for the sample RT/WR.

The adding of 10%CKD considerably decreases iron concentration by the precipitation of hydroxides which control iron solubility. Column with WR/RT+10%CKD shows its iron concentration considerably reduced to 0 mg/L after the first flush. This demonstrates the high capacity of CKD addition to suppress iron from solution. The sample representing the co-disposal of waste rock and tailings (WR+RT+10%CKD) reduces considerably the concentration of Fe (210 mg/L). This sample once again shows the good made of the addition of CKD in order to reduce the iron concentrations.

With the presence of RMB in the superior layer (RT+10%(CKD+RMB)/WR), a decrease of the iron concentration is observed when compared to sample without basic additive. When this layer is set under the waste rock (WR/(RT+10%(CKD+RMB)), the iron concentration falls near 0 mg/L value.

Sulfate concentrations are initially high for all columns after the first flush of water and rapidly decrease. Waste rock (WR) and the cover up of waste rock by reactive tailings (RT/WR) present the same sulfate concentration of 15 000 mg/L.

The co-disposal sample (WR+RT+10%CKD) reduces the first concentration of SO_4 to 8 140 mg/L.

The addition of alkaline material (10% CKD or (CKD+RMB)) in the overlay involves a higher sulfate concentration after the first flush compared to concentrations obtained only by acidic materials. This can be explained by the fact that in these samples, the sulfates come from the dissolution of accumulated products of corrosion and at the same time of the dissolution the arcanite present in the CKD. After the second flush, sulfate concentrations for the samples with alkaline materials decrease to values lower than those obtained for the acidic materials due to the precipitation of gypsum and/or ettringite. The sample with the CKD layer overlaying acidic materials without homogenization (10%CKD/RT/WR) presents also a higher sulfate concentration at the beginning of the experiment but the reduction is much slower as if the reactions were limited to the interface between the CKD and the reactive tailings. The sample with the double mass of waste rock (RT+10%CKD/2WR) shows high sulfate concentration which is 1.6 times that obtained with the sample with a single mass (RT+10%CKD/WR).

When the layer of alkaline materials is disposed under the waste rock as in samples (WR/RT+10%CKD) or (WR/RT+10%(CKD+RMB)), sulfate concentrations are similar to those obtained for the acidic materials. However, in this case, the dissolution of accumulated SO_4 products from waste rock is followed by the secondary precipitation of sulfate containing phases in the alkaline layer. The sulfate concentrations measured come mainly from the dissolution of arcanite.

According to the first results, the columns having the alkaline layer under the waste rock (WR/RT+10%CKD and WR/RT+10%(CKD+RMB)) reduce Al, Cu, and Zn concentrations compared to the columns without alkaline material. The same trend is observed with the co-disposal sample (WR+RT+10%CKD). Magnesium concentrations seem not influenced by the presence of alkaline materials. As for the results presented in section 4.2.2, alkali ions (Na and K) concentrations are increased in the presence of alkaline materials during the first 2 flushes due to the arcanite from the CKD and the residual sodium hydroxide from the RMB.

In short, the use of alkaline materials seems beneficial in order to maintain neutral pH conditions and to reduce the concentrations of iron and other metals such as Zn. The use of a liner composed of RT with 10%CKD under the waste rock gives immediately a neutral pH and reduces the concentrations of metal elements such as Fe, Al, Cu, and Zn in the leachate. To maintain these conditions in a long term basis, a cover composed of the same materials (RT with 10%CKD) should be placed over the waste rock.

The co-disposal of waste rock, tailings, and CKD (WR+RT+10%CKD) allows also to ensure a neutral pH and the reduction of metallic elements in solution. However, its effect seems to last less longer for reason difficult to explain.

The addition of RMB to CKD seems beneficial in order to reach more rapidly an increase in the pH conditions.

5 CONCLUSIONS

Leaching column tests have been carried out in order to evaluate the geochemical behavior of CKD and RMB as alkaline addition on leachate compositions when set up with acid producing tailings and waste rock. In order to accelerate the reactions, the fine materials were mixed with nonreactive sand to increase the permeability. Here, we did not take account of the capillary properties of materials. Fortin (2000) shown that the mixtures of fine materials used in this study have the qualities required in order to stay at a high degree of saturation, therefore to act like a capillary barrier in order to prevent pyrite oxydation.

The measurement of the pH and the chemical analysis of the leachates involve the following conclusions.

The addition of 10% CKD and 10% (CKD:RMB, 1:1) at reactive tailings is sufficient and effective to assure neutral pH conditions over more than 5.3 years of precipitation. The use of 10% RMB alone does not involve the neutralization of reactive tailings.

In order to reduce AMD from waste rock, the use of a liner composed of a mixture of reactive tailing with 10% alkaline materials (CKD or mixture of CKD and RMB) as a fine materials simultaneously with a cover layer of the same composition is advised. This disposition would make it possible to obtain a neutralization in the short run due mainly to the liner and to provide a longer-term neutralization starting from the cover. The mixture of CKD and RMB provides a better short term neutralization than the use of CKD alone.

6 ACKNOWLEDGEMENTS

We would like to extend our appreciation to M.Choquette, J.Frenette from Université Laval for the mineralogical work on Scanning electron microscop and X-ray diffraction. We would also like to extend our appreciation to M.Plante from Université Laval and L.Truchi from Université Sophia-Antipolis, France, for their part of the chemical analysis.

REFERENCES

Barrett, T.J., Cattalani, S., Chartrand, F. and Jones, P., 1991b. Massive sulfide deposits of the Noranda area, Quebec. II. The Aldermac Mine. Canadian Journal of Earth Sciences, v. 28, p. 1301-1327.

Brady, K., Smith, M.W., Beam, R.L., and Cravotta, C.A. 1990. Effectiveness of the use of alkaline materials at surface coal mines in preventing or abating acid mine drainage: Part 2. Mine site case studies. p. 227-241. In: *Proceedings, 1990 Mining and Reclamation Conference, April 23-26, 1990, West Virginia University, Morgantown, WV.*

Burnett, J.M., Bumett, M., Ziemkiewicz, P., and Black, D.C. 1995. Pneumatic backfilling of coal combustion residues in underground mines. In: *Proceedings, Sixteenth Annual Surface Mine Drainage Task Force Symposium, April 4-5, 1995, Morgantown, WV.*

Doye, I 1999. Caractérisation de résidus industriels dans le procédé d'entremêlement par couches pour neutraliser le drainage rocheux acide. M.Sc. École Polytechnique de Lausanne, Suisse.

Douglas, R.J.W. 1970. Geology and economic minerals of Canada. Department of Energy, Mines and Ressources, Canada.

Duchesne, J. and Reardon, E.J. 1998. Determining controls on element concentrations in cement kiln dust leachate, *Waste Management*, 18, 339-350.

Fortin, S., Lamontagne, A., Poulin, R., Tassé, N. 2000. The use of basic additives to tailings in layered co-mingling to improve AMD control. *Environmental Issues and Management of Waste in Energy and Mineral Production.* 549-556.

Gélinas, P., Lefebvre, R., Choquette, M., Isabel, D., Locat, J., Guay, R. 1994. Monitoring and modeling of acid mine drainage from waste rocks dumps. La mine Doyon case study. Final report presented to MEND Prediction committee. Rapport GREGI 1994-12.

Lawrence, R.W. 1990. Prediction of the behaviour of mining and processing wastes in the environment. *In Proc. Western Regional Symposium on Mining and Mineral Processing Wastes*, F. Doyle (ed). Soc. For Mining, Metallurgy, and Exploration, Inc., Littleton, CO.115-121.

MEF, 1995. Parc à résidus miniers Aldermac. Rapport de caractérisation par Consor Inc.. Ministère de l'environnement et de la faune. Jan 1995. 81 p.

MENVIQ 1989. Directives 019, Industries minières. 55p.

Perry, E.F. and Bradly, K.B. 1995. Influence of neutralization potential on surface mine drainage quality in Pennsylvania. *In: Proceedings, Sixteenth Annual Surface Mine Drainage Task Force Symposium, April 4-5, 1995, Morgantown, WV.*

Rich, D.H. and Hutchinson, K.R. 1990. Neutralization and stabilization of combined refuse using lime kiln dust at High Power Mountain. p. 55-60. In: *Proceedings, 1990 Mining and Reclamation Conference, April 23-26, 1990, West Virginia University, Morgantown, WV.*

Rose, A.W., Phelps, L.B., Parizek, R.R., Evans, D.R. 1995. Effectiveness of lime kiln flue dust in preventing acid mine drainage at the Kauffman surface coal mine, Clearfield County, Pennsylvania. p. 159-171. In: *Proceedings, Twelfth American Society for Surface Mining and Reclamation Conference, June 3-8, 1995, Gillette, WY.*

Savoie, A., Trudel, P., Sauvé, P., Hoy, L., Kheang, L. 1991. Géologie de la mine Doyon (région Cadillac). Rapport ET 90-05, Ministère de l'énergie et des ressources naturelles du Québec.

Skousen, J., Rose, A., Geidel G., Foreman, J., Evans R., Hellier, W. 1998. Handbook of technologies for avoidance and remediation of acid mine drainage. Morgantown, WV, USA: The National Mine Land Reclamation Center.

Wiram, V.P. and Naumann, H.E. 1995. Alkaline additions to the backfill: A key mining/reclamation component to acid mine drainage prevention. In: *Sixteenth Annual Surface Mine Drainage Task Force Symposium, April 4-5, 1995, Morgantown, WV.*

Tailings and Mine Waste '02, © 2002 Swets & Zeitlinger, ISBN 90 5809 353 0

Spatial surface flux boundary model for tailings impoundments

Maritz E. Rykaart & Del G. Fredlund
Department of Civil Engineering, University of Saskatchewan, Saskatoon, Saskatchewan, Canada

G. Ward Wilson
Department of Mining and Mineral Process Engineering, University of British Columbia, Vancouver, British Columbia, Canada

ABSTRACT: Rehabilitation of tailings impoundments is one of the most challenging aspects in mine closure, as not only does the potential for producing leachate pose a challenge to the rehabilitation designer, but also other aspects such as stability and settlement must be considered. The water balance of a tailings impoundment is unique in the sense that it usually hosts a pond that in turn causes a phreatic surface in the impoundment. The position of the phreatic surface defines the saturated and unsaturated zones in the impoundment, which of course varies spatially and temporally. Predictive modeling for this hydrologic system becomes difficult, as numerical models capable of analyzing the combined saturated/unsaturated zones are not adequately refined to accurately solve the flux boundary problem for infiltration at the surface of the tailings. This paper describes the development of a flux boundary model that enabled accurate modeling of the unsaturated zone in the tailings impoundment at Kidston Gold Mine, Queensland, Australia. The technique made it possible to accurately predict the spatial variation of infiltration to the tailings as a result of the presence of the phreatic table.

1 INTRODUCTION

Kidston Gold Mine is located 360 km southwest of Cairns in north Queensland, Australia. Mining operations commenced in 1984 and by December 1999, 3.2 million ounces of gold had been recovered from the Eldridge and Wises Hill pits. Mining is scheduled to cease early 2002. The climate is tropical with a distinct dry and wet season (November to April) when 80% of the average 702 mm precipitation falls. Annual potential evaporation averages 1651 mm (Rykaart 2001), resulting in a net negative climatic water balance.

1.1 *Kidston tailings impoundment description*

Tailings (68 Mt) was stored in an engineered tailings impoundment (310 ha surface area) between 1984 through to October 1997. Prior to 1991 the deposition technique was long-beach deposition at the back end of the impoundment with decant towers along the upstream edge of the main embankment. In 1991 the strategy was changed to one of sub-aerial deposition around the periphery of the impoundment with the final decant tower located in the back south-east portion of the impoundment where it is adjoined to a local hill, Paddy's Knob as illustrated in Figure 1 (Currey 1998).

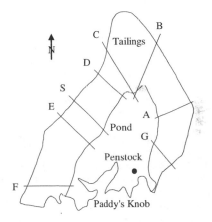

Figure 1. Simplified plan view of the Kidston tailings impoundment, showing the penstock location after 1991, and the piezometer section lines.

The 5.8 km long embankment wall, which encircles about 70% of the perimeter of the tailings impoundment is constructed from waste rock, and was progressively raised in five lifts (3 downstream and 2 centreline). The embankment height varies between 32 m at its highest point at the northern end, to less than 1 m in the south.

Seepage from the embankment (due to the presence of chimney & blanket drains) is collected via 2-3 m deep interception trenches cut down through a decomposed granitic layer. Seepage from the base of the impoundment is limited by the impervious nature of the underlying Einasleigh Metamorphics. Tests confirmed an average base permeability of 9.6 x 10^{-9} m/s (Gutteridge Haskins & Davey 1984, Rykaart 2001).

After October 1997 tailings deposition was changed to thickened tailings deposited in the mined out Wises Hill pit, and rehabilitation of the tailings impoundment started.

1.2 *Kidston tailings impoundment rehabilitation philosophy*

Kidston Gold Mine opted not to rehabilitate the tailings impoundment using the Australian Environment Protection Agency (EPA) best practice guidelines, which entails placing a soil cover to limit oxygen and water infiltration and act as a growth medium (EPA 1995). The closure plan called for direct vegetation of the tailings surface with native grass and tree species. Although uncommon, this method of rehabilitation is often used in South Africa (Blight 1989), and even recently in New Zealand (Mason et al. 1995). The philosophy behind Kidston's rehabilitation approach is based on economical and a couple of technical grounds. Firstly, should the vegetation be found to establish on the tailings surface, it would render an imported soil growth medium redundant. Secondly, the increased evapotranspiration rates are expected to result in a lowering of the tailings dam phreatic level, and ultimately in a reduction in the seepage rate from the impoundment.

2 PROBLEM STATEMENT

Kidston Gold Mine needed to prove that the closure alternative of direct vegetation to the tailings would not pose any environmental risk. To this effect Kidston initiated comprehensive research programs (Ritchie & Currey 2000), of which the understanding of the unsaturated zone water balance of the tailings impoundment is but one component. The following section describes the principles, which led to the development of the spatial surface flux boundary functions described in this paper.

2.1 *The spatial flux hypothesis*

A large saturated zone exists in the tailings impoundment due to the presence of the pool. The established phreatic surface has a shape that is governed by the tailings properties and the exit location is determined by the presence of drains in the embankment walls. If one thus considers a typical cross-section at any location through the tailings

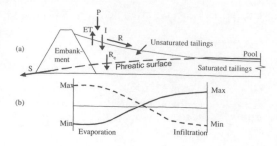

Figure 2(a). Typical cross-section through a tailings Impoundment, (b) Spatial distribution of surface fluxes of infiltration and evaporation.

dam there would be a unsaturated zone of tailings that varies in thickness from the embankment end to the pool end as illustrated in Figure 2(a).

The top tailings impoundment surface (beach profile) along this typical cross-section would be subject to all the usual water balance components of precipitation (P), evapotranspiration (ET), infiltration (I), runoff (R), recharge (Re), and seepage (S). It could however be expected that there would be a spatial variation in the magnitude of these components as we move between the embankment and the pool. The reason for this is the availability of moisture in the profile, which is governed by the presence of the phreatic level (Staley 1957, Blight 1997). Therefore, at a point close to the embankment we would expect evaporation to be a minimum, and as we move towards the pool the evaporation should increase until it reaches a maximum (potential evaporation) at the pool edge. Similarly we would expect infiltration to be a maximum close to the embankment, decreasing towards the pool. This is illustrated graphically in Figure 2(b). The present study illustrates an attempt to show what the spatial distribution of the surface fluxes are, thus paving the way for more accurate multidimensional unsaturated/ saturated flow water balance modeling.

3 SPATIAL SURFACE FLUX CALCULATION METHODOLOGY

Estimating tailings impoundment water balances has always been an important issue, be it for operational or closure water management. Most of the saturated zone water balance components are relatively well understood and can be estimated or measured with relative ease and with a high degree of confidence. However, the same cannot be said for the surface flux components above the unsaturated zone. The measurement of these fluxes is difficult, expensive and time consuming, and as a result engineers look towards numerical modeling to provide the answers. Important advances have been made in this regard, with the development of software such as SoilCover (SoilCover 1997), HELP (Schroeder et al. 1994), UNSAT-H (Fayer & Jones 1990), SWACROP (Feddes et al. 1984), HYDRUS (Simunek et al. 1998), and SWIM (Ross 1990), to name but the few most well known.

These models all attempt to calculate the surface flux components using numerous methods and assumptions. The most important single component is accurate calculation of evaporation. Numerous empirical and semi-empirical methods for calculating actual evaporation are available (Monteith 1965, Shuttleworth & Wallace 1985, Choudhury & Monteith 1988, Passerat De Silans et al. 1989). One mechanistic method for calculating actual evaporation is the modified Penman formulation as proposed by Wilson et al. (1994). A mechanistic approach was considered essential for this study. The only known model that currently uses the modified Penman formulation is SoilCover, and that makes it an appropriate tool to use.

Due to the detailed field data required for accurate use of a model like SoilCover, and the fact that it is only a one-dimensional (1-D) model, the surface flux boundary conditions are often oversimplified using coarse recharge numbers. It is common practice to solve these water balance problems using multidimensional saturated/unsaturated seepage analysis models like SEEP/W (GEO-SLOPE 1991) and MODFLOW (McDonald & Harbaugh 1998). These models do not allow for the calculation of the surface flux boundary conditions, as they are not coupled soil/atmosphere surface flux boundary models, but require some form of recharge input. To obtain this recharge value the modeler will calibrate towards a known parameter, mostly being a phreatic level, and as such the most suitable recharge value might not represent the surface flux boundary condition appropriately.

The authors suggest that by using the proposed conceptual model to determine the spatial surface flux boundary condition distribution, the tailings impoundment water balance can be accurately calculated based on actual calculated, not inferred infiltration rates. The spatial flux boundary functions are calculated using a conceptual model, and can be used directly as a surface flux boundary condition in multidimensional saturated/unsaturated seepage analysis models, effectively bridging the current gap between the two modeling systems.

4 CONCEPTUAL MODEL

Modeling the complete unsaturated profile of the tailings impoundment using SoilCover, proved to be difficult due to the temporal, and more important, the great spatial variability of the phreatic level. To overcome this problem a conceptual model of the generalized tailings impoundment cross-section was developed that would act as the basis for all Soil-Cover modeling. The conceptual model defines the geometry (boundaries) of a generalized tailings impoundment cross-section via a set of functions that describe the shape of the tailings impoundment beach profile and phreatic surface.

4.1 Beach shape function

The beach shape function was developed using the principles of particle segregation down a slope. Due to the hydraulic deposition of tailings, particle segregation takes place as the excess water flows to the pool. The coarsest particles settle out first, close to the embankment, while the finer slimes only settle in the pool base. Numerous researchers have studied this aspect, both in natural streams (Morris & Williams 1999) and more specifically on tailings impoundment beaches (Blight 1987).

The Kidston tailings impoundment surface was investigated and seven representative section lines, illustrated on Figure 1, were chosen and accurately surveyed. The slope of the beach profile proved to be on average 1%, and followed the classical exponential shape observed by Blight (1987) in his master profile studies. Figure 3 presents the measured beach slopes along the seven section lines together with the best-fit function to describe the beach shape. For comparison the measured profile of a gold tailings dam in South Africa as reported by Blight (1987) is included.

The function, which follows the principles of the tailings master profile (Blight 1987), is expressed as follows:

$$\frac{h_b}{y_b} = \left(1 - \frac{H_b}{X}\right)^{n_b} \tag{1}$$

where H_b (m) is the distance along the beach profile as measured from the embankment to the edge of the pool, and X is the maximum length of the beach profile (650 m for the generalized tailings impoundment cross-section at Kidston). The height of the beach

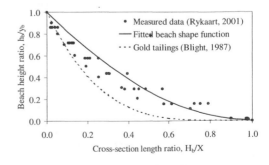

Figure 3. Fitted beach shape function for the measured beach profiles of the Kidston tailings impoundment.

surface above the pool level at any point along the beach profile is given by h_b (m), and the maximum height of the beach surface above the pool level, which is at the embankment is given by y_b (3.4 m for the generalized tailings impoundment cross-section at Kidston). The dimensionless shape constant, $n_b = 1.85$, and is based on the best fit with respect to the measured beach profiles.

4.2 Phreatic line shape function

Developing a function for the shape of the phreatic line was based primarily on observational methods. 42 Piezometers were installed along the seven representative section lines in the dam as illustrated on Figure 1, and the database of measured levels (spanning from 1997 to 2001) (Rykaart 2001) from the installed piezometers were used to develop a calibrated function that supports the shape well. This function is presented by:

$$\frac{h_p}{y_p} = \left(1 - \frac{H_p}{X}\right)^{-n_p} \qquad (2)$$

where X is as defined for Equation 1 and H_p (m) is the distance along the phreatic line profile as measured from the embankment to the edge of the pool. For the generalized Kidston tailings impoundment cross-section $H_p = H_b$. The depth of the phreatic level below the pool level at any point along the beach profile is given by h_p (m), and the maximum phreatic level depth below the pool level, which is at the embankment is given by y_p (10.0 m for the generalized tailings impoundment cross-section at Kidston). The dimensionless shape constant, $n_p = 4$, is based on the best fit with respect to the measured piezometer levels. Figure 4 presents how the shape function fits measured piezometer data from the seven section lines.

4.3 Combined (composite) conceptual model

Combination of the beach shape and phreatic line shape functions makes it possible to know exactly

what the depth to the phreatic surface would be at any point along the generalized tailings impoundment cross-section as illustrated on Figure 5. For the generalized Kidston tailings impoundment cross-section the maximum depth to phreatic level is 13.4 m. In Figure 5, $H = H_b = H_p$, $h = h_b + h_p$, and $y = y_b + y_p$. Since the units are non-dimensionalized this composite cross-section is representative of any location on the tailings impoundment.

5 MODELING METHODOLOGY

The conceptual model describes the generalized tailings dam cross-section. In order to calculate the spatial flux boundary functions this conceptual model must be solved by applying the 1-D coupled soil/atmosphere surface flux boundary model Soil-Cover. The section below describes how this can be accomplished.

5.1 SoilCover calibration

To conduct any modeling with a degree of certainty, model calibration is required. The SoilCover model for the results reported in this paper, was calibrated using the continuous data set of climatic, Bowen ratio, matric suction and phreatic level data measured at a single location on the tailings impoundment.

The data loggers were not fully operational for the entire monitoring period (1996-2001), and as a result three individual portions of uninterrupted data, which spans over a number of months were chosen to do the model calibration against.

Excellent data existed for the physical description of the model in terms of physical and hydraulic tailings properties, as well as tailings hydraulic conductivity. Good correlation's between modeled and measured tailings suctions at two depths (50 mm and 750 mm below surface) were obtained, although only by distributing all rainfall events over a 24-hr period. Using actual measured rainfall intensities

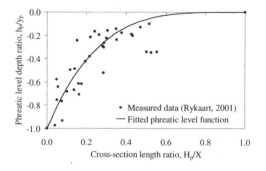

Figure 4. Fitted phreatic line shape function based on measured piezometer levels from the Kidston tailings impoundment.

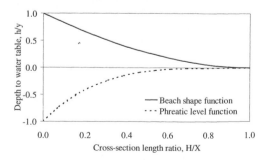

Figure 5. Combined (composite) beach- and phreatic line shape functions for the generalized Kidston tailings impoundment cross-section.

created significant instabilities in the modeling, and as a result was not used. The measured Bowen ratio data further supported the trends observed in the modeling.

The outcome of the calibration modeling was that SoilCover could be used with confidence to calculate the water balance in the upper unsaturated zone of the tailings impoundment (Rykaart 2001).

5.2 Solving the conceptual model using SoilCover

In order to conduct SoilCover modeling on the generalized tailings impoundment cross-section (650 m long) a decision was made to divide the section into 13 zones (each 50 m wide). An individual SoilCover model would then be run for each zone, and by integrating the computed surface flux boundary conditions over the entire tailings cross-section a good estimate of the cumulative result would be obtained. This approach thus allows for a two-dimensional (2-D) solution using the 1-D SoilCover model (Rykaart 2001).

5.3 Tailings properties for SoilCover modeling

Extensive laboratory testing, including 66 grain size distribution- and 25 soil water characteristic curve tests were done on the tailings material. The tailings varies from well graded sands (SW) (Unified Soil Classification System, Holtz & Kovacs (1981)), with an average D_{50} of 0.16 mm, to fine sands (ML), and an average D_{50} of 0.03 mm. The average specific gravity of the tailings is 2.74 (Rykaart, 2001).

The soil water characteristic curves indicated a saturated volumetric water content (θ_s) ranging between 34 and 56%, an Air-Entry Value (AEV) ranging between 1.5 and 12 kPa and a residual suction (ψ_r) ranging between 3.5 and 700 kPa (Rykaart, 2001). The data base of tested tailings properties was used to define three main tailings types for modelling purposes; coarse, intermediate and fine. Using a single averaged material type for the entire tailings impoundment cross-section would not be accurate as measured data had shown that the tailings becomes progressively finer as one moves away from the embankment towards the pool (due to natural particle segregation). The choice of three material types was also based on work done by Kealy & Busch (1971), where they modeled seepage from mill-tailings, and found that three tailings types becoming progressively finer between the embankment and the pool best describe the model.

The three soil water characteristic curves for these tailings types were respectively selected based on the 75-, 50- and 25%-tile values of the saturated volumetric water content, the Air-Entry Value and the residual matric suction and the values are listed in Table 1. The steepness of the curves caused modelling instability and these curves had to be flattened in the high matric suction range. Figure 6 presents

Table 1. Soil water characteristic curve properties for the three selected tailings types used in the SoilCover modeling.

Tailings type	θ_s	AEV	ψ_r
Coarse	38%	2.5 kPa	8.0 kPa
Intermediate	42%	3.2 kPa	10.0 kPa
Fine	44%	6.0 kPa	70.0 kPa

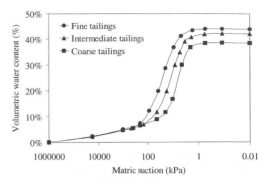

Figure 6. Soil water characteristic curves for the three selected tailings types used in the SoilCover modeling.

the soil water characteristic curves of the tailings as used in the SoilCover modeling.

Where the transitions between these tailings types down the beach profile would be was not firm, but would be determined upon the outcome of the actual modeling. Kealy & Busch (1971) used coarse tailings for the first 79 m of their 152 m long cross-section; 49 m of intermediate tailings and 24 m of fine tailings (i.e. moving from the embankment towards the pool). For the generalized Kidston tailings impoundment cross-section a final tailings zone ratio of 5:5:3 was adopted as a good transition for the tailings types. This means that for the first five of the 13, 50 m wide modeled zones, coarse tailings was used, the next five zones intermediate tailings, and the final three zones were modeled with fine tailings (250 m:250 m:150 m for this 650 m section).

To define the vertical saturated hydraulic conductivity on the tailings beach profile for each of the 13 zones, a theoretical function for saturated hydraulic conductivity was developed. This function was verified using measured field data as illustrated in Figure 7. The data consists of 29 laboratory saturated permeability tests, 62 Guelph permeameter tests, 8 double-ring infiltrometer tests and 14 rainfall simulator tests (Rykaart 2001). The resultant surface permeability function is described by the following expression:

$$k_s = 1.94 \times 10^{-5} \cdot e^{(-0.00977 \cdot H)} \qquad (3)$$

where k_s (m/s), is the vertical saturated hydraulic conductivity on the surface of the generalized Kidston tailings impoundment cross-section at any given point along the beach profile, H (m) is the dis-

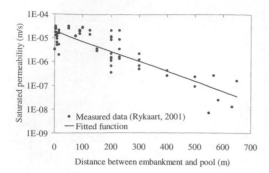

Figure 7. Vertical saturated hydraulic conductivity function for the beach profile of the generalized Kidston tailings impoundment cross-section.

tance along the beach profile as measured from the embankment, and e is the base of the natural log.

Vertical tailings profiles are not homogeneous, and the physical and hydraulic properties of each of the horizontal tailings layers can vary significantly. 27 Horizontal saturated conductivity (k_h) tests were performed on the Kidston tailings (Douglas Partners 1997, Earthtech Consultants 1999), with results varying over three orders of magnitude. This was confirmed through visual observation during trenching. Thin slimes bands encased by otherwise coarse tailings are found throughout all vertical profiles in the tailings Impoundment. These slimes bands are never continuous, as the deltaic depositional method of the tailings prevents this.

The average k_h/k_s ratio for the tailings is not constant, but for the purpose of the SoilCover modeling the ratio was set equal to 1. In surface flux boundary calculations, the surface properties have the greatest influence on the flow regime. This fact led to the decision that for the SoilCover modelling of the generalized Kidston tailings impoundment cross-section, homogeneous vertical layers would be considered. Since the model calibration was carried out using a homogeneous profile with good effect, it is believed to be a reasonable assumption. The random nature of the horizontal layering further supports the use of a homogeneous profile.

5.4 Overall tailings impoundment water balance

The use of the SoilCover model and procedure outlined above still presented a problem in the sense that there was no way to measure whether the integrated (composite) SoilCover solution was in fact correct. To overcome this problem a primary water balance of the entire tailings impoundment was carried out using available physical and climatic data. The water level in the pool was known since 1997, and this data together with survey data was used to calculate a stage curve for the tailings impoundment pool. Complete climatic data was available for the same period, which allowed for the determination of

total rainfall as well as the calculation of potential evaporation from the pond and tailings impoundment. Data for seepage rates from collection drains around the tailing impoundment were used to calculate an overall seepage loss, and together an estimated runoff value into the pond was calculated as being in the order of 42% (Rykaart 2001). This value was used as the guideline against which the SoilCover modeling had to be measured.

5.5 SoilCover climate data

The daily climatic data required for the SoilCover modeling presented in this paper consisted of a year of typical mean data for the Kidston site. This data includes minimum and maximum air temperature, minimum and maximum relative humidity, net radiation, windspeed, and rainfall. The typical data set was made up using continuous on-site weather station data from 1996, historic daily data from 1983, as well as regional records dating back 50 years (Rykaart 2001). The total annual precipitation used was 702 mm.

6 SPATIAL FLUX BOUNDARY FUNCTIONS

Solving of the 13 individual SoilCover simulations, using each of the chosen tailing types, as well as the final optimal composite solution, each presents different spatial flux boundary functions for the surface flux components. Figure 8 presents the dimensionless evaporation ratio (AE_r), which is defined as:

$$AE_r = \frac{AE_z}{AE_{max}} \qquad (4)$$

where AE_z = the individual zonal evaporation (mm), and AE_{max} = maximum individual zonal evaporation (mm). Essentially the evaporation ratio is equal to the actual evaporation (AE)/potential evaporation (PE) ratio (AE/PE), with AE_{max} = PE. From Figure 8 it is evident that the evaporation ratio is the least

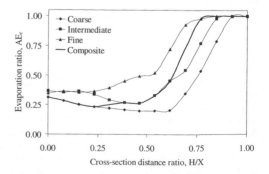

Figure 8. Spatial evaporation flux boundary functions for the generalized Kidston tailings impoundment cross-section.

close to the embankment and gradually increases to a value of 1 (where actual evaporation equals potential evaporation), near the edge of the pool. This is consistent with the proposed spatial flux hypothesis illustrated in Figure 2(b).

The overall evaporation ratio for cross-section comprising of only fine tailings exceeds that of the cross-section comprising only coarse tailings, and that can be explained by the increased capacity of the fine tailings to retain moisture, as well as the increased capillary suction.

It is interesting to note that the spatial evaporation ratio distribution is not linear, but remain almost constant in each case before rapidly increasing from a H/X ratio equal to ≈0.6 (390 m) for the fine tailings case profile, to a H/X ratio equal to ≈0.72 (468 m) for the coarse tailings case profile. The reason for this is the rapidly decreasing depth to the phreatic surface as a result of the two exponential shape functions illustrated in Figure 5. The coarse tailings evaporation ratio remain constant for longer due to the inability to replenish water from the water table, due to the lower tailings Air-Entry Value.

The net infiltration ratio (NI_r) is used to present the net infiltration (NI) data on the same basis as the evaporation ratio. The net infiltration ratio presented in Figure 9 is defined as:

$$NI_r = \frac{1 - (NI_z - NI_{max})}{-(|NI_{max}| + |NI_{min}|)} \qquad (5)$$

where NI_z = the individual zonal net infiltration for each of the 13 modelled zones (mm), NI_{max} = maximum individual zonal net infiltration (mm), and NI_{min} = minimum individual zonal net infiltration (mm). The net infiltration (NI) is defined as:

$$NI = P - R - ET \qquad (6)$$

where P = precipitation, R = runoff, and ET = evapotranspiration (or just evaporation if no vegetation is present).

The trends in Figure 9 again follow the spatial flux hypothesis presented in Figure 2(b), with the maximum net infiltration occurring at the embankment end, and the least happening at the pool end of the tailing impoundment. There is little difference between the coarse and intermediate tailings case profile curves, both showing a steep drop in infiltration from a H/X ratio equal to ≈0.75 (488 m). The fine tailings case profile indicate an overall lower net infiltration ratio, and drops steeply from a H/X ratio equal to ≈0.6 (390 m).

Combination of the spatial evaporation flux boundary function, and the net infiltration flux boundary function as presented in Figure 10, shows how these two spatial flux boundary functions compliment each other, and allow for a direct comparison with the hypothetical graph depicted in Figure 2(b).

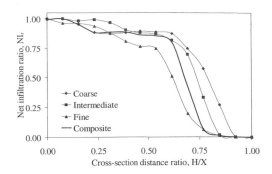

Figure 9. Spatial net infiltration flux boundary functions for the generalized Kidston tailings impoundment cross-section.

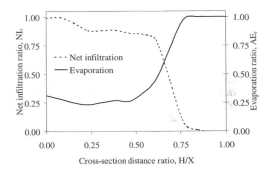

Figure 10. Combined spatial evaporation- and net infiltration flux boundary functions for the generalized Kidston tailings impoundment cross-section.

Table 2. Overall surface flux water balance based on the generalized Kidston tailings impoundment cross-section.

Tailings type	Runoff	Evaporation	Net infiltration
Coarse	51%	86%	-37%
Intermediate	32%	114%	-45%
Fine	20%	142%	-62%
Combined	40%	114%	-55%

The unsaturated zone water balance results for all the modelled cases are summarised in Table 2. All the data has been expressed as percentage of the total precipitation.

The negative net infiltration indicates a net negative water balance from the system, which in this case would imply the lowering of the phreatic level. It is important to note that the modeled cross-section reported on here is not vegetated, and as a result transpiration is not accounted for.

The spatial net infiltration flux boundary function for the generalized Kidston tailings impoundment cross-section presented here, can be used as the surface flux boundary condition in multidimensional saturated/unsaturated seepage analysis models such as SEEP/W or MODFLOW in the form of a net infiltration flux, q (mm/d):

289

$$q = \frac{NI}{t} \qquad (7)$$

where t = any selected time period (days for the data presented in this paper). Figure 11 present a step function showing how the monthly breakdown of the net infiltration flux data shown here can be used as a surface flux boundary condition in multidimensional seepage analysis models. The idea is that such a flux boundary function would be produced for each of the 13 zones, and these flux boundary functions would be the surface flux boundary conditions in multidimensional saturated/unsaturated predictive modeling. The Kidston tailing impoundment surface can thus be divided into 13 iso-zones (zones of equal net infiltration flux).

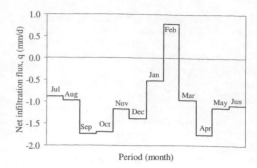

Figure 11. Monthly distribution of net infiltration fluxes aggregated over the generalized Kidston tailings impoundment cross-section.

7 SPATIAL FLUX BOUNDARY FUNCTION EVALUATION

In order to evaluate the validity of using the presented spatial flux boundary functions as surface flux boundary conditions in multidimensional saturated/unsaturated seepage analysis models, a comprehensive evaluation test was conducted. Essentially the aim was to develop a full 3-D model of the Kidston tailings impoundment and use the calculated spatial flux boundary function for each of the 13 zones as surface flux boundary conditions in the 3-D model. The modeled seepage rate from the tailings impoundment after a known period of time was then compared to actual field measured seepage rates from the Kidston site, and an excellent match was observed (Rykaart 2001). This evaluation confirmed that the flux boundary functions as presented in this paper do provide a suitable alternative.

8 CONCLUSIONS

The authors presented a conceptual model to allow the calculation of accurate unsaturated zone surface flux boundary conditions along a 2-D tailings impoundment cross-section, using the rigorous 1-D coupled soil/atmosphere surface flux boundary numerical model, SoilCover. The conceptual model is based fully on physical characterisation of the problem at hand, and can be applied and modified for any tailings impoundment with a spatially varying phreatic surface.

The real benefit of this work is in bridging the gap between calculating true surface fluxes, for use in multidimensional seepage analysis modeling. The spatial surface flux functions can directly be used as input in 2-D and 3-D saturated/unsaturated flow models, and eliminates the need for recharge guesswork and estimations.

The calculated spatial flux boundary functions confirm the spatial flux hypothesis presented in this paper. By calculating a calibrated spatial flux boundary function for a tailings impoundment, appropriate predictive modeling can be done, by merely manipulating the spatial flux boundary function according to the changed climatic conditions, without having to rerun the SoilCover model.

9 ACKNOWLEDGEMENTS

The authors wish to acknowledge the financial support from Kidston Gold Mines Limited, and their permission to publish the findings of this research. The Kidston Environmental Officer, Mr. Paul Ritchie and his assistant, Mr. George Ryan is thanked for their help in gathering a lot of the field data required in this study.

REFERENCES

Blight, G.E. 1987. The concept of the master profile for tailings dam beaches. *Proceedings of the International Symposium on Prediction and Performance in Geotechnical Engineering, Calgary, Alberta, Canada, 17-19 June 1987*: 361-365.

Blight, G.E. 1989. Erosion losses from the surfaces of gold-tailings dams. *Journal of the South African Institute of Mining & Metallurgy* 89(1): 23-29.

Blight, G.E. 1997. Interactions between the atmosphere and the Earth. *Geotechnique* 47(4): 715-767.

Choudhury, B.J., Monteith, J.L. (1988). A four-layer model for the heat budget of homogeneous land surfaces. *Quarterly Journal of the Royal Meteorology Society*, Vol. 114, pp. 373-398.

Currey, N.A. 1998. Tailings revegetation at Kidston Gold Mines. *Proceedings of the 2nd Annual Summit of Mine Tailings Disposal Systems, Brisbane, Australia, November 1998.*

Douglas Partners 1997. Factual report on tailings dam insitu testing Kidston Gold Mine. *Consultants report to Kidston Gold Mines Limited*, Edited by J. Thrupp & P. Carver. Project no. 21712, July.

Earthtech Consultants 1999. Report on piezocone testing Kidston tailings storage facility. *Consultants report to Australasian Groundwater & Environmental Consultants Pty Ltd*, Edited by C. Cooper & A. Middleton. Report no. MF1367, December.

Environment Protection Agency (EPA) 1995. Rehabilitation and Revegetation. *Module in series on: Best Practice Environmental Management in Mining*. Australian Federal Environment Department, June.

Fayer, M.L. & Jones, T.L. 1990. UNSAT-H version 2: Unsaturated soil water and heat flow model, *PNL-6779*, Pacific Northwest Laboratory, Richland, Washington, USA.

Feddes, R.A., Wesseling, J.G. & Wiebing, R. 1984. Simulation of Transpiration and Yield of Potatoes with the SWACROP-model. *9th Tri-annual Conference of the European Association of Potato Research (EAPR), Interlaken, Switzerland, July 2-6*.

GEO-SLOPE International Ltd. 1991. SEEP/W for finite element seepage analysis, Version 4 for Windows 95 and NT, *Getting Started Guide*, Calgary, Alberta, Canada.

Gutteridge Haskins & Davey 1984. Kidston Project: Tailings Dam Study. *Consultants report to Kidston Gold Mines Limited*. Original report, November 1983, Amended report, July 1984.

Holtz, R.D. & Kovacs, W.D. 1981. *An Introduction to Geotechnical Engineering*. Prentice-Hall, Inc., Englewood Cliffs, New Jersey, USA.

Kealy, C.D. & Busch, R.A. 1971. Determining Seepage Characteristics of Mill-Tailings Dams by the Finite-Element Method. *Report of Investigations 7477*, United States Department of the Interior, Bureau of Mines, January, 51 pp.

Mason, K.A, Gregg, P.E.H. & Stewart, R.B. 1995. Land reclamation trials and practices at Martha Hill Gold Mine, Waihi, New Zealand. *Proceedings of the 1995 PACRIM Congress: Exploring the Rim*, Auckland, New Zealand.

McDonald, M.C. & Harbaugh, A.W. 1988. MODFLOW, a Modular Three-Dimensional Finite Difference Groundwater Flow Model. *US Geological Survey*, Open-File Report 91-536, Denver.

Monteith, J.L. (1965). Evaporation and environment. *In The State and Movement of Water in Living Organisms, Symposium: Society of Experimental Biology*. Vol. 19, Edited by G.E. Fogg. Academic Press, San Diego, California, pp. 205-234.

Morris, P.H. & Williams, D.J. 1999. A Worldwide Correlation for Exponential Bed Particle Size Variation in Subaerial Aqueous Flows. *Earth Surface Processes and Landforms* 24: 835-847.

Passerat De Silans, A, Bruckler, L, Thory, J.L., Vauclin, M. (1989). Numerical modeling of coupled heat and water flows during drying in a stratified bare soil – comparison with field observations. *Journal of Hydrology*, Vol. 105, pp. 109-138.

Ritchie, P.J. & Currey, N.A. 2000. Tailing Dam Rehabilitation at Kidston Gold Mines. *Proceedings of ANCOLD 2000*, Cairns, Queensland, Australia, October 21-27, 2000.

Ross, P.J. 1990. SWIM – a Simulation Model for Soil Water Infiltration and Movement. *CSIRO Division of Soils*, Davies Laboratory, Townsville, Queensland, Australia.

Rykaart, E.M. 2001. *Spatial infiltration to the Kidston tailing dam*. Ph.D. Thesis, University of Saskatchewan, Department of Civil Engineering, Saskatoon, Saskatchewan, Canada.

Schroeder, P.R., Lloyd, C.M. & Zappi, P.A. 1994. *The Hydrological Evaluation of Landfill Performance (HELP) Model User's Guide for Version 3*, EPA/600/R-94/168a, USA.

Shuttleworth, W.J., Wallace, J.S. (1985). Evaporation from sparse crops – an energy combination theory. *Quarterly Journal of the Royal Meteorology Society*. Vol., 111, pp. 839-855.

Simunek, J., Huang, K., & van Genuchten, M.Th. 1998. The HYDRUS code for simulating the one-dimensional movement of water, heat, and multiple solutes in variably-saturated media, Version 6.0. *Research Report No. 144*, USA. Salinity Laboratory, USDA, ARS, Riverside, California, 164 pp.

SoilCover 1997. *SoilCover User's Manual*. Unsaturated Soils Group, Department of Civil Engineering, University of Saskatchewan, Saskatoon, Saskatchewan, Canada.

Staley, R.W. 1957. *Effect of depth of water table on Evaporation from fine sand*. Master of Science Thesis, Colorado State University, Fort Collins, Colorado, USA.

Wilson, G.W., Fredlund, D.G. & Barbour, S.L. 1994. Coupled soil-atmosphere modelling for soil evaporation. *Canadian Geotechnical Journal*. 31: 151-161.

Radioactivity and risk

Tailings and Mine Waste '02, © 2002 Swets & Zeitlinger, ISBN 90 5809 353 0

The benefits of the risk assessment to the mining industry

Tatyana Alexieva
SRK Consulting, Lakewood, Colorado, USA

ABSTRACT: Risk is present in every human endeavor. Business consists of the undertaking of risk for reward. But how can the level of risk be chosen to result in the best return? To answer this question we need to first identify the risks, then assess them, and finally decide how to manage them.

Tailings and mine waste disposal and storage is in many cases the critical component of a mine operation. It is often associated with environmental, human health, loss of life and loss of property risks. Not knowing what the risk is and by not managing it to an acceptable level exposes mine owners and operators to higher liabilities, higher insurance premiums, and a higher potential of a loss.

This paper gives a general overview of a risk management approach applied to tailings and mine waste disposal and storage. The basic components of the risk management process – risk analysis, risk assessment, and risk management, are defined, and a step-by-step procedure suggested. The benefits of adopting the risk management approach are illustrated through a number of case studies.

1 INTRODUCTION

Conducting a risk assessment for mining related projects has become a requirement and a standard practice in an increasing number of countries around the world. Many regulators now consider a mine facility that has not undergone a risk assessment to be a potential hazard and, most worrisome, an unknown hazard. The purpose of the risk analysis and risk assessment is to identify and evaluate the potential hazards and become proactive in managing the risks within acceptable levels.

Applying the risk management approach can be beneficial to the mining company at any stage of the mine development. For example, conducting a risk analysis could give an estimate of the reliability of the mine feasibility study, could identify the risks associated with the design or the operation of a waste disposal facility, or could single out the best option for closure of a mine facility.

SRK has provided a wide range of risk analysis and risk assessment services for many of our clients that facilitated the project optimization process and the decision making. This paper presents a general procedure for conducting a risk analysis, risk assessment and risk management. Some project examples are also included.

2 THE RISK MANAGEMENT PROCESS

Every mine manager knows that a consideration of the risks is incorporated into every step within the mine management process. Is the tailings storage facility stable during start-up, throughout its construction phases and operation, and after closure? Is it likely that it will fail during the design earthquake, and what would be the consequences of a failure? What is the risk of a loss of life, of environmental contamination, of production loss or a property loss? These are some of the questions that the mine managers have to answer prior to making a decision on what needs to be done next and how should the available budget be spent best. And although every mine manager considers the risks, some choose to make a subjective risk evaluation, based on their experience, while others prefer to go through a structured risk management process, which generally provides an objective evaluation of the risks and leads to a defensible decision.

The structured risk management process usually consists of risk analysis, followed by risk assessment and risk management. The stages in the structured risk management process are graphically presented in Figure 1 and described below.

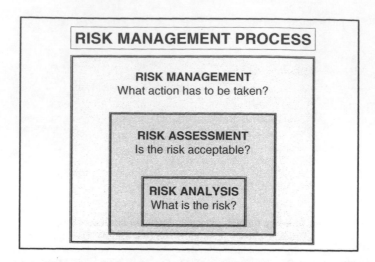

Figure 1. Graphical presentation of the risk management process.

2.1 *Risk analysis*

Risk is the product of the likelihood of a failure and the consequences of a failure. **Risk Analysis** is the process of evaluating the likelihood of undesired events, harm, or loss. The risk analysis consists of hazard identification, frequency analysis, consequences analysis and risk estimate.

2.1.1 *Hazard identification*

A hazard is defined as a set of circumstances that may cause adverse consequences. Hazard identification is carried out by asking the question "What will happen if …?" Examples of hazards are slope failure due to steep slope, contaminated seepage due to lack of seepage control measures, overtopping due to insufficient dam freeboard.

2.1.2 *Frequency/probabilistic analysis*

Frequency analysis is estimating the chance of occurrence of the identified hazard by answering the question "How likely it is…?". It is important to note that there is a difference between the terms "probability" and "frequency". Probability is a dimensionless number between 0 and 1, and is conditional and per event. Frequency is usually given in terms of number of occurrences of the hazard per unit time.

2.1.3 *Consequences analysis*

Consequence is the end result of a hazard, should it occur. For example, the consequences of a breach failure of the embankment of a tailings dam could be loss of life of the downstream inhabitants, property damage, and environmental contamination.

2.1.4 *Risk estimate*

The results of the hazard identification, frequency analysis, and consequence analysis are gathered to estimate the expected risks. Risk estimates can be performed using a variety of different methods. A preferred method by many engineers and scientists is the fault-event (cause-consequence) tree. The fault-event tree is a diagrammatic representation of the risk analysis process that incorporates its logic.

The risk estimate could be qualitative, semi-quantitative, or quantitative. In a qualitative risk estimate the hazards and the consequences are assigned descriptors such as "very likely", "unlikely" and "possible". The descriptors are arranged in terms of severity on a scale, such as 1 to 5. The semi-quantitative risk estimate allows the risks to be estimated in terms of probability while still based on assigned descriptors such as "very likely" and "possible" by having a probability value assigned to each descriptor. In a quantitative risk estimate the probabilities of occurrence of the hazards (basic faults) and the consequences of the hazards are either assigned or calculated using statistics and modeling in terms of probabilities.

The decision what method of risk estimate to use depends on the severity of the problem to be solved, the available information, and the particular project requirements. In general, a qualitative risk estimate provides a basis for comparison between options. The semi-quantitative method usually provides an order of magnitude accuracy of the probabilistic risk estimate and is preferred when insufficient information or time is available to conduct a fully quantitative risk analysis. The quantitative risk analysis is preferred when a comprehensive risk assessment is required by the regulators, when the accuracy of the risk estimate is essential for decision making, or when the risk analysis is used as a tool for conducting a cost-benefit analysis.

The main benefits of the risk analysis are as follows:

- Clearly identifies the main contributors to the risk;
- Facilitates easy sensitivity analysis: identifies the impact of one or a combination of proposed corrective measures on the risk reduction;
- Limits the subjectivity in the design process;
- Allows cost-benefit analysis by comparison between the cost involved in implementing the various alternatives and the corresponding level of risk;
- Assists in predicting the hazards and becoming proactive by developing a risk reduction program and prioritizing the required measures to fit within the budget constraints.

2.2 Risk assessment

Risk assessment is the process of deciding whether the estimated risks are tolerable. A decision on the tolerability of the risks can be made by comparison of the estimated risk with:

- Regulations criteria;
- Generally accepted risk levels;
- Appropriately established and negotiated risk limits in line with the particular conditions.

In a case where no regulations are available and no site-specific criteria are established, a risk of an environmental contamination in the order of 1 in 10,000 or less is generally considered acceptable within the industry. In some cases, especially where clean up or rehabilitation is required, the acceptable environmental risk levels are established with regard to the effect the environment would have on human health.

An example of risk criteria is shown in Figure 2, which presents the societal risk criteria curves prepared by ANCOLD (Australian National Committee on Large Dams). ANCOLD specified the ranges of acceptable, tolerable and intolerable risks as a relation between the probability of a failure and the expected number of fatalities due to the failure.

Figure 3, illustrating the probability of death for an individual per year of exposure is included for comparison and shows typical risks associated with the everyday life. In general, a probability of an involuntary loss of life of 1 in 1 million or less is considered acceptable.

2.3 Risk management

The risk management is an ongoing process of ranking and prioritizing in order to be proactive in controlling and managing the risks throughout the life of the mining facilities. Risk analysis and risk assess-

ment as part of the risk management process should be carried out periodically and particularly when changes such as increasing the design capacity of a TSF, adopting a new deposition method, or changing the embankment construction method, are expected.

The risk management is the decision-making stage that completes the logical thinking process followed in the risk analysis and risk assessment that gives an answer to the question "What is worth doing?" or "What action should be taken?". The main benefits of adopting the risk approach in managing the mine tailings and waste facilities are:

- The risk approach facilitates the optimization of the facility planning by selecting the most cost-effective way of satisfying the project requirements in terms of pre-agreed threshold levels of safety and environmental risk.
- The risk approach helps to identify the liabilities related to the project and to take appropriate measures to reduce the liabilities to an acceptable level. This serves as a reassurance to the shareholders regarding the value of their investment.
- The risk approach provides a justification to the relevant authorities and/or affected parties that the physical stability and environmental safety criteria for the facility are met. In some cases this may include a justification of a contaminant release by showing that the risk imposed to human health and the environment is within acceptable limits, as a "zero release" policy may not be always practically possible when managing a mine waste facility.
- The risk approach may assist in obtaining a reduction of the insurance premiums by demonstrating that the risks associated with the mine tailings or waste facility are known, managed, and controlled.

3 EXAMPLES OF RISK APPROACH APPLICATION

The following examples present some typical applications of the risk approach for managing mining related projects. All examples represent actual case studies performed by SRK. No project and client names are mentioned due to confidentiality issues.

3.1 Risk assessment to facilitate permitting

Following a number of tailings and other mine waste facility failures in the recent years, the mining industry is under an increased pressure from the regulators and the community to ensure the highest environmental standards are implemented in the design, construction and operation of all mine facilities. The regulators in many parts of the world now require

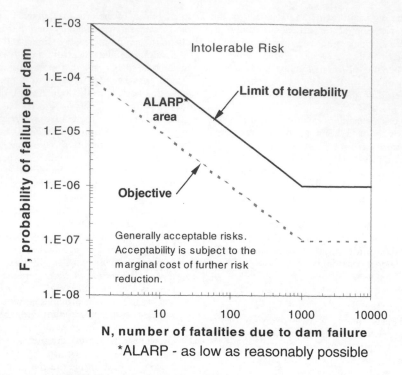

Figure 2. ANCOLD societal risk criteria (after ANCOLD, 1998).

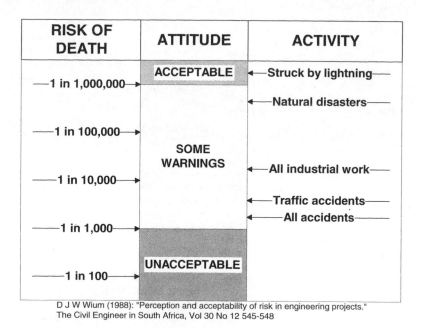

D J W Wium (1988): "Perception and acceptability of risk in engineering projects."
The Civil Engineer in South Africa, Vol 30 No 12 545-548

Figure 3. Risk of death per individual (after DJW Wium, 1988).

the adoption of Failure Modes and Effects Analysis early in the design phase of a project. A risk assessment in one form or another is also required when selecting a closure alternative or deciding on the scope and urgency of the appropriate remedial measures. Below are some examples of how the risk assessment was used to facilitate the permitting process.

3.1.1 *Permitting of a new TSF*
A tailings storage facility was proposed for a new mine development in an exceptionally sensitive environment. The focus of the suggested design was to limit the possibility of an adverse effect of the facility on the surrounding environment. Although a significant effort was made in achieving high environmental safety, both the regulators and the client were concerned about the degree of risk the facility potentially imposed on the environment.

A Failure Modes and Effects Analysis (FMEA) was conducted to evaluate the range or risks the proposed facility entails. FMEA is a structured risk analysis in which all failure modes are identified and ranked in terms of probabilities or with descriptors, according to the combined influence of their likelihood of occurrence and the severity of the consequences. The analysis included a review of all hazards associated with the construction, operation and closure of the facility. The probability of contamination of selected areas of interest was then estimated and related to the probability of a permanent loss of a number of endangered species. It was concluded that the probability of a permanent loss of 25% of the vegetation and the wild life within the potentially affected area, due to a potential failure of the facility, is in the order of 1 in 100,000.

The results of the conducted FMEA indicated that the probability of a failure of the TSF, and the consequences associated with it are within the boundaries, generally accepted by the regulatory authority.

3.1.2 *Selection of a mine remediation option*
In many cases when a number of closure alternatives are considered, selecting the optimum one could be a significant challenge. A risk assessment approach was used in selecting the optimum alternative for a long-term management of arsenic dust currently stored in the underground vaults of an inactive mine. Four alternatives for mine remediation were considered. Although the pros and cons of each alternative were identified and listed, without quantifying the risks associated with each option against a common benchmark, it was difficult to objectively weigh and compare the options. For this reason a new type of a risk assessment was implemented, which combines the engineering risk with the ecological and human health risks.

The objective of the engineering risk analysis was to estimate the probability of exceeding 1000 kg/yr

of arsenic loading to a local creek during and after the implementation of each considered option. The selection of 1000 kg/yr arsenic discharge to the creek was based on the findings of an ecological risk assessment study conducted previously, which concluded that an arsenic loading of more than 1000 kg/yr could have an adverse effect on human health. The results of the risk analysis showed that the risk associated with one of the options was significantly lower than the risk associated with the other three options. The conducted risk assessment served as a basis for comparison between the remediation alternatives and as a proof to the authorities and the public that the proposed measures will bring the risks to acceptable levels.

3.1.3 *Selection of a TSF closure option*
A TSF located in a tropical environment was approaching closure. Two main closure options were considered: a "wet" option and a "partial dry" option. The wet option involved a permanent pond over the entire tailings surface, which would inhibit acid formation. The partial dry option envisaged a small pond located away from the embankment and a permanent engineered dry cover over the exposed tailings surface. The wet cover option was the cheaper alternative, but there was a concern that the risks associated with the implementation of this option might not be acceptable.

Conducting a risk analysis was considered necessary to aid in assessing the levels of risk imposed by each option, and in finding the optimum solution. All hazards leading to potential failure modes were identified and incorporated into a fault tree. The fault tree was then extended into an event tree to predict the effect of the potential failure modes on the environment, the people and the property.

The results of the analysis clearly showed that the initially proposed "wet" option presents unacceptable risks, while the potential risks associated with the "dry" option were lower than generally acceptable. A sensitivity analysis was then conducted using the fault-event trees in search for a compromise between the two options.

Using the risk analysis approach provided assurance that the optimum solution associated both with acceptable risk and with lowest possible cost was found.

3.2 *Risk assessment to reduce liabilities*

All mines, like most commercial endeavors, create certain risks to people and the environment. The management of existing mines is faced with present and future environmental liabilities. By conducting a risk analysis and risk assessment, SRK has provided the management of a number of mines around the world with a comprehensive tool for decision-making. The results of the risk assessment have been successfully used as a logical and defensible framework for focusing relatively limited capital on projects where the spending has the largest impact on reducing corporate exposure.

3.2.1 *Preparation of a management plan*
The management of an existing mine was concerned about the potential consequences that a failure of any one of the mine facilities might have on the environment and on the people living downgradient of the mine. Although high standards of design and construction were adopted by the mine management, the risk analysis identified some potential hazards, which could lead to undesirable consequences. After a detailed review of the possible failure modes and their consequences, a management plan was compiled aiming at reducing the company liabilities to a minimum.

3.2.2 *Reduction of insurance premiums*
Keeping the risk of a loss of life due to a potential failure of the tailings dam was of a high priority for the management of large platinum mine. In the past, the mine had suffered the consequences of a failure of one of its tailings facilities which had resulted in a loss of life, loss of property and environmental contamination. To ensure that the risk of a failure of the tailings facilities is at a very low level, the mine management had initiated a risk assessment program. Part of the initiative was an extensive monitoring program. The results of the monitoring program were reviewed monthly and any deviations from the expected results were reflected in the risk analysis. If the obtained risks were higher than the pre-determined limits, remedial measures were immediately implemented. Being able to provide evidence that the risks are kept at a low level at all times helped to significantly reduce the insurance premium for the mine.

3.3 *Risk assessment to optimize operations*
Mining companies are under increased pressure by the regulators and the communities to ensure high levels of environmental and safety standards at their operations. Physical standards of design, construction, maintenance and closure are usually specified. However, these do not always address the possibilities of deficiencies and unexpected events that may arise during the various stages of the life of the mine facilities. For this reason, we use risk assessment methodologies to evaluate the possibility, probability and significance of events that may pose risk to the mine operations, the environment and the community. The mine management can then make well-informed and defensible decisions as to the best utilization of the available budget.

3.3.1 *Planning of environmental remediation*
A regulatory authority informed the management of large industrial site operations that they would face significant fines if the site did not meet the environmental standards within a specified time period. The site was suspected to be a source of contamination of the surface and ground water, of the soil and the air of the surrounding area. An initial audit of the site showed the complexity of the problem. Adopting a risk based approach for the design and scheduling of the necessary remedial activities was considered necessary.

The site was subdivided into nine areas according to levels of contamination and type of contaminants. The risks that each area imposes to other areas within the site and to the surrounding environment were estimated using a fault-event tree technique. In the process, the origin and the contaminant content of each source of contamination were examined. Based on the collected information remedial measures were designed to either eliminate or reduce the contamination to acceptable levels. Implementing all necessary remedial measures immediately was unrealistic. The fault-event tree was then used to conduct a sensitivity analysis to evaluate the effect of each remedial measure on the overall risk reduction. Based on the results of the sensitivity analysis, the remedial activities were ranked and prioritized to ensure maximum reduction of the risks of contamination at a minimum cost. As a result, the activities undertaken within the first year reduced the overall risk by 50%, while the associated costs represented only 20% of the total cost of the required remedial measures.

4 CONCLUSIONS

We do not live in a risk-free world. In the current competitive environment, taking just enough risk to bring the optimum benefit is of primary importance for achieving the best return on the investment.

By conducting a risk analysis it is possible to clearly identify the existing hazards, evaluate the effect of proposed changes to the design, or the effect of the implementation of suggested remedial measures. In addition, the probabilistic modeling widely used in the risk analysis process allows the implementation of less conservative designs, opposed to the old-fashioned deterministic approach.

Opposite to common thinking, the effort involved in conducting a risk analysis is usually negligible on the scale of the overall work involved in the design of the facility. All that is required is a good knowledge of the project and significant experience in conducting a risk assessment analysis. In most projects, all the data to conduct a risk assessment is readily available: a brainstorming session with the parties involved and an experienced "risk analysis" person is all that is needed to produce this excellent decision making tool.

Knowing the risks provides a basis for negotiations with the regulators and the public for the extent of the contingency plans and the range of measures that need to be taken to bring the risks to acceptable levels. The risk assessment approach demonstrates to the regulators, to the public, and to the shareholders that the risks associated with the mine facility of interest are known and are under control.

By implementing the risk assessment methods we have provided state-of-the-art solutions to our clients' problems. Using the risk approach has resulted in significant cost savings for the clients by:

- Making well-informed decisions on the best utilization of the available budget;
- Making defensible decisions on the amount of risk the company takes;
- Selecting the best alternative by conducting a cost-benefit or risk-cost analysis;
- Reducing the liability of the company and insurance premiums;
- Facilitating a "trouble-free" permitting process.

Although the term "risk" strikes fear and we try to avoid it by ignoring it, we know that there is risk in one form or another in all our endeavors. To be able to manage the risk and get the maximum benefits, we should understand it, assess it and manage it in a systematic manner.

REFERENCES

Australian National Committee on Large Dams, 1998, Guidelines for Design of Dams for Earthquakes.

Canadian Standards Association, 1991, Risk Analysis Requirements and Guidelines.

Department of the Army, U.S. Army Corps of Engineers, 1999, Risk-Basis Analysis in Geotechnical Engineering for Support of Planning Studies.

John M. Cyganiewicz, P.E. & Jon D. Smart, P.E., 20th Congress of ICOLD, U.S. Bureau of Reclamation's Use of Risk Analysis and Risk Assessment in Dam Safety Decision Making.

International Council on Metals and the Environment, 1998, Proceedings of the Workshop on Risk Management and Contingency Planning in the Management of Mine Tailings, Buenos Aires, Argentina.

Wium, DJW, 1988, Perception and Acceptability of Risk in Engineering Projects, The Civil Engineer in South Africa, Vol 30 No 12 545-548.

Tailings and Mine Waste '02, © 2002 Swets & Zeitlinger, ISBN 90 5809 353 0

Mobility tracing of radionuclides in the uranium tailings "Schneckenstein" (Germany)

T. Naamoun, D. Degering & D. Hebert
Institute of Applied Physics, T.U. Bergakademie, Freiberg, Germany

ABSTRACT: The gamma together with the alpha spectrometric measurements are the best tools for the resolving environmental problems having relation with radioactive contamination. In the current work, the equilibrium and desiquilibrium between most of elements of the uranium chain were studied. The activity of ^{234}U and ^{235}U were measured by means of the last mentioned method. The former mentioned one was used for the measurement of ^{230}Th, ^{226}Ra, ^{210}Pb, ^{227}Ac from ^{235}U chain and ^{40}K. Both analytical methods were applied for the ^{238}U activity determination. The results achieved by means of the two methods of measurement, are in close agreement. The (^{227}Ac/ ^{226}Ra) ratio indicate the secular equilibrium between ^{226}Ra and its mother nuclide ^{238}U. The ^{230}Th is found nearly in equilibrium with ^{226}Ra. The present condition of the tailing areas show a high stability on the part of ^{226}Ra, ^{227}Ac, (^{231}Pa) and ^{230}Th. The present results show that in the best case, no more than about 70 % of the total uranium content was removed during the ore processing. The (^{210}Pb/ ^{226}Ra) ratio shows a near equilibrium between the ^{210}Pb and ^{226}Ra. The loss of ^{232}Th during the mineral processing is probably not significant and its mobility is expected not to be important. The present condition does not favour the mobility of ^{40}K.

1 INTRODUCTION

Using Physical (gravitational and radiometric) and chemical (acid and mainly soda alkaline) procedures, uranium ores from the Schneckenstein area, Zobes, Niederschlema, Oberschlema and the area of Thuringia were treated in the uranium processing plant called "32 unit". The ore residue was discharged in special settling basins in the Schneckenstein site in the former uranium deposit. Thus the resulting tailings will contain material rich with radioactive elements with different ability to migration or "mobility". The gamma together with alpha spectrometry is an exemplary tool for such an investigation.

2 GENERAL DESCRIPTION OF THE SITE

The Schneckenstein site is located in the southwest of Saxony, in the Boda valley approximately 3 km from the village of Tannenbergsthal/county of Vogtland (Figure 1). The area of investigation is located in the subsidiary branch of the southern boda valley at an altitude of 740 to 815 m above sea level. It receives an average annual precipitation of 1050 mm. The bedrock is the Eibenstock granite covered with a weathered surface layer. Southwest of the site follows the contact zone with quartz-schist.

3 EXPERIMENTAL

3.1 *Field activity*

Four sediment cores were selected at different tailing sites by drilling four boreholes composing two boreholes in each tailing (Figure 2). The first two boreholes (GWM 1/ 96 and GWM 2 / 96) were drilled to the granite foundation in Tailing 2 (IAA I). The third borehole (RKS 1/ 96), Tailing 1(IAA II) was sunk to a depth of about eight meters. However, the granite foundation was not reached due to technical problems. The fourth borehole (RKS 2/ 98, Tailing1) is 12 m deep. The cores of 50 mm in diameter were cut into 1 m long slices and transported in argon filled plastic cylinders to avoid contact with ambient air.

3.2 *Sample preparation and analysis*

3.2.1 *Gamma spectrometry*
Before the beginning of the process of measurement with the gamma spectrometer, about 200 grams of soil material was taken from the bottom of each full casing and considered as a soil sample. each sample was dried in a cupboard below about 90 degree Celsius in order to prevent the loss of Pb. Each sample was filled into a gasproof measuring container of cylindrical shape. About two weeks-minimum rest were needed to attain the secular equilibrium be-

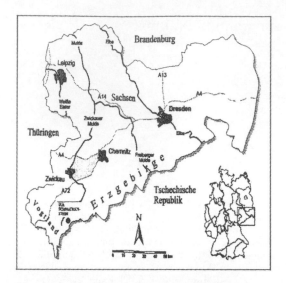

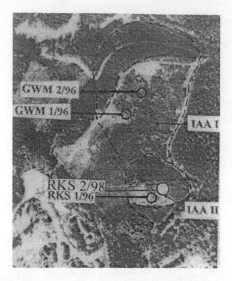

Figure 1. The area of investigation- Uranium tailings Schneckenstein.

Figure 2. Location of boreholes in the tailing sites.

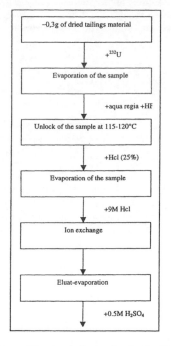

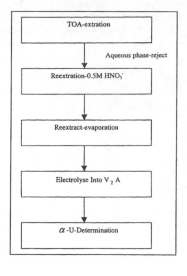

Figure 3. The schema of the chemical procedure for the Alpha spectrometric measurements.

tween radium and radon as well as it's daughter nu-clides (^{214}Pb and ^{214}Bi). In most cases about twenty four hours of counting were enough to receive an optimal gamma spectrum. The main component of the gamma spectrometer that was used for analysis is a p-type high purity Ge detector with 36% relative efficiency and 1.8 keV line width (FWHM) at 1.3 MeV. The detector was surrounded by a passive shielding of 9 cm lead, 2cm mercury and 4 cm elec-trolyte copper. Each gamma spectrum was analysed by comparing the count rate of a calibration source with the net peak count rate of the same measuring geometry. During the process of counting, the effect of self absorption was taken into account.

3.2.2 *Alpha spectrometry*

An alpha spectrometer with an energy range of 3Mev-7 Mev, a noise of 15 Kev (5.486 Mev) was used to determine the activities of three uranium nuclides (^{238}U (4,196 MeV), ^{235}U (4,39MeV) and ^{234}U (4,77 MeV)) in the tailings sediment. In order to accomplish this task, part of the gamma spectrometric samples were used. Before the beginning of the counting process, the sample passes through the following preparation stages (Nindel, 1988) illustrated in Figure3.

4 RESULTS

4.1 *Gamma spectrometric measurements*

4.1.1 ^{238}U *isotope*

In the borehole N°1, its content varies between 452 Bq/ kg in the first interval of depth and 2179 Bq/ kg in the last one (Figure 4). In the second core, it ranges from 326 Bq/ kg in the first sample and 3288 Bq/ kg in the sixth one. In the third one, it lies between 273 Bq/ kg in the first interval of depth to 8111 Bq/ kg in the fifth one. In the fourth borehole, it varies from 185 Bq/ kg in the next to last sample to 6193 Bq/ kg in the fifth one. Whereas, its mean concentrations are close to 1794, 1360, 2909 and 2506 Bq/ kg.

4.1.2 ^{226}Ra *isotope*

In the first core, its gross content ranges from 229 Bq/ Kg in the first interval of depth to 4483 Bq/ Kg in the next to last interval. In the second core, it varies between 339 Bq/ Kg in the first sample and 8866 Bq/ Kg in sixth one. In the third one, it lies between 335 Bq/ Kg in the first sample and 9121 Bq/ Kg in the fifth one. In the fourth core, it varies from 191 Bq/ Kg in the next to last interval of depth to 7674 Bq/ Kg in the fifth one. While, its mean concentrations are around 2960, 4251, 4216 and 3644 Bq/ kg.

On the other hand, in the first borehole, the (^{238}U/^{226}Ra) ratio differs from 0.2 in the last but one interval to 0.7 in the second one with a mean value of 0.6 (Figure 4). However, it exceptionally attains 2 in the first sample. In the second core, the mentioned ratio ranges from 0.3 to 1.0 with a mean value of 0.5. In the third one, it lies between 0.4 and 0.9 with a mean value of 0.7. In the fourth core, it varies from 0.4 to 1.0 with a mean value of 0.7.

4.1.3 ^{230}Th *isotope*

In the borehole N°1, its bulk concentration varies from 852 Bq/ Kg in the fourth interval of depth to 4028 Bq/ Kg in the (6.5–7.5 m) interval of depth but its amount is very weak and cannot be measured in the first one. In the second core, it differs from 480 Bq/ Kg in the first interval of depth to 9011 Bq/ Kg in the sixth one. In the third one, it ranges from 230 Bq/ Kg in the first interval of depth and 7537 Bq/ Kg in the fifth one. In the fourth core, it lies between 180 Bq/ Kg in the first sample and 6013 Bq/ Kg in the fifth one but it is undetectable in the two last two intervals of depth. Whereas, excluding the two last intervals of depth of the fourth borehole, its mean concentrations are around 2692, 4016, 3625 and 3069 Bq/ kg respectively.

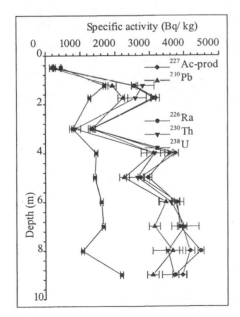 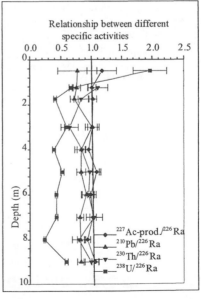

Figure 4. The specific activity change with depth and their relationships; Borehole N°1.

In addition, in the first borehole, the $(^{230}$Th/ ^{226}Ra) ratio is roughly 1.0 in most analysed samples with a mean value of 0.9, but it is near to 0.6 in the fourth interval of depth. In the second core, the mentioned ratio differs from 0.7 to 1.4 with a mean value of 0.9. In the third core, it ranges from 0.7 and 1.1 with a mean value of 0.8. In the last borehole, it lies between 0.5 and 0.9 with a mean value of 0.7.

4.1.4 ^{210}Pb isotope
In first core, its total mass lies between 178 Bq/ kg in the first sample and 3666. 4 Bq/ kg in the next to last sample. In the second core, it varies from 230 Bq/ kg in the first sample to 7519 Bq/ kg in the sixth one. In the third core, it differs from 343 Bq/ kg and 6673 Bq/ kg. In the fourth core, it ranges from 293 Bq/ Kg in the first interval of depth to 6941 Bq/ Kg in the seventh one. While, its mean concentrations are close to 2440, 3701, 3340 and 3055 Bq/ Kg respectively.

Furthermore, in the first borehole, the $(^{210}$Pb/ ^{226}Ra) ratio differs from 0.7 to 1.0 with a mean value of 0.8. In the second core, it ranges from 0.7 to 1.1 with a mean value of 0.9. In the third core, it lies between 0.7 and 1.0 with a mean value of 0.8. In the last one, it varies between 0.8 and 2.0. with a mean value of 0.9.

4.1.5 ^{227}Ac isotope
In the first core, its bulk content varies from 12 Bq/ Kg in the first interval of depth to 192 Bq/ kg in the last but one. In the second borehole, it differs from 22 Bq/ kg in the first interval of depth to 423 in the

sixth one. In the third borehole, it lies between 17 Bq/ Kg in the first interval of depth and 399 Bq/ Kg in the fifth one. In the last core, it ranges from 7 Bq/ Kg in the next to last interval of depth to 343 Bq/ Kg in the fifth one. Whereas, its mean concentrations are near to 137, 193, 180 and 157 Bq/ kg respectively.

Moreover, in the first core, the $(^{227}$Ac/ ^{226}Ra) ratio differs from 0.043 to 0.054 with a mean value of 0.047. In the second one, it ranges from 0.042 to 0.064 with a mean value of 0.046. In the third one, it lies between 0.040 and 0.050 with a mean value of 0.044. In the fourth core, it varies between 0.038 in the last but one interval of depth and 0.057 in the first one with mean value of 0.045.

4.1.6 ^{232}Th isotope
In the borehole N°1, its total amount ranges from 35 Bq/ kg in the fourth interval of depth to 63.3 Bq/ kg in the last but one interval (Figure 5). In the core N°2, it lies between 43 and 65 Bq/ kg. In the third one, it varies from 42 Bq/ kg in the first interval of depth to 70 Bq/ kg in the fourth one. In the last core, it differs from 35 to 70 Bq/ kg. In addition its mean concentrations are close to 50 Bq/ kg in the first and in the last core respectively and roughly 53 in the second and the third ones.

4.1.7 ^{40}K isotope
In the borehole N°1, its gross concentration differs from 834 Bq/ Kg in the sixth interval of depth to 1136 Bq/ kg in the next to last one (Figure 6). In the second core, it varies from 743 Bq/ kg in the sixth

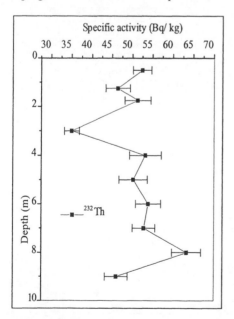

Figure 5. The specific activity change with depth of the ^{232}Th nuclide; Borehole N°1.

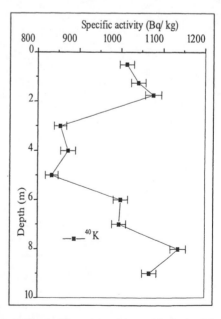

Figure 6. The specific activity change with depth of the ^{40}K nuclide; Borehole N°1.

interval of depth to 1290 Bq/kg in the second one. In the borehole N°3, it lies between 817 Bq/ kg in the fourth interval of depth and 1361 Bq/ kg in the sixth one. In the last core, it ranges from 612 Bq/ kg in the third interval of depth to 1297 Bq/ kg in the seventh one. While, its mean concentrations are 989, 1033, 1096 and 1017 Bq/ kg respectively.

4.2 *Alpha spectrometric measurements*

4.2.1 ^{238}U *isotope*
In the borehole N°1, its whole content ranges from 387 Bq/ kg in the first interval of depth to 2307 Bq/ kg in the last one. In the borehole N°2, it varies between 327 Bq/ kg in the first interval of depth and 3340 Bq/ kg in the sixth one. In the third core, it differs from 263 Bq/ kg in the first centimetres of depth to 8200 Bq/ kg in the fifth interval of depth. In the last core, it lies between 179 Bq/ kg in the last but one interval of depth and 4894 Bq/ kg in the sixth one. While, its mean concentrations are 1386, 1793, 3157 and 2543 Bq/ kg respectively.

4.2.2 ^{234}U *isotope*
In the first core, its total mass ranges from 397 Bq/ kg in the first interval of depth to 2237 in the last one. In the second one, it lies between 325 Bq/ kg in the first interval of depth and 3379 Bq/ kg in the sixth one. In the third borehole, it differs from 278 Bq/ kg in the first interval of depth to 8132 Bq/ kg in the fifth one. In the last one, it varies from 201Bq/ kg in the next to last interval of depth to 4852 Bq/ kg in the sixth one. Whereas, its mean concentrations

are roughly 1351, 1743, 3144 and 4467 Bq/ kg respectively. In addition, the mentioned nuclide is found in equilibrium with its predecessor ^{238}U in all boreholes.

4.2.3 ^{235}U *isotope*
In the borehole N°1, its entire content differs from 15 Bq/ kg in the first interval of depth to 85 Bq/ kg in the last one (Figure 7). In the second core, it ranges from 9 Bq/ kg in the first interval of depth to 111 in the sixth one. In the third one, it lies between 8 Bq/ kg in the first interval of depth and 250 Bq/ kg in the fifth one. In the last core, it varies from 3 Bq/ kg in the next to last interval of depth to 211 in the sixth one. In addition, its mean concentrations are close to 48, 30, 113 and 99 Bq/ kg respectively. Moreover, in the borehole N°1, the ($^{238}U/ ^{235}U$) ratio varies between 22 and 33 with a mean value of 29 (Figure8). In the second core, the mentioned ratio differs from 23 to 75 with a mean value close to 33. In the third core, it ranges from 21 to 47 with a mean value of 29. In the last borehole, it lies between 17 and 65 with a mean value roughly 30.

5 DISCUSSION

As expected, the uranium content (^{238}U) is considerably lower in the first intervals of depth than in the other ones in most tailing areas. It ranges from 0.002 % in the borehole N°3 to 0.003 % in the other ones. These results are in the sector of the uranium con-

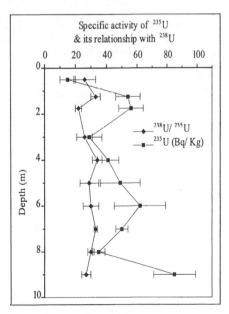

Figure 7. The specific activity change with depth of ^{235}U & its relationship with ^{238}U; Borehole N°1.

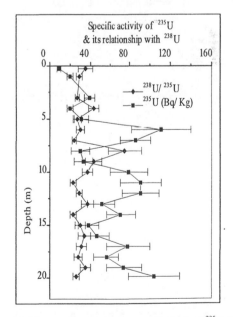

Figure 8. The specific activity change with depth of ^{235}U & its relationship with ^{238}U; Borehole N°2.

centration in rocks considered as trivial and without economic importance that contain less than 0.005 % of the already mentioned element (Wismut 1946-1955). In addition, as already illustrated in the first paragraph of this chapter, in the mentioned interval of depth, the tailings material is a mixture of a wide range of material types; from silt to cobbles. Therefore, this category of rocks apparently does not undergo any chemical processing and is used essentially in the revelation of the studied areas. Further, the uranium content in most other tailing parts is notably higher. The grain sizes of the tailing materials in the mentioned space is approximately uniform as results of grinding process that it is required for the exposure of the uranium minerals especially by alkaline leaching (Merritt, 1971). Moreover, as shown in Figures 9 and 10, the results achieved by means of the two methods of measurement, alpha and gamma spectrometry, are in close agreement.

Furthermore, according to Van Wambeke (1971), the (^{238}U/ ^{235}U) ratio has been gradually changed since the formation of the earth's crust. Thus, 2.0 10^9 years ago it was close to 26.9 but due to the difference between the decay rates of the two mentioned nuclides, $8*10^8$ years before the present it reached 72.2. Today it is roughly 21.7.

Moreover, it is of importance to point out that the ^{235}U nuclide is determined with a considerable relative error with regard to other uranium nuclides. This is due probably to the relative short time of measurement. Therefore, the above mentioned ratio is also calculated with an important standard deviation that must be taken into consideration.

On the other hand, the ^{227}Ac (half life 21.8 y), the daughter nuclide of ^{231}Pa, has a half life of $3.3*10^4$ y and belongs to the ^{235}U chain. Consequently, the (^{227}Ac/ ^{226}Ra) ratio reflects the geochemical processes rather than the (^{231}Pa/ ^{226}Ra) ratio. Further, the first mentioned ratio is a useful tool for the estimation of the equilibrium/ disequilibrium state between ^{226}Ra and its mother nuclide ^{238}U. Furthermore, the (^{227}Ac/ ^{226}Ra) ratio is found to reach nearly the best value of the activity ratio (0.046), considered as an indicator of the secular equilibrium between ^{226}Ra and its mother nuclide ^{238}U (Senftle et al., 1957) in most analysed samples. In addition, considering that the ^{235}U share is roughly to 0.711 % from the total uranium content (Grundy & Hamer, 1961), the mean value of the activity relationship of ^{238}U/ ^{235}U is around to 21.7. Thus, multiplying this (^{227}Ac/ ^{226}Ra) ratio by the above mentioned activity relationship of ^{238}U/ ^{235}U, the ^{227}Ac/ ^{226}Ra equivalent- unit of uranium is nearly to 1 in most tailing parts.

Moreover, assuming that Th tends to form carbonate complexes mainly in form of $[Th(CO_3)_5]^{6-}$ compound (Boyle, 1982) and presuming that the mobility of Th is considerably less than that of uranium, consequently its loss during the mineral processing is notably less than that of uranium. Therefore, the ^{230}Th is found nearly in equilibrium with ^{226}Ra nuclide in most analysed samples. Further, the little difference between the values of the ^{230}Th/ ^{226}Ra is apparently due to the chemistry of Th itself as already mentioned, as well as to the accuracy of measurement of this nuclide. Furthermore, under the alkaline leaching, the radium loss is not important since only between 3 and 5 % of its total content may be solubilized (Snodgrass, 1990). In addition, for ecological reason, Ra was probably fixated by the use of $BaCl_2$ instead of $BaSO_4$, since the Ba content is found to be very high in most tailing parts except in the first intervals of depth of the first and the second boreholes where it is under its detection limit. This is probably in relation with the non-treated heap material, whereas the sulphur content is found to be unimportant in most tailing areas. There-

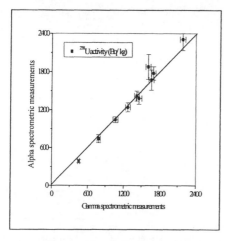

Figure 9. Comparison between the gamma & alpha spectrometric measurements; Borehole N°1.

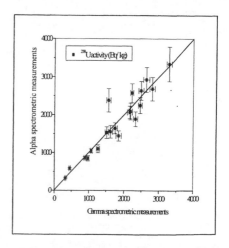

Figure 10. Comparison between the gamma & alpha spectrometric measurements; Borehole N°2.

fore, before the mineral processing, the ^{226}Ra nuclide was apparently in equilibrium with its mother nuclide 238U as well as with its predecessor ^{230}Th. Moreover, the present condition of the tailing areas show a high stability on the part of ^{226}Ra, ^{227}Ac, (^{231}Pa) and ^{230}Th. In addition, it is of importance to point out that the shortness of the used counting is probably the cause of the difference between the two (^{227}Ac/ ^{226}Ra) ratios in equivalent- unit of uranium since the ^{235}U was determined with a high error of measurement. Moreover, since the (^{227}Ac/ ^{226}Ra) ratio in equivalent- unit of uranium varies between 1 and 10 (Cherdynsev & Isabaev, 1955; 1957) for a secular equilibrium and for a high disturbance between ^{226}Ra and ^{238}U respectively, the calculated (^{227}Ac/ ^{226}Ra) ratio by means of the measured ^{238}U/ ^{235}U relationship shows unimportant displacement of

^{226}Ra with regard to its mother nuclide ^{238}U for most analysed samples. Therefore, the uranium series should be in equilibrium, and it is easy to conclude that the specific activity of ^{226}Ra reflects the uranium content. Consequently, the ^{238}U/ ^{226}Ra is a measure of the efficiency of the leaching.

Further, the processed ores in the borehole N°1 below 2m and in the borehole N°2 below 4m depth of the younger tailing shows a certain similarity as seen by the nearby constant and lower activity ratio ^{238}U/ ^{226}Ra compared to the profiles of the older tailing where a broad variation and an increase of this ratio occurs. This might be due to the improvement of the efficiency of the leaching procedure with time, since many procedures were experimented between 1953 and 1982 (IAEA, 1993). Therefore, the present results show that in the best case, no more

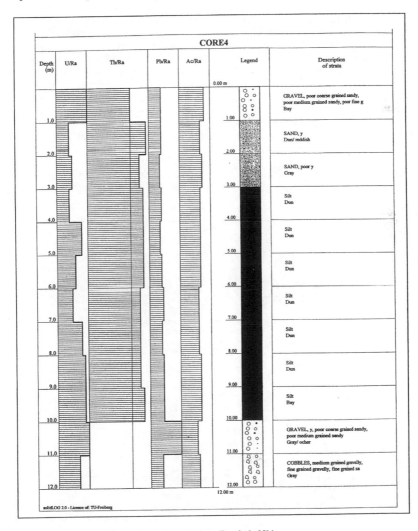

Figure 11. The change with depth of different physical parameters; Borehole N°4.

than about 70 % of the total uranium content was removed during the ore processing. Furthermore, the disturbance of the activity ratios between most of the nuclides in the first intervals of depth, excluding the revelation material, is probably due to the mixing of trivial and non-treated material with a processed one.

In addition, the low sulphur content in most parts of the tailing areas on the one hand, and the stability of Ra and the low loss of Th during the mineral processing on the other hand, emphasise the alkalinity of the leaching procedure.

On the other hand, the successor of the radioactive nuclide noble gas ^{222}Rn; ^{210}Pb with a half life of 22.3 y may be considered as the best tracer of the behaviour of its just mentioned predecessor. In fact, the (^{210}Pb/ ^{226}Ra) ratio show a near equilibrium between the ^{210}Pb and ^{226}Ra nuclides with a little dominance of ^{226}Ra (^{210}Pb/ ^{226}Ra < 1) mainly in the first intervals of depth. This is probably due to the ^{222}Rn diffusion from the tailings material with a wide range of grain sizes as well as with a lower degree of saturation and, consequently, a higher diffusion coefficient. Besides, its convective transport from the deeper to the upper layers mostly for the first intervals of depth under the effect of the heat change, is also a second origin of the ^{222}Rn and consequently of the ^{210}Pb depletion in the mentioned intervals. In addition, the large excess of ^{210}Pb (^{210}Pb/ ^{226}Ra = 2) in the interval of depth 10-11m of the borehole N°4 as, demonstrated in the Figure 11 is due to the enclosing of ^{222}Rn between two boundary layers with diffusion coefficients con-

siderably lower than that of the material of the above mentioned interval of depth where it prefers to move. Moreover, the near equilibrium between ^{210}Pb and ^{226}Ra confirms again the secular equilibrium state between most nuclides of the ^{238}U chain, i. e. their low loss during mineral processing as well as their stability under the present conditions of the tailings environment.

Concerning the ^{232}Th isotope, it is of importance to point out that its mobility is even lower than that of ^{230}Th isotope (IAEA, 1993). Therefore, its loss during the mineral processing is probably not important. Moreover, the selective extraction procedure shows that most of its content is held in the lattice structure of minerals. Consequently, its mobility is expected to be insignificant independent of the chemical conditions of the tailings environment. In addition, the current relationship between ^{232}Th and ^{238}U nuclides for most analysed samples (Figure 12) shows that the initial ^{232}Th/ ^{238}U ratio was roughly the same in the whole mineral product on the one hand as well as the increase of the leaching productivity with the time; from the first site represented with the third and the forth boreholes to the second one represented with the first and the second cores on the other hand.

With regard to the ^{40}K nuclide it is noteworthy to point out that the mobility of potassium is very low under all weathering conditions. Therefore, its loss during the mineral processing was probably insignificant. Also, the present conditions of the tailings environment do not favour its mobility.

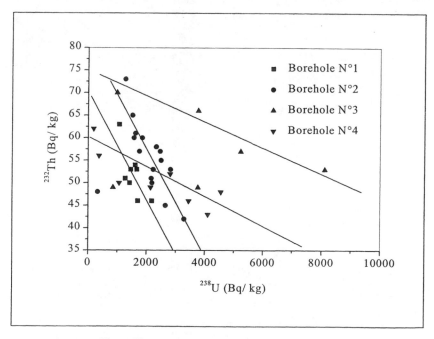

Figure 12. The relationship between ^{232}Th & ^{238}U nuclides for most samples.

Finally, assuming the relatively high mobility of uranium under the chemical conditions of the prospected areas as already demonstrated by means of the selective extraction as well as the clear migration of ^{222}Rn through the tailings material, these two nuclides must be taken into account in the radiation risk assessment of both contaminated areas.

6 CONCLUSIONS

The gamma together with the alpha spectrometric investigations prove their usefulness in the tracing of the mobility of radionuclides which is impossible to realise with other analytical methods.

REFERENCES

Boyle, R. W., (1982): Geochemical prospecting for thorium and uranium deposits, Develop. Econ. Geol., 16. Elsevier, Amsterdam.

Cherdyntsev, V. V. & Isabaev, E. A., (1955; 1957): Tr. 4-oi sessii komis. po opred. absolyutnogo vozrasta geol. formatsii akad. nauk SSSR otd. geol. geogr. nauk, Moskva.

Grundy, B. R. & HAMER, A. N., (1961): J. Inorg. Chem. 23, pp. 148/ 150.

IAEA., (1993): Uranium extraction technology. Technical reports series N°359: 110-111.

Merritt, R. C., (1971): The extractive metallurgy of uranium, Colorado school of mines research institute, Johnson Publishing Co., Boulder, CO, 576 p.

Senftle, F. E. et al., (1957): Comparison of the isotopic abundance of 235U and 238U and the radium activity ratios in Colorado Plateau uranium ores, Geochimica & Cosmochimica acta, 11, pp. 189/193.

Snodgrass, W. J., (1990): "The chemistry of 226Ra in the uranium milling process" The environmental behaviour of radium. Technical reports series N° 310, Vol. 2, IAEA, Vienna: 5-26.

Van Wambeke, L., (1971): The geology of uranium and thorium. In: Lesmo, R. (ed.). Report of the session Part II (1969), The geology of uranium and thorium.E.N.I. – Scuola Enrico Mattei (Italia).

Wismut, 1946-1955: Razrabotka mestarazdenia uranovi rud sovietska germanskim akcionernim obchestvom Vismuta za 1946-1955 godi.

Tailings and Mine Waste '02, © 2002 Swets & Zeitlinger, ISBN 90 5809 353 0

Radioactive tailings issues in Kyrgyzstan and Kazakhstan

R.B. Knapp, J.H. Richardson, N. Rosenberg, D.K. Smith & A.F.B. Tompson
Lawrence Livermore National Laboratory, Livermore, CA, USA

A. Sarnogoev
Center on Monitoring of Dangerous Natural, Man-Caused Processes and Rehabilitation of Tailings and Mining, Ministry on Emergencies and Civil Defence., Bishkek, Kyrgyz Republic

B. Duisebayev
National Atomic Company "KazAtomProm", Almaty, Republic of Kazakhstan

D. Janecky
Los Alamos National Laboratory, Los Alamos, NM, USA

ABSTRACT: Soviet era uranium mining and ore processing practices in Central Asia have left a nuclear legacy that threatens human health, promises severe and long-term environmental degradation, and retards economic development. We survey four sites in Kyrgyzstan (Mayluu-Suu, Kaji-Say, Ak Tuz, and Ming Kush) and two sites in Kazakhstan (Ulba Metallurgical Plant and Aksuek) that epitomize the situation. Mayluu-Suu, the site of greatest concern in Kyrgyzstan, has multiple tailings impoundments that are threatened by landslides, which could either push impoundments into the adjacent river or form dams and submerge them. The Ulba Metallurgical Plant in Kazakhstan is one of the oldest plants supporting the nuclear cycle. Operations have created a large quantity of waste, including actinides and beryllium, that is stored in retention basins. Some retention basins have sediment beaches, which act as sources for airborne contaminants. The clay-lining in some basins has cracked, producing contaminant plumes of unknown magnitude.

1 INTRODUCTION

1.1 *Background*

The early Cold War era saw the emergence of uranium as a strategic material and a boom for exploration and ore processing. This boom was especially intense in the former Soviet Union. It was during this time that Central Asia was discovered to be a uraniferous province of global significance (Fig. 1).

Early mining and ore processing activities both in the USA and the former Soviet Union were conducted under conditions and standards not accepted today. There was a sense of urgency in operations that were also conducted under a shroud of secrecy. Also, there was an incomplete knowledge of the impact of the radioactive materials on human health and a lower sensitivity for the environment. Consequently, standards of tailings and waste disposal were below those required to protect humans in the surrounding towns and region. Many tailings impoundments are precariously poised on river banks, poorly maintained, and close to population centers.

The disintegration of the Soviet Union brought a halt to many uranium mining and ore processing operations in Central Asia and a burgeoning recognition of the radioactive legacy of these operations. However, remedial actions are severely constrained by both depressed economies and a lack of information. Soviet era practices mandated that key information, including the content and structure of radioactive piles and impoundments, be keep confidential and stored outside of the region in a centralized location. Today, this information is inaccessible.

1.2 *Assessment considerations*

In this paper, we review the status of all of the radioactive mine tailings sites in Kyrgyzstan and two sites in Kazakhstan. In the absence of historical documents, we rely on our personal experience at each site; two of us (A. Sarnogoev and B. Duisebayev) have devoted our professional careers to these issues. We also use a World Bank (1998) study of the Kyrgyzstan impoundments.

In our qualitative assessment of the status, we consider the three exposure pathways: air, surface water, and groundwater. Air pathway exposure can be by radon, by direct gamma ray, and by the direct and indirect effects of radioactive dust particles. We assess the integrity of both radioactive impoundment covers and access restrictions (e.g., fences) and the to evaluate the air exposure pathway.

Surface water can be contaminated by runoff from the impoundments and by erosional entrainment of impoundment sediments. We assess impoundment cover and dam integrity and water diversion capabilities to evaluate this exposure pathway. Surface water exposure can be through ingestion of contaminated water or through ingestion of foods either irrigated or processed by contaminated water.

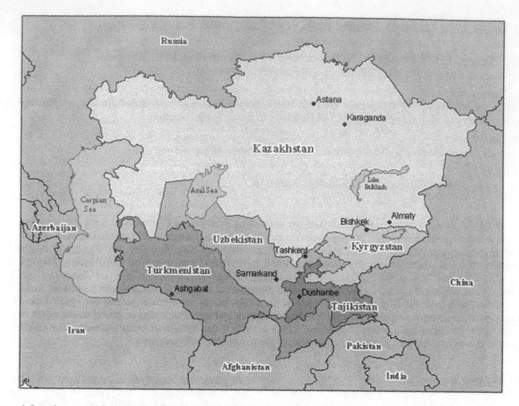

Figure 1. Location map of Central Asia and Kazakhstan and Kyrgyzstan.

The impact of radioactive tailings impoundments on groundwater is unknown in the region. We assess the groundwater pathway by examining the integrity of impoundment covers to prevent infiltration, by evaluating the current water content of the impoundments as an indicator of liner integrity. Groundwater exposure is by mechanisms similar to surface water exposure. The groundwater pathway is important in this semiarid and arid region because it comprises a significant fraction of the potable, irrigation, and industrial water used. Its impact is on a longer timescale than that of surface water, however. The time for surface water to move from point-of-contamination to point-of-use is much less because contaminated groundwater must both infiltrate and then slowly migrate, forming contaminant plumes. In general, this exposure pathway needs to be elevated in our recognition of its importance.

2 KYRGYZSTAN

2.1 Introduction

Kyrgyzstan was a major uranium producing region in the former Soviet Union with mining dating from the early 1950's; there are currently no active uranium mines. In total, there are five locations in Krgyzstan that have radioactive tailings impoundments. Three of these are former uranium mining sites: Mayluu-Suu, Kaji-Say, and Ming-Kush (Fig. 2). A fourth site, Ak-Tuz, is an active base and precious metal mine that has naturally occurring radioactive thorium as an uneconomic waste product. A fifth site, Kara-Balta, was a uranium ore processing center. We focus our attention on the four mining sites and the risks associated with their radioactive tailings impoundments.

In each of the sites discussed below, there is virtually no data regarding radioactive groundwater plumes. Evidence at each site strongly supports the existence of these contaminants plumes because the impoundments are dry, they have been poorly capped for significant periods of their lives, and the existence of linings is questionable.

2.2 Mayluu-Suu

Of all the sites in Kyrgyzstan, the uranium tailings impoundments at Mayluu-Suu pose the greatest threat to human health. Mayluu-Suu is the oldest and largest of the four sites and is composed of 23 distinct tailings piles. These radioactive piles are situated in steep ravines and on the banks of the Mayluu-Suu River and its tributaries. The Mayluu-Suu River feeds the Syr-Daria River, which is a major

314

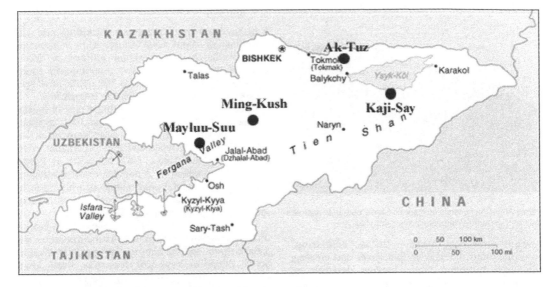

Figure 2. Location of the four Kyrgyzstan radioactive tailings impoundments.

source of irrigation water in the fertile Fergana Valley — the breadbasket of Central Asia.

Exposure via surface water pathways is the greatest risk at Mayluu-Suu. The ravines are highly susceptible to landslides and these landslides pose the greatest threat (Fig. 3). Over 200 landslides have occurred in the past 30 years at Mayluu-Suu. These have caused deaths of the local population and destruction of buildings and other property. With respect to the radioactive tailings piles, there are three possible consequences of a landslide:

1. damming of the Mayluu-Suu River and flooding of radioactive tailings piles,
2. "bulldozing" radioactive piles into the Mayluu-Suu River, and
3. no damage.

Large, active and potential landslides are poised above a narrowing in the Mayluu-Suu River (Fig. 4). Movement of these could easily dam the river and flood three upstream radioactive impoundments. If not removed quickly, then a radioactive lake could form behind the dam as the radioactive tailings impoundments flood and leak; landslide dams are notoriously unstable and a landslide dam retaining a large volume of radioactive water is not a favorable situation.

Radioactive pollution of the Mayluu-Suu River would be a catastrophe on a regional scale. 23,000 people live in the town of Mayluu-Suu just 3 km downstream of the impoundments. The fertile Fergana Valley is 20 km away; this has been a major agricultural area for more than three thousand years and it relies exclusively on irrigation water from the river. The Fergana Valley encompasses the countries of Kyrgyzstan, Uzbekistan, and Tajikistan and its products reach markets in neighboring Kazakhstan and Turkmenistan.

Several incipient landslides are poised above radioactive tailing piles (Fig. 5). Their release and

Figure 3. A major landslide behind a uranium tailings impoundment at Mayluu-Suu.

Figure 4. Potential landslide dam site at Mayluu-Suu.

315

Figure 5. Incipient landslides poised above potential dam site and uranium tailings impoundments at Mayluu-Suu.

downhill movement would result in "bulldozing" these piles into the Mayluu-Suu River and creating radioactive pollution plume moving downstream. Again, this threatens the town of Mayluu-Suu and the Fergana Valley.

It is very likely that there will be a major landslide at Mayluu-Suu in the next few years. Though there is uncertainty as to the consequences that this next landslide will cause, recent history shows that these landslides do impact humans (Fig. 6). In response to these threats, TACIS is funding rehabilitation of several of the impoundments of greatest risk at Mayluu-Suu. TACIS is a European Union organization whose mission is to aid countries of the former Soviet Union. Additional rehabilitation work is required to more fully secure the future of this locale.

Exposure via the air or direct contact with contaminated materials is also a possibility at Mayluu-Suu. Access is to each impoundment is open and domestic animals often graze on the surface, thereby exposure shepherds, the animals themselves, and consumers of animals products.

2.3 Kaji-Say

Kaji-Say has a single uranium mill tailings pile with a volume of about 150,000 m^3. This volume has been swelled by the continual addition of debris, some of which is radioactive contaminated equipment. Spot readings as high as 800 µR/hr have been recorded; the European standard is 100 µR/hr.

The Kaji-Say radioactive tailings pile is located on the banks of a wadi and less than 3 km upstream from Ysyk-Köl; Ysyk-Köl is a large fresh water lake of economic importance and the cornerstone of future development (Figs. 7-8). Ysyk-Köl is known across the continent as a popular resort destination and its environmental preservation is a high priority issue. For example, no motor craft are allowed on the lake to preserve water quality.

The hazard at Kaji-Say is the catastrophic erosion of the radioactive tailings pile, its incorporation into a flash flood, and the subsequent transport of radioactive sludge to Ysyk-Köl, less than 3 km downstream. This is a semi-arid climate with about 300 mm of rainfall per year. Most of the rain comes in two or so events that can cause flash floods in the wadi. The position of the tailings piles makes it susceptible to erosion by these extreme flow events.

Consequently, the primary human health risk is through surface water exposure. But the economic risk should not be ignored because pollution of the lake by radioactive materials would be a catastrophe.

Three actions are needed for rehabilitation at Kaji-Say: a dam downstream to prevent radioactive sludges from reaching Ysyk-Köl, erosion control at the base of the uranium tailings piles, and a specialized fence to prevent access by the local population to the tailings pile — a common occurrence. A specialized fence is required to prevent it from being an attractive item for theft.

A heretofore unconsidered exposure pathway, and hence unknown risk, is the degree radionuclides have penetrated into the groundwater when the ura-

Figure 6. Recent landslide at Mayluu-Suu that destroyed buildings, roads, school bus stops, and other infrastructure.

Figure 7. Down-slope side of the Kaji-Say uranium tailings pile, a temporary retaining dam, and Lake Ysyk-Köl in the distance.

Figure 8. Another down-slope side of the Kaji-Say tailings pile showing its unstable slope and position relative to the adjacent wadi.

nium tailings piles was uncovered. It is highly probable that percolation of radioactive solutions has occurred. What is not known is the magnitude of this percolation and the extent of any radioactive plume spreading toward Ysyk-Köl.

2.4 Ak-Tuz

Ak-Tuz is a site of active gold, silver, lead, and rare earth mining currently by a foreign owned company. Radioactivity in the mill tailings piles is associated with thorium, an element of no commercial value at Ak-Tuz.

There are four radioactive mill tailings piles at Ak-Tuz; two of them are still active. The active piles are owned by a foreign company and the two inactive piles are the responsibility of the Kyrgyz Republic. The Ak-Tuz tailings piles site above a tributary to the Chüy River on steep valley walls.

There have been two incidents related to the radioactive tailings piles at Ak-Tuz. In 1962, the dam to pile #2 (Fig. 9) broke and about 30,000 m³ of radioactive sludge entered the river below. Subsequently, much of this sludge was used for building materials by the local population. No exposure assessment has ever been done on the local populace and, because this event took place during Soviet time, the local populace has not been informed of its occurrence. This active impoundment remains unstable and a high risk.

Figure 9. The No. 2 tailings pile at Ak-Tuz is still active and poses a danger due to potential slope stability problems.

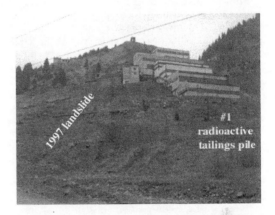

Figure 10. The operating plant sits above the inactive radioactive impoundment No. 1 at Ak-Tuz. Poor operating procedures resulted in a prolonged fluid leakage from a pipe, producing a landslide down the left-hand side of the impoundment.

Figure 11. The 1997 landslide at the No. 1 pile at Ak-Tuz, looking up.

317

The second event took place in 1997 when a leaking water pipe from active mining operations saturated the ground above tailings pile #1 (Fig. 10-11). A landslide occurred that swept through about one-quarter of the tailings pile, spreading it down the hill and across the upper reaches of the Chüy River tributary. Quick action reduced the downstream impact but no quantitative data has been collected.

The primary exposure pathway is via surface water and air/direct contact with radioactive debris. Slope stability issues of the impoundment dams and poor operation procedures make it likely that more impoundment failures will occur in the future. Like all of the other sites, the risk associated with groundwater exposure is unknown; though it is presumed high.

2.5 Ming-Kush

Ming-Kush is an isolated location that was created in 1955 exclusively for uranium mining and ore processing. The town associated with the mines was about 15,000 at its peak but it is now about 5,000 people who mostly engage in subsistence agriculture and animal husbandry.

There are four radioactive mill tailings piles at Ming-Kush. Tuyuk-Suu (Fig. 12) is the largest and also the impoundment with the highest risk. Its volume is about 300,000 m³ and peak surface readings are at 400 µR/hr. This pile was constructed across a valley with a perennially flowing stream. The stream is diverted around the pile by two bounding canals. The canals are poorly maintained — no maintenance for four years — and there are suggestions of losses in canal integrity at several places. In addition, there is evidence of erosion near one of the bounding canals at the downstream impoundment dam (Fig. 13).

The Tuyuk-Suu tailings impoundemnt also has an inadequate cover. This could lead to infiltration of water into the tailings pile and eventual collapse; monitoring wells drilled in 1987 revealed no water in the pile at that time.

In addition, there is no bounding fence to this pile — it was stolen — and animals freely graze on its surface. Efforts are required to re-fence the perimeter with less attractive material, to regularly maintain the canals, and to resurface the pile to prevent rain penetration.

The Taldybulak tailings impoundemnt at Ming-Kush has similar problems as the Tuyuk-Suu impoundment (Fig. 14). It was constructed across an ephemeral stream, whose flow was diverted beneath the impoundment. Maintenance of this culvert is lacking. Taldybulak is at a very remote location.

Two other uranium mill tailings piles at Ming-Kush warrant rehabilitation. The piles, named "D" and "K", were formed from slurries pumped to very remote locations. Evaporation of the liquid has lead

Figure 12. The Tuyuk-Suu tailings impoundment at Ming-Kush showing the variety of exposure pathways, including air/direct exposure and surface water exposure.

Figure 13. Front-side of dam at Tuyuk-Suu (Ming-Kush) showing erosion at right-hand side drainage.

Figure 14. Poorly maintained stream-water diversion culvert at the Taldybulak uranium tailings impoundment at Ming-Kush.

318

to the development of shrinkage cracks on the surface of the piles. These cracks are 30 – 40 cm wide, 3 – 4 m deep and run up to 10 m in length. They permit the penetration of rainwater deep into the tailings pile.

3 KAZAKHSTAN

3.1 *The Ulba metallurgical plant*

The Ulba Metallurgical Plant (UMP) is in the city of Ust-Kamenogorsk in northeastern Kazakhstan (Fig. 15). UMP is a large industrial complex located near city center. The plant was founded in 1949 to extract uranium from ore and supply enriched uranium and beryllium for atomic defense and naval reactors for the former Soviet Union. The plant represents one of the oldest facilities supporting the nuclear cycle in the former USSR and, as relayed by our hosts, is the world's largest nuclear fuels pellets producer. After the end of the Cold War, production of enriched uranium and metals diminished.

However in the last decade, demand for reactor fuel in Commonwealth of Independent States (CIS) has increased as has the need for beryllium (Be) and tantalum (Ta) metal in the civilian, defense, and aerospace sector. The UMP supplies most of 4.4% enriched ^{235}U reactor fuel used by the CIS and meets a strong international market for BeO, Be alloys, Be ceramics, Ta, and niobium (Nb).

Over nearly fifty years of continuous operations the plant has generated a significant quantity of legacy waste including actinides, beryllium, and metal by-products in solution (and eventually as solid precipitates). On-going practice is for the liquid effluent to be piped approximately three kilometers north of the plant and discharged into three large retention basins each with a capacity of between 25,000 m^3 and 30,000 m^3 of liquid (Fig. 16).

A fourth retention basin is presently under construction. The three operational basins are each filled with standing water (~ pH 8.0) and lined with clay

Figure 16. Piping system conveying liquid wastes from the UMP plant (3 km in background) to the retention basins.

Figure 15. Location of the Ulba Metallurgical Plant and the closed Aksuek uranium mine in Kazakhstan.

Figure 17. One of three actively used retention basins at the UMP plant in Ust-Kamenogorsk, showing existing discharges and the build up of precipitate wastes.

Figure 18. Closed retention basin "1-3" at the UMP plant showing the build up of precipitate wastes as a "beach" and the current construction of a protective clay cap. Water in this basin is now accumulated only from rain and snowmelt run-off.

and synthetic liners to attenuate infiltration of contaminated fluids. As the water evaporates, uranium and beryllium precipitates are created that settle out as sediments at the bottom of the basin. The sediments are kept saturated to prevent the airborne dispersal of particulates by controlling the level of water in the basin (Fig. 17).

One of the older discharge basins is leaking (Fig. 18). Standing water was let to completely evaporate out of basin "1-3" which resulted in the desiccation, cracking, and compromise of its clay and synthetic liners. Rainwater and snow melt subsequently allowed to collect in the pond, leak through the liner, and infiltrate to groundwater. Dissolved metals have subsequently appeared in groundwater samples produced from near-by monitoring wells.

The threat to a municipal water supply well-field for the city of Ust-Kamenogorsk situated downgradient from the disposal facility is real. In addition a sediment "beach" containing a thick pile (several meters) of uranium and metal precipitates (sediments) in the northeastern corner of the 1-3 retention basin is partially exposed to the atmosphere; only 30% of the beach is presently covered by a clay cap.

Over the long term, the basins will have a finite lifetime, either because precipitate build up will decrease their capacity or because the integrity of the lining system will ultimately diminish. The performance of the pond disposal system as a whole may also be susceptible to interruptions in the inflows and other operational problems that could occur at the plant. Because there is little new land for more ponds and because the plant as a whole is an extremely viable and productive enterprise, longer term disposal or treatment technologies for the waste stream that do not rely on the ponds may become of interest.

3.2 Aksuek

The Aksuek mine is located about 300 km northwest of Almaty, near the southwest tip of Lake Balkash (Fig. 15). It is an isolated location on the steppes of Kazakhstan and was discovered in the late 1940's; production started in 1957 . At Aksuek, uranium ore was mined underground from hard rock until the break-up of the Soviet Union, when all uranium mining and production activities were terminated. It currently crushes barite ore mined at a near-by location to produce a fine powder used in drilling fluids.

During uranium mining operations, excavated rock containing uranium but at concentrations below then commercial grade was disposed in a huge pile at the surface. Later, when technologies had improved, the uranium in this pile became commercial grade and leaching operations were conducted to extract more uranium. Readings at the pile's surface are typically at 100 µR/hr (international standards are 60 µR/hr) with maximum radioactivity at 500 µR/hr.

Figure 19. Collapse structure at Aksuek mine, providing ready egress for high levels of radon.

320

The hazards at Aksuek are two-fold. The greatest risk arises from wind-blown particles from the radioactive waste pile. During half of the year, the steppe wind blows strongly from northeast to southwest, which bring air-borne particles from the pile to the nearby (3-5 km) mining town (current population about 2,000). The wind blows in the reverse direction during the remaining half-year. There are no measurements of air-borne particles. The risk is restricted to the town and can not be quantified.

The second risk is from rain water percolating into the mine below and transporting radionuclides to the groundwater. This risk is not seen as significant. One of the mining tunnels collapsed during mining operations and created a crater at the surface (Fig. 19). Any rain (about 250 mm/yr) will penetrate into the mine workings through this collapse structure. But, in general, the mine is dry and the water table is greater than 900 m below the surface so any radionuclide transport is most like small and confined.

4 CONCLUSIONS

The Central Asian republics, especially Kazakhstan and Kyrgyzstan, inherited a plethora of radionuclide contamination issues upon receiving their independence. Many of these problems are associated with uranium mining activities and the disposal of mine tailings in precariously poised or poorly maintained impoundments. Their highly qualified technical staff are working diligently and under severe economic constraints to address the high priority issues.

The two main threats are the immediate exposure through airborne particles and catastrophic collapse of tailings impoundments into major river systems. However, the long-term threat and exposure through the groundwater system has been completely ignored and must be addressed.

REFERENCES

The World Bank, 1998, Hazard ranking and remedial action plan for the uranium mill tailings impoundments in the Kyrgyz Republic (draft).

Reprocessing, utilization, and treatment

Tailings and Mine Waste '02, © 2002 Swets & Zeitlinger, ISBN 90 5809 353 0

A brief discussion of the behavior of a loose iron tailing material under undrained triaxial loading

S. Tibana
Department of Civil Engineering, University of North Fluminense, Rio de Janeiro State, Brazil

T.M.P. de Campos
Department of Civil Engineering, Catholic University of Rio de Janeiro, Brazil

G.P. Bernardes
Department of Civil Engineering, UNESP-Guaratinguetá, Brazil

ABSTRACT: This work presents results of some laboratory tests carried out to assess the susceptibility to liquefaction of tailing dams built with mining waste. Three samples of industrial iron ore waste from two different mines were investigated. The experiments studied the instability under static loading of isotropically and anisotropically consolidated specimens. Monotonic triaxial tests under compression and extension were used. The instability line was found to be very close to the Ko-line in the p' x q space, and was independent of the shear strain rate. Despite some small differences in grain size distribution curve the wastes can be considered to be a single material. The instability line and the strength envelope define an area in the p'x q space where the specimens are not stable when drainage is not allowed. Some considerations of the stress strain behavior are presented and discussed.

1 INTRODUCTION

The undrained behavior of many tailing materials under monotonic and cyclic loading shows an instability line close to the Ko-line. Considering this behavior and that many tailing dams are built, and have their height increased by the upstream method, using mine residues, problems may be expected with their stability when the drainage system of the impoundment fails and the top flow line goes up. Some laboratories research carried out in Brazil (e.g., Tibana, 1997 & Tibana et.al., 1998a, 1998b & 1998c) shows that this collapse behavior occur at small levels of strain. The results presented here affirm that the collapse is independent of the strain rate. Results of undrained triaxial test on three samples of iron ore residues are also discussed.

2 TAILING CHARACTERIZATION

Results of tests on three samples of iron ore residues are presented; two from Samitri Mine (Samples A and B) and the other from the impoundment of the Fernandinho Dam (Sample C) collected after an accident in 1987.

The grain size distribution curve of all residues are presented in Figure 1. As observed in this Figure, all tailing samples are composed, predominantly, of fine sand and silt. The specific gravity of particles is higher than as usually found in soil, ranging from 3.68 to 4.17.

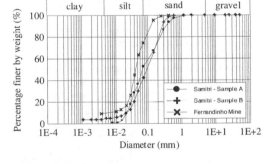

Figure 1. Grain size distribution.

The characterization tests are summarized in Table 1. The high specific gravity of the waste is probably dependent on the iron concentration. In addition to this, Tibana et.al. (1998a) showed that different fractions of one sample of iron ore residues may have different specific gravity. So, considering a grain size distribution in volume instead of by weigh it may show that the tailing is composed more by fine sand than by silt.

Table 1. Tailing characterization

Sample	G	Sand (%)	Silt (%)	Clay (%)
A	4,03	58	36	-
B	4,17	71	23	5
C	3,68	40	50	10

Analysis under the eletronic (MEV) and bi-ocular microscope confirm that the finer the fraction of the residue the more concentrated is the iron ore in it. The shape of the finer particles is like a plate while the fine sand fraction is prismatic. A volumetric grain size distribution shows that the tailing is predominantly composed of fine sand with prismatic shape so it will probably be more susceptible to liquefaction when the mass of the tailing is in a saturated and loose state. The Atterberg Limits tests showed that the silt and "clay" fractions are non plastic.

The maximum and minimum void ratios of each specimen are presented in Table 2. These tests follow the ASTM recommendations although the fine particles migrated to the side of mold. The high specific gravity of particles may justify the small range of the maximum and minimum void ratios.

Table 2. Specific gravity of residues

Sample	$e_{máx}$	e_{min}
A	0.7	0.33
B	0.96	0.55
C	1.24	0.68

3 TRIAXIAL TESTING PROGRAMME

Initially, sub-compaction and pluviation methods were used to prepare specimens but both methods proved not to be good for this material because of segregation of particles observed and low initial saturation (Tibana, 1997). So, tests were carried out to define the best procedures to prepare the specimens. It was concluded that it was best to pour the material in layers, into a mold partially filled with water. In this way the loose state of the specimens could be guaranteed without segregation and with higher water content. The main disadvantage is that with this procedure it is impossible to control the initial void ratio of the specimens making final interpretation of results more difficult.

All triaxial tests on isotropically consolidated specimens were performed at relatively low pressure consolidation stress levels, between 15kPa and 300kPa, despite Castro's (1969) procedures. In his work, Castro carried out tests using specimens consolidated to high stress levels to guarantee only a positive excess pore pressure. Castro's procedures have been used successfully by many researchers (e.g., Ishihara, 1983) to assessed the liquefaction of soils.

Monotonic triaxial tests performed on sample C concerned isotropically and anisotropically consolidated specimens shearing in compression and in extension paths. The displacement rate applied to the specimen during compression was 0.04 mm/min and 4.0 mm/min, while in the extension path it was 4.0

mm/min. The void ratio after isotropic consolidation with effective stress (15kPa - 300kPa) ranged from 0,64 to 1,33 for the sample C, which corresponds to relative densities of 1,08 and 0,36, respectively. Despite care during molding to maintain the loose state of the specimens the initial relative densities were around 50%. This high relative density may be explained by the procedures used to mold the specimens being different from those used to determine the limits void ratio. All tests with sample C were carried out with specimens of 1,5" diameter (Tibana, 1997).

Considering samples A and B which were prepared using the same procedures as for sample C, the void ratio of the specimens range from 0,51 to 0,61 and 0,88 to 0,94, respectively. The specimens had a diameter of 7,11cm and height ranging from 14,9 to 15,6. The strain rate to shearing the specimens was 0,75mm/min (Tibana, et.al. 1998c).

4 UNDRAINED TRIAXIAL TEST RESULTS

The undrained stress strain behavior of soils is dependent on the initial effective stress and void ratio of specimens. The higher the initial effective stress and initial void ratio, the higher is the contractive tendency and, consequently, the higher the excess pore water pressure in undrained conditions. Castro (1969) performing an undrained dead load triaxial tests under high initial effective stress shows that a loose specimen of sand can show softening behavior in a stress strain curve after rupture. At a large strain the effective stress of specimens is close to zero and the strain rate is very fast. As the initial effective stress decreases the stress-strain relationship changes and the curve shows a hardening after a drop in the deviator stress. The negative excess pore water pressure explains the hardening. It is important to state that independently of specimen molding the peak in the stress-strain relationship occurs, ie collapse, always at small levels of strain.

The results of preliminary CIU tests, to define the rupture envelope and the stress-strain relationship of samples C, A and B are presented in Figures 2, 3 and 4, respectively.

The tests on sample C were performed with no limited axial deformation while samples B and A it was stopped at about 10%.

Despite small differences between the samples, the dimensions of specimens and their strain rate, the stress-strain relationship were very similar. The collapse occured at small level of strain and before reaching maximum excess pore water pressure. For the effective stress range used in this work the deviator stress was not greater than 100kPa.

Some tests following the same procedures performed on sand from a deposit that had liquefied in

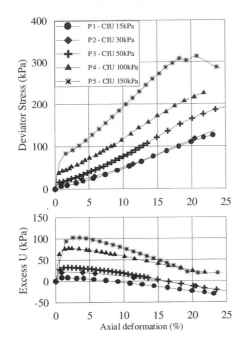

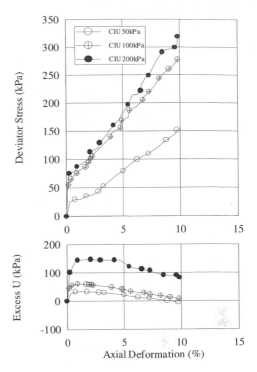

Figure 2. Results of CIU tests an Sample C.

Figure 4. Results of CIU tests an Sample B

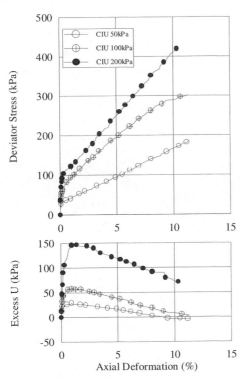

Figure 3. Results of CIU tests an Sample A.

Costa Rica did not show the same behavior. No tendencies to collapse behavior were observed although the effective stresses were approximately the same as for the iron ore residue tests.

Tests carried out to investigate the influence of strain rate on the stress strain curve were performed under shearing rates of 0,04 and 4,0 mm/min. The results show that the strain rate affects the stress-strain relationship only after reaching a deviator stress level where a sudden change in stiffness occurs. At low levels of deviator stress no influence can be observed in stiffness or strength. On the order hand, at large deformation levels the strength proved to be dependent on strain rate. The lower the strain rate the greater was the strength of specimens (Tibana, 1998b).

As shown in Figures 5, 6 and 7, initially, the stress paths tend to the left side, indicating positive excess pore water pressure, and after the collapse of the structure close to the maximum pore water pressure, the stress paths tend to the right side, indicating the tendency to develop negative excess pore water pressure. In Figure 5, for Sample C, it was possible to draw lines of axial deformation (0,5%, 1,0% and 1,5%) before the collapse. With samples A and B, however, the collapse happens at very small axial deformation (<0,2%), consequently, lines were not drawn.

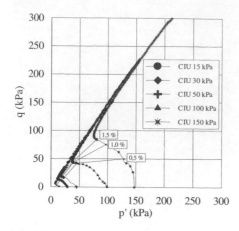

Figure 5. Effective stress path – Sample C.

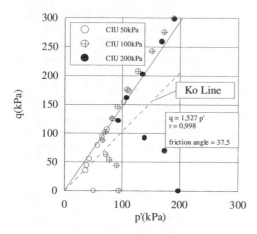

Figure 6. Effective stress path – Sample A.

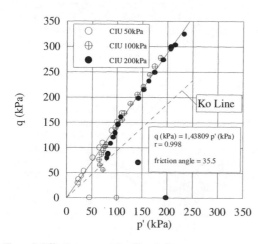

Figure 7. Effective stress path – Sample B.

Figures 6 and 7 show that collapse occurs close to Ko lines obtained by Jaky (1944).

Results of conventional undrained axial compression tests performed on anisotropically consolidated specimens are shown in Figure 8. The anisotropically consolidated samples follow the Ko path. After the consolidation step the valves of the cell were

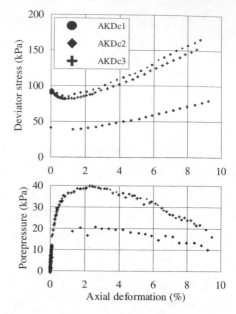

Figure 8. Stresses train curves of anisotropically consolidated specimens.

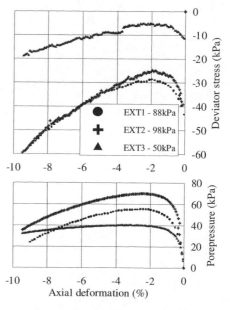

Figure 9. Results of CIU tests in extension.

closed so as not to allow drainage of the specimens while they were loaded. As soon as the specimens were loaded following a compression path their strength decreased, defined as a peak in the final deviator stress of consolidation. The pore water pressure reached its the maximum value at only 2% of axial deformation. After the decrease in deviator stress all the stress-strain curves show strain-hardening behavior.

Specimens in CIU tests following the conventional axial extension path (Figure 9) show stress-strain curves indicating a stiffer response of the tailing in extension than in compression. Peaks at small levels of axial strain (bellow 2,0%), indicated that the iron ore residues are also susceptible to collapse in extension. At about 2,0% axial deformation, the stress-strain curves change from strain-softening to a strain-hardening behavior. The pore water pressure reaches about 80% of initial effective stress and the maximums occur just shortly after the minimum deviator stresses.

5 DISCUSSION

Table 3 summarizes the parameters obtained from all CIU tests. The friction angles at 10% axial deformation were 35,5° for samples B and C and 37,5 for sample A. At collapse the friction angles range from 22,1 to 26,6.

For sample C the Ko line obtained by Jaky (1944) confirms the path followed automatically by the equipment with specimens consolidated anisotropically.

The three samples A, B and C have, approximately the same Ko-line and friction angle mobilized at collapse and at 10% of axial deformation.

The upstream method to increase dam height must not be done by transporting tailing as hydraulic fill. The dam body must be compacted to avoid positive pore water pressure.

The use of cyclone is not the best way to separate finer particles from other fractions because of differences in specific gravity that increase as the particle size diminishes. In that way the dam body can have finer particles that are more susceptible to liquefaction.

Table 3. Resume of parameters from CIU tests

Sample	M (15%)	ϕ' (15%)	ϕ' Collapse	Ko (Jaky) Ko= $1 - \sin\phi'$
A	1,527	37,5	26,6	0,39
B	1,438	35,5	24,7	0,41
C	1,439	35,5	22,1	0,42

6 CONCLUSIONS

Considering all the results presented here, it can be affirmed that iron ore residues investigated are susceptible to collapse at small levels of axial strains in compression and in extension.

Despite the differences between the three samples and the tests procedures used the results show that the tailings from different mines can be considered as a single material.

The fact that the Ko line is close to collapse envelope under undrained conditions is a evidence that residue in a loose state, in the impoundment and in the dam body, is unstable when loaded under undrained conditions. Consequently, any load monotonic or non monotonic, could trigger a rupture process followed by liquefaction.

REFERENCES

Castro, G. (1969). Liquefaction of Sands. Ph.D. Thesis. Harvard University, Cambridge, Ma.

Ishiraha, K. 1993. Liquefaction of Flow Failure During Earthquakes, Geotechnique, Vol. 43, No. 3, September, 351-415.

Tibana, S. 1997. Desenvolvimento de uma célula triaxial cíclica servo controlada e estudo da susceptibilidade à liquefação de um resíduo da lavra de mineração de ferro mineração. DSc. Thesis, PUC-Rio, Brazil.

Tibana, S. de Campos, T.M. P. & Bernardes de Paula, G. 1998a. Behavior of a Loose Iron Tailing Material Under Triaxial Monotonic Loading. Proceedings of 3rd International Congress on Environmental Geotecnics, September 7-11, Lisbon, Portugal.

Tibana, S., de Campos, T.M.P. & Bernardes, G.P. 1998b. Características de Resistência Não Drenada em Compressão e Extensão de um Resíduo de Mineração de Ferro. Prodeedings of XI Brazilian Congress on Soil Mechanics and Geotechnical Engineering, Brazil.

Tibana, S., Galvão T.C.B., Simões, G.F. & Carvalho, C.R.O. 1998c. Propriedades Geotécnicas de duas Amostras de Resíduos de Minério de Ferro da mina de Alegria. Prodeedings of XI Brazian Congress on Soil Mechanics and Geotechnical Engineering, Brazil.

New technologies and approaches

Tailings and Mine Waste '02, © 2002 Swets & Zeitlinger, ISBN 90 5809 353 0

Solar radiation on surfaces of tailings dams – effects of slope and orientation

G.E. Blight

University of the Witwatersrand, Johannesburg, South Africa

ABSTRACT: Evaporation from tailings surfaces and the growth of vegetation on the surfaces of rehabilitated tailings dams is directly dependent on the net solar radiation received by the surface. This depends on the latitude of the site, the season of the year and the cloud cover, as well as the slope angle of the tailings dam surface and its orientation. Evaporation not only controls accessibility to the tailings dam top surface, but evaporation from the slope of a ring dyke impoundment has an important influence both on surface shear stability and overall slope stability. This is especially the case in areas with arid or semi-arid climates. The paper presents sets of measured radiation data for the summer and winter solstices and the spring equinox at 26° South latitude. The method for calculation of incident radiation at any other latitude or season is also indicated.

1 INTRODUCTION

The water balance of a tailings dam is of prime importance for its operation and performance. In a cool, humid climate evaporation from tailings and water surfaces is relatively small and may play an unimportant role in the water balance. In a warm, arid or semi-arid climate, on the other hand, evaporation from the tailings surfaces and from decant and recycle ponds is usually the largest single source of water loss. Evaporation also plays an important role in slope stability. If the evaporation is small, phreatic surfaces will be higher and tailings strengths lower than if the evaporation is large. At closure of the facility, the surface of the impounded tailings will desiccate rapidly and sooner become accessible and trafficable if the evaporation is large than if it is small. On the other hand, high rates of evaporation, usually linked with low annual rainfall, will make it difficult to protect the tailings surfaces of a closed facility against erosion, by establishing a cover of vegetation.

The author has spent a number of years studying and quantifying the characteristics of evaporation from the surfaces of waste deposits (e.g. Blight and Blight et al, 1997, 1998, 1999, 2000, 2001). The purpose of this paper is to illustrate the influence of season and slope angle on potential rates of evaporation from tailings surfaces. The measurements presented here were made at a latitude of 26°S, but the results and supporting calculations can be used to estimate rates of evaporation in other latitudes and climates.

2 CHARACTERISTICS OF SOLAR RADIATION

Figure 1 shows successive positions and attitudes of the earth as it orbits the sun. The four positions shown are the two solstices and the two equinoxes. The diagram is labelled for the southern hemisphere. The important points to note are that because the axis of the Earth is inclined at 23½° to the plane of its orbit, the noonday sun is directly overhead the Equator at the two equinoxes and directly overhead either the Tropic of Capricorn or the Tropic of Cancer at the two solstices.

At noon at a particular spot on the Earth's surface, the solar power reaching the outer limit of the Earth's atmosphere is given by:

$$R_A = S_o(1 - \alpha)(\sin\phi\sin\delta + \cos\phi\cos\delta) \qquad (1)$$

Where R_A = net incoming radiation above the atmosphere (incoming minus reflected); S_o = solar constant (= 1380Wm^{-2}); α = planetary albedo or the reflectance of the Earth. According to various authorities (e.g. Robinson, 1966, Tyson & Preston-Whyte, 2000), α varies from 0.3 to 0.5 with 0.4 being a reasonable average; ϕ = latitude of the place under consideration; δ = declination of the sun which varies from 0° at the equinoxes to 23½° at the solstices. Assuming α = 0.4, $S_o(1 - \alpha)$ = 828Wm^{-2} which is the net incoming solar power at noon, on the Equator, at the equinoxes, or at the appropriate tropic (either Cancer or Capricorn) at the solstices. Figure 2 shows the variation of R_A with latitude at the equinoxes, the northern summer solstice and the southern summer solstice.

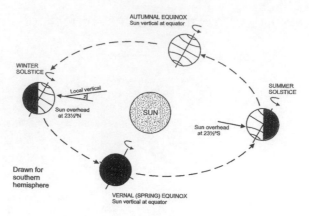

Figure 1. Earth's orbit about the sun (after Tyson & Preston-Whyte (2000)).

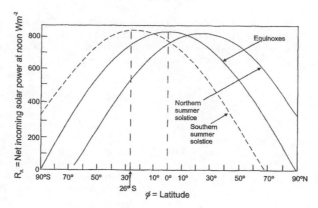

Figure 2. Solar power at noon at the solstices and equinoxes at outer limit of Earth's atmosphere for various latitudes.

The solar power can be converted into daily solar energy in MJm^{-2} by integrating the power over the length of the day from dawn to dusk. This can be done approximately by assuming that the power at a particular point varies through the day from sunrise to sunset in a parabolic fashion. For example the daily net solar energy at the outer limit of the atmosphere at the Equator at the equinox would be

$$2/3(828 \text{Wm}^{-2} \times 12\text{h} \times 3600\text{sh}^{-1}) = 23.85 \text{MJm}^{-2}$$

However, what is required for practical use is not the power or energy at the outer limit of the atmosphere, but the quantity available at the surface of the earth. The intensity of solar radiation will be reduced considerably as a result of absorbtion by the atmosphere, reflection from high clouds and dust particles and reflection from the surface of the Earth. According to Flohn (1969) these losses amount to about 50% for nominally cloudless days, leaving the balance available for heating the near surface air and soil and for evaporating water from the soil. This loss ratio will be revisited in section 5 of the paper.

3 MEASUREMENTS OF INCOMING SOLAR RADIATION AT THE EARTH'S SURFACE

The two main factors reducing the solar energy at the Earth's surface below that available at the outer limits of the atmosphere are the approximately 50% loss in the high atmosphere, referred to above, and interception by cloud cover. Figure 3 shows the results of 30 measurements of the relationship between R_i/R_A (where R_i is the incoming solar radiation at the Earth's surface) and the atmospheric clarity C. C is the extent to which the sun's rays reach the Earth's surface unimpeded by cloud, i.e. C = 1 corresponds to a cloudless sky (100% clarity) and C = 0 to one that is completely overcast by cloud. Each data point represents the result of a complete day's observations in which R_i was measured and C was assessed visually at hourly intervals throughout the day. The diagram shows that on average only 26% of R_A is received at the Earth's surface at the place of measurement (Johannesburg, South Africa, latitude 26°S, altitude 1700m AMSL). However, the ratio R_i/R_A for C = 1 will probably differ with latitude, al-

334

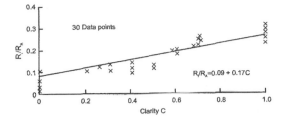

Figure 3. Effect of atmospheric clarity on solar energy received at Earth's surface as a fraction of net energy available at outer limit of atmosphere.

titude and climate. As will be seen later, for conditions of $C = 1$, R_i/R_A also differs with season and whether one is considering a spot measurement or an average over a day.

4 THE EFFECT OF SURFACE SLOPE AND ORIENTATION ON SOLAR RADIATION NORMAL TO A SURFACE

Figure 4 shows the effect of surface orientation on solar radiation normal to five surfaces throughout the day at the southern hemisphere spring equinox, as measured at latitude 26°S, altitude 1700m. The surfaces were oriented as follows: horizontal (O), and facing north (N), east (E), south (S) and west (W). Apart from the horizontal surfaces, the other four surfaces were all inclined at 30° to the horizontal, a common slope angle in geotechnical and tailings dam engineering. The curves in the upper diagram in Figure 4 show hourly values of solar power (Wm^{-2}) received normal to the five surfaces. Although measured on a cloudless day, it is apparent that more power was received before noon than after noon. (Compare the 11h00 peak for the east facing slope with the 14h00 peak for the west facing slope. Measurements of power for the two orientations should have peaked at the same number of hours before or after noon and have been the same.) The discrepancy is probably a result of varying concentrations of water vapour and air-borne dust in the atmosphere.

The lower diagram in Figure 4 gives the integrals of the power curves above (in MJm^{-2}). It is clear that for this latitude and season of the year, the north-facing slope received the most energy. (The numbers in parentheses to the right of the energy curves are the relative values of energy received over the day). The south-facing slope received less than 60% of the energy received by the north-facing slope.

Evaporation from a soil surface can be calculated by evaluating the terms in a surface energy balance given by:

$$L_e = R_n - G - H \qquad (2)$$

in which L_e is the energy utilized as latent heat of evaporation of water from the soil; $R_n = R_i(1 - a)$ and a = the reflectance or albedo of the soil surface; G is the energy absorbed in heating the near-surface soil; and H is the energy absorbed in heating the near-surface air. The cumulative value of L_e over a number of days is directly proportional to the cumulative value of R_n and, as shown in Figure 5, is about 80% of R_n. The value of the evaporation from the soil that corresponds to a given cumulative value of L_e is obtained by dividing L_e by the latent heat of evaporation of water which is $2.47MJ.kg^{-1}$ to give evaporation in kg of water per m^2, which is equivalent to mm. The figures to the immediate right of the cumulative curves in Figure 4 are evaporation values in mmd^{-1} assuming the albedo $a = 0.1$.

The rates of evaporation may not sound very impressive, but a north-facing 30° tailings slope 30m high and 1km long could be evaporating

$3.26lm^{-2}d^{-1} \times 2 \times 30 \times 1000 = 0.2Mld^{-1}$ (approximately),

while a 1km × 1km top surface could, on the same day, evaporate

$3.01lm^{-2}d^{-1} \times 1000 \times 1000 = 3Mld^{-1}$ (approximately).

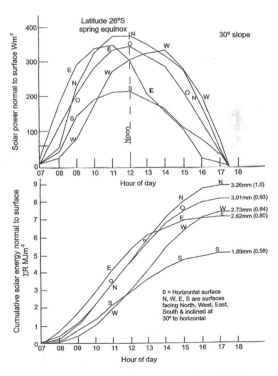

Figure 4. Solar power and daily energy received at Earth's surface at 26°S at spring equinox on 30° slope.

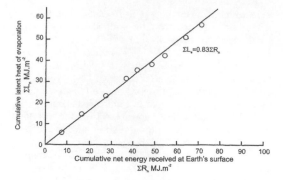

Figure 5. Relationship between cumulative net energy received at Earth's surface and latent heat of evaporation.

Figure 6 shows a corresponding set of curves for 45° surfaces at the southern hemisphere spring equinox. In this case the contrast between the north-facing slope and the other slopes is even greater. Figure 7 shows (above) the solar power curves for a day at the summer solstice at Johannesburg and (below) the solar power curves for a day at the winter solstice. The power curves for the autumn equinox should be very similar to those for the spring equinox and were not measured.

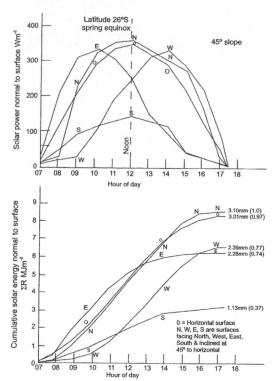

Figure 6. Solar power and daily energy received at Earth's surface at 26°S at spring equinox on 45° slope.

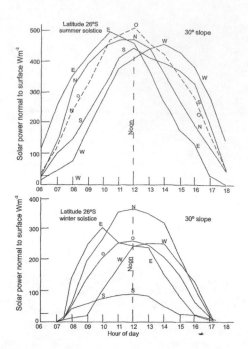

Figure 7. Solar power received at Earth's surface at 26°S on 30° slope at summer solstice (above) and winter solstice (below).

5 SOLAR ENERGY LOSSES IN THE ATMOSPHERE

Table 1 summarizes apparent losses of solar energy in passing through the atmosphere, for this study:

Table 1. Comparison of theoretical solar energy at outer limit of the atmosphere with measured solar energy at the Earth's surface (at 26°S, 1700m)

Noon peak at:	Calculated R_A	Measured R_i	R_i/R_A
Spring equinox	744 Wm^{-2}	340Wm^{-2}	0.46
Summer solstice	826	510	0.62
Winter solstice	538	260	0.48
Daily integration		Average	0.52
Spring equinox	23.85MJm^{-2}	14.01MJm^{-2}	0.59
Summer solstice	21.43	8.25	0.38
Winter solstice	15.48	5.86	0.38
		Average	0.45

Hence on the basis of the present fairly limited set of measurements, and an assumed value for α of 0.4, energy losses through the atmosphere average about 0.5 of the energy available above the atmosphere. Hence the loss ratio of 0.5 or 50% reported by Flohn is a good intermediate value to assume for purposes of estimation, although it is lower than the 70% loss apparent from Figure 3.

The data for Figure 3 were measured during a period of cloudy weather (unusual at Johannesburg) when the water content of the atmosphere was high. This probably accounts for the greater losses of energy in the atmosphere.

6 CALCULATING SOLAR ENERGY INCIDENT ON A SLOPE

Figure 8 shows the principle for calculating the solar power incident to a slope at noon:

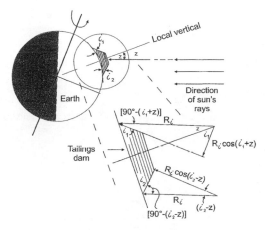

Figure 8. Principle for calculating solar power incident to a slope at noon considering latitude and zenith angle.

z, the zenith angle at noon is given by

$$\cos z = \sin\phi\sin\delta + \cos\phi\cos\delta \qquad (3)$$

For cloudless weather R_i is assessed as 50% of R_A, and for cloudy weather Figure 3 can be used. Referring to Figure 8, for a slope where the sun's rays at noon strike the surface at an angle

$$90° - (i_1 + z)$$

the incident solar power will be

$$R_s = R_i\cos(i_1 + z) \qquad (4a)$$

And for a slope where the noonday sun's rays strike the surface at

$$90° - (i_2 - z)$$

the incident solar power will be

$$R_s = R_i\cos (i_2 - z) \qquad (4b)$$

Example:

$\phi = 35°N$, $\delta = 0$, $i_1 = 30°$ (e.g. slope to N), $i_2 = 20°$ (e.g. slope to S).

Then $\cos z = \cos\phi$ and $z = \phi$.

$90° - (z + i_1) = 25°$ and $R_s = R_i\cos(65°) = 0.42R_i$; $90° - (i_2 - z) = 105°$ and $R_s = R_i\cos(-15°) = 0.97R_i$.

The power can then be integrated over the length of the day as shown in section 3 to give the daily incident energy.

7 CONCLUDING REMARKS

This paper has illustrated some seasonal effects for solar radiation on slopes having different angles and orientations. The quantitative measurements were made at 26°S latitude, and only the effects of slope angles of 0°, 30° and 45° were investigated. The results can be used, in conjunction with equations (1), (4a) and (4b) and Figures 2, 3 and 8 to make qualitative or quantitative assessments for other latitudes and slope angles, and for varying amounts of cloud cover.

REFERENCES

Blight, G.E. 1997. Interactions between the atmosphere and the Earth. *Geotechnique* (42) 715-766.
Blight, J.J. & Blight, G.E. 1998. Using the radiation balance to measure evaporation losses from the surface of the soil. *2nd Int. Conf. on Unsaturated Soils*: 327-332. Beijing, China.
Blight, J.J. & Blight, G.E. 1999. The microlysimeter technique for measuring evapotranspiration losses from a soil surface. *Geotechnics for Developing Africa*: 23-27. Netherlands, Balkema.
Blight, G.E. & Lufu, L. 2000. Principles of tailings dewatering by solar evaporation. *Tailings and Mine Waste '00*: 55-63. Netherlands: Balkema, 55-63.
Blight, G.E. 2001. Evaporation from wet and "dry" beaches of tailings dams. *Tailings and Mine Waste '01*: 33-39. Netherlands: Balkema.
Flohn, H. 1969. *Climate and Weather*. UK: Weidenfeld & Nicolson.
Robinson, N. 1966. *Solar Radiation*. Netherlands: Elsevier.
Tyson, P.D. & Preston-Whyte, R.A. 2000. *The Weather and Climate of Southern Africa*. South Africa: Oxford University Press.

Tailings and Mine Waste '02, © 2002 Swets & Zeitlinger, ISBN 90 5809 353 0

Stabilization of uranium in pitwaters using phosphate and coal tailings

S.C. Muller & T. Delaney
SRK Consulting, Fort Collins, Colorado

R. Ryan & J. Weber
SF Phosphate, Vernal, Utah and Rock Springs, Wyoming

S. Swapp, C. Eggleston & D. Nash
University of Wyoming, Laramie, Wyoming

ABSTRACT: Uranium and radium daughter products present potential exposure risks in abandoned or inoperative open pit waters and water filled tailings ponds. Wildlife, livestock and even human exposure have been documented as problematic throughout parts of Wyoming, Utah, New Mexico, Arizona and Texas. While backfilling and capping such impoundments reduces direct exposure risk, slugs of impacted groundwater can degrade aquifer water quality and discharge radionuclides to surface waters.

Water treatment is time consuming and economically prohibitive under current uranium economics. While this situation may change, the technical practicability of alternative stabilization technologies bears consideration. In an effort to optimize costs, the utilization of other mine wastes and tailings has been evaluated as a stabilization medium. Specifically, tests are being conducted through the support of the Abandoned Mine Lands Department (AML) of Wyoming Department of Environmental Quality (WDEQ) utilizing phosphate and coal waste and tailings. The Wyoming mining industry is being encouraged to participate in the pilot-testing phase. The nature of this research will initially focus on phosphate and coal tailings that are considered disposal problems at abandoned mines. This investigation will involve a statewide characterization of the uranium problem in pitwaters, column and settling testing results, and plans for pilot testing at impacted, open pits. The presentation will focus on Phase I data interpretation.

1 INTRODUCTION

Steffen, Robertson and Kirsten (SRK) Consulting, the University of Wyoming have commenced a research and development (R&D) project for the Wyoming Department of Environmental Quality (WDEQ) – Abandoned Mine Lands Division (AML) and the phosphate / coal industries. The purpose of this R&D project is to immobilize soluble uranium, radium and/or associated metals in uranium pitwater using phosphate or coal mine byproducts or wasterock. This stabilization technology may also prove beneficial for treatment of uranium in groundwater in-situ using the media as a reactive barrier or in surface water treatment cells prior to surface water discharge. For purposes of this investigation, we will refer to sorption media as Wasterock Attenuating Phosphate (WRAP) or Wasterock Attenuating Carbon (WRAC).

1.1 *Defining the problem*

AML has supported the backfilling of water-filled pits nationwide for some time. It has generally been assumed that a backfilled pit represents fewer risks to human exposure and ecological pathways. Recent uranium reclamation has favored the disposal of tailings in mined-out open pits, with special construction to eliminate excess pore-water pressure and provide groundwater pathways around the tailings. This approach removes the tailings from the active surface environment, reduces the chance of future human intrusion, controls the release of contaminants to acceptably low levels, and virtually eliminates any release of radon from the tailings to the atmosphere. The WDEQ normally recommends designing a clean fill to the water table with tailings deposition atop the clean fill to minimize impact of further dissolution of uranium and other metals.

Most of the uranium mines in Wyoming lie on lands administered by the Federal government, in particular, the Bureau of Land Management. In 1999, the U.S. Congress asked that the National Research Council assess the adequacy of the regulatory framework for hardrock mining on federal lands. The regulatory framework applied to hardrock (locatable) minerals--such as gold, silver, copper, and uranium--on over 350 million acres of federal lands in the western United States. To conduct the study, the National Research Council appointed the Committee on Hardrock Mining on Federal Lands in January 1999. This Committee surmised that if backfilling of mines is to be considered, it should be determined on a case-by-case basis, as was concluded by the Committee on Surface Mining and Reclamation (COSMAR) report (NRC, 1979). Site-specific conditions are too variable for prescriptive regulation.

Historical WDEQ records do show that certain abandoned or inactive uranium pits contained elevated levels of uranium, radium and associated metals of concern. The nature and extent of this problem has never been formally documented and would be incorporated as part of this proposed investigation. For those pits that have already been backfilled, it is unlikely that groundwater has been impacted due to design specifications that require clean backfill and segregation of wasterock above the groundwater interface.

Under a worst-case scenario, backfilled material could represent a concentrated source for uranium leaching especially if the pitwater chemistry is either acidic (< 5.5 s.u.) or moderately alkaline (>10.5 s.u.). These conditions could occur irrespective of irrespective of segregation technologies due to saturated uranium minerals and associated metals in the pitwalls. Groundwater gradients post-backfill typically change from an inward flow direction to a mound prior to returning to a pre-mine condition or gradient. The backfilled material may even impeded throughflow of groundwater is the fill material is finer or more tightly compacted than the original sediment. It is anticipated that the design specifications preferred by WDEQ have circumvented this possibility due to segregating backfill material. However, breaches in cover cap due to erosion or net-positive infiltration of precipitation can occur due to lack of vegetation. If a groundwater or surface water discharge problem were to develop in the future, a sorption media such a phosphate or coal waste might be utilized as a reactive barrier. Under such as scenario, the impacted groundwater could be directed through capture trench amended with the reactive media. Discharging surface waters with elevated metals could likewise be treated directly once detention times for reactivity are known with greater certainty.

The dynamics of uranium solubility is controlled by many factors with oxidation-reduction potential (ORP), pH and dissolved oxygen (DO) being critical to uranium stability. Iron abundance and valance state plays a significant role in microbial stabilization of uranium minerals. Organic carbon, iron species, phosphate, arsenic and sulfur in the water can complex uranium stability fields. Conceptually, uranium left in a pit wall is either oxidizing or in solution due to the aerobic prevailing conditions.

Exceptions may exist when high amounts of sulfide minerals are oxidizing contributing to a lower pH condition. While uranium can stabilize under these low pH conditions, other metals such as arsenic become mobile as pH decreases. Backfilling material to displace the pitwater should ultimately create an anaerobic or reducing condition that can be accelerated by natural, bacterial activity. The stabil-

ity of the uranium and associated radium daughter products in a backfilled pit will vary in accordance to the original pit water chemistry, the chemistry and physics of the backfill material. Other parameters will include the rate of bacterial activity that is controlled by microbial abundance, temperature and available nutrients, hydrologic conditions and caprock-protection from additional surface oxidation (i.e. surface water infiltration).

1.2 Advantages of retaining an open uranium pit

Water in many open pits (good and bad water alike) is largely accessible to wildlife although serious highwall hazards generally restrict human and livestock entry. Highwalls, unless gradually sloped or terraced, represent health and safety risks to humans and livestock. Often highwalls and pitwaters become habitat for wildlife serving as a refuge from the sun and wind. Occasionally the pitwall and pitwater may become favorable habitat for food sources and rearing young. Often wildlife species include sensitive raptors and bats if the open pit interconnects with underground workings.

Non-depleted, uranium ore bodies with pit access represent an asset for future resource development. Backfilling these pits diminishes the economics for future extraction. Treatment of waters prior to backfilling will ensure that groundwater is not degraded and may reduce the overall cost of future monitoring.

1.3 Wyoming phosphate and coal waste

Phosphate waste associated with the Phosphoria Formation of western Wyoming can represent a source for arsenic, selenium, cadmium and radionuclides if not properly managed. While Wyoming does not have the number of abandoned phosphate mines as the neighboring states of Idaho and Utah, such workings and wasterock represent a challenging, waste disposal scenario. Most Wyoming phosphate beds are steeply dipping requiring either underground mining or vertical stoping for extraction. Collapsed portals, highwalls and subsurface workings offer marginal void space for selective backfilling of the metal-laden wasterock.

Abandoned coal mines are widespread in Wyoming and can be found proximal or within 50 miles of most of the open pit uranium deposits. Waste rock generated from the prior mining includes "gob", scoria or cinder, fines, carbonaceous clays, pyritic and oxidized coal. Metals and anions of environmental concern are known to occur within several of the mining districts with wasterock elevated in selenium, sodium, arsenic and sulfur. Not all of the prior mining operations for coal extraction are amenable for local backfilling. Some of these coal workings represent erosion problems that can liberate metals from the wasterock.

It is uncertain how widespread the phosphate and coal wasterock with metals of concern might be throughout the State of Wyoming. During the uranium characterization Phase, we will attempt to better define this issue through sampling and review of mining records. Controlling metal or acid generating capacities of phosphate or coal wasterock would be most difficult at the actual mine sites without creating a RCRA-type or lined repositories. The abandoned and treated uranium pit offers a potential solution for this phosphate and coal wasterock disposal that is more resistant to metals migration.

1.4 *Research rationale and approach*

Prior investigators have demonstrated that uranium has an affinity to adsorb to phosphate ore and coal. Work recently conducted by the Oak Ridge Laboratory has shown that both phosphate and coal shown promise for stripping uranium out of solution. Activated carbon is commonly utilized in metal processing and more recently in water remediation projects. Work recently conducted by UFA Ventures at the Bunker Hill CERCLA site in Idaho (Chen, 1997) has shown success in the mitigation of soluble metals such as arsenic, copper and zinc with the mineral, Fluorapatite. The chief mineral associated with phosphate deposits in Wyoming and the neighboring state of Idaho and Utah is Fluorapatite.

Naturally occurring deposits of phosphate and coal often contain above background levels of uranium making associated waste products more expensive to stabilize on site. The potential for utilizing phosphate or coal byproducts and/or wasterock as a uranium-stabilizing agent could be mutually beneficial to the reclamation of uranium, phosphate and coal sites.

Methods for conditioning uranium tailings at mill sites to reduce harmful effects on the environment are directed toward recycling of the waste waters, precipitation and removal of radium from solution, and neutralization to precipitate heavy metals. The slurry is usually transported to an impoundment basin where the solid particles settle out, and the effluent is treated for removal of contaminants before discharge into a settling pond. Treatment ponds may contain low permeability liners where appropriate to control seepage. Barium chloride may be added to the tailings to precipitate dissolved radium-226 as radium-barium sulfate. Lime and limestone may be added to the tailings to raise the pH. Research is continuing in the industry into methods for further removing the contaminants during the mill process. The barium chloride treatment is neither effective for hexavalent uranium removal and/or the complexing of associated metals of concern.

Evidence supports that f phosphate and/or coal byproducts and waste can be used to stabilize uranium in solution. It is postulated that the attraction of uranium to these media will limit the potential for desorption or release back into groundwater or surface waters. What is in question are the kinetics and sorption sites to attract the uranium and associated metals. Uncertainty regarding the optimal material size or surface area, detention time, effect of catalyzing cations, complexing anions and organic carbon, field parameters (i.e. pH) will be investigated in this study. This program will also determine the optimal application process and the comparable economics to other stabilization methods. This research and development program has been design to answer these questions and to bring this new technology to the forefront of industry.

In terms of application of the stabilization media there are various possibilities, including:

- Reactive barriers for surface drainages or groundwater as French drains;
- Direct application to the pit through dispersion and slow settling methods;
- Pump and treat technology through the reactive media; and or,
- Admixing backfill media to stabilize metals from dissolution.

2 PROJECT GOALS AND OBJECTIVES

The project goals are to:

- To confirm the effectual use of phosphate and coal wasterock or byproducts as media for uranium and associated metal stabilization in pit water;
- To identify the nature and extent of uranium-contaminated pitwater in Wyoming;
- To determine the comparable economics of uranium stabilization using this technology to conventional technologies such as Barium Chloride treatment; and,
- To identify opportunities for a pilot scale or demonstration project that would be subsidized at least in part by the uranium, phosphate and/or coal industries.

If the above goals are met, the primary objective would be to bring this technology into use by industry for stabilization of uranium in pitwater expanding options for surface water use, uranium waste disposal and groundwater protection. An ancillary objective would be to demonstrate to AML, the State of Wyoming is model for environmental protection, waste reduction and mineral resource conservation.

3 PROJECT SCOPE

The project scope has been divided into three (3) discrete Project Phases, each spanning about six months over a year and one-half (1½) timeline. Each phased activity is dependent upon the results of the prior Phase and are stand-alone in respective reporting requirements. This will enable WDEQ-AML can assess each phased milestone as independent funding blocks for "go or no-go" decisions. Phase 1 will entail initial column testing and characterization of uranium pitwater and phosphate/coal wasterock exhibiting high metals accumulations. Phase 2 will test three actual pitwater samples and Phase 3 will be a pilot scale test at a uranium pit or effluent discharge site.

4 PHASE 1. INITIAL COLUMN TESTING AND PIT CHARACTERIZATION

The first Phase of the R&D project has been expanded from column testing utilizing phosphate and/or coal media to the characterization of the problem in water-filled, uranium pits. Since this database is non-existent to incomplete, WDEQ would like to identify the extent of the problem prior to implementing the cure. Problem phosphate tailings and coal with significant metal values will also be identified from the literature. To accomplish the characterization step will require identification of water-filled pits and access arrangements. While the primary focus will be on AML pits, Phase I will also include contacting uranium, phosphate and coal mine and impacted property owners.

Generally, there has been reluctance by uranium pit owners to participate in any public R&D program that might direct public attention to issues that they may have in water quality. Phase 1 will also consider a settling test should favorable results occur in the column tests. The primary goal of the R&D program is to identify favorable attenuating properties of either phosphate by-product or coal tailings. However, an ancillary goal is to develop an economical product that will encourage usage by the industry. To attain this secondary goal, amenability to application will be critical. Therefore, a Phase 1 column or settling test may reveal characteristics of the material that will facilitate alternative application methodologies such as broadcasting pelletized media or slurry spray.

4.1 Task 1-1. Column and settling vessel construction

The first step in the process is to determine the geochemical mechanisms that cause either coprecipitation or attenuation of the uranium, radium and associated metals species in pitwater. Variables such as pH, retention time, adsorption kinetics and resistance to desorption are the factors of primary interest during this first Phase of testing. Optimization of particle size, concentration and application rate, while important for commercialization, are of less importance than understanding the process works. To simulate conceptual conditions, a column or settling vessel has been chosen for testing in Phase 1.

The column that will be constructed will be of clear plastic and of a diameter of four inches (Muller, 1997). The length of the column that will be oriented vertically will be an arbitrary length of four feet. The column will be filled with tailings media of uniform size and type rendering the length factor of less importance than the surface area of the material. To promote uniform saturation the media size will be minus 10 mesh. To ensure a good hydraulic conductivity, the minimum size of the media will be above 100 mesh. To attain maximum saturation, the column will be filled from the bottom such that at least two thirds of the media will be completely saturated with the surrogate uranium pit water. Since low temperature thermodynamics is often enhanced at zones of evaporation, at least one third of the sample medium will remain unsaturated. Sample ports will be developed in the column to enable extraction of both liquid and solid without disturbing the column. This is necessary to observe the temporal aspects of precipitation kinetics and water parameters over time. Three columns will be prepared for simultaneous testing using: (1) a low pH pitwater, (2) a high pH pitwater and a (3) a neutral pH pitwater or control. Later these vessels will be utilized for replicate testing by the University of Wyoming.

4.2 Task 1-2. Dry sample analysis and preparation

The surface area of the phosphate or coal particle will be an important consideration for optimizing attenuation of uranium, radium and other metals of concern in the pitwaters. Initially the characteristics of the solid media will be optimized for passing fluid though a vessel such as the column. In this regard the distribution of solid size will be identified and likely optimized for water retention in a flow through vessel. Settling characteristics of the particles might be enhanced to facilitate exposure and detention if the particles are broadcasted onto the pitwaters. Specific gravity, attenuation sites and other considerations will be necessary to optimize the grain-size and composition for this purpose. Attempts will be made to develop a product that can be used directly in the pit. Therefore, size, shape and other parameters such as specific gravity and settling speed will be carefully monitored for reference in future applications.

Chemical and mineralogical analyses of the dry media will be characterized to ensure prior knowledge of any naturally occurring radionuclides, metals, potentially reactive cations such a phosphorus, arsenic, vanadium and iron. A full suite of oxides will be analyzed and the forms of sulfur will also be categorized.

4.3 Task 1-3. Surrogate uranium water preparation

To minimize the complexing affinities of other cations and anions to reaction kinetics, the initial pitwaters will be constructed at Energy Laboratories in Casper using a typical uranium concentration that would normally be above acceptable standards for drinking water. This surrogate pitwater would be prepared at a range of pH for application in three separate columns to determine the influence and change of pH on reaction times. Dissolved oxygen (DO) and oxidation-reduction potential (ORP or Eh in millivolts) will be calibrated to reflect field conditions and carefully monitored during the testing phase.

4.4 Task 1-4. Column testing and analysis

Prior to initiating the column testing with the uranium surrogates, the control column will be filled with distilled water for 24 hours, and the leachate will be analyzed to ascertain any soluble complexes under normal weathering conditions. Field parameters (re: pH, DO, ORP, Specific Conductance and iron species) will be analyzed to ascertain baseline reactions and solubility of cations and anions in the column. Key cations in the media (i.e. phosphorus) will be analyzed using unfiltered (re: total) chemistry.

As mentioned previously, three column tests would commence using acidic, neutral and alkaline pH-adjusted uranium surrogate effluents to the columns. Water sampling would occur on 8-hour intervals to determine potential changes in field parameters and retained for more detailed analysis of radionuclide constituents for 36 hours. The frequency would then be reduced to one per 24 hours for seven days. Solid media would also be extracted during these sample intervals and preserved by freezing to ensure that biological activities would not cause additional reactions prior to analysis. Microbial activity will be documented if noted.

Chemical and radiometric analyses of the solid media such as autoradiographs (i.e. X-ray film) will form the basis for corresponding mineralogical analysis of the samples. For example, if the solid media exhibits an apparent enrichment of radionuclides, then an autoradiograph will delimit the sites for detailed mineralogical test using the scanning electron microprobe (SEM) at the University of Wyoming and other selective techniques for mineralogical determination such as powder camera X-ray diffraction. In this manner, attenuating affinities such as mineralogical precipitation, chelation, cation exchange and other adsorption phenomenon can be discriminated.

As important as determining how, when and where attenuation might be occurring is the affinity for reversal of the process such as desorption or dissolution of the radionuclides. In order to better predict stability over a range of conditions post-treatment, the columns at the end of the testing period will be treated with another surrogate flush of slightly acidic or alkaline distilled water. In this instance the acidic column will be treated with t he alkaline bath and the alkaline column will be treated with an acidic bath to look at reaction extremes post-treatment. The neutral column will be treated with either acidic or alkaline solution depending upon the last pH of the lixiviate. If, for example, the last pH is 7.5, then the wash would be slightly acidic to bring the sample potentially back to a neutral range. The desorption testing will follow the same sampling sequence as the initial column tests.

4.5 Task 1-5. Pitwater access and permitting

Data on file with the AML division of the WDEQ shows uranium enrichment of pitwaters, however, much of this data is almost 10 years old or non-existent for some of the open pits. Further, numerous, water-filled open pits do not currently fall under AML and are in suspended or maintenance mode by the present owners or operators. The objective of this Task would be to identify, contact owners, and secure access release for sampling of all of the major uranium pits that have not been backfilled in the State.

4.6 Task 1-6. Pitwater sampling and analysis

While sampling pitwater might appear fairly straightforward, in deeper pits, there can be a high level of heterogeneity of water chemistry associated with limnological stratification. Shallow zones that are aerated by wind action may have significant algal blooms under sustaining pH conditions. Deeper zones with limited light might even develop anaerobic conditions particularly in the winter under icing conditions. Since uranium solubility is highly pH, ORP and DO dependent it is likely that some pits may exhibit stratification of uranium values. Therefore, obtaining representative samples requires some characterization of pit bathymetry, in-situ field parameter measurement and perhaps even selective sampling of limnological profiles for uraniferous samples. To achieve the goal of rapid characterization in the field, we plan to access the pitwaters in a small craft than contains a depth-finder (radar) for bathymetry, global positioning system (GPS) for replicate sampling verification and field parameter probes with sufficient cable to take discrete measurements at depth. A gross gamma probe with ample cable for underwater measurements will enable discrimination of "hot-zones" if any for select sampling or sample compositing with less mineralized waters.

Where background radioactivity appears elevated, samples will be collected to analyze chemical parameters including uranium and radium species, RCRA metals, key uranium complexing cations (Re:

P, As, V, Se, Fe^{+2}/Fe^{+3}), organic carbon and sulfate. This surface water quality data and applicable groundwater data from available wells will be compiled for future reference and potential testing during Phase 2 and 3.

4.7 Task 1-7. Replicate analysis and interpretation

The defensibility of the data generated during the testing program may be challenged by industry or other scientific researchers. To enhance the program's credibility and to enhance quality assurance/quality control (QA/QC) of the evaluation methods and results, replicate samples and testing will be undertaken by the University of Wyoming Technical Support Team (UWTST). In this manner, variance in data collection methods, sample-handling techniques, sample preservation and extraction procedures can be critically evaluated and if necessary challenged.

To minimize error during the testing Phases 2 and 3, sample splits of all water samples will be procured in accordance to a QA/QC plan. This plan will be developed prior to any column testing using actual pitwaters and will also address any recommended changes that the UWTST might have in the column testing and analysis methods being conducted by Energy Laboratories.

4.8 Task 1-8. Data reporting and preliminary interpretation

Upon completion of the testing phase of the column work and the collection of the pitwater water samples, there will be the need to report the interim results with interpretations and recommendations for Phase 2 testing. As part of our internal QA/QC process, we plan to generate a draft for review by both the WDEQ and UWTST.

5 PHASE 2. BENCH SCALE TESTING OF THREE PITWATERS

Phase 2 activities would commence upon reaching consensus on the results of the Phase 1 research and characterization. Potential modifications to the testing procedures might be employed based upon the results including the consideration of an alternative stabilization medium. If the results from Phase 1 do not support stabilization goals, then Phase 2 might even necessitate modification of the physical properties of the stabilization medium to enhance reaction time or to limit desorption affect.

However, assuming favorable or promising results from Phase 1 tests, recommended source waters from three distinct uranium districts would be undertaken to use a the applied effluent or settling media. This real case application would be necessary to understand any complexing effects of naturally occurring cations, anions and field parameters including microbial activity, temperature and seasonal variation.

5.1 Task 2-1. Selection of test waters and methods

The inherent goal of selecting different pitwaters by uranium district is to determine the relative effectiveness of the Phase 1 stabilization technology under diverse natural chemistries. For example, a selenium-rich, bicarbonate pitwater will likely react quite differently from an iron-rich, sulfate pitwater. More than one pitwater may be used in-lieu of multi-district pitwater if extreme variation exists that represents a category of likely pitwaters to be reclaimed in the future.

5.2 Task 2-2. Collection of samples

Samples would be collected from at least three pits in accordance to previously developed protocols. Since this will likely represent a different season (i.e. winter months), sampling may be undertaken through the ice that will likely form on the pitwater. In this manner, seasonal variation will also be evaluated when compared to data collected in the fall of 2001. Delay of the sampling program until the spring or summer 2002 would impact the timing of the pilot-scale investigation or Phase 3.

5.3 Task 2-3. Complete analysis of pitwaters

While it is anticipated that the Phase 1 testing and characterization program will be most complete, other parameters, in-situ measurement techniques and/or data needs might not be recognized until after review the Phase I data. For example, microbial interaction might be a critical enhancement to stabilization of the uranium or radium. Recognition of this potential attribute in Phase 1 might necessitate a higher order of analysis than previously conducted in Phase 1. Other information that might be necessary for planning or selection of the Phase 3 pilot investigation would be the composition of the bottom sludge of the pit or substratum.

This aspect of the analytical program can be important to understand naturally occurring attenuation processes in the specific pitwaters. This insight would also be useful in designing pilot applications that deal with direct amendments of the stabilizing media to the pitwater for testing for flocculation or clarification properties of suspended and dissolved constituents.

5.4 Task 2-4. Column testing and analysis

At this juncture, it is not known which sampling media (phosphate or coal by-product) or application approach (i.e. vessel through-flow versus settling directly in pitwater) will be optimal. There is also the possibility that one methodology will not work for the various categories (i.e. sulfate vs. bicarbonate) of uraniferous pitwater. To address these possibilities, this task must be flexible to allow for trying new methods as well as replicating previous methods for comparison with the actual pitwaters.

Further, thermodynamic modeling (Task 6) will precede the actual column or settling tests to establish some predicted results. In this manner the model sensitivities can be evaluated, modified or even recalibrated for a future tool in accessing cost and performance expectations for field scale operations.

5.5 Task 2-5. Replicate analysis and data validation

Multiple methods might be required to optimize conditions for uranium and/or radium attenuation. Therefore the methodologies employed for testing in Phase 2 might be some variation of previously employed methods. To ensure comparability with the Phase 1 data, identical controls suing the prior methodology of column testing would also be applied during Phase 2 in addition to variants that have evolved due to interpretation of the data.

5.6 Task 2-6. Data interpretation, modeling and reporting

Due to the complexity associated with working with actual pitwaters, established thermodynamic models such as MINTEQA2 or PHREEQQC and will be calibrated with previously generated data from Phase 1.Analytical data from the actual pitwater will be entered to determine critical phase relationships such as pH control.

The first level of modeling will actually proceed the Phase 2 testing to ensure that the field parameters are optimized for success in uranium and/or radium attenuation. Should complexing or competing analytes of field parameters exist in the natural waters, a separate test vessel utilizing additives determined from the modeling might be required to achieve optimal results.

6 PHASE 3. FULL SCALE PILOT TESTING

The pilot scale testing is proposed to evaluate reaction times, kinetics, desorption effects under actual field conditions. To achieve this goal, the Phase 2 results and institutional controls/ownership will likely guide the selection process. At this time it is not known if a classic water treatment or through-flow process through a vessel or cell will be the best method or if direct amendment is viable. Therefore, what we have planned is a combination test of both methods. To ensure that there are no negative impacts of the testing on the pit waters, isolation of the test waters from the main pit body would be optimal. In this manner, the pilot waters would be controllable should an off-site disposal of uranium-enriched byproduct become necessary due to regulatory guidelines.

Likely, a lined, constructed cell at the site would be the best precaution against dilution with pitwaters and/or creation of a licensed by-product. Because of this possibility, a permit analysis step has been added to Phase 3 to ensure that proper regulatory guidelines are followed and that there are not regulatory surprises that cause additional liabilities to WDEQ, AML, the owner/operators and/or the research entities. During this more cost-intensive Phase of activity, industry capital support will be easiest to obtain using the favorable results from Phase 1 and 2 investigations as a negotiating framework (Task 1). Further, WDEQ-AML Division would like to understand any additional permitting or compliance hurdles that might be required if the Pilot Program generates a byproduct that could require additional licensing. We therefore added another Task (2) to address this concern. Construction and operation of the Pilot Plant will likely require additional permitting requirements that will also be addressed prior to commencing field activities.

6.1 Task 3-1. Negotiations with mining companies

Preliminary discussions with Wyoming uranium companies indicates an interest yet a reluctance to participate in a testing program with WDEQ. It appears that the main concern is that if adverse water quality is detected then the companies may be subject to interim clean-up standards while the property activity is on hold due to depressed uranium prices. The one phosphate producer in Rock Springs has indicated an interest in securing other markets for phosphate byproducts. Several coal companies appear interested in off-site disposal of coal fines and carbonaceous shale.

AML sites may also have phosphate or coal feed-stock for uranium pitwater stabilization. Potential problems for disposal of waste rock at existing AML sites will be the potential for the deposits to contain background uranium, radium and other metals of concern such as selenium or arsenic (Re: "impacted wasterock"). A decision to include or not include "impacted wasterock" would likely follow the permit analysis and negotiations with uranium producers. Obviously, there are advantages of disposing of a "impacted wasterock" in an existing uranium pit, especially if the net effect is protection of human health and the environment. However, a current operator may need assurances that they will not increase their liability under such as scenario.

Phosphate and coal producers who could develop new markets for their product or byproduct might be willing to capitalize some of the pilot phase testing. A consortium of producers might be better than a single producer, so that such participants do not claim patent rights thus limit product availability. If we can attract the support of uranium producers, it would likely require certain assurances of confidentiality or liability protection that can only be granted by the WDEQ, DOE or the NRC. However, if a particular company has a known problem pit, the State could enter a cooperative agreement perhaps using the Wyoming Voluntary Remediation Plan (VRP) as a mechanism for implementation at a prospective Pilot Test site. Due to the number of variables associated with not knowing the specific treatment media and applicability of existing pits, a full financial assistance package will likely not be available until after the results of Phase I and II results become publicly available. Other potential sources for co-funding may include Federal entities such as EPA, Army Corps of Engineers or Department of Interior.

6.2 Task 3-2. Permitting for a pilot test and discharge

The regulatory compliance requirements for testing at an abandoned mine or licensed mine site may differ depending upon how the treatment technology is viewed by the regulators. Permitting may be required for the Pilot Test particularly if there is a discharge to a regulated waterbody even if it is an existing permitted pit. To minimize potential for future liability, certain permitting will be investigated. Potential for generating mixed waste, processed uranium, RCRA or TSCA waste will need to be assessed and a decision generated by the Wyoming Attorney General's office and/or EPA or DOE if the byproduct of stabilization could end up being regulated. In certain instances where the determination is uncertain or could take onerous amounts of time, operational precautions may be implemented such as disposing of all byproducts at an approved DOE repository. Permitting uncertainty and/or ambiguity may also influence the selection of the Pilot Test site and the nature of the testing program.

6.3 Task 3-3. Construction of the pilot plant

It is anticipated that minor surface disturbance would be associated with the Phase 3 Pilot Testing program. However, this assumption is predicated upon there being sufficient room at or near the pitwater level of the selected pit. Should water need to be pumped up the pit face or if a lined-cell requires construction, the pilot activities may require an elaborate construction and reclamation plan. For current budgeting purposes, it is assumed that existing infrastructure will be available at the site requiring little, if any, site preparation and reclamation activities. Byproduct would be disposed on site and all mechanical equipment would be skid-mounted or mobile.

6.4 Task 3-4. Pilot scale operations and testing

The most difficult element to cost at this time is that related to the actual Pilot Scale Testing since the instrumentation will evolve during the research and the testing will likely be constrained by site specific conditions, ownership and regulatory jurisdictions. While it is difficult to speculate all of the particulars associated with Pilot Scale Operations, certain objectives can be stated as follows:

- The Pilot Test should be conducted over a block of time significant enough to measure results. For example, if a settling test is undertaken, results may require measurement over a much greater period of time than that undertaken through a filtration vessel or extraction column.
- The Pilot Test should include the ability to compute the mass balance of the treated waters, losses due to evaporation or precipitation and volumes of applied media or reagents. Since pitwaters are in dynamic flux with surface and groundwater, the Pilot Test will likely not be of the scale to treat and analyze an entire pit waterbody.
- The Pilot Test should utilize to the greatest extent possible, readily obtainable stabilization media. The Pilot Test should consider overall economics during the Pilot Testing Phase to the extent possible without compromising the desired result. All of the Pilot Scale testing will need to be extrapolated to a full-scale application for costing purposes.
- Short- and long-term monitoring of the test results should be considered in order to ascertain "long-term effectiveness". To the extent physically possible, the monitoring program should address impacts over a period of years. This may necessitate consideration of a test repository or "monitored treated backfilled pit" (MTBP).
- To the greatest extent possible, the Pilot Testing should evaluate different forms of attenuating media to ascertain efficiencies in alternative distribution and detention methods. Treatment scenarios will likely vary if the future intended use of the pitwaters is elimination due to backfilling or retention as a surface water supply. Clean-up standards may also vary depending upon this intended future usage. It may be worthwhile to consider alternative Pilot Testing methods versus just one to address these differences.

We can assume that there will be three alternatives for treatment of uranium in pitwater:
(1) amendment of pitwater or backfill to protect groundwater;
(2) short-term treatment scenario to deal with influent production of uranium in surface waters; and

(3) long-term treatment scenario of surface waters such as design of a passive or continuous treatment system that can persist over long periods of time.

6.5 Task 3-5. Documentation of results

Each technical aspect of the Pilot Phase results will be documented in the utmost detail for future reproducibility. At this point, it is anticipated that this final technical document will include all of the prior Phase summaries and elements of Task 3-6 (Economic Analysis) and Task 3-7 (Highwall Reduction Analysis).

6.6 Task 3-6. Economic analysis and future risk

In order for any remedial technology to be deemed technically practical, it must be both technically feasible and within economic affordable. Generally "economically affordable" can be defined as part of the cost of extractable operations that are deemed profitable. Surface mining of uranium, at the ore grades found in Wyoming, have not paid the cost of reclamation in recent time. For those operators who are still in business, their biggest asset (re: open pit in a non-depleted orebody) is also their biggest environmental liability. Should the uranium market price not recover in the short-term, many of these operators will likely abandon their water-filled, open pits.

Can an open pit or potential asset be retained in its present form without being a risk to human health and the environment? If a technology exists to reduce the exposure from uranium, radon and carcinogenic metals from surface waters, perhaps non-depleted open-pits should stay open. If the pits must be backfilled to reduce potential risk to caving or steep highwalls, then one must ask the question about enhanced groundwater degradation and potential for groundwater discharge to surface waters.

If the proposed remedy (i.e. backfilling) presents yet another risk such as groundwater contamination, the cost of mitigating that other risk must be included in the economic formula. When local groundwater discharges to surface waters, the ensuing reclamation cost might be amplified. Therefore the cost of backfilling, compaction, regrading and reseeding cannot be the sole economic consideration for pit reclamation.

It is unlikely that uranium pitwater stabilization will not reduce highwall risks that are commonly the driving rationale for open-pit backfilling. However, stabilization technology could however eliminate the need for pit backfilling to prevent exposures to adverse water quality. Cleaner pitwater would equate to fewer negative groundwater and off-site surface water impacts.

If we calculate the treatment costs on the basis of uranium reduction in pit water and extrapolate the cost on an acre-foot of water, we can then calculate the beneficial value of the treated water. For purposes of comparison to backfilling, we can look at the cost of backfilling an open pit eliminating the direct receptor pathway of the pitwater to human health and the environment as discussed in Task 3-7 below.

6.7 Task 3-7. Highwall reduction analysis

The chief rationale for backfilling pits has been the risk of highwalls to human health and the environment. Without maintaining institutional controls such as fencing, trespassers, livestock and hoofed wildlife risk plunging to their deaths or risk injury from rockslides and failures. Pitwaters also represent a potential hazard to human health and the environment in the form of radioactive exposure, metals accumulation in shoreline vegetation and the risk of drowning.

Highwalls often serve to benefit wildlife, providing refuge, nesting and perches for raptors, bats, swallows, rodents, lizards and insects, some of which threatened, endangered or sensitive status. Pitwater also can benefit stock, wildlife and local residents.

To assess the "trade-offs", benefits and/or risks associated with retaining highwalls, the first question will be stabilization. If the pit walls can be stabilized, does it may sense to keep the pit open for future access of the uranium resource or as a beneficial use for stack and wildlife?

The scope of this task will be to assess various alternatives to backfilling that will alleviate the risks associated with highwalls. While the answer certainly has some site-specific aspects, there are several risk reduction measures that could be evaluated from the standpoint of general implementability and cost comparison. A few of the scenarios that would be investigated are:
(a) Highwall stabilization utilizing buttressing, benches, terracing and rock netting;
(b) Highwall reduction utilizing "cut-back" techniques;
(c) Highwall reduction through blasting; and,
(d) Highwall reduction by selective backfilling.

Economic comparisons will be developed using two actual pits to ascertain technical applicability and specific requirements.

7 CONCLUSIONS

Conclusions will be presented during the Tailings and Waste conference based upon the first Phase of testing.

REFERENCES

Adepoju, A.Y., P.F. Pratt, and S.V. Mattigod. 1986. Relationship between probable dominant phosphate compound in soil and phosphorus availability to plants. *Plant and Soil,* 92:47-54.

Altschuler, Z. S., B.S. Berman, and F. Cuttita. 1967. *Rare Earths in Phosphorites Geochemistry and Potential Recovery* U.S. Geol. Survey Prof. Paper 575B.

Chen, X.-B., 3. V. Wright, 3. L. Conca, and L. M. Peurrung. 1997a. Effects of pH on Heavy Metal Sorption on Mineral Apatite. *Environmental Science and Technology* , **31**:624-631.

Chen, X.-B., B. V. Wright, B. L. Conca, and L. M. Peurrung. 1997b. Evaluation of Heavy Metal Remediation Using Mineral Apatite. *Water, Air and Soil Pollution,* **98**:57-78.

Davis, A., M. V. Ruby, and P. D. Bergstrom, 1992. Bioavailability of arsenic and lead in soils from the Butte, Montana, Mining District. *Environmental Science and Technology,* **26**: 461-468.

Jackson, Henry M., Chairman, Senate Committee on Interior and Insular Affairs, *Coal Surface Mining and Reclamation, An Environmental and Economic Assessment of Alternatives,* Serial No. 93-8 (92-43), 143p., 1973.

Koeppenkastrop, D. and E.J. De Carlo. 1990. Sorption of rare earth elements from seawater onto synthetic mineral phases. *Chem. GeoL,* **95**:251-263.

Ma, Q. Y., Traina, S. and T. Logan. 1993. In Situ Lead Immobilization by Apatite. Environ. Sci. Technol. **27**:1803-1810.

Muller, S.C., 1997. Utilization of pyritic tailings to mitigate arsenic mobilization *in_*Tailings and Mine Waste '97: 509-516.

Nriagu, S., 1974. Lead orthophosphates, IV. Formation and stability in the environment. *Geochim. Cosmochim. Acta,* **38**:887-898.

Ruby, M. V., Davis, A. and A. Nicholson. 1994. In Situ Formation of Lead Phosphates in soils as a Method to Immobilize Lead. Environ. Sci. Technol. **28**: 646-654.

Saperstein, Lee W., "Environmental and Legislative Consequences of Strip Mining," Chapter 29, p.419-430, of Majumdar, S. K. *et al,* eds, Environmental Consequences of Energy Production, The Pennsylvania Academy of Sciences, Easton, Pennsylvania, 531p., 1987.

Wright, J. V., L. M. Peurrung, T. E. Moody, J. L. Conca, X. Chen, P. P. Didzerekis and E. Wyse. 1995. *In Situ Immobilization of Heavy metals it: Apatite Mineral Formulations,* Technical Report to the Strategic Environmental Research and Development Program, Department of Defense, Pacific Northwest Laboratory, Richland, WA, 154 p

Xu, Y. and F. W. Schwartz. 1994. Lead immobilization by hydroxyapatite in aqueous solutions. J. Contaminant Hydrology, **15**:187-206.

30 USC 1201 *et seq.,* The Surface Mining Control and Reclamation Act of 1977, Public Law 95-87, August 3, 1977.

Tailings and Mine Waste '02, © 2002 Swets & Zeitlinger, ISBN 90 5809 353 0

The tailings pond at the Milltown Dam, Montana, can be cleaned

G. Ter-Stepanian
Armenian Academy of Sciences, Yerevan, Armenia

ABSTRACT: Problems of storage of mine tailings and of threat of a catastrophe on bursting the tailings dams as that has arisen at the Milltown Dam in Montana can find an adequate solution through settling of suspension in tanks and obtaining hard sediments able to carry a load. An upwards directed suspension force is acting in suspensions that prevents the settling of solid particles under gravity in tailings ponds. This force manifests itself in decreasing the unit weight of solid particles; therefore the solid particles settle extremely slowly. Applying the seepage force directed oppositely to the suspension force the later can be destroyed, and the solid particles settle quickly. The solution of the problem of the Milltown Dam impoundment with 6.5 million cubic yards of toxic waste should be started as soon as possible because of unsatisfactory state of the dam and long time needed for cleaning the tailings accumulated during more than a century.

The rich copper deposits of western Montana started to be worked out as long ago as in 1864 at Butte, Mo. The mine tailings formed contain cadmium, arsenic, copper and other toxic mine waste. It was quite natural for those early days to throw down the mine tailings into rivers and then to the ocean. This practice continued too long. Since 1907 the last stop of mining waste became the Milltown Dam, six miles upstream from Missoula, a college town known as the Garden City.

Accumulation of toxic waste during this long time interval led to formation of a huge pile of toxic sludge containing 6.5 billion cubic yards of toxic waste, retained by an old, 19th century timber crib dam leaking through cracks. This insecure construction can be destroyed easily by an earthquake and cause an environmental catastrophe in the Clark Fork valley. The inactivity in this conditions is intolerable.

The solution of the problem must be environmentally effective and economically acceptable. The most fundamental solution of the problem consists in cleaning-up the accumulated sludge, in other words to separate the solid constituents from the water. As it is known the sludge does not settle in tailings ponds, and continues to be in the suspending state during centuries. The reasons of such behavior were unknown until now, and were considered as an immanent property of suspensions. In reality, it is the

result of action of a new force of Nature, discovered by the Author and called the suspension force (Ter-Stepanian 1998). This force is responsible for a number of enigmatic features in some types of landslides as landslides in quick clays (Ter-Stepanian 2000) and debris flows.

The Author had investigated into this problem and found the reasons why the mine tailings do not settle in tailings ponds. The reason is rather simple. Mine tailings are suspensions. Suspending is holding of solid particles in a fluid without touching. As the direct contact between solid particles does not exist, no forces can be transmitted from one solid particle to the other. As a result no effective stresses can be formed in a suspension, and transmission of forces is carried out through the fluid. High neutral pressure develops in suspensions due to the friction between the sinking solid particles and water. The neutral pressure and the piezometric head are high in lower layers and zero at the surface; consequently the hydraulic gradient is directed upwards. This gradient arises the suspension force, also directed upwards. By its size the volumetric suspension force J equals to the difference between unit weights of suspension and water, $J = \gamma_m - \gamma_w$.

Solid particles being submerged in water experience the known Archimedean or static uplift, which makes the unit weight of solid particles equal to the static (submerged) unit weight $\gamma_{st} = \gamma_s - \gamma_w$, where γ_s is the unit weight of solid particles. The suspension force J, being directed upwards, decreases the static unit weight γ_{st} of solid particles and makes it equal to the much lesser dynamic unit weight $\gamma_{dy} = \gamma_{st} - J$. Solid particles under action of feeble dynamic unit

For more information contact George Ter-Stepanian, Prof., Dr.Sc. (Eng.), Mem. Armenian Ac.Sc. 4 Sanford Dr., Shelton, CT 06484, USA. Fax: (203)922-8165.
E-mail: gterstepanian@hotmail.com

weight sink very slowly. The practice shows that the time needed for settling of mine tailings lasts decades and even centuries. Exploration of the tailings pond at the Milltown Dam will give invaluable data about state of tailings formed for almost a century.

The suspension force is related to the seepage force: both forces arise due to the friction formed due to relative movement of solid particles and water. In the first case the sinking solid particles move in relation to immovable water in tailings ponds, in the second case – the percolating water moves in relation to immovable walls of pores in the soil. Both forces are proportional to the hydraulic gradient and have flow nets. Hence follows that these forces are antagonists and being directed oppositely can destroy each other. The greater force prevails.

As the ores have quite different physical properties depending on type of rocks and the method of beneficiation used, a large range of water content included, a single method of economical and effective settling cannot be indicated. There is a number of auxiliary methods for increasing the seepage force, decreasing the suspension force, making the sediment dense, and so on that should be used additionally apart of the seepage force to find the optimum solution. Therefore laboratory and/or pilot studies of particular mining tailings are desirable. Usage of seepage force is necessary in all cases.

The settling of solid particles is conducted in tanks installed at concentration mills. Solid constituents are received in form of hard sediments able to carry a load, while fluid ones are solutions rich in toxic waste and reagents used for ore concentration. Hard sediments can be used as building, road or backfill materials while water solution can be recycled, and valuable components extracted.

The situation of February 1996 when ice jam 10 miles long and 14 feet thick built up stream and headed down the river taking away the upper 4 feet thick layer of mine tailings consisting of light harmless minerals as quartz, calcite and clay minerals is well known. That caused, as county officials say, a huge fish kill. According to the New York Times of May 7, 2001, that resulted a decline of 86% of young rainbow and brown trout and decrease of two thirds of the member of larger rainbow trout. Such cases can be easily repeated because the Clark Fork runs northwards. The ice in such rivers is taken up by river in the upper reaches earlier than in the lower ones because of more early setting in the warm weather and thawing of snow. However much more disastrous catastrophe would happen if the leaking 19th century crib dam will collapse, and the stream will take away the whole mine tailing mass, lower layers included. These layers contain the most toxic heavy minerals as cadmium, copper and arsenic, that will kill the biota in the river and enter the flood chain.

The advantage of using the tanks for cleaning-up of tailings at concentration mills creates the possibility to avoid the construction of the pipelines for transportation of the mine tailings to ponds, the construction of tailings dams and processing them on the site. The occupation of the valuable land for the storage of the toxic waste will become unnecessary. In the case of Milltown Dam the first tanks can be installed near the dam and cleaning up may be started in small scale without any large investment. Settling of solid particles will be performed using the seepage force. When the correct additional methods of settling the solid particles will be chosen from the proposed methods and their efficiency for the given tailings is proven new tanks may be added.

Natural separation of light an heavy minerals in the mine tailings pond at Milltown Dam can be used advantageously for extraction of useful minerals from the lower layers, applying the modern technology. In this case special precautionary measures will be needed for more effective obtaining of the valuable minerals from the waste. The waste should be extracted with minimum mixing.

REFERENCES

Ter-Stepanian G. 1998. Suspension force induced landslides. Proc., 8th Int. Congress Int. Assoc. Eng. Geol. and the Environment, Vancouver 1998.3:1905-1912.

Ter-Stepanian G. 2000. Quick clay landslides: their enigmatic features and mechanism. Bull. Eng. Geol. and the Environment. Berlin: Springer, 59(1):47 – 57.

Tailings and Mine Waste '02, © 2002 Swets & Zeitlinger, ISBN 90 5809 353 0

Consolidation of clay slurries amended with shredded waste plastic

N. Lozano
New Mexico State University, Las Cruces, NM

L.L. Martinez
Army Corps of Engineers, Albuquerque, NM

ABSTRACT: Large volumes of diverse waste tailings and dredged materials are generated yearly around the world by the mining and dredging industry. Waste slurries are typically fine-grained material with low permeability and rate of consolidation. These properties contribute to stability problems in their disposal and containment, which potentially lead to possible dangerous flow slides and environmental contamination. At the same time, the municipal solid waste stream of plastics has dramatically increased within the past decades. This study presents the result of consolidation tests that were performed by randomly mixing clay-plastic and placing the plastic strips in layers at mid-height of the slurry sample with 0%, 1%, 2% and 4% plastic by weight. In one sample, a wick drain was used instead of plastic strips. The total settlement, change in weight and volume of the admixed slurry were compared to the control sample. The results indicate that the clay slurry drained by a wick drain and 4% plastic layered contributed to the largest total consolidation and increase in the rate of consolidation. This indicates that the permeability of the slurry increased, which in turn contributed to faster pore pressure dissipation and consequent increase in shear strength. This increased rate of settlement of the material will contribute to the stabilization of mine tailings and dredge materials and to the beneficial reuse of waste plastic.

1 INTRODUCTION

The mining and dredging industries generate tremendous amounts of fine-grained saturated slurries. These deposits are composed of fine grained and saturated slurries with low permeability and rates of consolidation that develop high pore water pressures, low shear strength, and low bearing capacity. Therefore, understanding the physical and chemical behaviors of these freshly deposited materials is essential to a variety of engineering problems such as slope stability and pollution risks to the environment.

Several strategies and mechanisms such as: physical, vegetative, chemical, thickening, wick drains, and evaporation are available to enhance the surface stability of tailings (Johnston,1973; Dean and Havens, 1973). Although these methods do exist, they require complex practices of management from expert engineers in the field. These methods of stabilization can also be very expensive (Dean and Havens, 1973).

The failure to properly stabilize and contain surface mine tailing impoundments have resulted in slope failure, overtopping and liquefaction (Smith,1973). One of the worst failures occurred at Aberfan in South Wales on October 21, 1966 when a pit heap slide killed 144 people (Thomas, 1978). Although most tailing failures have not been as catas-

trophic, there have been cases where failures caused environmental pollution or the disruption of the surrounding ecology, communication systems and highways (Thomas, 1978). The dredging industry removes materials ranging from clean sands to weak cohesive clays (Bray, 1979).These dredged spoils primarily affect the surrounding ecology and environment. Also, the pungent smell of the fill material is very unique to dredged material (Bray, 1979). The methods for consolidating the slurries generated are by improved drainage methods, the increase of evaporation surfaces, and the preloading the soil slurries (Bray, 1979).

The total solid waste generation in the United States in 1995 was 208 million tons per year (EPA 1997). The plastic industry in the United States today is a $140-billion-per-year industry. That makes this industry greater than that of steel and aluminum combined (Ehrig, 1992). According to U.S. EPA, "The amount of plastic waste generated has been increasing by about 10% per year for the past 20 years." In 1995 the total plastic waste generated amounted to 19 million tons and only about 1 million tons were recycled. (EPA 1997). Twenty one percent of the plastic waste consisted of high-density polyethylene (HDPE) as containers of milk and water bottles (EPA, 1997) and about 2% was recycled (Khan, 1996). The fact that this plastic waste product is continually growing in volume opens up the door to be considered as a new engineering material to stabilize weak slurry impoundments.

2 SCOPE

The main objectives of this study are to enhance the stability of weak slurry deposits such as mine tailings and dredged materials and to identify and implement beneficial reuse of waste plastic, especially HDPE. Surface disposal impoundments of mine tailings and dredged materials contain diverse chemical contaminants that can adversely affect surrounding environments, ecological systems, and human beings upon penetration or breaching of their containment systems. It is of utmost importance that these impoundments remain stable to prevent hazardous contamination and disastrous flow slides. With the increased production of plastic containers, and their limited recycling possibilities, the opportunities for ingenious methods of reuse are at hand. Therefore, in this study the use of shredded HDPE plastic to stabilize fine-grained slurry impoundments such as mine tailings and dredge materials will be evaluated. The waste plastic will be used to increase the permeability and the rate of consolidation of the slurry during and after deposition of mine tailings and dredge materials. By dewatering these impoundments simultaneously as they are being deposited, the consolidation and settlement rates would be accelerated. The stabilization will contribute in preventing run out flow slides and possible environmental contamination

3 BACKGROUND STUDIES

Previous studies show that shredded plastic waste (shredded fibers) could be used to reinforce soils by increasing their shear strength and stiffness. (Edil and Benson, 1998) Shredded fibers also contribute in increasing the unconfined compressive strength and tensile strength of clays (Puppala and Musenda, 2000). Another preliminary investigation show that the discrete and randomly distributed fibers is the maintenance of strength isotropy and the absence of potential planes of weaknesses that can develop parallel to oriented reinforcement" (Schaefer, 1997). Another advantage of using shredded plastic waste in geo-technical applications is that these materials are generally resistant to biodegradation. Plastic fibers do not corrode and are resistant to biological and chemical agents such as fungi, insects, mineral acids, alkalis, dry and moist heat, and oxidizing agents (Jones, 1985). Shredded plastic wastes have the potential of becoming a beneficial re-use application in the geotechnical field. They can also save U.S. industries millions of dollars in landfill costs and at the same time minimizing environmental damage.

4 METHODOLOGY

4.1 Clay slurry

The main criteria to select the type of soil was to simulate the characteristics of mine tailings and dredge materials. The slurry was prepared using a medium plasticity clay from Chinle, Arizona with a specific gravity of 2.65. The sieve and hydrometer analysis determined that more than 85% of the soil was composed of soil particles smaller than 100 microns. The soil has a liquid limit of 37.5 and a plasticity index of 17. This clay has high cohesion, low permeability and high swelling potential. It was determined that this clay has a low permeability than soil slurries generated in mining and dredging operations. It was reasoned that if shredded plastic could increase the consolidation rate of the Chinle clay, then the same benefits could apply to soil slurries in the mining and dredging industries.

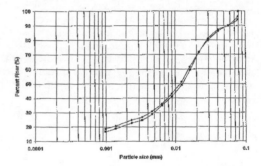

Figure 1. Particle size distribution of Chinle clay.

4.2 Plastic

Containers made of HDPE plastic for milk and water were shredded into 2 inch long by 0.25 inch wide strips. The plastics have a modulus of elasticity of 900 MP and a tensile strength of 500 MP. The density is approximately 0.95 g/cc. Strips of plastic were added to the clay in proportions of 0%, 1%, 2%, and 4% by weight. The plastics were amended to the slurry in a random manner for one set of tests and the other set had the plastic placed in layers within the clay slurry.

4.3 Design and construction of consolidometer

The standard consolidation apparatus found in soil mechanic laboratories was not adequate for testing the specimens with inclusions of plastic strips. To simulate the consolidation under field condition loads, and allow for the consolidation process to take place in a fully saturated environment, it was necessary to design and fabricate a consolidation device. The device consisted of a lever arm system with a weight hanger, load applicator and one end securely

Figure 2. View of consolidation equipment.

Figure 3. Consolidation cells, wick drain and plastic strips.

anchored to a firm support by a pin connection. The circular load applicator was placed on top of the cylindrical soil sample. Above the load applicators were placed two steel plates for the attachment of dial gauges to monitor the settlement. The dial gauges were attached to the steel plates with C clamps. The consolidation device was anchored to the floor slab and steel rails built for that purpose to avoid any displacement of the equipment once the testing was started (Figure 2).

4.4 Testing procedures

The first step towards starting the experiment was to grind and sieve the Chinle clay through sieve number 4. After taking into consideration the size of the inclusions, it was decided to use about 8 inches thick slurry in a six inch diameter cell. The amount of soil required to fill this volume was estimated to be about 5,000 grams.

The second step was to determine the initial moisture of the soils and afterwards the randomly mixed samples were prepared first. The sieved soil was mixed with the necessary amounts of plastic strips and water. The soil-plastic-water mixture was stirred until the soil was able to flow and had a homogeneous texture. The water content of the slurry was 40%. Two 12 inch long wick drains were placed along the inside wall surface of the mold across from each other. The wick drains were held secured to the wall of the mold by a circular steel wire that was placed about one inch from the bottom. This also helped to keep the wick drains out of the path of the load applicator. Finally, the slurry was poured into the consolidation molds and it was covered by a round piece of typar. During the sample preparation, all the components were weighed. Afterwards the load applicator was placed and left for 24 hours. Then weights were added every 24 hours until the total load reached 32 Kg. The internal mold depth was measured with a square ruler and at the same

time the dial gauges were used to record the displacement down to 1000s of an inch.

Once the consolidation and rate of settlement measurements were completed, the samples were drained, weighted, the moisture content and volume determined. The sample preparation for the layered system was very similar to the ones that were mixed, except the slurry was prepared without the plastic strips. First half of the slurry was poured into the mold followed by a plastic layer. The plastic strips were distributed as evenly as possible over the first layer of slurry. The other half of the slurry was poured over the plastic strip layer. Afterwards the steps were similar to the mixed sample (Figure 3).

5 RESULTS

The consolidation results for all the clay slurry specimens are tabulated in tables 1 and 2. The results include total settlement, changes in weight and volume, the time of 90% consolidation according to the Taylor method, the modified compression and recompression index and the coefficient of consolidation. The first test run conducted was on the clay slurry specimens mixed with specified percentages of shredded plastic fibers. The results for each mixed sample are tabulated below.

The second test run conducted was on the clay slurry specimens layered with specified percentages of shredded plastic fibers and one specimen containing a lateral filter drain at the midpoint of the sample. The results for the layered samples are tabulated above.

All clay slurry samples were tested and analyzed in the same manner and followed established procedures. Each test sample was prepared with different amounts of shredded inclusions and the inclusions were either randomly mixed or placed in layers with the slurry. After the consolidation tests were com-

Table 1. Mixed clay slurry samples

	Control	Plastic (%)		
		1	2	4
Max Sett.(in)	1.30	1.48	1.39	1.132
Weight Decrease %	5.32	2.338	5.131	7621
Volume Decrease %	14.8	14.8	14.4	13.6
T90-32 Kg (min)	96.04	2.25	4	6.25
Mod. Compression Index	0.1545	0.1632	.1513	0.14417
Modified recompresion index	0.0269	0.0307	.04613	0.02065
Coefficient of Consolidation in²/min	.00103	0.1104	0.0065	5.09E-3

Table 2. Layered slurry samples with shredded plastic and wick drain

	Filter Drain	Plastic (%)		
		1	2	4
Max Sett.(in)	1.47	1.35	1.67	1.98
Weight Decrease %	7.32	6.33	4.45	6.4
Volume Decrease %	16	14	17	19
T90-32 Kg (min)	6.25	25	3	2.25
Mod. Compression Index	0.17	0.16	.18	0.21
Modified recompresion index	0.0209	0.016	.014	0.0142
Coefficient of Consolidation in²/min	.0051	9.6E-4	9.2E-3	9.6E-3

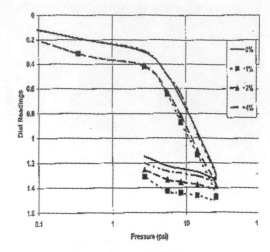

Figure 4. Consolidation curves for randomly mixed samples.

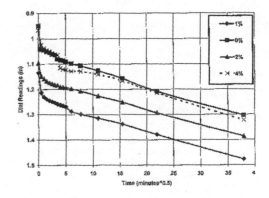

Figure 5. Time rate settlement curve for 32 kg.

pleted on all specimens, it was necessary to comparatively analyze the individual results according to the same standards and specifications in order to determine the significant change in behavior. By testing and analyzing the specimens in the same manner allowed for control and consistency of the produced results.

The initial and final weight of soil, water, plastic strips and cell for all clay slurry samples were determined and recorded. Also the initial and final height and volume of the samples were recorded. All slurry samples had 40% water content. It was necessary to identify these initial properties in order to determine the amount of settlement and changes in weight and volume for the individual samples.

5.1 Randomly mixed samples

The results of all the amended test samples were compared to the control sample in order to deter-mine the effect of the plastic strips on the consolidation and settlement rate of the clay. The total settlement of the control sample was 1.3035 inches, and a rebound of 0.1355 inches. (Figure 4). This sample also had a decrease in weight by 5.28% and a decrease in volume by 13.84%. The time required for 90% of consolidation for a load of 32 Kg by the Taylor method was 48.80 minutes (Figure 5). The analysis for the Casagrande method was not possible, because the primary consolidation did not occur during the 24 hour period. From the Taylor method the coefficient of consolidation was determined to be 1.03E-2 in²/min. The modified compression and recompression index were 0.15 and 0.0269.

All of the mixed samples achieved a larger total settlement compared to the control sample such as the total settlement increased by 11.69%, 9.93% and 6.24% for samples mixed with 1%, 2%, and 4% respectively. Also, the samples experienced reduction in their weight and volume due to decrease in water content and compaction of clay and plastic. The results indicate that the samples with 1% and 2% plastic strips had a greater total settlement and volume decrease than the mixed 4%, but it shows higher final moisture content. This appears to indicate that these samples were experiencing compaction between the clay and plastic and the pores were being collapsed. This is why there was an increase in settlement and a decrease in volume. The sample with 4% plastic experienced the least settlement and least decrease in volume, but the most decrease in weight. This indicates the sample with 4% plastic contributed to the increased drainage of water in the sample. The fact that the total settlement and reduction in volume did not increase indicates that the pores in this sample did not collapse while it was subjected to a load. It appears that the plastic provided drainage routes for the water to escape, but at the same time strengthened the soil mass. For randomly mixed samples, the sample with 4% plastic produced the best results in increasing the permeability of the clay slurry.

5.2 Layered sample

The total settlement of the clay slurry with layered plastic and wick drain increased by 16.17% for the wick drain, 2.48% for 1% , 31.34% for 2%, and 55.68% for 4% plastic in comparison to the control sample. The reduction in weight and volume of the layered samples in comparison to the control sample were as follows: wick drain was 38.54% and 15.77%, 1% was 19.94% and 1.76%, 2% 15.61% and 24.02%, and the 4% was 21.25% and 38.30%. These results indicate that the sample with 1% plastic experienced the poorest result. The layered 4% and 2% had the greatest increase in total settlement and volume reduction, but less than the wick drain (Figure 6).

The increase in total settlement and volume reduction can be attributed to the compaction of randomly open air spaces in the thick layer of plastic fibers. Although the layered 4% had the greatest total settlement and volume reduction, it drained out

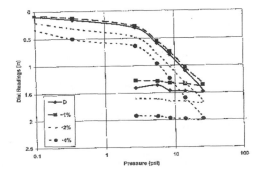

Figure 6. Consolidation curves for layered samples.

about 45% less water than the filter drain. The filter drain on the other hand had a significant increase in total settlement, weight reduction, and volume reduction. The filter drain appears to have undergone a very small settlement. Therefore, the wick drain appears to have contributed to faster consolidation and strengthening of the clay slurry.

In comparing both test runs, the mixed 4%, the filter drain, and the layered 4% produced the best results. The randomly mixed sample with 4% of shredded plastic provided faster drainage as indicated by the reduction in total weight. The layered sample with 4% plastic achieved the highest total settlement and decreasing in volume. The wick drain appears to be the best in providing faster consolidation indicated by the reduction in weight and volume of the sample.

There are advantages and disadvantages in mixing the plastic with the slurry or placing them in layers of certain thickness and width. When the slurry is mixed with the plastic strips, the inclusion will develop fiber type reinforcement. On the other hand, when the plastic is spread into layers of certain thickness and width, it will serve as a drainage path and lower the distance of travel of the draining water. The layered system might be easier to implement in the field as the shredded plastic and slurry are sequentially deposited in layers. Wick drains can do similar work, but it is not a waste material.

6 CONCLUSIONS AND RECOMENDATIONS

This research produced encouraging results for stabilizing saturated fine-grained slurries by amending it with waste plastic. The addition of waste plastic to slurry contributed in increasing the rate of settlement, decrease in weight and volume of the slurry specimens. The mixed 4%, layered 4%, and filter drain had the best overall effect on the stabilizing properties of the slurries. The increase in settlement was due to the ability of the plastic waste or filter drain to provide a drainage route for the water. As the water drained from the slurry mass, the weight, volume and pore water pressure decreased. This shows densification and strengthening of the slurry mass.

This approach may create a new way for reusing HDPE waste. If this new technique of stabilizing slurry masses with amended plastic HDPE waste can prove to be economical and effective, then a variety of applications could be introduced. Since this is the first time that this method of stabilization has been proposed, it is recommended that future studies be conducted to determine the consistency of these results. Although beneficial results were produced from three distinct amendment methods, the actual stabilizing properties enhanced should be analyzed. Future studies should focus on how dewatering, volume reduction, and increased rate of settlement contribute to the stabilization of the slurry masses.

7 ACKNOWLEDGEMENT

This study was carried out with the support of the New Mexico Alliance for Minority Participation (NMAMP) as part of their undergraduate research program in the laboratories of the Civil, Agricultural and Geological Engineering. We are very grateful to Mr. Wendell and the Manufacturing Center Technology for improving and assembling the consolidation apparatus.

REFERENCES

Bray, R.N. 1979. *Dredging: A Handbook for Engineers*. London: Edward Arnold.

Dean, K.C. and Havens, R. 1973. "Comparative Costs and Methods for Stabilization of Tailings" *Tailing Disposal Today,* 450-476.

Ehrig, R.J. 1992. *Plastics Recycling*. New York: Hanser.Edil, T.B. and Benson C.H. "Geotechnics of Industrial By-Products." *Recycled Materials in Geotechnical Applications*, ASCE 79 (1998): 1-18

Holtz, R.D. and Kovacs, W.D. 1981. *An Introduction to Geotechnical Engineering*. New Jersey: Prentice.

Johnston, C.E. 1973. "Tailing Disposal – Its Hidden Costs." *Tailing Disposal Today,* 762-815.

Jones, C.J.F.P. 1985. *Earth Reinforcement and Soil Structures*. London: Butterworths.

Khan, M.R. 1996. *Conversion and Utilization of Waste Materials*. Washington D.C.: Taylor &Francis.

Puppala, A.J. and Musenda, C. 2000 "Effects of Fiber Reinforce-ment on Strength and Volume Change Behavior of Expansive Soils" Transportation Research Board.

Smith, E.S. 1973 "Tailings Disposal – Failures and Lessons." *Tailing Disposal Today*, 356-376

Schaefer, V.R. 1997. Soil Improvement and Geosynthetic CommitteeReport. "Fiber Reinforced Soils" Ground Improve ment, Ground Reinforcement, and Ground *Treatment* (1997): 273-291

Thomas, L.J. 1978. *An Introduction to Mining*. Australia: Hicks Smith and Son, 1978

United States Environmental Protection Agency. 1997. "Municipal Solid Waste Factbook/Plastic Waste Components." http://www.epa.gov/epaoswer/non-hw/muncpl/factbook/internet/mswf/plastic.htm

The use of the powder marble by-product in the raw material for brick ceramic

G.C. Xavier, F. Saboya Jr. & J. Alexandre
State University of Norte Fluminense, Campos, Rio de Janeiro, Brazil

ABSTRACT: The decorative stone industries of marble and granites in north of Rio de Janeiro, Brazil, has as by-product a rock powder that has been causing great concerns regarding environmental and health issues. It has been blamed to cause severe lung disease. Thus, these activities have nowadays become an important object of research in order to find a suitable use of these solid wastes.

This work intends to discuss some technical aspects concerning the use of this material, which derives from the marble blocks cutting work, in the ceramic raw material (paste). The study has been carried out using clayey soils of the municipal district of Campos (RJ), where more than 130 ceramic industries are settled, favoring regional development.

Characterization laboratory tests were carried out and samples were conformed in a laboratory extruder, simulating the actual industry process. These samples were molded with different marble powder contents, fired at temperature raging from 750 to 950^{0}C and their mechanical properties were determined. The results depict the possibility of using this by-product in the paste composition of ceramic bricks for the use in civil engineering construction industry.

1 INTRODUCTION

Nowadays, there is a great concern amongst govern, contractors and environmentalist regarding the increasing amount of industrial waste (solid, liquid and gaseous). In Italva, district of Rio de Janeiro State, Brazil, great part of residential area is settled over landfills composed basically with marble powder. This by-product comes from the decorative stone industries and, after air-dried, it becomes a very fine material that can be easily inhaled by human being and animals as well, and it has been blamed as the reason of severe lung diseases among local people.

The decorative stone industries have as one of the main activities the sawing and polishing of rock blocks. This activity generates, per unity, a significant amount of mud-like waste, estimated in 32 to 40 m^3/day (sixteen trucks), composed essentially by water, lime, sand, marble and granitic powder.

The civil engineering construction industry is believed to be one of the most consumer of mineral resource generating, therefore, a great amount of solid waste. However, despite this scenario, the technical recycling is not enough explored in developing countries.

According to NITES, the amount of waste due to mining process of decorative stone can reach up to 40% of the total extracted volume. The non standard dimensions of the blocks, faults and piling of the pieces are amongst the main source of waste. However the poor composition of the abrasive mud, lack of calibration of the cutting tools and handling, are also important source of waste (Neves et all, 1998).

Silva (1998) and Freire Motta (1997), have shown that the amount of waste from cutting and sawing process in Brazilian decorative stone industries can easily reaches 20 to 25% of the total volume.

The Civil Engineering Laboratory of the State University of Norte Fluminense, in Campos, Brazil, has developed researches aiming basically the improvement of the brick ceramic for civil construction purposes. These studies comprehends also the optimization of the exploration of borrow pit areas and the use of the industrial waste in the raw material for brick ceramic manufacturing.

2 METHODOLOGY

The waste characterization tests were carried out at the Geotechnical Laboratory to determine the grading curve, index properties and specific gravity of the grains. Several powder marble contents were used in the raw material, in order to verify its influence on the properties of the brick ceramic, regarding its standard requirements for commercial use.

2.1 Grading curves

The grain size distribution of the marble powder is depicted in Figure 1, where it can be seen that 86% are of clay size and 14% are of fine sand size, indicating, thus, that the marble powder is a very fine clay-like material.

It can be seen that the grading curve of waste are rather uniform with uniformity coefficient C_u=1.7, showing very fine equivalent grain size, suitable for the use in raw material of clay brick.

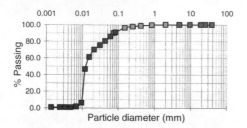

Figure 1. Grading curve of marble powder waste.

2.2 Specific density of the grains

The samples were dried at 110^0C and the specific gravity was determined by the picnometer method. The value obtained for marble powder was about 2.77g/cm^3 which is higher than expected for these material. This was believed to happen because of the presence of the iron powder (abrasive) in the mud.

2.3 Atterberg limits

As expected, the marble powder has shown no plastic behavior. However, it was found important to verify the influence of addition of powder in the clayey soil regarding the Atterberg Limits, once these values are very closely related to the optimum extrusion moisture content. This is also of crucial importance on the quality of the final product.

Figure 2 shows the Atterberg Limits of the raw clayey material as function of powder content. As expected, there is a marked decrease in the Plastic

Index until 10% of marble content remaining constant afterwards. The same behavior is observed for Liquid Limit.

2.4 Specific area

This test was carried out in order to verify the fineness of the marble powder, once it is a very important characteristic regarding the composition of the raw material for ceramic brick.

The Specific Surface value of the marble powder determined from Blaine Method was about 173m^2/kg which is fine enough for its use in ceramic raw-material.

2.5 Technological tests

For evaluating the mechanical behavior of the bricks, the samples were prepared by mixing clayey soil with different waste content of marble powder of 0, 5, 10, 15 and 20%. Mixing were made after 24 hours in oven to 110^0C and moisten up to extrusion water content, which is function of Liquid Limit (LL), according to the following expression given by Alexandre (1998):

$$W_{ext} = \frac{LL}{2} + 2\% \qquad (1)$$

It can be seen in Figure 3 that the value of W_{ext} decreases as the amount of waste in ceramic raw material increases, allowing a homogenization less moisten improving the mixing process and, thus, causing less linear variation during firing.

After this first step, the sample are then taken to laminator and to the extruder afterward. The final dimensions of the samples after extrusion are 13.0x2.7x1.7cm under a vacuum pressure of 26 inch of Hg. The samples were led to dry under temperature of 110^0C and fired at 750, 850 and 950°C in an oven with temperature rate of 5°C/min and naturally led to cooling down at room temperature.

2.6 Properties

After firing at selected temperature, the specimens were submitted to several tests in order to verify their technological properties. These tests are basi-

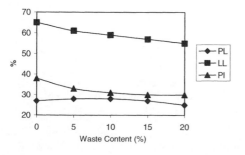

Figure 2. Variation of Atterberg limits with waste content.

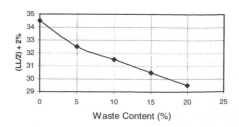

Figure 3. Extrusion water content against waste content.

cally, water absorption, specific gravity, porosity, linear variation and tensile strength on flexural test. These tests were carried out on triplicate, and described values represent their mean.

3 RESULTS

3.1 *Water absorption*

As depicted in Figure 4, the brick ceramic with 0% of powder marble has shown less water absorption for the temperature of 950^0C than of 850^0C. At 950^0C the water absorption increases up to 23% for those samples with 5% of waste and it reaches a minimum for samples with 10% of waste, increasing again with further waste content. At 750^0C, the minimum value takes place at 15% of waste content and a maximum at a 20% of waste content. Similar behavior is observed for samples fired at 850^0C. It is important to mention that, according to Brazilian Standards, the maximum values of water absorption is 22% (NBR 6480).

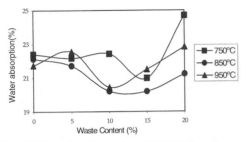

Figure 4. Water absorption against waste content.

3.2 *Linear shrinkage*

The linear variation of specimen dimensions is shown in Figure 5. It can be seen that the specimens fired at 850^0C and 950^0C have shown similar behavior, where a very small decrease of linear variation between 0 and 5% of waste content is noticeable. The Figure 5 indicates that the linear variation is apparently independent of waste content, despite the values found for 15% of waste.

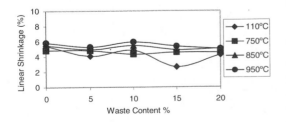

Figure 5. Variation of linear dimensions with waste content.

3.3 *Specific gravity*

As shown in Figure 6, a similar behavior is found for 750^0C and 850^0C curves where the values of specific gravity decreases as the waste content increases from 0% to 10% and increases slightly up to 15% of waste and then decreases slightly. The 950^0C curve shows and marked decreases as the waste content increases from 0% to 5%, reaches a maximum at 10% of waste content and then falls considerably for 20%. It is very interesting to mention that for 20% of waste content the specific gravity becomes independent of firing temperature.

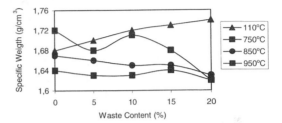

Figure 6. Specific weight with waste content.

3.4 *Flexural strength*

Figure 7 depicts the flexural strength of the prismatic samples. The values vary from 5.0 MPa to 11.3 MPa. The highest values are shown by those samples fired at 850°C with 20% of waste. The 750°C curve does not show any decrease in flexural strength with the increase of waste content. However, no further tests were carried out considering waste content higher than 20%.

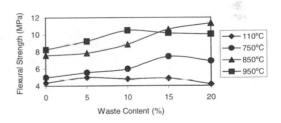

Figure 7. Flexural strength with waste content.

4 CONCLUSION

The marble waste can be easily handling without any direct kind of contamination. However mask are advisable to be used once the powder is very fine and can be breathed in by animals and mankind.

By analyzing the results shown before, it can be said that the use of 15 to 20% of waste content could be considered the best proportion. However the water absorption is one of the most critical properties for ceramic brick, and the values for 20% of waste content are higher than allowed for civil construction purpose. Thus, a raw-material composition of 15% of waste content fired at 850°C might be used in a industrial scale

REFERENCES

Neves, G.A.; Patrício, S.M.R.; Ferreira, H.C. e Silva, M.C. 1998. Utilization of granite waste in ceramic brick. 43° Brazilian Ceramic Congress *(in portuguese)*.

Silva, S.A.C. 1998. Characterization of granite waste. Potentiality of utilization in fabrication of soil-cement brick. MSc Thesis, Federal University of Espírito Santo, Vitória/ES, Brazil *(in portuguese)*.

Freire, A.S.; Mota, J.F. 1997. Potentiality of utilization of marble and granite waste. Revista Rochas de Qualidade, 123, Brazil *(in portuguese)*.

Alexandre, J. 2000. Analysis of raw material and composition of paste used in ceramic industry. DSc. Thesis, State University of Norte Fluminense – UENF, 1-170, Campos, Brazil *(in portuguese)*.

Tailings and Mine Waste '02, © 2002 Swets & Zeitlinger, ISBN 90 5809 353 0

Researches concerning the Purolite assimilation for use within the uranium separation-concentration Resin In Pulp process

E. Panturu, Gh. Filip, St. Petrescu, D. Georgescu, F. Aurelian & R. Radulescu
Research and Design Institute for Rare and Radioactive Metals, Bucharest, Romania

ABSTRACT : New strong anionite resins manufactured by the PUROLITE Company were studied comparing with the older AM resin, for U (VI) separation concentration from model solutions. Loading capacities for U (VI) were established using resins, A -500 and SGA- 600 respectively, followed by the study of loading capacity variation related to the recycling number.
Mechanical resistance determination was performed for SGA-600 and AM resins, showing superior characteristics for these resins and enabling their use in "Resin In Pulp" process.

1 INTRODUCTION

One of the most used procedures in uranium separation-concentration from leached uranium mineral pulps is the so-called "Resin In Pulp".

An ion exchange resin must have, beside a convenient selectivity and exchange capacity, a high mechanical resistance and average size granules superior to the meshes screen dimensions (which is used after the contacting phase at the resin separation).

Before the band filter appearance, this procedure was introduced in the '70 in order to eliminate the necessity for large filtering surfaces (on tambour filter), method used at that time to obtain clear uranium solutions.

These solutions were used to separate uranium by the ion exchange resin in fixed bed.

The Feldioara "R"- plant, which is processing the uranium domestic minerals through the alkaline leaching procedure, was designed and built in the '70. In the technological flow, the plant includes the uranium separation phase through RIP process.

The strong base anion exchange AM resin, manufactured in Russia, was taken into consideration in the plant designing and used afterwards from the beginning in the production process.

As the resin stock has been exhausted, a current problem is to choose an ion exchanger to replace the AM resin. In order to choose this new ion exchanger, a collaboration convention between the Romania Purolite firm, resin producer and The R & D Institute for Rare and Radioactive Metals - Bucharest was signed. The resin producer supplied the following types of resins for tests:

- Purolite A-500
- Purolite SGA-600

The researcher performed the specific characteristics study, comparing the Purolite resins to the AM one.

2 PUROLITE A-500 AND SGA-600 U (IV) LOADING CAPACITY

According to the product specification:
- the Purolite resin A-500 is a strong basic resin, macro porous type I special, with an excellent granule integrity and resistance to mechanical forces;
- the Purolite SGA-600 is a transparent gel, a strong basic anionite type I special, used to separate the uranium's anionitic complexes obtained through the alkaline or acid leaching processes.

2.1 *Purolite A-500 and SGA-600 (IV) loading capacity in "ideal" solutions*

The U (IV) loading capacity of the two resins was determined for the following anions content solutions:

$U = 0,66$ g/l
$CO_3^{2-} = 0,64$ g/l
$HCO_3^- = 1,41$ g/l
$SO_4^{2-} = 0,24$ g/l

The sorption determinations were performed according to Romanian standard STAS 9475/10-88. There were used: a column with 10 mm diameter, 5ml resin volumes, under the following conditions: continuous flow, effluent flow of 5 Bev/h.

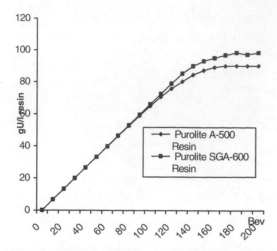

Figure 1. The U^{6+} sorption isotherms on purolite A-500 and SGA-600 resins.

The figure 1 represents data concerning the uranium's sorption on Purolite A-500 and SGA-600.

From the figure no.1 results that the U (VI) loading capacities of the tested resins in "ideal" solutions are:

- Purolite A- 500 = 90,49 gU/dm^3 resin
- Purolite SGA-600 = 98,85 gU/ dm^3 resin.

Purolite A-500 was eliminated from the next test due to its reduced U (VI) loading capacity and especially because of its macro porous form which in RIP sorption conditions, is leading to the resin's pores blocking by the fines minerals particles.

2.2 The U (VI) loading capacity of purolite SGA-600 in "model" solutions with ionic compositions similar to real ones

The results presented in figure 2 for the Purolite SGA-600 were obtained in the same conditions as those from the 2.1 paragraphs, using a sorption solution having the following composition:

U = 0,66 g/l
CO_3^{2-} = 14,4 g/l
HCO_3^- = 8,85 g/l
SO_4^{2-} = 10,03 g/l

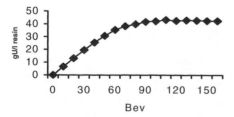

Figure 2. The uranium loading isotherm of purolite SGA-600 resin in solutions with ionic composition similar to the real ones.

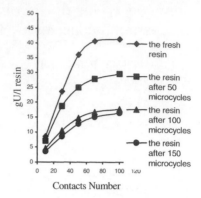

Contacts Number

Figure 3. The uranium loading capacity variations for purolite SGA-600 depending on the contacts number.

Table 1. The ion exchange capacity values of Purolite SGA-600 and IS resins.

Parameter	Resins	
	Purolite SGA-600	AM
Ion exchange capacity value (meq/ml)	1,04	1,04
Ion exchange capacity value (meq /ml)	3,97	3,97

2.3 The purolite SGA-600 U (VI) loading capacity variation related to the recycling number

The Purolite SGA-600 loading capacity variation depending on the recycling number (1-150) was studied.

The results obtained are represented in figure 3.

According to the experimental results illustrated in figure 3 the resin's loading capacity is decreasing to 71,42 % of it's initial capacity after 50 cycles, to 42,86 % after 100 cycles and 39,689 % after 150 cycles.

3 COMPARATIVE DETERMINATIONS OF PUROLITE AND AM RESINS MECHANICAL RESISTANCE

In order to determine their mechanical resistance two methods were used:

- the "granules breaking" apparatus;
- the determination of resins losses depending on the number of sorption – elution cycles.

3.1 The purolite SGA-600 and AM resistance determination with the "granules breaking" apparatus

The test was performed within Victoria Purolite firm laboratory and the results are included in table 2.

Table 2. The mechanical resistance of purolite SGA –600/3472 and AM resins

Parameter	Resin	
	Purolite SGA-600	AM
Breaking Force (kg)	0,858	0,508

According to the table No2. results, the breaking force necessary for the Purolite SGA-600 granules breaking is superior to that used for AM resin.

In order to determine the resin losses for Purolite SGA-600 and AM resins due to the osmotic shock, there were measured resin volumes lost through crumbling (having d< 425mm) depending of the recycling number.

3.2 The resin losses variation for purolite and AM resin versus the sorption-elution cycles number

According to the experimental results presented in the figure no.4 one can observe that the Purolite SGA-600 resin losses are comparable (lightly inferior) to AM resins.

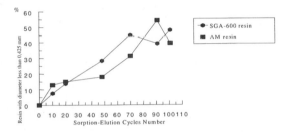

Figure 4. The resin losses for purolite SGA-600 and AM resin depending on the sorption-elution cycles number.

4 CONCLUSIONS

1. The Purolite A-600 resin compared to AM resin has a 10% bigger exchange capacity (meq/g) and a U (VI) loading capacity of 42,314g/l compared to 30,0g/l of AM.

2. The mechanical resistance of Purolite SGA-600 resin, that was determined through a break test, is 0,858 kg compared to 0,508 kg for AM resin. This increased mechanical resistance is seen as an important advantage when the resin is used in a high attrition media like the resin-in-pulp mixture.

3. Because of its superior characteristics, the Purolite SGA-600 resin provides a minimum 10% lower self-consumption, the elution (NaCl + Na$_2$CO$_3$) and precipitation (NaOH) reagents consumption is reduced with the same percentage (10%).

4. The Purolite SGA-600 resins fulfills all the technical and economical conditions to be successfully used as a substitute for the AM resin within the uranium separation-concentration through the RIP process.

5. Further sorption tests using Purolite SGA-600 in uranium-loaded pulps will be undertaken, in order to find the optimal sorption parameters and assess the real resin losses, before the industrial use.

REFERENCES

Golden, L.& Irving, J., 1972, Chemistry and Industry, 837-844.
Abrams, I., 11March 1975, Requirements for Ion-Exchange in Condensate Polishing, Diamond Schamrock Chemical Co.
David, B., Filip, Gh., Tataru, S., The behavior of the ion- exchange resin made in Romania under the uranium's hydrometallurgical conditions, Revista Minelor XIX, 4, 1968.
Researches concerning the using of the different kinds of ion-exchange resin made in Romania and in other countries in order to eliminate their import, ICPMRR File no.271/01160/1980.
Researches concerning the using of the different kinds of ion-exchange resins made in Romania and in other countries, ICPMRR File no.328/01527/1982.
Symposium Purolite, Victoria, Romania, 1-2 July 1998.
Slater, M.J., 1991 Principles of Ion Exchange Technology, Butterworst Heinemann, Oxford, 7, 54.

Tailings and Mine Waste '02, © 2002 Swets & Zeitlinger, ISBN 90 5809 353 0

Cu(II) separation and recovery from mining aqueous systems by flotation (DAF) using alkylhydroxamic collectors

L. Stoica, C. Constantin & O. Micu
University "Politehnica" Bucharest, Faculty of Industrial Chemistry, Romania

ABSTRACT: Metallic ions contained in mining waters are dangerous pollutants representing at the sometime potential sources of useful compounds. Among the known separation methods, those generally named "adsorptive bubbles separation methods" especially ion and precipitate flotation offer an optimum alternative to the classical methods. The present paper studies the behaviour of alkylhydroxamic acids (C_7-C_9) as unstudied collector and complexing reagents in ion and precipitate flotation. The surface tension properties of the collector reagents were investigated by surface tension isotherms. Complexing reagent function was studied by the formation of insoluble species with metallic ions (Cu) followed by their separation using ion and precipitate flotation. The influencing factors were studied: pH, molar ratio (c_c:$c_{Cu(II)}$), metallic ion concentration and foam processing. The colligend-collector interaction was investigated using IR spectrometry, thermal analysis, chemical elementary analysis and has established the nature of bonding in sublate which allows the evaluation of final product recovery. The separation yields (R>97%) the possibility of foam processing and finally metallic ion confirm the collector quality of the alkylhydroxamic acids studied. The good separation, efficiency, selectivity and possibility of foam processing represent the main characteristics of the proposed method.

1 INTRODUCTION

The presence of ion metallic in some aqueous technological systems at low concentrations (5 - 1000 mg L^{-1}) reclaims the study of separation method that permit both depollution and recovery of valuable compounds. The group of adsorptive bubble separation methods represents an alternative to the classical methods (Lemlich, 1972, Mavros, 1992). In this context the results obtained by applying ion flotation with all its variants is an alternative with large application possibilities. Some characteristics of flotation confer priority compared to other methods are:
- diversity of species (in nature and structure, likely to be separated);
- high efficiency of separation at low concentration 10^{-5} - 10^{-2} M of separable species;
- simplicity;
- rapidity;
- economically;
- flexibility and fiability of equipment;
- low consumption of reagents in order to form some insoluble species with advanced hydrophobicity and low specific weight.

In order to modify the surface properties of particles in aqueous systems containing metallic ions, favouring the interaction with the mobile phase, a surface tension substance named collector is introduced into the system. The collector has a heteropolar structure, with linear chain (C > 6) and a polar group, due to which it interacts with the insoluble hydroxospecies of the colligend M^{n+}, generally surface-inactive. In this manner, the product formed by the collector-colligend interaction-sublate becomes surface-active, being concentrated in foam, owing to the mobile phase. Flotation is a complex separation process, whose achievement depends on physical-chemical, hydrodynamic factors and interphase mass transfer. The study of the acid-base and of complexing equilibrium, correlated with the species structure and with the possibility of recovering of metallic ions from the foam, permits the inclusion of flotation within the group of high purification methods.

In this context the aqueous systems from the mining industry may contain metal ions (e.g. Cu(II), Co(II), Ni(II)), being pollutants and potential sources of useful compounds. Beside these sources, a series of economy branches generate impurifying metallic ions (chemical and electrochemical industry, ceramics etc.). The continuo diversification of technologies that generate aqueous systems impurified with metallic ions requires systematically studies for applying efficient methods, both for depollution and recovery of valuable compounds. In this way, the study of some new classes of collector reagents about the synthesis, characterisation, especially of the complexing groups, which confer surface tension properties, became a necessity. There are promising results with some organic reagents well-know in analytical chemistry and modified in mentioned way by alkyl chain grafting: alkyloxymes, alkylamines, alkylthiocarbamates (Stoica, 1998c).

2 MATERIALS AND METHODS

Hydroxamic acid is a known complexing analytical reagent for transitional metals cations. The alkylhydroxamic acids can be obtained by grafting radical alkyl to hydroxylamine, resulting two tautomeric forms (Stoica, 2000). The alkylhydroxamic acids having a weak acid character posses the capacity of forming mixed salt and donor bounding complexes (chelates) with mixed bonding (salt and donor) types. Starting from the collector functions in ionic and precipitate flotation, especially surface tension substance and in the same time complexing agent, this study presents the alkylhydroxamic acids behaviour (C = 7 - 9) in separation of Cu(II).

The collector characterisation was made by drawing the surface tension isotherms in collector-water system using the stalagmometrical method (Stoica, 1998b). Based on the obtained values of the surface tension the surface tension isotherms $\sigma = f(c)_T$ are drawing.

The quality of collector reagent has been established by flotation experiments (DAF variant) of "model" aqueous systems of Cu(II) on laboratory installation (Stoica, 1998a).

According to this purpose: $CuSO_4 \times 5H_2O$, high purity degree (stock-solutions $5.5 \times 10^{-2} M$) has been used to prepare work solutions; NaOH high purity degree, 0.1M and 2M solutions; alkylhydroxamic acids (C_7-C_9) $5.5 \times 10^{-2} M$ solutions.

The study of influencing factors (pH, molar ratio c_c:$c_{Cu(II)}$, metallic ions concentration, necessary air) by function %R = f(property) established the optima conditions for the Cu(II) separation.

The final Cu(II) concentrations was determine by atomic absorption spectrometry (PAY UNICAM SP 9).

The composition and structure of the chemical species (sublates) isolated in foam have been studied by chemical and physic-chemical methods: IR spectrometry (SPECORD M 80), reflection electronic spectrometry (SPECORD M 40), thermal analysis (E.PAULIK J.PAULIK ERDAY) and chemical elementary analysis (for N and Cu determinations).

3 RESULTS AND DISCUSSION

The tensioactive properties of alkylhydroxamic acids were studied by $\sigma = f(c)_T$ function. The experimental results indicate the decreasing of surface tension proportional with alkylhydroxamic acid concentra-

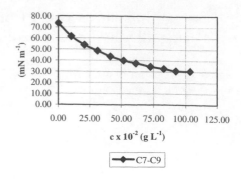

Figure 1. Surface tension isotherms of the studied collectors (18°C).

tion, which point out their surface activity. Figure 1 presents the descending aspects of $\sigma = f(c)_T$ curves. It can be observed the descending aspect of surface tension isotherms is more significant (at low concentrations).

The collector concentration is an important parameter of flotation process. When this value is high enough, the tensioactive ions have a stable thermodynamic arrangement having very closed hydrocarbon chains, faraway from water-surface, and its hydrophilic parts remaining into the water.

The heteropolar structure suggests a special behaviour in the flotation process: as a surface tension substance or as a complexing reagent, forming insoluble chelates with Cu(II) cations.

The transformation hydrophilic species $[M(H_2O)_x]^{n+} \rightarrow$ hydrophobic species could be achieved: directly by interaction with the collector in ion flotation process (I.F.), or indirectly by interaction with the collector, after insolubilisation as hydroxospecies in precipitate flotation process (PP.F.).

In ion flotation (I.F.) the collector-colligend molar ratio is stoichiometrical, while in precipitate flotation (PP.F.) is understoichiometral. For establishing the reagents behaviour in these two different processes, the order of collector addition has been studied: in the case of I.F. the sequencing was (I) collector in order to obtain metallic alkylhydroxamate, (II) pH regulator, while for PP.F. the sequencing was (I) precipitation reagent and pH regulator in order to obtain M(II)-hydroxospecies, (II) collector for emphasising their hydrophobia (Stoica, 1997, 1998a, b).

In this paper the influencing factors were studied for PP.F. variant.

pH is an important factor in the separation process; the optimum pH flotation value permits the optaining of maximum separation efficiency is proper for each collector-colligend system and it depends on: the nature of the existing species in the system, the mutual interaction and the sublate stability. Is well known that the optimum pH range of flotation is around the precipitated pH alkylhydroxamate and hydroxospecies (Sebba, 1962). Till now there is no information about the hydroxamates precipitating pH, and that is why the optimum pH must be established for each system.

The flotation experiments have been undertaken on Cu(II) 5.5×10^{-3} M solutions, pH range = 6.0 – 9.0 to different molar ratios, $c_c:c_{Cu(II)}$. The obtained results show an optimum pH value between 7.0 –9.0 when high values of separation efficiency are obtained (> 97 %) at understoichiometrical consumption, in the case of precipitate flotation (Fig. 2). The optimum flotation pH has the same value as precipitation pH of metallic ions because there is a competition between metallic alckylhydroxamate and hydroxides obtained. In interaction with Cu(II) at greater than 6.5, the collector (alhylhydroxamic acid) has tensioactive function.

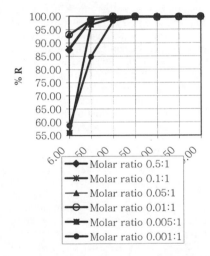

Figure 2. %R = f(pH) dependence at difference molar ratio $c_c:c_{Cu(II)}$ for the system Cu(II)-AH$_{7-9}$ in PP.F. variant

The optimum molar ratio $c_c:c_{Cu(II)}$ is important for establishing the separation conditions. An understoichiometrical consumption represents an advantage of the method, regarding the process price, generally. The experimental results regarding the optimum molar ratio are showed in Figure 3.

The optimum molar ratio $c_c:c_{Cu(II)}$ is 10^{-2} because the optimum pH range corresponds to form hydroxospecies.

The metallic ions concentration are important both for assessing the operating parameters, but especially for their recovery as valuable compounds. Our flotation experiments at optima pH values and optima molar ratios offers high separation efficiencies (92.7 - 99,9 %) in the range of 10 - 1000mg L^{-1} initial metallic ion concentrations (Tab. 1).

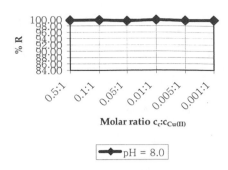

Figure 3. %R = f($c_c:c_{Cu(II)}$) dependence at optimum pH flotation for the system Cu(II)-AH$_{7-9}$ in PP.F. variant

These values prove both the tensioactive and ligand qualities of hydroxamic acid on the entire studied domain of concentrations.

The necessary air corresponds to a dilution ratio of on 3:1 and 5:1 respectively the working pressure being kept constantly at 4×10^5 N m^{-2}.

The optima conditions established for "model" aqueous systems have been checked for real samples. The concentration of Cu(II) in effluent was $c_{final} < 1$ mg L^{-1} after precipitate flotation process.

The study of collector-colligend interaction established in metallic ion flotation from aqueous systems represent an important stage in our research. The structure assessment of the concentrated species can explain the separation selectivity, the high separation yields and the ways of turning into account the valuable products. The type of colligend-collector interaction is interesting, especially for insoluble sublates of complexing coordination type with concentrations ranging in 10^{-2} - 10^{-6} M, where the ion-molecular flotation works, can not be isolated by applying common methods (precipitation, filtration, etc.). In this context the study of sublates of transitional metallic ions with alkylhydroxamic acids (C$_7$-C$_9$) becomes quite interesting. Regarding all this thinks it appears the importance of the knowledge of sublate structure in the recovering flotation of metallic ions as metallic hydroxamate or hydroxide. It was considered as a necessity to establish the optima flotation conditions

Table 1. Separation of Cu(II) at the different metallic ion concentrations (pH optimum = 8.0, molar ratio optimum $c_c:c_{Cu(II)}$ = 10^{-2}:1; flotation time = 5 min.; V$_{sample}$:V$_{water}$ = 3:1; p = 4×10^5 N m^{-2})

c_i (mg L^{-1})	pH	c_{final} (mg L^{-1})	%R
10	7.93	0.73	92.7
20	7.96	0.43	97.9
50	8.07	0.49	99.0
100	7.96	0.47	99.5
200	8.05	0.63	99.7
500	8.03	0.58	99.9
1000	8.09	0.65	99.9

of M(II) from dilute aqueous solutions as $M(OH)_nH_2O$, using alkylhydroxamic acid as collector, to study the mechanism of the interaction collector-M(II) for the establishing of the combination structure which is the sublate. According to this purpose, that sublates separated in optima pH conditions, stirring time, dilution ratio and different molar ratios $c_c:c_{Cu(II)}$ was subjected to chemical and physical-chemical analysis for establishing the composition and structure.

The elementary chemical analyses concerning the copper and nitrogen content suggests the formation of some proper chemical combination at molar ratios $c_c:c_{Cu(II)} = 1:1$ (Tab. 2).

The IR spectra data (Tab. 3) correlated with elementary analysis indicate the coordination to the metallic ion.

The conclusions of IR spectra are following:
- the presence of intense bands in the 1500 - 1700 cm^{-1} ranges assigned to the $\nu_{C=O}$ confirms the coordination of ligand by oxygen atoms;
- the bands in the 3100 - 3600 cm^{-1} ranges indicated the presence of the coordinated water;
- the band near 1100 cm^{-1} may be assigned to the Cu-O-H bending.

The absence of a RES signal (the absence of paramagnetisme) could be interpreted by obtaining of some dimmer groups with diamagnetic coupling spin. In acid medium the enol form is favoured for Cu(II); the dimmer species obtained in this conditions could be realised either by hydrogen bonds (O...H-N). For this complex isolated (Cu_2L_4), probable formulae (A) could be assigned. In neutral medium the complexation involve the formation of hydroxospecies having OH groups in bridges. The probable structure of $Cu_2(OH)_2L_2$ complex is (B) formulae.

Table 2. The elementary chemical analyses of sublates separated at different pH range and the same molar ratio $(c_c:c_{Cu(II)}= 1:1)$

Sample	% Cu calc./exp	% N calc./exp.	Chemical formulae
1	15.60/14.70	6.82/7.20	Cu_2L_4
2	25.19/24.95	5.51/5.85	$Cu_2(OH)_2L_2$

Where: Sample 1 is obtained at pH = 2,2 (copper alkylhydroxamate) and Sample 2 is obtained at pH = 7.0

Table 3. Characteristic infrared frequencies of AH, $Cu(OH)_2$ and isolated sublates and their assignments (Belamy, 1962, Nakamoto, 1970)

Sample	δ_{H2O}	$\nu_{C=O}$	δ_{Cu-O-H}
Literature	3100 - 3600	1500 - 1700	1100
AH		1500 - 1700	
$Cu(OH)_2$	3100 - 3600		1000 - 1100
1*	3100 - 3600	1550	1040
3*	3100 - 3600	1550	1030 - 1050

* The same conditions as well as Table 2

A probable formulae:

B probable formulae:

Supplementary information regarding collector-coligent interaction could be obtained from thermal analysis of the sublates isolated similar aspects reveal structural transformations of sublates which could be correlated with the probable formulae (A and B). For the sublates isolated at the same pH but understoichiometrical molar ratio suggest the existence of a mixture of alkylhydroxamate and hydroxospecies, not only the formation of a single one.

4 CONCLUSIONS

1. Flotation is a complex separation process, whose achievement depends on physical-chemical properties, hydrodynamic factors, and interface mass-transfer, correlated with the species structure.
2. Dissolved Air Flotation (DAF) is a depollutant separation process and its main advantages are: high efficiencies (%R > 95), rapidity, accessibility, the possibility of removal and recovery organic and inorganic species.
3. The present work established the optimum separation conditions for Cu(II) ions removal from aqueous systems by applying DAF technique, in PP.F. variant.
4. Alkylhydroxamic acids (C_7-C_9) have tensioactive properties and chelating action for the transitional metallic ions. These properties related by experimental researches demonstrate the possibility of using alkylhydroxamic acids as collector reagents in precipitate flotation (PP.F.) at understoichiometrical consumption $(c_c:c_{Cu(II)} \geq 10^{-2})$.
5. High separation efficiencies offer and the possibility of decreasing the metallic ions concentrations under the allowable limits and the recovery of valuable compounds.
6. The proposed method appears to be cost effective and environmentally acceptable.

REFERENCES

Belamy, L. J. 1962. *The infrared Spectra of Organic Molecules*. New York: Willey Interscience Publication.

Lemlich, R. 1972. *Adsorptive Bubble Separation Techniques*. New York, London: Academic Press.

Mavros, P. 1992. *Innovations in Flotation Technology*. Dordrecht, Boston, London: Kluwer Academic Publishers.

Nakamoto, K. 1970. *Infrared Spectra of Inorganic and Coordination Compounds*. New York: Willey Interscience Publication.

Sebba, F. 1962. *Ion Flotation*. New York: Elselveyer.

Stoica, L., Constantin, C. & Cioloboc, I. 1997. Alkylhydroxamic acids-collectors for metallic ions flotation from aqueous systems. *7th Balkan Conference on Mineral Processing, Vatra Dornei, 26-30 May 1997*: 346-351.

Stoica, L., Meghea, A. & Constantin, C. 1998a. Metallic ions separation and recovery from mining aqueous systems by flotation (DAF) using alkylhydroxamic collectors. In Sanches, M. A., Vergara F. & Castro S. H. (ed.), *Environment & Innovation in Mining and Mineral Technology; Proceedings of the IV International Conference on Clean Technologies for the Mining Industry, Santiago de Chile, 13-15 May 1998*:367-381.

Stoica, L., Meghea, A. & Constantin, C. 1998b. Cu(II) removal from aqueous systems by complexation with alkylhydroxamic acids and flotation. *1st International Conferince of the Chemical Societies of the South-East European Countries, Halkidiki, 1-4 June 1998*: 641.

Stoica L. 1998c. *Ion and Molecular Flotation*. Bucharest: Didactic and Pedagogic Publication.

Stoica L., Constantin C., Micu O. & Meghea A, 2000. New collectors for Co(II) and Ni(II) removal from aqueous systems by flotation (DAF). *Proceedings of the 8th International Mineral Processing Symposium, Antalya, 16-18 Oct*: 241-246.

Case histories

Pozo Azul tailings impoundment: Design modifications made to utilize a difficult site

Matthew L. Fuller
Olsson Associates, Denver, Colorado, USA

Robert L. Byrd, Joe Tagliapietra & Kenneth D. Ball
Breakwater Resources Ltd., El Mochito Mine, Honduras

ABSTRACT: The paper presents an interesting case history of the Pozo Azul Tailings impoundment at Breakwater Resources Ltd.'s El Mochito Mine in the department of Santa Barbara, Honduras. El Mochito is an underground lead-zinc-silver mine located on the eastern flank of the Santa Barbara Mountains of north central Honduras. The mine has been in operation since 1948 and is currently mining at a rate of 1,850 tons/day, yielding approximately 550,000 tons of tailings annually, which after mine backfill requirements are met is discharged to the Pozo Azul tailings disposal facility. Pozo Azul has a compelling history that has been fraught with technical challenges its entire life. Presented is a historical perspective of challenges faced by the tailings facility's designers and the innovative design modifications that have been implemented by the mine staff to overcome technical challenges associated with karst topography, seismic activity, changing environmental conditions, hurricanes and fluctuating metals prices.

1 BACKGROUND

El Mochito's first tailings facility, El Bosque, was commissioned circa 1969-70. El Bosque was constructed three kilometers east of the mill in a tributary to the Quebrada Raices, a major river flowing into Lago de Yojoa; a large inland, recreational lake. The embankment was constructed by the centerline raise method utilizing cycloned tailings underflow deposited over a clay starter dam. Tailings were disposed in the El Bosque impoundment for nearly twenty years until it was decommissioned in 1989-90.

During the late 1970's Rosario Resources, who at the time owned and operated the mine began developing plans for a new tailings disposal facility. A site was selected nine kilometers east of the mill in yet another tributary to the Quebrada Raices. The site was named for an artesian spring in the area called Pozo Azul.

2 POZO AZUL TAILINGS DISPOSAL FACILITY

The Pozo Azul site was at first look an ideal setting for a tailings disposal facility. A broad valley, with a relatively small drainage basin offering considerable storage volume by damming a narrow stream between two ridges located approximately 700-feet upstream of it's confluence with the Quebrada Raices. Because the site is located down gradient from El Bosque, tailings could be delivered by gravity to the facility by simply extending the slurry pipeline from El Bosque.

What turned out to be a drawback to the Pozo Azul site, however, is the geology. The valley is fault controlled and comprised of the Atima Limestone formation, overlain by deeply weathered lateritic soil. The Atima Limestone, which crops out throughout the valley, although massive, is also karstic and heavily jointed. The karstic limestone would prove later to be problematic from the standpoint of containment.

The initial design consisted of the centerline construction method utilizing a 100-foot high clay starter dam (crest elevation 1,847 feet), a graded filter blanket, a side hill decant intake and embankment construction using cyclone underflow. Construction of the starter dam, filter, decant intake discharge conduit and tailing slurry line extending from the El Bosque tailings impoundment was completed in early 1981 (Figure 1).

Following construction of the starter dam, cyclone separation of tailings at the dam was initiated with underflow being deposited on the dam for construction purposes and the overflow (slimes) being deposited in the impoundment.

During this period however, a decision was made to utilize a hydraulic backfill system within the mine to further the exploitation of the San Juan ore body. The hydraulic sand backfill operation diverted the tailings sands from the tailings embankment construction to mine backfill. This decision prompted the first of many design changes that would occur at Pozo Azul.

Figure 1. Clay starter dam being constructed in the Pozo Azul drainage, circa 1979-80. A large limestone outcrop can be seen in the valley invert.

To compensate for the loss of tailings underflow material for embankment construction purposes, a decision was made in 1982 to continue construction of the Pozo Azul embankment with earth fill materials similar to a water dam construction. This was carried out under contract during 1983 and 1984. The work included removal of underflow placed by cyclone separation on the downstream face of the original starter embankment and placing it on the upstream side of the dam. This action, in effect, created a crest of underflow tailings upstream of the starter dam at elevation 1,870 feet, for a total dam height of 123 feet above the valley invert. The downstream earthfill raise was not completed at this time, however, due to other problems encountered within the impoundment (Figure2).

Upon initial deposition into the Pozo Azul impoundment during 1981, significant leakage of tailings and slurry water occurred into the limestone formation (the impoundment basin was unlined).

Tailings were seen emanating from the "Pozo Azul" artesian spring located downstream of the impoundment on the Quebrada Jutal, a tributary to the Quebrada Raices.

Attempts to seal the impoundment area by subsurface grouting, and surface sealing with clay material were intermittently successful, for short periods of time. An early Rosario report indicates approximately 4 months in 1981 and 9 months in 1982 of tailings deposition into the Pozo Azul impoundment. Tailings deposition continued intermittently between Pozo Azul and El Bosque until late 1984 when disposal in Pozo Azul was discontinued entirely.

Meanwhile the El Bosque impoundment continued to operate as the primary tailings storage facility until 1989. In 1988, recognizing that El Bosque would soon reach its ultimate storage capacity, a feasibility study was performed to determine whether or not the Pozo Azul facility could be recommissioned.

Results of the feasibility study indicated that by utilizing a unique application of a geomembrane liner the existing tailings in the impoundment could be isolated from future tailings deposited over them. The design consisted of allowing the tailings in the basin to dry out during the dry season. Grading the surface of the tailings and installing a 30 mil polyvinylchloride (pvc) membrane directly over the tailings. Utilizing this design the problems of tailings loss through sinkhole development could be alleviated. The design included a series of finger drains below the geomembrane liner that drained into the side hill decant tower. These drains serve to intercept infiltration below the liner and evacuate it through the decant system, rather than allowing it to essentially "pipe" tailings into the underlying karstic formation (Figures 3, 4 and 5).

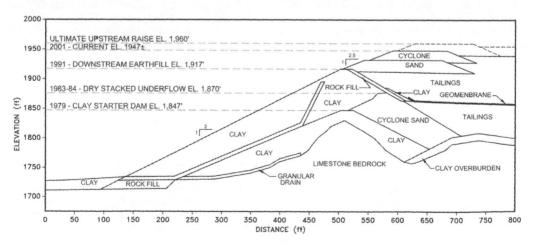

Figure 2. Typical cross section of the Pozo Azul embankment showing the various stages of its development.

Figure 3. Finger drains were constructed in the "dry" tailings bed below the geomembrane liner to intercept and evacuate infiltration below the liner.

Figure 4. A geomembrane liner was installed over the "dry" tailings bed and finger drains.

Figure 5. Completed installation of the liner system covering the bed of tailings.

Having overcome the impoundment containment issues associated with the facility an enlargement design for Pozo Azul was developed. A downstream earthfill raise that included a sand drain between the original starter dam and the downstream raise was designed. The downstream raise design (referred to as Phase I), which raised the dam to a total height of 167 feet (crest elevation 1,917 feet) was intended to accommodate another downstream raise (Phase II) to crest elevation 1,950 feet (total dam height of 200 feet) sometime in the future (Figure 6).

Figure 6. The downstream raise was actually initiated in 1983 when cycloned tailings were removed from the downstream slope and "dry stacked" upstream of the clay starter dam.

Figure 7. Construction the downstream raise to elevation 1,917 feet near completion.

Lining the Pozo Azul impoundment and constructing the Phase I dam raise would require approximately 18 months to complete. At this time the El Bosque tailings disposal facility was nearly full and would not accommodate an additional 18 months of tailings in its existing configuration. This prompted a feasibility study, which determined that El Bosque could accommodate an additional 18 months of tailings if the crest were raised by seven feet. A downstream, cycloned tailings dam crest raise combined with a small earth fill dike at the left abutment was designed and constructed at El Bosque in 1989 while the initial construction activities were underway at Pozo Azul.

Constructing the liner system raise at Pozo Azul was begun and completed during the dry season in 1989. The Phase I downstream raise took approximately 18 months to complete and was completed in 1991. Tailings disposal at Pozo Azul was initiated in 1990 and the unique liner design and installation proved to be a success (Figure 7).

Following on the successful re-commissioning of Pozo Azul, the El Bosque tailing disposal facility was decommissioned and put into closure and reclamation in 1991. Pozo Azul has served as the sole surface repository for tailings at El Mochito since 1990-91.

The geomembrane liner system was initially installed at Pozo Azul under an extremely tight schedule and intended to facilitate a rapid execution of work and minimize the time and expense required for re-commissioning the facility. As a result the liner was only installed over the tailings lying in the valley invert and did not extend up the slopes of the valley basin. Tailings being deposited in the impoundment would therefore rise above the upper limits of the liner in 1991-92 raising concerns of continued excessive leakage from the impoundment basin. Because of the relatively high costs associated with installing a geomembrane liner system throughout the basin a feasibility study was performed in 1991 to determine if an alternative side-hill liner system could be developed that would minimize the capital and/or operating expenses for tailings disposal at Pozo Azul.

A side-hill treatment alternative analysis was carried out for Pozo Azul in 1991. Through extensive field trials the study determined that tailing slimes deposited directly atop a geotextile filter fabric would blind off the openings in the fabric very quickly. It was found that by installing a geotextile filter fabric directly over the prepared ground surface, and depositing tailing slimes directly atop the geotextile that a filter cake could be developed rendering the geotextile essentially impermeable.

Managed tailings deposition could be utilized to assure the geotextile was covered quickly following installation in order to:

- Prevent degradation of the exposed geotextile from wildlife and vegetation;
- Facilitate draining the slimes deposited on the basin side slopes and;
- Develop a filter cake atop the geotextile as soon after its installation as possible.

Geotextile could be installed by mine staff field crews on an as needed basis as the tailings level in the impoundment rose. Thus a geocomposite liner system was designed for the basin side hill lying above the geomembrane. The geocomposite liner includes a prepared sub-grade, 16-ounce, needle-punched geotextile and managed tailings deposition. The geocomposite liner has subsequently been installed on an as needed basis as the level of the tailings rise in the impoundment since 1991.

Tailings disposal has continued at Pozo Azul since 1990. In 1995 residual tailings storage in the Pozo Azul impoundment had decreased to a point where it was infringing on the freeboard volume available for storm water storage. This is a critical element to impoundment operations and safety, because the impoundment was being operated without an emergency spillway. The design storm runoff was to be stored in the impoundment and discharged through the side hill decant structure.

In November 1995, after four rather uneventful years at Pozo Azul, Breakwater decided proactively to assess the situation relative to Pozo Azul operations. Because of the diminished storage capacity at Pozo Azul and the time required to design and construct a crest raise for the dam, Breakwater decided to develop an engineering design for raising the dam crest. In 1996 a downstream, earthfill embankment raise to crest elevation 1,950 feet (33 foot raise to a total dam height of 200 feet) (Phase II), which included an emergency spillway located at the left abutment was designed by Hydro-Triad, Ltd. An engineer's cost estimate for the project came in at $3.3 M for construction.

While the engineer was on site to present the Phase II design to mine staff, he recognized that a considerable quantity of whole tailings were being delivered to Pozo Azul, rather than being used underground for mine backfill. The engineer suggested that if the mine could assure a whole tailings run to Pozo Azul it might be feasible to construct an embankment raise by the upstream method using the underflow produced from cycloned whole tailings being delivered to Pozo Azul. Recognizing the capital costs savings associated with an upstream raise versus a downstream earthfill design, Breakwater commissioned Hydro Triad, Ltd., to perform a study to determine the feasibility of constructing an upstream raise at Pozo Azul.

The study determined that an upstream raise was feasible. Initially, a 15-foot upstream crest raise to crest elevation 1,932 feet was designed and included the following key elements:

- A drain internal to the embankment raise that keeps the underflow sand used to construct the raise dry, by evacuating residual water from cycloning and rainfall to the abutments.
- Achieving higher than 70% relative density by spreading and compacting cycloned tailings used for constructing the embankment crest.
- Strict operating criteria that prevents the release of tailings if a breach of the embankment were to occur due to liquefaction of the tailings underlying the embankment crest.
- Thorough QA/QC program and embankment monitoring program to assure the integrity of the facility during construction.

Construction of the initial, 15-foot upstream crest raise was initiated in 1997 and represented, yet again, a fundamental design modification to the Pozo Azul tailings embankment (Figure 8).

While the initial 15-foot crest raise was being constructed further engineering analyses were performed to determine the maximum height to which the embankment could be raised by the upstream

376

Figure 8. Initiating cycloning for the upstream raise in 1997.

method. A final design for the facility was developed that resulted in an ultimate crest elevation 1960 feet, representing a 43-foot upstream cyloned sand raise atop the original earth fill dam, equating to a total dam height of 210 feet.

Pozo Azul's upstream raise is currently (September 2001) being constructed. The embankment crest has been raised to elevation 1946 feet (dam height 196-feet), approximately. As a means of facilitating continued gravity-feed of tailings to Pozo Azul, the tailings delivery line from the mill, via El Bosque, has been reconstructed to a flatter slope. This allowed the pump station at the impoundment to be raised approximately 15 feet, which was necessary to allow flushing the line, or gravity feed to the impoundment in the event of power outages that occur periodically (Figure 9).

3 SUMMARY

The Pozo Azul tailings impoundment has experienced two significant design modifications since it was initially conceived in the late 1970's. Further-more two unique liner systems were utilized to assure the technical integrity of the impoundment and facilitate its continued use.

Initially a clay starter dam was constructed and cycloned tailings were deposited atop the clay starter dam in order to develop a centerline raise. During centerline construction, tailings slimes were being deposited in the impoundment. Two things occurred at this time, which changed the direction of Pozo Azul.

First, tailings slimes and supernatant deposited in the impoundment began to leak from the basin through the karst formations underlying the impoundment. Second, the mine began utilizing whole tailings for mine backfill thereby, depleting the source of construction material for a centerline raise at Pozo Azul.

To prevent continued loss of tailings and supernatant to the underlying karst formation a unique geomembrane liner application was designed that isolated the tailings previously deposited in the basin from those that would be deposited above the liner. The liner system was installed over "dry" tailings and included a series of finger drains under the liner that capture infiltration below the liner and routes it to the side hill decant tower, rather than allowing it to "pipe" " tailings into the underlying karstic formation which would ultimately undermine the liner.

As a result of the loss of whole tailings for construction the embankment design was modified from a centerline cycloned tailings dam to a downstream earth fill structure, similar to a typical water dam construction. Cycloned tailings deposited on the downstream slope were removed and dry-stacked on the upstream slope of the starter dam and served as an interim dam crest at elevation 1,877 feet (dam height of approximately 127-feet), while the downstream raise was constructed to crest elevation 1,917 feet (167-foot dam height).

Figure 9. An initial 15-foot upstream raise was completed at Pozo Azul in 1998.

Prior to the tailings and supernatant surface rising above the geomembrane liner a unique geocomposite side hill liner system was designed that include a prepared sub-grade, a geotextile filter fabric and tailings slimes deposited directly over the geotextile, which created a nearly impermeable barrier. This liner system could be installed on an as needed basis for approximately 20 percent of the cost of installing a geomembrane liner. This lining method was adopted in 1991 and continues today.

In 1995 yet another major design modification was made. Rather than spending an estimated $3.3 M to construct the Phase II downstream earth fill raise to crest elevation 1950 feet an upstream cycloned tailings embankment raise was designed. The upstream raise included the following key elements to assure its technical integrity throughout its operational and post-operational life:

- A drain internal to the embankment raise that keeps the underflow sand used to construct the raise dry, by evacuating residual water from cycloning and rainfall to the abutments.
- Achieving higher than 70% relative density of the cycloned tailings used for constructing the embankment crest by mechanical placement.
- Strict operating criteria that that prevents the release of tailings if a breach of the embankment were to occur due to liquefaction of the tailings underlying the embankment crest.
- Thorough QA/QC program and embankment monitoring program to assure the integrity of the facility during construction.

Ultimately the upstream construction will result in an embankment crest elevation of 1,960 feet (dam height 210-feet), which will accommodate tailings production through 2006. The upstream raise will extend the life of the facility by 3 to 4 years beyond what the original downstream embankment design would have accommodated.

Pozo Azul is an example of how an open-minded owner working together with an innovative engineer can implement unique ideas to reduce tailings handling capital and operating costs while maintaining the technical integrity of a challenging site.

REFERENCES

Hydro-Triad, Ltd., "El Mochito-Tailings Disposal," November, 1985.

Hydro-Triad, Ltd., "Mina El Mochito," July, 1988.

Hydro-Triad, Ltd., "Technical Specifications for El Mochito Mine, Construction of the Final Stage, El Bosque Tailings Embankment and Appurtenant Structure," August, 1988.

Hydro-Triad, Ltd. "El Bosque Tailings Embankment Enlargement Study," September, 1988.

Hydro-Triad, Ltd., "Pozo Azul Feasibility Study," December, 1988.

Hydro-Triad, Ltd., "Proposal: Mina El Mochito, Design Program for Permanent Spillway and Reclamation Plan, El Bosque Tailings Impoundment," March, 1990.

Hydro-Triad, Ltd., "Alternatives Evaluation Report, Impoundment Sidehill Treatment, Pozo Azul Tailings Storage Facility," March, 1991.

Hydro-Triad, Ltd., "Geotextiles Evaluation Report Impoundment, Sidehill Treatment, Pozo Azul Tailings Storage Facility, Mina El Mochito," July, 1991.

Hydro-Triad, Ltd., "Mina El Mochito, Pozo Azul Tailings Embankment Enlargement By the Upstream Cyclone Method, Engineering Design Report," March, 1997.

Performance of the Kennecott Utah Copper tailings embankment, Magna, Utah

Joergen Pilz
AMEC Earth and Environmental, Inc., Salt Lake City, Utah

Doug Stauffer
Kennecott Utah Copper, Magna, Utah

ABSTRACT: Kennecott Utah Copper has been operating their modernized tailing impoundment since 1997. Nearly four years of operational experience has been gained in hydraulic placement and compaction of cyclone sand tailings. During this period a variety of dozers and other equipment have been evaluated in tailings placement. Currently, rubber-tired equipment is used to manage and control tailings placement. The project design criteria with regards to compaction, percent fines content, and permeability/drainage are being readily achieved. During this initial operation period, the geotechnical performance has been monitored with a quality assurance testing program, and instrumentation consisting of piezometers, settlement plates, and survey measurements. This paper reviews tailings placement techniques, summarizes in place material properties, and presents an overview of the geotechnical performance of this state of the art embankment.

1 INTRODUCTION

The Kennecott Utah Copper Tailings Impoundments are located approximately 10 miles west of Salt Lake City, Utah along Interstate 80 and north of the town of Magna, Utah. These impoundments store the tailings from the world famous Bingham Canyon Mine, situated about 15 miles to the south. The impoundments are operated as the original south impoundment, which encompasses approximately 5,700 acres and the newer north impoundment, which encompasses about 3,400 acres. The south impoundment has been operational since about 1904 and has been constructed using the upstream method of construction. Operations transitioned from the south impoundment onto the north impoundment during the years 1996 to 2001. An active dewatering and reclamation program is underway on the south impoundment. This paper discusses primarily the north impoundment, which is considered to be a state of the art, modified centerline type of embankment, designed to meet all current seismic design criteria. The location of the impoundments and the surrounding vicinity are shown on Figure 1, Aerial Photograph.

2 DESIGN CRITERA

The north impoundment provides the storage capacity for continued copper mining activities through the year 2018 at tailings throughput of up to 176,000 tons per day. In addition to providing the necessary storage capacity, the north impoundment provides envelopment and buttressing to seismically vulnerable portions of the south impoundment. The north embankment has been designed to withstand a maximum credible earthquake (MCE) of Richter magnitude 7.0 occurring along the East Great Salt Lake Fault at a horizontal distance of 3 kilometers. The MCE peak horizontal ground acceleration is 0.52 g.

3 EMBANKMENT DESIGN AND CONSTRUCTION

Kennecott Utah Copper's mine and concentrator are located approximately 15 miles to the south of the tailings impoundment facility. Whole tailings are transported to the impoundment via gravity through a large diameter concrete pipeline. The whole tailings are then separated at two cyclone stations to produce a sand product used for embankment construction. The cyclone stations are located at the east and west abutment tie-ins with the south impoundment. From the cyclone stations, the sands are pumped through rubber lined pipelines to cyclone sand deposition cells. The cyclone overflow South Impoundment North Impoundment (slimes) are delivered via unlined 26-inch diameter steel lines to a header and spigot system located on the upstream perimeter of the cyclone sand embankment. The cyclone overflow deposition system is intended to develop uniform, flat beaches and provide continued dust control on the impoundment surface. Decant water is reclaimed through a floating pump decant barge system. Barge-mounted reclaim pumps recy-

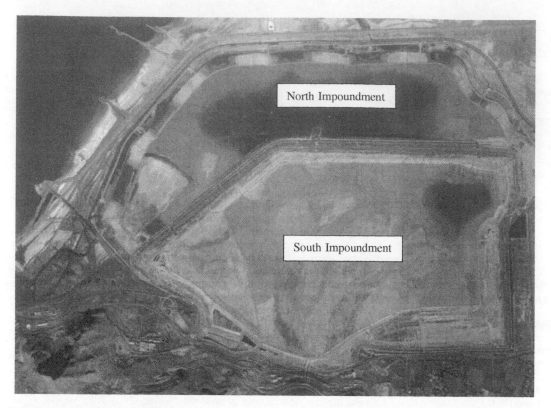

North Impoundment

South Impoundment

Figure 1. Aerial photograph of the tailings impoundment.

cle the water back through the cyclone stations as dilution water and back to the concentrator via a reclaim water system consisting of pipelines, canals, and pump stations.

Geologically, the impoundment is situated on a thick sequence of late Pleistocene lake bed clays. Although these clays provide an excellent hydrogeologic barrier between the natural groundwater system and tailings, the thick sequence of clays, sands, and silts represent a relatively "soft" foundation from a seismic design perspective. The cyclone sands provide such a seismically stable embankment. A detailed description of the embankment geotechnical design is provided in reference 1 (Ridlen, 1997). The entire embankment is underlain by a 3-foot thick blanket drain, consisting of 1.5-inch diameter slag drain rock surrounded by graded filters. The final design embankment will be triangular-shaped in cross section with a crest width of 100 feet, 235 feet high, and a footprint of about 1,100 feet. The final exterior slopes will be 4 horizontal to 1 vertical, while intermediate construction slopes are 3 horizontal to 1 vertical. The embankment will be constructed in three phases: phase 1 (Zone A) will be 90 feet high, phase 2 (Zone B) includes down stream cell construction from phase 1 and will be

150 feet high, and phase 3 (Zone C) will extend to the final design configuration.

4 MATERIAL CHARACTERISTICS

The Copperton whole tailings have approximately 65 percent passing the number 200 sieve (0.074 mm). Slurry feed from the Copperton concentrator flows through two 60-inch diameter concrete tailings lines at a solids content of about 40 to 43 percent (by weight). Specific Gravity of the solids is approximately 2.65. Cyclone sands, also termed underflow, constitute the coarse fraction of tailings after separation by hydrocycloning. Cyclone underflow contains less than 25 percent passing the number 200 sieve and averages about 20 percent fines. When discharged from the cyclones, the underflow contains approximately 65 percent solids (by weight) and is repulped down to 50 to 58 percent solids for distribution to the embankment construction cells. Cyclone Overflow (or slimes) constitute the fine fraction of tailings after being separated by hydrocycloning. Cyclone overflow contains approximately 95 percent passing the number 200 sieve and approximately 17 percent solids (by weight).

380

5 OBSERVATIONAL METHOD

An important aspect of the embankment construction is the use of the observational method to verify that embankment design parameters are being met and observe the geotechnical performance of the impoundment. The geotechnical performance characteristics include embankment and foundation pore pressures, settlement, seismic performance (accelerometers), density, gradations, and measurement of horizontal and vertical movements.

6 CYCLONE STATIONS

Each cyclone station consists of a feed system, cyclone clusters, underflow collection and distribution system, overflow collection and distribution system, and process water system. Whole tailings enter a concrete feed sump where they are diluted to about 25 percent solids content and are then pumped through three rubber lined slurry pumps to feed a circular arrangement of cyclones (termed a "cyclone cluster"). Each cyclone cluster consists of twenty-three, 20-inch diameter Krebs cyclones. A central distribution pot feeds the cyclones. Two circular launders are located beneath the cyclones to collect the overflow and underflow. Air-actuated knife gate valves operate the individual cyclones and adjust automatically to maintain a near constant cyclone entrance pressure of 15 psi. A single operator controls and monitors both the east and west cyclone stations from a computerized control room facility, located at the west cyclone station.

7 TAILINGS DISTRIBUTION SYSTEM

The cyclone underflow flows from the circular launder to a concrete collection sump where underflow is diluted to approximately 50 to 58 percent solids by weight to feed a pump system. The underflow pump system consists of 3 trains of 6 pumps in series, delivering 2,500 to 3,500 gallons per minute of underflow at a pressure of 500 to 800 pounds per square inch. To maintain sufficient pipeline velocity, the underflow pipe is 10 inches inside diameter rubber lined pipe. The underflow distribution pipeline extends as far as 25,000 feet from the cyclone station to the furthest point on the embankment. Two booster stations are used to maintain pressure and flow. Under adverse conditions and during three to four months in the winter, the whole tailings are delivered to the impoundment through a drop box system located at the east and west abutments. Maintenance of the cyclone sand distribution system is performed at this time.

Cyclone overflow is also distributed to the impoundment by pumping from each cyclone station.

Since the overflow comprises only about 17 percent solids, these pipelines consist of 40-inch diameter steel pipes. The overflow pipeline system is operated at less than 200 psi, for distances up to 25,000 feet. To maintain a uniform, wetted beach for dust control purposes, the cyclone overflow is delivered to the interior of the impoundment using 26-inch diameter steel "header and spigot" pipeline fitted with 3-inch diameter nozzles spaced at 20-foot centers. The overflow system discharges to the impoundment interior along a 4,400-feet segment for a period of 12 hours. After 12 hours, overflow is routed to a different 4,400-foot long section of the embankment such that the entire perimeter of the impoundment receives deposition at least once every 4 days.

8 EMBANKMENT CONSTRUCTION

Even at an operating capacity of 155,000 tons per day, material balance constraints prevent the entire embankment footprint from being raised at one time. Therefore, the embankment cross section is constructed in three zones. Currently, the 650 feet wide phase 1 embankment (Zone A) is being constructed. As underflow material becomes available due to the decreasing footprint of Zone A, Zone B will be constructed downstream of the current embankment. Likewise, as Zone A continues to narrow, Zone C sand placement will begin. The construction sequence and material balance assures that each successive zone catches up with the elevation of the previous zone to match the elevation of the impounded overflow and whole tailings. To achieve this balance, close monitoring of material placement and grades must be accomplished at all times.

The approximate 37,000 feet long embankment is divided into 16 segments or hydraulic deposition cells. Each cell is operated independently for underflow deposition and is defined with small (less than 5 feet high) dikes constructed of underflow sand. Each cell is approximately 2,200 feet long and varies in width from 650 feet near the base of the embankment to as narrow as 100 feet as the embankment is raised.

Construction of the North impoundment embankment is accomplished by hydraulically depositing underflow slurry into embankment cells and managing the newly deposited material with rubber-tired, self propelled equipment to prevent particle segregation, spread the sand into thin lifts, and assure that the material is placed to specified density criteria. Tracking is ongoing at all times while underflow is being deposited into a cell. If underflow is allowed to deposit for extended periods of time without equipment tracking, pockets of wet, loose, high fines content material may develop. The "slimes" pockets then require removal, mixing with cleaner sands and compaction to meet specification requirements. To maintain specification require-

Figure 2. Photograph of hydraulic cell.

ments, it is more economical to continually manage the material during deposition than to allow slimes pockets to develop. Figure 2 is a photograph of an active hydraulic cell.

9 EMBANKMENT PERFORMANCE

Oversight is provided by a full time staff of two to three senior materials technicians, a survey crew, Kennecott operations and quality assurance engineers, and by periodic independent review of the work. Material testing includes nuclear moisture density tests, sand cone density tests, proctor compaction tests and gradation analysis. All field test locations are recorded using a first order GPS (Global Positioning) survey system. The material tests are presented both in a tabular and graphical format in reports prepared once every two weeks. The quality control reports include tabular summaries of all the data acquired along with plots of test locations, trend line analysis of density, fines content, and passing/failing tests. The QC data are presented in Microsoft Excel® format, which are periodically archived and also used to present annual summaries of the embankment construction to the Utah State Engineers Office of Dam Safety.

Materials testing is performed within large test pit "trenches" excavated through the embankment about one week after deposition in a cyclone sand hydrau-

lic cell has ceased. Each test pit is about four to five feet deep and may represent about one month of deposition within an active cell. Initially, testing was performed during active tailings deposition, but this caused interference with placement operations. As operation of the cyclone sand deposition was gained, confidence in the placement techniques was developed. This confidence permitted this "confirmation" type of test program to be utilized.

The use of a GPS system to record the QC tests allows for a very accurate portrayal of the progress of the embankment construction, elevations, and test data. Accurately locating the QC tests provides good confidence that the large volumes of cyclone sands are being adequately tested in quantity, location, and areal extent. The GPS data are used to record not only the QC test locations, but also used for material balance calculations and payment of the earthwork contractor.

Cyclone sand placement is statistically evaluated for each cell within the embankment. The embankment specifications require that a "ten test" average not fall below 98 percent of the standard proctor density (ASTM D-698) and no individual test fall below 96 percent compaction. Figure 3 shows the running test averages for the hydraulic placement of cyclone sands during the year 2000. Likewise, the percent fines in the cyclone sands, which is closely related to the vertical drainage characteristics of the cells, is shown on Figure 4. Also shown on Figure 4

10 Test Average (Per Cell)

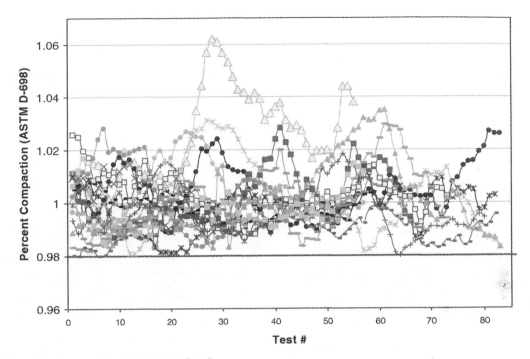

Figure 3. Ten test running average per hydraulic cell.

Year 2000 Density and Gradation Tests

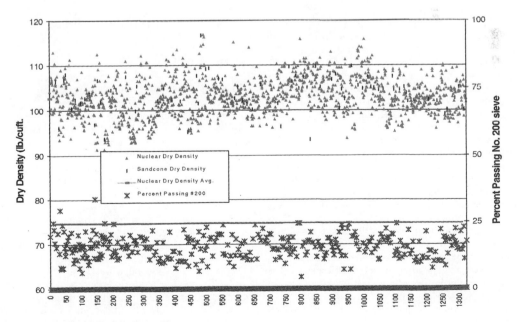

Figure 4. Dry densities and percent fines content.

are the dry density values versus the number of tests completed. The "cycles" observed within the grind are a function of season and ore type within the Bingham Mine.

10 SAND PLACEMENT EQUIPMENT

Cyclone sands are fed to the cells through a single HDPE feed line, which can be moved with the compaction equipment. During the initial 3 years of embankment construction, different equipment has been tested and utilized for sand management and compaction. Initially, standard D-8 size dozers were utilized. These dozers provided the required compaction and management of underflow segregation, but due to the abrasive nature of the tailings particles, under-carriage wear of the steel tracks and drive system was excessive.

In an attempt to reduce equipment maintenance and operation costs, rubber tracked "farm type" equipment was utilized for a trial period. Although meeting specification requirements for compaction and particle segregation, equipment availability was low (high down time) and maintenance costs were excessive. To remove the operating equipment from the abrasive, wet tailings, a modified backhoe was tested. A backhoe was fitted with a large compaction plate and the plate used to both manage tailings streams and densify the tailings by "pounding." This approach could not efficiently manage the large hydraulic cell areas (25+ acres) and high deposition rates (15,000 tons/day per cell).

The current equipment being used to manage tailings placement includes articulated four-wheel drive rubber-tired farm tractors. Each axle is fitted with duel tires. These units have proven to be the most economic equipment to operate. We believe that tailings management is a more appropriate term than "compaction," since this equipment does not track and compact the tailings in a manner identical to traditional earthwork placement. There are few moving parts exposed to the abrasive underflow sand and the tractors travel at speeds which can keep up with underflow deposition rates. Figure 3 shows operation of the rubber-tired tractor.

D-8 size steel track dozers are still utilized to construct containment dikes, access roads, and grade exterior slopes. However, these operations are performed in dry underflow sands and are not continuous. Component wear, especially track wear, is not as severe under these conditions.

Each of the tailings management units has been able to meet the specified density requirements. It is of interest to note the difference in overall embankment density produced with the different types of equipment. Following are the average compaction results for over 500 nuclear density tests measured while the various types of equipment were being utilized for underflow management:

Bull Dozer	108.9 lb/cf
Rubber Track Tractor	105.9 lb/cf
Rubber Tire Tractor	104.4 lb/cf

Results demonstrate that even the rubber-tire tractor compacts the underflow sand to about 100 percent of the Standard Proctor Density (ASTM D-698), which exceeds specification requirements of 98 percent.

11 PROCESS WATER RECLAMATION AND DISTRIBUTION

Water is decanted from each cyclone cell utilizing either weir boxes or portable diesel powered pumps, depending upon the hydraulic cell grades with respect to the impoundment. Water is allowed to pond at one location within each cell and is decanted or pumped to the impoundment interior. The location of the cell decant ponds are moved horizontally with each underflow deposition cycle to prevent pockets of tailings fines from accumulating.

Decant water is removed from the impoundment utilizing a floating pump barge facility. The floating barge is secured to two large trusses attached to fixed concrete anchors located along the slopes of the south tailings impoundment. The barge system consists of 20 smaller "sub-barges," each supporting vertical cantilever type electric pumps. The main barge holds a central manifold, which connects to large diameter HDPE pipelines for transport to a collection sump on the top of the south impoundment and delivery back to the process circuits.

12 DUST CONTROL

Dust control within the impoundment interior is accomplished efficiently by deposition of cyclone overflow. The impoundment beach is not allowed to dry for more than four days between deposition cycles. The impoundment surface is monitored daily by visual inspection from the perimeter roads and monthly by helicopter along a regular grid inspection. This system has operated efficiently throughout the initial four years of operation.

Dust control within the cyclone sand construction cells is considerably more difficult to manage. The excellent drainage characteristics of the sandy underflow tailings and the more intermittent deposition cycles permit the hydraulic cells to dry out quickly. To reduce pump and valve wear, half of the hydraulic cells are operated (intermittently) for several months at a time before switching over to inactive cells. During the inactive deposition cycle, a water sprinkler irrigation system is used to wet the surface of the inactive cells. The cyclone sand distribution system is converted to a sprinkler system during this inactive deposition period. Large (1,000 gpm) impact type sprinklers are installed at approximate 450-foot spacings along the centerline of the cells. Operation of each sprinkler is cycled to maintain a wetted surface. Embankment slopes are also treated with polymer and/or hydromulch applied using standard water trucks and hydromulch spray trucks. Hydromulch has been utilized where an area is dry, but will be disturbed again within a short time (<2 weeks). Polymer is utilized where longer-term dust control is desired. Vegetation is established where surfaces will be exposed for extended periods of time (>1 year), using a "drill and seed" approach to establish grasses. Access roads are treated using either water and/or a Magnesium Chloride solution. This system requires close oversight of weather forecasts and expected wind patterns and has been successfully operated for two years.

13 GEOTECHNICAL INSTRUMENTATION

A geotechnical instrumentation system was included in the original design to monitor the performance of the embankment, confirm that threshold values are not exceeded, and allow for optimization of the design as experience with its construction is gained. The geotechnical parameters monitored include foundation pore pressures, embankment drainage, and phreatic levels and settlements. As the embankment is raised, inclinometers will also be installed to measure horizontal deformations. The geotechnical instrumentation system is arranged along five principal geotechnical monitoring sections, as shown on Figure 5, Geotechnical Monitoring Locations.

Vibrating wire piezometers measure both foundation pore pressures and embankment phreatic levels. Settlements are measured using closed hydraulic settlement sensors fitted with vibrating wire piezometers. Settlements and pore pressures are plotted in conjunction with survey raise rates to assure that design foundation pore pressures are not exceeded. Instruments are monitored on a monthly to quarterly basis. Trend lines are plotted for changes in measured values of each instrument. A typical monitored embankment cross section is shown on Figure 6, Geotechnical Monitoring Cross Section.

To date, the geotechnical monitoring data has shown that the embankment has generally performed close to or exceeded the geotechnical design expectations. Drainage of the cyclone sand embankment through the blanket drain has exceeded design expectations. In general, embankment phreatic levels have been very low and essentially coincide with the drain blanket. As the embankment settles up to an estimated 20 to 25 feet, the drain blanket will become submerged. Flow will then become a function of piezometric head. A detailed study is currently underway to evaluate the phreatic surface response to cycles of active tailings deposition. The instrumentation data reveals that drainage of the hydraulic cells occurs within one week of inactivity of tailings deposition within a hydraulic cell. Such data will be used for future optimization of the drain blanket extension.

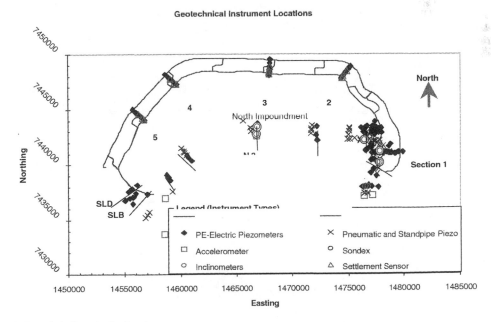

Figure 5. Geotechnical monitoring locations.

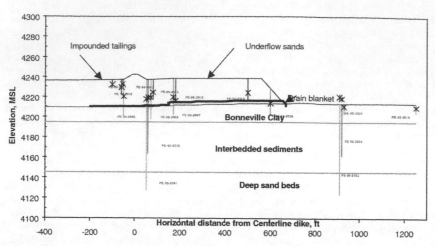

Figure 6. Geotechnical monitoring cross-section.

REFERENCES

Ridlen, P.W., Davidson, R. et al. 1997. *Geotechnical Design of Kennecott Utah Copper Tailings Impoundment Expansion, in Proceedings of the Fourth International Conference on Tailings and Mine Waste.* Rotterdam: Balkema Publishers: 741-752.

Wong, I., Olig, S. et al. 1995. Seismic Hazard Evaluation of the Magna Tailings Impoundment. *Environmental and Engineering Geology of the Wasatch Front Region* Publication 24.

Tailings and Mine Waste '02, © 2002 Swets & Zeitlinger, ISBN 90 5809 353 0

Performance of vertical wick drains at the Atlas Moab Uranium Mill tailings facility after 1 year

Michael E. Henderson & Jared Purdy
Olsson Associates, Denver, Colorado

Tracey Delaney
Independent Consultant

ABSTRACT: The Atlas Uranium Mill tailings facility is an unlined tailings disposal facility located on the banks of the Colorado River, near Moab, Utah. Perforated vertical wick drains were installed in 2000 to accelerate tailings consolidation and provide hydraulic relief prior to placing a final cover. This paper summarizes the performance of the wick drains, after nearly one year of operation.

1 INTRODUCTION

The Atlas Uranium Mill tailings facility is an unlined tailings disposal facility located on the banks of the Colorado River, near Moab, Utah. The facility was used for disposal of uranium tailings from 1956 to 1984, during which approximately 12 million tonnes were deposited in a sub-areal manner. The tailings impoundment covers approximately 53 hectares, at an average depth of about 20 meters, as shown on Figure 1.

The current closure plan for the tailings impoundment includes regrading the tailings surface and placement of a radon barrier. One of the concerns related to closure of the facility is that leachate could migrate from the tailings facility into the underlying groundwater system as the tailings consolidate following regrading and cover placement. This concern is primarily driven by the potential for increasing ammonia levels in the Colorado River adjacent to the tailings facility, should the groundwater discharge into the river system increase.

The second consideration was that under Federal rules, the surface of the tailings impoundment must

achieve 90 percent of the total consolidation before a final cap is placed. At the Atlas site, the cap was required to be placed by mid-2002, providing about two years to achieve the desired consolidation.

As a result of these factors, a program was undertaken in 2000 to evaluate options for dewatering the tailings, followed by detailed engineering design, permitting and construction.

2 FIELD INVESTIGATION

A detailed geotechnical and geochemical investigation was undertaken in early 2000 to characterize the physical and geochemical properties of the tailings. This investigation identified a zone of saturated slimes surrounded by a "donut" of non-saturated, consolidated sand.

The geotechnical investigation consisted of Cone Penetrometer testing (CPT) to collect in-situ geotechnical data of the tailings, and sample collection using a hollow-stem auger. ConeTec, Inc. (Salt Lake City, Utah) provided both truck and track-mounted CPT equipment for the field investigation. Agapito Associates (Grand Junction, Colorado) provided the auger drill rig. The CPT tests were conducted at 98 locations across the surface of the facility, and samples were obtained at 6 locations. Monitoring wells were installed at strategic locations in the tailings to measure water levels and sample tailings supernatant.

3 MODELING

A spatial model of the tailings mass and its physical and geochemcial properties was developed using mine planning software (e.g. Vulcan). A variably saturated groundwater model was developed using

Figure 1. Atlas tailings impoundment, looking east.

this data to evaluate dewatering options. Similarly, a geochemical transport model was developed to estimate changes in the geochemistry of the tailings mass resulting from dewatering options.

Consolidation was modeled using Sigma/W, from Geo-Slope International. Average total settlement was estimated to be approximately 2.0 meters after about 70 years.

Seepage was modeled using Seep/W, from Geo-Slope International.

4 EVALUATION

Following the field investigation and modeling effort, an evaluation was undertaken to select the appropriate technology which would reduce or eliminate consolidation-related seepage from the impoundment and which would speed up the rate of consolidation. The options evaluated included the do-nothing or baseline alternative, horizontal wells, vertical wick drains, vacuum-assisted vertical wick drains, and air sparging. Vacuum-assisted vertical wick drains were selected as the preferred method to reduce the amount of consolidation-induced seepage that could enter into the underlying groundwater system and which could meet the consolidation requirements. A more comprehensive discussion of the modeling and evaluation effort can be found in Henderson, 2000 and Henderson, 2001 (see references).

5 WICK DESIGN

It was determined that a triangular wick installation pattern would be the most effective means for dewatering the tailings due to a decreased flow path, when compared to a rectangular pattern. Spacing of the wicks was calculated using a wick design program provided by Nilex, Mebra, and a program developed by Dr. Jinchun Chai, MebraWin. The results of these programs were then checked with hand calculations.

The tailings material properties and specific site parameters required for the calculations included horizontal and vertical hydraulic conductivity, coefficients of horizontal and vertical consolidation, unit weight, compression index, initial void ratio, the coefficient M from the Cam-Clay equation, surcharge pressure, length of drainage path and time required to reach the requisite consolidation (see Table 1). The required material properties were determined from lab tests and the application of the CPT test results.

The results from the wick design programs indicated that the wicks be placed on 3-m centers in order to achieve 90% consolidation in 18 months under the design surcharge load.

Table 1. Required material properties and site parameters

Vertical Hydraulic Conductivity, k_v (m/s)	4.00E-9
Hydraulic Conductivity Ratio (k_h/k_v)	4
Unit Weight (kN/m^3)	18
Compression Index, Cc	1.37
Initial Void Ratio, e_o	2.08
M (from Cam-Clay)	1.35
Coefficient of Vertical Consolidation, c_v (m^2/s)	1.41E-7
Coefficient of Volume Compressibility, m_v (m^2/kN)	2.94E-3
Drainage Path (m)	Variable
Surcharge Pressure (kn/m^2)	Variable
Required Consolidation (%)	90
Time (months)	18

6 COLLECTION SYSTEM

In order to collect water expelled from the tailings impoundment through the vertical wicks a collection system was designed. The collection system consisted of horizontal wicks, a drainage collection trench, a central sump and a submersible pump.

Each vertical wick was tied into a horizontal wick. The horizontal wicks were then fed into the drainage collection trench.

The drainage collection trench consisted of a 4-in diameter perforated HDPE pipe surrounded by gravel and encased with a geotextile. The trench began at the east and west sides of the impoundment and sloped at a 1-2% grade towards the central sump.

7 WICK INSTALLATION

Approximately 360,000 meters of vertical wicks were installed from September to December 2000 by Nilex Corporation (Denver, Colorado). The wick was installed using specifically equipped track mounted excavators. The vacuum-assisted portion of the wick installation was installed, but has not been used to date. Figure 2 shows installation of the wick drains. Figure 3 shows settlement of the surface immediately following wick installation.

Figure 2. Wick rig installing vertical wicks at Atlas tailings impoundment.

Figure 3. Surface settlement immediately following wick drain installation.

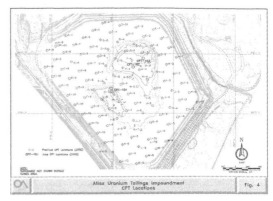

Figure 4. CPT test locations

8 SURCHARGE

Following the installation of the wicks, a surcharge load was applied. The design surcharge load consisted of sand of variable depths ranging from approximately 6-m at the center of the impound to approximately 2-m at the edges. Due to funding limitations, the full design surcharge load could not be applied. Consequently, the wicked area received from 20-65% of the design load, with the majority of the surcharge load applied to the outer region of the wicked area and only about 1 to 1.5-m in the center. The surcharge thickness over the slimes area is contoured at 1-ft (0.3048-m) in Figure 4. The surcharge loading outside of the slimes area was not contoured.

9 FIELD PERFORMANCE

A post-mortem CPT and survey program was performed in June 2001 to monitor performance of the wick drains. The CPT program was conducted at three points, near previously tested locations. The program was designed to evaluate the performance of the wicks over the 6-8 months of operation. Specific physical parameters being evaluated included density, pore water pressure changes and hydraulic conductivity.

The post-mortem survey program was designed to compare numerically predicted settlement with field performance. The CPT and survey program of June 2001 focused on three locations within the wicked area of the tailings impoundment. CPT-101 and CPT-103 were located midway between the center of the impoundment and the wick boundary on the west and east, respectively, CPT-102 was located just east of the center of the impoundment (See Figure 4). CPT locations 101 and 103 received between 2.5 to 3.5 meters of surcharge fill which equates to approximately 65% of the design fill. CPT-102 received approximately 1.5 meters of surcharge fill, which equates to about 25% of the design fill.

10 CALCULATIONS

Prior to comparing actual settlement to that predicted by the wick design programs, total consolidation settlement and the actual degrees of consolidation were calculated. Total consolidation settlement was calculated using the Skempton-Bjerrum method.

The CPT and survey program of June 2001 furnished actual values of settlement from which corresponding degrees of consolidation were calculated ($U=s/s_t$, where U = degree of consolidation, s = settlement and s_t = total consolidation settlement).

11 COMPUTER MODELING

Wick performance was again modeled using the program developed by Dr. Jinchun Chai entitled MebraWin. As well as determining requisite wick design patterns, the program can also calculate one dimensional consolidation settlement and its corresponding degree of consolidation with respect to time as a result of a given wick pattern.

Horizontal and radial consolidations are evaluated using Terzaghi's and Hansbro's theories, respectively. The corresponding degree of consolidation, U, is determined using Carrillo's relation: $1-U = (1-U_r)(1-U_z)$, where U_r is consolidation due to radial flow and U_z is consolidation due to vertical flow.

The settlement of the given material is calculated using the relationship of e-log σ', where e is void ratio and σ' is the effective stress increment. The effective stress increment, σ', is calculated from the degree of consolidation of the corresponding layer.

Using the material properties of the tailings that had been used in the development of the wick design, the impoundment was modeled in order to determine the predicted settlements and degrees of consolidation for the three CPT locations as a function of time and actual surcharge loads.

12 RESULTS

The values for consolidation settlement and their corresponding degrees of consolidation predicted by MebraWin were compared with the actual field data. These values, in turn, were compared to the case in which vertical wick drains were not utilized (See Figures 5 and 6). Table 2 presents these comparisons. It is important to note that 90% consolidation will not be reached in the requisite time period of 18 months due to the fact that the full design surcharge load was not applied.

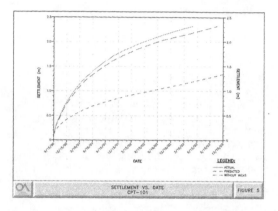

Figure 5. Comparison of actual settlement with predicted settlement – CPT-101.

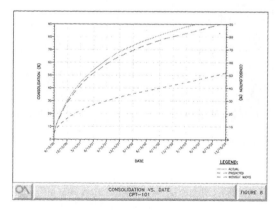

Figure 6. Comparison of actual consolidation with predicted consolidations – CPT-101.

Table 2. Comparison of settlement results

Location	Time (days)	Settlement (m)		
		Actual	Predicted	w/o Wicks
CPT-101	280	1.37	1.27	0.65
CPT-102	240	0.61	0.76	0.21
CPT-103	270	1.39	1.28	0.66

Table 3. Comparison of consolidation results

Total Consolidation Settlement (m)			
Location	Hand	Predicted	w/o Wicks
CPT-101	2.58	2.56	2.58
CPT-102	1.08	1.53	1.08
CPT-103	2.51	2.52	2.51

Degree of Consolidation (%)			
Location	Hand	Predicted	w/o Wicks
CPT-101	53.0	49.6	25.2
CPT-102	56.5	49.7	19.4
CPT-103	55.4	50.8	26.3

CPT-101 was used to present the graphical results of settlement and consolidation with respect to time, as it is most representative of the wicked region of the tailings impoundment.

Table 3 presents the total settlements and the present (as of June 2001) degrees of consolidation for each of the test locations for the current surcharge loading as calculated by hand and modeled using MebraWin.

For CPT-101, actual consolidation and settlement were 0.8% and 7.3% greater than predicted, respectively, and 105% greater than the case in which wicks were not utilized.

The curves corresponding to actual settlement and consolidation were interpolated for the period of June 2001 through December 2003 in order to determine the time required to achieve the requisite consolidation. It may be concluded from Figure 6 that a total of approximately 29 months will be required to reach 90% consolidation. Computer modeling results in a total period of approximately 32 months to achieve the same degree of consolidation. The increased time required to reach the requisite consolidation is the result of the application of partial surcharge loads.

The enhanced actual consolidation, as compared to modeled consolidation, is attributed to heterogeneities in the tailings. Silt and sand lenses allow for increased conductivity of pore fluids that cannot be modeled using MebraWin.

13 SUMMARY

The current closure plan of the Atlas Uranium tailings facility calls for regrading the tailings surface and the placement of a radon barrier. Two concerns associated with the closure plan were increased seepage to the underlying groundwater system and the requirement that 90% consolidation be achieved prior to the placement of the final cap (i.e. the radon barrier).

Following an evaluation of tailings dewatering options, it was determined that the most effective means of dewatering the tailings would be the usage of prefabricated vertical wick drains.

Data collected during field investigations provided the requisite information for the wick design. It was determined that the wicks be placed in a triangular pattern with 3-m spacing.

The installation of wick drains began in September 2000 and ended in December 2000. The full design surcharge load was not achieved due to funding limitations.

A post-mortem CPT and survey program was undertaken in June 2001 to monitor the performance of the wick drains. Following the program, the performance of the wick drains was compared to their predicted performance. The predicted performance was modeled using MebraWin and the actual surcharge load was used rather than the design load.

Results from the comparison show that the actual performance of the wicks is better than predicted. However, the required degree of consolidation, 90%, will not be achieved in the requisite time, 18 months, due to the placement of only a fraction of the design surcharge load.

REFERENCES

Geo-Slope International, Seep/W & Sigma/W, 1998, Calgary, Alberta ,Canada.

Henderson, Michael, E. (2000), "Tailings Dewatering at the Atlas Moab Uranium Mill Tailings Facility."

Henderson, Michael, E. (2001), "Closure of the Atlas Uranium Tailings Impoundment," Proceedings of the Eighth International Conference on Tailings and Mine Waste '01, A.A. Balkema, Roterdam, pp 123-136.

Tailings and Mine Waste '02, © 2002 Swets & Zeitlinger, ISBN 90 5809 353 0

Steep Rock Iron Mines: Dredging and draining of Steep Rock Lake and some of the effects after 45 years

V.A. Sowa
Jacques Whitford and Associates Limited, Vancouver, British Columbia, Canada

ABSTRACT: The largest undeveloped and the richest deposit of hematite iron ore on the North American continent was discovered in 1938 beneath Steep Rock Lake, near Atikokan, Ontario, Canada. Stimulus for development of the mine came during World War II when steel mills were facing a critical shortage of iron ore. Since the iron ore was beneath Steep Rock Lake, the development of the open pit mine required a massive water diversion scheme, including the diversion of the Seine River, draining of Steep Rock Lake, and dredging of a large quantity of lake bottom sediments to expose the iron ore. The volume of water pumped to drain the lake was 570 billion litres, and dredging removed 215 million cubic metres of lake bed overburden material. The mining project at the time was the largest mega project undertaken in Canada. The development of the project is described, including draining of the lake, dredging of the overburden, and some of the impacts of both the drainage of the lake and the disposed dredged material after 45 years.

1 INTRODUCTION

The richest undeveloped deposit of hematite iron ore on the North American continent at the time was discovered in 1938 beneath Steep Rock Lake, near Atikokan, Ontario, Canada. Atikokan is located as shown on Figure 1. Stimulus for the development of the mine came during World War II when steel mills were facing a critical shortage of iron ore. Increased steel production to supply the allied war effort coincided with attacks by Nazi U-boats on iron ore ships from South American mines. There was additional interest in the Steep Rock iron ore deposit because the deposit was such a high-grade ore that it could be charged directly into the blast furnaces. Development of the Steep Rock Iron Mines commenced during the war on an accelerated basis, authorized under Canada's War Measures Act.

Since the iron ore was beneath Steep Rock Lake, the development of the open pit mines required a massive water diversion scheme, including the diversion of the Seine River, draining of Steep Rock Lake, construction of various dams, tunnels, and other diversion works, and dredging of a large quantity of very soft lake bottom sediments to expose the iron ore. The mining project at the time was the largest mega project of its type undertaken in Canada.

Mining at the Steep Rock Iron Mines commenced in 1944 and ceased in 1979 after 35 years. In 1985 the mining company wished to abandon the mine and return the mine lease to the Province of Ontario (Crown). Prior to accepting the mining property, the Crown required an abandonment plan, an assessment of the condition of the water control structures and, including among other requirements, consideration of the impact of draining the lake and the disposal of the dredged material on the mine area.

Restoration of the diverted Seine River to its original channel was not possible. Firstly, the diverted Seine River could not be returned to its original channel because the river would then be flowing through 90 million cubic metres of disposed dredged materials, and the resulting increased sediment load could have serious environmental effects on the fish habitat downstream. Secondly, some of the diversion works, including some dams and drainage channels, were taken over by Ontario Power Generation as part of the water control facilities required for their coal-fired Atikokan Generating Station constructed after mining ceased. Consequently, most of the diversion scheme, including the water control structures, will need to operate in perpetuity. An evaluation of the condition of some of the water control structures, such as dams, prior to the transfer of the mine lease to the Crown has been presented elsewhere, (Sowa, et al. 2001a).

In order to understand the impact of dredging and drainage of Steep Rock Lake within the general area of the mine, it is necessary to place the mine development, construction of the water control structures, and the environmental changes in the correct chronological historical perspective. For this reason, a brief historical review of the four main mine development phases are described in the following section.

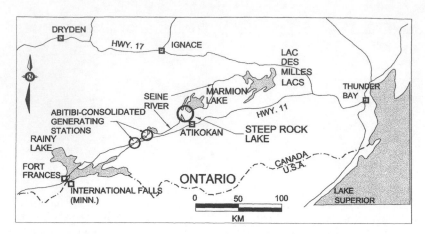

Figure 1. Location of Steep Rock Lake.

2 HISTORICAL BACKGROUND

2.1 *Phase 1 - 1929 to 1942 (Figure 2a)*

Phase 1 is the pre-development phase and Figure 2a shows the pre-development conditions in the area. The Seine River flows from Lac Des Mille Lacs to Marmion Lake, as shown on Figure 1, then to Steep Rock Lake and to Rainy Lake. In 1929 Boise Cascade (now Abitibi-Consolidated Inc.) constructed power dams on the Seine River, Figure 1, including the Moose Lake Generation Station, (Figure 2a). Steep Rock Lake forms an "M" shape, as shown on Figure 2a, with names assigned to each leg; West Arm, Middle Arm, East Arm, and Southeast Arm.

2.2 *Phase 2 – 1943 to 1950 (Figure 2b)*

To begin open pit mining in 1943 in the Middle Arm of Steep Rock Lake, it was necessary to isolate the Middle Arm and divert the Seine River around it. The route chosen was west of Marmion Lake, through Raft Lake and Finlayson Lake, then south through the West Arm of Steep Rock Lake, as shown on Figure 2b. Two control dams were constructed, the Raft Lake Dam to regulate Marmion Lake, and the Wagita Bay Concrete Dam to control flow into the West Arm. A third dam, the Narrows Dam was constructed in the Steep Rock Lake to separate the Middle Arm from the West Arm. The Moose Lake power plant was closed.

Isolation of the Middle, East and Southeast Arms from the Seine River allowed draining Steep Rock Lake by pumping and commencement of dredging. Approximately 15 million cubic metres of the clayey silt lake bed overburden was dredged from the Middle Arm and dumped into the West Arm. The Errington Mine in the Middle Arm was opened, and the first ore was shipped on October 3, 1944.

2.3 *Phase 3 – 1951 to 1954 (Figure 2c)*

During this period Steep Rock Iron Mines developed the Hogarth and Roberts open pit mines in the Middle Arm, Figure 2c. Dredging of the soft silty clay overburden from the Middle Arm began in December 1950 and the dredged material was dumped on the ice of the West Arm during the winter of 1950-51. A significant amount of the dredged clayey silt that was dumped on the ice was eroded during the spring melt of 1951 into the Seine River, and transported downstream as far as Rainy Lake. Both the Canadian and American residents along the Seine River and Rainy Lake were upset by the damaging effect of the dredged silt on the domestic water supplies and on fish. Requests were made to the International Joint Commission for an inquiry.

The solution to the dredge pollution problem was to isolate the West Arm of Steep Rock Lake to create a retention basin to allow settlement of the dredged clayey silt. Isolation of the West Arm was achieved in 1952 by constructing a series of dams to divert the Seine River to the west of the West Arm at shown on Figure 2c. The Reed Lake Concrete Dam was constructed and the Wagita Bay Concrete Dam was raised. Three earthfill dams, West Arm Dams No. 1, 2, and 3 were constructed to retain the dredged clayey silt overburden deposits, and to regulate the water levels. A tunnel was bored through the bedrock at the abutment of West Arm Dam No. 2 to provide an outlet from the West Arm.

2.4 *Phase 4 – 1955 to 1961 (Figure 2d)*

In 1953 Steep Rock Iron Mines Limited entered into an agreement with Caland Ore Limited to develop an open pit mine in the East Arm of Steep Rock Lake. The drainage basin contributing runoff into the East Arm is 65 km^2, and pumping the runoff

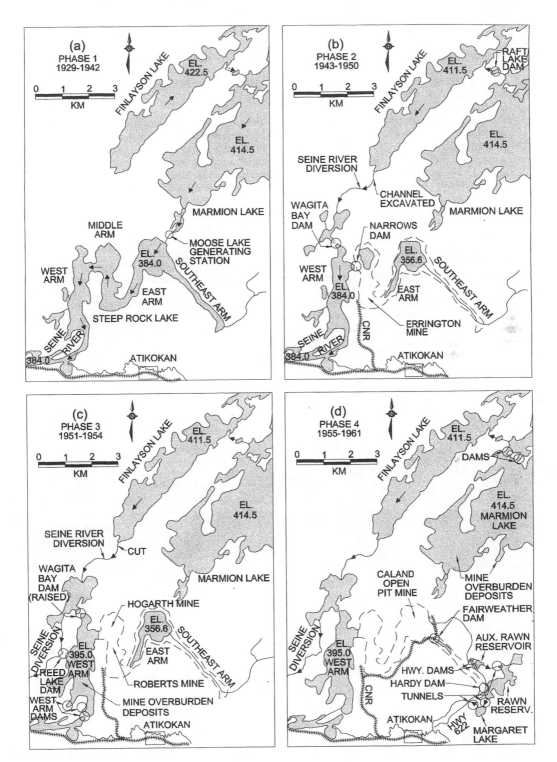

Figure 2. Historical development of Steep Rock Lake Mines.

from the bottom of the open pit would be a significant operating expense. To reduce pumping costs, dams, including Hardy Dam, channels, and tunnels were constructed as shown on Figures 2d and 3 to create the Rawn Reservoir. The Rawn Reservoir diverts the runoff from 40 km² of the East Arm drainage basin into the Atikokan River watershed. The Rawn Reservoir drains through a tunnel constructed to Margaret Lake, then from Margaret Lake through another tunnel under Highway 622 into the Atikokan River watershed. Three small creeks, damned by Highway 622 embankment fills, created the Auxiliary Rawn Reservoir which drains by tunnel into the Rawn Reservoir.

Caland Ore erected their plant facilities on the east shore of the East Arm. The Canadian National Railways (CNR) railway was extended, Figures 2d Railway (CNR) was extended, Figure 3, to the plant by crossing the dewatered Southeast Arm of Steep Rock Lake on a 27 m high earthfill embankment known as the Fairweather Dam. The embankment also acts as a dam to intercept runoff water flowing into the East Arm mining area. The intercepted water is pumped to the Marmion Lake watershed.

Caland Ore encountered the same type of clayey silt overburden covering the iron ore body that had caused Steep Rock Iron Mines so much difficulty. The southern part of Marmion Lake was converted into a retention basin and 120 million cubic metres of clayey silt overburden was deposited into the Marmion Lake retention basin.

2.5 Present situation

The coal-fired Atikokan Electrical Generating Station, owned by Ontario Power Generation, has taken over many of the dams and water control structures constructed for the mine as part of the water control system for the Atikokan Generating Station. Ontario Power Generation also pumps the water that collects upstream of the Fairweather Dam. The Auxiliary Rawn Reservoir was essentially eliminated in 1987 when a culvert was installed under one of the highway dams to drain the Auxiliary Rawn Reservoir into the Southeast Arm.

The Caland Ore plant buildings on the East Arm were turned into a sawmill after mining ceased. The sawmill is located 27 m below the original level of Steep Rock Lake. Other developments located on the old lake bed of Steep Rock Lake are a ski hill just upstream of the Fairweather Dam, a hydro line and transformer station, and a market garden operation just below the Hardy Dam beside the highway.

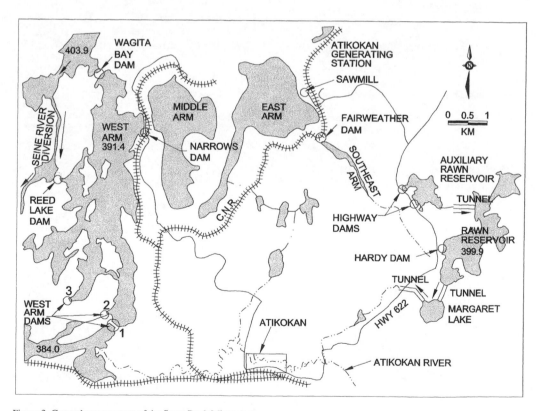

Figure 3. General arrangement of the Steep Rock Mines Area.

3 DRAINING STEEP ROCK LAKE

Draining Steep Rock Lake required assembling the largest fleet of pumping plant ever mounted on floating equipment in North America at the time, (Taylor, 1978). The total quantity of water pumped from the Middle and East Arms of Steep Rock Lake, with a surface area of 10 km^2, was 570 billion litres. Approximately 310 billion litres of water was pumped before the first mining of iron ore could start on the shallowest "B" ore body (Errington Mine). The pumping capacity of the pumps was three times the average daily water consumption of the City of Montreal at the time.

The pumping plant consisted of 14 centrifugal pumps, 610 mm in diameter, each with a capacity of 105,000 litres per minute and driven by a 500 H.P. electric motor. The pumps were mounted in pairs on specially constructed steel barges. The total capacity of the pumping plant was 2.1 billion litres per day.

The pumps were electric-powered and the Hydro Electric Power Commission of Ontario provided the electrical power supply. The was done by constructing a 110 KV electric transmission line, 193 km long, from Thunder Bay to Atikokan in a record time of 180 days under war time conditions.

Pumping commenced on December 16, 1943, and by January 4, 1944, all of the pumps were brought into service and the lake level started to drop at a rate of about 180 mm per day. By January 15, the lake level had dropped 4.0 m, by March 10, 13.8 m, and by April 13, it was down 21 m, to approximately the top of the lake bottom over the shallowest iron ore body, the Errington mine. By June, the lake level had dropped sufficiently to allow commencing the removal of the soft lake bottom material by dredging, and iron ore was first sighted in August, 1944. Mining of the iron ore commenced in late September, and the first trainload of twenty rail cars was loaded on October 3, 1944.

4 DREDGING AT STEEP ROCK LAKE

The initial dredging at Steep Rock Lake was handled by conventional size dredges, (Taylor, 1978). However, removal of a large quantity of lake bottom sediments above the Hogarth iron ore body required much larger dredging capacity. The dredging contract for removal of 38 million cubic metres of overburden was the largest single dredging contract ever let at the time. The dredge selected for the work was a 900-tonne ocean-going all electric dredge, the "Nebraska" that was used for the Kingsley Dam in Nebraska, and also for enlarging the Idlewild Airport in New York. The dredge, located in Boston, was dismantled, loaded onto almost 100 railway cars and transported to Atikokan, and re-assembled in

1950. The dredge was re-named the "Steep Rock", and started dredging on November 2, 1950.

The dredges at Steep Rock operated by standing on a "spud" or leg mounted at the stern, and swung back and forth by means of anchors attached to swing lines at the bow. The business end of the dredge was the suction end and the cutting head which were mounted on a "ladder" attached to the bow. The dredge was operated in a semi-circular motion, clearing a 12 m depth of overburden for each "set" of the dredge. The cutting head was driven by a 500 H.P. electric motor, rotating at 20 rpm, churning the overburden at the bottom of the lake into a slurry which was sucked through the 800 mm diameter suction pipe into the pumping system on the dredge. The huge pumps, powered by 6,000 H.P. electric motors, pumped the overburden through a 760 mm diameter floating discharge line at a rate of 2,300 cubic metres per hour into the West Arm of Steep Rock Lake.

The dredging proceeded well during the first few months of 1951 until spring break-up in 1951 when a major problem of pollution of the Seine River occurred caused by dredged material being washed into the Seine River. There was some discolouration of the Seine River in 1944 when the first small dredges began operation and the discolouration increased slightly when the second 380 mm dredge started up. However, with the commissioning of the giant "Steep Rock" dredge in 1950, the amount of clayey silt dumped into the river system increased twenty-fold. This circumstance, combined with the additional dredged material frozen on the ice in the West Arm which was released during spring thaw, resulted in a significant amount of dredged silt entering the Seine River. The solution for the silt control, was, as noted earlier, to turn the West Arm of Steep Rock Lake into a retention basin to allow more time for the silt to settle out.

When plans were made to open the Roberts Mine, Figure 2c, in the Middle Arm of Steep Rock Lake, another dredge was obtained, re-named the "Marmion", and commissioned on October 10, 1952. This dredge was slightly smaller than the "Steep Rock", with a displacement of 850 tonnes, a length of 46 m and a beam of 12 m. After the "Marmion" was launched, the two dredges worked on the Hogarth Mine until February 15, 1954 before moving both dredges to the Roberts Mine. The two huge dredges moved a combined 1.15 million cubic metres of overburden per month.

Dredging continued during the winter months, in weather as cold as - 40 to - 45 ^{0}C. In January, 1952, a ice-breaking tugboat was put into operation, and in spite of the 800 mm thick ice that formed during the extreme cold weather, the tug was able to keep the dredges free of heavy accumulation of ice. This was accomplished by keeping the tug constantly in motion around the dredges, breaking up ice before becoming too thick. Ice cakes that could not be handled by the tug or pumped by the dredges, were pushed to the shore and removed by dragline and bulldozer.

Dredging for the Roberts Mine continued from 1954 to 1962 when the last of the 37 million cubic metres of overburden was pumped and deposited in the West Arm of Steep Rock Lake. In total, 90 million cubic metres of overburden was deposited in the West Arm over a period of 18 years.

When the Caland Ore Company began preparing for mining in the East Arm of Steep Rock Lake, they acquired two dredges to remove 120 million cubic metres of overburden from the East Arm, (Whitman et al. 1962). Dredging in the East Arm started on April, 1955 and was completed on September, 1960. The southern end of Marmion Lake, isolated by a series of dams, was transformed into a retention basin to store the dredge spoil.

The two Caland dredges were even larger than used by Steep Rock Iron Mines in the Middle Arm. The two Caland Ore dredges were 915 mm diameter electrically powered suction dredges, each powered by a 10,000 H.P. motor, and mounted on hulls measuring 54 m long and with a 15 m beam. Each dredge was equipped with a discharge line which consisted of a 915 mm diameter floating discharge line supported by pontoons, and about 6.5 km of a 1067 mm diameter pipeline on the land. Two booster pumps, each powered by 10,000 H.P. motors were located on each line. Each dredge could move 2,700 cubic metres of dredged overburden when operating under average conditions. The static pumping head varied from 107 to 236 m.

The four large dredges, two operated by Steep Rock Mines, and two operated by Caland, were undertaking the largest earth-moving project in history at that time. The combined electric power required for operating four dredges and water pumps made the two mining companies, jointly, Ontario Hydro's largest industrial user of electrical power at the time.

The depth of overburden encountered by Caland Ore in the East Arm was considerably greater than that encountered by Steep Rock Iron Mines in the Middle Arm. In the East Arm the ore body was overlain by overburden with an average thickness of 100 m, plus 30 m of water.

Determination of the safe dredged slopes was a major problem. The varved clayey silt lake bottom material was a soft and sensitive soil that could be treacherous, and establishing safe dredged slopes was a challenge (Leggett, 1958, Whitman et al. 1962). If the selected dredged slopes were too steep, the mining personnel and mining operations could be put at risk from a slide. Alternatively, selecting flatter and more conservative slopes would incur a significant cost increase. The challenge was to establish the optimum slope which satisfied both criteria (Whitman et al. 1962). Some of the slopes were of substantial height, with one of the dredged slopes in the East Arm being about 120 m high.

The operation of the open pit mine while mining the iron ore also involved excavation of a substantial quantity of waste rock. Since the exposed dredged clayey silt slopes were subject to rapid erosion from surface water runoff, the slopes were protected with a facing of the waste rock. The maximum thickness of waste rock on the dredged slopes was 3 m.

5 SOME EFFECTS FROM DREDGING AND DRAINING STEEP ROCK LAKE AFTER 45 YEARS

5.1 General

Dredging and draining Steep Rock Lake had a significant influence on the lake, the mines, and the surrounding area, and this influence continues after mining ceased and the mines were abandoned. Some of the effects of dredging and draining Steep Rock Lake after about 45 years are considered in the following which includes future water management, and the impact on the West, the East, and the Middle Arms of Steep Rock Lake.

5.2 Effect of dredging and draining Steep Rock Lake on water management

Dredging and draining of Steep Rock Lake has significantly impacted on the past, present, and future water management as described elsewhere, (Sowa et al. 2001b). There are four main water control systems that require long-term operation and maintenance to ensure public safety, protection of existing developments, and environmental protection of the dredge spoil material impounded during the mining operation. The four water control systems are: (1) The Seine River Diversion around Steep Rock Lake; (2) Maintaining a water cover on the clayey silt dredge spoil deposited in the West Arm of Steep Rock Lake and in Marmion Lake, including maintaining the artificially high water levels in the West Arm, (3) Protection of facilities constructed on the old lake bed from rising water levels in the abandoned mines; and (4) The Rawn Reservoir diversion system constructed to reduce future pumping costs.

Pumping water from the mines terminated in 1979 when mining operations ceased. The abandoned mines, as deep as 300 m below the original lake level, are slowly filling with water and eventually will affect some of the water control structures and developments within the mine areas. There are numerous alternatives to manage the rising water level in the future, but the following are three main water management options:

(1) Protect all developments within the abandoned mines from flooding. This means maintaining the water levels artificially low to protect the lowest

development at elevation 356.6 m from the rising water level, which will require the installation and operation of large pumping capacity.

(2) Allow the water in the abandoned mines to rise to the level of the West Arm, Figure 3, at elevation 391.4 m and then flow out from the West Arm by gravity. The developments affected by the rising water will need to be moved or protected.

(3) Maintain a water level elevation between the above two limits of elevation 356.6 m and 391.4 m, with the amount of pumping based on a cost-benefit analysis at the time that a decision is required.

A final decision on the three future water level management options is not required until the water level in the East Arm reaches elevation 356.6 m in about 30 years in Year 2031, (Sowa et al. 2001b). In the meantime, the water control systems and the various water control structures must be maintained in a safe and reliable condition.

5.3 *Effect of dredging on the West Arm of Steep Rock Lake*

Following the contamination of the Seine River in 1951 from the dredged clayey silt, the West Arm of Steep Rock Lake was transformed into the West Arm Retention Basin in 1952. The creation of the West Arm Retention Basin required construction of new dams, including West Arm Dams No. 1, 2, and 3, and a tunnel as noted previously.

The West Arm of Steep Rock Lake, which once was up to 60 m deep, is almost completely filled with the deposition of 90 million cubic metres of dredged overburden. The present water level is 7.3 m above the original lake level, and typically the depth of water covering the dredged clayey silt overburden is about 0.6 to 1.8 m.

One of the West Arm dams, West Arm Dam No. 1, located as shown on Figure 3, is considered in the following. West Arm Dam No. 1, constructed in 1952, is a homogenous earthfill dam with the earthfill consisting of sand and gravel. A plan of the dam and typical sections are illustrated on Figure 4. The crest elevation of the dam is 396.2 m, the crest length is about 323 m, and the crest width is about 11 m. The maximum height of the dam is about 18 m. The average existing slope for the upstream and downstream slopes is about 2.5 horizontal to 1.0 vertical (2.5 H to 1.0V).

The water level behind West Arm Dam No. 1 has been up to elevation 395.0 m during dredging to provide adequate detention time to allow settlement of the dredged clayey silt material. After completion of dredging, the water level was later decreased to the current level of about 391.4 m. The elevation of 391.4 m provides sufficient water cover on the dredged clayey silt to ensure that wind erosion of the silt does not occur.

Evidence of different West Arm water levels is illustrated on the two cross-sections for West Arm Dam No. 1, Figure 4, which indicate the presence of wave erosion of the upstream slope. As can be seen there is a small concave erosion feature on the upstream slope at approximately the current water level in the West Arm. As noted previously, the water level in the West Arm was at a higher level during dredging, up to elevation 395.0 m, and there is a another concave erosion feature on the upstream slope at about elevation 393-395 m that was caused by wave action.

Dredged material deposited in the West Arm also has an effect on the groundwater levels in the West Arm Dam No. 1. Typical groundwater level results are illustrated on two sections on Figure 4 which also show the West Arm water level on the upstream side, and the tail water level on the downstream side. As can be seen from Figure 4, the groundwater level measured beneath the crest of the dam, within the earthfill or in the foundation soil, is quite low, essentially almost at the same level as the tail water at about elevation 384.0 m. It is reasonable to expect that the piezometeric pressure drop across the dam would be roughly linear which means that the groundwater level below the crest of the dam in the standpipes would be considerably higher than the observed elevation 384 m. A possible explanation for this behaviour is given in the following.

Examination of the sections on Figure 4 indicates that the upstream toe of the earthfill ends at about elevation 384 m, and the underwater ground surface then extends essentially horizontally. In contrast, the original ground contours shown on the plan, Figure 4, near the dam indicate that the original ground surface was as low as elevation 376 m. The difference between the two ground surface elevations is the thickness of dredged material that was dumped in the West Arm.

The presence of dredge spoil was confirmed from grain size analyses, Figure 5, on lake bottom samples taken at the toe of the dam. Also shown on Figure 5 are the results of grain size analyses on samples of varved clayey silt from the foundation soil below West Arm Dam No. 1. As can be seen, the clayey silt that was dredged from the Middle Arm and disposed into the West Arm Retention Basin at the toe of West Arm Dam No. 1 is very similar to the foundation clayey silt soil below West Arm Dam No. 1. The results shown on Figure 5 indicate that the dredged material consists of about 35 percent clay material, and the foundation clayey silt soil consists of about 25 to 33 percent clayey soil. Such clayey silt soils are relatively impervious.

The dredged material has covered the lake bottom and the upstream slope of the West Arm Dam No. 1 as an impervious blanket. As the dredged material is relatively impervious, only a small amount of seepage can occur into the dam. Since the sand and gravel earthfill is more pervious than the clayey silt, any water that does seep through the impervious

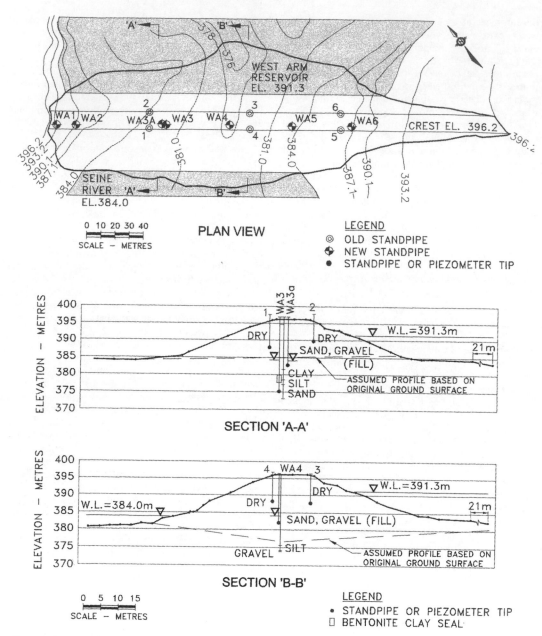

Figure 4. West Arm Dam No.1.

clayey silt blanket will drain more quickly to the downstream toe. This results in the low groundwater levels observed within and below the dam. This example illustrates the beneficial effect that the dredged material, acting as an impervious blanket, has on reducing the seepage and, consequently, in improving the stability of the dam.

In 1951 before the West Arm Retention Basin was created, some suspended silt from the dredge spoil in the West Arm entered the Seine River and caused concern about the quality of the drinking water and the effect on fish. After the creation of the retention basin in 1952, the quality of the river water improved, and the Seine River returned to being one of the best sport-fishing rivers in Northwestern Ontario (Taylor, 1978).

The West Arm supports a healthy fish population in spite of the 90 million cubic metres of dredged

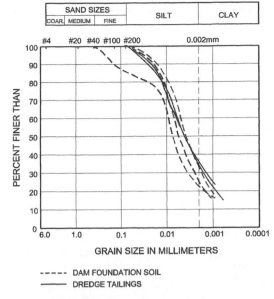

SAND SIZES | COAR. | MEDIUM | FINE | SILT | CLAY

Figure 5. Results of grain size analyses of dredged material and foundation soil.

material that has been deposited into the West Arm. Some fresh water is continuously allowed to spill over the Wagita Bay Concrete Dam at the north end of the West Arm, see Figure 3, to refresh the water in the West Arm. The most commonly found fish in the West Arm are bass, walleye, pike, white fish, and lake herring, (Nash, 2001).

The rate at which the water in the retention basins clarifies after dredging ceased appears to be relatively rapid. While no records were found for the West Arm retention basin, some observations have been reported for the Marmion Lake retention basin, (Whitman et al. 1962). During dredging of the East Arm and disposal into the Marmion Lake retention basin, make-up water for the dredging operation was taken at a point only 2,700 m from the closest dredge discharge point. Yet there was no problem with silt returning to the dredge pool, although there was a distinct discolouration of the return water. Of particular interest was the observation that the discolouration in the Marmion Lake retention basin almost completely disappeared within less than one year after completion of dredging.

5.4 Effect of draining the Middle and East Arms of Steep Rock Lake

The open pit iron ore mines were located in the Middle and the East Arms, and since the pumps were turned off in 1979, the pits are slowly filling with water, (Sowa et al. 2001b). The source of water is seepage from the bedrock slopes of the pits and precipitation. The Middle and East Arms are self-contained bodies of water and there is no other

source of fresh water. The water levels in both the Middle and East Arms are approximately the same at any one time. The open pits in the Middle Arm and the East Arm are not connected at the present time (2001) and will not be connected for some time.

Even though iron ore mines were operated in both the Middle and the West Arms, the ability to support fish in the two arms is quite different. Fish thrive in the East Arm, and a floating fish farm has been established in the East Arm since 1988, (Nash, 2001). Over the years, the fish farm operator has tried various species of fish but has been most successful with rainbow trout. The operator has harvested 100,000 kg of fish in a year.

In contrast there are no fish in the Middle Arm. Some research has been undertaken in an attempt to determine the cause for the lack of fish. Analyses of the water in the Middle Arm indicated that there are elevated levels of some metals in the water, but, individually, none of the metals are in sufficient concentration to adversely affect fish. The cause for the lack of fish in the Middle Arm has not been established, and further study will be required.

6 SUMMARY AND CONCLUSIONS

The rich deposit of iron ore at Atikokan was located below Steep Rock Lake. Development of the iron ore mine required the diversion of the Seine River, draining of Steep Rock Lake, dredging of a large quantity of overburden material to expose the iron ore, and the construction of 40 water control and water diversion structures, mainly dams and tunnels. The entire project was a massive undertaking. Water was a major factor during the development of the Steep Rock Iron Mines. Water was either redirected, drained, pumped, or used as medium to allow dredging to be used as an economical method for excavating the lake bed overburden. Continued management of the water in a safe manner will be required in the future.

Pumping to drain Steep Rock Lake was one of the largest operations of its kind and required assembling the largest fleet of floating pumping plant in North America at the time. Dredging was also a very major operation, requiring a fleet of large dredges. The main dredging contract was the largest single dredging contract awarded at the time.

The impact of draining the lake and dredging was significant during construction and the mining operations, and continues to be a major consideration after abandonment of the mines. The Seine River diversion must be maintained, a water cover will be maintained over the dredge spoil in the retention ponds, the facilities that have been constructed on the drained lake bed will need to be protected from the rising water in the open pit mines, and the Rawn

Reservoir must be maintained. These requirements still apply 45 years after the initial construction, and will continue to apply.

The environment was impacted considerably during mine construction and mine development, but there is evidence that the impact is not as severe and has been mitigated 45 years later. The Seine River is still one of the best sport-fishing rivers in the area. The West Arm retains 90 million cubic metres of dredge spoil, and yet supports a healthy fish population. Fish are thriving in the East Arm mine pit, which also includes a successful floating fish farm. No fish are currently present in the Middle Arm mine pit, but that may change in the future.

7 ACKNOWLEDGEMENTS

The author appreciated the opportunity to be involved with this project on behalf of the Ontario Ministry of Natural Resources, and is grateful to the Ministry for permission to publish this paper. The author also wishes to acknowledge the assistance received from Jacques Whitford and Associates with the production of the paper. The interpretations and opinions expressed in this paper are solely those of the author.

REFERENCES

Leggett, R. F. 1958. Soil Engineering at Steep Rock Iron Mines, Ontario, Canada. Proceedings of the Institute of Civil Engineers, London, Vol. 11, p. 169-188. Discussion, Vol. 13, p. 93-117.

Nash, T. J. 2001. Personal communication.

Sowa, V. A., Adamson, R. B., and Chow, A. W. 2001a. Abandonment and Reclamation of the Steep Rock Iron Ore Mine at Atikokan and related Hardy and West Arm No. 1 Dams. Proceedings of the 54th Canadian Geotechnical Society Conference, Calgary, Alberta, September 16-19, 2001.

Sowa, V. A., Adamson, R. B., and Chow, A. W. 2001b. Water Management of the Steep Rock Iron Mines at Atikokan, Ontario during the construction, operations, and mine abandonment. Proceedings of the 25th Annual Mine Reclamation Symposium, Campbell River, British Columbia, September 24-27, 2001.

Taylor, B. W. 1978. Steep Rock – The Men and the Mines. Quetico Publishing, Atikokan, Ontario.

Whitman, E. W., Reipas, S. A., and Hardy, R. M., 1962. Unusual problems in the development of the Caland mining operation at Steep Rock Lake. Presented at the Engineering Institute of Canada 75th Annual Meeting, Montreal, Quebec, June 12th to 15th, 1962.

Tailings and Mine Waste '02, © 2002 Swets & Zeitlinger, ISBN 90 5809 353 0

Community consultation in the rehabilitation of the South Alligator Valley Uranium Mines

P.W. Waggitt

Office of the Supervising Scientist-Environment Australia

ABSTRACT: Uranium was mined in the South Alligator Valley of the Northern Territory, Australia, between 1955 and 1964. The mines were not rehabilitated following mining. The valley was incorporated into the World Heritage listed Kakadu National Park in 1987. The Australian Federal Government carried out some hazard reduction works in the early 1990s. In 1996 the Gunlom Land Trust was granted the area under the Commonwealth's *Northern Territory Land Rights Act (1976)* but immediately leased the land back for continued use as a National Park. A condition of the lease is that all former mine sites and associated workings will be rehabilitated by 2015. The paper follows the development of an intensive and comprehensive consultation process involving all stakeholders. This has been a major challenge for all involved as the plan must be capable of satisfying the requirements of the regulators and supervising authorities, whilst still accommodating the needs of the Plan of Management for the park and the aspirations of the Traditional Owners. The paper ends with a discussion of lessons learned and a summary of the outcomes achieved and details of the program for the future.

1 INTRODUCTION

In the post war years there was a mad rush to locate and develop uranium resources for strategic purposes. Australia had a number of relatively high grade deposits which could be easily exploited and so exploration was intense in a number of areas of the country, nowhere more so than in the vicinity of the Pine Creek geosyncline. This area was the scene of a gold rush in the late 1800s and geologists flocked to the area in the 1940s to search for uranium. When deposits were located and exploited the environmental management practices left much to be desired, especially as rehabilitation was unknown. Also unacknowledged at the time was the notion of Aboriginal ownership of these lands. Since the mid-60s Aboriginal land rights have developed and as lands are handed back to Traditional Owners so there is a need to alleviate the impacts remaining from the earlier mining campaigns.

2 THE NATURAL ENVIRONMENT

The South Alligator Valley lies about 13° south of the equator with a tropical, rainy climate having a distinct dry season, A_w in the Köppen (1936) classification. The average annual rainfall is approximately 1200mm of which more than 90% falls between 1 November and 30 April ("the wet"). During the wet season the South Alligator River is fre-

quently in flood and access to, and around the valley, is often not possible for weeks at a time. Rainstorms during the "wet" can be very intense which increases the erosion risks. The soils of the area are nutritionally poor and physically fragile, once disturbed they are highly susceptible to erosion. For these reasons earthworks are normally only carried out in the "dry". Roads are not sealed and have to be extensively repaired and maintained each dry season.

Temperatures in the valley average about 21°C annually with maxima often around 40°C and minima rarely below 14°C.

The landscape is a relatively narrow valley passing through the Kambolgie sandstone of the Arnhemland escarpment, dominated by the edge of the escarpment and various outliers. The sides of the valley are steep and rise some 250 metres above the river.

The natural vegetation of the area is a dry schlerophyll savanna forest with the dry areas dominated by stands of *Eucalyptus* and *Acacia* species and *Pandanus* and *Melaleuca* in the lower lying and poorly drained parts.

3 BACKGROUND AND HISTORY

The deposit at Rum Jungle is perhaps the best known of the early uranium mines in the Pine Creek area, but it was followed soon after by a series of smaller mines that were located in the upper reaches of the South Alligator valley. The location of the valley is shown in Figure 1. The region was remote and the only acknowledged land use at the time was extensive cattle ranching in the bush land, which had begun around 1950. However airborne surveys, very unsophisticated by today's standards, did locate more than 50 radiological anomalies in the valley. Ground investigation revealed that there were indeed economically viable uranium deposits among these anomalies and mining began in about 1954/5. Details of the history of the local uranium industry in these times have been published elsewhere (Annabel, 1977; Fisher, 1968 &1988; Waggitt, 2000). The operations were very small by present day standards. The total production from all 15 mines between 1955 and 1964 amounted to 841.45 tonnes U_3O_8 (Waggitt, 2001).

The process of developing the mines was very simple with no Environmental Impact Statement or consultation process. The former was because there was no Environmental Impact Assessment (or equivalent) legislation in existence at the time; the latter beacuse the local population was considered to be only the pastoralist, although there were Aboriginal people present in the valley from time to time. There were no long-term Aboriginal camps in the area as it is regarded as "sickness country" by them and whilst people were allowed to transit the area, permanent settlement was not regarded as an option. However, Aboriginal people did get involved with the mines but there are few instances of real employment. In fact Aborigines had worked for the pastoralists in the region since the 1940s, primarily in cattle herding operations. In earlier times Aborigines had worked in mines in the region around Pine Creek as labourers. During the war years many of the area's Aborigines were compulsorily located in camps near towns, which kept them out of their country. In 1943 they were the only labour available and so kept the industry running. The mining industry restarted and the exploration opened up the country again. The development of some mines took place near or even within the boundaries of Aboriginal sacred sites, which caused great concern to some communities, but there was little consultation. Attempts by communities to seek the assistance of the Federal Government's Native Affairs Branch in stopping some developments came to naught (BHP, 1987).

Once the mining contracts had been completed there was no further market for the uranium and, although some exploration did continue into the mid-to-late 1960s, the operations together with the camps and associated facilities were allowed to run down. At the peak of operations there were two substantial camps, many kilometres of gravel roads, and mines at 13 locations, some of which were underground and some open cuts. The underground mines were mainly worked by open stoping. Other facilities included a crusher/battery, which separated out pitchblende by gravity methods. This was then hand sorted, placed in drums and exported to programs in the UK and the USA. There was also a small mill, which used solvent extraction technology to produce "yellow cake". The tailings from this mill were discharged to a flat area adjacent to the mill and on the flood plain of the South Alligator River. When the mining finished, sites were simply abandoned which was the accepted practice of the day. There were no rehabilitation bonds or environmental regulations at that time.

The environmental impacts from all this activity were surprisingly small with only one adit showing signs of acid rock drainage. However, the tailings were frequently inundated by floodwaters and washed into the river on numerous occasions (R.Fry, pers comm.). Throughout this time the Aboriginal people still had little say in what was happening on their traditional land.

3.1 The changing political scene

In 1974 Australia ratified the World Heritage Convention and went on to pass Environmental Impact and National park legislation in 1974/75. In 1976 the Commonwealth Government passed legislation giving recognition to Aboriginal Land Rights and offered the opportunity for Aboriginal Traditional Owners to lodge land claims to a tribunal in respect of lands with which they had established ties. In the case of the South Alligator Valley the people of the Jawoyn association considered just such a claim. In the meantime there was also a move to have land to the north of the area declared a National Park. Starting in 1979, stages 1 and 2 of Kakadu National Park were declared and subsequently inscribed on the World Heritage List as an area of outstanding significance on both natural and cultural heritage grounds. An area of the Jawoyn interests, comprising the two former pastoral leases of Gimbat and Goodparla, was earmarked to be incorporated as stage 3 of the Park.

Other political events at the time included the Northern Territory (NT) being granted self-government in 1978. This was significant since, as part of the process, the Federal government agreed to undertake some rehabilitation of former uranium mine sites in the NT. In 1987 the Federal government declared 65% of the two pastoral leases to be incorporated into Kakadu as stage 3 of the National Park, the remainder was declared a Conservation Zone and set aside for 5 years to allow for mineral exploration. Earlier studies had indicated that the area was highly prospective for a number of minerals including uranium, but particularly gold, platinum and palladium (OSS, 1987). Again the Traditional Owners were not really consulted extensively,

404

Alligator Rivers Region

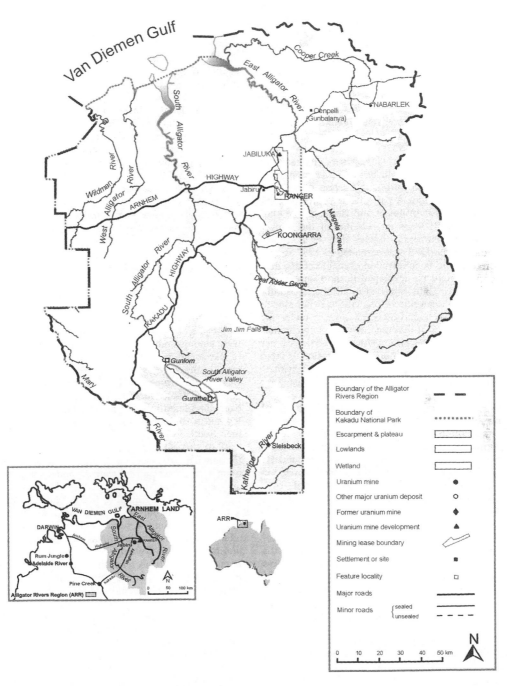

Figure 1. Location map.

although several sacred sites had been identified and registered in the valley. As the exploration proceeded the developers and the Federal Government authorities did consult with the Aborigines with growing frequency, usually through the Northern Land Council (NLC). The NLC had been set up under the auspices of the *Aboriginal Land Rights (Northern Territory) Act* 1976 to provide services and support to the land rights process including conducting negotiations with mining companies on behalf of Traditional Owners.

The South Alligator Valley lies within the boundaries of the Alligator Rivers Region and is thus within the operational area of the Office of the Supervising Scientist (OSS). This is an agency of the Commonwealth (Federal) Government charged with ensuring that there should be no adverse environmental impacts arising from any aspect of uranium mining in the region. A part of the Commonwealth Department of Environment and Heritage, the OSS was set up in 1978 as an outcome of the Ranger Uranium Environmental Inquiry (Fox et al, 1977). The remit of the OSS was extended in 1987 so that the office would be involved in oversight of any new mines that might be developed in the Conservation Zone. This oversight extended to all phases of mining, including exploration. The OSS was not able to consult directly with traditional owners under the protocols in place at that time, but relied on communicating through the NLC. This was obviously not the most effective means of communication as there was always a risk of misunderstanding or misinterpretation in either direction, but it was at least communication.

From 1987 until 1991 there was a controversial exploration program at Coronation Hill, the site of a former uranium mine (OSS, 1987). The Aborigines know the site as Guratba and part of the exploration licence/lease area lay within a sacred site boundary. Work was approved by the relevant authorities, but the Jawoyn community became divided over whether the mine should proceed or not and the whole situation developed into a national event. After an EIS had been submitted in draft form, commented upon and a final version presented and effectively accepted, the Government of the day created a Resource Assessment Commission to further assess the issue before any final approval to proceed could be made. Throughout the whole Coronation Hill saga the amount of consultation with Traditional Owners was increasing at each stage, although each side of the debate was possibly selective about who they consulted. The mining company had employed some Traditional Owners as labourers on the site, which further emphasised the division in the community.

In 1991 the Government made the decision to stop all future development programs in the Conservation Zone and incorporate all the area into Kakadu National Park. This was achieved later that year. The mining company immediately claimed compensation for the termination of their leases and a subsequent High Court decision was that some of the leases were in fact still valid and that compensation is due. This matter has still not been resolved at the time of writing.

The Jawoyn people began negotiating their land claim and options for the return of the land. The land claim was granted and in 1996 a lease was signed between the Gunlom Land Trust, the group of specific Traditional Owners for the area, and the Director of National Parks. The lease was immediately leased back as a Park for an initial period of 99 years. Amongst the clauses of the lease was a specific requirement that all former mining activity be rehabilitated. The lease specified that a Plan of Rehabilitation be agreed by December 2000 and that all subsequent works must be completed by December 2015. A whole new program of works was therefore required to achieve these objectives.

4 REHABILITATION HISTORY

After the signing of the NT self government legislation immediate efforts on uranium mine rehabilitation were concentrated at Rum Jungle (Kraatz & Applegate, 1992). The South Alligator Valley sites were amongst others to be considered later. A survey of the abandoned minesites on the Gimbat pastoral lease in the valley was undertaken by the Commonwealth Government's Housing and Construction Agency (1986). Once the Rum Jungle program had been completed in 1987, attention turned to the valley and a survey of radiological hazards at the former South Alligator mill site was undertaken by staff of the NT Department of Mines and Energy (Bromwich, 1987). The Commonwealth Government's Department of Primary Industries and Energy commissioned a survey of all the sites, which was then used to formulate a remediation strategy (Construction Group, 1988). The funds available for the works program were not sufficient for complete rehabilitation and so the concept of a hazard reduction program was developed. This entailed assessing hazards, primarily in relation to physical safety and radiological safety, at the sites and designing a program of works that would address all the issues. In 1990 two contracts were let by tender for programs of hazard reduction works. Again there had been very little consultation with Traditional Owners and the work was undertaken in the dry seasons of 1991 and 1992. This was between April and November in each year, the times when the area was freely accessible by road. The first phase was to clean up the former mill site and the associated mining village area and the second phase dealt with all the remaining mining and exploration sites identified as needing treatment. The work was undertaken by a local contractor, who went to considerable lengths to involve Traditional Owners in the program. This person had been involved with the Aboriginal workers at the Coronation Hill Joint Venture at the Guratba (Coronation Hill) project.

The hazard reduction works were uncomplicated, adits and shafts were to be sealed off to prevent public ingress, loose scrap metal and radiological materials were to be buried. One issue that arose on the sealing program was the discovery of colonies of bats in some old underground workings. The indications were that some of the bat species present were endangered and so a modification was made to the program with all bat-occupied shafts and adits being closed with grilles in such a way that the animals could still have access whilst preventing people entering. Throughout the works program OSS oversaw the erosion control and radiological protection aspects and park rangers assisted with revegetation supplies. When a site was completed a radiological check was made before the work could be "signed off". At the end of the program materials had been buried in 6 locations in the valley, usually close to the affected site.

There were no regulations in force at the time relevant to the design of low-level radioactive waste containments and so a local standard was derived. The general points were that sites should be above flood level, material should have a minimum of 1.5 metres of clean cover, erosion potential across sites should be minimized and radiation levels after completion of containments should be less than 0.15 μGhr^{-1} above background. Furthermore the sites would be inspected annually by OSS. Every year there would be an examination for erosion problems commencing 12 months after completion and every 3 years thereafter there would be a check on the radiological security of the sites. There were no permanent markers placed at any of the sites although their locations were plotted by OSS using a hand-held GPS and the results put on file at OSS and in the park headquarters as well as at the district ranger station.

5 THE NEW REHABILITATION PROGRAM

Once the lease had been signed in 1996 the clock began to run down in respect of the deadlines for the preparation of the overall plan. By this time it had become accepted practice that Aboriginal people had to be involved closely with every stage of works on their lands. Kakadu National Park is managed jointly by Parks Australia North and the Traditional Owners. The Plan of Management is drawn up jointly and it's implementation overseen by a Board of Management, which has a majority of Traditional Owner members. The first stage was to organise an information forum to advise Traditional Owners of what was going to happen and to set up a process for ensuring they were kept in the information loop and had their views taken into account and, most importantly, they had the opportunity to approve plans.

The initial information exchange meeting was held in October 1997. The organization of such a meeting took time with many smaller groups needing to be brought together and then logistics worked out for assembling people at one place in the valley. The location chosen was a former OSS field camp, which had been handed to Traditional Owners for use as a youth camp site and training facility. It was already clear that meeting inside offices was not a procedure which would encourage high attendance. Also representatives of several government agencies, both Commonwealth and NT, wanted to be present which created the risk of two groups developing, one predominantly European and one predominantly aboriginal.

The use of the open-air venue with two days put aside for talking relieved much of the stress for those unaccustomed to meetings. The format was made as informal as possible whilst still having a structure. The agenda was simple and consisted of an introduction, an explanation of the lease and the obligations of PAN presented by National Park staff known to the attendees. This was followed by an extended briefing by an OSS staff member on the past mining history and the present situation in the valley. This was the first major hurdle to be overcome. It transpired that whilst the Traditional Owners knew much about the sacred sites in the valley they were not familiar with the majority of the mining sites. The briefing included lots of photographs and discussions of locations expressed in terms of a site's position on the ground relative to a known sacred site. There were many questions from the local people which indicated they had little detailed knowledge of what mining activity there had been or of the potential environmental impacts, apart from the obvious visual impacts. Having to explain acid rock drainage was the first of many challenges for the non-Aboriginal speakers.

After a long information and discussion session there was a site visit to see two of the areas concerned. At the first site, Guratba, there were discussions about who could go on the site. Once the group had been selected it was apparent that there were subtle rituals being followed which the non-Aboriginal people were unaware of. For example, any stone that was moved for examination was treated with respect and returned to it's exact postion before the party moved on This was a very unusual concept for a group of "European" people, some of whom would normally think little about accidentally kicking a stone during a site visit.

At the second site, the El Sherana camp, there was an understanding about the problems of dismantling old buildings containing asbestos and sheds that looked unsafe. The issue of radiation hazards in the core shed was less easy to explain. It became apparent to the "experts" that communication was not happening as effectively as it might.

For many of the non-aboriginal people present this was their first experience of having to deal directly with the "clients" rather than through the NLC. The process was flawed without doubt. Some of the issues included: inappropriate language with too many long words and too much jargon, representatives from different organisations wearing similar uniforms, and so on. There were unrealistic expectations amongst some of the non-Aboriginal people that decisions would be made soon after what seemed like complete and logical explanations had been delivered. This revealed a poor understanding of how Aboriginal communities make decisions by consensus rather than by majority. This led to a sense of progress being too slow for some people. The problems continued after the meeting with the production of a summary record that was all words. The situation then stagnated as staff changes at PAN, the prime agency with carriage of the issue, resulted in the process virtually halting for several months. Also the death of a senior Traditional Owner meant that meetings had to be put off for a while.

Whilst this 'pause" was happening a trip by helicopter was arranged to enable one senior Traditional Owner to be present whilst the sites were photographed from the air. These pictures were then used to produce material to distribute to the community to show them what sites looked like. This operation was a great success and enabled some relationship building to begin. Time passed and, although technical studies were carried out on one site in the valley where acid rock drainage problems were occurring, progress was slow in terms of moving the process forward. It was not going to be possible to meet the deadline set by the lease at this rate. We obviously had the wrong process.

6 THE NEW PROCESS

A new approach was tried in 1999 with the idea of formalizing the consultation process through creation of a steering committee. A meeting was held at the Aboriginal community of Werenbun, about 50km north of Katherine. This was the permanent home of the majority of the Gunlom Trust members. The idea was to discuss, in an informal atmosphere, how we were all going to get the process back on track, especially how the concerns, opinions and aspirations of the Traditional Owners could be canvassed, discussed and addressed. After a slow start the various parties finally got to talking about the size and scope of the issue. The first agreement at the meeting was that Traditional Owners would be the majority of any committee that was formed. Who

else should be represented in the consultative process was a topic that was discussed at length. The final composition for the group was agreed as follows:

- Traditional Owners, a group selected by the community because they were custodians of sites and ceremonies within the affected areas and also some of them lived in the valley at times during the dry season. This group to be both men and women and to be drawn from the various communities, and to be the majority.
- PAN-The lessees and so the agency having responsibility to carry out the rehabilitation under the terms of their lease.
- NLC-Representing the interests of the Aboriginal community and providing them with specialist advice.
- OSS-technical advisers to PAN as well as having responsibility for uranium mining environmental affairs in the region.
- NTDME-The regulator of mining activity in the Northern Territory and so having some statutory obligations.

The other addition to the group would be the consultants appointed by any of the above parties to provide technical expertise to the issue of rehabilitation.

This group was formed and immediately decided to name itself the Consultative Committee as being the title that was most expressive of their primary function. This Consultative Committee then agreed to set a timetable for meetings and activities to try and ensure that the program would be completed in accordance with the deadlines set in the lease. The idea was that the technical experts could meet as often as they wished but at the agreed intervals progress reports would be presented to the whole group and decisions made as to the next step.

The style of operation was agreed to be a major gathering every 6 to 8 weeks with any member of the Aboriginal communities concerned being welcome to attend. These meetings would hear presentations from the experts and then discuss the ramifications of the information. The subjects of the presentations would be presented in information documents available beforehand. The style of the presentations was difficult to work out at first but great emphasis was put on the use of models, posters, pictures, diagrams and computer graphics. These techniques were very successful. For example, the lack of familiarity amongst the Traditional Owners with the types, capabilities and sizes of earthmoving machinery available meant that the project team had to be a little innovative in choosing a method of information transfer to explain what the proposals meant. At times the use of small models of machinery has been the best way to demonstrate options for earthmoving. There was also a need to alleviate fears about big machines causing greater disturbance to the sites and possibly exacerbating problems needing attention. The project team selected pictures and diagrams showing how smaller machinery had been used in similar situations before and causing the least possible disturbance.

Also the choice of venue was important as people had to feel comfortable with their surroundings before they would relax and discuss issues. Consequently conventional meeting rooms were not an option. However, this led to another problem. Having meetings in the open air at a shade house in the Ranger station or under trees at a campsite was fine in the dry season. But once the wet season approached it was essential to find venues that were sheltered and cool and where the group could be catered for easily.

A hotel resort at Katherine was able to provide a suitable shelter in the gardens, which seemed like the ideal place but as the days got hotter and more humid so it became more difficult for everyone to concentrate on the business in hand. For the next meeting, still during the wet, the Committee used a motel near the former Rum Jungle minesite and made use of their meeting room. There was more acceptance of this format and venue by this time, but the outdoor venues are still the first preference.

As time progressed there have been growing numbers of interested community members and their families who have wanted to find out about what is happening. Whenever possible meetings have been held within the valley area. This not only reinforces the links between the people and the land, but also enables site visits to compare the presentation with the reality to be undertaken quickly and easily. Also the consultative group members become more familiar with each other when camping in one location and sharing the process of discussion through meals as well as in more formal sessions. The build up of mutual trust and respect within the group has been the most gratifying and satisfying part of the process to date.

Two meetings were facilitated by an independent consultant, who also prepared a summary record of the event. This approach was tried as no one person was able to take notes without having to drop out of the discussions. Also it was felt that an independent "chair" would assist in developing the idea of transparency in communication. Throughout the sessions all outcomes, questions raised and points agreed were written up on a flip chart. Each page was photographed as it was completed and these photographs were compiled into the meeting record. In this way the community are confident that the record is accurate. Also they easily remember the context in which ideas were discussed or agreements reached. From time to time the Traditional Owners ask to have a private discussion so they can debate a point amongst themselves. This is often done with the assistance of an anthropologist, who is there to explain some of the finer details such as site locations in relation to registered sacred sites.

During the meetings a number of significant issues have become apparent which explain the earlier reluctance to deal with some of the problems. For example, cultural issues have been discussed more frequently and openly as the process has advanced. These have included the need to exclude women from discussions of some areas where there are sites sacred only to men. This has enabled progress to be made on the two sites concerned. Also consultants have been careful not to assign female staff when dealing with issues relating to these specific areas. A further issue is the notion that materials may not be allowed to be brought into, or taken out of, the boundaries of particular sacred sites. This has obvious implications for what options are feasible at certain sites, for example when dealing with the potential for backfilling open cuts. An innovative solution has to be developed in order to be able to provide a cover over a site to ensure that the required radiological standards can be achieved.

Also there are concerns about the disturbance to the land at a number of sites. This must be minimized and so smaller machines are preferred for earthworks and no drilling or blasting is permitted in the valley for fear of arousing malevolent forces. This last point creates a challenge as clean rock is likely to be required for capping material or erosion control works at several locations. At the moment the consultants' believe it would be possible to use hydraulic rock breakers or similar physical methods to obtain suitable materials. But obtaining consensus agreement on the "mining" method may be a very difficult task.

As time progressed it became necessary to negotiate an extension of 12 months to the original schedule, a reflection of the time required to advance each stage. Communities will not be hurried and the spacing of meetings enables people to relax before the next round of discussions and actions. Also out-of-session work such as field investigations and intra-community discussions can take place. Fieldwork is tasked to provide the data needed for the rehabilitation planning but also has to address issues raised at meetings. These usually relate to the safety of foodstuffs and water obtained from the valley area. These questions are raised frequently in discussion and each major food group, be it fish, meat, water or fruit and vegetables has been the subject of questions at one time or another. As a consequence sampling programs have been undertaken for food items in each of these groups and the results presented to the meetings. Test results have indicated that there are no radiological concerns in this regard at any location in the valley.

As part of the information transfer process a radiological protection seminar was organised for the Traditional Owners to explain what radioactivity is and how it relates to their everyday life. This was carried out by a specialist trainer and has certainly been very well received, so much so that the exercise is to be repeated for other communities in the region who are involved with uranium mining.

7 OUTCOMES

It is now expected that the rehabilitation plan for 80% of the sites will be completed on schedule by 31 December 2001. The remainder cannot be completed until other issues related to containment of low-level radioactive material, such as ore remnants and small patches of contaminated soil etc, have been resolved; an issue that requires input from federal authorities. The Traditional Owners are more confident that they understand what is being done in the rehabilitation process, and how and why it is being done. They can also see how the process of planning is being adapted to take account of their own ideas and aspirations for the future. The Gunlom Land Trust members are looking forward to seeing their plan being implemented and, hopefully, participating fully in the implementation as well as the long term stewardship of the sites within the overall Plan of Management for Kakadu National Park.

8 ACKNOWLEDGEMENTS

The author wishes to acknowledge the assistance of his colleagues in valuable discussions during the writing and for undertaking the peer review of this paper. These include associates in the World Heritage Branch and Parks Australia North within Environment Australia as well as from the Office of the Supervising Scientist. Special thanks are due to Ben Bayliss for assistance with the figure.

REFERENCES

Annabel R I. 1977. The uranium hunters. Rigby, Adelaide.
BHP. 1987. Aboriginal History Upper South Alligator Valley with Advice for Project Personnel. Coronation Hill Joint Venture. Unpublished manual.
Bromwich D. 1987. Radiological Hazards at the abandoned South Alligator Uranium Mill. Northern Territory Department of Mines and Energy Report OH 87/4. Darwin, NT, Australia.
Construction Group. 1988. Rehabilitation proposals for abandoned uranium mines in the Northern Territory Report 88/2, Department of Administrative Services, unpublished report.
Department of Housing and Construction SA/NT Region. 1986. Report on survey of abandoned mines on Gimbat pastoral lease Northern Territory. Department of Administrative Services, unpublished report.
Fisher, W.J. 1968. Mining practice in the South Alligator Valley in Proceedings of a symposium "Uranium in Australia" AusIMM Rum Jungle Branch 16-21 June 1968. AusIMM, Melbourne.
Fisher, W.J. 1988. A brief history of mining, exploration and development in the South Alligator river valley. WJ & EE Fisher Pty Ltd Darwin, unpublished report.
Fox, R.W., Kelleher, G.G. & Kerr, C.B. 1977. Ranger Uranium Environmental Inquiry-Second report. Australian Government Publishing Service, Canberra.
ICRP. 1990. Recommendations of the International Commission on Radiological Protection, Pergamon Press, Oxford.
Köppen W & Geiger R. 1936. Handbuch der Klimatologie. Bd. 1/M.C., Berlin.
Kraatz, M & Applegate, R.J. 1992. Rum Jungle Rehabilitation Project-Monitoring Report 1986-88. Technical Report Number 51, Conservation Commission of the Northern Territory. Darwin NT, Australia.
OSS. 1987. Supervising Scientist for the Alligator Rivers Region-Annual Report 1986-1987. Australian Government Printing Service, Canberra.
Waggitt, P.W. 1998. Hazard reduction works at abandoned uranium mines in the upper South Alligator valley, Northern Territory in Radiological aspects of the rehabilitation of contaminated sites, eds, Akber RA & Martin P, (1998) pub. South Pacific Environmental Radioactivity Association (SPERA), Christchurch, New Zealand.
Waggitt, P.W. (2000). The South Alligator Valley, Northern Australia, Then and Now: Rehabilitating 60's uranium mines to 2000 standards. in proceedings of the SWEMP 2000 Conference, Calgary, Canada. May30 – June 2, 2000. pub: Balkema.
Waggitt, P.W. (2001). Remediation of Abandoned Uranium Mines in the Gunlom Land Trust Area, Northern Australia. (IN PRESS) Proceedings of 8th International Conference on Environmental Management ICEM'01. Bruges, Belgium 30 September-4 October 2001. pub. American Society of Mechanical Engineers.

Tailings and Mine Waste '02, © 2002 Swets & Zeitlinger, ISBN 90 5809 353 0

Remediation of streamside tailings along Silver Bow Creek near Butte, Montana

W.H. Bucher, G. Fischer, B. Grant, A. Shewman & L. Cawlfield
Maxim Technologies, Helena, Montana

J.E. Chavez & T. Reilly
Montana Department of Environmental Quality, Helena, Montana

ABSTRACT: Mining wastes deposited along Silver Bow Creek downstream of Butte, Montana have posed a threat to human health and the environment for over 100 years due to the presence of heavy metals and arsenic in the wastes. Under the Comprehensive Environmental Response, Compensation and Liability Act (CERCLA), the State of Montana has undertaken the remediation of more than 2.5 million cubic yards of mine waste deposited along approximately 26 miles of Silver Bow Creek and its floodplain. During remediation, mine waste is excavated from the stream and floodplain, selected materials are treated with lime, and the waste is placed in local or regional repositories constructed for the project. The stream and floodplain are rebuilt to restore their natural function and maintain streambank stability during vegetation reestablishment. The long-term remediation objective of this source removal action is to attain groundwater and surface water quality standards in the floodplain and stream.

1 INTRODUCTION

This paper describes efforts to remediate a floodplain impacted by mine waste near Butte, Montana (Fig. 1). Impacts include stream channel and bank degradation, water-quality degradation, reduction of productivity of terrestrial plants, and impairment of the benthic macroinvertebrate community. Silver Bow Creek is a tributary to the Clark Fork River that typically has flows of about 20 cfs in the area near Butte. The stream is devoid of fish and most other aquatic life forms due to elevated concentrations of copper zinc, and other heavy metals in the water and stream sediments. The primary carcinogenic risk to human health is exposure to arsenic in soil and groundwater although some heavy metals exceed groundwater protection standards as well (MDEQ & EPA 1995).

The remediation of Silver Bow Creek is being undertaken by the Montana Department of Environmental Quality (MDEQ) under the Comprehensive Environmental Response, Compensation and Liability Act (CERCLA). The 26 mile-long floodplain of Silver Bow Creek from the west side of Butte to the Warm Springs Treatment Ponds has been designated the Streamside Tailings Operable Unit (SSTOU). The operable unit is divided in four subareas which are further subdivided into reaches, each about one mile long, as shown in Figure 2. The first mile of the floodplain is remediated and the next two miles are currently under construction. This paper describes the remedial design and remedial action undertaken by MDEQ and its engineering consultant, Maxim Technologies, Inc. (Maxim).

Figure 1. Location map.

2 BACKGROUND

The Silver Bow Creek/Butte Area was listed on the National Priorities List in 1983 by the United State Environmental Protection Agency (EPA) and remedial investigations began in 1984 by the State of Montana under an agreement with EPA. The owner of much of the Butte mining property, the Atlantic Richfield Company (ARCO), took over the remedial investigation in 1991 and produced a draft Remedial Investigation Report (ARCO 1995a). This investigation provided information for the baseline human health and ecological risk assessments (MDEQ 1994) and determined appropriate remedial actions through a feasibility study (ARCO 1995b). From the alternatives presented in the feasibility study, a remedy was selected by MDEQ and EPA and documented in the record of decision (MDEQ & EPA 1995).

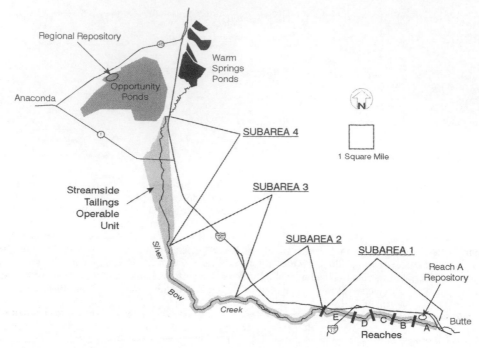

Figure 2. Site map of SSTOU.

2.1 Remedial investigation

The mine wastes located in the floodplain of Silver Bow Creek are impacting the environment through a number of pathways that are shown conceptually in Figure 3. The remedial investigation found that five heavy metals and arsenic were contaminants of concern at this site, and these contaminants were found primarily in three media: tailings/impacted soils, instream sediments, and railroad materials. Railroad materials are primarily waste rock from Butte mines used to construct portions of railroad beds that flank the stream through much of its course. Typical concentrations of these contaminants in these solid media in the first subarea of Silver Bow Creek are presented in Table 1. Although there are no numeric cleanup standards in Montana for these contaminants in solid media, concentrations of arsenic, copper and zinc are of most concern to human health or the environment.

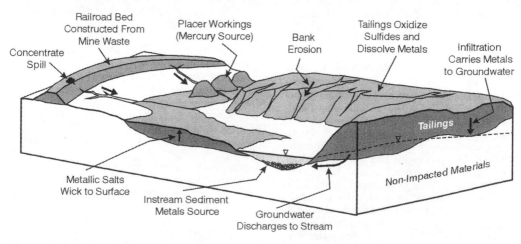

Figure 3. Conceptual contaminant migration model.

412

Table 1. Typical contaminant concentrations in solid media (mg/kg).

Medium	As	Cd	Cu	Pb	Hg	Zn
Tailings/ Impacted Soils*	278	7.8	739	540	2.1	2,400
Instream Sediments**	70.7	4.4	816	234	0.55	1,244
Railroad Materials*	375	2.0	1,081	587	NS***	687

* Median concentration from Subarea 1 (ARCO 1995a).
** Geometric mean concentration from Subarea 1 (ARCO 1995a).
*** NS - Not Sampled.

Table 2. Low flow concentrations of contaminants of concern in surface water (µg/L).

Station	As	Cd	Cu	Pb	Zn
SS-07	8.7	1.6	178	5.3	682
SS-08	5.3	1.5	192	13.5	771
SS-09	11.6	1.8	243	9.4	790
SS-10	14.5	2.5	322	15.2	860
Standard	18*	2.5**	9.3**	3.2**	120**

* Human health standard from Circular WQB-7 (MDEQ 1998).
** Chronic aquatic criterion evaluated at hardness of 100 µg/L from Circular WQB-7 (MDEQ 1998).

The remedial investigation also documented the presence of these contaminants in the surface water of Silver Bow Creek. Table 2 presents typical concentrations of the contaminants of concern for representative low flow conditions on Silver Bow Creek measured within Subarea 1 (ARCO 1995a).

The data in Table 2 are the geometric mean values for total metals concentrations for selected low flow events at four stations located in upstream to downstream order in the Subarea. The data suggest that copper, lead and zinc are particularly dangerous to aquatic organisms at their present concentrations. Mercury was not sampled in these data sets. The result of a risk assessment based on these data indicated sufficient risk that a remedy should be pursued (MDEQ 1994).

2.2 Feasibility study and record of decision

The feasibility study conducted by ARCO (1995b) presented seven site-wide remedial alternatives including no action. All the remedies were based on source removal activities and no analysis of water treatment options were included. The alternatives considered two treatment options for each of the solid media, in situ treatment and removal with varying degrees of these two treatment options appearing in the different alternatives. In situ treatment of materials consisted of lime amendment of tailings and impacted soils or railroad materials; it was not considered for instream sediments. Removal of material contemplated excavation and transport of material to local or regional waste repositories to be constructed for the project.

In their record of decision (ROD), MDEQ and EPA (1995) selected an alternative that treated or removed all materials. Tailings/impacted soils in the floodplain were required to be removed from the 100-year floodplain if they were in contact with groundwater, in situ treatment was impracticable (e.g. due to thickness), and evolution of the channel could cause materials to be eroded into the channel in the future. This decision required that the bulk of the tailings/impacted soils be removed and no in situ remedies have been applied to date. In addition, all instream sediment is to be removed under the record of decision.

The remedy required that all railroad materials that pose a risk to human health and the environment be excavated, treated or covered. In situ treatment or covers are appropriate for areas not likely to be eroded into the stream. Areas of direct contact with the stream are excavated, removed to repositories, and replaced with suitable material. Those railroad materials consisting of concentrate spills are to be removed to a Resource Conservation and Recovery Act (RCRA) Subtitle C landfill (ARCO 1997).

An essential component of the remedy was the reconstruction of the stream channel and floodplain as a naturally functioning system with the ability to migrate where infrastructure will not be threatened. This required a design of the channel that was consistent with sediment transport and geomorphic design principles (ARCO 1997).

3 ELEMENTS OF DESIGN

Two design packages have been produced for the project to date, one for Reach A and one for combined Reaches B and C (Fig. 2). Although there were several noteworthy differences in the two designs which are discussed in this section, most aspects of the designs were similar. The design for Reach A is described in detail in the Final Design Report – Reach A of Subarea 1 (Maxim et al. 1999), and an Addendum to the Final Design Report, Reaches B and C, Subarea One (Maxim 2001) provides information specific to the Reaches B and C design.

3.1 Dewatering

Working in the stream environment posed two major challenges, diverting the stream flow during construction and dewatering saturated material to allow removal and reconstruction. Surface water diversion was accomplished by different means in the two designs. In Reach A a short-term diversion channel was constructed to accommodate all flow during construction as well as high flows for a period after construction was complete. This allowed a cost savings in streambank construction methods because the banks did not have to withstand high flows during the period when vegetation was reestablishing. After vegetation is reestablished on the banks, the diversion channel will be removed and the banks should withstand high flows with allowable rates of deformation.

In Reaches B and C the existing stream channel was used to contain the stream flow while a parallel channel was constructed. This saved the cost of a short-term diversion channel but resulted in other complications: constraint of location of the new channel, need to construct channel crossings at certain points, and greater risk of out-of-bank flooding during construction. In addition, one portion of Reach C required installation of a temporary surface water diversion during construction because landowner constraints prevented establishment of a new channel alignment. This surface diversion was sized for flows expected during the construction season and consisted of a HDPE lined trapezoidal channel. It was less expensive to build than the diversion channel in Reach A.

Because a large portion of materials requiring removal were saturated by groundwater, dewatering was required to execute the clean-up and build the new streambed and streambanks. The groundwater dewatering systems consisted of sediment detention ponds, dewatering trenches, and sumps with the associated dewatering pumping systems. The sediment ponds were designed to increase retention times and reduce effluent turbidity. Each pond was designed to remove sediment from the longest trench that would discharge into it.

Groundwater dewatering trench locations, lengths, and depths were designed based on existing floodplain topography, a potentiometric map, and required material excavation elevations. Design information was also obtained from a pilot test conducted on a portion of the stream in 1997 (Maxim & Inter-Fluve 1998). In general, trenches were designed to dewater floodplain materials to an elevation two feet below the required excavation elevation. Dewatering trenches were routinely located along the new stream alignment to save space and obliterated as the new stream channel was built. Sumps consisting of upright HDPE pipe packed with washed gravel were constructed at the downgradient end of each trench. Pumps with a minimum capacity of 100 gpm per 100 feet of trench were required to remove the anticipated volume of water from each trench.

3.2 Mine waste excavation

Tailings and metals impacted soils were required by the *Comprehensive Work Plan* (ARCO 1997) to be removed to the "order-of-magnitude break" in metals concentration. Although not easy to define precisely, this depth was usually discernable in tailings/impacted soil concentration data obtained from test pits. Test pits were excavated on a rough 150-foot grid and samples were obtained at four inch intervals. Metals analysis were conducted using X-ray fluorescence methods and determinations of the base of tailings were made using these data. An elevation for the base of tailings was established from survey data and the base of tailings elevation surface was modeled to allow establishment of a 50-foot staking grid of the base of tailings. Typical tailings removal depths are two to three feet.

Because the base of tailings surface could be quite variable over short distances and because test pit data was limited to an approximate 150-foot grid, it was necessary to over-excavate the tailings/ impacted soils to ensure adequate removal of contaminants. The agencies set a goal of removing 90 percent of the tailings/impacted soils (as defined by the "order-of-magnitude break") with 95 percent confidence. An evaluation of the statistical properties of the excavation surface based on test pit data indicated that over-excavation of the tailings/impacted soils by six-inches would achieve this goal.

In Reach A, mine waste was to be excavated and transferred by haul trucks to the nearby repository. In Reaches B and C the waste was transferred by truck to loading areas where it was placed on railroad cars for removal to a regional repository.

3.3 Repository design

Two different approaches have been used for disposal of waste materials to date. In Reach A the design required a local waste repository with lime treatment of the waste materials; whereas, the Reaches B and C design employed a regional waste repository at the Opportunity Ponds where treatment of waste was not required. Waste removed during construction of Reach A was placed in a repository situated just outside the 100-year floodplain of Silver Bow Creek and adjacent to the construction area. Previous characterization studies provided information on the waste's properties including soil gradation, *in situ* moisture content and density, metals content, acid-base account and sulfur fractionation. Additional studies were completed to determine liming requirements for the waste and the leachate chemistry for both limed and unlimed waste (Grant *et al.* 2000).

The construction for the Reach A repository consisted of amending waste with lime, construction of a two-foot thick soil cap, and revegetation of the cap. Amendment of the waste was accomplished at the repository by spreading the waste in 18-inch lifts, applying and then incorporating lime followed by compaction of the amended waste in-place. At final build-out, the repository contained 150,000 cubic yards of waste. A typical cross-section of the repository waste is shown in Figure 4.

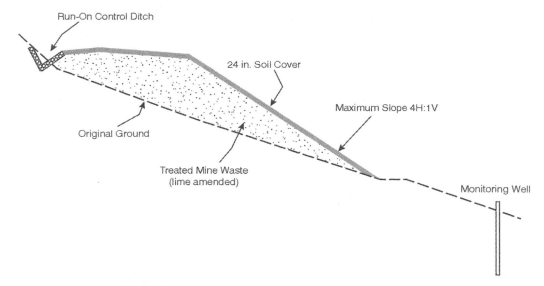

Figure 4. Typical repository section (Reach A).

For Reaches B and C, approximately 200,000 cubic yards of waste is being transported by rail to a regional repository at the Opportunity Tailings Ponds, located 15 miles from Reaches B-C (Fig. 2). The tailings ponds are 3,300 acres in area and no longer in operation. The existing waste in the tailings ponds exhibits similar characteristics to the wastes from Silver Bow Creek and the agencies have approved consolidation of waste at this location. Unamended waste is placed in two approximately five-foot thick lifts. A temporary six-inch thick cap of gravel covers the waste to prevent wind erosion of tailings while a final cap design is developed by ARCO.

3.4 Non-impacted materials

An objective of the remedial action is to reconstruct a naturally functioning stream channel and floodplain (ARCO 1997). In many areas, infrastructure development has resulted in an aggrading channel as sediment is deposited upstream of bridges and other constrictions. Restoring a more natural grade has required removal of these depositional areas, some as deep as six feet, and establishment of a more uniform channel grade. The lower channel elevation has also resulted in a lower floodplain elevation in these areas, often requiring excavation of non-impacted materials as well as contaminated materials to achieve the new topography. In addition, non-impacted materials are generally excavated as the new channel corridor is constructed.

The lowering of the floodplain that occurs in some portions of the project is a benefit in that it lessens the need for borrow material to rebuild the floodplain. However, the non-impacted materials that are required to be excavated are often too coarse to provide suitable vegetative backfill material and must be placed in areas with sufficient fill thickness that they will be covered with a lift of backfill materials suitable for vegetative growth. In some cases sufficient fill areas are not available and terraces have been constructed at the perimeter of the floodplain as well as permanent embankments outside the floodplain to accommodate these excess materials.

3.5 Railroad materials

Railroad materials consist of: 1) Contaminated materials that impact the remediated floodplain; 2) Contaminated materials that are in direct contact with Silver Bow Creek; and, 3) Ore concentrate spills. All three classes of materials have been encountered in Reaches A through C.

The first class of railroad material is railroad embankments within and outside of the SSTOU that are contaminated and have the potential for eroding arsenic and metals-contaminated sediment onto the remediated floodplain. Embankments within the SSTOU are being remediated by placing a granular rock cover over the contaminated material to dissipate rainfall and runoff energy and prevent migration of the underlying waste. Embankments outside of the SSTOU have the same erosional potential as embankments within the operable unit and these areas have been observed to produce contaminated sediment which eventually migrates to the Silver Bow Creek floodplain. Because these areas are outside this operable unit, sediment basins have been designed to retain runoff and entrained sediment from the off-site embankments pending remediation of these sites.

The second material class consists of railroad embankments whose contaminated materials are in direct contact with Silver Bow Creek. Typically these embankments are along the stream channel or form abutments for railroad bridges over Silver Bow Creek. Remediation for these areas requires removal of contaminated embankment material in direct contact with the stream and replacement with designed erosion protection. Protection typically consists of riprap armor designed to withstand 100-year flows.

Ore concentrate spills have very high concentrations of arsenic and metals and are found in small isolated areas along the railroad embankments. Remediation for this material also requires excavation of the material with adequate margins of excavation to assure complete removal. The material is then placed in Department of Transportation approved containers and shipped to an approved hazardous waste disposal site.

3.6 Stream reconstruction

Because the ROD required removal of contaminants from the bed and banks of the stream, reconstruction of the entire stream channel has been required to date. To comply with sediment transport and geomorphic constraints as well as to facilitate construction, the new channel has typically been constructed in a new alignment without reference to the original stream location. However, in some areas the channel location was fixed by infrastructure such as bridges and railroad embankments. In these areas a fixed or non-deformable channel has been constructed which will remain in place for all flows up to the 100-year recurrence event. The non-deformable channel design consisted of bank toes constructed from large rock, upper banks constructed from fabric encapsulated soil lifts, and bed material designed to resist scour in the 100-year flood event. Design concepts for stream reconstruction are presented in the *Final Conceptual Design Report* (Inter-Fluve & MEI 1998).

A deformable channel is the preferred construction method to allow development of a migrating stream which adjusts to the sediment and flow conditions. The concept for this part of the channel design was to build stream banks which would be initially stable under high flows but gradually deform after the vegetation on the banks and floodplain became well established (Maxim *et al* 1999). This concept was realized in different ways in the two designs. In Reach A, a diversion channel designed to accommodate the ten-year flood (about 610 cfs) was constructed to protect the channel and floodplain from high flows. This allowed construction of relatively inexpensive soil banks with a coir fabric protecting the face of the banks from low flows. Coir fabric is a coconut based product which gradually

degrades over a period of two to five years. In this time period the banks as well as the floodplain should establish sufficient vegetation to withstand erosion from low flows. The banks should also be sufficiently resistant to high flows to allow gradual erosion without catastrophic loss of floodplain materials. After a period of a few years, the diversion channel will be removed and the stream channel will begin to migrate naturally in response to higher flows.

A different approach was adopted for Reaches B and C. In these reaches, banks were built with coir fabric encapsulated soils which are designed to resist the 50-year flow. In addition, the outer banks of pools contain a coir fabric wrapped rock toe also designed to the same criterion. These construction techniques are illustrated in Figure 5. As the banks revegetate and the coir fabric degrades, these banks should become mobile at lower flows. Although this design approach resulted in greater costs for stream construction, a cost savings was realized by using the existing channel as the stream diversion channel rather than constructing a separate diversion channel.

Bed material for both designs was sized based on analysis of existing bed material gradations found in Silver Bow Creek and requirements of the sediment transport continuity analysis (MEI and Inter-Fluve 1997). The depth of the bed material, generally 1.5 feet, was based on a scour analysis for the design flow. Bed material was extended underneath the banks to compensate for the lack of alluvial sized materials in the floodplain backfill.

As part of the channel design, the bedform was designed to include riffle, run and pool sections. Run sections (Fig. 5) were relatively straight and contained a talweg approximately 0.5 feet deep which varied in position throughout the run. Pools were constructed at most stream bends. Pool construction was more complicated because the asymmetric section allowed for a deep pool and steep bank on the outside of the bend and a point bar on the inside of the bend (Figs 5, 6). A large rock toe was also constructed on the outside of the bend to limit the rate of stream migration. Riffle sections, which are relatively wide and shallow, were constructed at the tailout of the pools. Important geomorphic criteria employed in this design were that the channel bank-full capacity be between the one and two-year recurrence flow or about 140 to 210 cfs., and that the width to depth ratio of the channel should be between 10 and 15 for riffles and runs and between 7 and 12 for pools. This resulted in top widths for riffles and runs of 26 to 28 feet in this portion of the project. In addition, bend radii were required to be at least four times the channel width to prevent overly rapid bend migration rates (Inter-Fluve & MEI 1997).

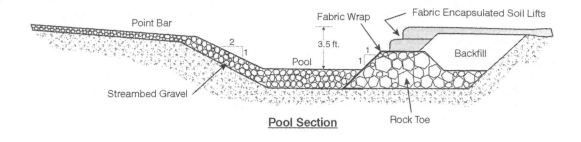

Pool Section

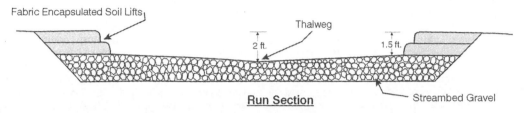

Run Section

Figure 5. Typical stream channel sections.

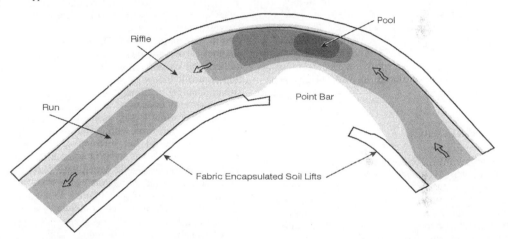

Figure 6. Typical pool plan.

3.7 Floodplain reconstruction

As part of the goal of attaining a naturally functioning stream and floodplain system, the floodplain has been rebuilt in a manner that allows lateral migration of the stream in most areas (MDEQ & EPA 1995). To accomplish this goal, a gently sloping floodplain has been constructed which drains towards the stream. Because the floodplain elevation has been lowered in many areas, it is often necessary to construct a new bank at the edge of the floodplain to meet existing terrain. This bank is constructed no steeper than 4H:1V to facilitate revegetation efforts and minimize erosion. Overbank terraces have been constructed in some areas, but these must accommodate a minimum of the five-year flow at their crest elevation.

Although no standing water wetlands have been constructed to date, the design concept allows for their construction if they are not likely to impact channel or overbank hydraulics. A particular concern with standing water wetlands is that the stream could avulse into these areas during high flows, potentially destabilizing the stream channel.

Reconstruction of the floodplain requires suitable materials for reestablishment of vegetation. Some of the non-impacted material remaining in the floodplain is not suitable for reestablishment of vegetation because of the coarse textures. Suitable backfill materials for vegetation reestablishment have been found within a few miles of the site and are imported as needed. Beginning with Reach B, the design requires that at least six inches of suitable backfill material be placed in all portions of the floodplains. In areas where the floodplain does not need to be backfilled because of the lowered floodplain elevation, this means that at least six inches of non-impacted materials must be removed to provide space for the suitable backfill.

Establishing a diverse, effective, and permanent vegetative cover on all reclaimed areas is an objective of the *Comprehensive Remedial Design Work Plan* (ARCO 1997). Vegetation is important to provide floodplain and stream channel stability as well. The vegetation design is based on the expected hydrology of the reconstructed floodplain which is generally classified in four hydrologic zones: wetland, transition, subirrigated and upland zones. Another consideration for revegetation is the soil texture. Coarse soils may be suitable for revegetation in wetland and transition zones but not drier zones. Hydrologic zones were initially based on groundwater elevations collected prior to construction. In practice, the significant changes in stream alignment and floodplain elevation changed the hydrology sufficiently that MDEQ decided to postpone revegetation planning until after construction and reevaluation of the hydrology. Revegetation consists of a seed mixes tailored to the hydrologic classes, shrubs and trees to be planted in specific areas, and plugs of selected sedges to be planted in wetland and transition zones.

4 CONSTRUCTION

Construction has occurred in several phases. Two general contracts have been let as well as smaller contracts for bridge construction and revegetation. The first general contract was for construction of Reach A, and a second general contract is underway for Reaches B and C. In general, construction in the floodplain has followed the following sequence:
- Construction of surface water diversions where required.
- Installation of groundwater dewatering systems.
- Removal of tailings/impacted soils to the repository.
- Excavation and placement of non-impacted materials to meet floodplain grading requirements.
- Excavation of the channel corridor.
- Construction of the new stream channel.
- Backfill of the floodplain.
- Diversion of the stream into the new channel.
- Revegetation of the floodplain.

4.1 *Reach A construction – 1999-2000*

Remedial Action construction for Reach A was implemented in 1999 and 2000. A competitive bid process resulted in selection of Jordan Contracting, Inc. (JCI) of Anaconda, Montana. Construction activities commenced in July 1999 and 1999 construction included completion of the diversion channel and structure, construction of the stream non-deformable channel under the Interstate bridges, excavation of tailings/impacted soils and placement of vegetative borrow between the Interstate bridges, preparation of a borrow area, and initial construction of the mine waste repository.

After a winter shutdown period, Reach A base contract items were substantially complete in July 2000. For Reach A, the total volume of tailings removed was 166,000 bank cubic yards from an area of approximately 36.6 acres, for an average depth of tailings of 2.8 feet. With space and extra lime available at the mine waste repository, additional tailings/impacted soils were removed from two areas within the Reach B floodplain, downstream of Reach A. Project completion was in December 2000. Construction cost, including the additional removal areas, was approximately $3,300,000.

The use of the short-term diversion channel for a major portion of Reach A where deformable banks were installed has worked well although the channel was expensive to construct and will require reclamation in the future. During flood events the diversion channel carries the peak flows and maintains the reconstructed channel flow at about 50 cfs. To date, this controlled flow method has resulted in no lost stream channel or floodplain materials. Upstream of the short-term diversion, a contractor initiative to pump Silver Bow Creek around a non-deformable reach rather than work with coffer dams in the channel was successful but instructive. The diversion channel alignment was steep in some areas and subject to erosion that could precipitate failure of the channel. Lining of the channel with PVC was improperly executed and nearly failed. In addition, sufficient pumping capacity was not always available or operable at the diversion, and at one point the work area was subject to some flooding.

The dewatering plan prepared for Reach A has worked well when implemented as intended. Initiatives by the contractor to postpone dewatering or use other methods have generally not proved effective. The required two-week dewatering period before tailings removal has proved effective in ensuring that the tailings will be desaturated sufficiently for excavation and subsequent handling at the repository.

A remedial action goal for Reach A was to remove 90 percent of tailings/impacted soils with 95 percent confidence. Removal verification sampling was conducted following tailings excavation, in accordance with the *Construction Quality Assurance Plan* (Maxim & RRU 1998) and the *Final Design Report* (Maxim *et al.* 1999). Verification sampling was conducted for record purposes only and it was not used to direct further excavation. At a point location, removal was considered to have been accomplished if four of the six metal constituents of concern were less than target concentrations. Using statistical properties of the pre-construction tailings depths database, approximately 37.1 percent of the verification samples could fail and still meet removal criteria. The measured failure rate for verification sampling in Reach A was 34.8 percent, slightly below the allowable failure rate. Reach A removal was based on over-excavating the base of tailings by six inches. Given the narrow margin of success, the over-excavation depth has been increased from six inches to nine inches for Reaches B and C. The use of statistical methods has allowed the setting of achievable removal goals for the site without resorting to after-the-fact "hot-spot" removal methods which are necessarily slower and more expensive.

Revegetation of Reach A was completed in 2001 under a separate contract. July rains alleviated concerns that the vegetation would succumb to drought and most areas have an adequate first-year cover. The contract included supplemental plantings supplied under the State of Montana's Natural Resource Damage Program, an additional source of funding outside of the State's remediation budget.

4.2 Reaches B and C construction – 2001-2002

Remedial action construction for Reaches B and C commenced in May 2001 and is scheduled to be completed in the spring of 2002. The construction contractor is Smith Contracting, Inc. (SCI) of Whitehall, Montana. SCI's base bid for the project was about $3,900,000.

As of September 2001, construction through Reach B was nearing completion. The plan for using the existing channel for the surface water diversion worked well except for a couple of high flow events when the stream overtopped its banks and flooded work areas. These events required a drying out period before construction could resume. This method of work also required that the contractor complete the channel construction before the tailings and non-impacted material be removed in the vicinity of the existing stream. In effect, this required the contractor to take two passes through Reach B with his excavation equipment. Stream construction methods are unfamiliar to the general contractors involved in work on the SSTOU to date, but the contractors have learned the construction techniques quickly.

Hauling mine waste on the railroad to the regional repository has worked quite well, especially since the railroad company has been able to load directly on its mainline without the need to construct rail spurs. The train typically transports 1600 cubic yards using 32 cars and it takes approximately one-half hour to load. One of the loadout areas was undersized due to space constraints and attempts will be made to maintain a minimum loadout area size in future designs. The unloading of waste at the regional repository site has been without problems although it requires the contractor to maintain a separate work crew at some distance from their other activities.

5 CONCLUSIONS

The design and construction of Streamside Tailings Operable Unit remedial action has been successfully initiated based on the two designs and construction contracts awarded in the first three miles of Silver Bow Creek near Butte, Montana. With construction nearing completion on the first two miles of stream, no major problems have been encountered to date with construction methods or results of the remedial action. The new stream channel is performing as expected, the floodplain revegetation has reestablished commensurate with expectations, and the repositories have successfully contained the waste. Although long-term monitoring has not commenced at this site, it is expected that it will show significant improvement to water quality as well as improvements in plant productivity in the terrestrial environment and benthic diversity in the aquatic environment.

Although construction costs of $2,500,000 per mile (excluding engineering and administrative costs) may appear expensive, they are considered cost-effective for the degree of remedy attained and should allow completion of the project within the projected budget.

REFERENCES

Atlantic Richfield Company (ARCO) 1995a. *Streamside Tailings Operable Unit, Draft Remedial Investigation Report.* Bozeman: Titan Environmental, Inc.
Atlantic Richfield Company (ARCO) 1995b. *Streamside Tailings Operable Unit, Draft Feasibility Study Report.* Bozeman: Titan Environmental, Inc.
Atlantic Richfield Company (ARCO) 1997. *Streamside Tailings Operable Unit, Silver Bow Creek/Butte Area NPL Site, Final Comprehensive Remedial Design Work Plan.* Bozeman: Titan Environmental, Inc.
Grant, B. G., Cormier, M. & Bucher, W.H. 2000. Alternatives for Mine Waste Relocation Repositories. *Proceedings of the Seventh International Conference on Tailings and Mine Waste.* Rotterdam: A.A. Balkema.
Inter-Fluve, Inc. & Mussetter Engineering, Inc. (MEI) 1997. Design Criteria for Silver Bow Creek, SSTOU, Subarea One, Remedial Channel Design. Bozeman: Inter-Fluve, Inc.
Inter-Fluve- Inc. & Mussetter Engineering, Inc. (MEI) 1998. *Final Conceptual Design Report for Silver Bow Creek, SSTOU, Subarea One, Remedial Channel Design.* Bozeman: Inter-Fluve, Inc.
Maxim Technologies, Inc. & Inter-Fluve, Inc. 1998. *Draft Pilot Test Construction Monitoring Report, Subarea 1 Remedial Design, Streamside Tailings Operable Unit, Silver Bow*

Creek/ Butte Area NPL Site. Helena: Maxim Technologies, Inc.

Maxim Technologies, Inc. & Reclamation Research Unit (RRU) 1998. *Construction Quality Assurance Plan, Remedial Action, Streamside Tailings Operable Unit, Silver Bow Creek/Butte Area NPL Site.* Helena: Maxim Technologies, Inc.

Maxim Technologies, Inc., Inter-Fluve, Inc., Reclamation Research Unit, & Bighorn Environmental, Inc. 1999. *Final Design Report, Reach A of Subarea 1, Streamside Tailings Operable Unit, Silver Bow Creek/ Butte Area NPL Site.* Helena: Maxim Technologies, Inc.

Maxim Technologies, Inc. 2001. *Technical Memorandum - Addendum to the Final Design Report, Reaches B and C, Subarea One, Streamside Tailings Operable Unit, Silver Bow Creek/Butte Area NPL Site.* Helena: Maxim Technologies, Inc.

Mussetter Engineering, Inc. & Inter-Fluve, Inc. 1997. *Channel Stability Analysis, Silver Bow Creek SSTOU, Subarea 1.* Fort Collins: Mussetter Engineering, Inc.

Montana Department of Environmental Quality (MDEQ)1994. *Draft Baseline Risk Assessment, Silver Bow Creek/Butte Area National Priorities List Site.* Helena: Camp, Dresser and McKee, Inc.

Montana Department of Environmental Quality (MDEQ) & United States Environmental Protection Agency (EPA) 1995. *Record of Decision, Streamside Tailings Operable Unit of the Silver Bow Creek/Butte Area National Priorities List Site.* Helena: Environmental Remediation Division.

Montana Department of Environmental Quality (MDEQ) 1998. *Circular WQB-7, Montana Numeric Water Quality Standards.* Helena: Planning, Prevention and Assistance Division.

A rapid response to cleanup – Gilt Edge Superfund Site, South Dakota

M.J. Gobla
U.S. Bureau of Reclamation, Denver, Colorado, USA

ABSTRACT: The Gilt Edge gold mine is an acid drainage site that has been put on an accelerated closure schedule. The mine ceased activities in 1999 when Dakota Mining Corporation declared bankruptcy forcing the State of South Dakota to immediately assume water treatment operations. Evaluation of conceptual closure plan options and cost estimates led the State of South Dakota to a decision to seek Federal assistance. The site has quickly moved into reclamation mode for the principal contamination source, the Ruby waste-rock dump. Designs and specifications for capping the Ruby waste-rock dump were prepared while Superfund listing was pursued. In October of 2000, mobilization of the first reclamation contractor began and by December the site was added to the National Priorities List. Capping the waste-rock dump will address a major acid drainage source. Water treatment requirements are expected to decline as conventional methods such as diverting clean water, backfilling, grading, capping, limestone neutralization, and revegetation are implemented. Acid seepage from underground workings, steep highwalls, and some pit backfills will remain. Major field trials of emerging technologies are nearing completion and some are showing promising results. Carbon reduction in a pit lake, and pyrite microencapsulation on simulated waste dumps, are showing initial success. Their application may minimize or eliminate the need for long-term active water treatment which has been a long sought goal for major acid rock drainage sites.

1 LOCATION

The Gilt Edge gold mine is located approximately 4 miles southeast of Deadwood, in the northern Black Hills of Lawrence County, South Dakota. The mine consists of patented mining claims (private ownership) and unpatented claims (USDA Forest Service lands). The mine is in Sections 5, 6, 7, and 8 of T4N, R4E within the US Geological Survey's Deadwood South 7 1/2 minute topographical quadrangle. Situated at elevations between 5,200 and 5,600 feet, the mine is located at the head of Strawberry Creek in mountainous terrain. Disturbed areas of the mine drain to Strawberry Creek, Hoodo Gulch, and Ruby Gulch. Surface water discharge from the mine ultimately reaches Bear Butte Creek near the town of Galena.

2 CLIMATE

Freezing temperatures are experienced from October through May. Temperature extremes vary from summer highs of 100 F to winter lows of -20 F. The frost depth is approximately 48 inches. The average annual snowfall measured at Deadwood is 97.1 inches of snow which is equal to approximately 10 inches of water. The site normally receives a total of between 25 and 30 inches of annual precipitation. Evaporation loss is approximately 18 inches per year.

3 MINE FACILITIES

The site consists of 4 open-pit mine excavations, a 2.2-million cubic-yard heap-leach pad, a 12-million cubic-yard waste-rock dump, topsoil sotckpiles, mine buildings, crusher system, water collection ponds, pump stations, a water-treatment plant, and 260 acres of disturbed lands. The principal mine buildings include an office, assay lab, and process building. Other facilities include electric power distribution lines and transformers, fuel storage and dispensing, ditches and culverts to route stormwater, and fresh water storage and distribution systems.

4 DISCOVERY

Mining activities at Gilt Edge followed shortly after the discovery and development of placer gold deposits on Whitewood Creek. This initial discovery led to the establishment of Deadwood in the summer of 1875. By 1876, prospectors spread out from Deadwood and made numerous discoveries of placer gold and hardrock deposits in the surrounding hills. The Gilt Edge and Dakota Maid claims were located on hardrock deposits while placer gold was found in Strawberry Creek and gold nuggets were found in Ruby Gulch. In the next few years additional discoveries of rich ores were made and many more claims were located. Important discoveries included the Anchor, Hoodo, Sunday, and the Oro Fino claims. The Anchor mine was the start of a famous fortune. Tom Walsh and a partner staked the claim but the partner left for Colorado. Walsh did not like working alone; after developing the mine he sold it in 1878 for $60,000. to some New York men. Walsh went to Colorado to split the profits with his partner. There he purchased the Camp Bird Mine and by 1902 he became one of Colorado's wealthiest men. His daughter Evalyn Walsh McLean, became famous as the owner of the Hope diamond.

5 UNDERGROUND MINING

The early mining activities at Gilt Edge extracted oxidized ores from shallow excavations. Recovery of the gold and silver was inefficient by the simple methods used. Crushing in an arrastre and panning soon gave way to stamp mills and amalgamation with better results. The Sunday mill was erected in Strawberry Gulch and ran from 1880 until 1894 (Waterland, 1991). Other amalgamation mills included the Dakota Maid mill erected in 1884, the Ruby mill erected in 1903, and the Rattlesnake Jack mill erected in 1913. Once the mines penetrated below the water table the simple milling techniques were not effective in processing sulfide ores and the operations ceased.

The cyanidation process was found to be effective on the sulfide ores. The Gilt Edge-Maid cyanide mill operated from 1904 until 1916. In 1935 the price of gold increased from $20. Per ounce to $35. per ounce prompting formation of the Gilt Edge Mines Inc. A new mill cyanide mill was erected and ran from 1937 until it was closed by the War Production Board in 1941.

6 OPEN PIT MINING

Cyprus, Amoco and other firms explored the site leading to identification of a gold deposit. In 1986 detailed engineering was initiated for the Gilt Edge Project. Construstion of the open-pit mine and cyanide heap leaching facilities were initiated in August 1987 under the name Brohm Mining Corporation. There have been a number of successive owners of the Gilt Edge surface mine including Lacana, Min Vin, and Dakota Mining Corporation. Mining of the Dakota Maid and Sunday pits occurred from 1988 until 1992. Waste rock from the two pits and spent

ore was placed into the Ruby Dump. Acid rock drainage from the Ruby Dump was first detected in 1993. The EPA and the State of South Dakota required Brohm to immediatly develop a mitigation plan to solve the acid drainage problem and increase the reclamation bond on the property.

7 SULFIDE DEPOSIT

High gold prices in the late 1980's led to identification of a low-grde sulfide gold deposit containing a resource of 1.7 million ounces. The project would have developed a 600-foot deep pit to extract 90 million tons of rock including 45 million tons of sulfide ore grading 0.038 oz/ton gold. The company spent about $5 million to identify the deposit and conduct detailed engineering studies. Permitting activities brought forth significant concerns regarding the environmental impacts associated with the proposed project. When gold prices declined the project was put on hold.

8 ANCHOR HILL PROJECT

Another gold deposit was found nearby at Anchor Hill. A large-scale mining permit for the Anchor Hill deposit was issued by the State of South Dakota on January 19, 1996. Anchor Hill consisted of Phase I on private land and Phase II on USDA Forest Service land. Mining of Phase I was initiated in May 1996 and was completed by August of 1997. During this same time, small pits were excavated into Langley Peak. The mining permit required the company to increase its reclamation bond from $1. million to $12. million. Having no other sources of funds, the company paid for the bond increase in increments as it mined the deposit. A percentage of the revenues received from gold sales from the Anchor Hill operations were sent to the State. The reclamation bond grew to $6 million by the end of Phase I.

Phase II of the Anchor Hill project required an Environmental Impact Statement. Approval came in November of 1997 when stripping of waste rock was initiated. Phase II would extract 5.07 million tons of ore and 6.34 million tons of waste rock to produce 120,000 troy ounces of gold (BOR, 2000). Additional drilling expanded the reserve to 160,000 troy ounces. In response to a lawsuit filed by the environmental activist organization Earthlaw, the Forest Service withdrew its approval without warning on February 18, 1998 and mining ceased. Dakota Mining Corporation went into negotiations with Earthlaw and the Forest Service. An agreement was reached, however, the company ran out of funds forcing it into bankruptcy. The company had expended all of its funds on water treatment during the negotiations and could not find backing to restart the mining operation. The State of South Dakota took over site operations in 1999.

9 REGIONAL GEOLOGY

The Gilt Edge mine is located in the northern Black Hills in an area intruded by Tertiary age volcanics. The site hosts many rock types and has a complicated geologic structure. The area includes volcanic stocks and laccoliths.

The Black Hills were uplifted by magmatic intrusions in late Cretaceous time about 60 to 65 million years ago. They form a northwest-southeast trending, ovla-shaped dome approximately 125 miles long by 60 miles wide. The core of the dome is composed of Precambrian metasediments with lesser volcanic and granitic rocks. Approximately 50 to 55 million years ago, the volcanics intruded into the Precambrian metasediments and the overlying Paleozoic and Mesozoic sedimentary rocks.

10 STRATIGRAPHY

The oldest rocks exposed in the Gilt Edge area are Precambrian amphibolite, metachert, and quartz-mica schist. Unconformably overlying is the upper Cambrian age Deadwood Formation composed of interbedded sandstone, shale, and carbonates that are locally metamorphosed to hornfels and quartzite by the Tertiary intrusions. The Tertiary volcanics are found in three units. They are hornblende-trachyte porphyry, alkali-trachyte porphyry, and quartz alkali-trachyte porphyry (MacLeod, 1986). Quaternary age sediments are found along creek bottoms and slope cover on the hillsides in the Gilt Edge area.

11 MINERALOGY

Historic undergrounding was conducted on mineralized vein structures and breccia zones. The principal minerals are coarsely crystalline quartz (SiO_2), and fine-grained pyrite (FeS_2) associated with microsocpic gold (Au). Visible gold was found in the Rattlesnake Jack mine, and gold nuggets up to 1.5 ounces in size were taken from Ruby Gulch in the early days of mining. Small amounts of arsenopyrite (AsFeS), chalcopyrite ($CuFeS_2$), malachite ($CuCO_3$), and chrysocolla ($CuSiO_4$) have also been encountered in the mine. Other minerals present include calcite ($CaCO_3$) at Anchor Hill, and Fluorite (CaF_2) in the deep sulfide deposit below the Dakota Maid and Sunday pits. Alterations of pyrite found in the Dakota Maid pit highwalls include copiapite ($Fe^{+2}Fe_4^{+3}(SO_4)_6(OH)_2.20H_2O$), gypsum ($CaSO_4.2H2O$), and other sulfates. A vein of maghemite (Fe_2O_3) is present in the Sunday pit. What appears to be conichalcite ($CaCu(AsO_4)(OH)$) has been discov-

ered in the Southeast Langley pit, however, the material has not been analyzed.

12 ACID ROCK DRAINAGE

Acid drainage sources at the Site include the Ruby waste-rock dump, Sunday pit, Dakota Maid pit, the spent ores on the heap-leach pad, road embankments, and disturbed areas. The underground mine workings also discharge acid drainage. Site characterization studies show that even the "oxidized ore" will eventually turn acid (Robertson Geoconsultants Inc., 2000). The Anchor Hill Pit has areas that are both acid generating and acid consuming. Acid drainage from the site typically has a pH between 1.8 to 3.5. As an example, one sample showed a pH of 2.85 with 240 mg/l Al, 0.51 mg/l As, 68 mg/l Cu, 270 mg/l Fe, and 4,200 mg/l sulfate.

13 WATER TREATMENT

Acid drainage is collected in membrane-lined ponds at the base of the Ruby Dump and in Strawberry Creek. Pumping facilities return the acid water back to the site where it is stored in the Dakota Maid and Sunday Pits. A 250 gpm water treatment plant is continuously operated to prevent releases of acid water from the site. The treatment plant utilizes sodium hydroxide solution as a reagent to form an iron hydroxide precipitate sludge. Heavy metals such as copper and cadmium are absorbed into the precipitate. Approximately 200 gallons/minute of treated water is discharged from the site. The sodium along with sulfate in the discharge water has caused some concern regarding TDS discharges. The EPA is currently evaluating conversion of the treatment plant to a lime reagent system rather than sodium hydroxide. Such a conversion would reduce the TDS in the plant discharge but sludge volume is expected to significantly increase. The water treatment system is assisted in the summer months by operation of evaporative sprayers reducing the inventory of acid water stored on the site.

14 CONCEPTUAL CLOSURE PLAN

The EPA asked the Bureau of Reclamation to prepare a conceptual closure plan for the Site. Work was initiated in November, 1999 and completed in February, 2000. The closure plan identified the major acid sources at the site, provided conceptual designs for closure and reclamation of individual mine features, and proposed a logical sequence and schedule to reclaim the site. Two options were compared, reclaim the Ruby dump in place or remove the entire Ruby dump and backfill the mine pits with the material. The first option was selected as its cost was estimated at $27 million verses $63 million for complete removal of the 20 million tons of waste rock from Ruby Gulch. The State of South Dakota, realizing that the $6. million bond was inadequate to close the Site, decided to ask the EPA to add the site to the National Priorties List and commence removal and remedial actions.

15 RAPID RESPONSE

For a major site, the Gilt Edge mine cleanup been progressing faster than any other Superfund project. The Bureau of Reclamation was asked to design and let a contract for construction of the Ruby Dump cap system as soon as possible. EPA appointed both an On Scene Coordinator to manage the Removal Actions and a Remedial Program Manager to begin planning the Remedial Actions. A number of consultants and in-house experts were called upon to perform various tasks.

Site sampling, drilling and groundwater investigations, site geochemical characterization, public coordination, design, preparation of feasibility studies, and other activities were conducted simultaneously rather than in sequence, shaving a year off the process. This required frequent meetings, conference calls, and site visits to keep everyone updated on the project progress. Most of these studies were completed by the end of the year. The cap design was finalized in September and contracting was started. Construction began in December 2000, the same month that the Feasibility Study was released (CDM Federal Programs Corporation, 2000). Early action Records of Decision were signed in 2001.

16 RUBY CAP DESIGN

Funding for the cap design was received in June, 2000. One problem with the site is a lack of clay-sized particles in the mine wastes. The Ruby waste rock is very porous and seepage at the toe of the dump increases rapidly in response to intense rainfall events. Flows can change from 25 gallons/minute to 200 gallons/minute in a few hours. Such a porous structure requires an impermeable cap. The concept report had identified two viable cap systems, a bentonite clay amended soil, or a high density polyethylene membrane. Final design selected a textured low density polyethylene membrane (LDPE), with a geocomposite drainage layer covered with 4 feet of granular material, 6 inches of topsoil, and vegetation. The 20-million- ton waste-rock dump was composed of many angle-of-repose slopes. A grading plan was established to relocate 1.3 million cubic yards of material in the dump. The proposed cap was subjected to large-scale shear box

testing at full saturation to ensure stability. It was decided that the membrane would be applied to a 3.5 to 1 slope. This required regrading the dump surface. This conservative slope was selected becaue of the wet climate. A major challenge will be holding the 6 inches of topsoil on slopes that are subjected to frequent freeze-thaw cycles and heavy snow loading. A system of drainage ditches will be built into the hillsides to intercept clean water runoff and direct it around Ruby Pond rather than continue to mix the clean water with the acid seepage. A system of drainage ditches will be placed over the cap to shed water to the hillside diversion ditch system. Once installed the cap is expected to reduce the dump seepage from it current range of 25 to 200 gallons per minute to a range of 0 to 20 gallons per minute.

17 INNOVATIVE TECHNOLOGIES

The EPA has also used the Gilt Edge Site as a full scale laboratory to test innovative technologies for use in mine closure. With assistance from EPA's Office of Research and Development in Ohio, two studies are underway.

The first is a $385,000. investigation into carbon reduction technology to treat acid mine drainage. The technology is a patented system provided by Green World Science Inc. of Boise, Idaho. The Anchor Hill Pit was filled with 70 million gallons of acid drainage from the site. Because the pH is too low for the bacteria in the system to work efficiently, it was raised to around 4.5 by adding hydrated lime to the pit lake. Next, approximately 60 tons of methanol and 130 tons of sugar along with some phosphorus are added to the pit. The technology uses chemical additions to promote the action of sulfate reducing bacteria in removing the metals from the water. The chemicals were added in the spring of 2001. After several weeks the pit began bubbling as it released large amounts of carbon dioxide. Metals are precipitating as sulfides on the floor of the pit lake. It is hoped that the reducing action of the bacteria will allow the pit to achieve surface water quality standards so it can be discharged without aditional treatment. Lab results have not yet been received.

The second test also involves the same technology, but rather than place it in an open reservoir of water, acid rock was placed into a lined pit to which the chemicals were added. This in-situ test was also implemented in the spring of 2001 and test results are expected soon.

18 MICROENCAPSULATION

Several organizations have been offering microencapsulation technologies for dealing with acid mine drainage. It was decided to conduct a side by side trial at Gilt Edge. A wooden frame was erected and minature versions of the Ruby Dump were constructed in ten membrane-lined cells. Some of the most acidic waste at the site (pH around 2.0) was placed inside the cells. A sump and drainage pipe leading to a collection bucket captures leachate from the piles. Two piles are untreated and serve as control. Two piles are treated with Keeco's silica microencapsulation chemicals, two piles are treated with the potassium permanganate developed by Dupont and now offered by the University of Nevada, Reno. Two piles are treated with phosphate chemicals called Envirobond, and two piles are simply neutralized with lime. Early indications are that at least some of the coatings appear to be working. The tests will continue for at least another year as longevity of the treatment is important.

19 SUBMARINE EXPLORATION

As part of the continuing site investigations, the Bureau of Reclamation pumped out, removed a dam, and re-opened the King adit and mapped the underground workings. The underground mine penetrates the sides and floor of the Dakota Maid pit and information was needed to plan the pit closure. A local mining company assisted in the underground safety evaluation that had to be completed prior to letting geologists and hydrologists into the adit and connecting workings. The 100 foot deep King shaft posed a bigger challenge. EPA brought in a remote controled submersible vehicle to explore the King shaft and lower mine workings. The device was able to descend the 100-foot shaft, and traverse most of the mine drifts including a square-set stope. A video recording was prepared from the camera feedback from the submersible unit. It was possible to confirm that the mine maps from the 1940's were accurate, and the top of the stope was intact and did not come near the floor of the Dakota Maid pit. The device saved the project a very costly underground exploration had conventional methods been applied. Hundreds of feet of mine workings, some as much as 150 feet away from the shaft were explored.

20 CONCLUSIONS

The Gilt Edge Mine Superfund Site has begun cleanup at an unprecedented rate. By dividing tasks amoung a number of organizations the project has quickly moved into construction. Regarding the Ruby Dump will be completed in 2001, and the capping system will be installed in 2002. This action will greatly reduce acid generation at the site, allowing closure to proceed for other parts of the mine that are currently needed for acid water storage.

The varied geology, wet climate, steep slopes, and presence of multiple surface and underground mine workings forms a complex site that requires careful planning to reclaim. Water treatment requirements will decline as remediation is implemented. While conventional methods such as excavating water diversions, sealing adits and pit floors, grading and backfilling mine wastes, capping, limestone neutralization, topsoil replacement, and revegetation can successfully mitigate the majority of the acid rock drainage problems at Gilt Edge, two difficult problems will remain: 1. Acid stormwater runoff from the mine highwalls. 2. Acid seepage from underground mine workings and backfills. Two emerging technologies, carbon reduction, and microencapsulation may soon emerge as new tools for closing acid mine sites.

REFERENCES

Bureau of Reclamation (2000) Closure Plan for Gilt Edge Mine, Lawrence County, South Dakota. U.S. Department of the Interior, Bureau of Reclamation, Technical Service Center, Denver, Colorado, 53 p.

CDM Federal Programs Corporation (2000) Draft Focused Feasibility Study for Gilt Edge Mine Site, Ruby Dump and Gulch Operable Unit 3 (OU3) Lawrence County, South Dakota. Contract No. 68-W5-0022.

MacLeod, R.J. (1986) The Geology of the Gilt Edge Area, Nothern Black Hills of South Dakota. Master's Thesis in Geology, Submitted to the South Dakota School of Mines and Technology.

Robertson GeoConsultants Inc. (2000) Geochemical Field Reconnaissance Survey of the Gilt Edge Mine, SD. Report 083001/3, Vancouver, BC, 31p.

Waterland, J. K. (1991) The Mines Around & Beyond. Grelind Printing Center, Rapid City, South Dakota, 424 p.

Tailings and Mine Waste '02, © 2002 Swets & Zeitlinger, ISBN 90 5809 353 0

A "no action" alternative that worked

R.B. Meade

Associate Professor, Department of Civil and Environmental Engineering, United States Air Force Academy, CO 80840

ABSTRACT: The Anvil Points site located in Colorado was used for research into oil shale processing from 1947 until about 1984. This paper describes the investigation into the stability of a tailings pile at the Anvil Points Facility, the recommendations made concerning the final disposition of the tailings, and the success of the "no action" alternative thus far.

1 BACKGROUND

The Piceance Basin in western Colorado contains an oil-bearing rock known as the Green River Formation. The oil locked within the rock has potential to free the United States from dependence on foreign oil. The price of crude oil has governed the exploitation of the oil shale. In the early 1970's the economics for development of oil shale seemed favorable and the United States government provided funds to investigate oil shale processing technologies. The Anvil Points Facility in Garfield County, Colorado became an active site for the research into oil shale processing.

The Green River shale is a calcareous rock, more marlstone than shale that contains entrapped hydrocarbon called kerogen. When the oil shale is crushed and heated, the entrapped waxy kerogen melts and the droplets are collected at the base of the heated column. With the oil comes fluid called "retort water" that must be disposed. The retort water is produced at about the same yield as the oil. The retort water is green in color and has a foul smell. The water and oil are separated by gravity and retort water can be collected in ponds and encouraged to evaporate by recirculating the water through a spray nozzle located above the pond surface.

2 THE ANVIL POINTS FACILITY

2.1 *The site*

The Anvil Points Facility (APF) is located a few kilometers west of Rifle Colorado and a few kilometers north of Interstate Route I-70. The terrain is relatively flat in the vicinity of the interstate that runs along the Colorado River and rises steeply onto ridges that terminate at the base of a high plateau,

the Roan Cliffs. Two unusual uplifted points, the Anvil Points, cap the cliff. The ridge containing APF is one of several parallel ridges oriented on a general north-south axis that are bounded by deep ravines that contain low volume streams. These streams are subject to flash flooding. The stream that bounds APF on the east is Sherrard Creek. Sherrard Creek drains south into the Colorado River. A view of the facility looking north along Sherrard Creek toward the Anvil Points is shown as Figure 1.

The ridge and ravines are formed from layered sedimentary rocks known as the Wasatch Formation. The plateau was formed by the Green River Formation. The shale was extracted from mines located high up on the sides of the plateau. The mines had some openings visible from APF. Oil shale was trucked down to the APF processing site.

APF consisted of a retort facility, some laboratory buildings and shale "tailings" piles. Casting or shoving the tailings over the cast edge of the ridge formed the pile. The installation was a research site and oil shale from various places in the world was brought to the site for processing. During the oil crisis of the 1970's the pace of research at the site increased sharply. Many tons of local Green River oil shale were processed at the site using the Parahoe process.

The end of the oil crisis in the early 1980's caused the processing operations at the site to cease. At that time, the facility was operated under the Department of Energy (DOE). DOE wanted to evaluate the stability of the tailings pile and assess its effect on the local environment. Among the issues to be resolved were the geomechanical stability of pile and options for treating or moving the pile if necessary. This case study discusses the geomechanical stability only.

I was the project engineer for the study. This paper is my recollection of the project. The opinions

Figure 1. Anvil points facility looking north.

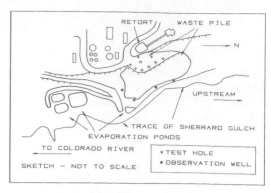

Figure 2. Layout of the processing plant and pile.

Figure 3. Composite photo of pile (1982).

expressed in this paper are personal and may not represent the views of DOE or the U.S. Army Corps of Engineers (COE), my employer at the time of the study.

2.2 *The pile*

The tailings consist of processed shale called spent shale, and unprocessed shale in the form of fines called raw shale. The raw shale fines were produced by the crushing operation needed for the shale processing. The fines were unsuitable for use in the retort methods of oil production.

Spent shale and raw shale were disposed separately. The tailings pile is the product of many years of dumping. The pile was placed during three periods of research (Garland & al.1982). From 1947 to 1955 about 543 kN (61 tons) of spent shale and 400 kN (45 tons) of raw shale were dumped in the Sherrard drainage basin. From 1964 to 1968 another 400 kN (45 tons) of raw shale and 158 kN (18 tons) of spent shale were added. From 1973 to 1978 substantial quantities of local oil shale were retorted at the site using the Parahoe process. About 490 kN (55 tons) of raw shale and 1059 kN (119 tons) of spent shale were disposed above Sherrard Creek.

The spent shale pile abutted and somewhat overlapped a raw shale pile located immediately north of the spent shale. The relative position of these piles with respect to the retort and Sherrard Creek is shown in Figure 2. Both of these piles were draped on the eastern slope of the ridge with Sherrard Creek flowing in the ravine below. A composite photo of the pile taken in 1982 is shown in Figure 3.

The Sherrard Creek streambed had been relocated in the vicinity of the pile. The purpose of the relocation of the stream was to provide space for a bench. The bench provided room to construct several small ponds to hold process water. The bench also served as a stable platform. A low berm of tailings had been constructed on the bench at the toe of the pile. The original stream channel ran beneath the bench.

3 THE PLAN

3.1 *The office study*

The COE began a study for DOE to assess the state of the pile and make recommendations regarding the future treatment or disposal of the pile. The study began in 1981 with the examination of existing information about the tailings. The mechanical properties of spent shale produced by the Tosco and Parahoe processes had been previously examined by the COE (Townsend & Peterson, 1979).

One of the anticipated methods of disposal was movement of the spent shale to a fill disposal site. In addition to storage for spent shale, raw shale could be disposed as a fill. The COE had begun a study of design methodology for the creation of surface storage fills in the vicinity immediately west of the Anvil Points Facility (Strohm & Krinitzsky, 1983).

The design methodology project and the tailings pile stability project were undertaken concurrently by the COE. Each of these studies reviewed the existing body of information (as of 1981) concerning oil shale processing. Oil shale is a highly variable material whose processing strategy depends on both the yield of oil expected and the in situ characteristics of the shale. Two popular oil-recovery processes were the Tosco process and the Parahoe Process. The character of the spent shale is strongly influenced by the oil-recovery process used.

The primary difference in the Tosco and Parahoe process was the particle size of the oil shale. The Tosco process used oil shale crushed to a sand size particle and the Parahoe process used a fine gravel size. The spent shale in each process was a friable material with a platey particle shape and was black in color. The tailings pile consisted of mostly spent shale from the Parahoe process but the research activities of the facility caused some quantities of foreign shale to be dumped in the pile. The author does not know the oil extraction process used on the foreign shale. A priority in the field investigation was to examine the diversity of spent shale and to determine the nature and effects of segregation produced by differing raw materials, operating practices, and weathering.

3.1.1 *A fire in the raw shale fines pile*

In spring of 1979, oil was seen seeping from the base of the raw shale pile. A fire had developed within the raw shale causing oil to be released. TOSCO Corporation (1982) discussed the origin of the fire in a report to the Bureau of Mines. No conclusions were presented for the cause of the fire. Spontaneous combustion within the raw shale was considered unlikely. The report suggested that spent shale may have been inadvertently dumped onto the raw shale. The Tosco Report noted that at times the boundaries between the two piles could not be easily distinguished. An infrared photograph taken at sunrise in June 1979 showed hot spots at isolated locations in both the raw shale fines pile and the spent shale pile. The fire was extinguished using water sometime after the summer of 1979.

3.1.2 *First impressions*

I accompanied Mr. W. Strohm (COE) on a site visit in 1981 to collect some samples of the Wasatch Formation to evaluate the stability of the native materials would underlying any proposed disposal fill. The pile stood out as a lumpy black mantle shrouding the eastern-facing slope of the facility. During the visit steamy vapors were visible above the crest of the pile in the early morning hours. The surface materials on the pile were at the state of limiting equilibrium. A gust of wind would send a thin sheet of spent shale sliding down the pile surface. Such movement was very localized. The movement was nearly invisible due the uniform black color of the pile. The movement was first noticed when a barely audible tinkling-ringing sound was heard that accompanied the sliding. Hushed sounds were noticeable due to the lack of industrial activity at this remote location.

3.2 *Slope stability*

The primary geomechanical issue was the stability of the pile. Assessment of slope stability required obtaining a sound understanding of the geometry of the tailings pile and the engineering properties of the tailings. Borings were planned for the crest region of the pile, the toe of the pile, surface samples from the sloping portions of the pile and a topographic survey of the pile to prepare cross sections for analysis.

Previous studies had examined the geotechnical properties of the spent shale (Townsend & Peterson, 1979). These studies had used tailings recovered directly from the APF retort and shipped in barrels to the COE lab for testing. In the pile the tailings were end-dumped or shoved over the edge of the slope so considerable size segregation and stratification parallel to the slope were expected. To capture these features a variety of sampling techniques were used.

The pile was about 47-m high, about 180-m long and rested at the slope angle of about 38 degrees. The crest of the pile was level with the APF retort about elevation 1817-m. The toe of the pile was about elevation 1770-m on a bench above Sherrard Creek.

4 THE FIELD INVESTIGATION

4.1 *The borings*

Borings performed through the crest and toe of the tailings pile included Standard Penetration Tests (SPT) that obtain 38.1 mm (1.5-inch) diameter sample taken with a 50.8mm (2-in) nominal diameter split spoon, 76.2 mm (3-in) nominal diameter samples, and samples using a "continuous sampler." The continuous sampler is a tube that is inserted inside a lead hollow-stemmed auger. A continuous column of material is obtained. Bag samples of materials on the slope were also taken.

The field investigation began in November 1982. The first two borings performed were those on the toe of the pile and then borings were made on the crest of the pile. Three layers of materials were present at the toe.

Spent shale extended from the surface (elevation 1766.3-m) to a depth of 1.5-m (elevation 1764.8-m). Brown moist dense silty sand extended from 1.5-m to about 5.5-m. The sand had been worked to form a bench above Sherrard Creek. Some spent shale was found mixed with the sand at a depth of 4.3-m. The interface between the sand and the native materials was inferred at a depth of about 5.5-m. The native materials were weathered Wasatch Formation. From a depth of 8.2- m to 11.3-m the rock was cored using air to remove cuttings. An observation well was set at the shale-sand interface on completion of the boring. The boring was dry at completion of drilling.

A second boring was completed at the toe of the pile. This boring was located on a low berm of spent shale. Again three strata were encountered. Spent shale was present from the surface (elevation 1774.5-m) to a depth of 3.6-m (elevation 1770.9-m). Below the shale was clayey sand with some gravel and occasional pieces of spent shale overlying the Wasatch Formation. This boring was dry at completion. These two borings concluded the first day's drilling.

The next morning the drill rig was moved to the crest of the pile (elevation 1815-m). The first boring at the crest used a continuous sampler to recover a core of the spent shale. This boring was planned to extend through the spent shale and penetrate the native slope materials, but it was not to be. The shale brought up in the sampler was hot, too hot to handle. Everything had to be re-thought.

The pile rarely retained a cap of snow throughout the winter. The black color of the tailings and the intense ultraviolet radiation seemed to fully explain the lack of snow cover.

The first boring at the crest had used the continuous sampler to obtain a core of material. Two 1.5-m (5-ft) core runs were made. The continuous sampler was steaming when it was raised from the hole. The continuous sampler about 60 percent recovery. It was not clear if the material was initially loose and became packed in the sampler or some of the sample had fallen during removal of the sampler from the hole.

The second continuous sample was hotter than the first. It also had about 60 percent recovery. A third continuous sampler run was made and the material seemed hotter yet. It was decided to drive an SPT sample at the 4.57-m depth. The blow count was 6. When the split-spoon was pulled from the hole, the shoe of the split-spoon had become blue with heat instead of the normal metallic gray. The split-spoon itself was too hot to be handled with gloves. The drill crew and the field engineer had no thermometer available, so we could only estimate the temperature.

Eventually, our testing revealed temperatures in excess of 200°C at a depth of 3-m below the surface of the pile. We pulled out the augers and watched them steam as we placed them against the drill rig.

The drill crew was a contract crew who had expected a routine investigation. Everyone was reluctant to handle the hot tools. Boring #3 was terminated at a depth of 5.5-m.

That evening, I purchased a candy thermometer from a grocery store to make some temperature readings in the tailings.

4.2 Salvaging the plan

The next morning the candy thermometer was used to obtain the temperature in the borings. The crew tried to complete more borings into the heart of the pile. Perhaps the hot tailings were a local condition. The intent was to map the hot materials within the pile. Four more borings were attempted through the pile. The work proceeded slowly because of the need to cool the tools. It was not considered appropriate to use water to cool the tools for fear of ruining the samples. The tools were allowed to air-cool. SPT testing was alternated with the continuous sampler in borings #4 and #5. Borings #6 and #7 used SPT testing at 1.5-m (5-ft) interval. The borings were terminated when the heat of the samples and tools became oppressive. The borings ranged in depth from 7.8-m to 13.4-m. Only boring #6 penetrated into the underlying Wasatch Formation. The boring had a surface elevation of 1817.2-m and the interface between the spent shale and the native soils occurred at a depth of 10.5-m (elevation 1806.7-m).

I returned to COE Geotechnical Laboratory in Vicksburg, Mississippi with a number of jar samples of the tailings and began to re-assess the situation.

The tailings were undergoing some sort of smoldering combustion. The fine combustion product was a powdery gray ash similar to the remains of a charcoal fire. The powder was loose and appeared to have a very high void ratio because it seemed to have extremely low weight. It was feared that the use of a drive sample such as the SPT would not yield a representative sample. The powder would compact in the spoon or fall out unless either a trap or flap valve was used. Moreover, some method of estimating the strength of the tailings was needed. A cone penetrometer could provide strength data.

The extreme heat in the pile would destroy a conventional electric cone penetrometer. The strain gages will become inoperative at temperatures in excess of 50°C. A custom electric cone was built to survive the high temperature. It was desired to measure temperature as well as the conventional cone parameters of tip resistance and side friction. The temperatures had seemed incredible and some reliable system of measurement was needed.

The interior of the custom cone housed a thermocouple and left little room to add gages to measure both side friction and tip resistance. Some compromise was necessary to create a cone for this situation. The tailings were known to be loose and cohesionless from the November 1982 fieldwork. Material of this type has little side friction. Therefore, the custom cone was designed to measure tip resistance and temperature.

The strain gage system to measure tip resistance was built of high temperature-resistant gages bonded with a high temperature-resistant epoxy. This system could tolerate heat that measured 240°C. A Teflon-coated 4-conductor cable was used in the hope that the cable and its insulation could resist melting and sticking to the interior surfaces of the drill rods. A thermocouple wire with asbestos, fiberglass and silicon insulation was used to measure temperature. The thermocouple was measured with a pyrometer capable of measuring temperatures as high as 650°C. The temperature of the drill rods, the pile surfaces, and recovered samples were read using a digital heat probe thermometer.

The electric cone was manufactured at the Waterways Experiment Station and calibrated to measure load in pounds based on the tip resistance strain gages within the cone. The cone had a 10-sq. centimeter projected area and a 60-degree cone tip. The thermocouple sensor was placed against the inside of the friction sleeve portion of the cone.

The custom cone was compared to a conventional electric cone used by the Bureau of Reclamation. Both cones were used to penetrate medium stiff soil on the grounds of the Federal Center in Denver, Colorado. The results were encouraging and the custom cone was taken to the Anvil Points site where it was operated using the Bureau of Reclamation cone penetrometer rig. The rig was designed to push either a mechanical cone or an electrical cone. This ability proved to be vital.

At the site the custom cone experienced some difficulty in providing a consistent measurement when it seemed that moisture in the form of hot vapors affected the strain gage. The electric cone functioned reasonably well until the strain gages that measured tip resistance were destroyed by heat in excess of 260°C. The temperature measuring capability of the cone was never impaired. The mechanical cone was used to measure both side friction and tip resistance.

4.3 Borings again

Finally, SPT testing and sampling was conducted at selected locations within the pile to supplement the cone data. All of the sampling and testing was compiled to produce an understanding of the range of materials present and form a general notion of the stability of the pile.

5 THE NATURE OF THE PILE

The spent shale was loose silty sand composed of friable particles. The interior of the pile was hot due to some type of smoldering combustion. The typical

SPT value was less than 10. The cone penetrometer data was indicative of meta-stable sand.

A representative profile from the crest of the pile is shown as Figure 4. The data is a composite of all of the sampling and testing for boring #6. Hotter and looser materials were recorded at other locations. Boring #6 was selected because it was the only boring to penetrate the full depth of the pile. Boring #6 was located near the center of mass of the pile.

The tip resistance is shown in Figure 4 as the jagged vertical profile. Two columns of temperature data are shown on the right of the figure. The temperature shown in the extreme right column was the temperature measured using the custom cone. The column containing data in the form 7 − 70 shows the SPT blow count as the number preceding the hyphen and the temperature measured within the split spoon sample as the number after the hyphen. For the example shown, 7 is the SPT blow count and 70°C was the temperature in that split spoon sample.

The heat generated by the combustion within the pile affected the temperatures in the Wasatch Formation and native slope materials that underlay the pile. Ground water temperature in an observation well installed in boring #8 at the base of the pile on the bench above Sherrard Creek registered ground water temperatures of 22° C. This temperature is about 5.6° C warmer than temperatures recorded in nearby wells that also were place in the bench above Sherrard Creek.

Conventional slope stability analysis required a confident estimate of the pile geometry, material parameters such as unit weight, and strength properties. Each attempt at investigating the pile painted a more complex model. The principal difficulty was the prognosis for the combustion. Combustion reduced both body forces acting within the pile and the strength of the spent shale. The combustion produced a moisture flux within the pile.

The pile had made the native soils warm and induced evaporation of ground water. The water vapor moves up through the pile by convection. Near the surface of the pile the water vapor either condenses on the particles or evaporates into the atmosphere.

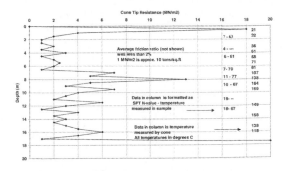

Figure 4. Temperature, cone, & SPT data boring #6.

431

Vapor was constantly rising from the pile. This steam was particularly noticeable in the early morning hours and in cold weather. Consider the well placed in boring #6 located at the top rear of the pile. The bottom of the standpipe well was always dry. The top portion of standpipe was always wet with condensate. The drilling tools raised from hot holes had water running down the surface of the drill rods while the sample never displayed any free water.

The hot pile seemed to wick up water from the ground water table and this water vapor was often observed at the top of the borings, both coming from open holes and those containing a standpipe. Steam could be seen issuing from cracks at the crest of the pile when the air temperature was cool. The absence of free water within the pile indicated that surface water does not move downward through the pile to the groundwater.

The combustion was causing a continuous change in spent shale properties and acting as the engine in a water flux phenomenon in the pile. Conventional slope stability analysis was not practical. It was clear that the pile was currently stable at the angle of repose of the tailings. This angle of repose was approximately 38 degrees. The cone data indicated that the interior of the pile was meta-stable. Therefore, some triggering agent such as an influx of water or severe winds might create instability in the pile. Barring some triggering action the pile may remain intact.

The pile was an unpleasant combination of troubling conditions. The pile interior contained meta-stable zones. The combustion created the obvious effects of heat and produced decomposition of the tailings. The vapors might contain volatile compounds that could be harmful if breathed. The ash qualities of the decomposed tailings would provide a fugitive dust problem if the surface armor of the pile were disturbed. Still, recommendations had to be made.

6 THE RECOMMENDATIONS

The pile was stable at the time the recommendations were proposed. Any change to the pile could create hazards for the personnel working the pile. The pile might become destabilized by natural causes. The pile certainly would be destabilized by any attempts to alter the geometry or extinguish the combustion.

6.1 The alternatives

The alternatives presented to DOE were:
1 no action (monitor the pile)
2 treat the pile (put out the combustion and stabilize the pile)
3 move the pile (deal with the combustion during the move).

Risks were involved in each alternative. With no action the pile may deflate due to combustion. Flash flooding in Sherrard Creek could threaten the toe of the pile. High winds could scale off the armor and permit erosion.

Each of the other alternatives required extinguishing the ongoing combustion. Using water could create potentially undesirable leachate and erode the pile. The process of using water to extinguish the combustion might require washing the tailings into the old ponds at the toe of the slope. Then, the material would need to be dried back and compacted in a disposal fill in some location other than adjacent to the present location above Sherrard Creek.

The report by Strohm and Krinitizsky discussed numerous issues regarding disposal fills. Many questions needed resolution. For example, what was the effect of retort water in the compaction process? What locations for the proposed fill were most appropriate? How should the material be moved (conveyor, truck)? Apparently, DOE opted to manage a stable pile with an uncertain future rather than run the risks associated with the other alternatives.

6.2 The no-action alternative works

The pile has been stable (no movement or collapse) for over 17 years. The other alternatives were more costly and had more severe risks. The situation seems to be no worse for wear after 17 years. Still, the pile has no certain destiny. The nature of the tailings defied traditional analysis methods. The pile may be stable indefinitely. If predictability is a requirement, then the no action alternative is insufficient. If monitoring (with an action plan is feasible) then the no action alternative has a future.

7 LESSONS LEARNED

In 1984 when I made the three recommendations I believed that the pile would have to be moved. The 'no action' alternative was akin to the null set in my mind. I included it for completeness. To the author's surprise the 'no action' alternative, now 17 years later, seems to have been the best choice.

Meta-stable does not equate to unstable. Meta-stable materials can be in equilibrium indefinitely in the absence of triggering agents. A follow-on study of the effects of combustion completed after my report seemed to confirm my belief that the pile would move towards instability (Lavery, 1985). Somehow the combustion must have ceased allowing the pile to remain meta-stable.

Luck seems to have played a significant role. Extreme events such as floods, wind storms, and earthquakes could have made the "no action" alternative a disaster. Such "no action" alternatives always run risks. The magnitudes of "no action" risks are difficult to estimate; as are the consequences should they occur.

The wisdom of the "no action" alternative is clear in hindsight. The other alternatives remain available today. In fact, the no action alternative could be continued. The passage of time has kept all the alternatives viable.

REFERENCES

Garland, T.R, Wildung, R.E, & Zachara, J.M., 1982. "A Case Study of the Effects of Oil Shale Operations on Water Quality I, History of Operations and Hydrological Controls," PNL-SA-9716, Pacific Northwest Laboratory.

Lavery, P.L., 1985. "Determination of Particle Deterioration and Collapse Susceptibility of Retorted Oil Shale Piles Caused by Potential Internal Combustion," Final Report DOE/LC/11045-2072.

Meade, R.B., 1984. "Development of a Geotechnical Stabilization Plan for the Existing Retorted Shale Pile, Anvil Points, Colorado" Final Report DOE/LC/10870-1568.

Strohm, W.E., and Krinitzsky, E.L., 1983. "Methodology and Concepts for the Design of Su face Storage Fills at the Anvil Points Oil Shale Retort Facility, Final Rep ort DOE/PC/42202-1542.

TOSCO Corporation 1982. "Data Summary of Anvil Points Raw Shale Waste Pile Combustion," Interim Report, Contract No. J02755001, Bureau of Mines, U.S. Department of the Interior, Washington, D.C.

Townsend, F.C. & R.W. Peterson. 1979. "Geotechnical Properties of Oil Shale Retorted by the Parahoe and TOSCO Processes,"Final Report, U.S.Army Engineer Waterways Experiment Station, Technical Report GL-79-22, Vicksburg, MS, BuMines OFR 47-81 Contract No. HO262064.

Mercury spill response, clean-up, and assessment, Choropampa, Peru

K.J. Esposito, T.A. Shepherd & M.C. Meyer
Shepherd Miller, Inc., Fort Collins, Colorado USA

N.R. Cotts
Minera Yanacocha S.R.L., Cajamarca, Peru

ABSTRACT: In June 2000, mercury being transported from mine to market was accidentally spilled over 41 Km of a highway in rural Peru. The mercury was a byproduct of gold produced by Minera Yanacocha, S.R.L. (MYSRL), Peru's largest gold producer, at its mine complex near Cajamarca, in the Andean highlands of northwestern Peru. In response to the spill, MYSRL marshaled a massive and comprehensive emergency response program that addressed all aspects of the environmental, human health, and social impacts that resulted from the spill. This paper describes the technical aspects of the impacts of the mercury spill, the emergency response taken, and the results achieved. This mercury spill was a uniquely challenging event from which much was learned. It is to the credit of all involved, MYSRL, the government officials and agencies, their respective consultants and advisors, and the local citizens, that the situation was so effectively and efficiently addressed.

1 INTRODUCTION

This case history summarizes the activities undertaken by Minera Yanacocha S.R.L. (MYSRL) in response to the mercury spill incident that occurred near the towns of San Juan, Choropampa, and Magdalena, Peru, on June 2, 2000.

Located 375 miles north of Lima in the Peruvian Andes, the MYSRL mine complex is situated at an elevation of 14,000 feet above sea level. The mine site is 30 miles north of the city of Cajamarca. MYSRL is the largest gold producer in Latin America. Currently, Minera Yanacocha consists of four open pit mines--Carachugo, Maqui Maqui, San Jose Sur and Cerro Yanacocha. Mercury is produced as a by-product of gold production.

1.1 *Spill history*

On June 1, 2000, a transport truck owned and operated by RANSA, a contract carrier for MYSRL, was loaded at the mine with nine small containers of elemental mercury (Hg), each containing approximately 180-200 kg of mercury. In addition, ten empty chlorine gas cylinders were loaded onto the forward portion of the semi-trailer. As the truck was preparing to leave the minesite for Lima, the MYSRL warehouse foreman noticed that the RANSA driver appeared to be ill, and directed that he be examined at the Yanacocha medical clinic. The clinic advised the driver that he was unfit to drive the truck, and the driver returned to Cajamarca for the night. Early the next morning, June 2, 2000,

the driver returned to the minesite with his supervisor to pick up the truck and begin the journey to Lima. The driver left the minesite, traveled to Cajamarca (where he stopped at the local RANSA office), and then continued the trip along the road between Cajamarca and the Pan American Highway. Approximately 50 kilometers from Cajamarca along the road to the Pan American Highway (at kilometer marker 155, Km 155), one of the chlorine gas cylinders became dislodged, and, before it fell off of the trailer, apparently disrupted one or two of the mercury containers from their original positions. Elemental mercury later was detected in the area of Km 155, near the point where the chlorine gas cylinder fell off the trailer, and was also detected at 25 additional points along the highway between Km 155 and the village of Magdalena (Km 114), where the truck was finally parked on the evening of June 2. It was later determined that approximately 151 kg of mercury was lost along this 41 km stretch of highway. MYSRL personnel first received word of the spill on the morning of June 3, when the mine manager then on duty received a call from an acquaintance who lives in the town of Choropampa, reporting that a substance appearing to be mercury was being collected on the highway by local citizens. The manager responded by immediately sending personnel to the area to evaluate the situation. After talking to a number of local citizens, the MYSRL personnel concluded that an unknown amount of mercury had been spilled and that most of the spilled mercury in Choropampa had been collected local citizens from the highway.

At this point, a second response team from MYSRL was dispatched to Choropampa, to locate and clean up any visible mercury remaining on the highway. On its way to Choropampa, this team encountered the RANSA driver and his supervisor, who were attempting to retrieve the chlorine gas cylinder that had fallen from the truck, and who reported that they were unaware of any mercury spill. The second response team then encountered the initial MYSRL personnel, who reported that the mercury in Choropampa appeared to have been collected by the local people. The second response team then proceeded to Magdalena, where the truck was parked. The team noted that the mercury containers on the trailer were in disarray, and at least one container was open. At this point the police in Magdalena had sealed off the truck from public access. Following the arrival of a district attorney from Cajamarca, the MYSRL response team was allowed to clean the truck and the surrounding area. The response team returned the collected mercury to the minesite.

In cooperation with MYSRL, the authorities in Magdalena sent an ambulance with a loud speaker to Choropampa, to warn the local people of the toxic nature of mercury and to encourage the people to return any collected mercury to the local medical post. By the evening of June 3, a third MYSRL response team had been dispatched to attempt to recover mercury from local citizens. This effort met with little success. At this time, MYSRL and RANSA personnel jointly prepared a press release, for publication on Monday morning, June 5, warning of the toxicity of mercury and asking that it be returned. It was jointly decided that this press release should be issued by RANSA, rather than MYSRL, so as to dampen speculation that the mercury contained gold or otherwise held great value. For the next few days, MYSRL, RANSA and local authorities focused their efforts on persuading local citizens to return mercury that had been collected. On Friday, June 9, MYSRL was informed by public health authorities in Cajamarca that mercury-related health problems were being reported by local citizens. At this point, MYSRL mobilized a comprehensive crisis management team to manage the situation. Teams were then organized to marshal and coordinate the resources and expertise needed to manage each aspect of the situation.

1.2 Response activities

The response activities undertaken by MYSRL can be grouped into the following categories, which are discussed in greater detail in later sections of this document:

Public Health: Immediately after becoming aware that health issues were associated with the spill event, MYSRL initiated an intense program to assist the public health authorities in identifying and treating affected individuals. Local individuals came in contact with the spilled mercury as they collected, stored, or utilized the mercury in their homes. MYSRL undertook a search to locate and retain both local medical doctors and world-renowned experts in the field of mercury toxicity who were willing to help. In addition, MYSRL provided assistance to the Cajamarca Regional Hospital, and to local medical clinics, in the form of medicines, nurses, ambulances, laboratory assistance, database compilation, support personnel and supplies. MYSRL also initiated long-term health monitoring and education programs.

Home Remediation: MYSRL worked closely with the Peruvian Directorate General of Environmental Health (DIGESA) and the Peruvian Civil Defense Authority to identify which homes had unsafe levels of mercury present in the indoor air. When unsafe concentrations of mercury were identified, the residents of these homes were temporarily relocated in MYSRL-provided housing until their homes could be cleaned to levels that were safe for long-term occupancy. Following the identification of contaminated homes and the delineation of highly contaminated areas within each home, a remediation program was initiated to ensure that each home was safe for long-term occupancy.

Identification of Mercury Spill Locations: The identification of locations where mercury was spilled from the truck was conducted in two distinct phases. Initially, visible mercury was located with a thorough program of walking the entire 41 km and visually searching for mercury droplets in the road or roadside. The visible mercury inspection identified 16 locations along the highway were MYSRL implemented immediate efforts to collect all of the visible mercury. After the visible mercury was located, a second phase was conducted including a soil sampling and analysis program to find areas where there was elevated mercury in the roadside soils that could not be identified visually.

Road and Roadside Remediation: Mercury was collected and removed from road and roadside areas which were identified as contaminated. Mercury or mercury contaminated soils were collected by sweeping, vacuuming, and/or excavation of the contaminated area. All of the excavated material was transported to, and isolated in, a heap leach facility at the MYSRL Mine Complex. The areas of asphalt contaminated by spilled mercury were either excavated and replaced or repaved with a new layer of asphalt, effectively entombing the mercury in-place.

Environmental Monitoring: A comprehensive plan for environmental monitoring was also initiated to determine if mercury from the spill was being mobilized into the environment. The environmental monitoring program included the collection of water, sediment, and ecological samples (plants, soils, insects, and fish). Environmental samples were collected in streams and rivers near the known spill sites where the highest likelihood of contamination would be found. In addition, environmental samples were also collected at outlying locations in the drainage basin to gain an understanding of the natural variability and natural background concentrations of mercury to be expected in the area of the spill.

Mercury Mass. Balance: The development of a mercury mass balance has been extremely important to provide an accurate evaluation of the fate of the mercury that was spilled. A mercury mass balance was initially developed at the outset of the project and was revised as more data and information became available. Iterative revisions of the mercury mass balances, provided an increasingly accurate account of mercury potentially remaining in the environment or in possession of residents.

Human Health and Ecological Risk Assessments: The human and ecological risk assessments are still in process and are expected to continue into 2002 to fully assess the potential for long-term risk. Sampling in support of the human and ecological risk assessments was implemented in August of 2000. Both terrestrial (soil, plant, and insect) and aquatic (benthic invertebrate and fish) samples were collected to: 1) document baseline mercury conditions in the overall spill area prior to any potential mercury migration, 2) provide data for calculation of site-specific mercury transfer factors between ecological components, and 3) to allow for monitoring of mercury dynamics over time. Follow-up sampling was conducted in the fall of 2001 at essentially the same locations in order to evaluate if there was mercury movement in the environment, and to assess the potential for long-term risk. In addition to these efforts MYSRL has provided technical assistance to government agencies in understanding the risk assessment process and helping these agencies explain the potential risk and the risk assessment process to the local communities.

2 PUBLIC HEALTH AND MEDICAL ASSISTANCE

The protection of public health was, and continues to be, the primary objective of MYSRL's response to the mercury spill. MYSRL undertook a range of public health and medical assistance programs and actions, in cooperation with local and national health authorities and medical providers. Many of these actions were designed to address the immediate needs of the affected individuals and communities. Other actions were designed to provide on-going protection and prevention.

Initial efforts following the mercury spill focused on locating spilled mercury that had been collected by local individuals, and warning the public of the dangers associated with mercury exposure. It became apparent soon after the spill that a significant amount of mercury had been collected by local residents and taken into their homes. Elemental mercury can cause adverse health effects when it vaporizes and is inhaled by individuals. Many of the homes in the affected communities are poorly ventilated and can become quite hot during the afternoon increasing mercury volatilization. Additionally, a few individuals reportedly boiled mercury in their homes, mistakenly hoping to recover gold from it (in fact, the mercury contained no gold).

The focus of the initial medical response occurred in Choropampa, where the first cases of intoxication became known and patients with symptoms were first identified. The first step was to diagnose and begin treatment of those individuals who had been exposed to mercury vapors and who were experiencing symptoms of mercury intoxication. Individuals complaining of symptoms were processed through the local medical clinics and, in some cases, transported to the Cajamarca Regional Hospital for treatment. Others with lesser symptoms were treated on an outpatient basis from the local medical clinics. MYSRL immediately provided ambulances to assist with the transport of patients to Cajamarca, and provided nurses to assist both in the local medical clinics and in the Regional Hospital. MYSRL engaged a physician from Lima full time between June and September 2000, interacting with local doctors and assisting in the diagnosis and treatment of patients. The doctor worked with outside experts to identify and evaluate exposed persons, as well as develop treatments for each category of patient. The medical doctor also compiled and maintained a comprehensive patient database, which provided valuable data about the treatment and recovery of affected individuals. MYSRL obtained and shipped to Cajamarca the drugs needed to provide the most up-to date treatment available.

Because it was evident that exposure had occurred primarily in homes, relocation of exposed individuals from those homes, along with other family members, became a top priority. MYSRL arranged for the lodging for family members of those admitted to the Cajamarca Regional Hospital. MYSRL, in cooperation with DIGESA, facilitated a program of public health screening based on urine analysis of the potentially exposed population. Voluntary urine analysis of the general population in the Choropampa area was undertaken at the end of June to screen for mercury exposure. This program was designed to identify all exposed individuals, even those without symptoms. Voluntary urine analysis screening for the general population was conducted on two additional occasions, in early July in Choropampa and in September in San Juan. In addition, urine analyses were used throughout the summer to screen potential patients and assess exposure of individuals and families, and will continue to be a primary diagnostic tool to assess exposure in the future. MYSRL commissioned the Centre De Informacion Control Toxicologica (CICOTOX) to set up a urine and blood analysis laboratory in Cajamarca, using state-of-the art equipment provided by MYSRL from its mine site laboratory. This laboratory operated between early July and mid-October, and provided quick turn-around analysis for diagnostic screening of potential exposure cases. All patients treated for mercury intoxication at the Cajamarca Regional Hospital and in village medical clinics have successfully completed their course of treatment.

2.1 Long-term human health monitoring and case management

Based on health monitoring and treatment results, MYSRL's medical experts are confident that there are no long-term health issues associated with exposure to elemental mercury for a short duration, particularly at the relatively low levels seen in this case. Once exposure to elemental mercury has ended, health effects from that exposure also are expected to end in a matter of weeks. Experts from the Peruvian Ministry of Health have agreed with this assessment. Thus, no future long-term or latent health effects are anticipated among those individuals exposed to mercury from this incident.

With respect to future exposures, the environmental and public health data collected by MYSRL shows that remediation of the spilled mercury has been successful in lowering potential outdoor and indoor exposure levels to internationally accepted safe levels. Thus, future exposure to humans is unlikely, unless individuals still have mercury in their possession and re-expose themselves or others in the future. Should future exposures related to this incident be identified, however, MYSRL will provide appropriate assistance to address the specific situation. MYSRL believes that the local and regional medical agencies are equipped with the knowledge and training to adequately address future problems that may develop. No further general public health screening is planned. Medical monitoring will continue for a period of two years for individuals who experienced relatively higher levels of exposure, women who were pregnant while exposed, and young children who were exposed. While no long-term health impacts are expected for any of these individuals, this continued periodic monitoring will provide a level of comfort for those perceived within the community as having higher levels of risk. MYSRL also provided training and on-call support to local and regional health care professionals with respect to long-term mercury exposure management. To offer additional reassurance to those exposed and to their families, MYSRL has also purchased 5 years of medical insurance for each of the exposed individuals, covering any future medical treatment determined to be necessary as a result of exposure to mercury.

3 REMEDIATION ACTIVITIES

This section summarizes the remediation activities that occurred in both affected homes and along the road. The objective of these remedial activities was to restore indoor and outdoor environments to conditions that posed no short-term or long-term human health or environmental risk. Data collected throughout the remediation process show that these efforts have been successful.

3.1 Homes

A key concern of the MYSRL remediation team was to identify those homes where mercury collected from the spill sites was brought into the home and stored (or in some cases spilled). Once such homes were identified, efforts were undertaken to prioritize and schedule those homes for cleanup, and, to clean, replace, or otherwise eliminate or recover mercury from these contaminated houses, including contaminated furnishings, personal effects, and stored food. The following sections describes each of these activities, as well as the verification procedures used to assess the success of the clean-up effort.

3.1.1 Assessment of indoor air mercury vapor levels
MYSRL developed a cooperative sampling program with DIGESA to conduct indoor air monitoring surveys designed to identify all homes that required potential evacuation and remediation. MYSRL established appropriate protective indoor air clean-up standards (Table 1) based on an extensive review of worldwide public health exposure targets and standards. These standards are consistent with U.S. Environmental Protection Agency standards for long-term habitability.

Table 1. Indoor air mercury concentration action levels

Action Level	Air concentration (mg/m³)	Recommended Action
Level 1	> 3000 ng/m³	Immediate evacuation is required. Cleaning is required.
Level 2	Between 1000 and 3000 ng/m³	Evacuation is recommended for exposures exceeding four weeks. Cleaning is required.
Level 3	Between 300 and 1000 ng/m³	No evacuation is required, but additional cleaning is warranted if point sources of mercury can be found and removed.
Level 4	< 300 ng/m³	No further cleanup is required and monitoring will be for assurance that no new sources of contamination enter the home/structure.

3.1.2 *Mercury vapor analyzers*

DIGESA and MYSRL initially performed indoor-air surveys in Choropampa using the Jerome mercury vapor analyzer. The Jerome mercury vapor analyzer measures mercury vapor in the air and is the standard occupational monitoring device used to identify mercury vapor in industrial settings. Mercury vapor analyzers measure the vapor that is released when metallic mercury volatilizes; higher concentrations of metallic mercury release higher concentrations of vapor in the area, resulting in higher mercury vapor readings. The Jerome mercury vapor analyzer has a relatively high detection limit (300 ng/m^3), and, because of its measurement technology, is prone to interference from other airborne elements (dust, ammonia) commonly present in indoor air in rural Peru. Interference by ammonia from urine, during the initial Jerome surveys, proved to be a problem in several homes where domestic animals were present.

Because of the high detection limits and problems with interference, MYSRL acquired a number of Lumex mercury vapor analyzers for use in indoor air monitoring. Lumex analyzers have sufficiently low detection limits and are not affected by positive interference, allowing MYSRL to more accurately measure mercury levels in the monitored homes. In contrast to the Jerome mercury vapor analyzer, the Lumex RA915+ mercury vapor analyzer is a sophisticated optical spectrophotometer, which selectively measures only mercury vapor in the air. Lumex mercury analyzers have a detection limit of approximately 2 ng/m^3. The first Lumex analyzer arrived on site on June 29, and due to its versatility, six additional analyzers were purchased between July 1 and September 1, 2000.

3.1.3 *House Survey and Mapping*

The utility of the initial house surveys was limited because they provided insufficient information about the specific locations in each home that would require the most intense remedial effort. Therefore, MYSRL implemented a subsequent house-mapping program designed to direct the actions needed to assess and clean up elevated indoor air concentrations of mercury. Prior to remediation, MYSRL personnel performed detailed surveys of the home interior to identify those specific locations contributing to high room concentrations. In addition to the homes initially identified with elevated mercury levels, MYSRL ultimately surveyed all of the accessible homes in Choropampa, San Juan, Magdalena, and the surrounding communities. In every home where elevated concentrations were detected (Action Levels 1-3 on Table 1), a detailed house map was drawn to direct the remediation crews to the areas requiring the most immediate attention. By November, 2000, MYSRL had surveyed 2814 homes in Choropampa, San Juan, Magdalena, and the outlying communities, had identified 115 homes in need of some level of remediation, and had completed remediation in all 115 of those homes.

3.1.4 *House remediation*

House remediation activities were conducted in two specific phases relative to the presence of visible and non-visible mercury.

Visible mercury in homes was immediately removed before detailed air surveys using mercury vapor analyzers were conducted. Remediation crews entered the homes and recovered any visible mercury by placing contaminated materials in sealed bags and immediately moving them to the mine area. Following removal of all visible mercury some homes still required additional remediation.

Identification of non-visible mercury contamination in homes provided a much greater challenge than the visible component. MYSRL implemented a detailed house-mapping procedure using the Lumex analyzer to document the site-specific sources of elevated mercury vapors (e.g. soil, clothing, food items, furniture, etc.). The clean-up options available for use by the remediation teams ranged from ventilation to excavation of contaminated floors and walls.

3.1.5 *House verification*

MYSRL's primary objective throughout the spill response was to ensure the protection of human health within the affected areas. Thus, it was essential to develop a reliable verification procedure to ensure that house remediation was effective and that safe conditions for long-term occupancy existed within the affected homes. The verification procedure that was developed and implemented resulted very often, in an iterative process requiring several episodes of cleaning before a home could be confirmed clean. After the areas of identified contamination were removed, the area was allowed to remain undisturbed for at least 12 hours. A follow-up Lumex survey was then performed to determine if the clean-up had been effective. The original locations of contamination were resurveyed to determine if all of the contaminated material (floor or wall, etc.) was cleaned. If elevated mercury vapor concentrations were still observed, additional remedial action was performed. However, if the original location appeared to be clean, another Lumex survey of the home was completed and secondary (or tertiary) contaminated areas were identified and delineated for additional clean up. This process of remediation followed by Lumex survey was continued until contaminated locations could not be identified and mercury vapor concentrations were less than 0.0003 mg/m^3 (Level 4) everywhere within the home. Once confirmation was complete, the home was inspected again by DIGESA, which certified it as acceptable for long-term habitation.

3.2 Roads

The initial assessment of the roads focused on identifying all areas where visible mercury was present, either on the asphalt road surface or along sides of the road. The initial visual assessment consisted of 137 people, divided into 30 groups of 4 people each, with each group walking a 3 kilometer section of highway four times in search of visible mercury on the asphalt road surface or on the roadside dirt surface. This effort resulted in the identification of 16 spill sites (shown in Figure 1).

MYSRL implemented a subsequent systematic roadside sampling program to determine if additional areas of dilute, non-visible mercury also existed along the road. Soil sampling, analysis, and mercury vapor monitoring approaches were used to identify all areas affected by non-visible mercury.

Roadside soil sample collection was initiated during early July. Systematic samples were collected from both sides of the road between Km 155 and Magdalena. When elevated concentrations of mercury in roadside soils were detected, additional samples were collected at closer intervals (between 10 and 50 meters depending on the site) to more clearly define the contaminated area and direct clean-up activities. In addition, site delineation was conducted with the Lumex mercury vapor analyzer. All nine additional non-visible mercury spill sites (Figure 1) that were identified were remediated using the same procedures as the original 16 sites.

In order to make timely decisions regarding mercury levels in soils, fast and accurate turn-around of soil analysis was required. MYSRL tasked a US commercial laboratory with setting-up a field laboratory in Cajamarca to provide timely analysis of soil samples.

3.2.1 Roadside remediation

MYSRL established a clean-up standard of 1000 ppb for mercury contaminated roads and roadside soils. The 1000 ppb (1 ppm) target cleanup level is considered very conservative in light of US EPA soil clean-up standards, which range between 10 and 400 ppm, depending on the environment and the potential exposure pathways. The primary remedial measure at contaminated roadside sites was removal of soils containing mercury concentrations above the clean-up targets. Contaminated roadside soil was typically removed either by sweeping, hand digging, or using vacuums. Heavy equipment (backhoes and bobcats) was also used when hand digging and vacuuming was insufficient. Several innovative clean-up techniques (such as water washing and forced volatilization) were also evaluated on the roadsides, but these techniques did not prove as effective as excavation. Similar to house remediation, roadside remediation was an iterative process. Initially, the least intrusive cleaning option, such as sweeping and

vacuuming, would be implemented to remove the most highly contaminated soils at shallow depths. The area would be resurveyed with the Lumex, and any remaining areas with elevated readings would be cleaned more aggressively (i.e., digging and excavation). This process of Lumex survey followed by remediation would be repeated until acceptable Lumex readings were obtained. After a site was confirmed clean with the Lumex, several confirmation soil samples were collected for total mercury analysis and one composite soil sample was collected from each site for Toxicity Characteristic Leaching Procedure (TCLP) analysis.

3.2.2 Asphalt remediation

There were two locations where the asphalt road surface contained significantly elevated concentrations of mercury such that remediation was required. The two areas correspond to the 1.6 kilometers of asphalt through the town of Choropampa and the asphalt surfaces near 12 of the spill sites. Asphalt remediation was accomplished either by removal or by entombment (chip-sealing).

MYSRL made an immediate decision to remove and replace 1.6 kilometers of asphalt on the road through the town of Choropampa. This decision was made due to the presence of finely disseminated visible mercury beads widely scattered on the entire road surface. The mercury was either dispersed by citizens carrying mercury away from the pool that was discovered in front of the medical post, by vehicle traffic or both. The finely dispersed mercury was contributing to elevated air levels near the road, which was affecting homes in close proximity to the road. Asphalt removal through Choropampa began on June 15 and the old asphalt was transported to the Yanacocha Mine Complex for safe disposal and a new asphalt surface was installed in mid-October.

Elevated mercury vapor readings were identified around 12 of the spill sites totaling approximately 3.6 kilometers of roadway. MYSRL decided to chip-seal the non-visible mercury contaminated asphalt areas rather than remove them due to the rapidly approaching wet season and the possibility that asphalt removal might contaminate the roadside soils that had just been cleaned. The chip-sealing procedure does not preclude future removal, if required, based on the results of the continuing environmental sampling and the ecological risk assessment.

3.2.3 Roadside results

Nearly all areas of roadside remediation achieved MYSRL's clean-up goal of 1000 ppb mercury, nevertheless, all sites were remediated to mercury levels that are well below the US EPA standard. TCLP testing was also used to evaluate the mobility of any mercury that remained on the roadside. TCLP extraction is a relatively aggressive procedure that indicates the quantity of metal that can be potentially

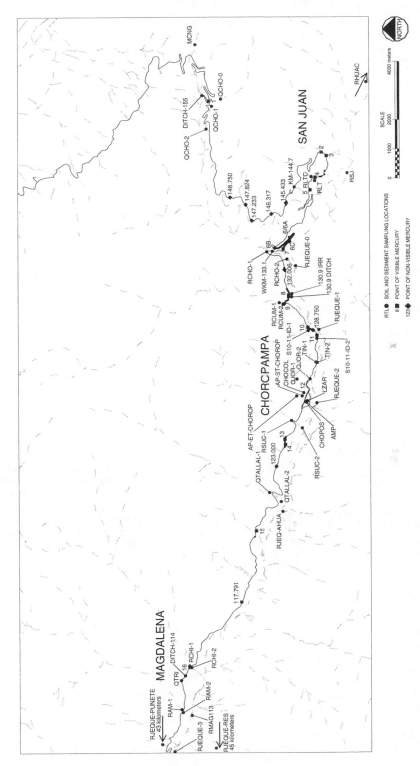

Figure 1. Spill locations and environmental sampling points.

441

leached into the environment from hazardous materials. This procedure was chosen to determine whether the low concentrations of mercury remaining in the soils following clean up could be mobilized into the environment. The very low TCLP test results show that post-remediation conditions (all results were less than 0.1 mg/L) are chemically stable and protective based on the US standard. For comparison of the TCLP results, the US EPA action level (that defines material as hazardous waste) for TCLP extractable mercury is 0.200 mg/L.

The post-remediation sampling data shows that the roadside soil remediation was successful and has restored these areas to safe conditions. However, seventy-two permanent concrete sediment traps were installed around the known spill sites for added safety. These traps were designed to detain any mercury-containing sediment that may potentially be mobilized from a spill site prior to reaching a tributary stream. The traps were also designed such that accumulation of stagnant water would be minimal. The sediment traps are monitored weekly and cleaned as required such that sediments do not overtop the traps. Sediment within the traps is sampled for total mercury on the same schedule as the water sampling events (described below). Following the first significant rainstorm after being installed, one of the sediment traps operated as designed and intercepted mercury that was mobilized from a previously undiscovered location. Results of from the remaining sediment traps shows mercury concentrations ranges between 26 ng/g and 2070 ng/g with an average of 472 ng/g. These concentrations are well within the range of background mercury concentrations found in roadside soils and do not indicate movement of spilled mercury from sites.

4 ENVIRONMENTAL MONITORING

MYSRL implemented a comprehensive environmental monitoring program during the week of June 15, knowing that data would be required not only to assess immediate impact potentials and to direct remediation plans, but also to support a long-term ecological and human health risk assessment. The overall objective of the monitoring program was to provide a comprehensive, quantitative characterization of the situation, so that the most efficient use of available resources could be made to achieve timely and effective mitigation of impacts. This section describes the methodologies and environmental sampling used to characterize conditions immediately after the spill.

Water and sediment sampling locations included: 1) up gradient locations unaffected by the spill, 2) major drainage crossings near the known spill sites, 3) the Choropampa town water supply, 4) several locations along the Jequetepeque River, and 5) 45 kilometers downstream of Magdalena at the Gallito Ciego Reservoir.

The purpose of the program was to characterize baseline conditions and to provide on-going monitoring data to determine whether mercury was being mobilized into the environment. This program also provides baseline data for conduction any long-term monitoring, if determined necessary based on the results of the ecological and human health risk assessments.

4.1 Water and sediment sampling

The water and sediment sampling program was designed such that samples would be collected weekly for the first month following the spill, or after significant rainfall. Sampling frequency was less intense during the subsequent dry months and more frequent at the onset of the wet season. Initial samples were sent to a local Peruvian laboratory for analysis. While the analytical results of the sediments were acceptable, all of the reported mercury concentrations in water samples were below the labs detection limit of 400 ng/L. Subsequent water/sediment analyses were shipped to a US laboratory to achieve lower detection limits.

Water samples were also analyzed for dissolved mercury, silver, arsenic, and total suspended solids in the laboratory. Field measurements included pH, electrical conductivity, temperature, and estimates of flow rates. Unfiltered water samples were collected for total mercury determination, whereas samples filtered through a 0.45 μm filter were analyzed for dissolved mercury. Ultra-clean sampling procedures were followed and the samples were not preserved with acid prior to shipment to the laboratory due to the potential for mercury contamination from the acid. Each soil and sediment sample sent to the US for analysis was sieved and particles less than 2 mm were set aside for digestion and analysis. Utilizing only smaller particles provides a more conservative determination of mercury concentrations.

4.2 Environmental monitoring results

The water and sediment sampling data has been compiled for of greater than one year since the spill. These results show relatively constant low concentrations of mercury, which indicates that no mercury has been mobilized away from the spill sites into the environment. This section presents the water and sediment quality results, evaluates data trends, and presents an estimate of local background concentrations.

4.2.1 Water

Total mercury concentrations in the collected water samples ranged between 0.11 and 1282 ng/L, with the highest mercury concentration in water collected 5 kilometers upgradient of the site of the first spilled mercury (sampling point MCNG). This concentration likely represents the naturally occurring background mercury concentrations that can be expected in this basin. The average dissolved concentration for all water samples collected from upgradient locations is 7.9 ng/L. The average total mercury value from these same locations is 26 ng/L. Since these locations are uphill and/or upgradient from the first spill location, these average concentrations represent background conditions expected in the Jequetepeque basin. For comparison, the average dissolved concentration for all sampling sites in the Jequetepeque basin (average of dissolved mercury values is 52 ng/L and the average total concentration average of total mercury values in is 74 ng/L. These average concentrations are conservative averages (higher than expected) because, in all calculations, the concentrations that were reported as less than the detection limit were assumed to be at the detection limit. These average concentrations are very similar to the known background concentrations discussed above, and indicate that all of the observed mercury concentrations within the water in the Jequetepeque basin result from natural sources. For comparison, the US EPA action levels for drinking water is 2000 ng/L, and the acute, and chronic aquatic values for protection of aquatic life are 1800, and 940 ng/L, respectively.

The occasional elevated concentrations can be attributed to natural sample variability or contamination (either in the field or in the laboratory), due to the fact that systematic trends are not evident in the data. Although there are several sampling locations that show slightly decreasing mercury concentrations between June and October, there are no sampling locations that showed a systematic increase in mercury concentration (or concentrations above apparent background) that would indicate mercury contamination from the spill. The decreasing concentrations in some locations between June and October are very low concentrations (background) and are likely a result of the natural flow dynamics of the basin, such as decreasing flow observed during the dry season, rather than a reflection of impact from the spill. These water quality data indicate that there has been no movement of mercury from the spill into the natural watercourses of the Jequetepeque basin.

4.2.2 Sediment

Sediments were collected along with water samples and were analyzed for total mercury and percent solid. The sediment concentrations range between 4 and 3388 ng/g on a dry weight basis.

Three background sediment sampling sites are upgradient of the first spill location in the same sites as background water samples. The average total concentration for all sediment samples collected from these upgradient locations is 404 ng/g, and range between 8 and 1390 ng/g. The highest background sediment mercury concentrations were found in the same location as the elevated concentrations in water, sampling site MCNG. Sediment samples at this location are among of the highest concentrations observed in the sediments in the basin, ranging between 661 and 1390 ng/g. In addition to the upgradient locations that represent background conditions, the mercury concentrations in the Gallito Ciego Reservoir (approximately 45 kilometers downstream of Magdalena) were as high as 2080 ng/g. Due to the time required to move sediments this large distance (especially during the dry season), and the fact that the elevated concentrations in the downstream samples are higher than any sediment concentrations found near spill sites, the elevated sediment concentration at the reservoir is likely natural and unrelated to the spill. From observations, the reservoir collects a significant degree of fine sediments from the upper Jequetepeque basin, so elevated mercury concentrations in the sediments are expected due to the affinity natural mercury has for fine sediments.

The sediment data shows that all of the sediments within the natural watercourses in the Jequetepeque basin are within background concentrations described above and well below comparable regulatory standards (although there is no standard for mercury in stream sediments, the US EPA standard for soils of 10,000 ppb or 10 ppm). As with the water samples, there is no obvious trend in sediment concentrations at any particular sampling point. The sediment data indicate that after the mercury from the spill is not moving into the environment via the streambeds.

5 MERCURY MASS BALANCE

The development of a mercury mass balance has been an extremely important element of the work performed to provide an accurate evaluation of the spilled mercury. A mercury mass balance was developed at the outset of the project. The mass balance was revised as more data and information became available to provide an increasingly accurate account of mercury potentially remaining in the environment or in the possession of residents.

5.1 MYSRL mercury mass balance

MYSRL initiated the development of a mercury balance reconciliation immediately after the spill event occurred. The most significant variables are the uncertainty in the quantities of mercury that were collected by the population, and the fate of that mercury. The mercury mass balance was revised a final time by MYSRL in late August based on the data and information collected up to that time. This revised balance provided a reasonable estimate of the distribution of mercury at the time of the spill and at the end of August relative to known locations and estimated spill quantities. The revised mercury balance was based on data and information collected at the site during the extensive environmental and remedial investigations that began in early June 2000. Observations from the initial response period (June 3-9) provide important data regarding the amount of mercury present in the environment at the time of the spill. The mercury recovered from the population and from the remediated spill areas represents the most significant recovery quantities, and provides very good data upon which spill locations and quantities were estimated. Other aspects of the mercury balance are based on the environmental data collected and on calculations made using scientific methods and information about the chemical and physical properties of mercury as well as the fate of mercury in the environment. MYSRL's revised mercury balance concluded that approximately 9 kg of mercury (from the 151 kg spilled) might remain in the environment. The remaining mercury is either widely dispersed at low concentrations in the environment or in the possession of private citizens (which is the most likely case, as the environmental sampling has not shown any potential sources that have not been remediated).

5.2 Independent auditor mercury balance review

Consulcont S.A.C. (a Peruvian environmental consultant) served as an independent auditor for the Ministry of Energy and Mining (MEM) to evaluate the effectiveness and accuracy of remedial activities and assessments conducted by MYSRL. In July, 2000 Consulcont S.A.C. was critical of the initial mercury balances provided by MYSRL and thus questioned the effectiveness of the remedial activities conducted. As a result of these findings, a second independent auditor was brought in by the MEM to evaluate the revised MYSRL mercury balance and Consulcont findings. The second auditor evaluated Consulcont's findings and MYSRL's overall assessment procedures and mercury balance. Both auditors met for a week in November 2000, and came to agreement on the amount of mercury that potentially be remaining in the environment. Their findings represent a refinement or restatement of MYSRL's revised mercury balance that was based on different analytical approaches and additional data available in November. Although many of the assumptions and calculations were different from those used by MYSRL, the mercury balance developed jointly by the auditors concluded that less than 6 Kg of mercury potentially remains in the environment. This is in good agreement with the conclusions reached in MYSRL's revised mercury balance.

The revised mercury balances, developed by MYSRL and the independent auditors, indicate that a small amount of mercury (6 to 9 kilograms) is still potentially unaccounted for. On-going environ-

mental monitoring and the ecological and human health risk assessments will provide the data to confirm that mercury remaining in the environment poses no short-term or long-term risk.

6 RISK ASSESSMENT

As part of the assessment of intermediate and long-term risks to humans and the terrestrial and aquatic ecosystems within the Jequetepeque basin, MYSRL initiated a risk assessment process early in the remediation response effort. Baseline data on mercury concentrations in soils, waters, sediments, and plant, insect, and aquatic invertebrates and fish tissue were collected prior to the first wet season in order to establish baseline conditions and to allow for calculation of site-specific transfer of mercury between ecological components. The sampling program was also designed to detect if any mercury had already been mobilized within the ecosystems surrounding the spill locations, prior to the initiation of the first wet season. Initial work was also directed at understanding ecosystems and receptors at potential risk to mercury in the environment, identifying potential routes of mercury exposure, and conceptualizing the fate and transport of mercury around the spill locations. Additional risk assessment efforts including providing technical support to local and national governmental officials regarding 1) the risk assessment process, 2) help in educating the local populations on the potential risks associated with mercury in the environment, and 3) helping interpret data collected by governmental groups.

Although initial screening-level risk assessments concluded that the potential risk to humans and ecological receptors was quite low after the successful implementation of remediation efforts, additional sampling was conducted a year after the initial sampling effort to confirm these findings. The second year sampling was conducted at essentially the same locations as the earlier sampling and was conducted to confirm that if there was residual mercury in the environment, that this mercury had not been further mobilized in the environment, and that it does not pose significant risk to humans or ecosystems. Results from this sampling are not yet available, so final conclusions on the long-term risk potential cannot be made. MYSRL will utilize these findings to decide what if any future monitoring of the ecosystems is required.

7 SUMMARY

7.1 General

- MYSRL was fully committed to the complete remediation of the mercury loss incident, using the most stringent international standards for the protection of human health and the environment.

- MYSRL retained some of the leading experts in the world on mercury health effects, appropriate treatment protocols, cleanup standards, remediation methodologies, and environmental monitoring and risk assessment.
- MYSRL estimates it will spend between US$9-11 million dollars on medical treatment and support, home remediation, soil and asphalt remediation, environmental monitoring, studies and risk assessments, and expert fees.
- Further it currently estimates expenditures of US$4 million for individual compensation, community compensation, health insurance for impacted individuals, and government fines.
- MYSRL has reviewed and improved all of their hazardous materials handling procedures.

7.2 *Public health*

- All patients treated for mercury intoxication at the Cajamarca Regional Hospital and in village medical clinics have completed their course of treatment.
- MYSRL's medical experts are confident that there are no long-term health issues associated with exposure to elemental mercury for a short duration, particularly at the levels seen in this case.
- MYSRL will implement a long-term public health monitoring program, and will provide health insurance for each impacted individual, for a period of five years, covering any medical condition diagnosed as related to exposure to mercury.
- Finally, MYSRL has agreed to provide funding to staff the medical posts in Choropampa and San Juan for a period of two years.
- All of the homes in Choropampa, San Juan, Magdalena, and outlying rural areas where elevated levels of mercury were found have been cleaned to EPA standards for long-term habitability. MYSRL surveyed 2814 homes in Choropampa, San Juan, Magdalena and the outlying communities, and of those, 115 homes required some level of cleaning.
- MYSRL, independently and in cooperation with regional and local health care agencies, has undertaken a variety of public health education programs focused on, but not limited to, mercury exposure education.

7.3 *Environmental impacts*

- The mercury was spilled on asphalt and soils along the highway, during the dry season allowing the vast majority of the mercury to be collected or to evaporate. Of the 151 kg of mercury spilled along the road, MYSRL now estimates that approximately 142 kg have been collected or have volatilized, leaving a balance of 9 kg remaining in soils or collected by other individuals.
- MYSRL has implemented a comprehensive environmental monitoring program, covering soils, water courses, and sediments both upgradient and downgradient from the places along the road where mercury has been found. To date, no mercury contamination has been detected in any stream or sediments.
- All soil and asphalt remediation has been completed. Permanent sediment traps have been installed at the known spill locations along the road.
- Now that remediation has been completed, MYSRL's focus has shifted to a long-term human health and ecological risk assessment. This risk assessment will focus on human health and potential risk to terrestrial and aquatic ecosystems and will be based in part on the company's estimates of the amount of mercury possibly remaining in the environment following the remediation work.

445

Physical and geochemical characterization of mine rock piles at the questa mine, New Mexico: An overview

S. Shaw, C. Wels & A. Robertson
Robertson GeoConsultants, Vancouver, BC, Canada

G. Lorinczi
Molycorp Inc., Questa Division, Questa, NM, U.S.A.

ABSTRACT: The Questa molybdenum mine is located in the Sangre de Cristo mountains in Taos County, northern New Mexico. Currently, the mining operations consist of underground block caving; however, between 1965 and 1983 lower grade molybdenum ore (0.185% MoS2, 74 million tonnes) was recovered by open pit mining methods, with some 320 million tonnes of waste rock produced. A comprehensive physical and geochemical characterization of the mine rock piles has been carried out at the site over the last few years to evaluate the current conditions and predict the future conditions. This paper describes the results of a drilling program completed as part of a larger, site-wide characterization program. The drilling program was carried out in phases with the objective of characterizing the geochemical and physical properties of the mine rock at surface and at depth and the oxygen and temperature variations with depth in the piles. Testing included (i) geochemical testing of drill cuttings (moisture content, paste pH/EC, ABA, leach extraction and forward acid titration testing); (ii) physical testing of mine rock samples (grain size analysis, moisture retention, permeability), and (iii) *in-situ* monitoring of temperature and oxygen/carbon dioxide in 10 boreholes. The results of the characterization study indicate that the majority of the mine rock, where it has been tested, is potentially acid generating, but, after more than 20 years, the mine rock may not have reached a mature state of oxidation and acid mine drainage. The particle size and moisture content of the mine rock is quite variable and the *in-situ* temperature and oxygen monitoring suggest that there is on-going sulfide oxidation and advective airflow (chimney effect) within the rock piles. The results of this rock pile characterization and monitoring have been used to calibrate an air transport and ARD production model of the mine rock piles (see Lefebvre et al., 2002).

1 INTRODUCTION

The Questa molybdenum mine is located 5.6km east of the town of Questa in Taos County, New Mexico in a region with a long history of mining (Figure 1). The mine site is located on the south facing slopes of the north side of the Red River Valley between an east-west trending ridgeline at about elevation 3,200m and State Highway No. 38 adjacent to the Red River at about elevation 2,440m. An aerial view of the site is shown in the photograph in Figure 2.

Figure 1. Location map.

The mine encompasses three main tributary valleys (Capulin Canyon, Goathill Gulch and Sulphur Gulch from east to west). Mine disturbance has been most extensive in Sulphur Gulch and Goathill Gulch in which the open pit mine and the active subsidence zone (from underground block cave mining) have been developed, respectively. The locations of the major mine surface facilities (open pit, mine rock piles, the cave zone, and buildings) are shown on Figure 3. The mine rock piles are very high extending vertically from the Red River, at elevation ~2.440m to about elevation 2,930m, resulting in some of the highest mine rock piles in North America. The height of these rock piles, and their relatively shallow depth (above natural terrain) results in unusual conditions of air and moisture movement through the piles. Changing air and moisture conditions have significant implications with respect to long term oxidation and acid mine drainage from these rock piles. This paper describes some of the physical and geochemical characterization studies that have been completed on these mine rock piles.

The climate on the mine site is semi-arid with mild summers and cold winters. The average month-

Figure 2. Aerial photograph of the Questa Mine Facilities.

ly temperature is below freezing for five months of the year (November through March) and the long-term average monthly temperature is approximately 4.2°C. The mine is located in an area of high relief, and the distribution pattern of precipitation and net infiltration is complex. There is a general trend of increasing net infiltration with increaseing elevation, largely a result of the difference in snowpack at different elevations (see Wels et al., 2002).

Deeply incised valleys transect the mine site and surrounding area. Many of the side slopes of these incised valleys are very steep with grades typically between 2:1 and 1.4:1 on the numerous scree slopes and near vertical where rock outcrops are seen. North of the Red River many of the valleys contain areas of weathered, hydrothermally altered and highly erosive rock that are typically referred to as hydrothermal scars. These scars are located both on and off the mine site and occur as a result of high rates of erosion resulting from a combination of erosive susceptibility of the weathered mineralized rock in which they form, the steep slopes and lack of vegetation. Periodic debris flows are generated from these scars during high rainfall periods, and have resulted in the development of large debris fans (silty and sandy, boulder-rich gravels). These fans form as alluvial/colluvial infill in the lower reaches of these tributary valleys. Large debris flows periodically have reached the Red River and have caused damming in the river to form alluvial terraces and meadows at locations such as Fawn Lakes and the town of Red River. These scars and their associated debris fans are comprised of oxidizing sulfide (predominantly pyrite) enriched rocks and are a natural source of significant loads of ARD to the Red River (RGC, 2001a).

A complex geological history characterizes the Questa area and a great deal has been written on the geology in the district (Johnson & Lipman, 1988; Schilling, 1956). The basement rocks in the region consist predominantly of Precambrian metamorphosed ocean floor volcanic, granitic and gabbroic intrusive rocks, schists and quartzites (Czamanski et al., 1990). The area has experienced periods of mountain building (Reed et al., 1984; Carpenter, 1968) and volcanic activity (Carpenter, 1968; Lipman, 1988). Andesitic flows overlain by rhyolitic welded ash flow tuff approximately 26 Ma can be seen in outcrop, along the ridgeline at the crest of the hydrothermal scars between the Red River and the Cabresto Canyon to the north. This volcanism was associated with the extension of crustal rocks and incipient Rio Grande rift formation. The Questa caldera collapse features (Reed et al., 1983; Lipman, 1988) concomitantly with the eruption of tuff. It is along the collapse-related fractures that mineralizing intrusions (porhpyritic stocks, sills and dikes) were emplaced from which hydrothermal processes evolved (Roberts et al., 1991; Czamanske et al., 1990).

Molybdenum in the Questa system occurs as hydrothermally deposited molybdenite (MoS_2) and is most often associated with quartz veins and magmatic hydrothermal breccia proximal to the tops of granitic intrusions (approximately 1 to 1.5 km below the surface). The resulting Climax-type porphyry molybdenum deposit is associated with a fluorine-rich hydrothermally altered system (Molling, 1989;

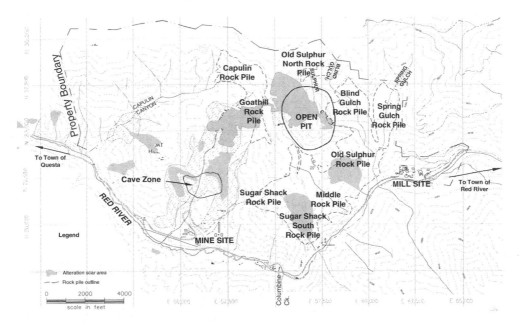

Figure 3. Questa mine site map.

Carten et al., 1993). The economic mineralization exists as two separate and discontinuous zones of high grade (>0.2% by weight) molybdenite mined by underground block cave methods, and as a vein stockwork system mined by open pit mining methods.

Associated with the hydrothermal mineralization processes are the alteration halos and vein zones, including pyrite veining that in places has resulted in pyrite concentrations of 2 to 3% by volume (such as in the areas where the hydrothermal scars are evident). Long term oxidation of the pyrite in these areas has resulted in pronounced jarositic staining (yellow), and chemical weathering of the host rocks. In areas where there is less pyrite (<1%) the surficial rocks are typically stained by minerals such as goethite, hematite, ferrimolybdite and manganese oxides, and chemical weathering of the matrix in the host rocks is less severe.

There are four primary lithological units in which open pit mining was conducted (see Figure 4). These include (oldest to youngest) the andesite flows, andesite-to-quartz latite porphyry flows, rhyolite tuff, and aplite porphyry. The andesite and prophyritic aplite units comprise the bulk of mined material. Due to the alteration of rocks in the area (both hydrothermal and atmospheric weathering processes), field classification of units is sometimes difficult and the term 'mixed volcanics' subsequently has been used as a classification term in the mine rock piles. Typically, the 'mixed volcanics' are composed of the andesite porphyry rock type with minor amounts of other units such as rhyolite dikes

or tuff. The aplite porphyry in the mine rock piles is easily distinguishable from other units as it is relatively competent, leucocratic rock with a 0.1 to 0.5 mm groundmass and little visual evidence of sulfide oxidation processes and physical weathering. The northern portion of the pit has been mined through a hydrothermal scar (Sulphur Gulch Scar) and the upper portions of this scar can still be seen above the pit wall. Much of the mined rock from this area is weathered scar material.

Open pit mining between 1968 and 1985 produced some 320 million tonnes of mined rock, end dumped into various steep valleys adjacent to the open pit. The piles typically extend vertically over 490m and are on the order of 110m thick.

2 INVESTIGATIONS

2.1 Drilling and sampling

One of the investigation programs completed on site was a drilling program to determine how conditions in the rock piles vary with depth. Geochemical and physical laboratory tests were completed on samples collected from drill cuttings. In conjunction with the laboratory tests, the boreholes were instrumented. These instruments allowed an assessment of the in-situ characteristics (temperature, O_2, CO_2,) with depth and over time.

Phase 1 of the drilling and sampling program was completed in July and August 1999, and Phase 2

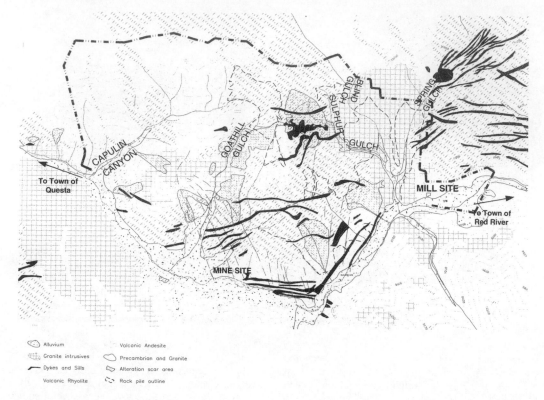

Figure 4. Geology of the Questa Mine area.

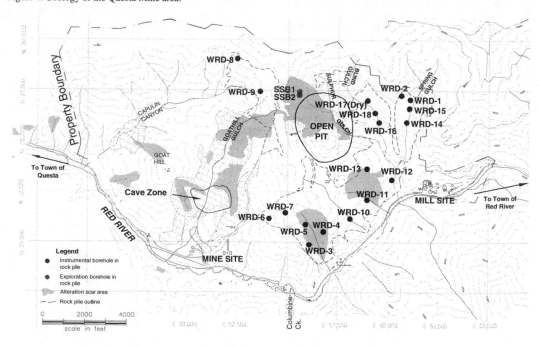

Figure 5. Borehole location map.

was undertaken between August and November 2000. Phase 1 consisted of nine boreholes in various mine rock piles and two boreholes in the Sulphur Gulch Scar. Phase 2 consisted of nine additional boreholes, primarily in piles not drilled during Phase 1 and in areas where additional information was required. The locations of these boreholes are shown on Figure 5.

Drilling was completed with an AP-1000 hammer drill (also referred to as a Bekker Hammer rig) supplied by Layne Western Drilling Company. The rig was equipped with a 15.2cm (6-inch) inner diameter by 22.9cm (9-inch) outer diameter casing. The casing was advanced by a top mounted diesel hammer and samples recovered by pumping compressed air downward to the cutting head through the annulus between the inner and outer casings. Cuttings were returned through the inner casing and routed through a cyclone and sample splitter before being collected in a bucket. This type of drill rig is used to minimize the geochemical disturbance, caused by crushing and/or the use of water to return samples.

Samples were taken at the end of each 1.5m (5 foot) drilling interval. Each sample was logged describing lithology, mineralization, grading, degree of alteration and oxidation, with emphasis on sulfide (pyrite) and carbonate contents. Field paste pH, conductivity and moisture contents were measured on each 1.5m sample. An example of the depth profiles for these field parameter results from selected boreholes is provided in Figure 6.

Sub-sets of the samples collected were submitted for geochemical and physical testing. Geochemical testing was completed by Canadian Environmental and Metallurgical Inc. in Vancouver, Canada and physical tests were completed by Advanced Terra

Testing in Lakewood, Colorado, USA., AMEC Inc. Laboratory in Phoenix, Arizona, USA and M.D. Haug and Associates Laboratory in Saskatoon, Saskatchewan, Canada.

2.2 Geochemical testings

The majority of geochemical testing consisted of static laboratory tests including rinse pH and conductivity measurements, modified acid base accounting (ABA), multi-element ICP analyses, forward acid titration tests and leach extraction tests. Additional information can be found in RGC Report No. 052008/10 (RGC, 2000c).

Rinse pH and rinse conductivity analyses were completed in the lab using the same procedure as was used for the field paste pH and conductivity measurements. These analyses, in part, served as a QA/QC measure correlate closely with the field data. The modified ABA test was used to determine the balance between acid producing minerals (sulfides) and acid consuming minerals (predominantly carbonates) present in the sample. The results were used to classify samples as to their potential for acid generation and acid neutralization. The results of the ABA analyses, while variable, show some classification based on lithology. The 'mixed volcanics' samples (predominantly andesite), located primarily in Capulin, Goathill, Sugar Shack West, Sugar Shack South, Middle and Old Sulphur rock piles, typically classified as potentially acid generating. The aplite samples, located primarily in the Spring Gulch and Blind Gulch rock piles, were either non acid generating or uncertain with respect to acid generating potential. A lithological unit described as black andesite (typically chloritized or hornfelsed) located in Blind Gulch and on the surface of Old Sulphur and

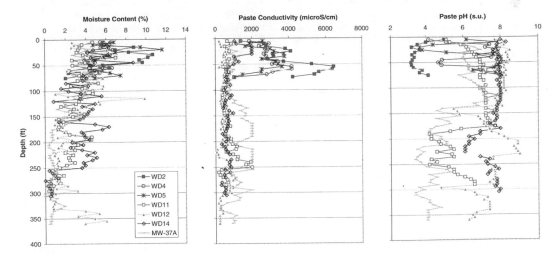

Figure 6. Results of the field paste parameters and moisture content with depth in selected boreholes.

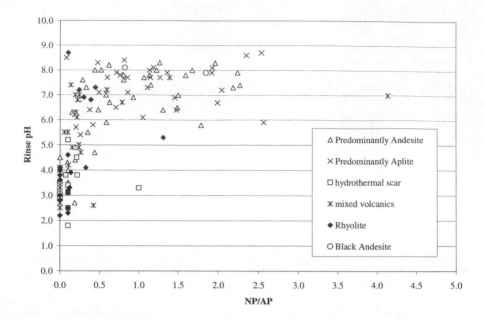

Figure 7. Neutralization potential: Acid production potential ratio versus rinse pH.

Middle rock piles were also largely non acid generating or uncertain with respect to acid generation potential. The fourth primary lithology, rhyolite, had highly variable ABA results.

While geochemical characteristics change with depth (Figure 6), they also appear to be primarily dependent on lithology. Figure 7 shows the variation ABA results with respect to lithology. This figure shows the ratio of neutralization potential (NP) to acid production potential (AP) versus the paste pH for waste rock samples and indicates that, while there is variability, the majority of material has the potential to generate acid (i.e. with an NP/AP ratio less than 3), but that only that material with NP/AP ratios less than ~1 has become acidic.

The forward acid titration tests consisted of the addition of sulfuric acid (H_2SO_4) to a sample while measuring the pH of the paste. It was used to qualitatively determine the acid neutralizing capacity of the material and the buffering capacity available at different pH conditions. The amount of acid required to reach each pH interval is dependent on the amount of neutralization present. 'Buffering' occurs where a significant amount of H_2SO_4 is added before the pH drops, thereby producing a 'step-like' curve in a plot of pH versus volume of acid added. Within the pH range of 5.5 to ~7.0, carbonate minerals in the sample dissolve and neutralize acidity. Between the pH range of 3.0 to 3.7, iron oxyhydroxides (FeOOH) will buffer acid and at even lower pH values (i.e. below ~3), aluminosilicate minerals such as the feldspars in the sample will dissolve and buffer

the added acid. The results of this test for a sub-set of samples from Questa are shown in Figure 8. The Black Andesite, Rhyolite and Aplite samples show some carbonate buffering capacity (in decreasing order) while the Mixed Volcanics and Scar samples are buffered only by aluminosilicates and possibly iron oxyhydroxides.

The leach extraction test used on the Questa samples was the EPA 1312 test procedure modified to utilize a leachate reagent of de-ionized water acidified to pH 5.5 (Meteoric Water Mobility Test (MWMT) reagent) to represent rainwater. This test provides an assessment of the amount (by weight) of soluble material in a sample (RGC, 2000a). Figure 9 shows the sulfate concentration in the leachate versus depth in various drill cuttings. The soluble sulfate content in the samples is greater closer to the surface, where secondary minerals have formed as a result of sulfide oxidation and subsequent acid neutralization. Figure 10 shows the relationship of soluble copper concentration versus depth. Again, the soluble copper content is greatest nearer the surface. Significant concentrations of copper are not seen in the leachate at pH values above ~4.0 (Figure 11). This pH dependency is typical of most metals commonly associated with sulfide deposits (with the exception of molybdenum, which is mobile in alkaline pH and immobile in acidic conditions).

The metal concentrations in the leachate therefore are largely dependent on pH, and in different pH ranges different metals are mobile. While ARD typically has a pH below ~4.5 and has high metal con-

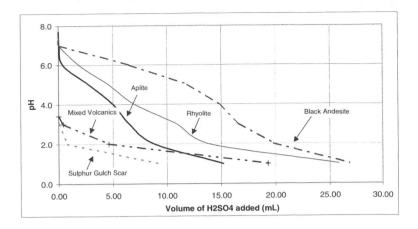

Figure 8. Forward acid titration test results.

centrations, buffered ARD may also have elevated concentrations of certain elements such as Mo, Zn etc. It is important to note that the leach extraction tests use a significantly higher liquid to solid ratio than occurs in the field. Therefore, the leach extraction concentrations are not directly comparable to expected field porewater concentrations (i.e. leachate concentrations are typically more dilute than porewater conditions). Geochemical speciation modeling is required to relate the leachate concentrations to expected *in-situ* concentrations (RGC, 2000b).

2.3 *Physical testing*

The physical testing of mine rock samples included grain size analysis, moisture content, shear strength, permeability tests, Standard Proctor compaction tests and moisture retention (soil water characteristic curve). The mine rock at the Questa mine can be grouped into four different classes (i) well-graded (finer) mixed volcanics; (ii) poorly-graded (coarser) aplite/black andesite; (iii) fine-grained erosion material (typically gap-graded); and (iv) very coarse rubble material (predominantly aplite/black andesite). Figure 13 shows the typical ranges of particle size distributions for these four classes of mine rock (based on 52 mine rock samples).

The particle size distribution has a strong influence on other physical characteristics such as permeability (to both air and water), water retention capacity, and strength. In general, the finer-grained mine rock showed a higher moisture retention capacity and lower saturated hydraulic conductivity than the coarser-grained mine rock (see RGC, 2000a, 2000c). These characteristics determine how air and water enters and migrates through the piles. Coarse rock at the base of piles (a product of end dumped pile construction) combined with coarse layers in the

interior of the pile produces a chimney effect for air movement through the rock piles. Finer materials developed due to weathering near the surface of the piles (see erosion material in Figure 12) inhibit infiltration and also provide moisture retention such that more water can be removed by evapotranspiration, reducing the net influx and therefore ARD seepage.

2.4 *Field instrumentation and monitoring*

Borehole instrumentation was completed for all 9 boreholes drilled during the Phase 1 program and one of the monitoring well drillholes (MW-37A) drilled in the mine rock piles during the Phase 2 program. The objectives of the instrumentation program were to enable monitoring of internal temperature, oxygen and carbon dioxide concentrations in the mine rock piles, and to sample any water that may be encountered in the boreholes.

The instrumentation was strapped to a 25mm (1-inch) diameter, Schedule 40 PVC casing with slip joint couplings. The casing is required to support the instrumentation, at the correct intervals, and provide a frame onto which the instrumentation and protecting layers could be strapped. The bottom 3m (10 feet) of each casing was slotted to allow the sampling of any water encountered at the base of the hole. The thermistor wires (for temperature) and nylon pore gas sampling tubes (for O_2 and CO_2 measurements) were taped to the outside of the casing prior to insertion into the borehole. The instrumented bundle was then lowered into the borehole and the holes were backfilled to within ~9m (30 feet) of surface with silica sand. The upper 9m of each drillhole was backfilled with drill cuttings and a concrete slab was poured on the surface.

Pore gas sampling tubes consist of 6.4mm O.D. by 4.3mm I.D. polyethylene tubes. Sampling ports consist of an open tube covered with a filter fabric to

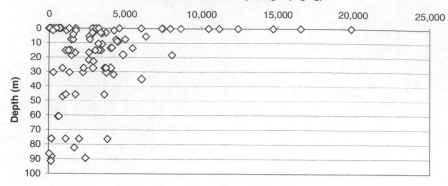

Figure 9. Sulfate content versus depth in drill cuttings.

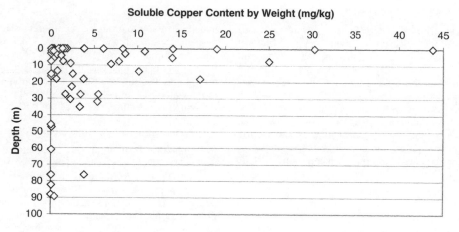

Figure 10. Copper content versus depth in drill cuttings.

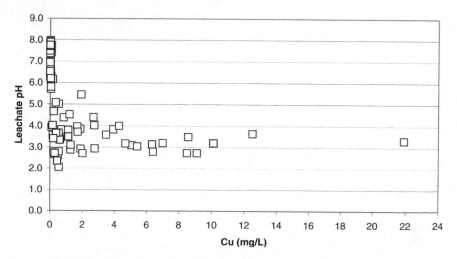

Figure 11. Copper concentration versus leachate pH in drill cuttings.

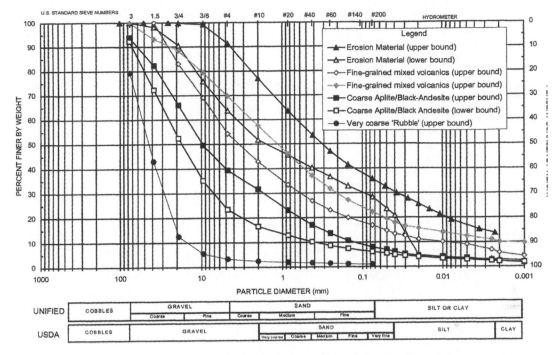

Figure 12. Upper and lower bounds of particle size distributions for Questa mine rock (based on 52 samples).

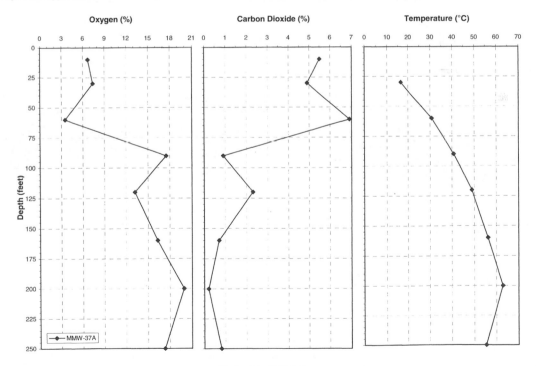

Figure 13. Temperature, O_2 and CO_2 concentrations with depth in MW-37A.

455

prevent clogging. Thermistor strings consist of a series of stainless steel jacketed thermistors linked to a pre-wired, weather proof surface terminal through a heavy duty PVC coated multi-pair cable. Thermistor cables were assembled and calibrated by RST Instruments (manufacturer) prior to shipment to the site.

The instrumented boreholes are monitored monthly for pore gas concentrations as well as *in-situ* temperature. Pore gas monitoring is completed using a Nova Analytical Systems Inc. (NOVA) Model 309BCWP portable O_2 and CO_2 Analyzer. To extract gas from various levels at the interior of the mine rock piles, the Nova Analyzer is connected to the sampling tubes via plastic quick-release unions and air is pumped across the detector in the analyzer.

Temperature measurements are collected from the weather-proof 10-channel surface terminals mounted on steel support frames at the surface of each borehole. This control box is the external extension of the thermistor string (RST Instruments Ltd.). Temperature measurements are performed with an Omega Engineering Model 865F, 2252 ohm digital thermometer. The thermometer and the thermistors were both pre-calibrated by the manufacturer.

An example of the results of the temperature, oxygen and carbon dioxide measurements over a one year period (Y2000) for boreholes MW-37A and WRD-5 are provided in Figures 13 and 14 respectively. Figure 13 provides the results with depth in the deepest borehole (MW-37A) and Figure 14 shows the seasonal trends in a representative bore-

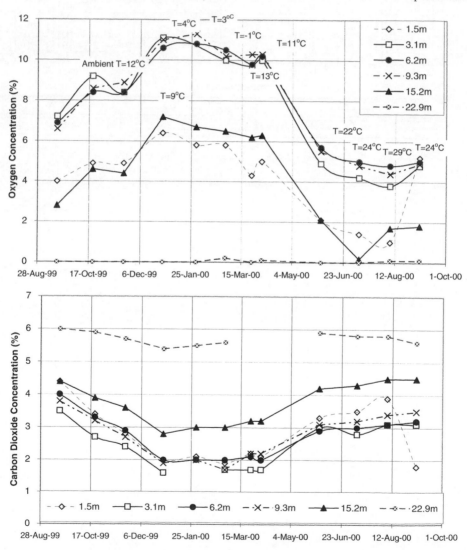

Figure 14. Seasonal trends in O_2 and CO_2 concentrations in WRD-5.

456

hole (WRD-5). From these results it is apparent that, for almost the entire depth of the hole, the oxygen concentrations are reduced but oxidation would not be oxygen limited. The elevated temperatures indicate on-going oxidation. Temperature differentials between in the inside of the pile and the ambient outside temperature result in a chimney effect drawing air into the rock pile near the toe of the pile. This convective effect is greatest in winter when the temperature contrast is greatest resulting in a clear pattern of seasonal increase and decrease in the oxygen content. CO_2 is produced by carbonate neutralization of the oxidation products (ARD) and its concentration also fluctuates seasonally depending on the rate at which it is flushed from the rock pile by air movement.

Similarly, occasional monitoring of barometric pressure changes with depth is done in the boreholes using a system linking three differential pressure transducers (Setra Model 239) installed at specific levels and a surface barometer (Setra Model 270). The transducers use the same pore gas sampling tubes to access the desired depths of measurement. A Campbell Scientific CR10X datalogger records in-situ differential air pressure and ambient barometric air pressure (at surface) in 2-min intervals and stores summary statistics (hourly and daily min-max values). The transducer-barometer system including data logger is portable and can be moved between the various boreholes. It is powered by a solar panel and rechargeable lithium batteries. Preliminary readings with this apparatus suggested that the air pressure within the pile responded very quickly to outside changes in air pressure. This response would be indicative of very high air permeability in the pile. However, a quantitative analysis of the air pressure data was not possible because changes in ambient (outside) air pressure were not much greater than the sensitivity of the surface barometer.

The pore gas and temperature monitoring data suggest that there is advective transport of air into the mine rock piles. Such advective air flow may also result in redistribution and/or removal of water from the piles in vapor form. This "drying" effect is likely to be most pronounced in the interior of the pile where the pore air can "heat up" thus significantly increasing its water holding capacity. The potential for vapor transport and its effect on the overall water balance of the mine rock pile has been evaluated with the help of the air flow model TOUGHAMD (Lefebvre et al., 2001a; 2001b; 2002).

3 CONCLUSIONS

The results of the characterization studies to date indicate that the majority of the mine rock, where it has been tested, is acid generating. The data further

indicates that elevated temperatures within the mine rock piles (due to on-going oxidation) and the high elevation gradient (rock piles are up to 490m high) result in high rates of air flow within the rock piles (chimney effect). As a result the rock piles are not oxygen limited and will continue to oxidize for a long time. The oxidation has produced leachable oxidation products in the rock piles. Associated with the air movement, there is moisture movement which effects the heat transfer in the pile as well as the moisture migration, and hence leachable contaminant migration and deposition, within the upper portions of the pile. The potential for long term ARD from the rock piles will result from a combination of all these effects.

4 ACKNOWLEDGMENTS

This paper presents the results of investigations and studies funded by Molycorp Inc. and could only have been completed with their large and dedicated commitment to understanding the physical and geochemical processes that are active on the Questa mine site. The authors of the paper would like to acknowledge the extensive contributions by both Molycorp personnel, in particular David Shoemaker, Bruce Walker and Anne Wagner; the other consultants who have participated in the development of this characterization program, notably Ralph Vail; and the constructive comments during the development of work plans from personnel from both New Mexico Mining and Minerals Division and Environmental Department.

REFERENCES

Carpenter, R.H. (1968). Geology and Ore Deposits of the Questa Molybdenum Mine Area, Taos County, New Mexico. *Ore Deposits of the United States, 1933-1967 vol. 2*: 1328-1350. AIME Graton-Sales.

Carten, R.B., White, W.H. and Stein, H.J. (1993). High-grade Granite-related Molybdenum Systems: Classification and Origin. Kirkham, R.F., Sinclair, W.D., Thorpe, R.I. and Duke, J.M. (ed.), *Mineral Deposit Modeling: Geological Association of Canada, Special Paper 40:* 521-554.

Czamanske, G.K., Foland, K.A., Hubacher, F.A. and Allen, J.C. (1990). The 40Ar/39Ar Chronology of Caldera Formation, Intrusive Activity and Mo-ore Deposition near Questa, New Mexico. *New Mexico Geological Society Guidebook, 41st Field Conference, Southern Sangre de Cristo mountains, New Mexico:* 355-358. 1990.

Johnson, C.M. and Lipman, P.W. (1988). Origin of Metaluminous and Alkaline Volcanic Rocks of the Latir Volcanic Field, Northern Rio Grande Rift, New Mexico. *Journal of Mineralogy and Petrology*: 1988.

Lefebvre, R., Hockley, D. Smolensky, J. and Gelinas, P., (2001a). Multiphase Transfer Processes in Waste Rock Piles producing Acid Mine Drainage, 1: Conceptual Model and System Characterization. *Journal of Contaminant Hydrology, Special Edition on Practical Applications of Coupled Process Models in Subsurface Environment, in print.*

Lefebvre, R., Hockley, D. Smolensky, J. and Lamontagne, A., (2001b). Multiphase Transfer Processes in Waste Rock Piles producing Acid Mine Drainage, 2: Applications of Numerical Applications. *Journal of Contaminant Hydrology, Special Edition on Practical Applications of Coupled Process Models in Subsurface Environment, in print.*

Lefebvre, R., Lamontagne A., Wels C. and Robertson, A. (2002). Assessment of ARD Production and Water Vapor Transport in Mine Rock Piles at Questa Mine, New Mexico. *Proceedings of the Tailings and Mine Waste 2002,* these proceedings.

Lipman, P.W. (1988). Evolution of Silicic Magma in the Upper Crust: the Mid-Tertiary Latir Volcanic Field and its Cogenetic Granitic Batholith, Northern New Mexico, USA. *Transactions of the Royal Society of Edinburgh: Earth Sciences, vol. 79*: 265-288.

Meyer, J. and Leonardson, R. (1990). Tectonic, Hydrothermal and Geomorphic Controls on Alteration Scar Formation Near Questa, New Mexico. *New Mexico Geological Society Guidebook, 41st Field Conference, Southern Sangre de Cristo mountains, New Mexico*: 417-422. 1990.

Molling, P.A. (1989). Volcanic, Plutonic, Tectonic and Hydrothermal History of the Southern Questa Caldera, New Mexico. *University of California, Santa Barbara PhD Dissertation:* 28.

Reed, J.C. Jr., Lipman, P.W. and Robertson, J.R.M. (1983). Geological Map of the Latir Peak and Wheeler Peak Wildernesses and Columbine-Honeo Wilderness Study Area, Taos County, New Mexico. *U.S. Geological Survey Miscellaneous Field Studies, Map MF-1570B:* Scale 1:50,000.

Roberts, T.T., Parkison, G.A. and McLemore, V.T. (1990). Geology of the Red River District, Taos County, New Mexico. *New Mexico Geological Society Guidebook, 41st Field Conference, Southern Sangre de Cristo mountains, New Mexico*: 375-380. 1990.

Robertson GeoConsultants Inc. (2000a). Progress Report: Results of Phase 1 Physical Waste Rock Characterization, Questa Mine, New Mexico. *Report prepared for Molycorp Inc.:* June 2000.

Robertson GeoConsultants Inc. (2000b). Interim Mine Site Characterization Study, Questa Mine, New Mexico. *Report prepared for Molycorp Inc.:* November 2000.

Robertson GeoConsultants Inc. (2000c). Interim Mine Site Characterization Study, Questa Mine, New Mexico. *Report prepared for Molycorp Inc.:* November 2000.

Robertson GeoConsultants Inc (2001a). Integrated Geochemical Load Balance for Straight Creek, Sangre de Cristo Mountains, New Mexico. *Report No. 052008/13 prepared for Molycorp Inc.:* January 2001.

Schilling, J.H. (1956). Geology of the Questa Molybdenum (Moly) Mine Area, Taos County, New Mexico. State Bureau of Mines and Mineral Resources New Mexico institute of Mining and Technology, Campus Station, Socorro, New Mexico, *Bulletin 51*: 87.

Wels C., Loudon, S. and Fortin, S. (2001). Infiltration Test Plot Study for Waste Rock at Questa Mine, New Mexico. *Proceedings of the Tailings and Mine Waste 2002, these proceedings.*

Assessment of store-and-release cover for Questa tailings facility, New Mexico

C. Wels, S. Fortin & S. Loudon
Robertson GeoConsultants, Vancouver, B.C., Canada

ABSTRACT: A test plot study has been initiated to study the performance of a store-and-release cover consisting of alluvial soil over tailings proposed for final closure of a large tailings facility in New Mexico. The test plot study consists of two closed lysimeters covered with 0.23m (9") and 0.60m (24") of alluvial soil (silty gravel) over backfilled tailings and an instrumented deep *in-situ* tailings profile with a 0.28m (11") alluvial cover with existing grass/shrub vegetation. All test plots were instrumented with temperature/suction sensors and moisture content sensors to monitor moisture movement in the cover/tailings profile. The lysimeters are free-draining and outflow is collected and monitored continuously using a tipping bucket. The lowest rate of net infiltration (20mm or 6% of total precipitation) during the first year of monitoring was observed in the deep *in-situ* cover/tailings profile. Net infiltration into the unvegetated lysimeter test plots were 55.9mm (17.3%) for the 0.23m alluvial cover and 117mm (36%) for the 0.60m alluvial cover. Initial calibration of a soil atmosphere model (SoilCover) to the test plot data suggested that the lower rates of net infiltration into the *in-situ* profile are a result of (i) higher evapotranspiration due to the presence of vegetation and (ii) lower vertical hydraulic conductivity of the *in-situ* tailings (relative to the backfilled tailings in the closed lysimeters). The observed rates of net infiltration for the unvegetated, backfilled lysimeters may therefore significantly overestimate cover fluxes for long-term (post-closure) conditions. The calibrated model should be used to predict cover performance for a range of climate conditions (e.g. wet year *vs* dry year) and cover design parameters (e.g. cover thickness).

1 INTRODUCTION

Molycorp Inc. owns and operates a large tailings facility located near the village of Questa, New Mexico (Figure 1). Over the last 33 years a total of nearly 100 million tons of tailings from the Questa Molybdenum Mine have been discharged into this facility covering a total surface area of about 260 ha (640 acres) (as of 1997). The tailings originate from a hydrothermally altered molybdenum porphyry deposit of volcanic origin and are produced at the Questa mine, located 8 km to the east of the Questa tailings facility.

Final closure of the facility will require covering of the tailings with locally available alluvial soils (silty gravel) after final closure of the facility (RGC, 1998). This soil layer would prevent erosion of the tailings (primarily by wind), provide a growth medium for revegetation and, in conjunction with the underlying tailings, represent a water storage (or "store-and-release") cover that would reduce infiltration into the deeper tailings profile.

This test plot study focuses on the function of the proposed alluvial cover to control net infiltration and ultimately seepage from the tailings impoundments. A store-and-release cover typically consists of a

well-graded soil layer (or multiple soil layers) that stores precipitation during wet periods and releases the moisture back to the atmosphere via evapotranspiration during dry periods. The net effect is a significant reduction, or elimination, of net percolation (also called "cover flux"), and ultimately seepage from the tailings impoundment. The semi-arid climate conditions experienced at Questa, New Mexico favor the use of the store-and-release cover over the more conventional water barrier ("low permeability") cover. While the dry climate generally improves the performance of the store-and-release cover (due to high rates of evaporation coupled with low precipitation) it often compromises the performance of a water barrier cover due to problems with desiccation (e.g. Swanson et al., 1997; Bews et al., 1997). Store-and-release covers have been designed for several mine sites in arid or semi-arid climates (e.g. PTI and WESTEC, 1996; GSM, 1995; and O'Kane et al., 1998).

The main advantages of the store-and-release cover compared to other alternatives considered (low permeability and capillary barrier covers) are:

- a store-and-release cover will require the least long-term maintenance and has the lowest potential for long-term degradation (and ultimately failure);

- the effectiveness of a store-and-release cover is expected to increase in time as the vegetation matures to the desired climax vegetation (e.g. sagebrush and juniper); and

- the proposed store-and-release cover can be implemented for a fraction of the cost of more complex engineered cover types.

Molycorp has implemented a test plot study at the Questa tailings facility to evaluate net infiltration through the proposed store-and-release cover (RGC, 2000a). The test plot study was initiated to collect site-specific data of cover performance for the purpose of final design. Specifically, the test plot study was designed to meet the following objectives:

- measure climatic conditions at the site;
- measure *in-situ* material properties (characteristic curves); and
- calibrate a soil-atmosphere model in order to predict net infiltration.

It is important to recognize that the primary objective of the test plot study was not to measure actual cover fluxes *per se* but instead, to calibrate a soil-atmosphere model for known (measured) boundary conditions. The ultimate goal is to use the calibrated soil-atmosphere model to predict with a high degree of confidence the performance of the storage cover for conditions relevant to final closure (i.e. deep, unsaturated tailings profile; mature vegetation; range of climatic conditions). These final closure conditions cannot, of themselves, be duplicated in a field trial of only a few years, but instead have to be simulated.

2 SITE DESCRIPTION

The Questa tailings facility is located west of the village of Questa in northern New Mexico (Figure 1). The tailings facility lies in an alluvial plain at an elevation of about 2320m amsl, bordered by the Sangre de Cristo Mountains to the east and the Guadalupe Mountains to the west. The tailings were impounded in two deeply incised valleys (so-called "arroyos") behind two earth fill dams (Dams 1/1C and Dam 4, respectively). Tailings are currently discharged behind Dam 4.

The test plots for this study were constructed on the tailings beach between Dam 1 and Dam 1C (Figure 1). Dam 1 was constructed in 1966 as an earthfill dam across an arroyo and was raised by downstream construction using earthfill. In 1975, "Old" Dam 1C was constructed of cycloned sand 200m upstream of Dam 1. In 1981 "New" Dam 1C was constructed at its present location (Figure 1). This 'new' dam replaces the old dam, which is contained within the tailings deposited behind "New" Dam 1C (see Wels

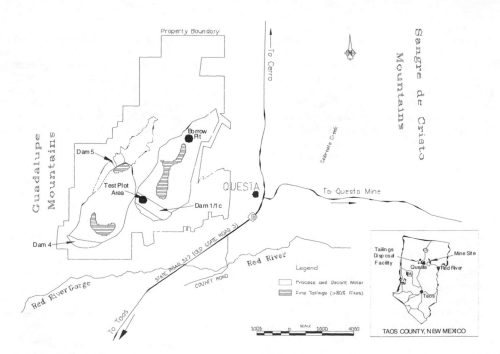

Figure 1. Location map of Questa tailings facility.

460

et al., 2001 for more details on the discharge history).

A detailed physical and geochemical characterization of the Questa tailings was carried out as part of the development of the Revised Closure Plan for this facility (RGC, 1998). Briefly, the Questa tailings are geochemically similar mixtures of aplite and andesite tailings with low-moderate sulfide content (0.5–1.5% pyrite). The predominantly andesitic tailings have higher sulfide-sulfur contents and a higher potential to generate acid than the slimes tailings or aplite tailings. Despite the oxidation potential of sulfide minerals, the Questa tailings are currently not acid generating. The surface tailings, which have been most susceptible to oxidation processes, have remained grey in color and are consistently circum-neutral with respect to paste pH. The field and laboratory data suggest that the buffering capacity of the tailings is sufficient to maintain circum-neutral pH in the tailings pore water (Wels et al., 2000).

A total of 32 tailings samples were analysed for standard grain size analyses (ASTM D422) to characterize the spatial distribution of the tailings at the surface of the impoundments according to size fraction. The tailings were subdivided into three classes:

i.   coarse tailings with <50% fines content (where fines constitute silt and clay sized particles with a grain diameter smaller than 0.075 mm (# 200 mesh));
ii. intermediate tailings with 50%< fines content <80%; and
iii. fine tailings with >80% fines.

This survey suggested that the coarse tailings represent about 2/3 of the total surface area. The fine tailings comprise only about 12% of the present tailings surface area (see Wels et al., 2001 for additional details).

The climate of the study area is semi-arid. Annual precipitation at nearby Cerro averages about 310mm (12.2 inches) with much of this precipitation occurring as summer thundershowers (on average 43% of total precipitation occurs from July to September). The summers are moderately warm with maximum daily temperatures around 27°C (81°F). The winters are long with temperatures dropping below freezing almost every night from October through to April. However, typically clear skies bring sunshine during most days with temperatures rising to above the freezing point. During the winter much of the precipitation falls as snow. Nevertheless, a significant snow pack rarely develops due to intermittent snowmelt and/or sublimation. Based on frost data collected by the U.S. Weather Bureau at Cerro a growing season of 120 days is average for the study area. As expected for this semi-arid climate, the potential evaporation rates far exceed precipitation rates during all months on record. The annual pan evaporation is estimated to be about 1715mm (67.5 inches).

3 DESIGN OF TEST PLOT STUDY

The test plot study was designed to evaluate factors controlling the performance of the cover and the net infiltration to the tailings, i.e. material properties, climate conditions and cover thickness (RGC, 2000a). A total of three test plots were constructed in the beach area between Dam 1 and Dam 1C of the Questa tailings facility (Figure 1). This tailings area has been covered for about 25 years, which allowed the establishment of a mature grass/shrub vegetation on the cover material. An initial reconnaissance survey indicated that the interim cover placed historically in this area was quite variable (ranging in thickness from 0.23m (9") to greater than 0.60m (24")). The test plot location finally selected (in the eastern portion, see Figure 1) was deemed most representative of the requirements for the deep in-situ test plot TP-1, i.e. an existing shallow alluvial cover with mature grass/brush vegetation.

The three test plots were constructed and instrumented to measure the performance of three different combinations of cover thickness and vegetation development:

Test Plot #1: Existing 0.28m (11") thick alluvial cover with mature grass/shrub vegetation overlying in-situ sandy tailings (very deep tailings profile);

Test Plot #2: 0.23m (9") thick alluvial cover with no vegetation overlying back-filled sandy tailings (~2.5 m deep tailings profile); and

Test Plot #3: 0.60m (24") thick alluvial cover with no vegetation overlying back-filled sandy tailings (~2.5 m deep tailings profile).

Test plot #1 is most representative of post-closure steady-state conditions, i.e. with mature vegetation established on the alluvial cover overlying undisturbed, hydraulically placed tailings. This test plot is designed as an open system, with the monitoring instrumentation was installed into the existing cover and tailings profile (without prior excavation). Any excavation of the cover and tailings profile would destroy the root system of the vegetation, which is considered a vital component of this cover system, as well as the 'natural' soil structure and density of the hydraulically placed tailings. Test plot #1 will be used to calibrate the soil-atmosphere model (including effects of vegetation on evapotranspiration) against measured changes in soil moisture and soil suction over time. The calibrated soil-atmosphere model of test plot #1 will be used to predict the net

461

flux through a mature cover system for post-closure steady-state conditions.

Test plots #2 and #3 represent free-draining "lysimeter plots", in which the rate of net percolation through the cover and shallow tailings profile is monitored directly and leachate is collected at the base of the lysimeter (Figure 2). Test plots #2 and #3 will be used to assess the influence of cover thickness on net percolation (initially without the influence of transpiration by plants). Both test plots will allow calibration of the soil-atmosphere model not only against observed suction/moisture content in the profile but also against measured net fluxes observed discharging at the base of the lysimeters.

Two sets of sensors were installed in the three test plot profiles. *In-situ* temperature and soil matric suction is measured using a thermal conductivity type sensor CS 229 (Campbell Scientific, 1998). The volumetric moisture content is measured indirectly using the capacitor-type sensors Enviroscan™ distributed by SENTEK Environmental Technologies, Adelaide, Australia. (SENTEK, 1997).

3.1 Construction and instrumentation

The construction and instrumentation varied between the deep, relatively undisturbed profile (test plot TP-1) and the closed lysimeters (test plots TP-2 and TP-3). For test plot TP-1 an access ditch was excavated along one side of the test plot to a depth of approximately 2.7m. The access ditch was used to determine *in-situ* material properties, collect representative samples, and to provide access to the revegetated, undisturbed tailings profile for lateral installation of the suction-temperature sensors.

For construction of the lysimeter test plots TP-2 and TP-3, thick walled (19mm) HDPE tanks with a diameter of 2.4m and a height of 2.3m were used (Figure 2). For each test plot, a large pit was first excavated into the existing tailings profile, and the tailings logged, sampled and stockpiled. A lysimeter tank was then lowered into the excavation and backfilled with the stockpiled tailings. An attempt was made to reproduce the same layering and *in-situ* properties (moisture content and bulk density) observed during excavation. A 50mm diameter discharge pipe was installed at the base of the tank to allow free drainage. This discharge pipe connects to a manhole where outflow is monitored continuously using a tipping bucket and a data logger.

After backfilling the lysimeter and construction of the discharge/manhole system, the alluvial cover material was placed in a single lift without compaction using the backhoe. The alluvial cover was placed over a foot print area of about 6.0m x 6.0m, or about eight times the foot print area of the lysimeter in order to minimize boundary effects. The alluvial material used for cover construction on the lysimeter test plots (TP-2 and TP-3) was taken from

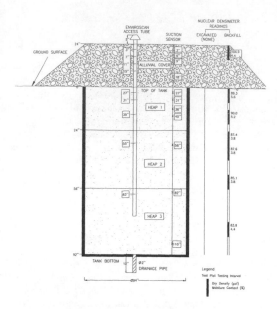

Figure 2. As-Built profile of test plot TP-3.

a local borrow pit with alluvial soil deemed representative of the alluvial soils proposed for final cover of the tailings facility (see Figure 1 for location).

A fully automated meteorological station was set up in the test plot area to evaluate atmospheric boundary conditions for future numerical model calibration. The weather station consists of an all-weather rain gauge, relative humidity/temperature probe, wind sensor, and net radiometer. In addition, a Bowen Ratio system was installed to measure actual evapotranspiration. More details on the construction and instrumentation of the test plots are provided by Wels et al. (2001).

3.2 Material testing

Representative samples of tailings and alluvial cover material were taken during construction of the test plots and submitted for laboratory testing to determine hydraulic properties and soil water characteristic curves (see RGC, 2000b for details). The geotechnical characterization included grain size analysis, moisture content, standard proctor compaction, permeability, and soil water characteristics (i.e. moisture retention). The alluvial cover material represents a fairly well-graded silty/clayey gravel (GM-GC) with a fines content ranging from 7.8 to 10.9%. In contrast, the tailings encountered at the test plots represent a poorly graded silty sand (SM) with a fines content ranging from 12.4% to 48.2%. The tailings used for the test plot study are representative of the "coarse" tailings, which cover about 2/3 of the total tailings area at the Questa tailings facility.

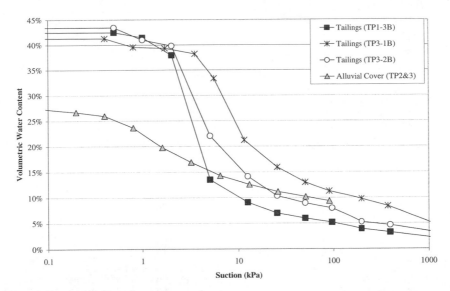

Figure 3. Soil Water Characteristic Curve for test plot samples.

Figure 3 shows the soil water characteristic curves (SWCC) determined for representative samples of the tailings and the alluvial cover material. The (coarse) tailings have a higher porosity and a steeper SWCC than the alluvial soil. The air entry value (AEV) of the tailings is slightly higher (~1 to 5 kPa) compared to that of the alluvial material (<1 kPa). The large range of particle sizes (from gravel to clay size) in the alluvial cover material is responsible for the low AEV and the gradual decline in moisture content with increase in suction (i.e. successive drainage of increasingly smaller pore spaces with increasing suction).

4 1ST YEAR MONITORING RESULTS

The monitoring results of the test plot study for the first year of monitoring (August 2000 – July 2001) are summarized in Table 1. The climate conditions for the first year of monitoring were typical for the region with a total precipitation of 323mm (12.7 inches) and potential evaporation of 1120mm (44.1 inches). Potential evaporation was calculated from local climate data, including daily average air temperature, relative humidity, net radiation, and wind speed (Penman, 1948).

During the first year of monitoring no outflow was recorded at the two lysimeters. However, detailed measurements of volumetric water content allowed the calculation of net infiltration into the cover/tailings profile. This was done by integrating the volumetric water content profiles at the start and end of the first year of monitoring. The rates of net infiltration (precipitation minus actual evapotranspiration) ranged from 75.7mm (23% of precipitation) for TP-1 to 101.8mm (31.7% of precipitation) for TP-3. The net infiltration for test plot TP-2 was intermediate at 92.7mm (28.7% of precipitation). Note that the rate of net infiltration at the surface is still significantly higher than the rate of deep percolation (or "cover flux") into the deeper tailings profile, in other words the test plots are still wetting up and have not yet reached (pseudo) steady-state conditions. This is evident in the lack of outflow at the base of the two lysimeters. A good estimate of deep percolation will only be attainable after several years of monitoring, including direct measurements of lysimeter outflow over time.

Figure 4 shows the time trends of volumetric water content for the deep, undisturbed test plot profile (TP-1). The daily precipitation is also shown for comparison. The first significant wetting occurred after a heavy precipitation period in mid-August when 48mm (1.9 inches) of precipitation occurred in 2 days. In the near-surface layers (top 0.38m) of the cover/tailings profile the volumetric moisture con-

Table 1. Summary of 1st year test plot monitoring.

	TP-1	TP-2	TP-3
Cumulative precipitation (mm)	--------- 322 ---------		
Cumulative Potential Evaporation (mm)	-------- 1120 --------		
Net Infiltration (mm)[1]	20.5	57.4	117.9

Notes:

1. Change in storage calculated from measured volumetric water content profiles

463

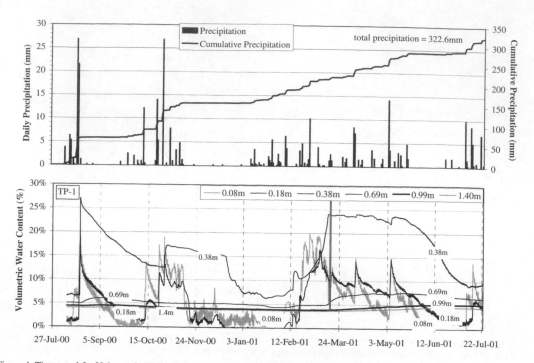

Figure 4. Time trend for Volumetric Water Content for selected depths in test plot TP-2, showing precipitation for comparison.

tents exhibited a very rapid increase (during the rainfall period) followed by a gradual decline due to subsequent evapotranspiration and drainage. During the winter months (Nov 2000 – Feb 2001) the soils in the test plots froze to a depth of ~0.5-0.6m (RGC, 2001a). The freezing of the soil is clearly recognizable in Figure 3 by the sudden drop in the volumetric water content (the sensors measure only liquid water content).

A combination of thawing of the upper soil layers and precipitation during the early spring period resulted in renewed wetting of the upper cover/tailings profile (late February to early March, 2001). During the latter portion of the spring period the alluvial cover experienced repeated wetting and drying in response to precipitation events (see sensors at 0.07 and 0.18m depth, Figure 4). These wetting-drying cycles maintained very wet conditions in the uppermost layers of the in-situ tailings profile (e.g. at 0.38m depth); however, they did not result in significant wetting of the deeper tailings profile (>0.69m depth). Drier conditions prevailed between May and July 2001 resulting in drying of the alluvial cover and the upper layers of the tailings profile (Figure 4). By the end of the first year of monitoring the water content in the deeper tailings profile (>1.0m) still remained very close to the initial (residual) water content.

Figure 5 shows the time trends of volumetric water content for the two lysimeter test plots TP-2 (upper panel) and TP-3 (lower panel). The general sea-

sonal pattern of moisture contents is similar to that observed in the in-situ profile of TP-1; however there are several important differences. First, the water contents in the lysimeter test plots (in particular in TP-3) did not show as fast a decline in water content during dry periods as observed in TP-1. This is a result of the absence of vegetation. In the unvegetated lysimeter test plots evaporation at the soil surface is the only removal mechanism of soil moisture. In contrast, the existing vegetation in TP-1 is able to remove soil moisture deeper from the soil profile by root uptake and plant transpiration in addition to evaporation. Second, the wetting front progressed much deeper into the tailings profile of the lysimeter plots compared to the deep in-situ tailings profile. For example, a clear increase in moisture content was observed in both lysimeters at around 1.25m in response to the spring rains compared to no significant response in TP-1 (compare Figures 4 and 5). The faster wetting of the lysimeter plots is believed to be a result of the loose structure of the backfilled tailings compared to the in-situ tailings profile, which is strongly layered due to hydraulic placement. The backfilling likely increased the vertical hydraulic conductivity of the tailings material (despite the similarity in bulk density). The increased supply of soil moisture in the upper profile (due to lack of plant transpiration) may have also contributed to the faster wetting in the two lysimeter test plots.

464

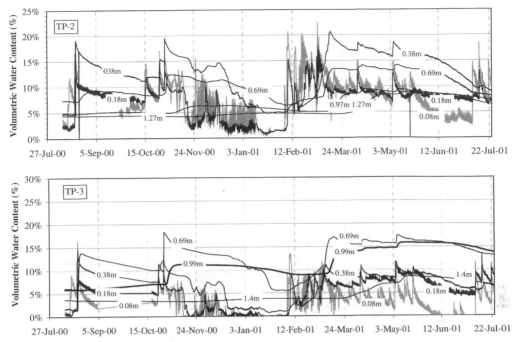

Figure 5. Time trend of Volumetric Water Content for selected depths in test plots TP-2 and TP-3.

Figure 6 compares suction and moisture content profiles in all three test plots for selected dates during the first year of monitoring. The depth of the alluvial cover is shown in these profiles for reference. The suction and moisture content profiles clearly illustrate the deeper wetting that occurred in the lysimeter test plots TP-2 and TP-3 compared to in the *in-situ* profile TP-1. The volumetric moisture content in the alluvial cover was generally significantly lower than in the underlying tailings despite similar, if not lower, suction values (Figure 6). This is a result of the much coarser soil texture and hence lower porosity of the cover relative to the tailings (Figure 3). The lower porosity of the alluvial cover also results in a deeper wetting front for the case of a 0.60m thick cover (up to 1.4m wetting in TP-3) compared to the case of a 0.25m cover (~1.0m and < 0.7m in TP-2 and TP-1, respectively). These observations highlight the fact that the tailings have a much better storage capacity than the alluvial cover.

The volumetric water content values near the base of the lysimeters TP-2 and TP-3 started to show a small increase towards the end of the first year of monitoring (Figure 7 and 8). However, they are still much lower than the field capacity of the tailings. These observations are consistent with the lack of any outflow from the lysimeters recorded in the manhole. Based on the observed rates of net infiltration it may take another 1-2 runoff seasons (or more)

before there will be any measurable discharge from the base of the lysimeters.

4.1 *Initial soil atmosphere modeling*

The soil atmosphere model SoilCover (Geoanalysis 2000 Ltd., 2000) was used to simulate the transient soil moisture conditions in the tailings test plots. For each test plot, model runs were carried out using local climate data collected at the primary met station over the period of August 8, 2000 to February 19, 2001. Most of the initial calibration runs, however, focused on the period August 8 to October 8, 2000 to avoid complications with snow cover and soil freezing. The goal of this initial calibration was to get a general agreement of the model predictions with observed trends in volumetric water content and soil suction.

For each test plot a SoilCover model was calibrated using a trial-and-error approach (RGC, 2001b). The model requires the input of both atmospheric data as well as material properties for the cover/tailings. In general, the climate data were well known (measured on site) and were used as fixed input to the model. The only exception was the potential evaporation (which is calculated internally by the model based on climate data). Initial calibration runs (using unadjusted climate data) suggested that SoilCover overpredicted actual evapotranspiration during days of high rainfall. This discrepancy is

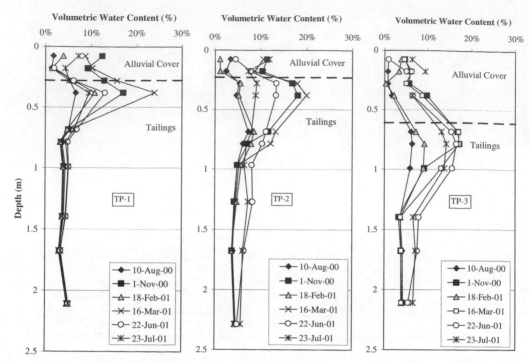

Figure 6. Volumetric Water Content profiles for TP-1, TP-2 and TP-3.

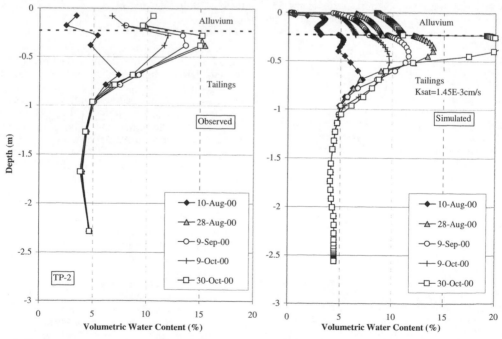

Figure 7. Simulated and observed depth profiles of Volumetric Water Content at TP-2.

likely a result of the fact that SoilCover assumes rainfall to be uniformly distributed across a given day, whereas in reality heavy rain showers generally occur in the late afternoon or evening when potential ET rates are sharply declining. A much better fit with field measurements were obtained when assuming no evapotranspiration for days with precipitation greater than 5mm (0.2 inches).

In this initial calibration the emphasis was placed on varying the hydraulic conductivity of the material rather than on varying the SWCC. This approach was chosen because the SWCC data (from the lab and field) are considered more reliable than estimates of hydraulic conductivity (uncertainty in extrapolation of lab measurements to the field). Hence the SWCC used as input to the model were derived from the laboratory data and the *in-situ* measurements (RGC, 2001a). The key variable to be determined in the calibration process was the saturated hydraulic conductivity. In all cases, the initial estimates were taken from representative laboratory measurements. In subsequent analyses the initial guesses were adjusted to either increase or decrease the advance of the wetting front in the alluvial cover and/or tailings.

Figure 7 compares simulated and observed depth profiles of volumetric water content for selected dates in test plot TP-2. The hydraulic conductivity assumed for the "calibrated" model were $3.8*10^{-3}$ cm/s and $1.45*10^{-3}$ cm/s for the alluvial cover and underlying tailings, respectively. These calibrated values fall well within the range of Ksat values de-

termined for these materials in the laboratory (RGC, 2001b). The match between observed and simulated suctions and in particular VWC is very good in light of the few numbers of parameters that had to be adjusted to obtain this fit.

TP-1 differs from TP-2 and TP-3 in that a grass/shrub vegetation is present. The influence of vegetation on net infiltration was simulated using the vegetation option in SoilCover. For modeling purposes it was assumed that plant transpiration occurs only on days where the daily maximum temperature is 7°C (44°F) or greater. This assumption is believed to most closely represent actual site conditions (see RGC, 1997). The other vegetation parameters (root zone depth and leaf area index) were determined by sensitivity analyses. A root zone depth of 0.28m (i.e. the depth of the alluvial cover) and a leaf area index of 1 gave the best fit to the observed field data (RGC, 2001b).

Figure 8 shows the simulated and observed depth profiles of volumetric water content for selected dates in test plot TP-1. The hydraulic conductivity of the alluvial cover had to be increased by a factor of 10 relative to the calibrated value for TP-2. The higher Ksat of the existing interim cover at TP-1 is likely due to the presence of vegetation providing root holes and reduced compaction. The hydraulic conductivity of the deeper tailings profile (below 0.46m depth) had to be reduced by an order of magnitude (down to $1.45*10^{-4}$ cm/s) in order to reproduce the very slow observed advance of the wetting front into the deeper tailings profile. This lower (ver-

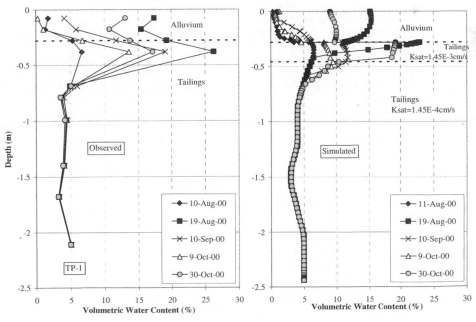

Figure 8. Simulated and observed depth profiles of Volumetric Water Content at TP-1.

tical) hydraulic conductivity is consistent with field observations made during test plot construction, which indicated thin layers of finer tailings (silty sand). While of limited vertical extent (a few inches) these layers of finer tailings may be very effective in reducing the vertical hydraulic conductivity of the tailings.

5 CONCLUSIONS

This paper describes the design and installation/instrumentation of cover performance test plots designed to study the performance of a store-and-release ("water storage") cover for final closure of the Questa tailings facility. Initial results for the first year of monitoring suggest that a 0.28m (11") thick alluvial cover with mature vegetation is more effective in reducing net percolation into the deeper tailings profile than either a 0.25m (9") or 0.60m (24") thick alluvial cover without vegetation. The field data support earlier modeling results, which suggested that the tailings are an important component of the store-and-release cover (RGC, 1997). The high porosity of the tailings combined with reduced vertical permeability (due to the presence of thin layers of finer-grained tailings) resulted in storage of the incoming precipitation just below the alluvial cover. Much of this soil moisture was removed from the soil profile during drier periods by evapotranspiration.

Initial calibration of the soil-atmosphere model SoilCover to the test plot data suggested that (i) the existing grass/shrub vegetation (with a 0.28m deep root zone and a LAI=1) in TP-1 significantly improves the removal of moisture from the cover and underlying tailings profile, and (ii) the vertical hydraulic conductivity of the *in-situ* tailings is significantly lower than that of the backfilled tailings (and laboratory estimates). These modeling results imply that the observed rates of net infiltration for the barren, backfilled lysimeters may significantly overestimate cover fluxes for long-term (post-closure) conditions. Work is currently in progress to predict long-term cover performance for a range of cover design parameters (e.g. cover thickness) and climate conditions (e.g. wet year vs dry year).

6 ACKNOWLEDGEMENTS

The test plot study was funded by Molycorp Inc. The authors of the paper would like to acknowledge the extensive contributions by Molycorp personnel, in particular Anne Wagner, Harvey Seto and Armando Martinez, and the constructive comments during the development of work plans from personnel from both New Mexico Mining and Minerals Division and Environmental Department.

REFERENCES

Bews, B.E., M. O'Kane, G.W. Wilson, D. Williams, and N. Currey (1997) The design of a low flux cover system for acid generating waste rock in semi-arid environment *Proceedings of 4th Int. Conference of Acid Rock Drainage p. 747-762*, Vancouver, B.C.

Campbell Scientific (1998), 229 Soil Water Potential Probe Manual , 15 p.

Geoanalysis 2000 Ltd. (2000), SOILCOVER User's manual, May 2000.

Golden Sunlight Mining (1995) Evaluation of proposed capping designs for waste rock and tailings facilities at Golden Sunlight Mine, Volume 5 Hard Rock Mining Permit and Plan of Operations, prepared by Schafer and Associates, Bozeman, Montana.

O'Kane, M., D. Porterfield, M. Endersby, and M.D. Haug (1998) The design and implementation of the field test plots at BHP Iron Ore, Mt. Whaleback – A cover system for an arid climate, Preprint 98-70, Society for Mining, Metallurgy and Exploration.

PTI and WESTEC (1996) Materials handling plan for Twin Creeks Mine, Humboldt County, Nevada, USA. Prepared for Santa Fe Pacific Gold Corporation by PTI Envir. Services, Boulder, Colorado and WESTEC, Reno, Nevada.

Robertson GeoConsultants Inc. 1997. Study of groundwater flow and tailings seepage near Questa, New Mexico. Report No. 052002/1 prepared for Molycorp Inc., October 1997.

Robertson GeoConsultants Inc. 1998. Questa tailings facility – Revised closure plan. Report No. 052004/1 prepared for Molycorp Inc., April 1998.

Robertson GeoConsultants Inc. (2000a) Work Plan for Storage Cover Test Plot Study, Questa Tailings Facility, New Mexico, RGC Report 052008/3 prepared for Molycorp Inc., January 2000.

Robertson GeoConsultants Inc. (2000b). As-Built Report – Storage Cover Test Plot Study, Questa Tailings Facility, New Mexico, RGC Report 052010/5 prepared for Molycorp Inc., November 2000.

Robertson GeoConsultants Inc. (2001a). Quarterly Monitoring Report - Storage Cover Test Plot Study, Questa Tailings Facility, New Mexico (January 9, 2001 – April 9, 2001), RGC Report 052010/7 prepared for Molycorp Inc., July 2001.

Robertson GeoConsultants Inc. (2001b). Initial Soil Atmosphere Modeling for Storage Cover Test Plot Study, Questa Tailings Facility, New Mexico, RGC Report 052010/8 prepared for Molycorp Inc., August 2001.

Sentek Pty Ltd. (1997), EnviroSCAN Hardware Manual, Version 3.0, 63 p.

Swanson, D. A., S.L. Barbour and G. W. Wilson (1997) Dry site versus wet site cover design, in: *Proceedings of 4th Int. Conference of Acid Rock Drainage p. 1595-1609*, Vancouver, B.C..

Wels, C., Shaw, S. and M. Royle (2000). A Case History of Intrinsec Remediation of Reactive Tailings Seepage- for Questa Mine, New Mexico. Proceedings from the 5th International Conference on Acid Rock Drainage (ICARD 2000), Denver, Colorado, Volume 1, pp. 441-458.

Wels, C., M. O'Kane, and S. Fortin (2001) Assessment of Water Storage Cover for Questa Tailings Facility, New Mexico. In Proceedings of the 9th Annual Conference of the American Society for Surface Mining Reclamation, Albuquerque, New Mexico, June 2001.

Tailings and Mine Waste '02, © 2002 Swets & Zeitlinger, ISBN 90 5809 353 0

Factors influencing net infiltration into mine rock piles at Questa mine New Mexico

C. Wels, S. Loudon & S. Fortin
Robertson GeoConsultants, Vancouver, B.C., Canada

ABSTRACT: From 1965 to 1983 large-scale open pit mining at the Questa mine, located in the Sangre de Cristo Mountains, New Mexico, produced over 300 million tonnes of mine rock, which was end-dumped into several steep valleys adjacent to the open pit. As a result, the mine rock piles are in general at angle of repose and vary in aspect and elevation (2,440m – 3,050m). Results of a detailed drilling program indicated significant variations in moisture contents between different rock piles (Shaw et al, these proceedings). Detailed monitoring of local climate conditions at six sites and measurements of net infiltration into the mine rock profile using four lysimeter test plots at three sites was carried out to evaluate the influence of local microclimate and physical mine rock properties on net infiltration. Initial results for the first year of monitoring indicated large variations in net infiltration. The local micro-climate at the high elevation site, in particular the development of a snow pack during the winter months, was found to result in significant net infiltration (~30% of annual precipitation). In contrast, warmer air temperatures combined with strong winds prevented the development of a continuous snow pack at the mid and low elevation sites resulting in higher actual evaporation and much reduced net infiltration (<6% of annual precipitation). The physical properties of the mine rock (particle size distribution, moisture retention capacity, Ksat) also influence net infiltration but to a lesser extent than the local microclimate.

1 INTRODUCTION

The Questa molybdenum mine, owned and operated by Molycorp Inc., is located 5.8 km east of the town of Questa in Taos County, New Mexico. From 1965 to 1983 large-scale open pit mining at the Questa mine produced over 300 million tonnes of mine rock, which was end-dumped into various steep valleys adjacent to the open pit (Figure 1). As a result, the mine rock piles are typically at angle of repose and have long slope lengths (up to 600m) and comparatively shallow depths (~30-60m). These conditions have large implications with respect to long-term oxidation and acid mine drainage from these rock piles and require definition and evaluation before appropriate mine closure measures can be developed.

Molycorp Inc. initiated a comprehensive characterization program in 1998 for the mine rock piles in order to provide a basis for the development of a closure plan (see Shaw et al., these proceedings). This characterization program included field reconnaissance and sampling of mine rock, physical and geochemical testing in the laboratory, as well as test plot and numerical model studies (RGC, 2000c). This paper focuses on the infiltration test plot study, which was initiated as part of this comprehensive characterization program (RGC, 2000a).

The principal objective of the infiltration test plot study is to collect site-specific data for estimating the net infiltration rate into the mine rock piles (i.e. rate of infiltration into the mine rock to a depth where it is no longer available for evapotranspiration). These estimates will provide the basis for estimating seepage from the rock piles and developing a water and load balance for the mine site.

2 SITE DESCRIPTION

The Questa mine site is located on the south facing slopes of the Sangre de Cristo Mountains in the middle reach of the Red River Valley (Figure 1). The mine rock piles cover a surface area of about 275 ha and extend vertically from just above the elevation of the Red River (~2,470m amsl) to about elevation 2,990m amsl, resulting in some of the highest mine rock piles in North America.

The physical and geochemical characteristics of the Questa mine rock are described in Shaw et al. (these proceedings). Briefly, the Questa mine rock can be divided into four broad categories (i) well-graded (finer) mixed volcanics; (ii) poorly-graded (coarser) aplite/black andesite; (iii) fine-grained erosion material (typically gap-graded); and (iv) very coarse rubble material (predominantly aplite/black

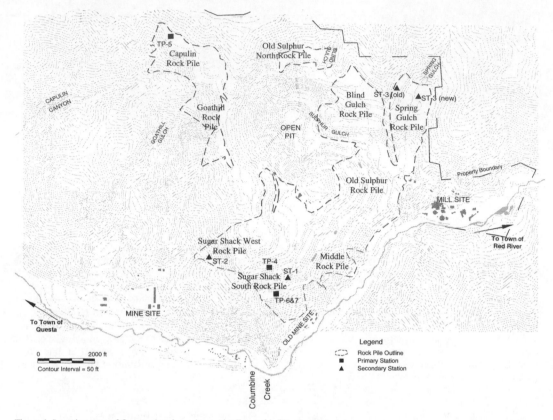

Figure 1. Location map of Questa showing mine rock piles and infiltration test plots.

andesite). Typically, the 'mixed volcanics' are composed of hydrothermally altered and highly weathered andesitic rock (including "scar material"). The erosion material represents mixed volcanic material, which was mobilized upslope and re-deposited further downslope during heavy rainstorm events (RGC, 2000b). This erosion material typically forms a very thin "crust" (~0.3-0.6m) in the up- and mid-slope areas but can be several meters thick near the toe of the rock slopes. The aplite porphyry and the black andesite are relatively competent, coarse rocks with little visual evidence of sulfide oxidation processes and physical weathering.

The climate is semi-arid with mild summers and cold winters. The average monthly temperature is below freezing for five months of the year (November through to March). The long-term average annual precipitation at the mill site (located at the base of the mine site) is approximately 400mm (15.8 inches). Lake evaporation (large free-water surface) and actual evapotranspiration (land surface including transpiration from vegetation) on site are estimated to be 1000mm (39.4 inches) and 400mm (15.8 inches), respectively (RGC, 2000c).

3 DESIGN OF TEST PLOT STUDY

In general, the amount of net infiltration into mine rock piles is controlled by (i) local climate conditions (precipitation and potential evaporation); (ii) physical properties of the mine rock (e.g. grading, moisture retention, permeability); and (iii) geometry and structure of the rock pile (e.g. slope angle, layering).

The key climate parameters influencing net infiltration (i.e. precipitation and potential evaporation) can be expected to vary significantly at a high-relief site such as the Questa mine. Long-term climate data from regional weather stations suggest a general increase in precipitation with elevation (approximately 43mm for every 90m increase in elevation, Wels et al, 2001). In addition, the duration of snow cover and the depth of snowpack generally increases with higher elevations, which may also significantly increase the amount of net infiltration (RGC, 2001b). Potential evaporation also shows a relationship with elevation (i.e., a decrease with increase in elevation) at a regional scale. However, potential evaporation is influenced by a variety of climate parameters (air temperature, relative humidity, net radiation and

470

Table 1. Summary of instrumentation at primary and secondary stations.

| Location | Station ID | Material | Lysimeter Test Plot Instrumentation | | | In-situ Monitoring | Met Station Instrumentation (dedicated) | | | |
			Soil Suction Sensors	Soil Moisture Access Tube	Drain flow Monitoring	Soil Moisture Access Tube	PREC	RH & T	W SP	NET R
Primary Stations										
upper bench on Sugar Shack South (near WRD-5)	TP-4	well-graded (finer) mine rock	10	1	yes	1	X	X	X	X
top of Capulin (near WRD-8)	TP-5	well-graded (finer) mine rock	10	1	yes	1	X	X	X	X
lower bench on Sugar Shack South (near WRD-3)	TP-6	poorly-graded (coarse) mine rock	-	1	yes	N/A	X	X	X	X
	TP-7	1ft of finer sediment over poorly-graded (coarse) mine rock	-	1	yes	N/A				
Secondary Stations										
midslope on Sugar Shack South (near WRD-4)	ST-1	poorly-graded (coarse) mine rock covered with fine sediment layer	N/A			5	X	-	-	-
midslope on Sugar Shack West (near WRD-6)	ST-2	well-graded (finer) mine rock covered with sediment layer	N/A			5	X	-	-	-
plateau area in Spring Gulch (near WRD-1)	ST-3	well-graded (coarser) mine rock	N/A			5	X	-	-	-

Legend N/A = not applicable
X = to be installed
- = not to be installed
5 = number of sensors/instruments

Abbreviations: PRECIP = Precipitation (tipping bucket w/ snowfall adaptor)
RH & T = Relative humidity and temperature (RH/T sensor)
W Sp = Wind Speed (Wind Monitor)
NET R = Net Radiation (net radiometer)

wind speed) and as such can be expected to deviate from this regional trend depending on local site conditions (e.g., aspect, ground cover, location along slope).

The physical properties of the mine rock near the surface are also expected to influence the amount of net infiltration. Field reconnaissance and laboratory testing indicated that the particle size distribution and associated hydraulic properties (soil water characteristic curve (SWCC) and hydraulic conductivity function) of the Questa mine rock varied significantly (see RGC 2000b, Wels et al., 2001). Hence, the rate of net infiltration can also be expected to vary significantly among those material types.

Finally, the geometry of the mine rock pile, in particular slope angle and layering, may also influence the amount of net infiltration. The major difference between infiltration on a sloped surface and a flat surface is the potential for greater surface runoff and capillary break effects on the slope resulting in lateral flow parallel to the slope. These factors would tend to result in lower rates of net infiltration along the slopes relative to the flat top of the mine rock piles.

The test plot study was designed to evaluate all three factors controlling net infiltration, i.e. climate conditions, physical material properties and rock pile geometry. A total of three primary stations and three secondary stations were instrumented (see Figure 1 for locations). Table 1 summarizes the instrumentation at the various locations. At each primary station, an instrumented lysimeter and a detailed weather

station were set up. The lysimeter tank, which has a diameter of 2.4m and a depth of 2.3m, was placed in an excavation and backfilled with mine rock material from that location (Figure 2). The base of the excavation was prepared as a conical depression (about 5-20cm dip over the radius of excavation) to

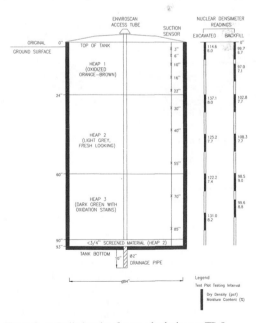

Figure 2. As-Built drawing for test plot lysimeter TP-5.

471

Table 2. Summary of climate data (July 23, 2000 – July 23, 2001).

	TP-4	TP-5	TP-6	TP-7
Cumulative precipitation (mm)[1]	380.0	395.5	462.3	
Cumulative Potential Evaporation (mm)[1]	1213.7	931.7	1143.8	
Lysimeter Outflow (mm)	0.055	115.92	25.90	0

Notes:

1. Missing data were patched with data from nearest weather station where required.

allow the tank to develop a slight slope towards the 50mm diameter drain in the center of the tank. The lysimeters are free-draining and outflow is monitored continuously using a tipping bucket and a data logger (see RGC, 2000d for more details).

Note that construction of such lysimeters was only feasible on flat surfaces. Attempts were made to install moisture content sensors into the angle-of-repose rock slopes (at the secondary stations) in order to measure in-situ moisture contents. However, these sensors did not provide reliable field data (Wels et al, 2001). Therefore, the net infiltration for sloped surfaces will have to be modeled using a soil-atmosphere model calibrated using the soil monitoring data from the lysimeter test plots and climate data from the secondary stations.

The four lysimeter test plots at the primary stations were designed to allow a detailed assessment of net infiltration of four different scenarios (see Figure 1 for location):

- well-graded (finer) mine rock material at mid-elevation (El. 2820m) (TP-4 at upper bench of Sugar Shack South rock pile);
- well-graded (finer) mine rock material at high-elevation (El. 2990m) (TP-5 on top of Capulin rock pile);
- poorly-graded (coarse) mine rock material at low elevation (El. 2660m) (TP-6 at lower bench of Sugar Shack South); and
- poorly-graded (coarse) mine rock covered with 0.3m (1ft) of fine-grained sediment material at low elevation (TP-7 at lower bench of Sugar Shack South).

Note that the lysimeter test plots were not revegetated. For more details on the design, construction and instrumentation of these test plots the reader is referred to Wels et al. (2001).

4 1ST YEAR MONITORING RESULTS

Table 2 summarizes the monitoring results at the four lysimeter test plots for the first year of monitoring (August 2000 – July 2001). The first year summary data indicate a very large difference in lysimeter outflow ranging from insignificant or no outflow

for TP-4 and TP-7 to 115.8mm (4.56 inches) in TP-5. The differences in total precipitation and potential evaporation was comparatively small (<25% between any two stations) and showed no clear correlation with the outflow data (Table 2).

These summary statistics indicate that factors other than precipitation and potential evaporation such as material properties and/or other climatic factors must influence net infiltration into the mine rock piles. A detailed analysis of the climate and lysimeter data (including regression analysis and time trend analysis) was carried out to determine the factors controlling net infiltration into the mine rock.

4.1 Climate data

Figure 3 shows scatter plots of daily precipitation measured at the primary stations TP-5 and TP6&7 versus those measured at TP-4. The results of a linear least square regression analysis are also shown (see solid line). The 1:1 line is shown for comparison (dashed line). The scatter plots suggest relatively uniform rates of precipitation across the mine site. The linear regression explains between 75-82% of the variation between any two sites.

The high elevation site (TP-5) received systematically lower precipitation during the few mid to high precipitation events (>10mm/day); however the opposite trend was observed for low precipitation events (Figure 3). In general, the low elevation site (TP6/7) matched precipitation rates observed at the mid-elevation site (TP-4) very well (Figure 3). However, TP6/7 received significantly more precipitation during some of the winter snowfall events resulting in higher total precipitation for the year (Table 2). In summary, the first year monitoring data do not support the general trend of increasing precipitation with higher elevation observed for the regional climate stations (Wels et al., 2001).

Figure 4 shows scatter plots of daily potential evaporation at the primary stations TP-5 and TP6&7 versus those measured at TP-4. Note that potential evaporation was calculated using the Penman method, which requires daily average air temperature, relative humidity, net radiation, and wind speed (Penman, 1948). The scatter plots indicate a very good correlation between the mid and low elevation

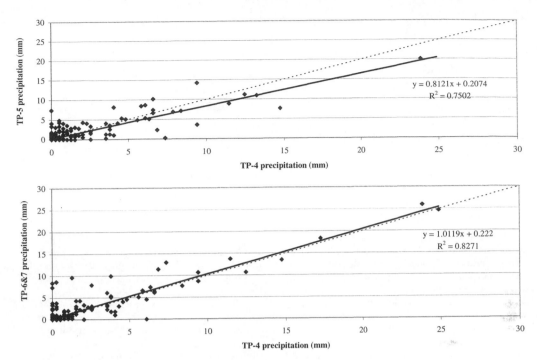

Figure 3. Comparison of daily precipitation at the three primary stations.

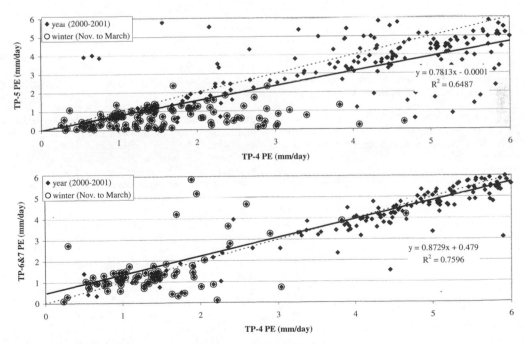

Figure 4. Comparison of potential evaporation at the three primary stations.

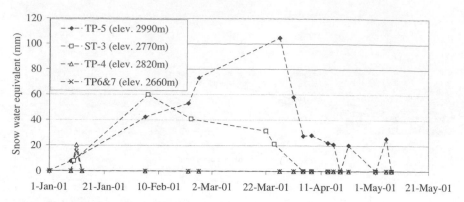

Figure 5. Results of snow surveys at primary and secondary stations (in water equivalents).

sites on Sugar Shack South (TP-4 and TP-6&7). In contrast, the high elevation site on Capulin (TP-5) recorded significantly lower potential evaporation rates, in particular during the winter months (circles). The lower potential evaporation rates at the high elevation site are predominantly a result of lower air temperature combined with lower net radiation (due to a continuous snow pack for most of the winter months, see below).

Snow surveys were carried out at the primary and secondary stations between January and May 2001 to estimate the accumulation and ablation of the snowpack across the mine site. Figure 5 shows the measured snow water equivalents (average of 10 point values per site) for the spring of 2001. Note that some snow accumulation occurred in November 2000 (not shown). However, most of this snow had disappeared by early January 2001, when the first snow survey was taken. The snow survey data indicate significant differences in snowpack development across the site. As expected the largest snowpack was observed at the high elevation site (TP-5) with a peak snow water equivalent (SWE) of 107mm (4.2 inches) in late March 2001. Some, albeit smaller, snowpack development was also observed at the secondary station ST-3 (Spring Gulch) (see Figure 1 for location). At the other three primary stations on Sugar Shack South (TP-4 and TP-6/7) the winter snowfall did not result in any significant snowpack accumulation (presumably due to the warmer temperatures/higher net radiation on these south-exposed slopes). These differences in snowpack development were found to have a strong influence on the rate of net infiltration into the mine rock piles (see below).

4.2 Test plot response

Figure 6 shows the time trends of soil suction at selected depths for the first year of monitoring in test plots TP-4 and TP-5. Soil suction was measured in these lysimeter test plots using a thermal conductive

ity type sensor (Campbell Scientific Inc. model 229). Similar measurements could not be taken in test plots TP-6 and TP-7 due to the coarse nature of the backfilled mine rock in those lysimeters (Wels et al., 2001).

The time trends in soil suction illustrate the difference in the wetting of the two mine rock profiles. In test plot TP-5, the heavy rains in August 2000 (78.7mm fell between August 13-29) resulted in a rapid wetting of the top 0.75m (note steep decline in soil suction at 0.76m depth around late August). The wetting of the deeper mine rock profile of TP-5 occurred more gradual. By mid-December the entire mine rock profile had "wetted up" and soil suction had reached values near saturation (<1 kPa). Note that the soil suction remained near saturation throughout the spring runoff and early summer 2001, except for some drying in the near-surface layers (<0.30m).

In test plot TP-4, the heavy precipitation in August and October-November 2000 also resulted in a significant reduction in soil suction (wetting) in the near-surface (e.g. at a depth of 0.15m, Figure 6). However, the near-surface layers did not remain near saturation for any prolonged period of time, but instead, increased again within days after precipitation to higher suctions. The drier near-surface conditions resulted in a much slower wetting of the deeper mine rock profile. For example, at 0.76m depth a significant decrease in soil suction was only observed in mid-November. In the deeper mine rock profile (>1.40m below surface) the suction actually increased temporarily during the winter months suggesting a slight drying trend. The soil suction near the base of the lysimeter (at 2.24m depth) at the end of the first year of monitoring was still >70kPa, i.e. well below saturation.

The total outflow observed at the base of test plots TP-4 and TP-5 is consistent with the soil suction trends shown in Figure 6. In test plot TP-4 no discharge was observed in the first 12 months of

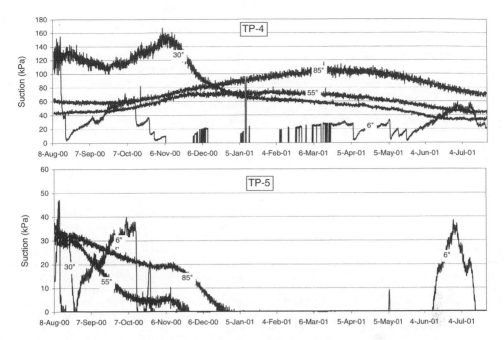

Figure 6. Time trends of soil suction in lysimeter test plots TP-4 and TP-5.

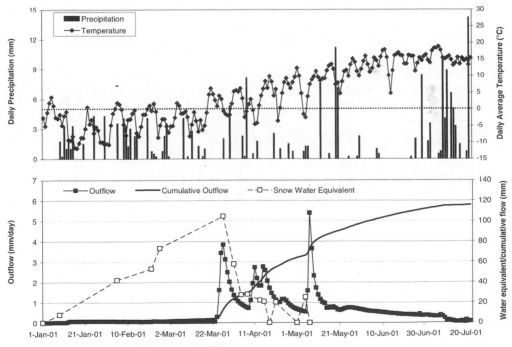

Figure 7. Outflow from lysimeter test plot TP-5 for the period Jan. 1 – July 23, 2001 (lower panel). Relevant climate parameters are shown for comparison.

monitoring (Table 2). In contrast, the total (cumulative) outflow collected from test plot TP-5 was 116mm (or ~30% of total precipitation, Table 2).

Figure 7 shows the daily and cumulative outflow from test plot TP-5 for the period January 1 – July 20, 2001 (lower panel). The snow water equivalent of the snowpack in this area is shown for comparison. Daily precipitation and average air temperature at this site are also shown (upper panel). The first significant outflow from TP-5 occurred in mid-March following several days of air temperatures above the freezing point (Figure 7). A detailed analysis of hourly outflow rates and climate data suggested a lag time of 24-36 hours between days of significant snowmelt and/or precipitation and increased outflow at the base of the lysimeter (RGC, 2001c). The outflow from the lysimeter gradually declined to very low rates (<0.254mm/day) after final depletion of the snowpack in mid-May. Note that the heavy precipitation events in June-July did not produce any significant outflow from the lysimeter. The data strongly suggest that the vast majority of net infiltration into the mine rock profile occurred during the snowmelt period.

Figure 8 shows the outflow data for test plot TP-6, located at the low elevation site. Outflow from test plot TP-6 was significantly lower (33mm) and more delayed than at TP-5 (c. Figures 7 and 8). This result was somewhat unexpected considering that test plot TP-6 was backfilled with much coarser, more permeable, mine rock than TP-5 (Wels et al., 2001). The much lower outflow rates are a result of higher storage requirements in the coarser mine rock and/or higher evaporative losses (relative to TP-5) resulting in reduced rates of net infiltration.

The main difference in climate conditions between the high and low elevation sites was the magnitude and duration of snowpack development (Figure 5). The lack of any snowpack at TP-6 likely allowed significant evaporation to occur from the mine rock surface throughout the winter and spring period. In contrast the development of a snowpack at TP-5 prevented evaporation until late into the spring season while, at the same time, delivering a large amount of melt water (up to 104mm of water equivalent) during a very short snowmelt period (~3 weeks). The amount and duration of a snowpack is believed to have a major control on the rate of net infiltration into the Questa mine rock piles.

Note that no outflow was observed in test plot TP-7 which only differed from TP-6 in the presence of a thin (0.3m) layer of finer-grained erosion material on top of the coarse-grained mine rock (Table 2). A comparison of these two test plots suggests that even a thin layer of finer-grained material can reduce the rate of net infiltration (under these climate conditions).

5 CONCLUSIONS

This paper describes the design and initial monitoring results of lysimeter test plots designed to study the net infiltration into mine rock piles at the Questa

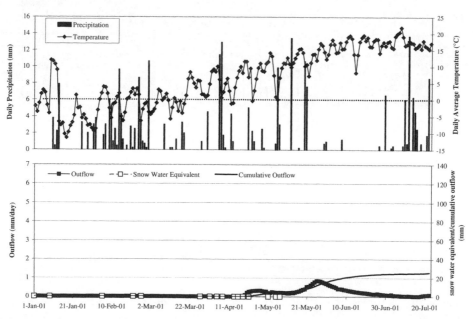

Figure 8. Outflow from lysimeter test plot TP-6 for the period Jan. 1 – July 23, 2001 (lower panel). Relevant climate parameters are shown for comparison.

mine. Initial test plot data collected during the first year of monitoring illustrate the large variability in infiltration that might be expected under field conditions on the mine rock piles. The lysimeter test plot located at the high elevation site (El. 2990 m amsl) wetted up within 5 months and experienced 114mm (4.5 inches) of outflow during the spring runoff period. In contrast, the lysimeter test plot at the mid elevation site (El. 2820 m amsl) did not show any outflow within the first year despite similar physical properties of the mine rock profile. The lysimeter test plot located at the low elevation site (El. 2660m amsl) showed an intermediate response with some outflow (38.1mm (1.5 inches)) late in the spring runoff period.

The much higher rates of net infiltration at the high elevation site are attributed to the development of a continuous snowpack. The snowpack prevents any evaporation from the mine rock surface during the winter and spring months. In addition, melting of the "ripe" snowpack during spring runoff results in prolonged, high rates of infiltration (e.g. 104mm of snow water equivalent melted in ~21days).

This finding is consistent with initial soil atmosphere modeling carried out for the Questa mine rock piles which indicated that the rate of net infiltration is most sensitive to the amount of "effective precipitation" (i.e. snowmelt and rainfall) occurring during the spring runoff period (RGC, 2001a).

The physical properties of the mine rock, in particular in the upper 0.3m, were also found to have an influence on net infiltration, albeit to a lesser extent than the climate conditions. The presence of a thin (0.3m) layer of well-graded (finer) erosion material, commonly found on the slopes of the Questa mine rock piles, significantly reduced the rate of net infiltration at the low elevation site (no outflow has yet been recorded).

The results of this infiltration study are consistent with field observations which suggest significantly lower rates of seepage from the mine rock piles at mid to low elevations (e.g. Sugar Shack South) compared to those at high elevation (e.g. Capulin and Goathill). The spatial differences in local microclimate and net infiltration will have to be taken into consideration when developing a water and load balance for the Questa mine site.

Modeling work is currently in progress to calibrate the soil-atmosphere model SOILCOVER (Geo -Analysis 2000 Ltd., 2000) using the observed test plot data (RGC 2001b). Once calibrated, this soil atmosphere model can be used to assess the influence of local climate conditions, material properties and slope angle on the rate of net infiltration.

6 ACKNOWLEDGEMENTS

The test plot study was funded by Molycorp Inc. and could only have been completed with their large and dedicated commitment to understanding the physical and geochemical processes that are active on the Questa mine site. We would like to acknowledge the extensive contributions by both Molycorp personnel, in particular Geyza Lorinczi, David Shoemaker, and Anne Wagner; and the constructive comments during the development of work plans from personnel from both New Mexico Mining and Minerals Division and Environmental Department.

REFERENCES

Geo-Analysis 2000 Ltd. (2000), SOILCOVER User's manual, May 2000.

Penman, H.L. 1948. Natural evapotranspiration from open water, bare soil and grass, *Proc. Royal Soc. London Ser.* A 193: 120-145.

Robertson GeoConsultants Inc. 2000a. Work Plan for Waste Rock Water Balance Study, Questa Mine Site, New Mexico, *RGC Report 052008/4 prepared for Molycorp Inc., February 2000.*

Robertson GeoConsultants Inc. 2000b. Progress Report: Results of Phase 1 Physical Waste Rock Characterization, Questa Mine, New Mexico. *RGC Report 052007/4 prepared for Molycorp Inc., June 2000.*

Robertson GeoConsultants Inc. 2000c. Interim Mine Site Characterization Study, Questa Mine, New Mexico, *RGC Report 052008/10 prepared for Molycorp Inc., November 2000.*

Robertson GeoConsultants Inc. 2000d. As-Built Report – Infiltration Test Plots for Mine Rock Piles, Questa Mine, New Mexico, *RGC Report 052008/11 prepared for Molycorp Inc., November 2000.*

Robertson GeoConsultants Inc 2001a. Initial Soil-Atmosphere Modeling for Mine Rock Piles, Questa Mine, New Mexico. *RGC Report 052008/14 prepared for Molycorp Inc., January 2001.*

Robertson GeoConsultants Inc 2001b. Progress Report on Water Balance Analysis for Mine Rock Piles, Questa Mine, New Mexico. *RGC Report 052008/17 prepared for Molycorp Inc., February 2001.*

Shaw, S., C. Wels, A. MacG. Robertson and G. Lorinczi 2002. Physical and Geochemical Characterization of Mine Rock Piles at the Questa Mine, New Mexico: An Overview, *9th International Conference on Tailings and Mine Waste '02:* Rotterdam: Balkema.

Wels, C., M. O'Kane, S. Fortin and D. Christensen 2001. Infiltration Test Plot Study for Mine Rock Piles at Questa Mine, New Mexico. *Proceedings of the 9th Annual Conference of the American Society for Surface Mining Reclamation, Albuquerque, June 2001:* 195-209. New Mexico: The American Society for Surface Mining and Reclamation.

ARD production and water vapor transport at the Questa mine

R. Lefebvre
INRS-Géoressources, Quebec Geoscience Centre, Quebec, Qc, Canada

A. Lamontagne
Experts Enviroconseil Inc., Quebec, Qc, Canada

C. Wels & A.MacG. Robertson
Robertson GeoConsultants Inc., Vancouver, B.C., Canada

ABSTRACT: Numerical simulations of acid rock drainage production in the mine rock piles at the Questa Mine were performed to (i) identify the main processes responsible for the observed present-day conditions, such as temperature and oxygen, and (ii) provide an estimate of the significance of gas phase humidity transfer on the water balance of the pile. Numerical simulations show that the mine rock pile geometry on mountain slopes favors strong thermal air convection which results in evaporation of pore water leading to a reduction in the moisture content of the mine rock. Conceptual simulations indicated that "air drying" could be a significant component of the water balance for the mine rock pile. Numerical simulations considering benches also indicated significant redistribution of water vapor with vaporization of pore moisture in the lower portion of the pile and condensation in its upper portion. However, water balance calculations indicated that the net vapor loss from the entire rock pile is relatively small, accounting for only about 2 % of net infiltration.

1 INTRODUCTION

Acid rock drainage (ARD) is caused by oxidation of pyrite-bearing rocks. The rate of pyrite oxidation is controlled by a complex combination of biochemical processes influencing reaction kinetics, as well as by coupled physical processes including the movement of air and water within piles (Figure 1).

Most mine rock piles are partially saturated and contain sufficient pore water to sustain pyrite oxidation. In contrast, the movement of air (and oxygen) within rock piles is often sufficiently slow to control pyrite oxidation which consumes oxygen. Heat production also occurs since pyrite oxidation is strongly exothermic (11.7 MJ/kg pyrite). The release of heat drives temperature up, as high as 70 °C in places (Gélinas et al., 1992). In high permeability material, this rise in temperature leads to thermal air convec-

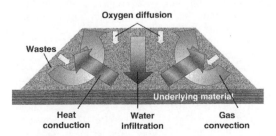

Figure 1. Conceptual model of physical processes acting within waste rock piles (after Lefebvre et al., 2001a).

tion within the piles. This process is a much more efficient oxygen transport mechanism than diffusion and it sustains higher global oxidation rates. Water infiltration through the material picks up soluble components to form an acidic leachate containing sulfate and metals. The quantitative representation of these processes requires numerical simulation.

A comprehensive physical and geochemical characterization of the mine rock piles has been carried out at the Questa mine over the last few years to evaluate the potential for current and future ARD production (Shaw et al., these proceedings). As part of this study, the numerical simulator TOUGH AMD was used to model ARD in the Questa mine rock piles. Table 1 summarizes the processes represented by TOUGH AMD (Lefebvre, 1994, Lefebvre et al., 2001a, 2001b). TOUGH AMD is based on the general multiphase simulator TOUGH2 (Pruess 1991).

This paper focuses on the effect of water evaporation within mine rock piles due to thermal gas convection. This process results from the entry of relatively cold air within the pile. Heating of this air flowing through the rock pile increases its water content. Part of the pore water then evaporates and is carried away in the gas phase. This process could lead to partial drying of waste rock locally. This process was of interest for the Questa mine rock piles since the site is located in a relatively dry climate and gas convection was believed to be important (Shaw et al., these proceedings).

Table 1. Processes simulated by TOUGH AMD.

System Phases and Components	

The system includes two fluid phases (liquid and gas) and four components (water, air, oxygen in air, and heat) in both phases. The state of the system is fixed by 4 primary variables: fluid pressure, water saturation, oxygen mass fraction in air and temperature.

Multiphase Fluid Flow

A multiphase formulation of Darcy's Law represents the simultaneous flow of gas and liquid phases under flow potentials including the effects of fluid pressure, temperature and density. Fluid pressures include capillary pressure, which is a function of water saturation. Relative permeability for each fluid phase is also related to water saturation.

Heat Transfer

Conduction (Fourier's Law), fluid flow (gas and liquid), and gaseous diffusion of heat are simulated. Latent heat related to liquid vaporization and condensation is considered. The heat stored in solids is considered as well as the heat in fluids. A semi-analytical method is used to calculate conductive heat loss to impermeable confining units.

Gaseous Diffusion

Diffusion of all mass components in the gas phase is modeled using Fick's Law and an effective diffusion coefficient for a partially water-saturated porous media.

Pyrite Oxidation Kinetics

A reaction core model is used to calculate the pyrite oxidation rate with first order kinetics relative to oxygen concentration. The reaction consumes oxygen and produces heat according to pyrite oxidation thermodynamics. Pyrite remaining and sulphate production are tracked.

Dissolved Mass Transport

Dissolved mass transport of the sulphate produced by pyrite oxidation is considered by a simple mass balance.

2 SITE DESCRIPTION

The Questa molybdenum mine, owned and operated by Molycorp Inc., is located in the Sangre de Cristo mountains in Taos County, northern New Mexico. From 1965 to 1983 large-scale open pit mining at the Questa mine produced over 297 million tonnes of mine rock, which was end-dumped into various steep valleys adjacent to the open pit. As a result, the mine rock piles are typically at angle of repose and have long slope lengths (up to 600 m), and comparatively shallow depths (~30-60 m). Molycorp Inc. has initiated a comprehensive physical and geochemical characterization program of the mine rock piles to assist in the development of suitable closure measures (Robertson GeoConsultants Inc., 1999a, 1999b, 2000). The climate on the mine site is semi-arid with mild summers and cold winters. Mean annual precipitation at the mill site is about 400 mm. Average net infiltration into the mine rock piles is estimated to range from 30 to 90 mm/yr (Robertson GeoConsultants Inc., 2001).

This paper focuses on the Sugar Shack South (SSS) rock pile which contains about 18 million m^3 of mine rock covering a surface area of approximately 1200 m by 450 m with a maximum thickness (in the valley center) of little over 100 m. Figure 2 shows a section through an edge of the Sugar Shack South mine pile in which three instrumented boreholes were installed in 1999. The rocks in the pile are mixed volcanics with grain sizes ranging from a d_{10} of 0.005-0.15 mm to a d_{90} of 25-42 mm.

Figure 3 shows temperatures and oxygen concentrations in three monitoring boreholes in the SSS pile (Robertson GeoConsultants Inc., 2000). WRD-3 is located on a bench in the lower half of the pile whereas WRD-4 and WRD5 were drilled through

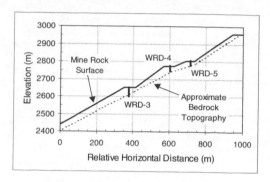

Figure 2. Section through the Sugar Shack South mine pile showing the location of instrumented boreholes (after Robertson GeoConsultants Inc., 1999a).

upper benches (Figure 2). WRD-3 shows a very small increase in temperature and almost atmospheric oxygen concentration. These conditions indicate that the material is less reactive in the upper portion of the pile in this area as indicated also by the paste pH and descriptions of the material drilled there. WRD-4 and WRD-5 instead exhibit higher temperatures that are steadily increasing with depth and that are diagnostic of high pyrite oxidation rates over the entire pile thickness.

Despite the inferred high reaction rate, oxygen concentrations remain high within the SSS pile. We conclude from these conditions that oxygen supply is important relative to consumption in the SSS pile and that oxygen is provided by strong lateral thermal air convection upslope of the pile. The next section on numerical simulation provides more details on the processes leading to the observed conditions in the pile.

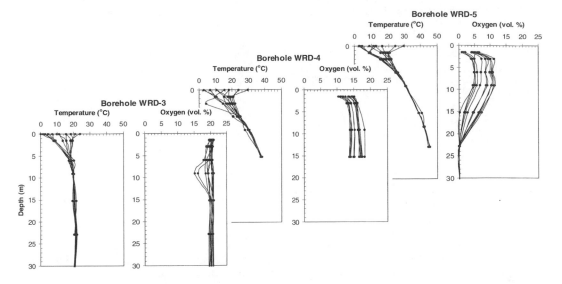

Figure 3. Monthly conditions observed in the SSS pile in three monitoring boreholes in year 2000 (Lefebvre et al., 2001).

3 NUMERICAL SIMULATIONS

3.1 Properties and conceptual model

Two different numerical models were used in this study: 1) a model with a simplified geometry was developed to investigate whether or not the process of moisture transfer in the gas phase could be significant; and 2) a more detailed model with a geometry more representative of the actual SSS pile was used to numerically reproduce the observed conditions in this pile. This second model allowed the identification of the processes acting in the SSS pile and provided the basis to estimate the importance of moisture transfer in the gas phase in the water balance of this pile.

Table 2 summarizes the material properties used in the two numerical models. As indicated in the table caption, the "bulk of material" and "boulder layer" properties represent the values used in the "best case" simulation of the SSS pile whereas the "conceptual model" properties were used in the model using a simplified geometry of the mine rock piles. Compared to other sites, the Doyon Mine in Canada and the Nordhalde pile in Germany, the material in the Questa mine rock piles was found to have an intermediate reactivity, a high permeability, and a peculiar geometry due to the mountain slope setting of the site (Lefebvre et al., 2000).

Figure 4 shows the numerical grid for the first model using a simplified geometry of the mine pile which is not taking into account the benches along the slope of the pile. This initial model was based on partial data provided by the site characterization

program and it was based only on the early monitoring data. On this basis and with this simplified geometry, it was found necessary to use four layers with increasing permeability from 10^{-9} m^2 at the surface of the pile to 5×10^{-8} m^2 at its base in order to numerically reproduce the general conditions observed in the SSS pile. At the time, this steady increase in permeability with depth was believed to generally represent the conditions of the site. Further observations showed the actual field conditions to be different from this idealized model.

Figure 5 shows the representation of the SSS pile with the two-dimensional vertical irregular grid used for the second set of simulations. The grid takes into account the presence of benches as well as the main changes in waste rock thickness. As before, the

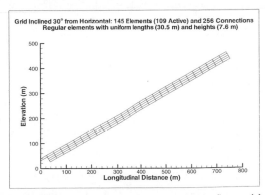

Figure 4. Rectangular numerical grid used for the first model with a simplified mine rock pile geometry.

481

Table 2. Physicochemical properties of the mine rock pile materials. The "bulk of material" and "boulder layer" properties represent the values used in the "best case" simulation of the SSS pile whereas the "conceptual model" properties were used in the model using a simplified geometry of the mine rock piles.

Properties	Symbols and units	Bulk of Material	Boulder Layer	Conceptual Model
Properties of the mine rock material				
Pyrite mass fraction in solids	w_{py} dim.	0.035	0.035	0.07
Solids density	ρ_s kg/m^3	2723	2723	2740
Porosity	n dim.	0.31	0.31	0.33
Average water saturation	S_w dim.	0.38	0.21	0.35
Humid global density	ρ_b kg/m^3	2059	2006	2235
Properties related to the pyrite oxidation rate				
Volumetric oxidation constant	K_{Ox} s^{-1}	1.0×10^{-7} *	1.0×10^{-7}	2.5×10^{-7}
Diffusive/Chemical total times	τ_d / τ_c dim.	2.5	2.5	2.5
Properties related to fluid flow				
Residual water saturation	S_{wr} dim.	0.14	0.14	0.14
van Genuchten "m" factor	m dim.	0.25	0.45	0.25
van Genuchten "α" factor	α Pa^{-1}	0.002	0.0035	0.0036
Horizontal permeability	k_h m^2	3.5×10^{-9}	1.0×10^{-8}	1×10^{-9} to 5×10^{-8}
Vertical permeability	k_v m^2	3.5×10^{-10}	1.0×10^{-8}	1×10^{-9} to 5×10^{-8}
Effect. vertical air permeability	k_a m^2	3.0×10^{-10}	0.9×10^{-8}	variable
Water infiltration rate	q_i m/y	0.100	0.100	0.095
Properties related to heat transfer				
Average thermal conductivity	K_{th} W/m·°C	1.65	1.65	1.65
Heat capacity of solids	c_{ps} J/kg·°C	837	837	837
Properties related to gas diffusion				
Effective oxygen diffusivity	D_e m^2/s	2.87×10^{-6}	3.65×10^{-6}	3.20×10^{-6}

*: $K_{Ox} = 1.5 \times 10^{-10}$ s^{-1} for low reactivity material

model grid was oriented at a 30° angle with the horizontal.

Table 2 presents the properties assigned to the two SSS material types considered for the second model: 1) the bulk of the waste rock consists of layers of relatively fine and intermediate size mine rock which can have a lower reactivity locally; and 2) a coarse boulder layer is found at the base of the pile due to material segregation caused by free dumping. Initial parameter values were determined from site characterization data and later modified in the calibration phase of the simulations. The parameters varied in the simulations to take into account the uncertainties in field values were 1) the magnitude and anisotropy of permeability; 2) the oxygen volumetric oxidation constant; 3) water infiltration; 4) the presence and permeability of the "boulder layer" at the base of the pile; and 5) the presence and distribution of low reactivity rocks.

Figure 6 illustrates the distribution of materials used in the second model representing the SSS pile. The bulk material is present over most of the model but it has a lower reactivity in the lower part of the pile near WRD-3. A more permeable boulder layer is represented at the base of the pile.

The boundary conditions used in the second model are: 1) yearly cyclic surface temperature variations (10 °C sinusoidal changes around a mean value of 14 °C); 2) hydrostatic gas pressure profile with depth; 3) atmospheric oxygen concentration; and 4) fixed water saturation at the surface imposing a constant water infiltration of 100 mm/y. For the purpose of this modeling exercise free drainage is assumed along the base of the rock pile (fractured bedrock). Also, conductive heat loss is represented from the base of the pile to the underlying bedrock.

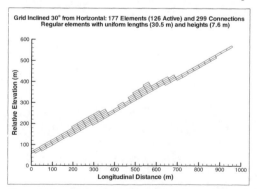

Figure 5. Numerical grid with benches used in the 2nd model representative of the Sugar Shack South mine rock pile.

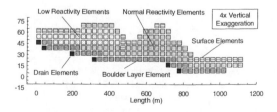

Figure 6. Material distribution and boundary conditions used for the "best case" of the Sugar Shack South simulations.

Similar conditions were used for the simplified model except that surface temperature was fixed at 10 °C rather than varied.

3.2 Simplified model results

Figures 7 and 8 illustrate the results of the simplified model. A vertical exaggeration is used in figures showing simulation results for better viewing. Figure 7 shows temperature and gas velocity on the top graph, whereas oxygen mass fraction and fluxes are presented on the bottom graph. This figure shows that a strong penetrative gas convection pattern is reaching the base of the pile and that a higher gas velocity is found in the more permeable basal layer. A steadily increasing temperature profile is obtained with the highest temperatures found near the base of the pile as observed in the monitoring boreholes at the site. Also, enough oxygen is brought in by gas advection to maintain high oxygen concentrations in the pile. The temperatures and oxygen concentrations simulated are thus in the right range compared to field data (Figure 3). The temperature distribution is found to be affected by water vaporization which is cooling the lower third of the pile whereas water condensation heats the upper third of the pile.

Figure 8 presents temperature and water vapor transfer fluxes on the top graph, with liquid water saturation and fluxes on the lower graph. This figure

supports the concept that major water vapor transport is physically possible under the simulated conditions: 1) the water vapor fluxes are in the same range as liquid water fluxes, and 2) water saturation is reduced very significantly in the core of the pile by vaporization from 0.35 to 0.24. These results thus show considerable drying of the core of the pile due to water vapor transport in the gas phase. Other simulations not presented here even reached total material drying locally when using either lower initial water saturation or lower water infiltration.

Although conceptually interesting, these initial simulations may not be representative of actual field conditions. First, the physical conditions presented in the previous figures are achieved for a modeling period limited to 10 years. This time frame was selected to obtain relatively fast simulation answers, and finally with the expectation that pseudo-steady state conditions could be reached in the system within that time frame. Our latest results show that such stable conditions may be reached only after a 20 to 25 year period (Figure 11). The second important limitation of these initial simulations is the simplified geometry representing the system. Benches can have an important effect on gas flow patterns as demonstrated by our latest simulation results. Finally, the data available on which to base the values of the material physical properties were not extensive at that time.

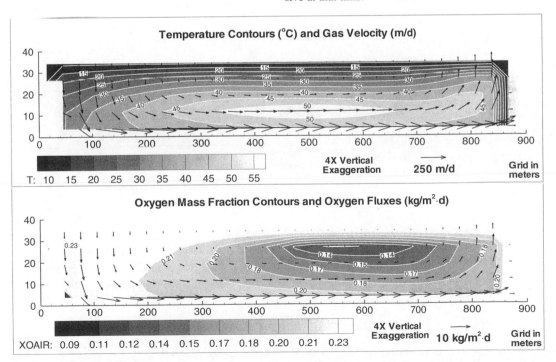

Figure 7. Initial simulation results with a simplified rectangular grid: temperature and gas velocity (top) and oxygen mass fraction and fluxes (bottom). The grid makes a 30° angle with the horizontal.

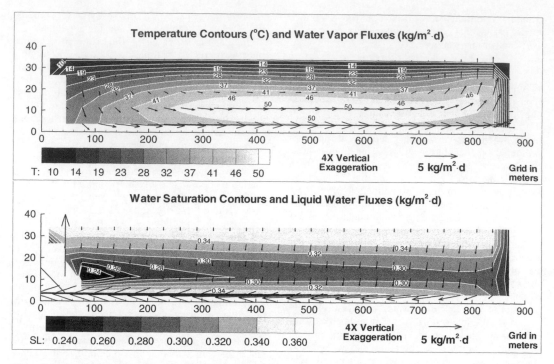

Figure 8. Initial simulation results with a simplified rectangular grid: temperature and water vapor fluxes (top) and liquid water saturation and fluxes (bottom). The grid makes a 30° angle with the horizontal.

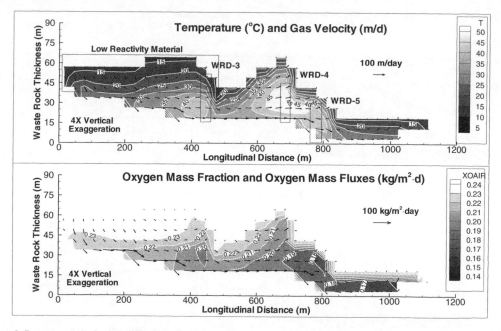

Figure 9. Best case results for Sugar Shack South with low reactivity material: temperature and gas velocity (top) and oxygen mass fraction and flux (bottom). The grid makes a 30° angle with the horizontal. The approximate locations of boreholes WRD-3, WRD-4 and WRD-5 are shown by rectangles (see also Figure 2).

484

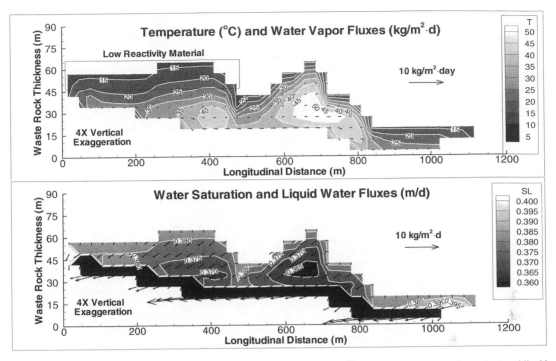

Figure 10. Best case results for Sugar Shack South with low reactivity material: temperature and water vapor fluxes (top) and liquid water saturation and fluxes (bottom). The grid makes a 30° angle with the horizontal.

These initial conceptual simulations still demonstrate that 1) water vapor transfer can be an important component of a mine rock pile water balance, 2) temperature in a mine rock pile can be significantly affected by the vaporization and condensation of water, and 3) an inclined geometry favors the development of strong lateral thermal gas convection.

3.3 Sugar shack south model results

The "best case" results for the SSS model is illustrated by Figures 9 and 10. To obtain these results, low reactivity material was assumed to be present in the lower portion of the pile as observed in the top portion of the drill log for WRD-3. This low reactivity material reduces heat production and leads to temperatures more representative of observations. Also, since less oxygen is consumed in the lower part of the pile, more oxygen reaches the upper portion which shows temperature and oxygen values close to observed levels (Compare Figures 9 and 3). As seen in Figure 10, the water vapor fluxes are much lower in magnitude than liquid water infiltration even though there is significant drying of the material as shown by the local water saturation reductions.

For the SSS pile, the main properties found to influence ARD were material permeability and reactivity. Permeability controls thermal gas convection

bringing the atmospheric oxygen required to sustain pyrite oxidation. Reactivity determines if the oxygen brought in the pile is consumed fast and thus close to the edges of the pile, or rather slowly so that it can penetrate deeper within the pile.

Figure 11 shows the evolution through time of the simulated average conditions for the entire pile for the Sugar Shack South under its present-day configuration. Results are shown for parametric runs including 1) a homogeneous case, 2) an anisotropic case where horizontal and vertical permeabilities are different, 3) a boulder layer case where a very permeable basal layer is added, and 4) the best case where low reactivity material is taken into account. Figure 11 shows temperature, remaining pyrite mass fraction, volumetric oxidation rate, and oxygen mass fraction. A summary of the main conditions obtained for these cases is also compiled in Table 3.

Figure 11 provides an overview of the general simulated conditions resulting from a given set of properties. Also, the figure gives an indication of the time required to reach pseudo steady-state conditions under which temperature, oxygen mass fraction and oxidation rate remain about constant. Such conditions can be achieved after a relatively short period in some cases (less than 10 years for the Homogeneous Case) but they may just be reached after 20 to 25 year simulation for all the other cases. On all these

Table 3. Approximate average conditions for the Sugar Shack South parametric simulation cases.

Simulation Case	Maximum Temperature (°C)	Maximum Oxidation Rate (kg/m³·y)	Maximum Oxygen Mass Fraction	Pyrite Mass Fraction After 25 years	Local Minimum Water Saturation
Homogeneous	45	1.15	0.21	0.75	0.30
Anisotropic	34	0.42	0.18	0.90	0.39
Boulder Layer	36	0.48	0.20	0.89	0.38
Best Case	31	0.37	0.21	0.87	0.36

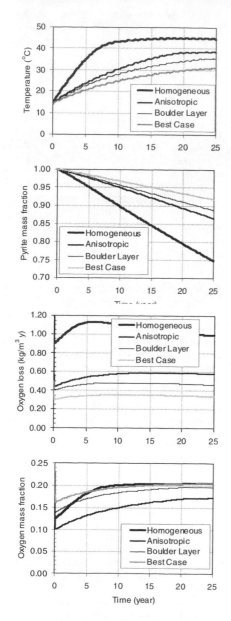

Figure 11. Average conditions through time for the different simulated conditions for the SSS mine pile.

figures, small cyclic variations can be seen, especially for temperature. These are related to the cyclic surface temperature imposed as boundary condition.

The initial homogeneous case is the one reaching the highest temperature, oxidation rate and oxygen mass fraction. This is due to the higher permeability and reactivity used for this case compared to the other ones. All other cases (anisotropic, boulder layer, and best case) exhibit similar behaviors because the permeability and reactivity of the materials used in these cases are nearly the same. The lower average temperature and reactivity of the base case is caused by the inclusion of very low reactivity material for a large part of the pile. The homogeneous case also shows the most important reduction in water saturation compared to other cases because gas convection is more important with the conditions used in that model.

4 DISCUSSION

4.1 Controls on gas flow and oxidation rate

A comparison of the conditions obtained from initial simulations using the simplified rectangular grid with the results of the more detailed grid with benches provides an indication of the effects of benches on gas flow. The presence of benches is seen to strongly influence gas flow and temperature patterns. In the initial rectangular grid, a single large gas convection cell formed from the base to the top of the system. For the irregular grid, there are instead two main gas convection cells in the lower and upper halves of the pile.

In the initial grid, layers of increasing permeability had to be introduced to obtain a convection pattern that could cover the entire system and provide oxygen everywhere. In the final irregular grid, gas convection can be triggered either with homogeneous or anisotropic material and even without the presence of a basal boulder-type material. In other words, the onset of strong gas convection patterns may be controlled more by the geometry of the system, especially the sloping of the pile and the presence of benches, rather than by a specific distribution of materials or specific values of permeability for these materials.

Even though the benches do not represent large irregularities at the scale of the Questa waste rock piles, they are preferential gas entry and exit pathways. The horizontal portions of benches reduce the gas flow section and the naturally upward hot gas flow tendency is enhanced by these surfaces. Also, the toes of the slopes starting upslope of these horizontal surfaces provide preferential air entries. This behavior shows the importance of keeping a representative geometry of the system in the numerical grids.

4.2 *Gaseous and liquid moisture transfer*

The initial simulations with a rectangular grid showed that significant drying could occur due to water vapor transfer of humidity under these simulated conditions. Local drying of some cells could even occur when small initial water saturations were used in these simulations. However, for the "best case" simulations shown here, we find that humidity transfer in the gas phase reduces water saturation within the pile but only intercepts a small proportion of the water infiltration. This reduction in water saturation agrees with the observed trends in mass water content reduction with depth in the piles (Shaw et al., these proceedings). The reduction in saturation is quite variable depending on the simulation cases considered but no case with the irregular grid has led to the complete drying of the pile. A water balance was performed for the best case on the liquid and vapor phase water transfer in the entire pile. This calculation shows that only 2% of the infiltrating liquid water is removed from the pile by water vapor transfer (assuming 100 mm/yr net infiltration). The relative contribution of water vapor loss to the overall water balance of the mine rock pile would be proportionally higher for a lower net infiltration.

The process of moisture transfer in the gas phase and its effect on the removal of water from a waste rock pile has some practical implications on ARD production from waste rock pile. In the Questa mine rock piles, very few places show leachate production at the toe of piles. There was thus a possibility that partial drying of water infiltration could reduce or at least delay the production of acidity from the pile. If this process had been very important, it would not have been advisable to cover the site and thus reduce this helpful process. The hypothetical effect of a cover on the SSS pile is discussed further by Lefebvre et al. (2001).

Moisture transfer in the gas phase also has implications on the evaluation of ARD production using temperature profiles. Figure 7 showing the results of the simulations with the simplified grid indicates that the temperature in a pile can be significantly modified by latent heat effects related to evaporation and condensation of water vapor within a pile. When these processes are strong, the temperature near the entry of a pile can be reduced by the vaporization of water: the pile then acts as a cooling system. On the contrary, in the upper portion of the pile, if water condenses following a reduction in temperature, heat is released by this process and contributes to an increase in temperature in this portion of the pile. If oxidation rates were determined in a mine rock pile based on temperature profiles without taking into account latent heat effects, this would lead to an underestimation of heat production and pyrite oxidation in the lower part of a pile whereas an overestimate would be made instead for the upper portion of the pile.

Questions remain regarding the prevalence of moisture transfer and latent heat effects in waste rock piles. Previous simulation studies had identified potential water saturation reductions due to this process in other sites in quite different contexts at the Doyon Mine in Canada and for the Nordhalde in Germany (Lefebvre et al., 2001b). Given the variability in material properties, it may be very difficult to identify the effect of this process based on measurements of water content in mine rock. Still, in favorable contexts, such as dry climates and sites with strong thermal convection, it may be important to determine the effect of "air drying" on the water balance of a mine rock pile.

Field observations using isotopic tracers support the concept of internal vaporization as a significant process within mine rock piles. Sracek (1997) obtained samples of pore waters in the Doyon pile using suction lysimeters and analysed their Deuterium and [18]Oxygen content. The isotopic data suggested that this water had been subjected to internal evaporation and condensation.

The magnitude of internal drying due to water vapor transport in mine rock pile may be related to the heterogeneity of the material. The conceptual model of mine rock piles with dipping layers of coarse and fine grained material presented by Wilson et al. (2000) would favor vapor transport. In such a model, water infiltrates down fine layers whereas gas is transported upward in coarse layers. Such a system would promote vapor exit from piles because the temperature in the coarse layers could remain relatively high all the way up the surface of the pile.

5 ACKNOWLEDGMENTS

Molycorp Inc. is gratefully acknowledged for supporting the work presented in this paper and for granting permission to publish it.

REFERENCES

Gélinas, P., Lefebvre, R., and Choquette, M., 1992. Characterization of acid mine drainage production from waste rock dumps at La Mine Doyon, Quebec. *Second Int. Conf. on Environm. Issues and Manag. of Waste in Energy and Mineral Prod.*, Calgary, Sept. 92.

Lefebvre, R. 1994. *Caractérisation et modélisation numérique du drainage minier acide dans les haldes de stérile* [In French: Characterization and modeling of acid mine drainage in waste rock piles]. Ph.D. thesis, Faculty of Science and Engineering, Laval University, Quebec, 375 p.

Lefebvre, R., Hockley, D., Smolensky, J., and Gélinas, P., 2001a. Multiphase transfer processes in waste rock piles producing acid mine drainage, 1: Conceptual model and system characterization. *Journal of Contaminant Hydrology*, special edition on *Practical Applications of Coupled Process Models in Subsurface Environments*, in print.

Lefebvre, R., Hockley, D., Smolensky, J., and Lamontagne, A., 2001b. Multiphase transfer processes in waste rock piles producing acid mine drainage, 2: Applications of numerical simulations. *Journal of Contaminant Hydrology*, special edition on *Practical Applications of Coupled Process Models in Subsurface Environments*, in print.

Lefebvre, R., D. Hockley, and C. Wels, 2000. Le comportement des haldes de stériles [In French: The behavior of waste rock piles]. *NEDEM 2000, Colloque sur la recherche de méthodes innovatrices pour le contrôle du drainage minier acide*, Sherbrooke, Québec, October 3-5, 2000, 2-23 to 2-32.

Lefebvre, R., Lamontagne, A., and Wels, C., 2001. Numerical simulations of acid rock drainage in the Sugar Shack South rock pile, Questa Mine, New Mexico, U.S.A. Proceedings, *2nd Joint IAH-CNC and CGS Groundwater Specialty Conference, 54th Canadian Geotechnical Conference*, Sept. 16-19, 2001, Calgary, Canada, 8 p.

Pruess, K., 1991. *TOUGH2 - A general-purpose numerical simulator for multiphase fluid and heat transfer*. Lawrence Berkeley Laboratory LBL-29400, 102 p.

Robertson GeoConsultants Inc., 1999a. *Interim Report: Questa Waste Rock Pile Drilling, Instrumentation and Characterization Study*. Interim Report 052007/1 Prepared for Molycorp Inc., September 6, 1999, 13 p. plus attachments.

Robertson GeoConsultants Inc., 1999b. *Progress Report on Questa Waste Rock Investigation: Workplans for Geochemical and Physical Characterization*. Report 052007/2 Prepared for Molycorp Inc., November 1999, 20 p. and attachments.

Robertson GeoConsultants Inc., 2000. *Interim Questa Mine Site Characterization Study*. Report 052008/10 Prepared for Molycorp Inc., November 2000, 80 p. and attachments.

Robertson GeoConsultants Inc., 2001. *Progress Report on Water Balance Study for Mine Rock Piles, Questa Mine, New Mexico*. Report 052008/17 Prepared for Molycorp Inc., February 2001, 28 p. and attachments.

Sracek, O., 1997. *Hydrogeochemical and isotopic investigation of acid mine drainage at Mine Doyon, Quebec, Canada*. Ph.D. Thesis, Faculty of Science and Engineering, Laval University, Quebec.

Wilson, G.W., Newman, L.L., and Ferguson, K.D., 2000. The co-disposal of waste rock and tailings. *ICARD 2000, Fifth International Conference on Acid Rock Drainage*, Society for Mining, Metallurgy, and Exploration (SME) of AIME, Denver, Colorado, May 21-24, 2000, 789-796.

Shaw, S., C. Wels, A. MacG. Robertson and G. Lorinczi 2002. Physical and Geochemical Characterization of Mine Rock Piles at the Questa Mine, New Mexico: An Overview, *9th International Conference on Tailings and Mine Waste '02*: Rotterdam: Balkema.

Tailings and Mine Waste '02, © 2002 Swets & Zeitlinger, ISBN 90 5809 353 0

Selection of a water balance cover over a barrier cap – a case study of the reclamation of the Mineral Hill Mine dry tailings facility

R.J. Frechette
Shepherd Miller Inc. (SMI) Fort Collins, CO, USA

F.W. Bergstrom
Amerikanuak Inc. (AKI) Gardiner, MT, USA

ABSTRACT: The Mineral Hill Mine in Jardine, Montana was an underground gold mine operated most recently from 1989 to 1996. Final reclamation of the site is in progress and is expected to be complete by 2002. Processing involved milling, cyanide leaching in tanks, and Merrill-Crowe recovery of metals. Tailings were disposed dry in a lined facility following vacuum filtration (disk filter).

The original plan for reclamation of the tailings was to use a clay cap to produce a barrier to infiltration and seepage. In the evolution from reclamation concept to final closure design and implementation, a vegetative water balance cover (VWBC) was chosen instead of the clay cap. The VWBC received all required State and federal agency approvals and was installed during 2000 and 2001.

This paper provides a case study of the optimization process that led to development of the VWBC concept and site-specific reasons why a VWBC was selected instead of the clay cap for this project.

1 INTRODUCTION

1.1 *Mine history summary*

The Mineral Hill gold mine is located near Jardine, Montana, approximately 5 miles northeast of Gardiner and the west gate of Yellowstone National Park (Fig. 1).

Beginning in the late 1800's, various owners and operators conducted mining and processing activities including arsenic production (1920's to 1940's) (Amerikanuak, 2000). Contemporary underground mining and processing of gold ore occurred from 1989 to 1996. Processing involved milling the ore and leaching the material in tanks using a dilute sodium cyanide solution. The process tailings were subsequently dewatered using a disk filter and then trucked to the lined tailings storage facility (TSF) for permanent disposal. At its peak, the mine processed approximately 300 tons per day (TPD) of gold-bearing ore.

1.2 *Tailings facility description*

The TSF is a synthetic-lined, shallow basin feature that drains to a lined catchment area at the south end of the TSF called the seepage collection pond (SCP). The TSF was designed and constructed in stages upon a glacio-fluvial terrace above the Bear Creek drainage, which is approximately a mile south of the mineral processing facility (Fig. 2). Bear Creek joins the Yellowstone River three miles downstream from the site.

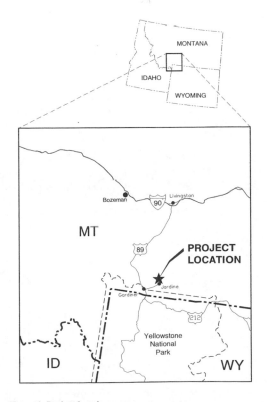

Figure 1. Project location map.

Each stage of the TSF is lined with a synthetic geomembrane placed upon a compacted, low permeability soil liner. A central channel drain was installed over the liner that directs leachate from the TSF pile and runoff from the active tailings face to the SCP. Slotted drain pipes were placed within the channel drain to convey the effluent (leachate and runoff) to a concrete structure at the toe of the TSF starter embankment. The concrete structure was designed to allow sedimentation from the drain pipes prior to discharging effluent to the SCP.

Tailings were dewatered using a disk filter to produce a filter cake with a nominal 18 percent moisture content. The tailings were then loaded into haul trucks and transported from the mill area to the TSF. At the TSF, the tailings were end-dumped onto the pile in layers and compacted to minimize precipitation absorption and to maintain trafficability. During operations, compaction monitoring was routinely conducted to verify compliance with permitting requirements.

The final operational configuration of the TSF is comprised of 14 acres of lined surface and contains approximately one million tons of tailings. The pile had a maximum height of approximately 100 feet and side slopes of 3:1 (horizontal:vertical) nominally. The north end of the pile (active face), had a

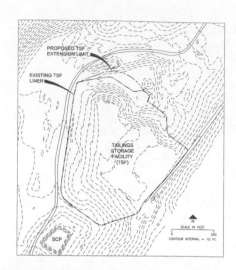

Figure 3. Post-operational tailings pile configuration (after Shepherd Miller, 2000).

steeper slope of approximately 1.5 to 2:1 since the operation was terminated prematurely and final sloping was not completed at that time (Fig. 3).

2 ORIGINALLY APPROVED TSF RECLAMATION PLAN

2.1 Clay cap description

During project permitting, a clay cap was designed and approved for TSF reclamation to prohibit infiltration and seepage through the tailings pile. As permitted, the cap would be constructed by incorporating bentonite into the upper 12 inches of tailings, and covering that surface with 3 inches of gravel followed by 12 inches of topsoil.

During final closure planning, the owner initiated additional engineering evaluation of the cap to address material compatibility and drainage issues. As a result of this process, a tentative revised clay cap design was contemplated that involved the following components from bottom to top (Fig. 4):

- 13 inches of compacted clay
- 12 inches of crushed/screened gravel
- geosynthetic fabric layer
- geosynthetic clay liner (GCL)
- 12 inches of topsoil

The lowermost clay layer was provided as an infiltration reduction layer. The gravel layer was provided as a drainage layer. The geosynthetic layer was provided for a cushion between the GCL and gravel. The GCL is the initial barrier to infiltration and intended to also retard deeper root penetration. The topsoil was for reclaiming the surface to produce a revegetated landform.

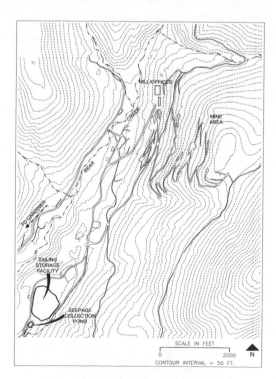

Figure 2. Mine site configuration (after Shepherd Miller, 2000).

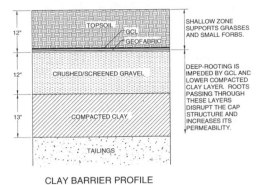

CLAY BARRIER PROFILE

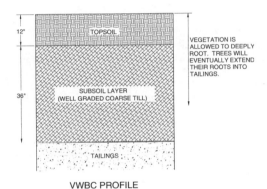

VWBC PROFILE

Figure 4. VWBC profile compared with clay barrier profile.

2.1.1 Clay cap design review

An independent engineering review of the proposed reclamation design was conducted for the owner for project valuation prior to implementing closure.

The review uncovered some technical flaws in the tentative design and concluded that the plan was not economically attractive. Based on the cap design modeling, the proposed clay cap would continuously maintain a saturated condition within the clay layer. This condition would result in at least two undesirable consequences: elevated seepage rates (even during periods without precipitation), and an increased probability of slope stability failure. The long-term need to control establishment of deep rooting woody species was also unattractive from a low maintenance / permanent closure perspective.

Given the 3:1 final side slopes, the saturated clay layer would be subject to instability. This condition would be exacerbated by seasonal snowpack loading that would add to the two-plus feet of permanent surcharge loading by the overlying gravel, geosynthetics and topsoil layers. With these driving forces, localized slope failures could be expected seasonally

that, over time, would compromise the effectiveness of the cap.

The designer's estimated cost of the reclamation cap was approximately $2 million. For a relatively small dry tailings facility such as the TSF, this cost ($3.28/sq ft) is considered to be fairly high and is comparable to closure of a wet tailings facility (Frechette, 1995). Material sourcing factors contributed significantly to the high cost estimate for the reclamation. Available on-site borrow materials include glacio-fluvial till materials and salvaged topsoil. The proposed clay layer would need to be obtained from an off-site source. The lowermost gravel layer could be processed from on-site sources.

3 ENGINEERING OPTIMIZATION AND VWBC CONCEPTUAL DESIGN

3.1 Comparison between clay cap and VWBC concepts

Because of the uncertainty in the water quality of effluent that may drain from the pile over time, it was considered desirable to implement a closure design with a prospect of eliminating infiltration/seepage over time. The saturated clay cap element of the original design, although providing a low permeability media, nonetheless would result in a finite ongoing seepage flow which would reflect seasonal and annual precipitation fluctuations.

A VWBC has the potential to seasonally store infiltrated meteoric water and deplete the stored water via evapo-transpiration processes. By implementing a soil profile thickness that optimizes available pore space for fluid storage relative to both the precipitation expectations (seasonal or annual), and plant rooting depth, a water balance can be maintained. This means that the VWBC can temporarily store, and eventually eliminate, as much precipitation as it receives without contributing to seepage through the profile. The active portion of the profile is normally defined as the lowermost boundary of significant root penetration. Beneath this point, moisture extraction can be fairly minimal and moisture build-up would eventually result in downward seepage migration (Fig. 5).

3.2 Frost heave and root penetration

Freeze-thaw activity and root penetration can be critical processes to consider for a clay cap that relies on maintaining low permeability. A protective layer is needed to mitigate these processes. Otherwise, disruption of the clay barrier may occur, resulting in a significant increase in infiltration/seepage. Freeze-thaw processes are of course limited to specific climate regions generally defined by latitude or elevation constraints. At the Mineral Hill site, expected frost penetration is four feet.

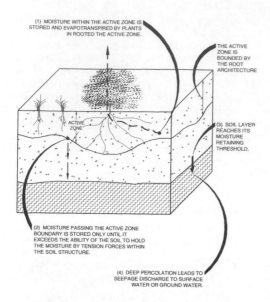

(1) MOISTURE WITHIN THE ACTIVE ZONE IS STORED AND EVAPOTRANSPIRED BY PLANTS IN ROOTED THE ACTIVE ZONE.

THE ACTIVE ZONE IS BOUNDED BY THE ROOT ARCHITECTURE

ACTIVE ZONE

(3) SOIL LAYER REACHES ITS MOISTURE RETAINING THRESHOLD.

(2) MOISTURE PASSING THE ACTIVE ZONE BOUNDARY IS STORED ONLY UNTIL IT EXCEEDS THE ABILITY OF THE SOIL TO HOLD THE MOISTURE BY TENSION FORCES WITHIN THE SOIL STRUCTURE.

(4) DEEP PERCOLATION LEADS TO SEEPAGE DISCHARGE TO SURFACE WATER OR GROUND WATER.

Figure 5. VWBC water storage and movement schematic.

However, extreme years with low snowpack and extended periods of extreme cold could result in much deeper frost penetration, particularly in coarse-grained soil with a high thermal conductivity. The local Water District office suggests an extreme frost depth design value of six feet (Gardiner-Park County, 1999).

Root penetration is even more of a concern than frost heave since rooting depths can far exceed the potential depth of frost penetration. Douglas fir which populate the undisturbed hillside above and below the TSF are expected to cover the reclaimed TSF over the long term. Typical rooting depths for mature Douglas fir are in the order of 10 feet. This depth range significantly exceeds a reasonable clay cap thickness.

A VWBC is relatively insensitive to freeze-thaw mechanics and typically performs better the deeper the roots penetrate because this extends the active zone for moisture holding and extraction (Fig. 5).

3.3 Slope stability

Slope stability can be a concern where material shear strength properties are exceeded by the stresses applied on the system. Clays, in particular, are sensitive to their environment and their shear strength can vary considerably under different conditions. For instance, the friction angle and cohesion in a dense, unsaturated clay may be in the order of 32 degrees and 10 psi, respectively. However, under saturated conditions with confinement that can generate excess pore pressure, the friction angle for the same clay material can decrease to zero. In such a reduced strength state, a clay that was once part of a stable cover when

stable cover when unsaturated, could render the cover unstable when saturated. A slope stability check was conducted on the clay barrier cap for the TSF side slopes under saturated conditions. Under this condition, which may be expected any time, and particularly during the snowmelt period, the clay barrier is likely to be unstable.

A VWBC typically employs the available soils on a site and does not rely on clay soils to function properly. As such, for Mineral Hill, glacio-fluvial till will comprise the majority of the reclamation cover thickness. This material possesses a moderate to high shear strength that will maintain slope stability under the range of site conditions expected (e.g. during temporary saturation), and also under seismic loading. In addition, over time, the slope stability associated with a VWBC typically improves due to maturation of the vegetation. The roots and fibrous biomass structure of vegetation create reinforced zones within the soil that have a higher shear strength than the soil alone. As the density of the biomass increases, this shear strength also increases (Gray and Sotir, 1996). The density can increase in two ways, either by root diameter increasing or by including a larger number of same-size roots. Both of these can occur as vegetation establishes and matures. For the Mineral Hill TSF, the proposed VWBC provides adequate slope stability for short through long-term, and significantly enhances slope stability over the previously considered clay cap.

3.4 Construction materials and cost

In addition to the superior performance in the areas described above, the use of a VWBC instead of a clay cap for the TSF at Mineral Hill provides substantial cost savings. The cost savings can be directly attributed to elimination of off-site borrow sources and manufactured materials. The VWBC total proposed thickness was four feet in comparison to the clay cap, which was either two feet (originally), or three feet (as modified subsequently). Of the multiple cap layers in the modified clay cap, the clay, gravel and GCL layers would either be derived from off-site sources or require processing. These three layers comprise roughly two-thirds of the total clay cap structure and represent over 80 percent of the associated capital cost. Maintenance costs of the clay cap, for repair and vegetation control, would be considerably greater than expected for the VWBC, which is largely maintenance-free.

4 VWBC INSTALLATION

4.1 Construction

The TSF footprint was expanded to the north to contain additional tailings from historic operations to consolidate management of the site reclamation. The

VWBC was installed over the regraded TSF throughout the expanded footprint. The final design of the VWBC incorporated three feet of subsoil and one foot of topsoil (Fig. 4). The subsoil was borrowed from the area immediately adjacent to the north end of the TSF. The subsoil consisted of the glacio-fluvial till material which was comprised of relatively well-graded mixtures of gravel, sand and silt. The material included only a minor clay fraction and occasional cobble to boulder-sized material. Placement of the subsoil was directly upon compacted tailings utilizing scrapers.

The topsoil was derived from material salvaged during site disturbance. In addition, wood planar shavings were imported from a local supplier to supplement the existing topsoil organic content. The topsoil was placed in one uniform lift at an approximate thickness of 12 inches using scrapers.

The TSF was regraded and covered with the first half (18 inches) of the VWBC subsoil layer during 2000. The remainder of the subsoil layer (another 18 inches) and the topsoil layer (12 inches) were placed in 2001. Following topsoil placement, the reclaimed surface was seeded with the project seed mix, and 5,000 10-cubic inch Douglas fir tubelings.

4.2 Cost savings

The cost of reclaiming the TSF, including the expansion of the footprint and hauling historic tailings to containment, was approximately $600,000. This is less than one-third of the cost estimated for the clay cap that was previously contemplated and equivalent to a very reasonable $1/sq ft.

5 SUMMARY

Although not applicable for all climate conditions or all site conditions, a VWBC is typically an appropriate solution in semi-arid climates. At sites where granular soil materials are prevalent as the available reclamation material, a VWBC is normally the most cost-effective choice for a cover. It can be a particularly attractive alternative to a clay cap in an application where frost penetration is a concern. In the case of Mineral Hill, use of a VWBC at the TSF enabled the development and implementation of a closure cap that reduces infiltration, maintains slope stability and provides a productive landform with beneficial use. The performance of a VWBC improves over time as the vegetation establishes and matures - this is true for both the infiltration/seepage and slope stability aspects of performance.

Use of a clay cap at this project would result in slope stability concerns, would not eliminate seepage, and would cost at least three times the cost of a VWBC to implement. Over time, the clay cap's expected performance would decline as the clay layer integrity deteriorates due to root invasion, periodic frost heave and at least localized slope failures that are likely to occur during heavy snowpack melt periods.

REFERENCES

Amerikanuak, Inc. 2000. *TVX Mineral Hill Mine Consolidated Closure Plan*. March (Revision 2).
Frechette, R.J. 1995. "Reclamation of Saturated Tailings Impoundments, New Remediation Technology in the Changing Environmental Arena." Proceedings of the 1995 Annual SME Meeting, Denver, Colorado, March 1995.
Gardiner-Park County Water District. 1999. Application for Water Service.
Gray, D.H. and R.B. Sotir. 1996. *Biotechnical and Soil Bioengineering Slope Stabilization: A Practical Guide for Erosion Control.*
Shepherd Miller, Inc. 2000. Appendix 6 of TVX Mineral Hill Mine Consolidated Closure Plan. March (Revision 2), *Mine Development Recontouring and Tailings Reclamation Plan, Mineral Hill Mine, Gardiner, Montana. February.*

Hard rock mine closure under modern rules – TVX Mineral Hill Mine case study

F.W. Bergstrom
Amerikanuak, Inc., Gardiner, Montana, USA

R.J. Frechette
Shepherd Miller, Inc., Ft. Collins, Colorado, USA

ABSTRACT: The underground TVX Mineral Hill Mine has been reclaimed in accordance with strict regulatory requirements, and will now provide critical winter range for the northern Yellowstone elk herd and other wildlife.

Portals and raises were plugged and dumps were fully recontoured, reestablishing original slope contours and forming forest clearings. Soil amendment with planer shavings promotes a fungal-based microfauna/flora which promotes long-term Douglas fir re-establishment. Fall plantings of ten-cubic inch Douglas fir tublings achieved 90%+ survival.

Arsenic is removed from mine drainage by zero-valent iron prior to discharge in an infiltration field. A second mine drainage is now a local water supply.

The tailings pile water balance cap controls percolation, enhances revegetation success, and minimizes maintenance by supporting Douglas fir forest. Remaining seepage is biologically treated to protect a zero discharge evapotranspiration bed.

Following full reclamation, the site will be donated to the U.S. Forest Service. Historic mine camp structures and site geological / mining information will be preserved for cultural and educational purposes.

1 INTRODUCTION

1.1 *Type area*

TVX Mineral Hill Mine was an underground gold mine, which produced gold doré and arsenic trioxide from ironstone and quartz vein ores. The mining district evolved as a series of ventures in and on Mineral Hill, operating intermittently between 1882 and 1996. The most recent redevelopment was in 1988, when the current cyanide leach plant was constructed under a Montana Metal Mine Reclamation Act Operating permit. All disturbances associated with the redevelopment were regulated via that permit. In addition, two unreclaimed historic (pre-law) tailings ponds were to be reclaimed.

Following the decision (1999) to permanently close and reclaim the Mineral Hill Mine, the reclamation plan, originally approved in 1986, was reviewed by the owner and its contractor (cumulatively referred to as the company) for constructability and long-term operability and maintenance requirements. Several improvements to the plan were found which would enhance closure performance in the near and long-term, and reduce long-term maintenance requirements. In fact, cost savings were realized through implementing superior designs.

Closure land use goals are 1) wildlife habitat, 2) water quality and soil protection, and 3) recreation.

The site is included in the critical winter range of the northern Yellowstone elk herd, as well as that of wolves, grizzly bears, bison, deer, and many other species. These goals are best achieved via creation of a robust vegetation cover.

Critical to this goal is the tailings cap. While the original plan would have limited revegetation to shallow rooting herbaceous species, the revised plan promotes their establishment along with woody deep rooting species, especially Douglas fir.

Modifications to the approved closure plan consist of:

- Change from bentonite clay amended tailings cap with shallow overburden to a thick (four foot) water balance cover
- Underground backfill of shallow stopes and near surface zone of all adits was added
- Reclamation of some historic – non-permit – workings that presented hazards to public safety
- Removal of contaminated subsoils under one historic tailings pond and importation of topsoil
- Full recontouring of development rock piles at the various portals on the mountain slope rather than partial recontouring, which made revegetation of the steep coarse pile toe extremely difficult
- Construction of a buried high density polyethylene (HDPE) pipeline to carry one of two mine drainages to a stream outfall and local water sup-

ply, rather than constructing a permanent water-tight adit plug
- A simple ferrihydrite system for removal of dissolved arsenic at the second – much smaller – mine drainage, which was ignored in the original plan
- Residual seepage from the TSF treated in a biological treatment system, rather than assuming zero seepage following capping.

To assure coordinated and effective management of the post mining land use, the owner intends to donate the 556-acre private land holding to the U.S. Forest Service Gallatin National Forest. More effective elk winter range management will result from elimination of this private inholding in the National Forest. Should administrative/legislative processes obstruct the donation, a private sale would be sought. That is, it is not within the business strategy or interests of the company to act as land manager of reclaimed mine lands. Such is better accomplished by land management agencies or other private parties.

Few if any mines have been successfully closed by the operating companies under the Montana Metal Mine Act. Unfortunately, a number of operations have – for financial reasons – terminated business, bonds have been foreclosed, and the state is or will be performing the closures. TVX Mineral Hill Mine is a successful closure performed by a responsible and financially sound owner/operator in a cost effective manner consistent with the approved bonding level.

Project reclamation is broken into two main components discussed in the next two sections; 1) the mine and facilities area, and 2) the tailings area.

2 MINE AND FACILITIES AREA

TVX Mineral Hill Mine is located 5 miles northeast of Gardiner, Montana at the historic mining camp of Jardine (Figure 1).

Mineral Hill was mined via a series of adits, with ore extraction largely by cut and fill drift mining. These hillside portals are associated with development rock piles, including the 450, 610, 750, 1050, 1200, and 1300 Levels. Other surface disturbances include, four ventilation raises, a waste pass, roads, the mill building, crusher building, and office/shop building (Figure 2).

Portals were backfilled to stable ground, a concrete bulkhead was cast to preclude reentry, and the development rock piles were excavated and pushed upslope to cover the portal pad and cut slope, reestablishing the original contours. Forest clearings were created providing expanded edge habitat. Although the 1986 approved reclamation plan only re-

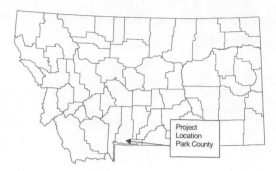

Figure 1. Project location.

quired partial recontouring, the coarse grained (boulder) angle-of-repose pile toe would have been difficult to successfully topsoil and revegetate. By constructing benches via uphill casting with a CAT 330 excavator and steep push roads for D6 and D8 CAT dozers, these piles were recontoured. Due to the short pushes, costs were $1.72 per cubic yard moved or $1,400 per acre.

In-situ reclamation of these materials was made possible via evidence showing acid rock drainage (ARD) would not become a problem. Multiple borings were completed into each rock pile. Samples were collected and each hole was completed as a suction lysimeter. Unsaturated water samples were thus obtained. These samples were all of basic pH – in the 8 to 9 SU range. Dissolved metals content, including arsenic were less than one part per million. Isolated core samples showed net acid generating potential, but the net chemistry of the piles was consistently basic. Mineral Hill development rock piles are composites of historic and modern development rock. Thus, they are up to 100 year old kinetic ARD tests. Based on these data, MDEQ concurred with in-situ reclamation.

Forest soils have thin A horizons, which are destroyed during salvage operations. In order to reestablish surface soil organic content above one percent, planer shavings were incorporated into the top six inches. Prodgers (2000) suggests a fungal dominate soil microfauna/flora is desirable for long-term propagation of Douglas fir (*Pseudotsuga menziesii*). Slow decomposition of wood cellulose is forecast to help promote long-term fungal growth and facilitate nutrient cycling. Planer shavings are also devoid of bark, which could be a vector for noxious weed seed importation. Nitrogen consumption was managed through planting freshly inoculated red clover for nitrogen fixation. A 2000 season seeding rate of 60 pounds per acre has resulted in excellent first year ground cover; however, the seeding rate was reduced in 2001 to 20 pounds per acre to help limit dominance of aggressive species.

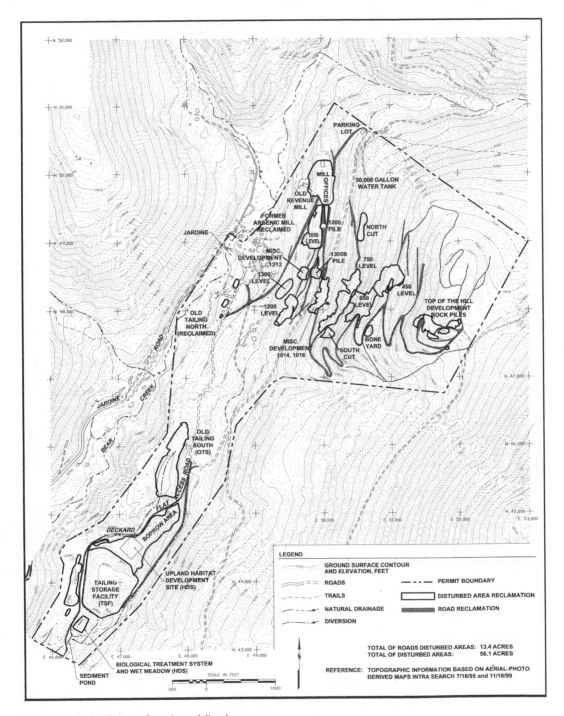

Figure 2. Mineral Hill site configuration and disturbance areas.

Figure 3. Gooseberry plant grown from local seed at end of first growing season.

Ten-cubic inch tubelings of snowberry (*Symphporicarpus occidentalis*), gooseberry (*Ribes cereum*), and woods rose (*Rosa woodsii*) were grown from seed collected on-site and planted at roughly 400 per acre. 2001 survival of these and same sized Douglas fir tubelings was greater than 90% following Fall 2000 planting (Figure 3).

Rubber rabbitbrush (*Chrysothamus nauseosus*), also grown from local seed, showed remarkable first growing season growth (Figure 4), illustrating the desirability of local seed procurement and nursery propagation to 10-cubic inch tubelings. The remarkable survival results in a jump start to reclamation at an affordable price ($1.00 per planted tubeling).

Cut/fill access and exploration roads were easily refilled using the CAT 330 excavator. Forest soil, stockpiled in the fill berm, required no fertilization; however, a light chemical fertilization was added along with the seed mix, and greater than 90% ground cover was achieved in the first growing season (Figure 5).

3 TAILINGS

Tailings reclaimed include the modern tailings storage facility (TSF) and the Old Tailings South (OTS), an historic pond reclaimed as a condition of the Montana Metal Mine Operating Permit. The second historic tailings pond, the Old Tailings North, was reclaimed during modern operations.

3.1 *Tailings storage facility (TSF)*

The TSF is a dry facility: that is, tailings were dewatered, then placed and compacted by heavy equipment. The TSF was constructed within a small ephemeral channel, which has been easily re-routed around the reclaimed facility.

Details of the 14 acre TSF cap can be found in a separate paper within this volume (Frechette, 2002), and this paper will limit discussion of the TSF to the

Figure 4. Rubber rabbitbrush plant at beginning (top) and end (bottom) of first growing season.

advantages realized by installing a water balance cover. The water balance cover consists of three feet of sand and gravel soils placed over compacted tailings and topped with one foot of salvaged topsoil. The native topsoil has roughly 2 to 3 percent organic matter, and has proven to be fertile. Planer wood shavings were applied at 100 cubic yards per acre

Figure 5. Exploration road recontoured with revegetation at end of first growing season.

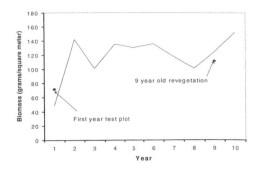

Figure 6. Clipable biomass in grams per square meter predicted by EDYS with field results obtained from first year test plots and nine year old TSF revegetation.

along with a grass/forb seed mix, followed by a Fall planting of 5000 Douglas fir tubelings (nursery grown from locally collected seed). Seeding in mid-July led to germination and roughly 30% ground-cover by September.

Establishment of vegetation is critical to the function of the water balance cover. The deep soil column (4 feet) acts to store precipitation, which the vegetation seasonally removes by evapotranspiration. With time, root diameter and density will increase, improving cap performance. Very importantly, a clay cap design would require removal of deep rooting plants. The water balance design eliminates this requirement.

Design modeling using the proprietary Ecological Dynamics Simulation (EDYS) model (McLendon, 2002) showed the relative efficiencies of both water barrier (clay) and water balance cap designs. Over the first 20 year period, the clay cap drained an average 0.47 gallon per minute (gpm), while the water balance design drained an average 0.14 gpm. Above-ground biomass, as predicted by EDYS, becomes established quickly. First year test plots and a nine year old reclaimed tailings slope verified the biomass predictions (Figure 6).

Subsequent to re-design and re-permitting, results were obtained for the OTS reclamation (discussed below). Quantitative measures of ground cover have not been obtained, but visual comparison to first year test plots shows significantly greater biomass on the OTS (Figure 7). As a result, the water balance cap is predicted to reduce seepage dramatically in the first year following reclamation.

3.2 Old tailings south (OTS)

Reclamation of the OTS (pre-Montana Metal Mining Act) was incorporated into the 1986 approved reclamation plan. This six-plus acre site was a shallow (less than 20 feet) wet tailings pond, which had never been capped, and was a source of windblown tailings to adjacent land and leachate to groundwater. Suction lysimeters installed at the base of the

Figure 7. OTS revegetation at end of first growing season.

tailings revealed marginally unsaturated conditions. pH was circum-neutral to slightly alkaline and dissolved iron was less than 1 ppm to 30 ppm.

A total of 110,000 cubic yards of material was excavated from the OTS and hauled to a prepared expansion of the TSF. All tailings were removed and from one to three feet of subsoil was also removed, using a total arsenic cutoff of 100 ppm. Following regrading for drainage and slope reduction, one foot of topsoil salvaged from the TSF and an offsite pit was placed, amended with composted horse manure and inorganic fertilizer, and seeded. 8,000 10-cubic inch sage and rubber rabbitbrush seedlings were planted. A shallow depression was planted with willow and quaking aspen tubelings. All tubelings were nursery grown from locally collected seed. Following the first growing season, none of these plants were found to have expired, although their location within the dense planted grasses and volunteer annual weeds hindered identification (Figure 7).

4 WATER MANAGEMENT

Three water streams must be managed. First, is the residual seepage from the TSF; second, is the Crevice (610 Level) exploration adit; and third, is the Mineral Hill Mine workings, which report to the lower most portal.

4.1 TSF seepage

The TSF double-underliner and French drain system collect and export seepage to a buried biological treatment system (BTS). Following project shut down and prior to construction of the BTS, seepage was collected in a pond, treated by high pressure (1000 psi) reverse osmosis (RO), and discharged to Bear Creek – a tributary of the Yellowstone River within Yellowstone National Park. Costs of $1.00 per gallon were initially realized for RO treatment, but following system optimization, costs were re-

duced to $0.30 per gallon. High RO cost (direct and indirect) and the need for a sizable facility to house a pre-treatment pH adjustment and settling system and associated tankage, made RO treatment an operational and financial option of last resort. A lower cost long-term alternative was necessary until such time as the cap reduced seepage to inconsequential levels.

A biological treatment system (BTS) was designed and constructed, with the goal of controlling toxicity to phreatic plants located in a constructed "wet meadow" evapotranspiration (ET) bed. The system is zero discharge in that all seepage is evaporated or transpired.

Treatment was designed and constructed in four stages (Figure 8).

- An anoxic limestone drain imparts alkalinity
- An aeration and settling stage precipitates iron
- An anaerobic sulfate reducing bacteria (SRB) stage removes metals and some sulfate
- A final aerobic bacteria stage was intended to remove manganese.

A contingency upland ET bed was constructed for wet years (EDYS modeling predicted 1 in 20 years), when excess capacity may be required. Because the upland site is not lined, a provision was installed to mix the treated seepage with Crevice Adit drainage water such that the water would meet all groundwater standards prior to discharge to the upland ET bed, thus assuring groundwater discharge compliance.

The system is sized to treat up to one gpm seepage. This is consistent with the maximum seepage rate predicted by EDYS modeling (Figure 9).

The BTS performed well at the bench scale, but following placement of the OTS tailings on the TSF expansion, seepage quality changed in an unpredicted way. While OTS lysimeter data showed dissolved iron less than 30 ppm and the TSF seepage dissolved iron had consistently been less than 100 ppm, the TSF seepage dissolved iron rose to 700 ppm immediately upon placing the OTS tailings, and has remained between 500 and 700 subsequently. The OTS tailings were placed directly on new liner,

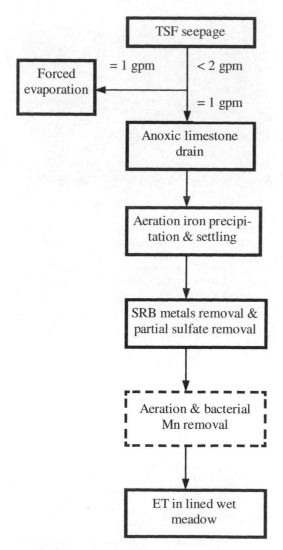

Figure 8. BTS system.

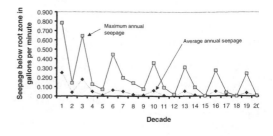

Figure 9. EDYS predicted TSF seepage as ten year means over 200 years.

and do not drain through any of the previously placed modern tailings. This additional iron load and associated acidity exceeded the buffering capacity of the limestone drain, necessitating the installation of a temporary caustic soda feed.

Commissioning difficulties are common in biological systems. These issues are discussed in Wildman (1993). The advantage of the buried tank/gravity driven BTS is that it can be retrofitted for changed conditions, maintained, and otherwise modified easier than a constructed wetland. In this case the wet meadow serves only to ET water. Carbon is added to the system as ethanol, reducing required residence time and allowing for indefinite carbon feed – as opposed to solid carbon sources in a constructed wetland.

The contingency of forced evaporation was incorporated into the closure plan to consume seepage in excess of the 1 gpm BTS treatment capacity and in the event of commissioning difficulties. Operating costs for forced evaporation are $0.25 per gallon, including blowdown disposal.

Seepage from the uncapped TSF declined to two gpm, as of fall 2001. With the establishment of cap vegetation during the 2001 growing season, pile recharge will begin to decline. EDYS seepage predictions verification is forecast for the 2002 growing season.

4.2 Mine drainage

Two Mineral Hill portals drain to surface; the Crevice Exploration Adit Portal (610 Level) and the 1300 Level Portal.

4.2.1 Crevice adit
The Crevice Adit was driven in the mid 1990's to access an exploration play on Crevice Mountain, some three miles distant. Water inflow intersected at the Palmer Creek Fault stopped advance at 6000 feet. The Crevice Adit is connected to the Mineral Hill Mine workings in only one location via a dry crosscut. Thus, the Crevice Adit is not hydraulically connected to Mineral Hill workings.

The original reclamation plan was to install a watertight plug. Construction would have been expensive, long-term maintenance requirements would be difficult to quantify, and other beneficial uses of the water would be precluded.

Water quality of Crevice Adit drainage is excellent. Most metals are below detection levels with dissolved arsenic at 20 ppb. Thus all drinking water quality criteria are met, and this water has been developed into the company water supply.

The reclamation plan was revised to include buried high density polyethylene (HDPE) pipelines, which carry the adit drainage to Bear Creek for discharge (Figure 10). No treatment is necessary for this high quality water. A separate pipeline carries

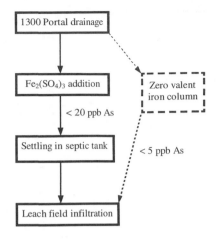

Figure 10. Crevice Adit drainage system.

water from a developed water supply well within the adit to a 50,000 gallon storage tank and associated distribution system to the company Jardine structures. A third HDPE pipeline carries some of the adit drainage on demand to the upland ET bed for pre-dilution.

Operation of this system necessitates continued underground access to the wellhead located 1500 feet underground. Adit rock quality is excellent and only minimal long-term ground control will be required. The portal, as well as several other minor historic portals and an historic partially collapsed raise have been closed with bat grates. These bat grates have been employed where underground workings can be left open without significant risk of collapse subsidence. The Crevice Adit is the only adit associated with the Montana Metal Mine Act Operating Plan that will not be backfilled. All other openings closed with bat grates are pre-law and minor in scale.

4.2.2 1300 Adit
The multiple levels of the Mineral Hill Mine all drain by gravity to the historic and collapsed 1300 Level. Drainage from the collapsed portal is collected in a French drain system and transported to a treatment building by buried HDPE pipeline (Figure 11). Dissolved arsenic is the only constituent of concern. All other metals and nutrients meet state groundwater quality criteria. Arsenic is precipitated via co-precipitation of iron, which is metered into the stream in the form of ferric sulfate. This method of adsorption on ferrihydrite is U.S. EPA's Best Demonstrated Available Technology (BDAT) for arsenic removal (Twidwell, 2001). Precipitated ferrihydrite is settled in a buried vault and discharged via a drainfield.

Test work on a retrofitted ferric sulfate system with a zero valent iron treatment system (Dipankar,

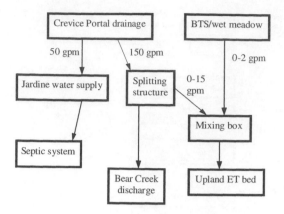

Figure 11. 1300 Level drainage/treatment system.

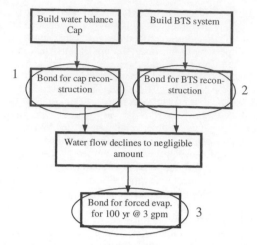

Figure 12. TSF cap and seepage control system bonding components.

2000) has shown very encouraging results. Dissolved arsenic is reduced from nominally 700 ppb to less than the 5 ppb detection limit without production of a solid, thus the treated water can be directly discharged without settling.

5 RECLAMATION BOND

Under the 1986 approved operating permit – as updated in 1997 – TVX Mineral Hill Mine was bonded for reclamation with the Montana Department of Environmental Quality (MDEQ) for $7.7 million, $2.05 million of which was for short-term (3 year) water treatment. The approved plan assumed complete draindown of the TSF, and no long-term water treatment requirements. When the company reassessed the reclamation plan and conducted the EDYS modeling, it became apparent this was – in the short-term – an optimistic assumption. Permitting the reclamation plan modifications resulted in preparation of a new EIS. During that process – and in light of recent mining company bankruptcies – MDEQ increased the bond to cover 100 years of water treatment.

MDEQ bonded incrementally for each component of the cap and seepage control systems. First, the cap itself was bonded, should reconstruction be required. That is, if the cap fails to stem percolation into the pile, money is available to the state to reconstruct, assuring a functional cap (Figure 12). Second, MDEQ holds bond for reconstruction or replacement of the BTS system, assuring a functional treatment system. Third, MDEQ bonded for a contingency treatment plan – should the cap and BTS systems fail – calculated as the net present value of a 100 year discounted cash flow. This final component of the bond is termed the "water treatment bond". It is a

contingency, which results in an increase of $3.14 million over the 1997 water treatment bond.

Actual closure costs for the TVX Mineral Hill Mine are consistent with the 1997 bond estimate. Reassessment of the 100 year water treatment portion of the bond has been requested as part of the bond release request, encompassing 2001 reclamation work. With cap and treatment system operational data in hand, it is anticipated a useful case study will be established for future bond calculations.

6 POST CLOSURE PERMITS

Although a goal of the TVX Mineral Hill Mine closure is to establish wildlife habitat and minimize any associated maintenance, some residual maintenance and operation will be required – especially as relates to water management. As long as water emanates from a mining facility, a permit will likely be required. Remaining permits will consist of:

- Montana Pollutant Discharge Elimination System permit for 1) the Crevice Adit discharge to Bear Creek and 2) the 1300 Adit discharge to groundwater
- U.S. EPA Underground Injection Control permit for the 1300 Adit discharge to groundwater
- MDEQ Montana Metal Mining Act Operating Permit

The entity which owns the property in the future would comply with these permits, including very limited monitoring.

7 PROPERTY DISPOSITION

The company proposes to donate the fully reclaimed property to the U.S. Forest Service (USFS) Gallatin National Forest. A total of 556 acres of patented surface and mineral rights would be donated, along with 194 acres of mineral rights, currently in public surface ownership. The process has been facilitated by USFS involvement in the closure planning and execution process, allowing their administrative hazardous materials review to run concurrent with company operations. No environmental risks have been identified.

The donated lands would contribute to the coordinated management effort for the northern Yellowstone elk winter range and migration corridor. Local recreation management and opportunities would also be enhanced.

The donation process involves legislative and administrative steps. The USFS wishes to preserve cultural and historic features of the Jardine mining district, including the historic Red Mill (ruin), which was built in 1898 during an earlier Mineral Hill operation, along with several other smaller structures. Over 100 years of mining and geologic records are available. Funding for this management plan is contained in authorizing legislation for the land donation. A donation agreement is then required between the USFS and the company. USFS plans would require public funding, but importantly, all closure related costs are being born by the company.

Should the donation process fail to reach a successful conclusion, a private sale would be made.

8 SUMMARY

The TVX Mineral Hill Mine was operated intermittently between 1882 and 1996. Final closure commenced in 1999. Reassessment of the 1986 Montana Metal Mine Reclamation Act Operating Permit suggested modification should be made to improve results, reduce maintenance requirements, and potentially reduce unnecessary costs. Several omissions in the original plan were rectified, where early planning had missed drainage issues.

Three drainages are managed by a variety of biological, inorganic, or simple drainage measures. Soils were amended with wood planer shavings, composted horse manure, and inorganic fertilizer, as appropriate. Results obtained exceeded expectations during the first growing season, confirming that the water balance cap is a suitable and superior technology for capping the dry tailings facility.

The result is successful reestablishment of wildlife habitat, water quality and soil protection, and recreation with minimal ongoing maintenance requirements and associated costs. Some continuing permit compliance requirements exist, but these are minimal and unavoidable. The future property owner will either be the USFS or some private party, depending on the processes involved and timeframes realized.

REFERENCES

Dipankar, C. 2000. Removal techniques of arsenic from groundwater. Council of Scientific and Industrial Research, Government of India, Jadavpur University, Calcutta, February, 2000.

Frechette, R. 2002. Selection of a water balance cover over a barrier cap – case study of the reclamation of the mineral hill mine dry tailings facility. *Proc. Tailings and mine waste '02, Fort Collins, Colorado, January 27-30, 2002.* Rotterdam: Balkema.

McLendon, T., W. M. Childress, C. L. Coldren, F. Bergstrom, R. Frechette. 2002. Evaluation of Alternative Designs for a Water-Balance Cover Over Tailings at the Mineral Hill Mine, Montana, Using the EDYS Model. *Proc. Tailings and mine waste '02, Fort Collins, Colorado, January 27-30.* Rotterdam: Balkema.

Prodgers, R. A. 2000. Evaluating organic amendments for revegetation, *Proc. 2000 Billings Land Reclamation Symposium.*

Twidwell, L. 2001. Technologies appropriate for removing arsenic, selenium, or thallium from mine and waste waters, *Proc. Mine Design, Operations & Closure Conference 2001*, Montana Tech of the University of Montana, April 8-12, 2001.

Wildman, T., Brodie, G., and Gusek, J. 1993. Wetland design for mining operations. BiTech Publishers Ltd, Canada.

Evaluation of alternative designs for a water-balance cover over tailings at the Mineral Hill Mine, Montana, using the EDYS model

Terry McLendon, W. Michael Childress, Cade L. Coldren & Rick Frechette
Shepherd Miller, Inc. Fort Collins, Colorado 80525 USA

Frank Bergstrom
Amerikanuak (AKI), Gardiner, Montana 59030 USA

ABSTRACT: TVX Mineral Hill, Inc. (TVX) is in the process of closing the Mineral Hill Mine in southern Montana, one requirement of which is to revegetate the tailings storage facility (TSF). We used the EDYS ecological model to simulate 1) above- and below-ground development of the plant communities and 2) water dynamics of the TSF cover design, under three topsoil variations, three precipitation regimes, and three revegetation scenarios.

The simulations indicate that all three topsoil variations, under all three precipitation regimes, should result in a productive and species diverse grassland community establishing on the site in the first two years. Plant community development and water use continued to increase during the 200-year simulation.

The 6-inch topsoil design resulted in 87% of the precipitation being returned to the atmosphere by evapotranspiration (ET) in the first 10 years under the normal precipitation regime, compared to 91% with the 12-inch topsoil design. Under normal precipitation, 12 inches of topsoil, and Douglas fir plantings, drainage decreased to 0.04 gpm, or about 0.4% of the precipitation received, when averaged over 200 years.

1 INTRODUCTION

TVX Mineral Hill, Inc. (TVX) is in the process of closing the Mineral Hill Mine. One goal in this closure operation is to revegetate the 14-acre tailings storage facility (TSF). The revegetation plan used to meet this goal should provide for 1) surface stabilization of the site, 2) minimize water movement through the tailings, and 3) provide a vegetative cover that is favorable for wildlife habitat and livestock use.

The first step in the development of the cover design was to contour the tailings to form a landscape surface compatible with the surrounding landscape. This contoured landscape consisted of three primary areas: 1) a north-facing 3:1 (horizontal:vertical) slope on the north quarter of the TSF, 2) a west-facing 3:1 slope on the west 40% of the TSF, and 3) a relatively flat (0-5% slope) top on the southeast one-third of the TSF (Figure 1). A fourth area included in the design was a portion of the liner without tailings, along most of the edge of the TSF. Once contoured, a 3-ft layer of sand and gravel was placed over the entire TSF, including the perimeter portion of the liner.

A water balance cover was evaluated as the primary method of preventing infiltration of water through the tailings. Initially, three design variations were evaluated relative to topsoil treatments: 1) 6 inches of topsoil placed over the 3-feet of sand and gravel, 2) 12 inches of topsoil placed over the sand

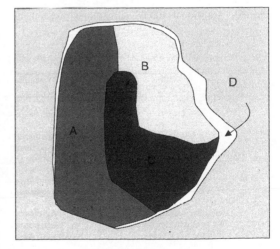

Figure 1. Map of vegetation zones in the EDYS simulation of the Mineral Hill Tailings Storage Facility.

and gravel, and 3) 6 inches of topsoil over the sand and gravel and benonite mixed with the upper 6 inches of tailings (6% benonite, 94% tailings).

Critical to the evaluation of these three cover design variations was a dynamic analysis of the effects of the variations on the successional development of the revegetated plant community. At a minimum, a dynamic analysis must include the ecological effects of variations in precipitation, the continuous development of the plant community (above- and below-ground), and the water balance dynamics influenced by the variations in substrate and the interactions between the developing plant community and the impacts of daily precipitation events. Static-state plant models are not appropriate. The plant community characteristics change on a daily, monthly, and annual scales, and each of these types of changes have significant impacts on water dynamics.

The EDYS (Ecological DYnamics Simulation) model was used to evaluate these cover design variations. EDYS is a spatially-explicit, mechanistic, computer model that is used to simulate plant community development (above- and below-ground) over time, the responses of ecological systems to environmental stressors, and the hydrological dynamics related to ecosystem dynamics (Childress and McLendon 1999, Childress et al. 1999, Childress et al. 2001). It has been applied to revegetation, land-use planning, and ecological responses to environmental stressors by the US Army Corps of Engineers, Natural Resource Conservation Service, National Park Service, US Forest Service, USAF Academy, US Marine Corps, CSIRO-Australia, and several mining companies (McLendon et al. 2000a, McLendon et al. 2000b).

The initial evaluation of the three topsoil treatments resulted in the 12-inch topsoil treatment being selected. Additional EDYS simulations were then conducted to evaluate several revegetation scenarios. Each of these scenarios was evaluated on the basis of vegetation development and water use over 20- and 200-year periods. In addition, potential disturbance of the plant communities from drought, fire, and heavy elk grazing were also evaluated on the basis of impacts of these stressors on vegetation recovery and water dynamics.

2 SITE DESCRIPTION

The Mineral Hill Mine is located in southwestern Montana, approximately 5 mi. northeast of Gardiner, Montana. Elevation of the site is approximately 6500 ft and average annual precipitation is approximately 16 inches. The site is in a transitional area between big sagebrush shrubland and Douglas fir forest, with the sagebrush primarily located on upland flats and the Douglas fir on the slopes.

The TSF site includes approximately 14 acres of tailings placed over a synthetic-lined basin. Maximum depth of the tailings is approximately 100 ft. Established stands of Douglas fir woodland exist along the northern and western edges of the TSF, with revegetated grassland along the eastern edge, and a aspen-cottonwood community transitioning to a sagebrush community along the southern edge.

Table 1. Design seed mix and local woody plant species for tailings storage facility.

Common name	Scientific name	Seeding rate (lb/ac)	Biomass (gm/m^2)
Douglas fir	Pseudotsuga menziesii		0.05
Big sagebrush	Artemisia tridentata		0.1
Green rabbitbrush	Chrysothamnus parryi		0.1
Wood's rose	Rosa woodsii		0.1
Red-osier dogwood	Cornus stonolifera		0.1
Snowbrush	Ceanothus velutinus		0.1
Oceanspray	Holodiscus discolor		0.1
Beardless wheatgrass	Agropyron ineme	7	0.784
Western wheatgrass	Agropyron smithii	10	1.120
Slender wheatgrass	Agropyron trachycaulum	6	0.672
Mountain brome	Bromus marginatus	10	1.120
Big bluegrass	Poa ampla	5	0.560
Canada bluegrass	Poa compressa	1	0.112
Green needlegrass	Stipa viridula	8	0.896
Indian ricegrass	Oryzopsis hymenoides	6	0.672
Idaho fescue	Festuca idahoensis	8	0.896
Alsike clover	Trifolium hybridum	2	0.224
White yarrow	Achillea millifolium	2	0.224
Heartleaf arnica	Arnica cordifolia	trace	0.100
Sulfur buckwheat	Eriogonum umbellatum	trace	0.100
Silky lupine	Lupinus sericeus	trace	0.100
Cicer milkvetch	Astragalus cicer	2	0.224

* Seed deposition rate only in randomly selected 5% of cells.
Note: All woody species are assumed to disperse naturally throughout the Area.

3 APPLICATION OF EDYS TO THE TSF COVER DESIGN

The 14-acre TSF was divided into four topographic types for the EDYS application: 1) a north-facing 3:1 (horizontal:vertical) slope covering the northern 25% of the TSF, 2) a west-facing 3:1 slope covering the western 40% of the TSF, 3) a relatively flat (0-5% slope) top covering the southeast one-third of the TSF, and 4) the perimeter toe of the TSF, which consists of the edge of the liner with no tailings, but covered with the 3-ft of sand/gravel and the topsoil (Figure 1). Each of the four areas was divided into 4 m x 4 m cells, for a total of 3,907 cells. Each cell had its appropriate elevation, slope, aspect, and tailings depth entered in the EDYS spatial grid.

EDYS divides the soil profile for each cell into 13 layers, with physical and chemical variables parameterized for each layer based on site-specific data. In this application, the characteristics of each of these layers varied, depending on which topsoil scenario (6-inch, 12-inch, 6-inch + bentonite) was used.

A 32-year historic daily precipitation regime was constructed using data from Jardine (adjacent to the mine site) and Gardiner (5 miles from the site). This constructed daily precipitation data set was used as the "normal" precipitation regime in the simulations. The 10-yr simulations used the first 10 years of the constructed data set. The 20-yr simulations used the first 20 years of the data set. The 32-year data set was repeated (i.e., Year 33 = Year 1, Year 65 = Year 1, etc.) to simulate a 200-year period. It is important to note that EDYS simulates precipitation events on a daily basis. This allows the simulation of "pulse" events (e.g., rapid snow melt at the end of the winter or a wet rainy period following spring snow melt), which are often very important in determining hydrological dynamics of a natural system.

The simulations began with no plant community on the TSF. Instead, an initial seed bank was placed in the surface soil layer uniformly across the TSF surface. This simulated the drilling of the revegetation seed mix (Table 1). The first-year plant community consisted of the plants that germinated and established from this seeding, plus some natural invasion of Douglas fir and big sagebrush. EDYS

Table 2. EDYS simulations of end-of-growing season aboveground biomass (g/m^2) of the north-facing revegetated plant community on the Tailings Storage Facility with 6 inches of topsoil, over a 10-year period.

Species	Year									
	1	2	3	4	5	6	7	8	9	10
Douglas fir	t	t	t	t	t	t	1	1	2	4
Big sagebrush	t	t	t	t	t	t	t	t	t	t
Rabbitbrush	2	1	1	1	1	1	1	1	1	1
Wild rose	t	t	t	t	t	t	t	t	t	t
Red-osier dogwood	t	t	t	t	t	t	t	t	t	t
Snowberry	t	t	t	t	t	t	t	t	t	t
Oceanspray	t	t	t	t	t	t	t	t	t	t
Beardless wheatgrass	5	11	10	8	10	12	9	9	12	12
Western wheatgrass	4	10	10	7	7	8	7	6	6	6
Slender wheatgrass	2	4	5	6	7	11	10	9	10	9
Mountain brome	5	8	8	8	11	14	9	5	6	7
Idaho fescue	4	8	10	12	18	24	21	17	19	18
Big bluegrass	2	4	5	8	12	18	22	33	51	65
Canada bluegrass	1	1	1	2	2	2	1	2	2	2
Green needlegrass	6	11	12	18	24	33	21	25	30	26
Indian ricegrass	3	5	8	9	15	22	18	15	17	18
Alsike clover	2	3	3	3	3	2	1	1	1	1
White yarrow	2	5	5	7	7	6	3	3	2	1
Heartleaf arnica	1	1	1	1	1	1	1	1	t	t
Sulfur buckwheat	1	1	1	1	1	1	1	1	1	t
Silky lupine	t	1	1	1	1	1	1	1	1	t
Cicer milkvetch	1	2	2	1	1	1	1	1	1	t
Total trees	t	t	t	t	t	t	1	1	2	4
Total shrubs	3	3	3	3	3	2	2	2	2	2
Total grasses	32	62	69	78	106	144	118	121	153	163
Total forbs	7	13	13	14	14	12	8	8	6	4
Total aboveground	42	78	85	95	123	158	129	132	163	173

Note: t = trace

simulated the development of this plant community over time as the "grassland" community. Two other revegetation scenarios were also simulated. The "Douglas fir" community consisted of the same seeded grassland community, plus Douglas fir seedlings planted across the TSF landscape the first year at a rate of 400 trees/acre. The "sagebrush" community consisted of the seeded grassland community plus big sagebrush seedlings planted at a rate of 400 shrubs/acre.

Two sets of simulations were conducted. The first set evaluated the three topsoil treatments, over a 10-year period and under the grassland revegetation scenario. The purpose of these simulations were to determine if the 12-inch topsoil depth or the bentonite treatment provided significant improvement over the 6-inch topsoil depth. Endpoint criteria for these simulations were 1) development of a sustainable plant community and 2) minimal drainage into the tailings.

The second set of simulations evaluated the 12-inch topsoil treatment under a set of environmental stressors over a 200-year period. The stressors in-

cluded 1) shifts in precipitation regime (75-300% of normal), 2) fire, and 3) heavy grazing by elk. End-point point criteria for these simulations were 1) development of a sustainable plant community, 2) minimal drainage into the tailings, and 3) minimal surface soil erosion.

4 SIMULATION RESULTS

4.1 Evaluation of topsoil variations

4.1.1 Aboveground plant dynamics
The EDYS simulations indicate that a productive and species diverse plant community should rapidly develop on the TSF cover. With 6 inches of topsoil and under a normal precipitation regime, the first-year community will be dominated by grasses, but with significant amounts of forbs and shrubs (Table 2). Standing crop biomass should more than double by the second year, and then gradually increase through Year 10. Grasses will continue to dominate the community throughout this period, but scattered Douglas fir saplings and trees will slowly become

Table 3. Comparison of EDYS simulations of end-of-growing season aboveground biomass (g/m^2) values of the revegetated plant communities on the four areas of the TSF with 6 inches and with 12 inches of topsoil, at the end of 10 years under a normal precipitation regime.

Species	6-inches Topsoil					12-inches Topsoil			
	North	West	Flat	Toe		North	West	Flat	Toe
Douglas fir	6	6	7	0		9	9	9	0
Big sagebrush	t	t	t	t		t	t	t	t
Rabbitbrush	1	1	1	1		1	1	1	1
Wild rose	t	t	t	t		t	t	t	t
Red-osier dogwood	t	t	t	t		t	t	t	t
Snowberry	t	t	t	t		t	t	t	t
Oceanspray	t	t	t	t		t	t	t	t
Beardless wheatgrass	12	12	12	10		12	12	12	11
Western wheatgrass	6	6	6	7		7	7	7	8
Slender wheatgrass	9	9	9	10		9	9	9	10
Mountain brome	7	6	6	8		7	6	6	8
Idaho fescue	17	17	17	21		18	17	17	22
Big bluegrass	68	67	68	22		61	60	61	23
Canada bluegrass	2	2	1	2		2	2	2	3
Green needlegrass	26	25	25	36		28	27	27	37
Indian ricegrass	18	18	18	24		19	19	19	25
Alsike clover	1	1	1	1		1	1	1	1
White yarrow	1	1	1	2		1	1	1	2
Heartleaf arnica	t	t	t	t		t	t	t	t
Sulfur buckwheat	t	t	t	t		t	t	t	t
Silky lupine	t	t	t	1		t	t	t	t
Cicer milkvetch	t	t	t	1		t	t	t	t
Total trees	6	6	7	0		9	9	9	0
Total shrubs	2	2	1	2		2	2	2	2
Total grasses	165	162	162	140		163	159	160	147
Total forbs	3	3	2	5		4	3	3	4
Total aboveground	176	173	172	147		178	173	172	153

Note: t = trace

important constituents. Forbs will remain in the community, but only as secondary species. Shrubs will remain minor constituents, probably as widely-scattered individuals or clumps. The EDYS simulations of 700 g/m² (6200 lbs/ac) of aboveground biomass by Year 10 is for standing crop biomass, which includes perennial portions such as crowns and basal stems of perennial grasses, and not annual production only.

There was little difference in plant community dynamics or standing crop biomass among the three areas that were over tailings (north-facing, west-facing, and flats). The flat should be slightly more productive and the west-facing area slightly less productive than the north-facing area (Table 3). The higher potential production on the flats is the result of increased moisture because of less slope, therefore less runoff. The west-facing slope is slightly less productive because the west aspect results in slightly more solar energy being received and therefore more evaporation. After 10 years, the flats should have slightly more bluebunch wheatgrass and

big bluegrass than the other areas. Idaho fescue and green needlegrass should be slightly more abundant on the north-facing slopes.

The EDYS simulations indicate that the plant community developing on the toe areas (topsoil and sand directly on top of the liner along the edges of the TSF) will be different in both production and composition from the areas over the tailings (Table 3). Standing crop biomass will be about 15-20% less on the toes than the other parts of the TSF because of the thinner profile, and therefore lower water holding capacity. The plant community on the toes should have more green needlegrass, Indian rice-grass, and yarrow and less big bluegrass than the communities over the tailings. Although different from the communities over the tailings, the plant community on the toes would still be a productive and diverse plant community, providing favorable forage and habitat to wildlife and livestock.

The 12-inch topsoil treatment increases aboveground biomass compared to the 6-inch addition, but only by 2-4% (Table 3). The estimated plant-

Table 4. EDYS simulations of end-of-growing season root biomass (g/m²) of the north-facing revegetated plant community on the tailings disposal area with 6 inches and 12 inches of topsoil, over a 10-year period.

Substrate layer	Thickness (mm)	Year									
		1	2	3	4	5	6	7	8	9	10
6-inch topsoil											
1 Topsoil	25	6	15	21	26	33	42	44	47	53	59
2 Topsoil	25	6	16	22	28	36	45	48	51	58	63
3 Topsoil	50	5	12	17	23	30	39	41	45	52	57
4 Topsoil	50	4	11	15	20	27	35	37	40	47	51
5 Sand	100	4	9	14	18	23	31	33	36	39	43
6 Sand	200	3	7	11	14	19	27	28	31	35	38
7 Sand	300	1	3	4	7	10	13	15	18	21	23
8 Sand	300	t	1	3	4	7	9	10	13	15	16
9 Tailings	150	t	1	1	1	2	3	5	7	9	9
10 Tailings	350	t	t	t	t	1	1	2	3	6	9
11 Tailings	500	0	0	t	t	t	1	1	3	5	10
12 Tailings	500	0	0	0	0	0	0	t	t	t	t
13 Tailings	500	0	0	0	0	0	0	t	t	t	t
Total	3050	29	75	108	141	188	246	264	294	340	378
12-inch topsoil											
1 Topsoil	25	7	17	23	27	35	43	45	48	54	59
2 Topsoil	50	9	22	29	37	47	60	62	67	75	82
3 Topsoil	75	6	15	21	28	36	47	49	52	60	65
4 Topsoil	150	6	15	21	27	36	48	50	55	62	67
5 Sand	100	5	10	14	17	22	29	30	32	35	37
6 Sand	200	4	8	11	14	19	24	25	28	30	31
7 Sand	300	1	3	5	6	10	13	14	16	19	19
8 Sand	300	1	2	3	3	6	8	9	11	13	14
9 Tailings	150	t	t	t	1	1	2	3	5	7	6
10 Tailings	350	t	t	t	1	1	1	2	4	7	8
11 Tailings	500	0	0	t	t	t	t	1	1	3	7
12 Tailings	500	0	0	0	0	0	0	t	t	t	t
13 Tailings	500	0	0	0	0	0	0	t	t	t	t
Total	3200	39	92	127	161	213	275	290	319	365	395

Note: t = trace

available water holding capacity (field capacity) of the profile with 12 inches of topsoil is 517 mm, compared to 498 mm with 6 inches of topsoil. This is an increase of only 4%.

The 12-inch treatment had more of an effect on species composition than it did on production (Table 3). The 12-inch treatment resulted in proportionally more Douglas fir and all species of grass except big bluegrass, and less big bluegrass than the 6-inch treatment. Forb production was the same on both treatments and shrub production was slightly higher on the 12-inch topsoil. These results suggest that succession to a Douglas fir forest would probably occur faster with the 12-inch topsoil than with the 6-inch topsoil.

4.1.2 Belowground plant dynamics

The EDYS simulations indicate that a few roots will reach the upper part of the tailings by the end of the first growing season, but that significant amounts will not reach the tailings until the seventh or eighth year (). By Year 10, we estimate that 61% of the total root biomass will be in the topsoil (0-150 mm depth) on the 6-inch treatment, 32% in the subsoil (150-1050 mm depth), and 7% in the tailings. With 12 inches of topsoil, we estimate that 69% of the root biomass will be in the topsoil (0-300 mm depth) after 10 years, 26% in the subsoil (300-1200 mm depth), and 5% in the tailings. Therefore, the 12-inch topsoil treatment will reduce the amount of roots in tailings material. This would be an important consideration if there was a concern about bioaccumulation of metals contained in the tailings, and subsequent surface deposition.

Neither slope nor aspect appear to have major impacts on root architecture, at least for the first 10 years. The thinner profile on the toes, and the less productive plant community associated with it, resulted in less root biomass than on the profiles over the tailings. The 12 inches of topsoil over tailings

contained 273 g of roots (Table 4), compared to 211 g of roots over the toes. The 900 cm of subsoil, under 12 inches of topsoil, over the tailings contained 101 g of roots, compared to 107 g in the subsoil over the toes.

4.1.3 Water dynamics

There are four sources of water loss from the TSF cover: evaporation, transpiration, runoff, and drainage. Evaporation is water loss directly from surfaces to the atmosphere, and EDYS separates evaporation by source, i.e., leaf surface of the plant canopy, soil surface (including the litter layer), and snowpack. Transpiration is evaporative water loss through plants. In most other models, evaporation and transpiration are combined into evapotranspiration (ET). However, the dynamics of the two sources can be very different, especially when there are significant changes in vegetation over time, therefore they are modelled separately in EDYS. Runoff is overland movement of water from the site. Drainage is percolation of water through the profile, past the rooting zone. In this application of EDYS, drainage is movement of water past the modelled profile into the tailings.

The EDYS simulations indicate that total annual evaporation loss will decrease over time (Table 5), primarily because of the development of the plant community canopy and therefore the shading of the soil surface. However, this rate of decrease is not equal among the three components of evaporation. Evaporation from the soil surface decreases relatively uniformly over time in response to canopy development, but evaporation from the plant canopy increases over time since there is more leaf area to intercept precipitation.

Transpiration water loss increases over time (Table 5) as the plant community becomes more productive (Table 2). Surface runoff also decreases over time because of greater structural development of

Table 5. EDYS simulations of water dynamics (m³) on the tailings disposal area with a 6-inch topsoil cover design, over a 10-year period under a normal precipitation regime.

Year	Precip	Evaporation			Transpiration	Surface Runoff	Drainage
		Canopy	Soil	Snow			
01	20,494	173	4,399	2,588	5,716	446	238
02	23,274	591	5,064	3,014	8,770	456	579
03	18,606	421	3,389	4,967	7,799	207	189
04	21,033	852	4,708	4,521	8,545	191	0
05	24,477	1,106	4,563	3,947	10,865	207	496
06	24,414	1,723	3,085	3,022	13,838	137	210
07	11,575	911	1,588	2,973	9,812	47	111
08	20,349	1,028	2,661	1,671	11,382	80	562
09	23,211	2,142	1,797	3,479	14,743	58	880
10	21,096	2,478	914	1,889	15,040	43	710
Sum	208,529	11,425	32,168	32,071	106,510	1,872	3,975 (0.20 gpm)
%	100.0	5.5	15.4	15.4	51.1	0.9	1.9

the plant community. Greater structural development increases the amount of plant biomass at the soil surface, which decreases the rate of surface flow and allows for greater infiltration. Estimation of these changes in the water dynamics of a vegetated system is a major advantage that EDYS provides over most other models that simulate runoff.

Annual drainage is only partly dependent on development of the plant community. It is also a function of the amount of annual precipitation received and the available water storage capacity of the profile and the underlying substrate (the tailings in the case). In wet years (e.g., Years 2, 5, 6, and 9), drainage generally increases because of the increased amount of water entering the profile. However, the higher drainage values in Years 8-10 are also caused, in part, by the increase in water content of the profile (Table 6).

The percentages of annual precipitation presented in Table 5 do not sum to 100. This is because of the lag time caused by storage in the profile and snowpack (Table 6). Moisture stored in the snowpack is lost each year to evaporation from the snow, some melting during the winter, and a large pulse of moisture during spring snow melt. Therefore a significant amount of the annual water budget is transferred from one year to the next as snowpack (e.g., 1740 m^3 in the first year, which is over 8% of the annual precipitation).

Table 7 presents a comparison of the summaries of the 10-year water dynamics on nine simulation scenarios. Under all three precipitation regimes, there is little difference in drainage between the 6-inch topsoil and the 6-inch topsoil plus bentonite treatments. The 12-inch topsoil treatment reduces drainage compared to the 6-inch topsoil treatment.

Table 6. Water balance on the tailings disposal area with a 6-inch topsoil cover design under a normal precipitation regime, based on EDYS simulations of water dynamics (Table 5).

Year	Balance 1 Jan	Annual Precip	Annual Losses ET	Annual Losses Runoff	Annual Losses Drainage	Precip - Loss	Storage on 31 Dec Profile	Storage on 31 Dec Snowpack
01	3,063	20,494	12,876	446	238	6,934	8,257	1,740
02	9,997	23,274	17,439	456	579	4,800	13,057	0
03	13,057	18,606	16,576	207	189	1,634	13,776	915
04	14,691	21,033	18,626	191	0	2,216	15,290	1,617
05	16,907	24,477	20,481	207	496	3,293	19,588	612
06	20,200	24,414	21,668	137	210	2,399	20,959	1,640
07	22,599	11,575	15,284	47	111	3,867	18,424	308
08	18,732	20,349	16,742	80	562	2,965	21,152	545
09	21,697	23,211	22,161	58	880	112	20,470	1,339
10	21,809	21,096	20,321	43	710	22	19,393	2,438
11	21,831							

Note: Both annual and end of year values are in m^3 of water.

Table 7. Comparison of EDYS simulated water dynamics on the tailings disposal area under different cover designs and precipitation regimes.

	Precip	Evaporation Canopy	Evaporation Soil	Evaporation Snow	Transpired	Surface Runoff	Drainage
Wet Regime							
6" topsoil	230,474	16,778	30,950	38,559	114,981	1,873	5,717
12" topsoil	230,474	17,939	29,343	37,777	120,943	1,740	4,173
Bentonite	230,474	16,686	31,086	38,679	113,531	1,875	6,226
Normal Regime							
6" topsoil	208,529	11,425	32,168	32,071	106,510	1,872	3,975
12" topsoil	208,529	12,644	29,470	31,571	115,114	1,699	2,942
Bentonite	208,529	11,352	32,342	32,129	104,933	1,876	3,964
Old Design	208,529	12,348	30,066	31,698	111,625	1,743	10,819
Dry Regime							
6" topsoil	189,466	9,271	32,410	30,865	95,834	1,818	3,907
12" topsoil	189,466	10,451	30,137	30,381	105,086	1,648	2,536
Bentonite	189,466	9,223	32,559	30,886	95,007	1,818	3,964

Values are in m^3 of water over a 10-year period.

511

Under a normal precipitation regime, this reduction is about 1000 m³ over the 10 years, or 26% less than the 6-inch rate. Under a wet regime, the reduction is 27% and 36% under the dry regime. Under each precipitation regime, the 12-inch topsoil increases transpiration and canopy evaporation and decreases evaporation from the soil and snow surfaces.

Short-term variations in drainage are often important for water management planning. The values presented in Tables 5 and 6 are annual totals and those presented in Table 7 are 10-year totals. EDYS calculates all water dynamics on a daily basis and these values are also available. Daily drainage through the TSF cover will vary significantly throughout the year and from year to year. In the first 4 years, the profile is relatively dry, with a large capacity to hold water (Table 6). This will minimize drainage on an annual basis. However, some drain-age will occur following short-term inputs of relatively large amounts of water (Figure 2). In the second year of the simulation, 579 m³ of water drained through the profile (Table 5), but most of this drainage occurred in 7 days in August and September (Figure 3). Most of the drainage in Year 9 also occurred in 7 days, but the 7 days were scattered between April (snowmelt) and October (Figure 4). Almost 25% of the annual drainage occurred on one day in May.

4.2 Evaluation of revegetation scenarios

The topsoil evaluations (Section 4.1) indicated that the 12-inch topsoil treatment resulted in a more effective cover design than the 6-inch topsoil treatment. Based on these results, further EDYS simulations to evaluate revegetation scenarios and effects

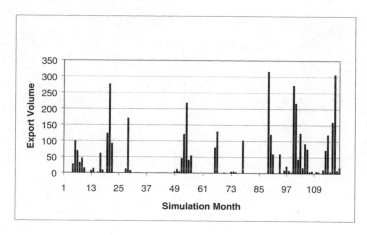

Figure 2. Estimated monthly drainage values (m³) through the TSF cover with 6 inches of topsoil.

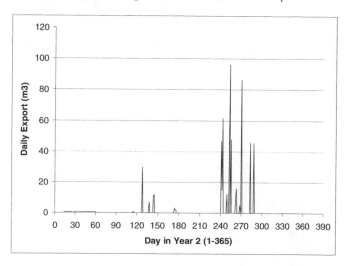

Figure 3. Estimated daily drainage values (m³) in year 2 through the TSF cover with 6 inches of topsoil.

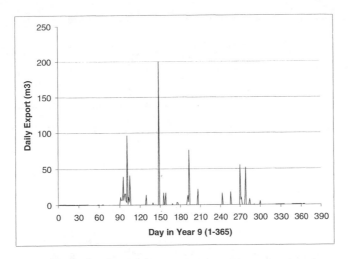

Figure 4. Estimated daily drainage values (m³) in year 9 through the TSF cover with 6 inches of topsoil.

of long-term precipitation fluctuations and environmental stressors used to 12-inch topsoil design only.

Three revegetation scenarios were evaluated: grass, Douglas fir, and sagebrush. The grass scenario was the seeded scenario used in Section 4.1, i.e., grasses and forbs drilled into the topsoil, with Douglas fir allowed to invade naturally. The Douglas fir scenario used the same grass-forb seed mix, but also included the planting of 400 Douglas fir seedlings per acre in the first growing season. The sagebrush

Table 8. EDYS simulations of drainage (gpm) through the TSF cover under three revegetation scenarios, assuming a normal precipitation regime.

Year	100% Precipitation		
	Grass	DgFir	BSage
01	0.108	0.108	0.108
02	0.074	0.080	0.080
03	0.220	0.181	0.181
04	0.000	0.000	0.000
05	0.146	0.008	0.008
06	0.128	0.282	0.053
07	0.281	0.783	0.560
08	0.065	0.045	0.086
09	0.684	0.328	0.480
10	0.751	0.487	0.622
11	0.039	0.018	0.018
12	0.118	0.093	0.093
13	0.077	0.000	0.000
14	0.217	0.000	0.000
15	0.283	0.095	0.095
16	0.089	0.081	0.081
17	0.296	0.141	0.141
18	0.169	0.000	0.000
19	0.064	0.000	0.000
20	0.147	0.000	0.000
Annual Mean	0.188	0.136	0.130

Cover design for Grass, DougFir, and Sagebr scenarios = 12 inches of topsoil over 36 inches of sand/gravel.

scenario was similar to the Douglas fir scenario except that big sagebrush seedlings were planted (400/acre) rather than Douglas fir seedlings. A 20-year simulation was used to compare these three scenarios, using the normal (average) precipitation regime.

Average annual drainage during the first 20 years was predicted to be 0.136 gpm under the Douglas fir scenario and 0.130 gpm under the sagebrush scenario (Table 8). Both were substantially less than the 0.188 gpm under the grass scenario.

These drainage differences among scenarios were the result of differences in vegetation development caused by the three revegetation scenarios (Table 9). Tree biomass develops most rapidly under the Douglas fir revegetation scenario and shrub biomass under the sagebrush scenario. Grass production was consistently higher in the Douglas fir scenario than in the grass scenario. This somewhat surprising result was probably the result of available moisture being greater in the upper soil profile under the Douglas fir scenario. This was because of more snow catchment by the greater tree biomass and less water lost by evaporation. The combination of high tree and high grass biomass resulted in relatively high transpirational water loss and a corresponding lower drainage.

4.3 Evaluation of effects of precipitation fluctuations

We tested the sensitivity of the cover design to fluctuations in precipitation by increasing and decreasing the amount of daily precipitation in the simulations, while keeping all other input parameters the same as in the normal precipitation scenario. We evaluated the grass, Douglas fir, and sagebrush revegetation scenarios by 1) decreasing the daily

Table 9. EDYS simulations of vegetation dynamics (g/m^2) on the TSF cover under various revegetation scenarios, under the normal precipitation regime.

Year	Grass Scenario				Douglas Fir Scenario				Sagebrush Scenario			
	Tree	Shrb	Grss	Forb	Tree	Shrb	Grss	Forb	Tree	Shrb	Grss	Forb
01	0	4	32	9	4	4	31	10	0	6	32	10
02	t	3	48	11	8	3	53	13	t	5	53	14
03	t	4	66	16	13	3	80	21	2	6	82	22
04	3	4	83	18	22	3	107	25	5	5	109	26
05	8	4	97	19	35	3	138	27	12	8	143	29
06	30	4	111	20	66	2	148	28	31	8	152	29
07	52	4	122	21	104	2	182	30	55	9	187	31
08	88	4	119	21	156	2	214	31	90	9	221	32
09	140	4	128	22	228	2	216	30	142	10	224	31
10	237	4	154	22	345	2	278	30	238	11	289	32
11	311	4	159	24	393	2	291	30	306	11	301	31
12	357	4	162	24	433	2	324	29	351	12	335	31
13	397	4	166	23	473	2	334	29	387	12	345	31
14	474	4	170	25	548	2	397	29	461	12	408	30
15	534	4	163	24	604	2	361	29	516	12	372	30
16	580	4	172	25	650	2	387	29	560	13	398	30
17	694	4	190	25	749	2	435	29	658	13	447	30
18	764	5	198	24	806	2	451	29	716	14	463	30
19	802	5	176	25	842	2	436	29	751	14	446	30
20	852	4	181	25	886	2	463	29	792	14	474	30

precipitation value from the normal (100%) precipitation regime by 25% (i.e., the 75% precipitation regime) and 2) increasing the normal daily precipitation by 25% (i.e., the 125% precipitation regime).

A 25% reduction in precipitation over 20 years would substantially reduce infiltration through the cover, under all three revegetation scenarios (Table 10). Under the grass scenario, mean annual infiltration decreased from 0.188 gpm with 100% precipitation, to 0.016 gpm with 75% precipitation. The number of years with net infiltration also decreased, from 19 to 10. Under the Douglas fir scenario, mean annual infiltration decreased from 0.136 gpm with 100% precipitation, to 0.010 gpm with 75% precipitation. Number of years with net infiltration decreased from 14 to 5. Under the sagebrush scenario, mean annual infiltration decreased from 0.130 gpm with 100% precipitation, to 0.010 gpm with 75% precipitation. Number of years with net infiltration decreased from 14 to 6.

A 25% increase in precipitation over 20 years increased simulated infiltration through the cover on all three revegetation scenarios, but the percent increase in infiltration was greater than the 25% increase in precipitation (Table 10). Under the grass scenario, mean annual infiltration increased from 0.188 gpm with 100% precipitation, to 0.832 gpm with 125% precipitation. Under the Douglas fir scenario, mean annual infiltration increased from 0.136 gpm to 0.395 gpm, and under the sagebrush scenario it increased from 0.130 gpm to 0.354 gpm.

All three revegetation scenarios continued to provide substantial reduction in infiltration through the cover as their respective plant communities contin-

ued to mature (Table 11). When averaged over 200 years, the Douglas fir scenario resulted in slightly less infiltration than the sagebrush scenario, and both were substantially more effective than the grass scenario. With both the 100% and 125% precipitation regimes, the sagebrush scenario was most effective in reducing infiltration for the first 30 years and the Douglas fir scenario was most effective after 30 years. Average annual infiltration with the Douglas fir scenario was 0.014 gpm with 75% precipitation, 0.041 gpm with 100% precipitation, and 0.086 gpm with 125%, or 0.2%, 0.4%, and 0.6% of annual precipitation, respectively. With 125% precipitation, average annual infiltration never exceeded 0.2 gpm on the Douglas fir scenario after the first 10 years.

4.4 Evaluation of disturbance impacts

Sensitivity of the design to disturbance events was evaluated by simulating the effects of fire and heavy grazing by elk. The Douglas fir revegetation scenario was used for the fire simulation and the sagebrush scenario was used for the elk grazing simulation. For both disturbances, the normal precipitation regime was used and the results were compared to the dynamics of the same scenario without the disturbance.

4.4.1 Fire

In EDYS, ecological effects of fire are simulated by designating when a fire event is to occur, or this can be randomly generated, and which cell the fire begins in, which can also be randomly selected. Once the cell is designated, the fire either stops in the sin

Table 10. Effect of changes in precipitation regime on simulated drainage (gpm) through the TSF cover under three revegetation scenarios.

Year	75% Precipitation			100% Precipitation			125% Precipitation		
	Grass	DgFir	BSage	Grass	DgFir	BSage	Grass	DgFir	BSage
01	0.000	0.000	0.000	0.108	0.108	0.108	0.260	0.260	0.260
02	0.000	0.000	0.010	0.074	0.080	0.080	0.145	0.155	0.155
03	0.050	0.033	0.033	0.220	0.181	0.181	0.390	0.344	0.344
04	0.000	0.000	0.000	0.000	0.000	0.000	0.124	0.155	0.085
05	0.000	0.000	0.000	0.146	0.008	0.008	0.456	1.056	0.350
06	0.002	0.000	0.000	0.128	0.282	0.053	1.254	1.395	1.127
07	0.026	0.000	0.000	0.281	0.783	0.560	2.565	1.775	1.634
08	0.000	0.000	0.000	0.065	0.045	0.086	0.259	0.098	0.160
09	0.000	0.000	0.000	0.684	0.328	0.480	2.340	0.887	1.065
10	0.116	0.023	0.023	0.751	0.487	0.622	3.267	1.024	1.153
11	0.000	0.000	0.000	0.039	0.018	0.018	0.243	0.093	0.092
12	0.000	0.000	0.000	0.118	0.093	0.093	0.378	0.176	0.176
13	0.000	0.000	0.000	0.077	0.000	0.000	0.327	0.000	0.000
14	0.031	0.000	0.000	0.217	0.000	0.000	0.469	0.000	0.000
15	0.048	0.056	0.056	0.283	0.095	0.095	1.897	0.111	0.111
16	0.000	0.022	0.023	0.089	0.081	0.081	0.306	0.097	0.097
17	0.019	0.067	0.067	0.296	0.141	0.141	0.969	0.238	0.238
18	0.025	0.000	0.000	0.169	0.000	0.000	0.380	0.000	0.000
19	0.004	0.000	0.000	0.064	0.000	0.000	0.225	0.035	0.035
20	0.005	0.000	0.000	0.147	0.000	0.000	0.385	0.007	0.008
Mean	0.016	0.010	0.010	0.188	0.136	0.130	0.832	0.395	0.354
Amt/Ppt	0.002	0.001	0.001	0.018	0.012	0.012	0.060	0.029	0.026

Table 11. EDYS simulation of mean annual drainage (gpm) per decade through the TSF cover over a 200-year period, under various revegetation scenarios, at different precipitation regimes.

Year	75% Precipitation			100% Precipitation			125% Precipitation		
	Grass	DgFir	BSage	Grass	DgFir	BSage	Grass	DgFir	BSage
001-010	0.019	0.006	0.006	0.246	0.230	0.218	1.106	0.715	0.633
011-020	0.013	0.015	0.015	0.150	0.043	0.043	0.558	0.076	0.076
021-030	0.059	0.071	0.071	0.276	0.130	0.129	1.107	0.192	0.188
031-040	0.016	0.022	0.022	0.225	0.050	0.051	0.565	0.118	0.124
041-050	0.023	0.010	0.010	0.166	0.013	0.014	0.311	0.033	0.050
051-060	0.068	0.036	0.036	0.219	0.065	0.066	0.503	0.113	0.113
061-070	0.049	0.016	0.016	0.183	0.043	0.050	0.341	0.067	0.070
071-080	0.040	0.008	0.010	0.107	0.018	0.023	0.216	0.038	0.045
081-090	0.031	0.002	0.009	0.094	0.013	0.014	0.194	0.042	0.044
091-100	0.056	0.031	0.032	0.154	0.060	0.067	0.361	0.083	0.087
101-110	0.012	0.002	0.002	0.040	0.009	0.010	0.221	0.024	0.029
111-120	0.011	0.000	0.005	0.041	0.002	0.006	0.399	0.017	0.024
121-130	0.021	0.029	0.025	0.077	0.052	0.058	1.224	0.068	0.072
131-140	0.003	0.002	0.009	0.121	0.010	0.012	1.420	0.024	0.043
141-150	0.001	0.000	0.000	0.291	0.000	0.000	1.119	0.003	0.009
151-160	0.006	0.022	0.024	0.635	0.046	0.051	0.912	0.054	0.066
161-170	0.001	0.000	0.002	0.839	0.004	0.005	0.822	0.010	0.016
171-180	0.000	0.000	0.000	0.574	0.000	0.000	0.122	0.001	0.016
181-190	0.001	0.017	0.019	0.469	0.035	0.040	0.092	0.036	0.043
191-200	0.000	0.000	0.001	0.159	0.000	0.004	0.023	0.010	0.017
Ann Mean	0.022	0.014	0.015	0.253	0.041	0.043	0.581	0.086	0.088
Amt/Ppt	0.003	0.002	0.003	0.023	0.004	0.004	0.043	0.006	0.007

gle cell because there is insufficient fuel to sustain the fire, or it spreads to adjacent cells. Which cells it then spreads to across the landscape is dependent on the fuel load in each cell, the fuel conditions, and the landscape characteristics.

For the TSF simulations, we selected a 100-year burn cycle, with the first fire starting early in the development of the revegetated plant community. This was done because this should be the most damaging time, from the standpoint of successional development of a woody-plant dominated community, for a burn to occur. Year 19 was the driest year in the first 20 years, so it was selected as the year of the first burn. The second burn occurred in Year 116, the driest year about 100 years after the first burn. No post-fire re-planting was simulated. Only natural recovery occurred.

Overall, two fires within 100 years had little effect on infiltration dynamics (Table 12). Annual infiltration, averaged over 200 years, was 0.05 gpm on the burned TSF, compared to 0.04 gpm of the unburned TSF. There was a discernable fire effect for 50 years following the first fire, but this decadal increased was never more than 0.012 gpm per year. The second fire had a greater initial effect, increasing the average annual infiltration in Years 111-120 from 0.002 to 0.034 gpm, and its impact also lasted about 50 years.

The primary effect of the fires on the vegetation was to reduce the dense crown biomass of grasses. The trees were not dense enough for a crown fire to develop, therefore only some individual trees were killed by the fires. The productivity of the remaining trees increased in response to the short-term decrease in competition from grasses. Effects of the second fire on standing crop biomass of grasses was evident for 80 years, whereas the effects on trees lasted less than 50 years.

4.4.2 Elk grazing

Heavy elk grazing was simulated by increasing elk herbivory from 0.004 lbs forage/ac/day under baseline conditions, to 0.126 lbs/ac/day under heavy grazing. This higher rate equals a resident elk stocking rate of 4 elk per section, year-round.

Over 200 years, simulated elk grazing increased infiltration from 0.04 gpm at light grazing to 0.06 gpm with heavy grazing (Table 12). Elk grazing had its greatest impact early in the 200-year period, when the plant community was shifting from a grass-dominated to a tree-dominated community. The revegetated cover was most sensitive to grazing during the first 50 years. Protection of the site (e.g., by fencing) for the first 10-20 years would decrease its sensitivity substantially.

The primary effect of elk grazing was to decrease the amount of grasses and forbs (Table 13). In time, this resulted in an increase in woody species because of a decrease in competition from grasses. These vegetation dynamics explain the effect on infiltration dynamics. Early in the 200-year period, elk grazing decreases grass production. Lower productivity of grasses results in less water transpired, and this lower water use is not equaled by the increased wa-

Table 12. EDYS simulation of the effect of fire and elk grazing on mean annual drainage (gpm) per decade through the TSF cover over a 200-year period, under the normal precipitation regime.

Years	Drainage			
	Unburned	Burned	Light Grazing	Heavy Grazing
001-010	0.230	0.230	0.218	0.257
011-020	0.043	0.043 Fire	0.043	0.101
021-030	0.130	0.142	0.129	0.228
031-040	0.050	0.056	0.051	0.100
041-050	0.013	0.016	0.014	0.041
051-060	0.065	0.073	0.066	0.095
061-070	0.043	0.048	0.050	0.057
071-080	0.018	0.013	0.023	0.020
081-090	0.013	0.015	0.014	0.034
091-100	0.060	0.061	0.067	0.080
101-110	0.009	0.011	0.010	0.014
111-120	0.002	0.034 Fire	0.006	0.026
121-130	0.052	0.079	0.058	0.062
131-140	0.010	0.028	0.012	0.011
141-150	0.000	0.008	0.000	0.017
151-160	0.046	0.068	0.051	0.043
161-170	0.004	0.010	0.005	0.001
171-180	0.000	0.000	0.000	0.003
181-190	0.035	0.046	0.040	0.019
191-200	0.000	0.014	0.004	0.000
Annual Mean	0.041	0.050	0.043	0.060
Amt/Ppt	0.004	0.005	0.004	0.006

Table 13. EDYS simulation of effects of heavy elk grazing on the vegetation dynamics (g/m²) on the TSF cover, under the big sagebrush revegetation scenario and the normal precipitation regime.

Year	Light Herbivory				Heavy Herbivory			
	Trees	Shrub	Grass	Forbs	Trees	Shrub	Grass	Forbs
01	0	6	32	10	0	6	31	10
02	t	5	53	14	t	8	36	12
03	2	6	82	22	1	9	27	11
04	5	5	109	26	4	10	44	14
05	12	8	143	29	12	12	33	14
06	31	8	152	29	37	14	52	19
07	55	9	187	31	64	16	58	22
08	90	9	221	32	102	18	71	26
09	142	10	224	31	173	22	81	36
10	238	11	289	32	285	27	133	41
11	306	11	301	31	341	31	144	41
12	351	12	335	31	387	36	159	43
13	388	12	345	31	424	38	164	43
14	461	12	409	30	503	41	186	44
15	516	12	372	30	562	46	184	41
16	560	13	398	30	606	50	188	41
17	658	13	447	30	719	60	225	39
18	716	14	463	30	787	67	235	37
19	751	14	446	30	824	69	211	34
20	792	14	474	30	873	75	221	33
50	1,793	27	913	30	2,403	237	300	16
100	2,841	111	1,479	30	4,042	487	402	14
150	3,100	234	1,992	30	4,492	918	356	11
200	3,258	422	2,452	30	4,677	1,461	290	12

water use by woody plants since most of the woody plant productivity is from Douglas fir, which has a relatively high water use efficiency. Over time however, the production of Douglas fir and shrubs increases to the point that its water use more than compensates for the lower water use by grasses. Therefore, within 70 years, infiltration in the heavily-grazed scenario is less than in the lightly-grazed scenario (Table 12).

5 CONCLUSIONS

The EDYS simulations indicate that all three cover variations (6-inch topsoil, 12-inch topsoil, and 6-inch topsoil plus bentonite in the upper tailings) will result in a productive and diverse plant community. In all three cases, the community should be relatively productive (130-150 g/m²) by the second year, and standing crop biomass should almost triple by the tenth year. The plant communities will still be grass-dominated by Year 10, but some Douglas fir will have established and over time the site will become dominated by Douglas fir.

The 12-inch topsoil depth resulted in 25% less infiltration than the 6-inch depth (0.147 and 0.199 gpm, respectively), averaged over the first 10 years under a normal (average) precipitation regime. In addition, less root biomass would be present in the tailings with the 12-inch topsoil depth than with the 6-inch depth. This could be a significant consideration if bioaccumulation of metals might be a source of concern. The bentonite treatment did not improve the performance of the cover.

Based on these results, whether 6 inches or 12 inches of topsoil are used in the cover design is largely a question of economics. Both will meet the vegetation criteria of providing surface stability and favorable forage and habitat for wildlife and livestock. The 12-inch layer will reduce drainage, but will cost more to install than the 6-inch layer.

Planting either Douglas fir or sagebrush seedlings in addition to a grass and forb seed mix will decrease infiltration, compared to planting the seed mix alone. Planting sagebrush seedlings will result in slightly less infiltration for the first 30 years than planting Douglas fir seedlings (0.130 and 0.134 gpm, respectively), but planting Douglas fir will result in less infiltration thereafter. The EDYS simulations indicate that the proposed cover design will provide the required reduction in infiltration through the tailings over a wide range of precipitation fluctuations and the revegetated plant communities should provide an effective vegetation cover even when subjected to drought, fire, and heavy grazing by elk.

Two important points can be made from this study concerning the design of water balance covers. First, short-term hydrologic dynamics may be as important, or sometimes more important, than annual

517

averages in determining the effectiveness of these covers. Short-term pulse events, such as spring snow melt or short periods of high rainfall, accounted for most of the infiltration through this cover. These pulse events would have been ignored if only annual or average daily precipitation values had been used, in which case, the use of these average values would have resulted in a significant underestimation of drainage through the tailings.

Secondly, changes in vegetation over time have a critical impact on the performance of water balance covers. These vegetation changes include plant succession, responses to normal environmental fluctuations (e.g., precipitation, grazing), and periodic disturbance events. If these vegetation dynamics are not accounted for in a design evaluation, the evaluation will either misrepresent the performance of the design or the design will have to be over-designed in order to provide sufficient protection.

REFERENCES

Childress, W.M. and T. McLendon. 1999. Simulation of multi-scale environmental impacts using the EDYS model. Hydrological Science and Technology 15:257-269.

Childress, W.M., D.L. Price, C.L. Coldren, and T. McLendon. 1999. A functional description of the Ecological DYnamics Simulation (EDYS) model, with applications for Army and other Federal land managers. US Army Corps of Engineers Research Laboratory Technical Report 99/55. Champaign, Illinois. 42 p.

Childress, W.M., C.L. Coldren, and T. McLendon. 2001. Applying a complex, general ecosystem model (EDYS) in large-scale land management. Ecological Modelling (In press).

McLendon, T., W.M. Childress, and C.L. Coldren. 2000a. Two-year validation results for grassland communities at Fort Bliss, Texas and Fort Hood, Texas. Technical Report SMI-ES-019. Shepherd Miller Inc. Fort Collins, Colorado. 79 p.

McLendon, T., C.L. Coldren, and W.M. Childress. 2000b. Evaluation of the effects of vegetation changes on water dynamics of the Clover Creek watershed, Utah, using the EDYS model. Technical Report SMI-ES-020. Shepherd Miller Inc. Fort Collins, Colorado. 56 p.

Author index

Ahmann, D. 13, 21
Aiken, S.R. 101
Alexandre J. 357
Alexieva, T. 295
Aurelian, F. 361
Ayres, B.K. 163

Bain, J. 213
Ball, K.D. 373
Barnekow, U. 113
Barrera, S. 87
Belem, T. 139
Benzaazoua, M. 139
Bergstrom, F.W. 489, 495, 505
Bernardes, G.P. 325
Biggar, K.W. 263
Binega, Y. 39
Bledsoe, B.P. 17
Blight, G.E. 333
Blowes, D. 21, 213
Boger, D.V. 129
Bonifazi, G. 27
Boshoff, J.C.J. 121
Boyce, S. 173
Büchel, G. 67
Bucher, W.H. 411
Buffington, R. 173
Bussière, B. 139
Byrd, R.L. 373

Carlson, K. 21
Castendyk, D. 181, 189
Cawlfield, L. 411
Chanasyk, D.S. 263
Chavez, J.E. 411
Childress, W.M. 505
Christensen, D. 163
Clayton, C.R.I. 45
Clements, W.H. 23
Coldren, C.L. 505

Cole, W.J. 101
Constantin, C. 365
Cotts, N.R. 435
Cutaia, L. 27

Dagenais, A.M. 139
de Campos, T.M.P. 325
Degering, D. 303
Delaney, T. 339, 387
Doye, I. 271
Duchesne, J. 271
Duisebayev, B. 313
DuTeau, N. 21

Esposito, K.J. 435

Figueroa, L. 21
Filip, Gh. 361
Fischer, G. 411
Ford, K.L. 245
Fortin, S. 459, 469
Foster, S. 53
Frechette, R.J. 489, 495, 505
Fredlund, D.G. 283
Fuller, M.L. 373

Garbarino, J. 13
Geletneky, J.W. 67
Georgescu, D. 361
Gipson, A.H., Jr. 93, 101
Gobla, M.J. 421
Gong, W. 249
Grant, B. 411

Harrington, J.M. 251
Hassani, F. 205
Hathaway, E. 53
Hebert, D. 303
Henderson, M.E. 387
Hossein, M. 205

Jakubick, A.T. 113
Janecky, D. 313
Jewell, P. 181, 189
Julien, P.Y. 17

Knapp, R.B. 313
Kowalewski, P.E. 173

Lamontagne, A. 479
Lanteigne, L. 163
le Roux, G.J.R. 35
Lefebvre, R. 479
Linkov, I. 53
Lorinczi, G. 447
Loudon, S. 459, 469
Lozano, N. 351
Lutze, W. 249

Macalady, D.L. 3, 13
Martens, P.N. 149
Martinez, L.L. 351
Massacci, P. 27
McLendon, T. 505
Meade, R.B. 427
Meyer, J. 13
Meyer, M.C. 435
Micu, O. 365
Moellerherm, S. 149
Muller, S.C. 339

Naamoun, T. 75, 217, 231, 303
Naeth, M.A. 263
Nuttall, H.E. 249

O'Kane, M. 163
Ouellet, J. 205

Panturu, E. 361
Paul, M. 67, 113
Petrescu, St. 361

Pilz, J. 379
Pinto, M. 87
Purdy, J. 387

Radulescu, R. 361
Ranville, J. 23
Reardon, K. 21
Reilly, T. 411
Richardson, J.H. 313
Robertson, A. 447, 479
Roselli, I. 27
Rosenberg, N. 313
Rust, E. 45
Ryan, R. 339
Rykaart, M.E. 283

Saboya, F., Jr. 357
Sarnogoev, A. 313
Sego, D.C. 263
Sellgren, A. 155

Shackelford, C.D. 3, 9, 21
Shaw, S. 447
Shepherd, T.A. 435
Shewman, A. 411
Shnorhokian, S. 205
Silva, M.J. 263
Smith, D.K. 313
Smyth, D. 213
Sofra, F. 129
Somot, S. 205
Sowa, V.A. 393
Spink, L. 213
Stauffer, D. 379
Stoica, L. 365
Sugatt, R. 53

Tagliapietra, J. 373
Ter-Stepanian, G. 349
Tibana, S. 325
Tompson, A.F.B. 313

Vermeulen, N.J. 45
Vos, H.P. 101

Waggitt, P.W. 403
Walker, M. 245
Walqui Fernandez, H. 93
Watson, C.H. 17
Weber, J. 339
Wels, C. 447, 459, 469, 479
Westall, J. 13
Whitlock, L. 155
Wildeman, T.R. 9, 21
Wilson, G.W. 283
Woods, S.L. 3, 21

Xavier, G.C. 357

Yarar, B. 197

Printed and bound by CPI Group (UK) Ltd, Croydon, CR0 4YY

25/10/2024

01779528-0001